THE INSTITUTION OF CIVIL ENGINEERS

Pre-failure deformation behaviour of geomaterials

Editors
R. J. Jardine, M. C. R. Davies, D. W. Hight, A. K. C. Smith,
S. E. Stallebrass

Thomas Telford

Géotechnique subcommittee for the Symposium in print:
Dr R. J. Jardine, Imperial College of Science, Technology and Medicine, London
Professor M. C. R. Davies, University of Dundee
Dr D. W. Hight, Geotechnical Consulting Group, London & visiting Professor, Imperial College, London
Dr A. K. C. Smith, Fugro Limited, Hemel Hempstead, Hertfordshire
Dr S. E. Stallebrass, City University, London

Published by Thomas Telford Publishing, Thomas Telford Ltd, 1 Heron Quay, London E14 4JD.
URL: http://www.t-telford.co.uk

Distributors for Thomas Telford books are
USA: ASCE Press, 1801 Alexander Bell Drive, Reston, VA 20191-4400
Japan: Maruzen Co. Ltd, Book Department, 3-10 Nihonbashi 2-chome, Chuo-ku, Tokyo 103
Australia: DA Books and Journals, 648 Whitehorse Road, Mitcham 3132, Victoria

First published 1998

A catalogue record for this book is available from the British Library

ISBN: 0 7277 2642 0

Typeset by Keytec Typesetting Ltd, Bridport, Dorset
Printed and bound in Great Britain by Henry Ling Limited, Dorset

Contents

Introduction

R. J. JARDINE*

The experimental and theoretical characterization of the pre-failure deformation properties of geomaterials has developed enormously over the past 15 years. Intensive studies have been carried out at many centres, and the new research results have made an important impact on practice, particularly in the UK and in Japan. Recognizing these developments, a *Géotechnique* Symposium in Print (SIP) on Pre-failure Deformation Behaviour of Geomaterials was held at the Institution of Civil Engineers, London, on 4 September 1997. This volume brings together the reprinted *Géotechnique* SIP papers, seven new papers, and the written discussion submitted to the ICE.

The September SIP was the successful outcome of an idea first suggested in March 1995 during the ICE Conference on Advances in Site Investigation Practice. Professor Michele Jamiolkowski was keen that the British contingent on the recently formed ISSMFE Technical Committee TC-29 should organize a conference in London to follow up the first International Symposium on Pre-failure Deformation Properties, which had been held the year before in Hokkaido, Japan. Concerned that a full International Conference on the same theme might not be viable so soon after Hokkaido, we hit upon the idea of an International Symposium in Print, published and produced by *Géotechnique*, and sponsored and promoted by both the BGS and the ISSMFE (through TC-29).

Long discussions with Professor Roy Butterfield, who was Chair of the *Géotechnique* Advisory Panel (GAP), led to the format adopted. An international Call for Papers was issued which emphasized the need for submissions of the highest scientific quality. The *Géotechnique* review process was applied rigorously and an additional issue of the journal was produced in August 1997 which was devoted entirely to the SIP. A GAP SIP subcommittee was formed to oversee the work. Joining Professor David Hight and me on the SIP subcommittee were Professor Michael Davies, Dr Andrew Smith and Dr Sarah Stallebrass. We were assisted very efficiently by Pam Skantham in the key initial stages of the project and in the later stages by Sharon Iliffe and Lesley Wilson from the ICE.

More than 80 papers were considered, reviewed and re-reviewed with help from the GAP and many colleagues in the UK and overseas. The 19 papers finally published in August 1997, and the seven additional papers and discussions presented in this volume, offer an excellent summary of the latest research into pre-failure deformation behaviour and its practical application. The SIP was truly international, with many overseas delegates attending and just over half of the papers originating outside the UK. The breadth of material allowed the presentations to be grouped under the three main headings of 'Measurements in the laboratory and field', 'Characterization and modelling' (in two parts) and 'Applications, design and case histories'. Each of these sessions was introduced by an SIP subcommittee member, and Professor Michele Jamiolkowski wound up the lively discussion. His final summary, published at the end of this Volume, includes an interesting personal 'shopping list' for the next phase of research, development and application in the field of geomaterial pre-failure deformation properties.

*Imperial College, London.

Part one

Symposium in print

Pennington, D. S., Nash, D. F. T. & Lings, M. L. (1997). *Géotechnique* **47**, No. 3, 391–398

Anisotropy of G_0 shear stiffness in Gault Clay

D. S. PENNINGTON,* D. F. T. NASH* and M. L. LINGS*

The inclusion of anisotropic stiffness parameters in sophisticated constitutive models necessitates their determination in the field and laboratory. Hitherto, anisotropy of small-strain stiffness of clays has occasionally been examined in the laboratory by measurements on specimens sampled at different orientations. The authors have developed a device to propagate and receive horizontal shear waves with both vertical and horizontal polarization through 100 mm triaxial samples. The device incorporates bender elements embedded in the pads of a horizontal belt. In tests on both reconstituted and natural samples of Gault Clay, each of the transmission velocities $V_{s(vh)}$, $V_{s(hv)}$ and $V_{s(hh)}$ has been measured during stress path tests. The ratio $G_{0(hh)}/G_{0(vh)}$ has been shown to be highly dependent on stress state. Laboratory data on natural samples show results broadly consistent with in situ results.

KEYWORDS: anisotropy; laboratory equipment; stiffness.

Pour pouvoir inclure des paramètres de rigidité anisotrope dans des modèles constitutifs perfectionnés, il faut les déterminer sur le terrain et en laboratoire. Jusqu'à présent, on a parfois examiné l'anisotropie de la rigidité d'argiles aux petites contraintes en mesurant des échantillons prélevés à diverses orientations. Les auteurs ont mis au point un dispositif qui permet la propagation et la réception d'ondes de cisaillement horizontales avec polarisation verticale et horizontale dans des échantillons de compression triaxiale de 100 mm. Le dispositif comprend des éléments de flexion noyés dans les plaquettes d'une bande horizontale. Dans des essais de parcours de contrainte sur des échantillons reconstitués et naturels d'argile de Gault, on a mesuré les vitesses de transmission $V_{s(vh)}$, $V_{s(hv)}$ et $V_{s(hh)}$. On a montré que le rapport $G_{0(hh)}/G_{0(vh)}$ dépend en grande partie de l'état de contrainte. Les résultats obtenus en laboratoire sur des échantillons naturels concordent assez bien avec les résultats obtenus sur le terrain.

INTRODUCTION

The initial shear modulus G_0 is now widely considered to be a fundamental soil stiffness property. It is material to practical geotechnical problems, both in earthquake engineering and in the prediction of soil structure interaction. Simpson *et al.* (1996) have shown that anisotropy of G_0 can have a significant influence on predicted deformations due to tunnelling in stiff clays. That shear stiffness is anisotropic is not surprising, since sedimentation produces an inherently cross-anisotropic fabric. This has been widely investigated in geophysical field tests (e.g. White *et al.*, 1983; Sully & Campanella, 1995; Butcher & Powell, 1995), but there is now a need to develop techniques to measure appropriate anisotropic parameters in the laboratory.

The anisotropy of an elastic material is frequently reduced from the fully anisotropic description, given by 21 independent material properties, to cross-anisotropy, which is represented by five independent elastic constants E_v, E_h, ν_{vh}, ν_{hh} and G_{vh} (Love, 1927). Shear deformations in the horizontal plane are governed by the dependent shear modulus G_{hh}, where

$$G_{hh} = E_h/2(1 + \nu_{hh}) \quad (1)$$

At very small strains, soils are believed to behave as elastic materials, so $G_{0(ij)}$ represents the initial elastic stiffness of any σ_{ij}' against ε_{ij} curve. Some, but not all, of the 'elastic' parameters used in the more elegant constitutive models (e.g. Atkinson & Stallebrass, 1991; Simpson, 1993) can be obtained at slightly larger strains from triaxial stress path tests. There is an obvious need to determine the $G_{0(ij)}$ parameters within the same specimen.

Laboratory determination of G_0 is frequently made from the measurement of shear wave velocity in resonant column devices or using the bender element method, introduced to soils testing by Shirley & Hampton (1977). These techniques are based on conversion of the velocity of propagation of the shear wave V_s to G_0 though the equation

Manuscript received 16 October 1996; revised manuscript accepted 19 February 1997.
Discussion on this paper closes 1 November 1997.
* Department of Civil Engineering, University of Bristol.

$$G_{0(ij)} = \rho V_{s(ij)}^2 \tag{2}$$

where ρ is the bulk density. With resonant column devices only one shear wave velocity (usually $V_{s(vh)}$) can be measured in each test. Similarly, where bender elements have been used in oedometers and triaxial cells, they are generally used to measure shear wave velocity propagated in one direction with one polarity. In the triaxial cell the bender elements have invariably been mounted in rigid end platens to evaluate $G_{0(vh)}$ (e.g. Dyvik & Madshus, 1985), and sampling at different orientations and testing under isotropic stress states have been used to evaluate $G_{0(hh)}$ (Simpson *et al.*, 1996). Additional benders mounted through the rigid wall of an oedometer have been used by Jamiolkowski *et al.* (1995) to investigate the interrelationship of $G_{0(vh)}$ and $G_{0(hh)}$ during one-dimensional loading or unloading. While these tests enable $G_{0(vh)}$ and $G_{0(hh)}$ to be determined on the same sample, the equipment lacks the advantage of the stress control possible in the triaxial cell.

Much experimental data has been published on the cross-anisotropy of sands reconstituted in the laboratory (e.g. Roesler, 1979; Stokoe *et al.*, 1985, 1995; Bellotti *et al.*, 1995) but relatively little has been written about clays. However, using their instrumented oedometer, Jamiolkowski *et al.* (1995) show that the following expressions for G_0, proposed by Hardin & Blanford (1989), apply to clays:

$$G_{0(vh)} = S_{vh} F(e)(\mathrm{OCR})^k p_a^{(1-nv-nh)} (\sigma_v')^{nv} (\sigma_h')^{nh} \tag{3a}$$

for the shear modulus perpendicular to the bedding plane, and

$$G_{0(hh)} = S_{hh} F(e)(\mathrm{OCR})^k p_a^{(1-2nh)} (\sigma_h')^{2nh} \tag{3b}$$

for the shear modulus on the bedding plane.

In these expressions S_{ij} are non-dimensional material constants reflecting the soil's fabric, $F(e)$ is a voids ratio function, k is an empirical function dependent on the Plasticity Index of the clay, p_a is a reference stress, and nv and nh are empirical stress exponents. Jamiolkowski *et al.* (1995) evaluated the constants for a number of clays and showed that $k = 0$.

However, the scope of that work is limited by equipment problems in following reloading paths reliably and the inability to control stress ratios in the samples. The determination of $G_{0(vh)}$ and $G_{0(hh)}$ during tests along controlled anisotropic stress paths is clearly desirable, and this paper describes a new device for use on triaxial specimens. Some preliminary results from tests on both natural and reconstituted Gault Clay are presented and compared with a field geophysical study.

LABORATORY EQUIPMENT

Triaxial cell bender arrangement

The tests reported here have been conducted in a modified 100 mm stress path cell of Imperial College design. Four sets of bender elements have been incorporated in the cell in order that waves can be propagated and received in orthogonal pairs both axially and diametrically across the mid-height section, as shown in Fig. 1. The four measurements performed are

(*a*) $V_{s(vh)b}$: wave propagating vertically with horizontal polarization at the same azimuth (horizontal direction relative to north) as waves propagating from the horizontal 'bender belt'

(*b*) $V_{s(vh)}$: wave propagating vertically with horizontal polarization at the azimuth perpendicular to the waves propagating from the horizontal 'bender belt'

(*c*) $V_{s(hv)}$: wave propagating horizontally with vertical polarization

(*d*) $V_{s(hh)}$: wave propagating horizontally with horizontal polarization (this wave coincides with the plane of isotropy in a cross-anisotropic medium, where equation (1) describes the relationship between elastic constants).

Since in a cross-anisotropic material $V_{s(vh)} = V_{s(hv)}$, the taking of three such measurements introduces redundancy and the ability to check that the soil is cross-anisotropic.

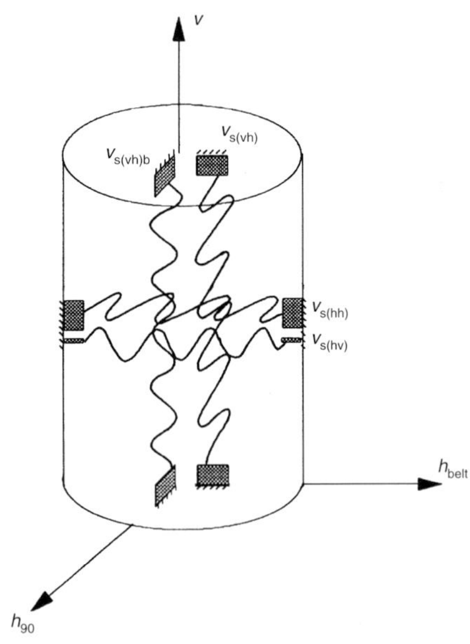

Fig. 1. Orientations of bender elements in the triaxial samples

Mounting of the bender elements

Bender elements are made from high-impedance piezoceramic material, and therefore need perfect insulation in the triaxial cell. Design and assembly techniques were based on those described by Dyvik & Madshus (1985) and particular attention was paid to shielding all electrical connections. Pairs of horizontal benders were arranged orthogonally and potted in pads as shown in Fig. 2. The horizontal benders protrude 3 mm into the sample through slots cut in the membrane. The pads are linked though pin joints to a horizontal belt instrumented to measure changes in diameter and hence the wave path-length.

Operation of the bender elements

A programmable function generator was used to generate waveforms to energize the driver elements. A range of driver signals based on sine pulses, and including the skewed pulses introduced by Jovičić *et al.* (1996) were used. Signals from the receiver elements were interfaced though a charge amplifier to provide a significantly higher input impedance than the digital storage oscilloscope, thereby increasing efficiency. Typical signals received from the bender belt are shown in Fig. 3. The accuracy of the shear wave velocity measurements was estimated to be better than ±1%.

SAMPLES AND THEIR PREPARATION

Samples of Gault Clay were retrieved from the research test site at Madingley near Cambridge. The site has been well characterized by Kay & Parry (1982), Marsland & Powell (1990), Butcher & Lord (1993) and others. Of particular interest to this study is a detailed in situ seismic shear wave study (Butcher & Powell, 1995), where various shear wave measurement techniques provide profiles of $V_{s(vh)}$, $V_{s(hv)}$ and $V_{s(hh)}$ as a function of depth.

The Gault is a grey, heavily overconsolidated clay deposit of the late Cretaceous geological period, which has been folded and eroded since deposition. At Madingley it is approximately 40 m thick, and is a very stiff to hard, finely fissured high-plasticity silty clay with a typical $CaCO_3$ content of 30%. High-quality samples were obtained from a depth of 7–8 m using the BRE 250 mm diameter thin-wall over-cored push sampler. At this depth K_0 and OCR were estimated to be 2·1 and 30 respectively (Butcher & Lord, 1993).

The sample azimuth (relative to the field cross-hole seismic tests conducted at the site) was determined, and preserved though sample preparation so that the bender belt could be attached along the same azimuth. Samples were extruded in a high-humidity room, trimmed by hand and finally sized using a bandsaw. During specimen preparation, suctions were monitored using a suction probe (Ridley & Burland, 1996), and were observed to increase by less than 20% from 120 kPa, an indication of negligible additional disturbance.

Reconstituted samples were prepared by remoulding intact material at a moisture content of 1·8 times the liquid limit with the addition of

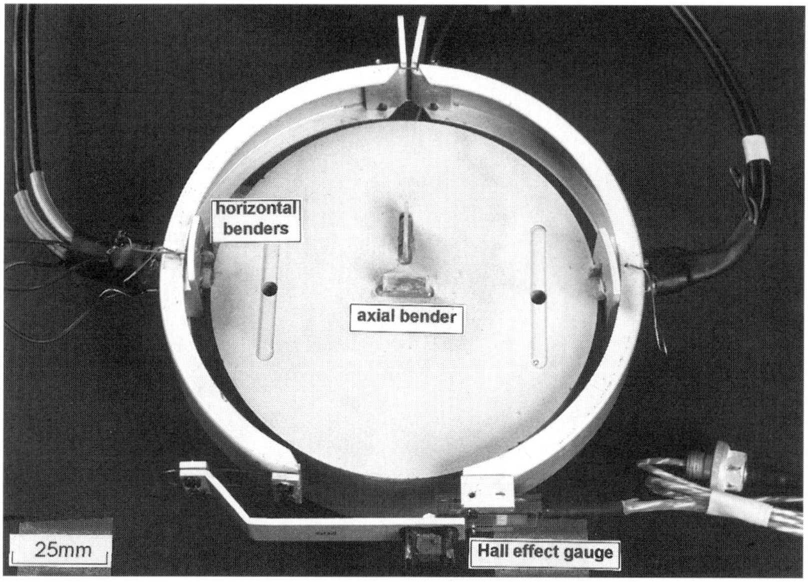

Fig. 2. Arrangement of bender elements within top 100 mm top platen and radial belt pads

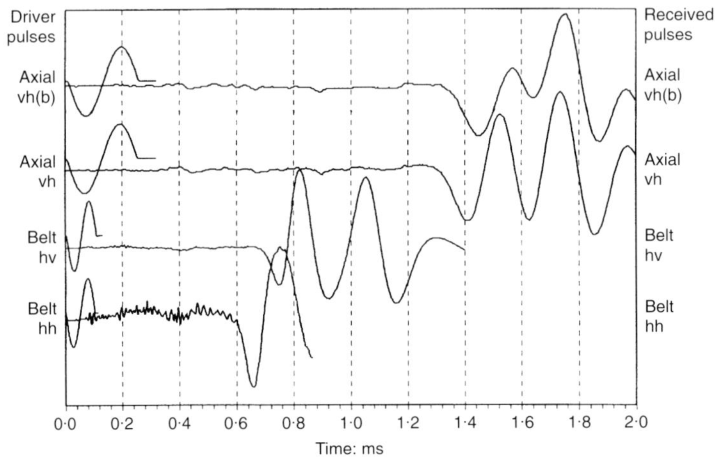

Fig. 3. Typical suite of driver and received pulses for axial and radial belt bender elements

deionized water. The resulting slurry was de-aired and one-dimensionally consolidated to a vertical effective stress of 165 kPa.

LABORATORY RESULTS AND THEIR INTERPRETATION

Reconstituted Gault Clay

The potential benefits of the bender belt were first explored on a reconstituted sample, since its preparation should have produced a cross-anisotropic material. The sample was tested in two stages.

Firstly, it was subjected to incremental isotropic loading and unloading, leading to the e against p' history shown in Fig. 4(a). Fig. 4(b) shows the corresponding $G_{0(ij)}$ against p' relationships for the sample established at the end of primary consolidation after each increment. The $G_{0(\mathrm{vh})}$ and $G_{0(\mathrm{vh})\mathrm{b}}$ results were virtually coincident and similar to the $G_{0(\mathrm{hv})}$ results and have been omitted for clarity. The results show different $G_{0(\mathrm{hv})}$ against p' relationships for the normally consolidated loading paths and overconsolidated (unload–reload) paths, as found previously by Viggiani & Atkinson (1995). The $G_{0(\mathrm{hh})}$ against p' results are similar, but consistently higher than the $G_{0(\mathrm{hv})}$ results. This is further illustrated in Fig. 4(c), where the ratios $G_{0(\mathrm{vh})}/G_{0(\mathrm{hv})}$ and $G_{0(\mathrm{hh})}/G_{0(\mathrm{hv})}$ are plotted against p'. Despite OCR varying by up to one order of magnitude, both these ratios remain essentially constant. This intuitively reasonable result for isotropic loading and unloading fits with the proposition that the degree of anisotropy is a consequence of the fabric of the clay, resulting from its one-dimensional deposition.

After returning to an isotropic stress of 100 kPa (an OCR of 6), a second stage of testing was undertaken. While maintaining p' constant, the sample was subjected to two cycles of anisotropic stress reversal between $\sigma_{\mathrm{h}}'/\sigma_{\mathrm{v}}' = 0{\cdot}4$ and 2·3. Fig. 5(a) shows the results, and as expected $G_{0(\mathrm{vh})}$ and $G_{0(\mathrm{hv})}$ had similar values, varying significantly less than $G_{0(\mathrm{hh})}$. At a stress ratio of 0·4, $G_{0(\mathrm{hh})}$ results are similar to $G_{0(\mathrm{hv})}$ results, whereas they indicate a stiffness 80% higher than the $G_{0(\mathrm{hv})}$ results at a stress ratio of 2·3.

Natural Gault Clay

Measurements of $G_{0(ij)}$ have been made on a natural sample. After initial isotropic consolidation under the in situ vertical effective stress, they have been brought back to the estimated in situ K_0 of 2·1 by increasing σ_{h}' while maintaining constant σ_{v}'. At various points during consolidation the procedure was halted and the pore pressure across the sample allowed to equalize before $G_{0(ij)}$ readings were taken; results for one of the tests are shown in Fig. 5(b). Since changes of void ratio were minimal, the variations of $G_{0(ij)}$ result primarily from stress changes. Throughout this test the two $G_{0(\mathrm{vh})}$ results increased by only 6%, and remained essentially equal to each other, indicating minimal variation of fabric with azimuth. In contrast, $G_{0(\mathrm{hv})}$ and $G_{0(\mathrm{hh})}$ were 45 and 265% higher than $G_{0(\mathrm{rh})\mathrm{b}}$ and increased by 5 and 26% respectively.

In order to enable comparisons to be drawn between the natural and the reconstituted tests, the $G_{0(ij)}$ data described above have been normalized by the voids ratio function

$$F(e) = e^{-1{\cdot}3} \tag{4}$$

proposed by Lo Presti (1995). These values are

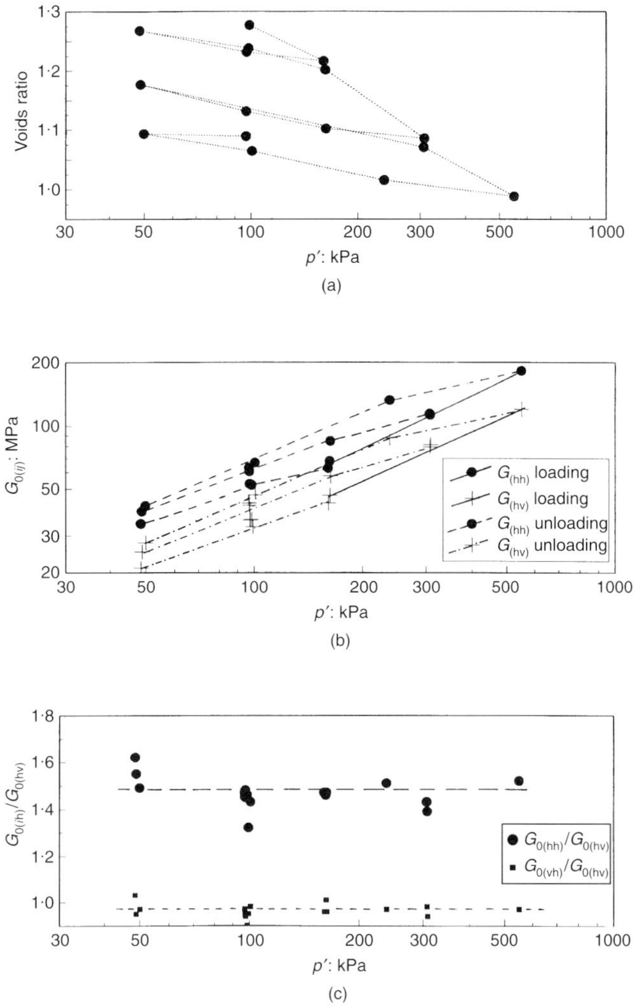

Fig. 4. Reconstituted sample: (a) voids ratio plotted against effective confining stress; (b) $G_{(ij)}$ plotted against effective confining pressure; (c) $G_{(ij)}/G_{(hv)}$ ratios

plotted against effective consolidation stresses $\sigma_v{}'\cdot\sigma_h{}'$ and $\sigma_h{}'\cdot\sigma_h{}'$ in Fig. 6. The values of S_{ij} and *ni* (from equations (3a) and (3b)) deduced from this plot are listed in Table 1.

This preliminary analysis of the data confirms broad agreement with the relationships implied by equations (3a) and (3b), although there has not yet been sufficient testing to confirm the voids ratio function. The significant differences between the reconstituted and natural material are probably due to differences in fabric, ageing and chemical bonding.

Comparisons with field geophysical data

Profiles of shear wave velocities obtained at Madingley by Butcher & Powell (1995) are reproduced in Fig. 7, together with the results from bender element tests on two natural reconsolidated samples. The $V_{s(vh)}$ results were obtained using a seismic cone while the $V_{s(hv)}$ and $V_{s(hh)}$ results were from cross-hole measurements using axial and rotary hammers. The accuracy of the velocity measurements is of the order of ±1% (Butcher, 1997), which is similar to that of measurements in the laboratory. There is broadly good agreement between the results, and as with the laboratory data, the in situ $V_{s(hh)}$ data lie substantially above the $V_{s(vh)}$ and $V_{\hat{s}(hv)}$measurements.

It is noticeable that the in situ $V_{s(hv)}$ and $V_{s(vh)}$ profiles are significantly different from each other, and that the laboratory values show similar but smaller differences. While the evidence from the

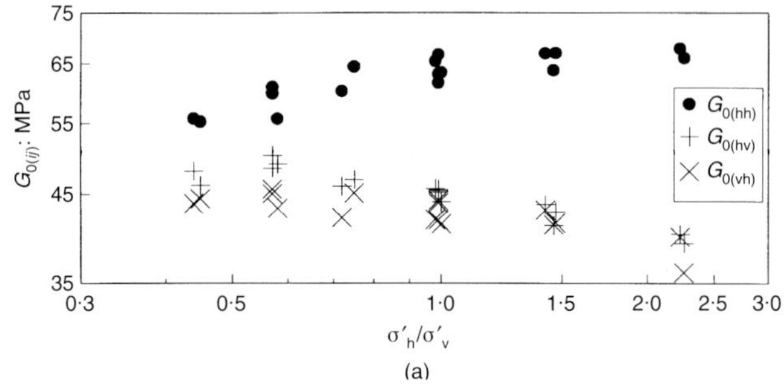

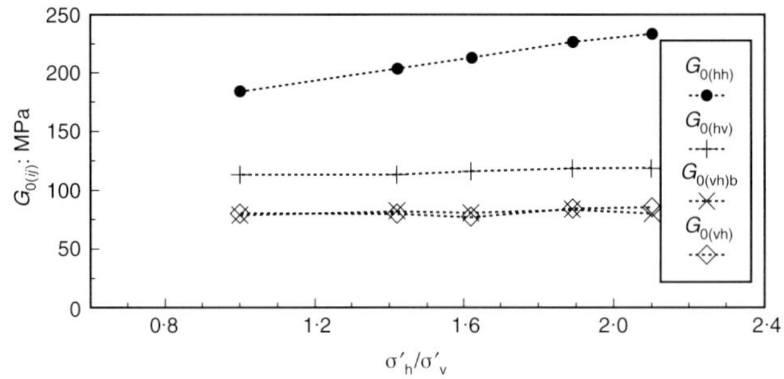

Fig. 5. (a) $G_{0(ij)}$ plotted against σ'_h/σ'_v for anisotropically reconstituted sample (OCR = 6, constant p' = 100 kPa, voids ratio = 1·09); (b) $G_{0(ij)}$ plotted against $\sigma_h'\sigma_v'$ for natural sample (OCR > 30, constant σ_v' = 120 kPa, voids ratio = 0·84)

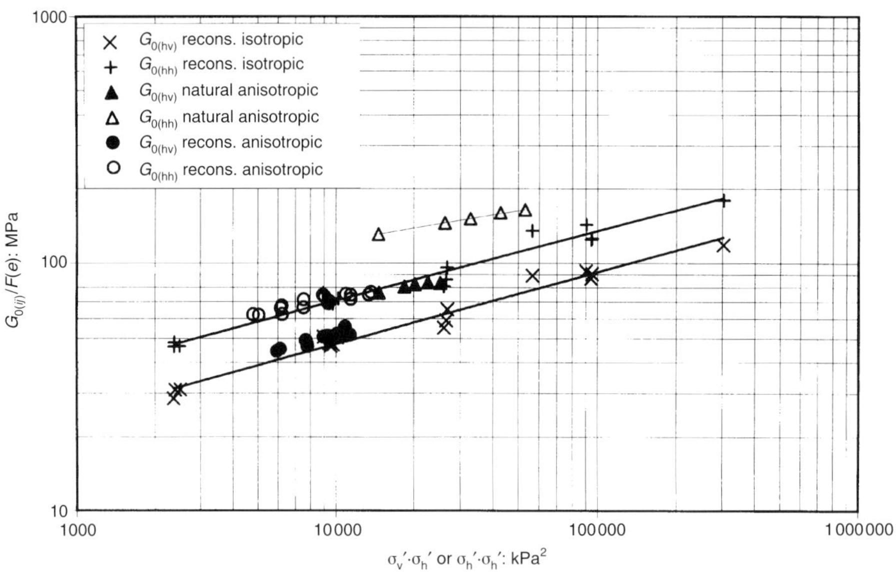

Fig. 6. Normalized shear moduli plotted against effective consolidation stresses for reconstituted and natural samples following isotropic and anisotropic stress paths

Table 1. Anisotropy stiffness constants from triaxial bender tests

Test	S_{hv}	S_{hh}	nv	nh	OCR	e
Reconstituted isotropic	3·4	5·4	0·30	0·28	1–11	1·09–1·30
Reconstituted anisotropic	3·6	5·5	0·30	0·28	6	1·09
Natural	16·8	25·4	0·16	0·18	30	0·84

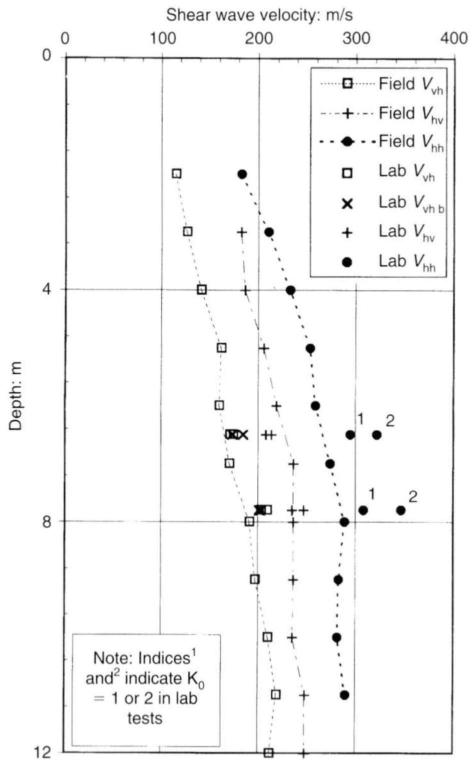

Fig. 7. Comparisons between field $G_{0(ij)}$ plotted against depth, and laboratory measurements, on two natural samples at $K_0 = 1$ and 2 (the numbers 1 and 2 indicate $K_0 = 1$ and 2 in lab)

laboratory test on reconstituted clay under axisymmetric stress conditions presented above indicated essentially cross-anisotropic fabric, neither the field nor laboratory tests on natural materials confirm perfect cross-anisotropy. This is possibly due to layering. The vertically propagated waves have to pass through each layer, whereas the horizontally propagated waves will tend to travel along layers of highest stiffness, hence $G_{0(hv)}$ values are consistently higher than $G_{0(vh)}$ values. A possible explanation of smaller differences between laboratory $V_{s(hv)}$ and $V_{s(rh)}$ compared to in situ values could lie in consideration of equations (3) which suggest that $G_{0(vh)}$ measured in situ would be influenced by variations of horizontal stress with azimuth. Since the field $V_{s(vh)}$ readings were taken with their plane of excitation perpendicular to the direction of propagation of the cross-hole $V_{s(hv)}$ (Butcher, 1997), it is plausible that a proportion of the observed differences are due to variation of in situ lateral stress with azimuth. Bearing in mind the complex stress history of the area, such azimuthal stress variations are not unexpected, and have previously been suspected in the Gault clay elsewhere in Cambridgeshire (Dalton & Hawkins, 1982).

CONCLUSIONS

The following conclusions can be drawn from the data presented.

(*a*) A novel bender belt has been developed which makes possible the determination of initial shear stiffnesses $G_{0(hv)}$ and $G_{0(hh)}$ on one specimen in the triaxial cell.
(*b*) $G_{0(ij)}$ is inherently anisotropic owing to the depositional fabric of a clay, and the natural material is significantly more anisotropic than the reconstituted material.
(*c*) Testing has shown that $G_{0(ij)}$ (and thus the ratio $G_{0(hh)}/G_{0(hv)}$) is highly dependent on stress state.
(*d*) Laboratory data and field geophysical data are broadly consistent, but uncertainties over the applicability of cross-anisotropic assumptions for $G_{0(ij)}$ require further investigation.
(*e*) Variations of in situ stresses with azimuth have implications for the interpretation of in situ tests. Indeed a combination of laboratory and field shear wave data may be useful for the assessment of K_0.

ACKNOWLEDGEMENTS

The authors are grateful to the EPSRC for sponsoring the research project described, and to the Building Research Establishment for providing the clay samples. The successful design of the laboratory equipment would not have been possible without the willingness to share ideas of Jim Bennel, University of Wales at Bangor, Vojcan Jovičić and colleagues at City University, London, and Steve Ackerley, Imperial College, London. Mike Pope has been invaluable in the manufacture and development of the devices, Tim Hetherington has assisted in the laboratory testing and Andrew Ridley, Imperial College, made the suction measurements.

REFERENCES

Atkinson, J. H. & Stallebrass, S. E. (1991). A model for recent history and non-linearity in the stress–strain behaviour of overconsolidated soil. Proc. Computer methods and advances in Geomechanics, (ed Cairns), pp 555–560.

Bellotti, R., Jamiolkowski, M. & Lo Presti, D. C. F. (1995). Anisotropy of small strain stiffness in Ticino Sand. *Géotechnique* **46**, No. 1, 115–132.

Butcher, A. P. (1997). Private communication.

Butcher, A. P. & Lord, J. A. (1993). Engineering properties of Gault Clay in and around Cambridge. *Int. Symp. Hard Soils and Soft Rocks, Athens* (eds Anagnostopolos *et al.*), pp. 405–416. Rotterdam: Balkema.

Butcher, A. P. & Powell, J. M. M. (1995). The effects of geological history on the dynamic stiffness in soils. *Proc. 11th European Conf. on Soil Mechanics, Copenhagen*, Vol. 1, pp. 1.27–1.36.

Dalton, J. C. P. & Hawkins, P. G. (1982). Fields of stress—some measurements of the in-situ stress in a meadow in the Cambridgeshire countryside. *Ground Engng* **15**, No. 4, 15–23.

Dyvik, R. & Madshus, C. (1985). Lab measurements of G_{max} using bender elements. In *Advances in the art of testing soils under cyclic conditions* (ed. V. Khosla) (Proc. of session sponsored by the Geotechnical Engineering Division in conjunction with the ASCE Convention in Detroit, Michigan), pp. 186–196. New York: ASCE.

Hardin, B. O. & Blanford, G. E. (1989). Elasticity of particulate materials. *J. Geotech. Engng Div., ASCE* **115**, No. 6, 788–805.

Jamiolkowski, M., Lancellotta, R. & Lo Presti, D. C. F. (1995). Remarks of the stiffness at small strains of six Italian clays. In *Developments in deep foundations and ground improvement schemes* (ed. Balasubramaniam *et al.*) pp. 197–216. Rotterdam: Balkema.

Jovičić, V., Coop, M. R., & Simic, M. (1996). Objective criteria for determining G_{max} from bender element tests. *Géotechnique* **46**, No. 2, 357–362.

Kay, J. N. & Parry, R. H. (1982). Screw plate tests in Gault Clay. *Ground Engng* **15**, No. 9, Sept. 22–30.

Lo Presti, D. C. F. (1995). General report: measurement of shear deformation of geomaterials in the laboratory. *Proc. Int. Symp. on Pre-failure Deformation of Geomaterials* (eds S. Shibuya, T. Mitachi & S. Muira), Vol. 2, pp. 1067–1088.

Love, A. E. H. (1927). *A treatise on the mathematical theory of elasticity*, 4th edn. Cambridge University Press.

Marsland, A. & Powell, J. J. M. (1990). Pressuremeter tests on stiff clays and soft rocks—factors affecting measurements and their interpretation. *Proc. 24th Annual Conf. of Engineering Group of the Geology Society. Field testing in engineering geology.* Special publication issue 6 (eds F. G. Bell, J. C. Cripps & M. G. Culshaw), pp. 91–110.

Ridley, A. M. & Burland, J. B. (1996). A pore water pressure probe for the insitu measurement of a wide range of soil suctions. In *Advances in site investigation practice* (ed. C. Craig), pp. 510–520. London: Thomas Telford.

Roesler, S. K. (1979). Anisotropic shear modulus due to stress anisotropy. *J. Geotech. Engng Div., ASCE* **105**, GT7, July, 871–880.

Shirley, D. J. & Hampton, L. D. (1977). Shearwave measurements in laboratory sediments. *J. Acoust. Soc. Am.* **63**, No. 2, 607–613.

Simpson, B. (1993). Development and application of a new soil model for prediction of ground movements. In *Predictive soil mechanics* (eds G. T. Houlsby & A. N. Schofield) (Proc. Wroth Memorial Symposium, Oxford.), pp. 628–643. London: Thomas Telford.

Simpson, B., Atkinson, J. H. & Jovičić, V. (1996). The influence of anisotropy on calculations of ground settlements above tunnels. *Proc. Geotechnical Aspects of Underground Construction in Soft Ground*, The City University, London, pp 591–595.

Stokoe, K. H., Hwang, S. K., Lee, J. N.-K. & Andrus, R. D. (1995). Effects of various parameters on the stiffness and damping of soils at small to medium strains. *Proc. Int. Symp. on Pre-failure Deformation of Geomaterials* (eds S. Shibuya, T. Mitachi & S. Muira), Vol. 1, pp. 785–815.

Stokoe, K. H., Lee, S. H. H. & Knox, D. P. (1985). Shear moduli measurements under true triaxial stresses. *Proc. session sponsored by the Geotechnical Engineering Division in conjunction with the ASCE Convention in Detroit, Michigan*, pp. 167–185. New York: ASCE.

Sully, J. P. & Campanella, R. G. (1995). Evaluation of in situ anisotropy from crosshole and down hole shear wave velocity measurements. *Géotechnique* **45**, No. 2, June, 267–282.

Viggiani, G. & Atkinson, J. H. (1995). Stiffness of fine grained soil at very small strains. *Géotechnique* **45**, No. 2, June, 249–265.

White, J. E., Martineau-Nicoletis, L. & Monash, C. (1983). Measured anisotropy in Pierre Shale. *Geophysical Prospecting* **31**, No. 5, 709–726.

Heymann, G., Clayton, C. R. I. & Reed, G. T. (1997). *Géotechnique* **47**, No. 3, 399–405

Laser interferometry to evaluate the performance of local displacement transducers

G. HEYMANN,* C. R. I. CLAYTON* and G. T. REED*

The authors have developed a Fabry–Pérot laser interferometer system for displacement measurement, which is compact and can be used as a stand-alone local strain measuring system. Interferometry operates on the principle of interference of a coherent light source and therefore provides an absolute standard of measurement. The purpose of this development is to investigate the extent of the linear-elastic range of geomaterials in triaxial stress space, and to obtain values of stiffness which can subsequently be compared with those obtained in other ways, both in the laboratory (using resonant column or bender elements) and in the field (using field geophysical techniques). The interferometer is also suitable as a high-accuracy instrument for calibration of other linear displacement transducers. This paper describes the instrument configuration and evaluates the performance of a commercial LVDT calibrated against the interferometer.

KEYWORDS: deformation; laboratory equipment; laboratory tests; stiffness.

Les auteurs ont mis au point un interféromètre à laser Fabry-Pérot de dimensions réduites qui peut servir indépendamment à mesurer les contraintes locales. L'interférométrie repose sur le principe de la perturbation d'une source de lumière cohérente et offre donc un étalon de mesure absolu. Les auteurs visaient dans leurs travaux à examiner le domaine linéaire-élastique de matériaux dans l'espace de contrainte triaxiale et à obtenir des valeurs de rigidité qui puissent être comparées plus tard aux valeurs obtenues par d'autres moyens en laboratoire (éprouvettes à fléchir) et sur le terrain (techniques géophysiques).

INTRODUCTION

Local displacement transducers are now commonly used to measure the small-strain behaviour of geomaterials. These transducers measure the relative displacements of positions on soil samples remote from the ends, thereby eliminating errors introduced by apparatus compliance, bedding and principal stress rotation near the sample ends. The ideal local displacement transducer should be accurate over a large measurement range and, furthermore, it should not be affected adversely by the cell fluid or cell pressure, or be damaged at large soil strains. Despite the large numbers of commercially available transducers such as LVDTs and proximity transducers, as well as a number of purpose-made local strain devices (Burland & Symes, 1982; Jardine *et al.*, 1984; Clayton & Khatrush, 1986; Clayton *et al.*, 1989; Tatsuoka, 1988; Goto *et al.*, 1991), no single device satisfies all these requirements. Different resolutions and accuracies are reported for these instruments, as shown in Table 1.

An instrument should be calibrated using a reference system having an accuracy 3–10 times better than the accuracy of the instrument being calibrated (Sydenham, *et al.*, 1989; Doebelin, 1983; Collett & Hope, 1983; Sydenham, 1982). Table 1 shows that the performance of local displacement transducers has surpassed the calibration capabilities that are commonly available in research and commercial laboratories and clearly the need has developed for a cost-effective, high-accuracy calibration instrument. Previous workers have apparently used a Michelson interferometer to calibrate LVDTs and proximity transducers (e.g. Lo Presti *et al.*, 1993). However, no details have been given with regard to the apparatus configuration, laser performance and signal interrogation

Manuscript received 27 October 1996; revised manuscript accepted 17th January 1996.
Discussion on this paper closes 1 November 1997.
* University of Surrey.

Table 1. Characteristics of local displacement transducers

Instrument	Reported range: mm	Reported accuracy: μm	Reported resolution: μm	Calibration instrument	Reference
Electrolevel gauge	—	±2	—	—	Burland & Symes (1982)
	15	—	< 1	LVDT with accuracy of 0·2 μm	Jardine *et al.* (1984)
Hall effect gauge	2·5	±6	< 1*	Micrometer with 2·54 μm resolution	Clayton *et al.* (1989)
	7	±30	—		
Local deformation transducer (LDT)	0·2†	0·09‡	0·12	Micrometer with 1 μm resolution	Goto *et al.* (1991)
Proximity transducer	5	±2	1	Steel slip gauges with accuracy of ±0·2 μm	Hird & Yung (1989)
LVDT	2	1	0·3	—	Scholey *et al.* (1995)

* Hall effect gauge resolution reported by Clayton & Khatrush (1986).
† LDT range of 1·5 mm has been reported by Tatsuoka (1988).
‡ LDT accuracy calculated from reported strain accuracy of 10^{-6} and gauge length of 90 mm.

system and therefore no judgement could be made on the performance of these calibration systems.

INTERFEROMETER CONFIGURATION AND PERFORMANCE

A Fabry–Pérot interferometer (Fabry & Pérot, 1897) has been developed at the University of Surrey as a local displacement instrument for triaxial measurement of geomaterial behaviour at ultra-small strains. The interferometer is also suitable as a high-accuracy instrument for calibration of other linear displacement transducers. The layout is shown schematically in Fig. 1. An 8 mW helium–neon (HeNe) laser light source with a wavelength (λ) of 0·6328 μm is used. The laser has an output power stability of ±2·5% and a longitudinal mode spacing of 438 MHz. The light is launched into a single-mode optical fibre with a core diameter of 4 μm and numerical aperture of 0·12 by way of a laser-fibre pigtail assembly. The coupling efficiency of the pigtail assembly is of the order of 55%. As the light exits the fibre it is passed through a cylindrical graded index (GRIN) lens with a length of 4·3 mm and a diameter of 1·8 mm, which collimates the light (divergence angle less than 0·3°). The collimated light subsequently impinges on a partial mirror with a diameter of 12·5 mm and a thickness of 1 mm. It is machined with a surface flatness of 1 μm and a parallelism of better than 5 arc minutes. One side of the partial mirror has a dielectric coating which reflects 60% of the light and transmits 40%, while the other side has an anti-reflectance coating. The light transmitted through the first mirror is projected on to another partial mirror with similar geometric and optical properties. This second partial mirror is fixed to a silicon photodetector with an active area of 15 mm^2. The two partial mirrors are positioned parallel to each other to form the Fabry–Pérot cavity.

Multiple reflections occur between the two partial mirrors, resulting in the well-known Fabry–Pérot transfer function shown in Fig. 2 (see, for example, Born & Wolf (1965)). The sharpness of the peaks is quantified as the full width at half maximum, which is the width between two points

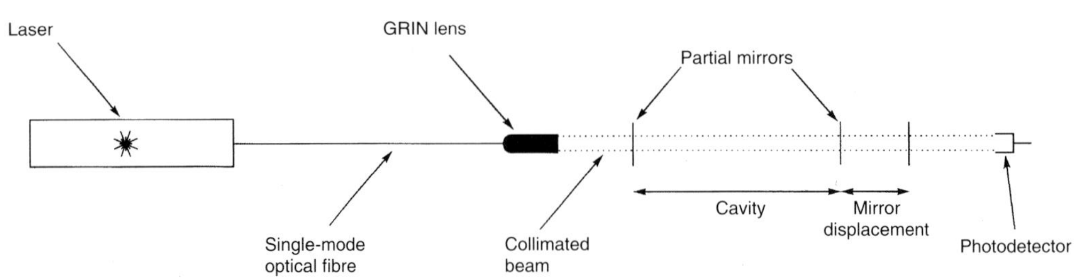

Fig. 1. Schematic layout of Fabry–Pérot interferometer

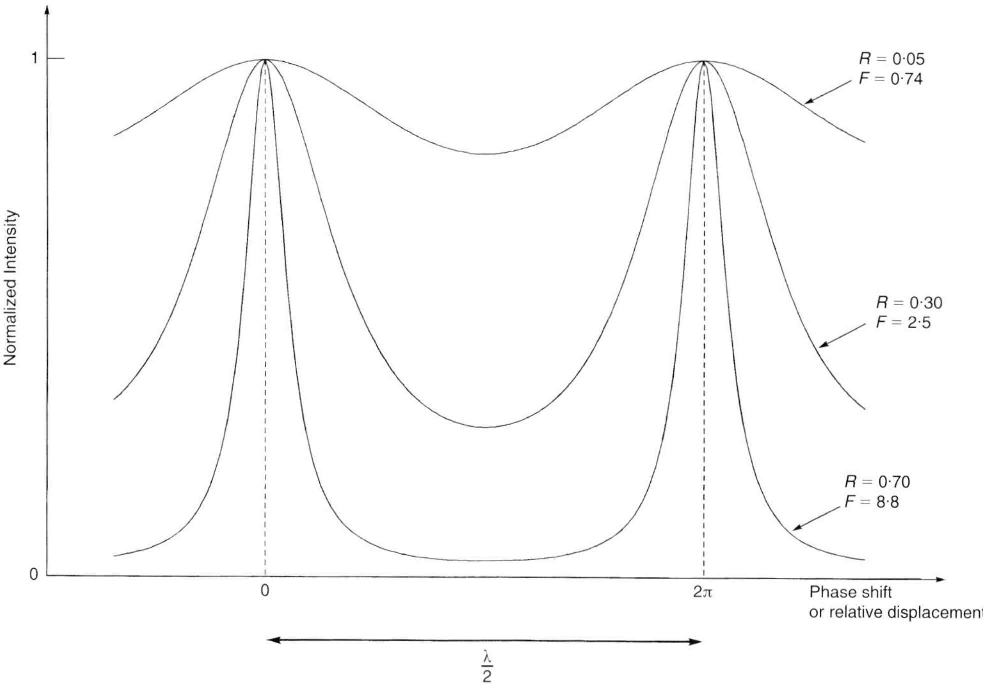

Fig. 2. Characteristic Fabry–Pérot transfer functions

on either side of a maximum where the intensity has decreased to half its maximum value. The ratio of the separation of adjacent peaks and the full width at half maximum is called the finesse (F) of the interferometer and is a function of the fraction of the incident light reflected by each mirror (R):

$$F = \frac{\pi\sqrt{R}}{1 - R} \tag{1}$$

The normalized intensity may be calculated, for any relative movement of the two mirrors, in terms of the finesse F and the phase shift ϕ:

$$\text{Normalized intensity} = \frac{\text{Output intensity}}{\text{Input intensity}} = \frac{1}{1 + \left(\frac{2F}{\pi}\right)^2 \sin^2\left(\frac{\phi}{2}\right)} \tag{2}$$

Figure 2 illustrates the fact that the relative mirror displacement required to induce a 2π phase shift (one peak to the next) is $\lambda/2$ and therefore only a function of the light wavelength. Both high- and low-finesse interferometers have displacement measurement applications. High-finesse interferometers have the advantage that the peaks can be identified with high accuracy and are therefore particularly appropriate to be used in association with nulling techniques, where an induced displacement is countered by returning the mirror to zero phase shift using some mechanical means such as a piezo actuator. Low-finesse interferometers are particularly suited to the technique of fringe counting of a digital signal, and can be used effectively for large-range displacement measurements (several millimetres). Fig. 3 shows the measured normalized output intensity of the interferometer developed by the authors, as well as the theoretical transfer function. The two mirrors forming the cavity were mounted on to an invar plate to minimize any temperature effects. One mirror was kept stationary while the other was displaced using a piezo actuator with a resolution of 0·004 μm.

The accuracy of measurements using fringe counting relies on the phase stability of the peaks and is therefore dependent on the accuracy with which the wavelength is known, as well as on any fluctuation of the wavelength. The wavelength of the light exiting the GRIN lens was measured by an optical spectrum analyser with an accuracy of ±0·0001 μm. The peak amplitude of the spectrum was at 0·6328 μm and the full width at half maximum was 0·0014 μm. Wavelength fluctuations of gas lasers occur if changes in temperature, barometric pressure or relative humidity occur. Collett & Hope (1983) reported that for helium–neon gas lasers, a change of one part per million of the wavelength will occur for a temperature change of 1°C, a pressure change of 0·3 kPa or a

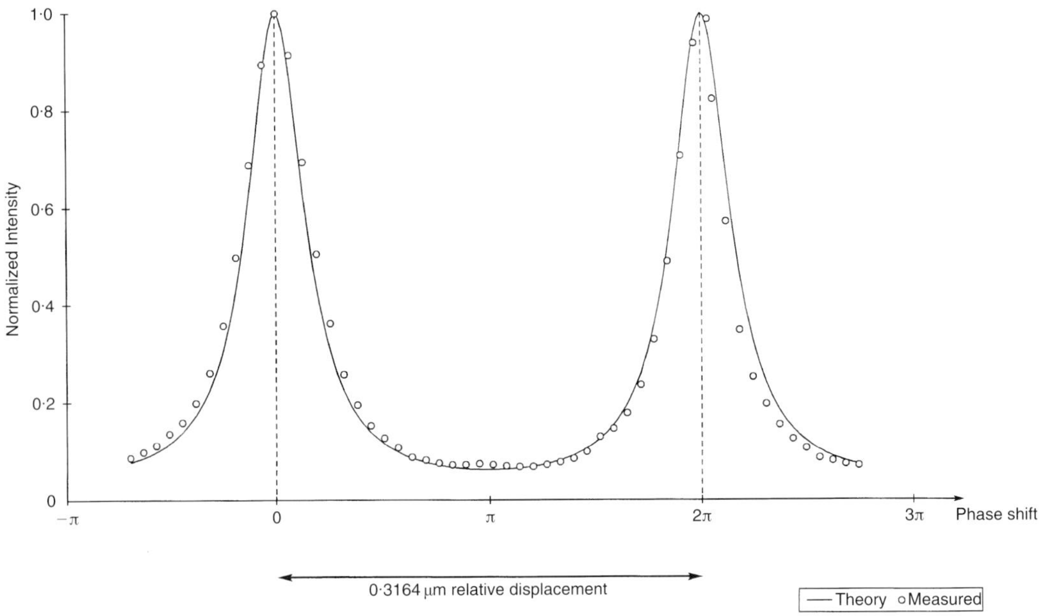

Fig. 3. Theoretical and measured interferometer output

change in relative humidity of 30%. When the interferometer was used to calibrate LVDTs, the procedure took only a few minutes and was carried out in a temperature-controlled laboratory. Changes in temperature, pressure and relative humidity were therefore within the limits which would cause a change of wavelength of one part per million.

Interpolation between peaks may introduce additional errors, as the Fabry–Pérot transfer function has to be known accurately. Effects which may introduce errors into the function, such as uncertainties in the mirror reflectance and fluctuation of laser output power, must be assessed in order to quantify the accuracy of the interferometer when used with interpolation techniques.

LINEAR TRANSDUCER CALIBRATION

Ideally a local displacement transducer should have high accuracy and a large measurement range. These two characteristics are generally in conflict, as higher accuracy normally results in a smaller measurement range. A convenient way of improving the performance of electronic linear transducers is to modify the signal conditioning. This often takes the form of increasing the amplification of the output signal, thereby increasing the output sensitivity. The level to which the output signal can be amplified is limited by the high-frequency electronic noise, as both the signal and the noise are amplified. High-frequency noise is induced by electromagnetic sources such as the amplifier circuit components and external electronic devices often found in the laboratory. This is in addition to low-frequency noise, which may be caused by temperature changes or amplifier drift. The high-frequency noise may be reduced by shielding or by using filtering techniques. It is important to note, however, that increasing the output sensitivity or reducing the noise does not improve the accuracy. Judgement on the accuracy can only be made after the instrument has been calibrated against a suitably accurate calibration device.

The performance of a commercial LVDT with a range of ±5 mm, used for triaxial local strain measurements, was evaluated. Two calibration ranges were used. Firstly, the LVDT was calibrated over the full range using a micrometer with a resolution of 2·54 μm. Secondly, in order to evaluate the limit of the LVDT's performance, it was calibrated over a smaller range using the interferometer. A 12 bit analogue to digital (A/D) converter which integrated the signal over a period of 30 ms was used in both cases. This type of converter is particularly suitable for typical soil mechanics test applications. The result of the calibration using the micrometer is shown in Fig. 4 as the measurement error with respect to a linear regression. The resolution of the A/D converter was 2·7 μm and the digital output signal was stable, indicating that the noise level was less than the resolution. If the error is assumed to represent a Gaussian distribution and the accuracy is taken as the 95% confidence level

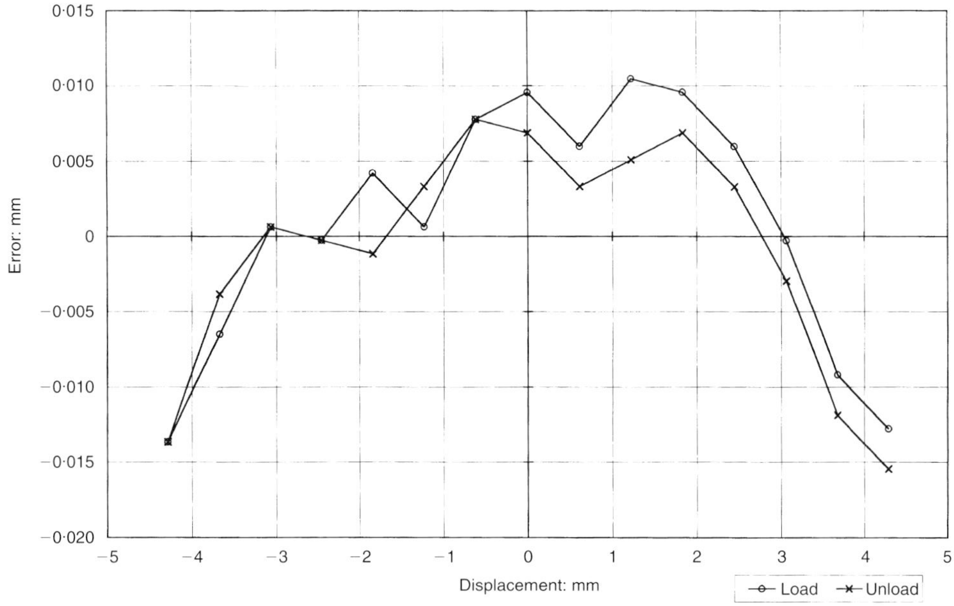

Fig. 4. LVDT calibration for ±5 mm range

(±2 standard deviations), the accuracy may be calculated as ±16 μm.

The arrangement for the LVDT calibration using the interferometer is shown in Fig. 5. Both the LVDT and interferometer were mounted on a translational stage used for routine calibration of linear displacement devices. The stage slides on precision linear bearings, and possible roll or pitch errors have been neglected. The micrometer handle was rotated by a geared motor at a rate of one revolution per hour. This was necessary, given the slow A/D converter. Hand rotation of the micrometer handle would be satisfactory if a high-speed converter were used. The amplification of the demodulated output signal was maximized while keeping the amplification of the high-frequency noise within acceptable levels. At this high amplification, the range of the LVDT was ±50 μm and the resolution of the A/D converter 0·012 μm. Fig. 6 shows the error between the measured displacement and a linear regression. For a 95% confidence level, the accuracy may be calculated as ±0·13 μm.

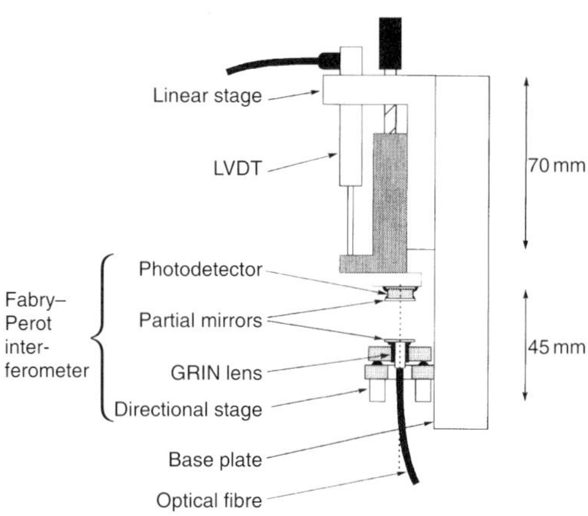

Fig. 5. Calibration apparatus

Comparison of the results from the two calibrations shows that the ratio of the accuracy of the LVDT to the measurement range is 0·32% for the large calibration range and 0·26% for the smaller calibration range. Additional calibrations were carried out for a number of ranges smaller than ±5 mm and in all cases the accuracy was between 0·2% and 0·5% of the range; this compares favourably with the manufacturer's specification of 0·5%. This small variation indicates that the LVDT is the limiting component in the system.

The above results show that a compromise is required between accuracy and measurement range. This may be addressed by using high amplification for small-strain measurements and reducing the amplification at larger strains. This strategy is acceptable for most testing programmes as the accuracy requirements are normally less stringent at larger strain levels. It may, however, present a problem for cyclic testing programmes where small stress excursions are undertaken at various levels of strain.

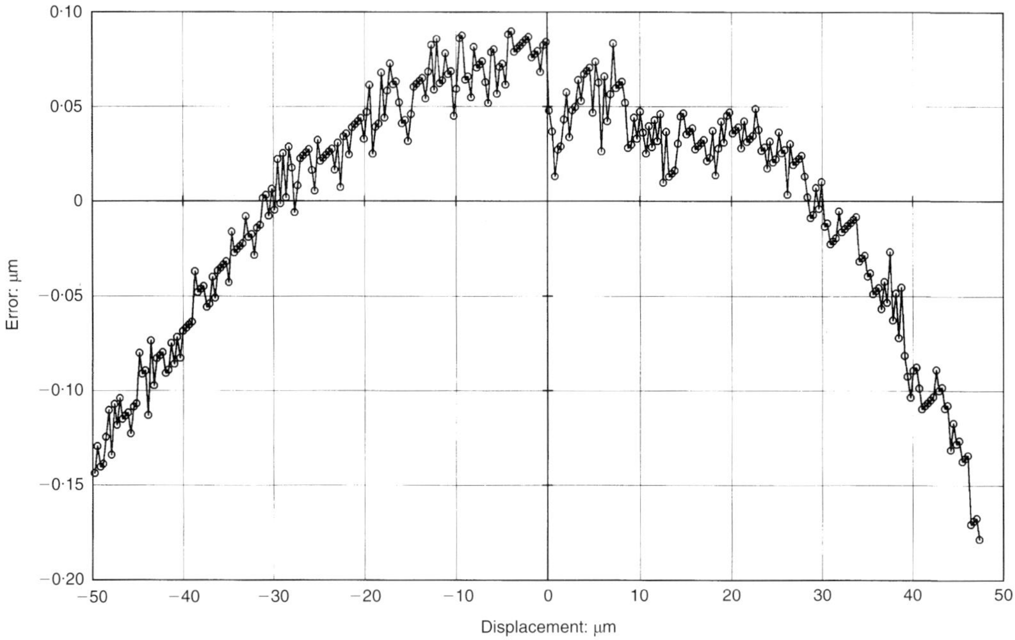

Fig. 6. LVDT calibration for ±50 µm range

DISCUSSION AND CONCLUSIONS

A Fabry–Pérot laser interferometer has been designed and constructed, and has been used with great effect to carry out calibrations of an LVDT intended for local strain measurement. This experience has demonstrated that the interferometer is sufficiently compact and robust to be used as a means of directly (i.e., without the use of LVDTs or Hall effect devices) measuring the ultra-small-strain stiffness behaviour of soil and weak rock specimens. Such testing is currently under way at the University of Surrey.

Many previously reported accuracies for local strain measurement devices have been better than the accuracy of the calibration system. This is unacceptable. The measurement of displacement with an interferometer depends only on the coherence of the light used, and therefore gas laser interferometers can provide a much higher accuracy than more conventional calibration systems.

Calibrations of the LVDT have shown that the system accuracy (including power supply stability, modulation/demodulation and analogue to digital conversion) is inversely proportional to the measurement range. Improved signal conditioning of electronic displacement transducers does not, of necessity, improve the accuracy of the system output. Accuracy can only be determined once the transducer system is calibrated against a suitably accurate reference system.

ACKNOWLEDGEMENT

The authors gratefully acknowledge the support received from the Engineering and Physical Sciences Research Council and the University of Pretoria.

REFERENCES

Born, M. & Wolf, E. (1965). *Principles of optics. Electromagnetic theory of propagation, interference and diffraction of light.* 3rd edn. London: Pergamon.

Burland, J. B. & Symes, M. (1982). A simple axial displacement gauge for use in the triaxial apparatus. *Géotechnique* **32**, No. 1, 62–65.

Clayton, C. R. I. & Khatrush, S. A. (1986). A new device for measuring local axial strains on triaxial specimens. *Géotechnique* **36**, No. 4, 593–597.

Clayton, C. R. I., Khatrush, S. A., Bica, A. V. D. & Siddique, A. (1989). The use of Hall effect semiconductors in geotechnical instrumentation. *Geotech. Testing J.* **12**, Mar., 69–76.

Collett, C. V. & Hope, A. D. (1983). *Engineering measurements.* Bath: Pitman.

Doebelin, E. O. (1983). *Measurement systems—application and design.* McGraw-Hill.

Fabry, C. & Pérot, A. (1897). Sur les franges des lames minces argentées et leur application à la mesure de petites epaisseurs d'air. *Annales de Chimie et de Physique* **12**, 459–501.

Goto, S., Tatsuoka, F., Shibuya, S., Kim, Y. & Sato, T. (1991). A simple gauge for local small strain measurements in the laboratory. *Soils Fdns* **31**, No. 1, 169–180.

Hird, C. C. & Yung, P. C. Y. (1989). The use of proximity transducers for local strain measurements in triaxial tests. *Geotech. Testing J.* **12**, No. 4, 292–296.

Jardine, R. J., Symes, M. J. & Burland, J. B. (1984). The measurement of soil stiffness in the triaxial apparatus. *Géotechnique* **34**, No. 3, 323–340.

Lo Presti, D. C. F., Pallara, O., Lancellotta, R., Armandi, M. & Maniscalco, R. (1993). Monotonic and cyclic loading behaviour of two sands at small strains. *Geotech. Testing J.* **16**, No. 4, 409–424.

Scholey, G. K., Frost, J. D., Lo Presti, D. C. F. & Jamiolkowski, M. (1995). A review of instrumentation for measuring small strains during triaxial testing of soil specimens. *Geotech. Testing J.* **18**, No. 2, June, 137–156.

Sydenham, P. H. (1982). Standardisation of measurement—fundamentals and practice. In *Handbook of measurement science* (ed. P. H. Sydenham), Vol. 1, pp. 49–94. Chichester: Wiley.

Sydenham, P. H., Hancock, N. H. & Thorn, R. (1989). *Introduction to measurement science and engineering.* Chichester: Wiley.

Tatsuoka, F. (1988). Some recent developments in triaxial testing systems for cohesionless soils. In *Advanced triaxial testing of soil and rock*, ASTM STP 977, pp. 7–67. American Society for Testing and Materials.

Zdravkovic, L. & Jardine, R. J. (1997). *Géotechnique* **47**, No. 3, 407–437

Some anisotropic stiffness characteristics of a silt under general stress conditions

L. ZDRAVKOVIC* and R. J. JARDINE*

Practical geotechnical problems invariably involve rotations of the principal stress axis direction and changes in the relative magnitude of the intermediate principal stress. To assess stiffness characteristics under such conditions requires a hollow cylinder apparatus (HCA) with independent controls and accurate measurements of the component stresses (σ_z, σ_r, σ_θ and $\tau_{z\theta}$) and strains (ε_z, ε_r, ε_θ and $\gamma_{z\theta}$). This paper describes how the anisotropy of a dense quartzitic silt was explored over the non-linear range by conducting tests on 254 mm diameter HCA samples, and on smaller triaxial specimens. It is shown that the stiffness characteristics depend strongly on the stress path direction and major principal stress axis inclination. K_0 consolidated samples show a particularly stiff response to torsional shear straining and the silt's compliance matrix is non-symmetrical and non-unique, even at small strains. The cross-coupling terms are different from those expected conventionally. For example, $\tau_{z\theta}$ stress changes can affect ε_z, ε_r and ε_θ strongly and the soil tends to compress vertically when subjected to either torsional shear straining or uniaxial lateral loading. The patterns of anisotropy are altered by overconsolidation.

KEYWORDS: anisotropy; stiffness; silt.

Dans la pratique, les problèmes géotechniques font invariablement intervenir la rotation de l'orientation des axes des contraintes principales et la variation de l'importance relative de la contrainte principale intermédiaire. L'évaluation des caractéristiques de rigidité dans de telles conditions exige l'emploi d'un appareil à cylindre creux à commandes indépendantes et la mesure précise des composantes de la contrainte (σ_z, σ_r, σ_θ et $\tau_{z\theta}$) et de la déformation (ε_z, ε_r, ε_θ et $\gamma_{z\theta}$). L'article décrit l'étude de l'anisotropie d'un limon quartzeux dense dans la gamme non linéaire à l'aide d'essais effectués sur des cylindres creux d'un diamètre de 254 mm et sur de plus petites éprouvettes de compression triaxiale. Les résultats montrent que les caractéristiques de rigidité sont fortement influencées par la direction des parcours de contrainte et par l'inclinaison prédominante des axes des contraintes principales. Les éprouvettes de sol consolidé K_0 manifestent une réponse particulièrement rigide à la déformation de cisaillement en torsion, et la matrice d'élasticité du limon est asymétrique et non unique, même à de petites contraintes. Les couplages réciproques diffèrent de ceux auxquels on pourrait normalement s'attendre. A titre d'exemple, des changements de la contrainte $\tau_{z\theta}$ peuvent fortement influencer ε_z, ε_r et ε_θ, et le sol tend à se comprimer verticalement quand il est soumis à une déformation de cisaillement en torsion ou à une charge latérale uniaxiale. Les modes d'anisotropie sont modifiés par la préconsolidation.

INTRODUCTION

Soil's particulate nature leads to predominantly non-linear and inelastic stress–strain behaviour. Strength and stiffness are also usually anisotropic because of: non-spherical particle shapes; directionally dependent particle contact sets developed at the time of formation; and the usually anisotropic subsequent stress history.

Recent research which allowed new insights into geomaterial stiffness is described by King (1970), Tatsuoka & Shibuya (1991), Jardine (1995), Lo Presti (1995), Belloti *et al* (1996) and others. Non-linearity has been studied extensively in locally instrumented triaxial tests and anisotropy has been investigated, usually at very small strains, using geophysical techniques. But few studies have been made, under general stress conditions, of anisotropy over the non-linear range. It remains unclear how these facets of behaviour can be understood or modelled.

Hight & Higgins (1995) and Jardine (1995)

Manuscript received 30 October 1996; revised manuscript accepted 18 April 1997.
Discussion on this paper closes 1 November 1997.
* Imperial College, London.

emphasized that practical geotechnical problems invariably lead to: (i) a highly non-linear response involving a wide range in strain level and (ii) markedly anisotropic stress changes, involving rotations (α) of the σ_1 axis direction to values between 0° and 90°, and changes in the relative magnitude of σ_2 (leading to the parameter b varying between 0 and 1·0). Triaxial tests, where α can only be 0° or 90° and $b = 0$ or 1·0, cannot model these conditions. Conventional direct, simple and torsional shear testing changes b and rotates the σ_1 axis, but special measures are required to monitor the stress conditions and measure the strain response accurately. A hollow cylinder apparatus (HCA) can offer such an opportunity, provided it is capable of: achieving acceptably uniform distributions of the stresses defined in Fig. 1(a); fully independent controls over, and measurements of, σ_z, σ_r, σ_θ and $\tau_{z\theta}$; and accurate local measurements of the strains ε_z, ε_r, ε_θ and $\gamma_{z\theta}$.

The large Imperial College (IC) HCA, which tests 254 mm high samples with an outside diameter of 254 mm and 25·4 mm wall thickness, meets these criteria (Hight *et al* 1983). By applying independent inner and outer cell pressures (p_i and p_o), a vertical load (W) and a torque (M_T), as shown in Fig. 1(a), this HCA allows full control of the component stresses σ_z, σ_r, σ_θ and $\tau_{z\theta}$ and hence α and b; local strain instruments measure all of the component strains ε_z, ε_r, ε_θ and $\gamma_{z\theta}$. Zdravkovic (1996) has recently improved the resolution and stability of the local strain sensors, as summarized in Appendix 1, and developed a new computerized control system.

This paper presents data from experiments, performed in the IC HCA and a stress-path triaxial cell, which explored the non-linear stiffness anisotropy of a dense artificial silt consolidated under K_0 conditions. Following Hight *et al* (1983), the stress states developed in the HCA tests are described by four independent variables, t, p', b and α, as defined in the Notation. The invariant deviatoric stress measure J may be used as an alternative to t, which signifies the largest deviatoric stress developed on any plane. The relationship between J and t depends only on b; they are identical when $b = 0{\cdot}5$ and $J = 2t/\sqrt{3}$ when $b = 0$.

Stiffness parameters

Tatsuoka & Shibuya (1991) showed that the stiffness of reconstituted granular soils is unlikely to be linear or elastic once the imposed strains exceed $\approx$ 0·001%. Jardine (1992, 1995) described the subsequent response in terms of successive yielding events, which were termed Y_1 to Y_4. Following the end of the elastic range ('Y_1 yielding') larger strain cycles show clearly hysteretic (rather than non-linear elastic) behaviour, with energy being consumed at the particle contacts. Plastic strains start to develop, or become markedly more pronounced, once a second relatively-low threshold strain is exceeded in a process termed Y_2 yielding, which probably marks the onset of particle slippage. The Y_3 and Y_4 events are associated with yielding at larger strains and are not the main subject of this paper. Following Jardine (1985), Al-Tabbaa & Wood (1989), Atkinson *et al* (1990), Jardine *et al* (1991) and Simpson (1992), the small strain behaviour of such soils is likely to be best described in terms of multiple yield surfaces, some of which move with the current stress point and develop as the stress state changes. Nevertheless, elastic behaviour is commonly assumed at very small strains and it is convenient when reporting behaviour at larger strains to retain some elastic stiffness terms to label specific directional (albeit non-linear and non-elastic) stiffness quantities.

When considering the response of an elastic material to general stress changes, one way of characterizing stiffness is to compute octahedral shear, G_{oct}, and bulk, K', stiffnesses at given deviatoric and volumetric strains, where:

$$K' = \frac{\Delta p'}{\Delta\varepsilon_{vol}} \tag{1}$$

$$G_{oct} = \frac{\Delta J}{\Delta\varepsilon_d} \tag{2}$$

While J and ε_d are usually defined in terms of principal stresses and strains, ΔJ and $\Delta\varepsilon_d$ are more conveniently calculated* from HCA measurements as:

$$\Delta\varepsilon_d = \frac{1}{\sqrt{3}}\sqrt{\begin{aligned}&4(\Delta\varepsilon_z^2 + \Delta\varepsilon_r^2 + \Delta\varepsilon_\theta^2) + 3\Delta\gamma_{z\theta}^2 - \\ &- 4(\Delta\varepsilon_z\Delta\varepsilon_\theta + \Delta\varepsilon_z\Delta\varepsilon_r + \Delta\varepsilon_r\Delta\varepsilon_\theta)\end{aligned}}$$

$$\Delta J = \frac{1}{\sqrt{3}}\sqrt{\begin{aligned}&\Delta\sigma_z^2 + \Delta\sigma_r^2 + \Delta\sigma_\theta^2 + 3\Delta\tau_{z\theta}^2 - \\ &- \Delta\sigma_z\Delta\sigma_\theta - \Delta\sigma_z\Delta\sigma_r - \Delta\sigma_r\Delta\sigma_\theta\end{aligned}}$$

If the material's stiffness response is stress-path dependent and non-linear, then different types of test performed on identical specimens will give systematic sets of $K' - \Delta\varepsilon_v$ and $G_{oct} - \Delta\varepsilon_d$ curves,

* Note that ΔJ and ε_d must be evaluated by substituting the changes in stress and strain components defined with respect to local origins. So, for example, ΔJ can be positive in tests where σ_1, σ_2, and σ_3 change, even if the overall value of J remains constant. This is possible as J is essentially a scalar distance from the origin of stress space, whereas ΔJ is a scalar distance from the current stress state.

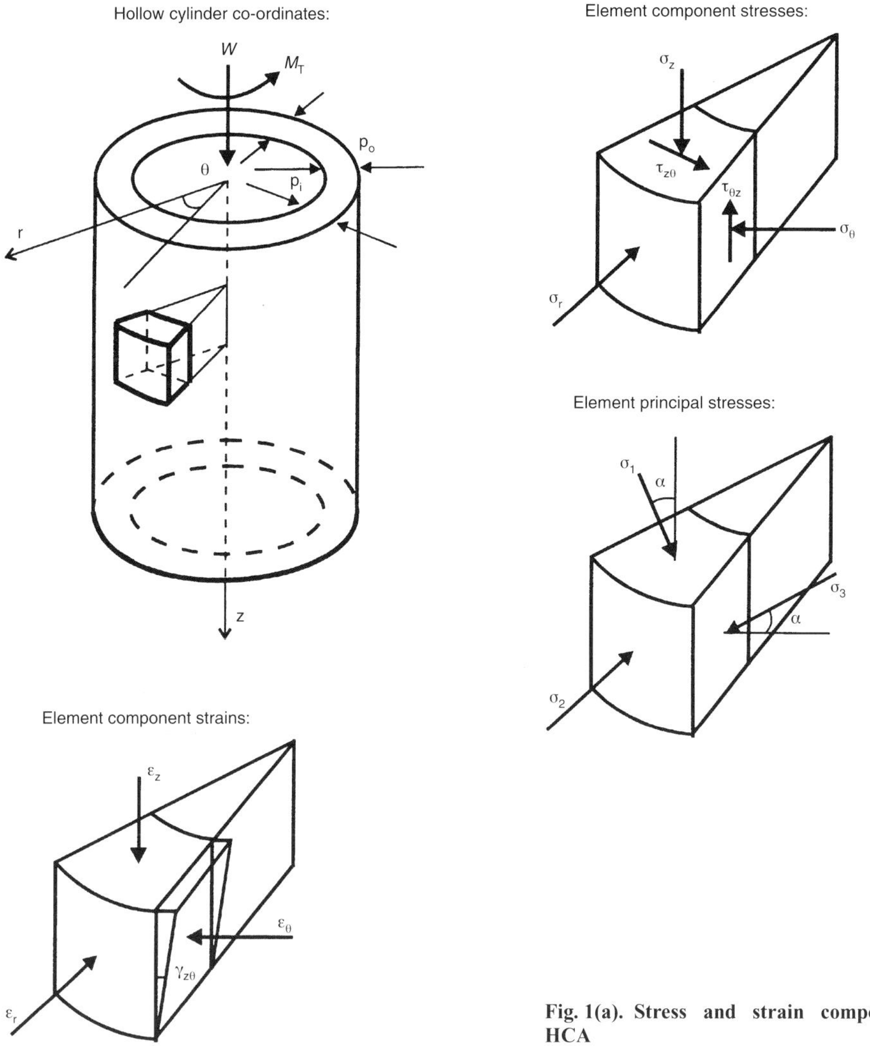

Fig. 1(a). Stress and strain components within the HCA

as proved for anisotropic Bothkennar and Queenborough clays by triaxial tests directed from a K_0 stress state at various inclinations ($\tan\theta = 2\,dt/dp'$) by Smith (1992) and Smith *et al* (1992). HCA testing is made more complex by the two additional independent variables (α and b), making it necessary to choose how t and p' should vary with b and α.

With a generally anisotropic elastic medium an alternative approach is to describe stiffness anisotropy by the compliance matrix defined in equation (3), which has 36 terms. Note that the unique compliance coefficients apply equally to loading and unloading stress increments and that cross-coupling compliance terms may exist between any pair of stress and strain components.

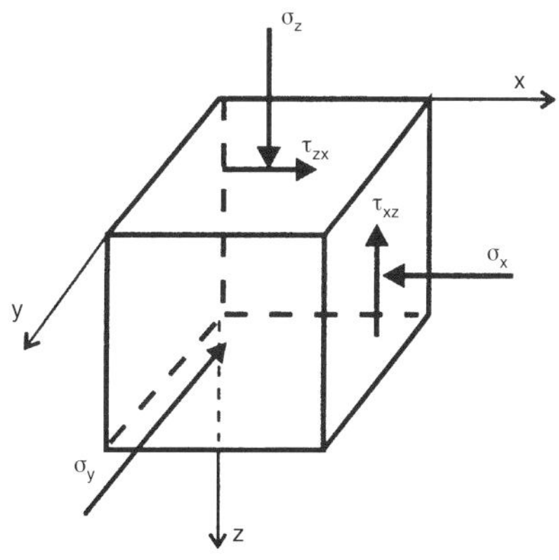

Fig. 1(b). Cartesian coordinates

$$\begin{pmatrix} \Delta\varepsilon_x \\ \Delta\varepsilon_y \\ \Delta\varepsilon_z \\ \Delta\gamma_{xz} \\ \Delta\gamma_{yz} \\ \Delta\gamma_{xy} \end{pmatrix} = \begin{pmatrix} C_{11} & C_{12} & C_{13} & C_{14} & C_{15} & C_{16} \\ C_{21} & C_{22} & C_{23} & C_{24} & C_{25} & C_{26} \\ C_{31} & C_{32} & C_{33} & C_{34} & C_{35} & C_{36} \\ C_{41} & C_{42} & C_{43} & C_{44} & C_{45} & C_{46} \\ C_{51} & C_{52} & C_{53} & C_{54} & C_{55} & C_{56} \\ C_{61} & C_{62} & C_{63} & C_{64} & C_{65} & C_{66} \end{pmatrix} \begin{pmatrix} \Delta\sigma_x \\ \Delta\sigma_y \\ \Delta\sigma_z \\ \Delta\tau_{xz} \\ \Delta\tau_{yz} \\ \Delta\tau_{xy} \end{pmatrix} \tag{3}$$

Because many soils have horizontal bedding, their initial structure is often assumed to give stiffness properties that do not vary in the horizontal plane. Considering the case where axes x and y denote two orthogonal principal stress directions in the isotropic (horizontal plane), and axis z denotes the vertical as shown in Fig. 1(b), equation (3) reduces to give a compliance matrix with 16 terms. Conventionally, six of the cross-coupling terms are taken to be zero, as shown in equation (4).

$$\begin{pmatrix} \Delta\varepsilon_z \\ \Delta\varepsilon_x \\ \Delta\varepsilon_y \\ \Delta\gamma_{zx} \end{pmatrix} = \begin{pmatrix} \frac{1}{E'_z} & -\frac{\nu'_{xz}}{E'_x} & -\frac{\nu'_{yz}}{E'_x} & 0 \\ -\frac{\nu'_{zx}}{E'_z} & \frac{1}{E'_x} & -\frac{\nu'_{yx}}{E'_y} & 0 \\ -\frac{\nu'_{zy}}{E'_z} & -\frac{\nu'_{xy}}{E'_x} & \frac{1}{E'_y} & 0 \\ 0 & 0 & 0 & \frac{1}{G'_{zx}} \end{pmatrix} \begin{pmatrix} \Delta\sigma'_z \\ \Delta\sigma'_x \\ \Delta\sigma'_y \\ \Delta\tau_{zx} \end{pmatrix} \tag{4}$$

Love (1927) and Pickering (1970) consider that the matrix should be symmetrical, leading to equation (5), which reduces to five the number of independent constants. Usually these are the drained Young's modulus in the vertical direction, E'_z ($= E'_v$); that in the horizontal direction, E'_x ($= E'_y = E'_h$); the Poisson's ratio linking the strains in each horizontal direction to a vertical direct strain, ν'_{zx} ($= \nu'_{zy} = \nu'_{vh}$); the Poisson's ratio for a strain in one horizontal direction due to a horizontal direct strain imposed at 90°, ν'_{xy} ($= \nu'_{yx} = \nu'_{hh}$); and the modulus of shear deformation in the vertical plane, G_{zx} ($= G_{vh}$). The Poisson's ratio for vertical strains due to a horizontal direct strain is usually calculated from equation (5).

$$\nu'_{hv} = \frac{E'_h}{E'_v}\nu'_{vh} \tag{5}$$

G_{hh}, the shear modulus relating shear strains developed in the horizontal plane to shear stresses imposed in the same plane is given by equation (6):

$$G_{hh} = \frac{E'_h}{2(1+\nu'_{hh})} \tag{6}$$

When analysing axially symmetric bodies such as an HCA sample, it is more convenient to use cylindrical co-ordinates (equation (7)) in which the z, r and θ subscripts are substituted for z, x and y respectively. As before, the 16-term compliance matrix is usually reduced to 10 coefficients by assuming that there are no cross-coupling terms between the normal strains and $\Delta\tau_{z\theta}$ (i.e. $C_{14} = C_{24} = C_{34} = 0$). If the soil is transversely isotropic then the coefficients for r and θ may be interchanged, reducing the matrix to six terms. Assuming symmetry leads to $\nu'_{rz} = (E'_r/E'_z)\nu'_{zr}$ from equation (5) and leaves only five independent parameters.

$$\begin{pmatrix} \Delta\varepsilon_z \\ \Delta\varepsilon_r \\ \Delta\varepsilon_\theta \\ \Delta\gamma_{z\theta} \end{pmatrix} = \begin{pmatrix} \frac{1}{E'_z} & -\frac{\nu'_{rz}}{E'_r} & -\frac{\nu'_{\theta z}}{E'_\theta} & 0 \\ -\frac{\nu'_{zr}}{E'_z} & \frac{1}{E'_r} & -\frac{\nu'_{\theta r}}{E'_\theta} & 0 \\ -\frac{\nu'_{z\theta}}{E'_z} & -\frac{\nu'_{r\theta}}{E'_r} & \frac{1}{E'_\theta} & 0 \\ 0 & 0 & 0 & \frac{1}{G'_{z\theta}} \end{pmatrix} \begin{pmatrix} \Delta\sigma'_z \\ \Delta\sigma'_r \\ \Delta\sigma'_\theta \\ \Delta\tau_{z\theta} \end{pmatrix} \tag{7}$$

Hardin (1978), and Hardin & Blandford (1989) have shown that with soils the anisotropic compliance terms depend on the current, anisotropic, effective stress components. Transverse isotropy is only likely for horizontally-bedded soil in which: the horizontal stresses (σ_x and σ_y, or σ_r and σ_θ) have always remained practically equal; and σ'_1 has always acted vertically or horizontally. Hollow cylinder tests typically transcend these limitations, as do the stress conditions developed in most practical situations. Tests which involve $\Delta\sigma'_r \neq \Delta\sigma'_\theta$ are likely to develop anisotropy in the horizontal plane; this anisotropy is axially symmetric, but it does not coincide with the orthogonal cartesian co-ordinates (see Fig. 1(b)).

It is also difficult to apply the cross-anisotropic elastic approach to cover non-linear soil behaviour. Tests in which only one stress component is varied can show how the direct compliance, C_{ij} ($i = j$), and relevant cross-coupling, C_{ij} $(i \neq j)$, or 'Poisson's ratio' terms vary with the direct strain component, or with ε_d and ε_{vol}. But they provide little information as to how the other direct compliance terms are affected by the stress changes experienced. Combined monotonic static and geophysical, or unload–reload, tests which measure the very small strain stiffness values during larger stress excursions, show that the mutual ratios of the current tangent compliance terms ($G_{z\theta}/E'_z$ etc.) and load-to-unload stiffnesses change continuously; see for example Tatsuoka & Shibuya (1991) or Porovic (1995). Clearly, the compliance matrix becomes non-unique and any particular compliance matrix can only be associated with a fixed initial effective stress point (and soil state) and apply only to very small monotonic strain excursions from that point.

Undrained tests

Equation (7) can be manipulated to relate drained and undrained behaviour. In uniaxial undrained triaxial tests, the axial strain predicted from equation (7) must match that expected from the conventional undrained stiffness expression, $\Delta\varepsilon_z = \Delta\sigma_z/E_u$, leading to:

$$\frac{\Delta\sigma_z}{E_u} = \Delta\varepsilon_z = \frac{1}{E'_z}\Delta\sigma'_z - 2\frac{\nu'_{rz}}{E'_r}\Delta\sigma'_r \tag{8}$$

If the undrained pore pressure changes developed under triaxial conditions are characterized in terms of Skempton's parameter A, with $B = 1{\cdot}0$, then manipulation of equation (8) leads to:

$$\frac{E'_z}{E_u} = \frac{E'_z}{E'_r}2A\nu'_{rz} + (1 - A) \tag{9}$$

Anisotropically consolidated soil invariably develops different E_u values over the non-linear range (at equal strains) in compression and extension tests; see for example Jardine (1995).

In simple undrained triaxial compression σ_z increases while σ_r remains constant. Assuming that $\Delta u/\Delta\sigma_z$ $(= A)$ is positive but less than 1·0, then σ'_z increases while σ'_r falls. In this sense the vertical effective stiffness, E'_z, response matches that for loading, while the radial effective stiffness, E'_r, response corresponds to that for unloading and equation (9) can be rewritten as:

$$\frac{E'^{\,load}_z}{E^{load}_u} = \frac{E'^{\,load}_z}{E'^{\,unld}_r}2A\nu'_{rz} + (1 - A) \tag{10}$$

where $E^{load}_u = E_u$ (compression).

In the same way simple undrained triaxial extension gives rise to a fall in σ'_z (unloading) and an increase in σ'_r (loading), provided $0 < \Delta u/\Delta\sigma_z < 1{\cdot}0$. Equation (9) can then be rewritten for triaxial extension conditions, as:

$$\frac{E'^{\,unld}_z}{E^{unld}_u} = \frac{E'^{\,unld}_z}{E'^{\,load}_r}2A\nu'_{rz} + (1 - A) \tag{11}$$

where $E^{unld}_u = E_u$ (extension).

As noted before, equation (7) assumes that torsional shear tests should cause no change in the three effective normal stresses or strains, and so the undrained and drained values of $G_{z\theta}$ should be equal.

When interpreting undrained tests from locally instrumented HCA or triaxial samples it is important to recognize that, while water is not crossing the sample boundaries, local volume changes can take place. Numerical analyses by Zdravkovic & Potts (1997) show that end restraint and other effects lead to internal pore water redistribution. It was estimated theoretically, and proven experimentally, that compressive volume changes of up to $0{\cdot}1\varepsilon_1$ can develop in the centres of IC-HCA samples during the initial (contractant) stages of undrained tests on HPF4 silt. Such volume changes generally become less significant as strains grow. Equations (8) to (11) remain valid despite the local drainage, as constant volume conditions are not assumed. It is only necessary to measure the effective stresses correctly, with pore pressures being fully equalized within the sample.

As shown later, equations (10) and (11) allow undrained test data to be used to compute and compare the ranges of E'_z and E'_r values developed during loading and unloading events.

DESCRIPTION OF TESTING PROGRAMME

The silt tested was an artificial, quartzitic rock flour termed HPF4 which has elongated and angular particles and the index properties summarized in Table 1. Further details are given by Zdravkovic (1996), who also describes how dense samples were formed in the HCA by pluviating a slurried silt-water mixture through water-filled, membrane-lined, double moulds. The samples had initial void ratios of $0{\cdot}65 \pm 0{\cdot}03$, corresponding to a relative density of 88%, and were saturated using elevated back pressures. Similar techniques were used to form 38 mm diameter triaxial samples for tests in a Bishop and Wesley stress path cell.

The tests considered here were performed on samples consolidated under K_0 conditions to OCR = 1·0 or 1·3, before being sheared following a range of undrained and drained effective stress paths, as detailed in Table 2. The tests were all computer controlled with fine changes being applied in σ_z, σ_r, σ_θ and $\tau_{z\theta}$ to follow the desired stress paths. Frequent transducer readings were taken so that stress conditions could be updated smoothly, keeping approximately constant strain rates when required and taking account of changes

Table 1. Index properties of HPF4 silt

e_{min}	e_{max}	I_L: %	I_P: %	Specific gravity	Particle size: mm	Mineralogy
0·563	1·332	27	0	2·65	$d_{10}=0{\cdot}005$ $d_{50}=0{\cdot}04$ $d_{90}=0{\cdot}1$	100% quartz

Table 2. Summary of tests

Test code	Brief description	Values of b and α
TC1·0, TE1·0	Undrained triaxial tests in compression and extension, starting from A in Fig. 2, OCR = 1·0	$b=0$ and $\alpha=0$ for all tests, except extension experiments when $\sigma'_r > \sigma'_z$
TC1·3, TE1·3	As above, starting from B in Fig. 2, OCR = 1·3	As above
*Mb*stage	Drained HCA test at point A in Fig. 2; OCR = 1·0; t and p' constant	b changing from 0 to 0·5; $\alpha=0$
*Rα*stage	As above	$b=0$; α changing from 0° to 90°
M0	Undrained HCA test from OCR = 1·0, t increasing to failure from A	$b=0{\cdot}5$; $\alpha=0$
$M0_{b=0{\cdot}3}$	As M0	$b=0{\cdot}3$; $\alpha=0$
M90	As M0, but in extension	$b=0{\cdot}5$; α jump rotating from 0° to 90° when $\sigma'_\theta > \sigma'_z$
D0	Uniaxial drained HCA test, with σ'_z increasing from B in Fig. 2, OCR = 1·3	$b=0$; $\alpha=0$
E0	As D0, but with σ'_θ increasing from B	b increasing from 0 to 1·0, then reducing; α jump rotating from 0° to 90° when $\sigma'_\theta > \sigma'_z$
G0	As D0, but with $\tau_{z\theta}$ increasing from B and with undrained conditions	b increasing from 0 to 0·3; α rotating from 0° to 32°
QA0	Undrained HCA constant t test starting from B in Fig. 2	$\Delta b=0{\cdot}5$; α increasing with $dt/d\alpha = 0$ kPa/°
QA3·5	As QA0, but with t increasing with α	$\Delta b=0{\cdot}5$; α increasing with $dt/d\alpha = 3{\cdot}5$ kPa/°
QA−3·5	As QA0, but with t decreasing as α increases	$\Delta b=0{\cdot}5$; α increasing with $dt/d\alpha = -3{\cdot}5$ kPa/°

in sample geometry. Where ageing periods are referred to, these extended for 12–24 h, allowing the ε_1 creep rates to fall to 0·05%/day or less. All perturbing stages were controlled to run at a deviatoric strain rate $d\varepsilon_d/dt \approx 5\%$/day, which was sufficient to ensure fully equalized pore pressures. Zdravkovic (1996) provides more experimental details and describes further experiments which involved more complex stress histories.

Consolidation paths

The HCA and triaxial tests described in Table 2 all started with compression from slurry following the path shown in Fig. 2 to point A ($t = 75$ kPa, $p' = 200$ kPa, $K_0 = 0{\cdot}5$), where the samples were allowed to age. The overconsolidated samples were swelled to point B (OCR = 1·3, $t = 50$ kPa, $p' = 167$ kPa, $K_0 = 0{\cdot}57$) as shown in Fig. 2, and aged again before shearing. Fig. 3 illustrates the typical $e - \log\sigma'_v$ characteristics over the same stress range; note the substantial creep experienced at A and B.

Triaxial tests

Undrained compression and extension triaxial tests were performed from both OCRs. As shown later, the silt showed: strong anisotropy, both contractant and dilatant phases of behaviour, and steeply non-linear stiffness characteristics.

HCA tests at OCR = 1

Stiffness parameters measured at OCR = 1 are often uncharacteristic of behaviour at higher OCR's (Jardine *et al* 1984; Jardine, 1995) and Zdravkovic's M and R series tests were not intended, primarily, to study stiffness. Nevertheless, useful and interesting data will be presented from five types of HCA tests on normally consolidated samples:

(*a*) The tests termed *Mb*stage involved changing the three normal effective stresses as shown in Fig. 4 under drained conditions; b gradually increased from 0·0 to 0·5, while t, p' and α remained constant. Ageing was then allowed.

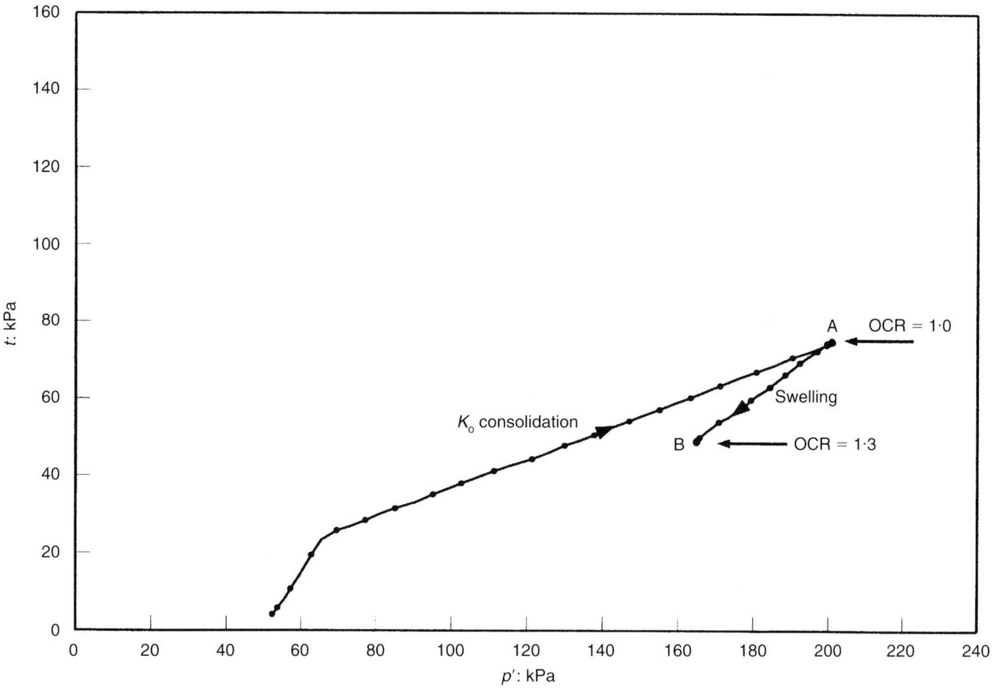

Fig. 2. K_0 consolidation paths for OCR = 1·0 and 1·3

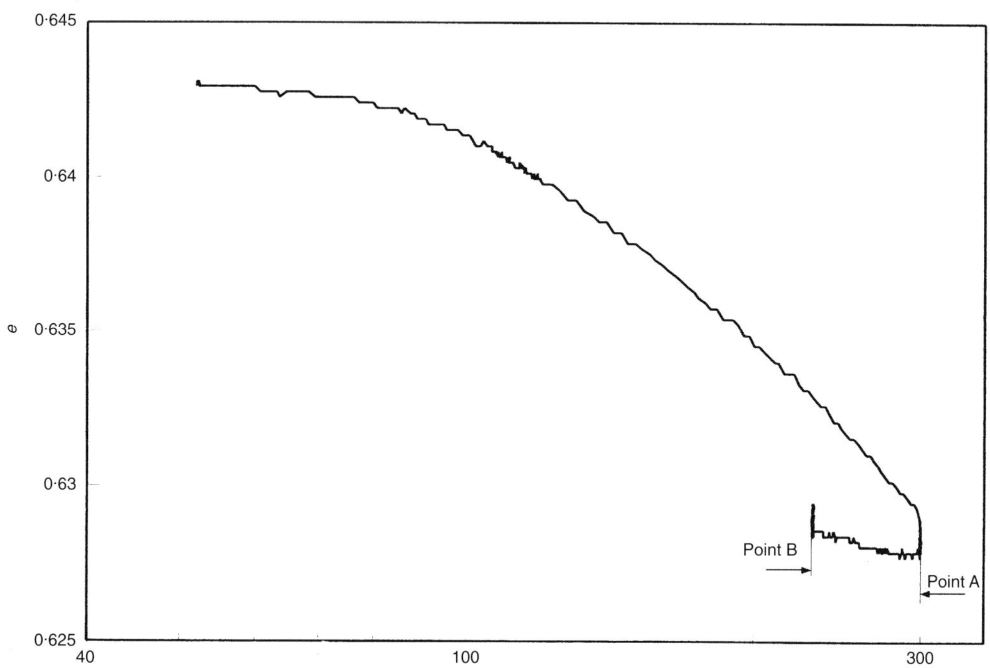

Fig. 3. Non-linearity of K_0 compression curve in the e–log σ'_v plane

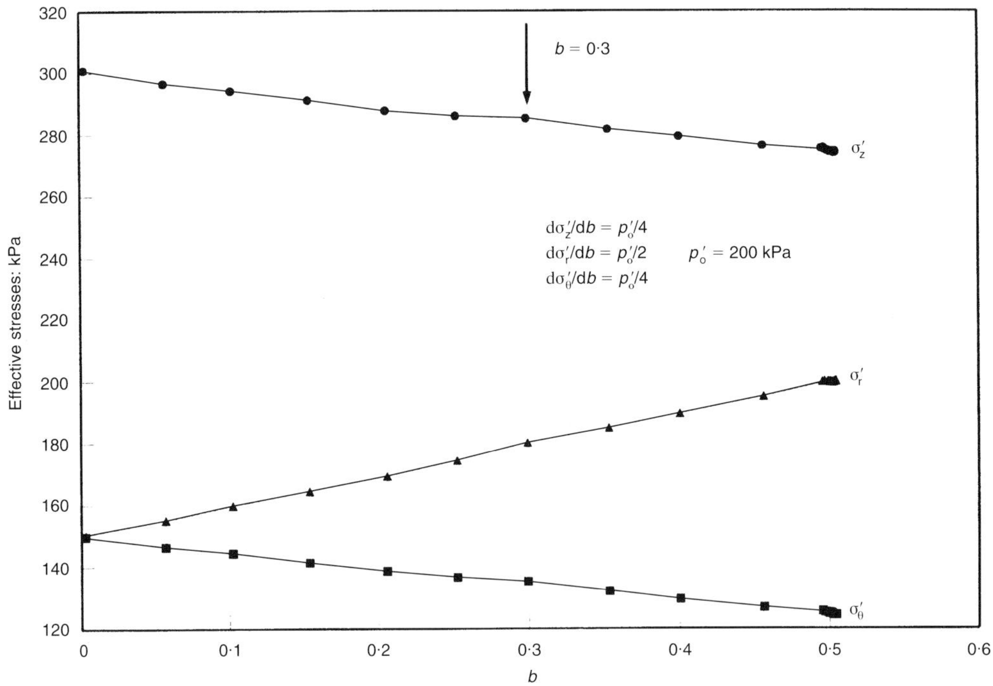

Fig. 4. Normal stress changes during *Mb*stage tests

(*b*) The tests termed *Rα*stage consisted of changing the stresses under drained conditions as shown in Fig. 5, so that α gradually rotated from zero to 90°, while keeping the same t, p' and b values. Ageing was allowed after reaching the end of σ_1 axis rotation.

The remaining three types of test were all from the 'M-series' and involved undrained shearing following drained '*b*-stages'. The mean total stress, p, was held constant, but pore pressures could change during shearing.

(*c*) Test M0 involved increasing $(\sigma_z - \sigma_\theta)/2$ to produce an 'active' undrained failure, while keeping $\alpha = 0$ and $b = 0{\cdot}5$. To do this σ_z increased and σ_θ reduced by equal amounts $(d\sigma_z/d\sigma_\theta = -1)$, while σ_r was constant.

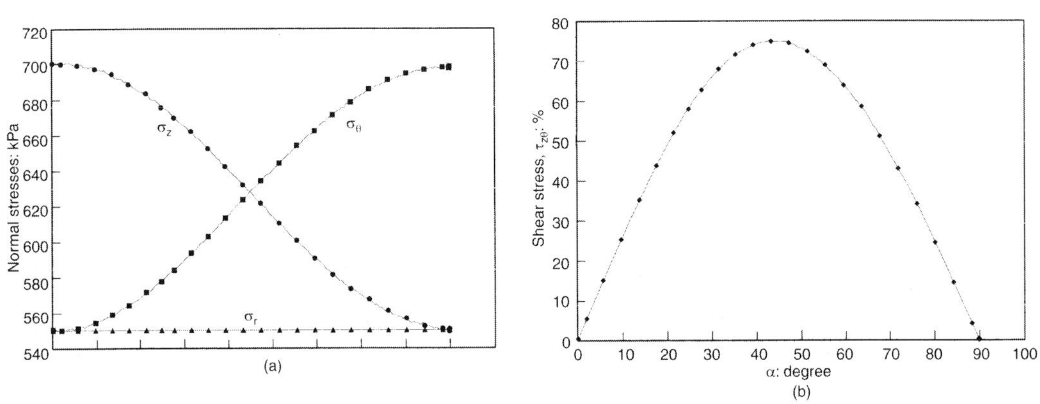

Fig. 5. Normal and shear stress changes during drained *Rα*stage tests

(*d*) Test $M0_{b0\cdot3}$ was identical to Test M0, except that the prior '*b*-stage' stopped at $b = 0{\cdot}3$ and the normal stresses were varied to keep $b = 0{\cdot}3$ throughout ageing and shearing.

(*e*) In Test M90 the stress controls were the same as in M0, except that σ_z was reduced, keeping $b = 0{\cdot}5$ and $d\sigma_z/d\sigma_\theta = -1$; $(\sigma_z - \sigma_\theta)/2$ fell continuously from point A in Fig. 2, to reach an eventual undrained passive failure with σ_1 acting circumferentially ($\alpha = 90°$).

The M-series tests involved $b > 0$ in order to model plane strain conditions, which normally lead to $0{\cdot}3 < b < 0{\cdot}5$ at failure. Comparisons with triaxial data could reveal how *b* affects stiffness, yielding and strength. Zdravkovic (1996) performed multiple *Mb*stage, *Rα*stage and M-series experiments involving similar undrained unloading stages to M90. The close match seen between the stiffnesses measured provided clear evidence of the HCA data's experimental repeatability. The effective stress paths for these tests are presented later in Fig. 25.

HCA tests at OCR = 1·3

The main purpose of the HCA experiments at OCR = 1·3 was to investigate stiffness anisotropy at small to moderate strains. The tests included:

(*a*) Those in which only one stress variable was changed, allowing the terms defined in equation (7) to be evaluated directly.

(*b*) Undrained experiments in which *t*, *b* and *α* could be varied together in different combinations; these tests match more closely the conditions developed in, for example, an undrained foundation or slope.

(a) HCA tests at OCR = 1·3 *involving uniaxial stress changes*

Two drained HCA uniaxial experiments, D0 and E0, and one undrained uniaxial experiment, G0, were performed on overconsolidated samples. Their effective stress paths are specified in Fig. 6.

(i) D0 consisted of loading σ'_z, while keeping σ'_r and σ'_θ constant, so that E'_z $(= E'_v)$ and ν'_{zr} $(= \nu'_{vh})$ could then be evaluated directly as functions of strain; see equation (7). The values of *α* and *b* remained zero throughout.

(ii) E0 involved increasing σ'_θ while keeping σ'_z and σ'_r constant, allowing E'_θ (initially equal to $E'_r = E'_h$), $\nu'_{\theta r}$ $(= \nu'_{r\theta} = \nu'_{hh})$ and $\nu'_{\theta z}$ (initially $= \nu'_{rz} = \nu'_{hv}$) to be determined as functions of strain; *α* remained 0° and *b* gradually increased from zero to 1·0 while σ'_θ remained less than σ'_z. Once $\sigma'_\theta > \sigma'_z$, *α*

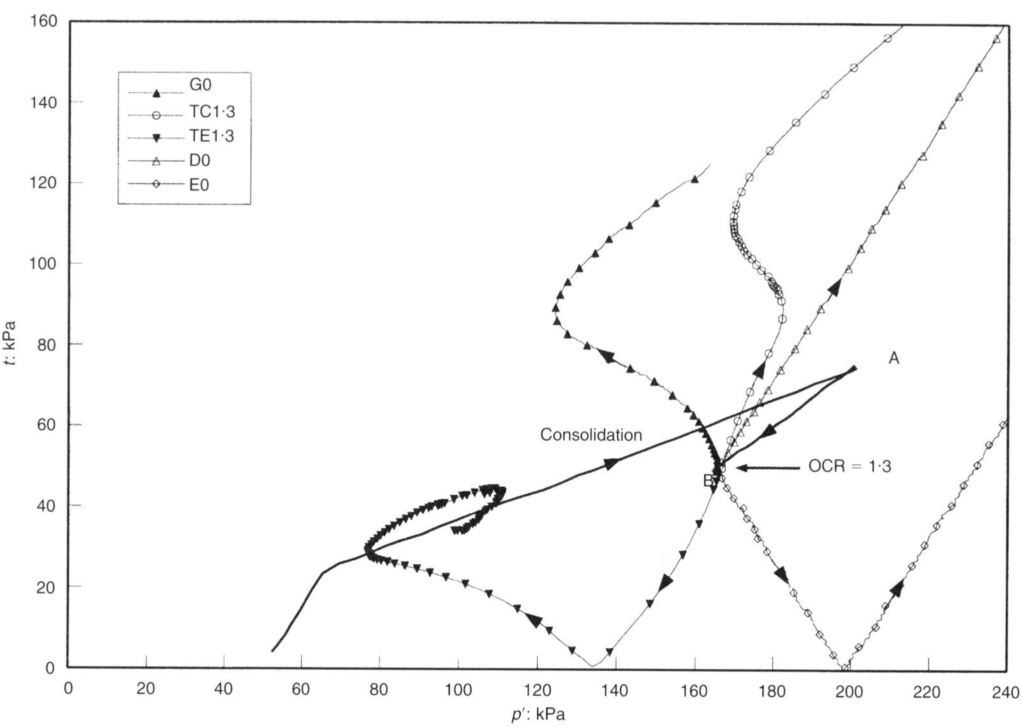

Fig. 6. Effective stress paths in t–p' plane from uniaxial tests at OCR = 1·3

jumped to 90° and b gradually decreased from 1·0 towards $\approx 0{\cdot}2$ at failure.

(iii) In Test G0, only increments of $\tau_{z\theta}$ were applied, showing how the parameter $G_{z\theta}$ ($= G_{vh}$) varied with shear strain; α increased monotonically, reaching 32° before developing an undrained failure, while the test conditions gave $b = \sin^2 \alpha$.

(b) HCA tests at OCR = 1·3 *involving polyaxial stress changes*

Three undrained HCA tests were performed on OCR = 1·3 samples in which changes in t were combined with incremental α rotations, such that the $\mathrm{d}t/\mathrm{d}\alpha$ ratio was kept constant during shearing. As with the M-series tests at OCR = 1, the aim was to mimic the conditions developed in practical plane strain problems. However, in this case σ_2 was varied incrementally to keep the changes in σ_2 halfway between those of σ_1 and σ_3. That is:

$$\Delta b = (\Delta\sigma_2 - \Delta\sigma_3)/(\Delta\sigma_1 - \Delta\sigma_3) = 0{\cdot}5$$

The above criterion led to $b \sim 0{\cdot}3$ at failure.

The effective stress paths followed during the experiments are shown on Figs 7 and 8; the second diagram also shows the anisotropic envelope of the Phase Transformation Points†, identified by Zdravkovic from her M-series experiments. The stress components applied to achieve the desired paths are plotted as functions of α in Figs 9–11.

(i) In QA0 $\mathrm{d}t/\mathrm{d}\alpha$ was zero, keeping the same three-dimensional (total stress) Mohr circle while rotating the pole monotonically about the circle's origin. While the early stages of QA0 and G0 (described above) involved similar stress changes, the tests' effective paths diverged once $\alpha > 10°$ (see Fig. 8).

(ii) QA3·5 involved increasing t while rotating α to keep $\mathrm{d}t/\mathrm{d}\alpha = 3{\cdot}5$ kPa/°, giving a stress path falling between QA0 (or G0) and D0 or TC1·3.

(iii) QA–3·5 was the mirror image of QA3·5: t decreased while holding $\mathrm{d}t/\mathrm{d}\alpha$ at $-3{\cdot}5$ kPa/°. The stress path followed fell between QA0 (or G0) and E0 or TE1·3.

† Phase Transformation being the effective stress point where the soil's undrained response changes from being contractant to dilatant, with $\mathrm{d}p'/\mathrm{d}\varepsilon_d = 0$.

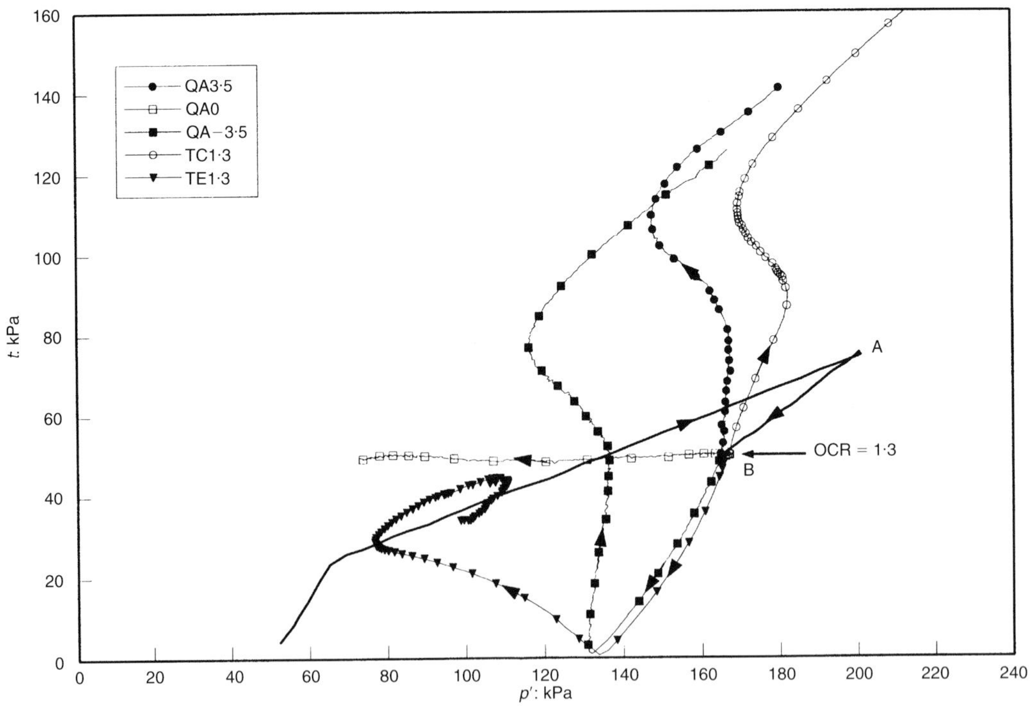

Fig. 7. Effective stress paths in t–p' plane from polyaxial tests at OCR = 1·3

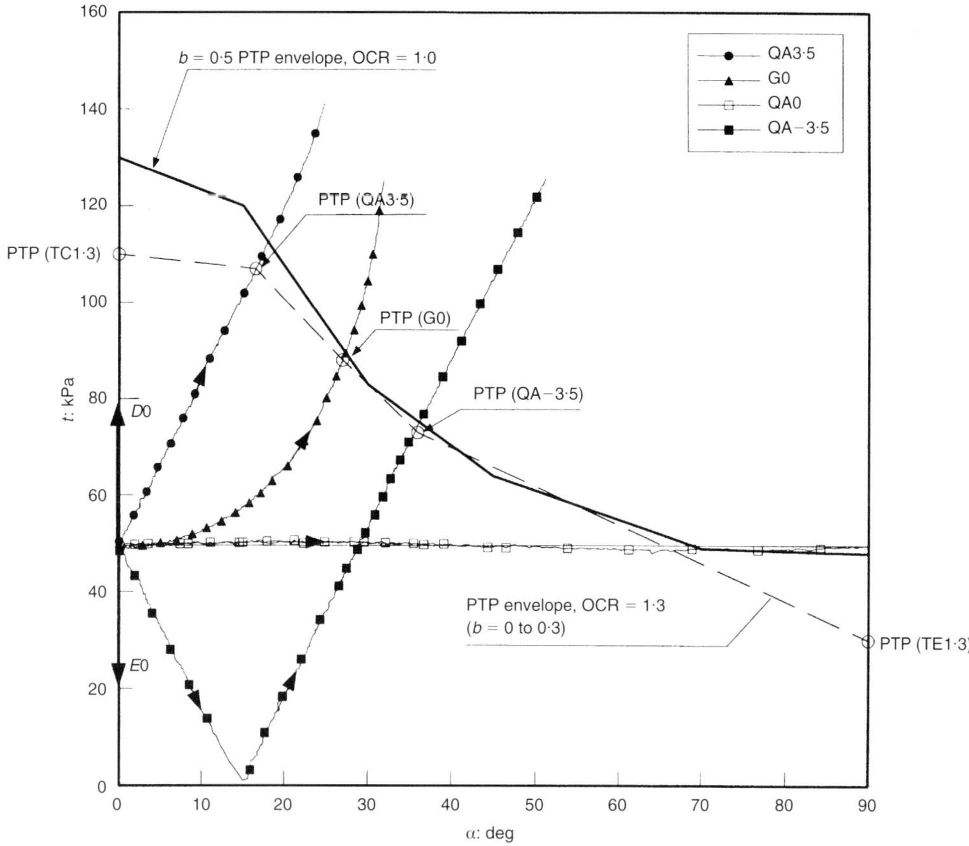

Fig. 8. Effective stress paths in t–α plane from uniaxial and polyaxial tests at OCR = 1·3 (Note: PTP signifies Phase Transformation Point)

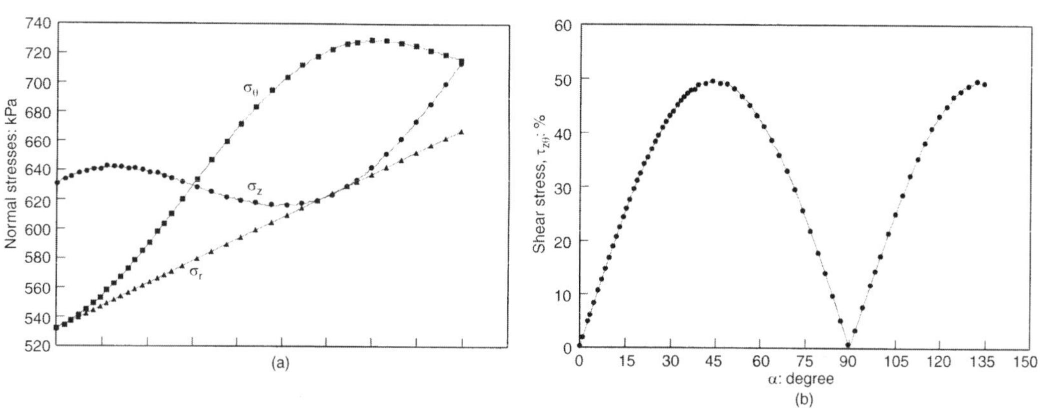

Fig. 9. Normal and shear stress changes for test QA0

RESULTS AND INTERPRETATION

The overconsolidated experiments are described first, as they aid the interpretation of the OCR = 1 tests.

Uniaxial tests at OCR = 1·3

Dealing first with the undrained triaxial tests, it can be seen from Fig. 6 that the compression and extension samples behaved in completely different

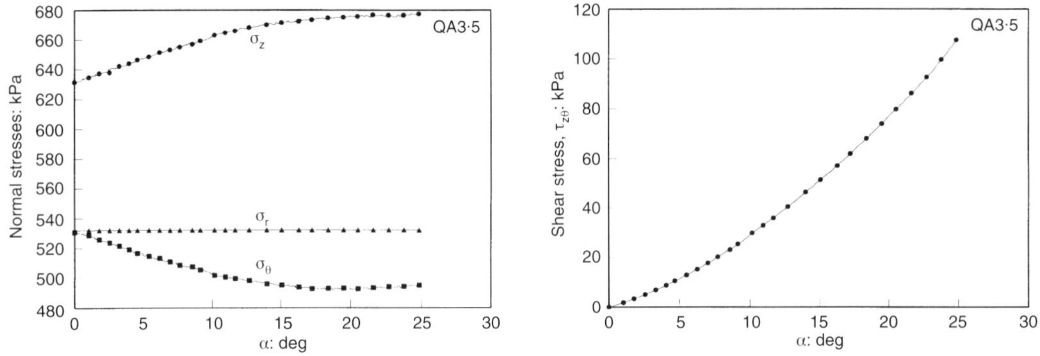

Fig. 10. Normal and shear stress changes for test QA3·5

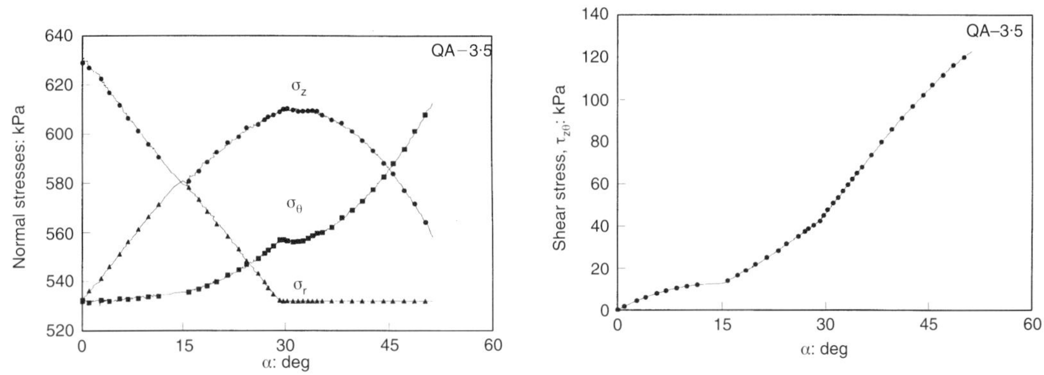

Fig. 11. Normal and shear stress changes for test QA–3·5

ways. The compression test (TC1·3) followed an initially almost linear effective stress path (with p' increasing and Skempton's parameter $A = 0{\cdot}18$) which continued until a form of large-scale yielding (termed Y_3 here) developed, after which the soil's response changed abruptly to become contractant, with p' reducing. The contractant phase ended when the Phase Transformation Point (PTP, or Y_4 yield point) was reached after which the sample exhibited strongly dilatant behaviour, climbing towards a high ultimate deviator stress ($t = 350$ kPa) without developing any clear shear band.

The extension test (TE1·3) developed a similar initial effective stress path direction (now with reducing p' and $A = 0{\cdot}18$) which curved progressively to the left, after passing through a poorly defined Y_3 yield zone, and showing increasingly contractant behaviour. This trend continued after reaching the $t = 0$ axis as the sample was extended (with t increasing again and $\alpha = 90°$) until a PTP or Y_4 yield point was reached and the sample started to show dilatant behaviour. The PTP values of t and p' were far lower than in compression.

The strong anisotropy of the effective stress path response was evident in all the undrained experiments, as shown in Figs 6 and 7, and by the envelopes‡ of PTP points plotted as functions of α in Fig. 8. Zdravkovic (1996) discusses these features further; this paper concentrates on the silt's small strain stiffness anisotropy.

(a) Anisotropy of octahedral shear stiffness, G_{oct}

A first impression of the stiffness characteristics at OCR = 1·3 is given in Fig. 12. The curves of secant octahedral shear stiffness, plotted against ε_d, for the five 'uniaxial' experiments appear to tend towards a plateau at small strains. It is difficult to define the linear initial 'elastic' stiffness accurately as the linear range (i.e. for strains within Y_1) appears to end before $\varepsilon_d = 0{\cdot}002\%$, as found by others. However, shear stiffness anisotropy is clear

‡ The derivation of the PTP envelopes from the full set of M-series tests is described by Zdravkovic (1996); note that no distinction is made here between the b values applying in different tests.

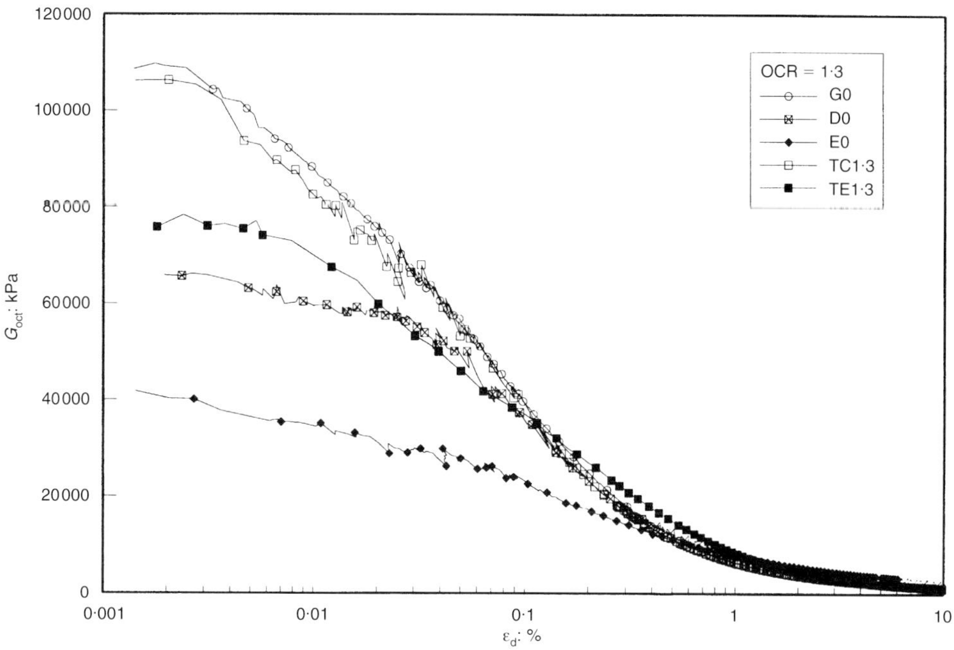

Fig. 12. Octahedral shear stiffness at OCR = 1·3 from uniaxial tests

at small to moderate strains, with a ratio of almost three between the highest and lowest initial G_{oct} values:

(i) The data confirm trends noted for low OCR soils by Jardine (1995): shear stiffness is at near maximum in undrained compression (TC1·3) and is smaller in undrained extension (TE1·3).
(ii) Drained tests involving changes in both t and p' (D0 and E0), develop far larger 'cross-coupling' shear strain components which lead to lower octahedral shear stiffness than undrained experiments.
(iii) The softest response is that of drained uniaxial horizontal loading test (E0).
(v) High torsional shear stiffness is seen in test G0.

(b) Relationships between individual strain components

All strain components were measured locally in the HCA experiments§ and Fig. 13 shows some interesting features from G0. It is evident that the shear strain component $\gamma_{z\theta}$ dominated during the test's early stages and that ε_r and ε_θ remained negligibly small. However, the sample clearly developed significant axial strains from a relatively early stage (with ε_d around 0·04%), after which the incremental ratio $d\varepsilon_z/d\gamma_{z\theta}$ became $\approx 0{\cdot}5$. Note that the axial compression coincided with σ'_z falling due to excess pore pressure generation; see Fig. 6.

The relationships between the strains in test G0 are completely different to those expected from equation (7). While changes in $\tau_{z\theta}$ are predicted to have no effect on any of the normal strains, and hence the effective stresses, actually $C_{14} \neq 0$, $C_{24} \neq 0$ and $C_{34} \neq 0$. We note also that the converse did not apply: uniaxial normal stress changes caused no change in $\gamma_{z\theta}$ in tests D0 and E0 (i.e. $C_{41} = C_{42} = C_{43} = 0$): the cross-coupling compliance matrix terms linking ε_z and $\gamma_{z\theta}$ are not symmetrical. The marked change seen in the strain increment ratio at $\varepsilon_d = 0{\cdot}04\%$ is interpreted as Y_2 yielding, showing the plastic nature of behaviour of the stress–strain response.

The strain components developed in test D0 are displayed in Fig. 14. It can be seen that $\varepsilon_r/\varepsilon_\theta \approx 1{\cdot}0$ at small strains, proving initial transverse isotropy in triaxial compression‖, with $\varepsilon_r/\varepsilon_z$ also holding approximately constant, see Fig. 16. Note

§ Only the axial strains were measured locally in the triaxial test; G_{oct} was computed assuming $\varepsilon_r/\varepsilon_z = -0{\cdot}5$.

‖ Note that ε_r and ε_θ diverged at moderate strains ($\varepsilon_z > 0{\cdot}1\%$), possibly due to bulging, but reconverged to reach the same value at larger strains.

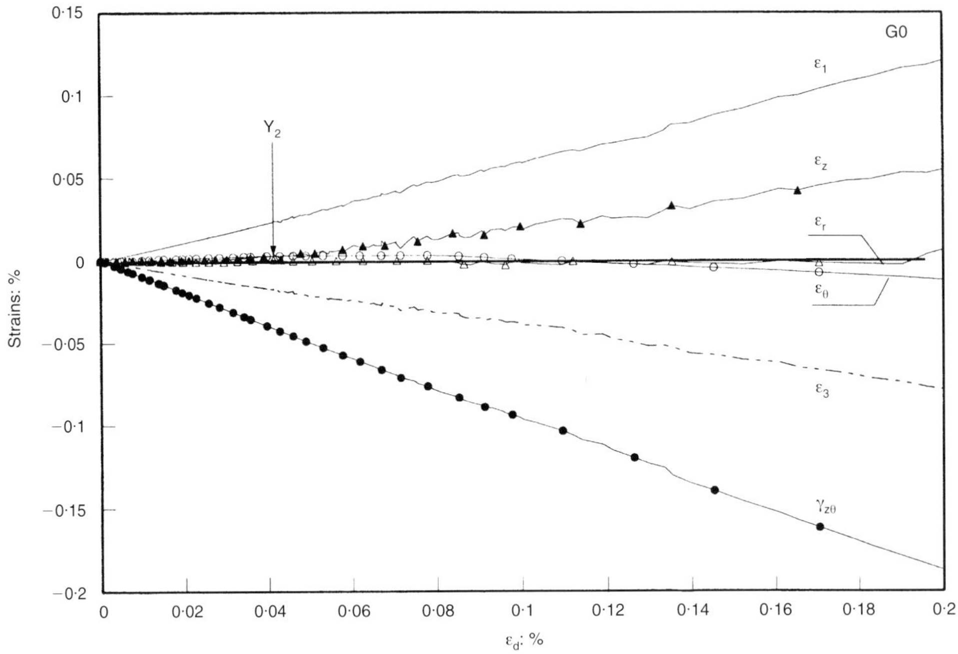

Fig. 13. Effects of strain cross-coupling in test G0 (torsional shear at OCR = 1·3)

that $\varepsilon_r/\varepsilon_z$ was far more negative than would be expected conventionally. The equivalent (positive) Poisson's ratio ν'_{zr} was $\approx 0{\cdot}46$ when $\varepsilon_d < 0{\cdot}04\%$, then grew (reaching $\approx 0{\cdot}58$ at $\varepsilon_d \approx 0{\cdot}13\%$) before falling back to an approximately constant value (0·5), until marked dilatant behaviour commenced at the Phase Transformation (or Y_4) Point when $\varepsilon_d \approx 1\%$. It is suggested that the changes in $\varepsilon_r/\varepsilon_z$ at $\varepsilon_d \approx 0{\cdot}04\%$ and $\varepsilon_d \approx 0{\cdot}13\%$ reflect Y_2 and Y_3 yield points. The effective stress paths of tests D0 and TC1·3 ran almost parallel over their initial portions, confirming that only small volume changes were developing within test D0 prior to the PTP (see Fig. 6).

Figures 15 and 16 also show the strains and strain ratios developed in test E0. The data for $\varepsilon_r/\varepsilon_\theta$ follow a broadly similar trend to that of $\varepsilon_r/\varepsilon_z$ in test D0, although there is more scatter at small strains. A slight change appears to develop at $\varepsilon_d \approx 0{\cdot}04\%$, followed by a steady change to more negative values from $\varepsilon_d \approx 0{\cdot}13\%$. The ratio $\varepsilon_z/\varepsilon_r$ fell steadily from 0·35 at $\varepsilon_d \approx 0{\cdot}003\%$ to reach a minimum of $\approx 0{\cdot}18$ when $\varepsilon_d \approx 0{\cdot}13\%$, and then remained relatively constant. The latter of the two strain limits might be interpreted as the Y_3 yield point; no clear Y_2 point can be seen. The trends for $\varepsilon_z/\varepsilon_r$ are remarkable. If interpreted in terms of conventional elastic theory, the data would indicate a negative Poisson's ratio ν'_{rz}: the sample was compressing vertically as a result of loading on one horizontal axis. This strongly anisotropic feature is the main cause of the low G_{oct} values developed in test E0 (see Fig. 12). Another conclusion is that this result is also incompatible with the assumption of a symmetrical compliance matrix: equation (5) cannot hold, as ν'_{rz} and ν'_{zr} have opposite signs, even at relatively small strains.

(c) Anisotropy expressed by 'directional' stiffness terms

Figure 17 shows the stiffness data from the uniaxial tests at OCR = 1·3 in terms of the directional parameters defined in equation (7). Also shown are the TC1·3 and TE1·3 curves and a trace for $G_{r\theta} = G_{hh}$ which has been calculated by substituting the E'_r and $\nu'_{r\theta}$ data from test E0 (for a range of ε_d) into the axially symmetric form of equation (6). This figure summarizes the stiffness anisotropy in terms of the directional components defined in equations (6) and (7).

The relationship between the drained and undrained vertical stiffness terms is also interesting. Considering compression first, and substituting (from TC1·3 and D0) the observed undrained values of A and E_u^{load}, $E_z'^{load}$ and ν'_{rz} into equation

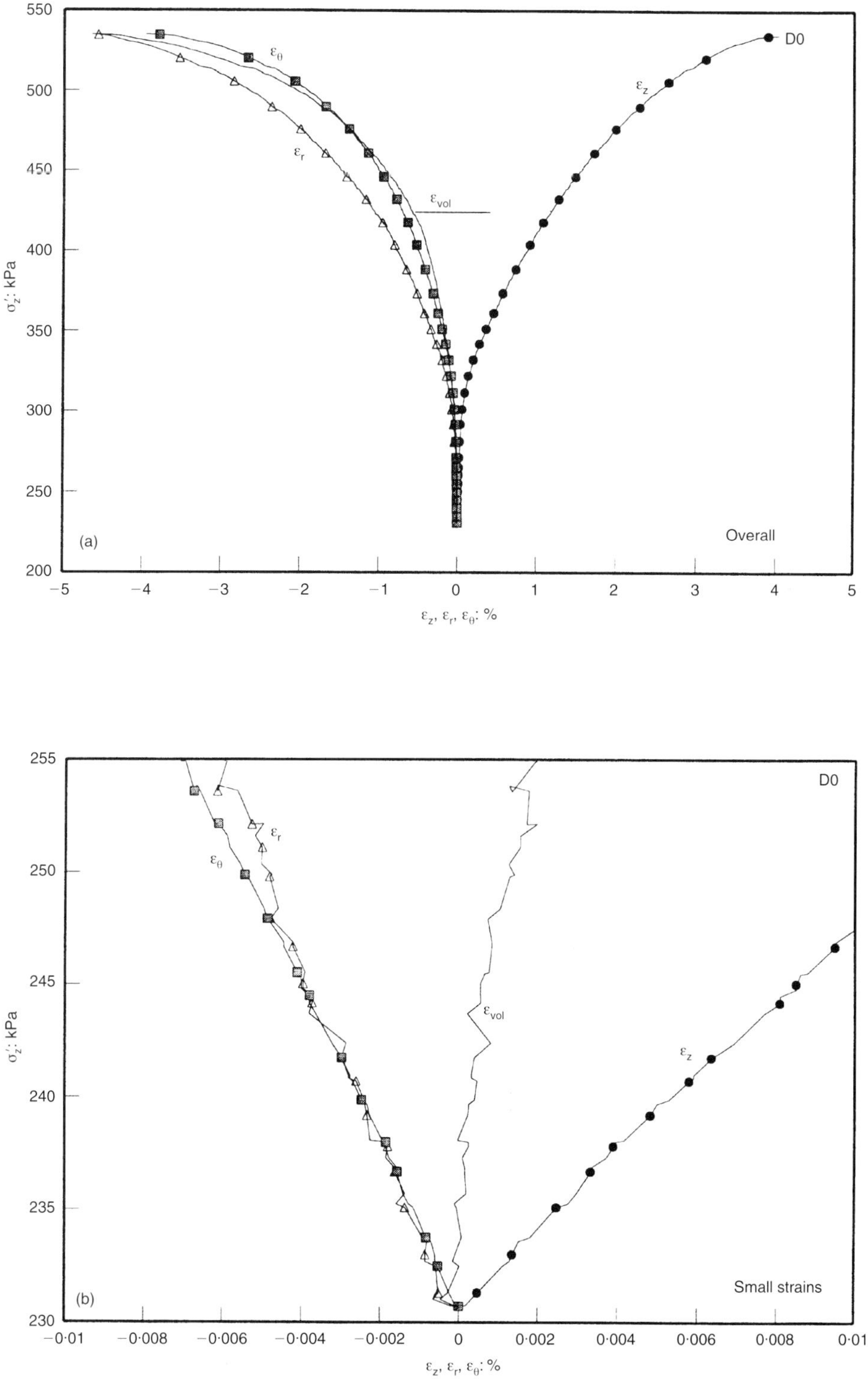

Fig. 14. Strain components developed during drained direct vertical loading from OCR = 1·3 (test D0)

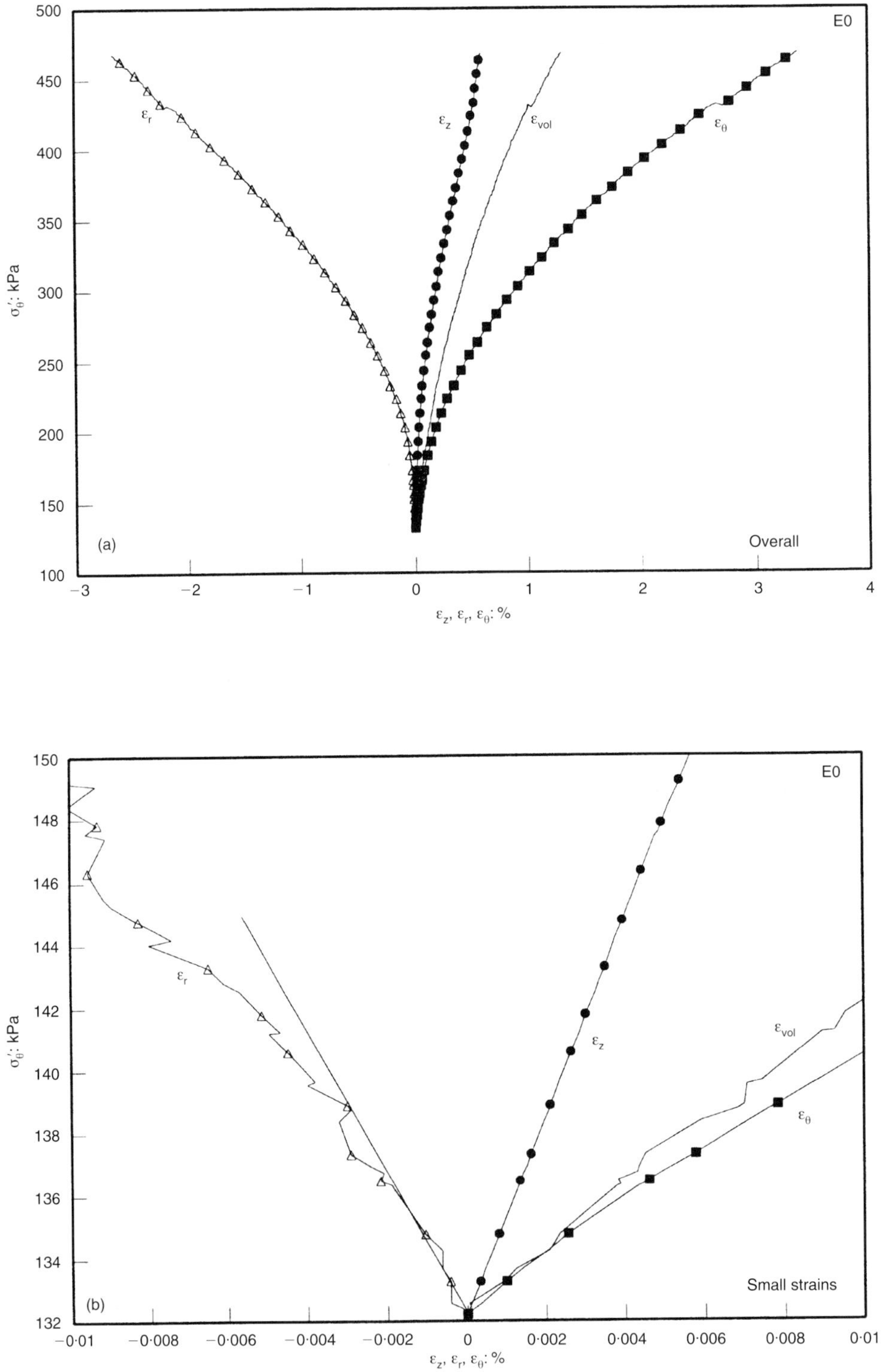

Fig. 15. Strain components developed during drained direct horizontal loading from OCR = 1·3 (test E0)

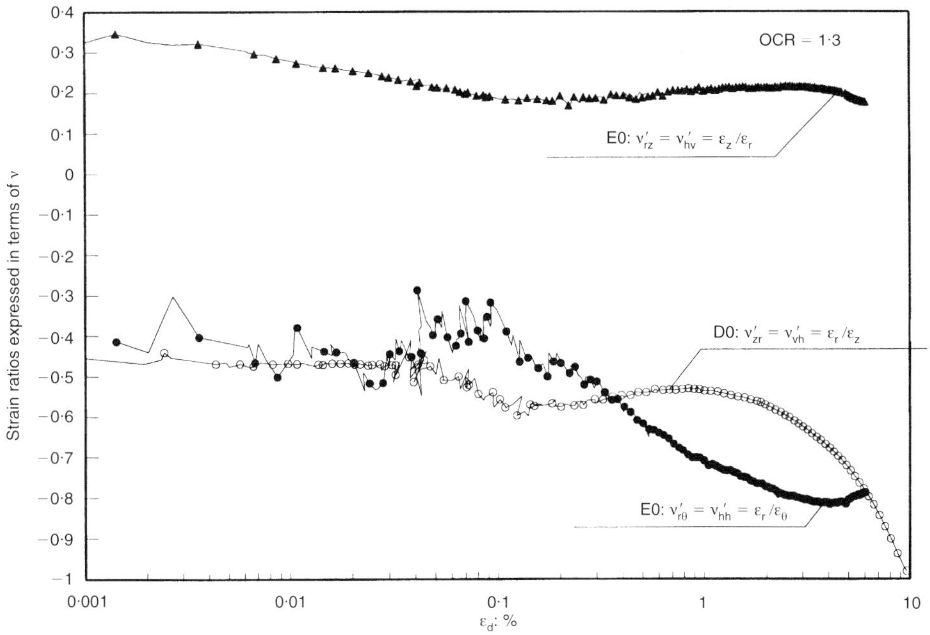

Fig. 16. Strain ratios calculated from tests D0 and E0

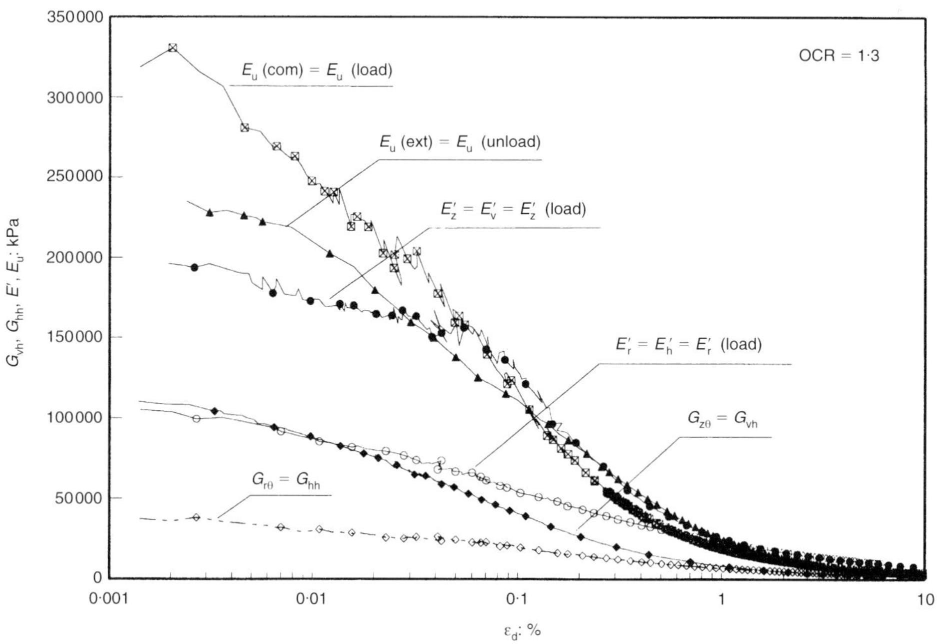

Fig. 17. Uniaxial and torsional shear drained stiffness of K_0 consolidated silt at OCR = 1·3

(10) allows us to predict values for E'_r in unloading; comparison with those measured in uniaxial horizontal loading (i.e. E'^{load}_r) will show whether the radial stiffnesses are unique functions of strain.

In the same way, substituting into equation (11) values (from TE1·3) for A and E^{unload}_u, and taking ν'_{rz} from D0 and E'^{load}_r from test E0, allows us to predict values for E'_z in unloading, which can be compared with the loading values ($E_z'^{load}$) from test D0. As summarized in Table 3, this approach indicates that, even at relatively small strains, E_r' (or E_θ') is approximately 25% stiffer in undrained unloading than in uniaxial drained loading, while E_z' is approximately 25% smaller in undrained unloading than in drained loading.

Noting that the calculations are relatively insensitive to the exact values of ν_{rz}', we may conclude that the terms in equation (7) are non-unique, with coefficients that depend on the sign of the loading change, as well as the strain level, even at relatively small strains, and even when there is no torsional shear loading.

Results from polyaxial tests at OCR = 1·3

As mentioned earlier, the undrained polyaxial HCA experiments QA3·5, QA0 and QA−3·5 (described in Table 2) involved simultaneous changes in $\tau_{z\theta}$, σ_z, σ_θ and σ_r. Fig. 7 illustrated the effective stress paths followed, Figs 9–11 reported the total stresses applied.

The soil's stress–strain response is described in Figs 18–20 and Table 4. The data reinforce three earlier conclusions:

(*a*) The soil compresses vertically in response to loading in $\tau_{z\theta}$: as in G0, axial compression

Table 3. Summary of directional stiffness parameters from tests at OCR = 1·3

Strain level: ε_d		0·003%	0·01%
Measured stiffness: MPa	$E_u^{loading}$ (TC1·3)	285	245
	$E_u^{unloading}$ (TE1·3)	230	210
	$E_z'^{loading}$ (D0)	190	170
	$E_r'^{loading}$ (E0)	100	90
Measured	ν_{rz}' (E0)	−0·3	−0·3
	ν_{zr}' (D0)	0·46	0·46
	$\nu_{r\theta}'$ (E0)	0·4	0·4
Inferred from equations (10) and (11)	$E_z'^{unloading}$ (11)	151	137
	$E_r'^{unloading}$ (10)	134	145
Ratios	$E_r'^{load}/E_r'^{unld}$	0·75	0·62
	$E_z'^{load}/E_z'^{unld}$	1·26	1·24

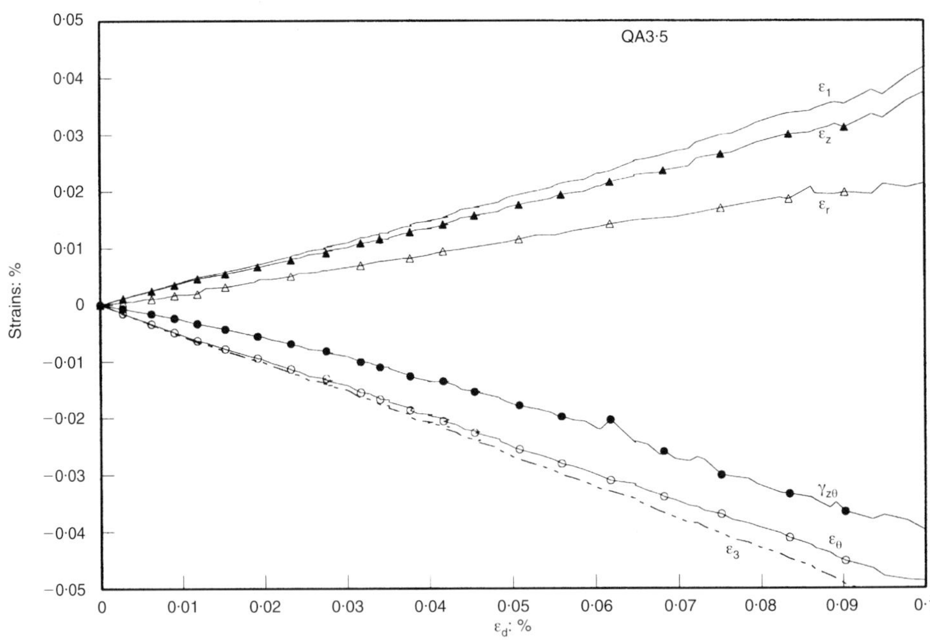

Fig. 18. Strain components for QA3·5 shearing stress path (undrained test from OCR = 1·3)

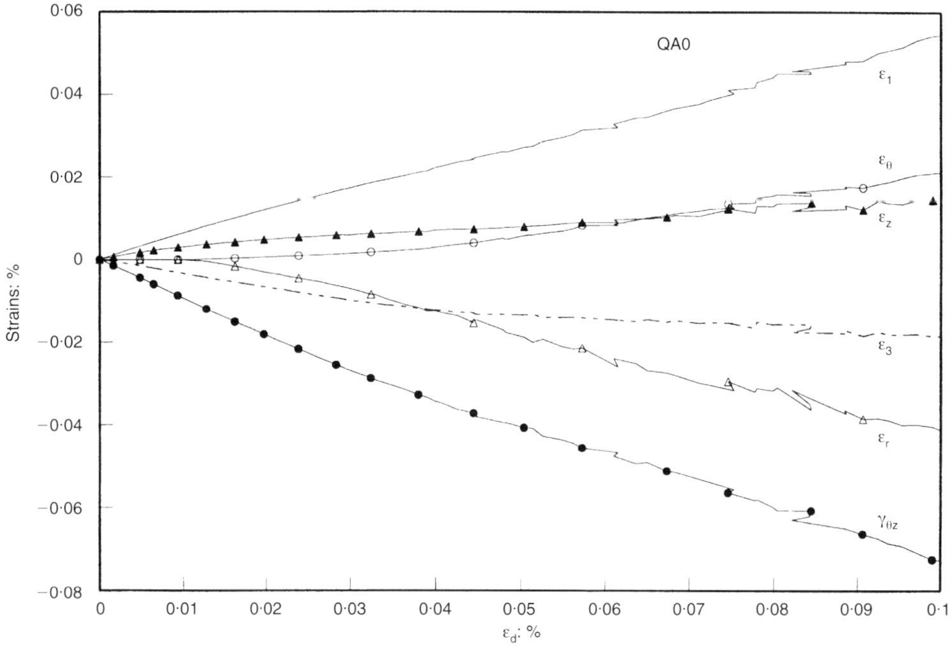

Fig. 19. Strain components for QA0 shearing stress path (undrained test from OCR = 1·3)

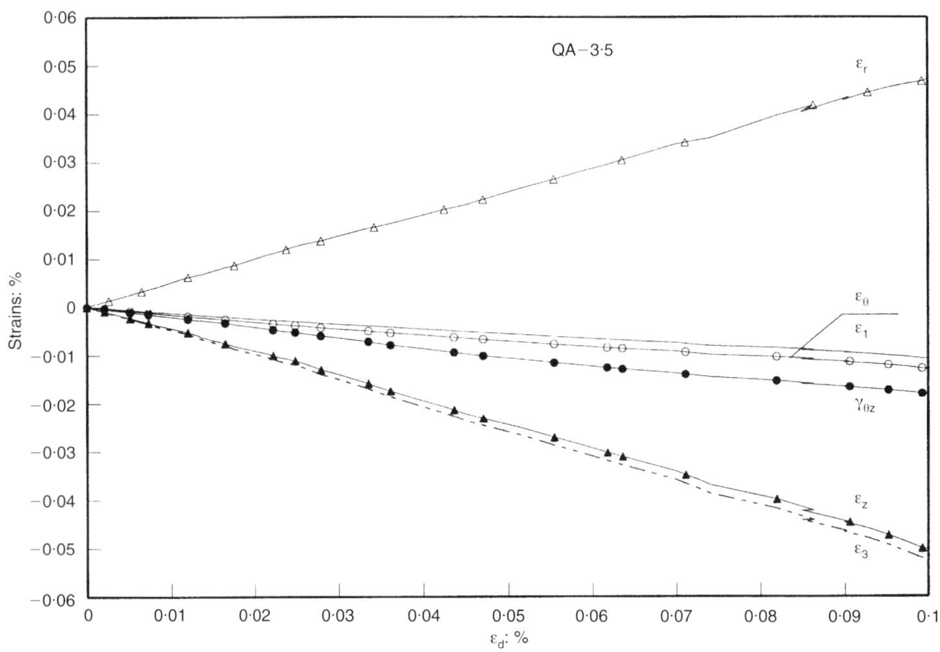

Fig. 20. Strain components for QA−3·5 shearing stress path (undrained test from OCR = 1·3)

Table 4. Summary of straining patterns seen at small to moderate strains in polyaxial tests at OCR = 1·3

Test	Stress path	Straining pattern
QA3·5	Loading both $\tau_{z\theta}$ and $(\sigma_z - \sigma_\theta)/2$; t increasing	Torsional distortion combined with strong axial compression and circumferential extension
QA0	Loading $\tau_{z\theta}$, unloading $(\sigma_z - \sigma_\theta)/2$; t constant	Torsional distortion combined with weak axial compression and circumferential extension
QA–3·5	Loading $\tau_{z\theta}$, unloading $(\sigma_z - \sigma_\theta)/2$; t decreasing	Minor torsional distortion, marked axial extension and strong radial compression

took place in QA0, even though $(\sigma_z - \sigma_\theta)/2$ was reducing.

(*b*) The soil's axial response is stiffer in compression than extension: the absolute values of $\Delta(\varepsilon_z - \varepsilon_\theta)/\gamma_{z\theta}$ are far smaller in the 'active' test QA3·5 than that in the corresponding 'passive' case QA–3·5.

(*c*) Comparing the curves of $G_{z\theta}$, calculated as $\Delta\tau_{z\theta}/\Delta\gamma_{z\theta}$, from tests G0, QA0, QA3·5 and QA–3·5 shows that changes in $(\sigma_z - \sigma_\theta)/2$ have little effect on $\gamma_{z\theta}$ (see Fig. 21).

Figure 22 compares the G_{oct} characteristics of the three polyaxial tests with those of TC1·3 and TE1·3, confirming that:

(*a*) The response was non-linear when $\varepsilon_d >$ 0·002%.

(*b*) Combining significant changes in $\tau_{z\theta}$ and $(\sigma_z - \sigma_\theta)/2$ leads to a softer octahedral response than either type of stress change when applied uniaxially: the curve for QA3·5 falls below those of G0 (only $\tau_{z\theta}$ changing) and TC1·3 (triaxial compression). The G_{oct} trace for QA–3·5 falls well below those of G0 and TE1·3 (triaxial extension). These results reflect the non-symmetrical cross-coupling between $(\varepsilon_z - \varepsilon_\theta)$ and $\tau_{z\theta}$.

(*c*) As noted above and shown in Fig. 21, the torsional shear response is relatively stiff. As a result, test QA0, which principally involved loading in $\tau_{z\theta}$ at its start, gave practically identical G_{oct} values to G0 and TC1·3.

Figure 23 explores how G_{oct} depends on the combinations of $\Delta\tau_{z\theta}$ and $(\Delta\sigma_z - \Delta\sigma_\theta)/2$ (or t and α). The stress path direction is expressed (from the incremental Mohr's stress circle) as the orientation of $\Delta\sigma_1$ with respect to the vertical, $\alpha_{\dot\sigma}$, which is calculated as:

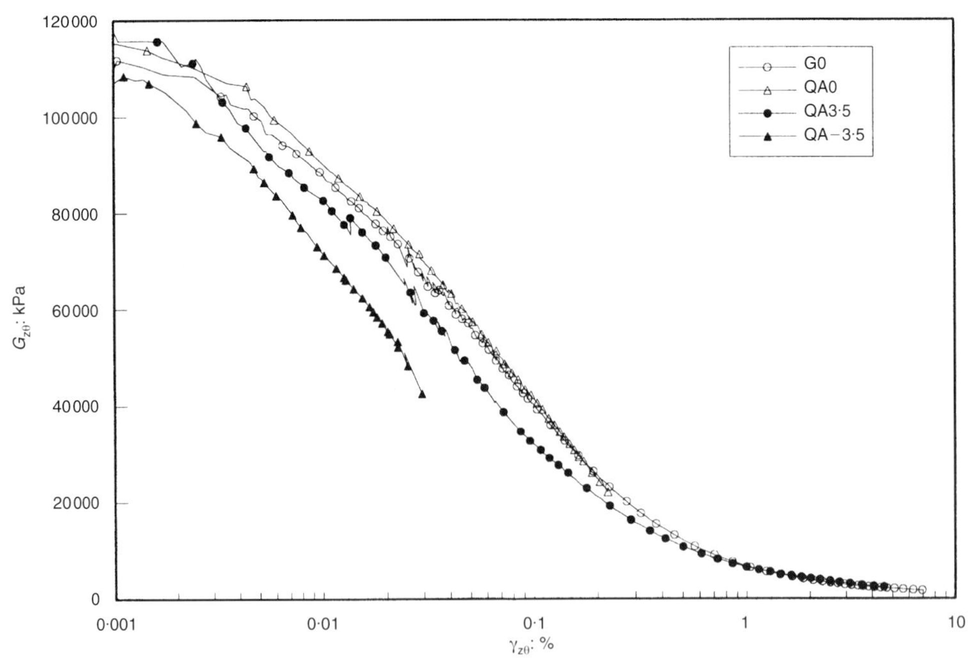

Fig. 21. Torsional shear stiffness for undrained tests from OCR = 1·3

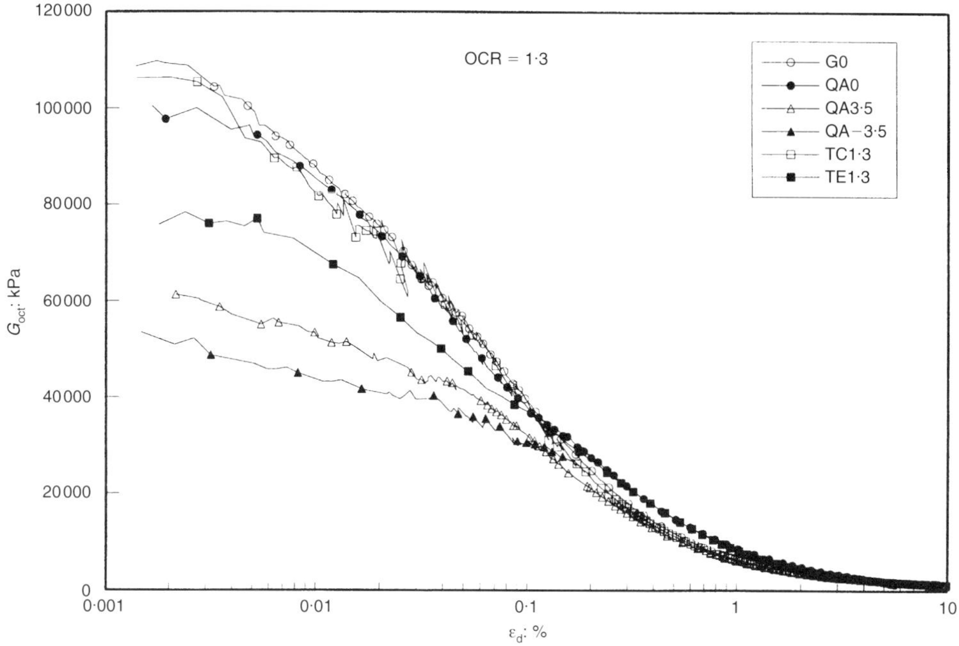

Fig. 22. Octahedral shear stiffness at OCR = 1·3 from undrained tests

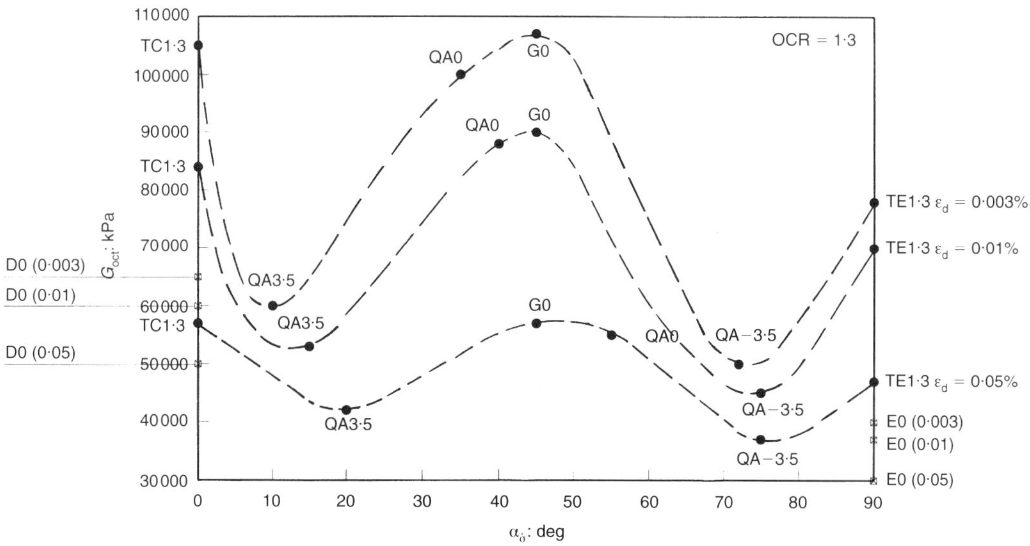

Fig. 23. G_{oct} plotted against angle of incremental stress change for different strain levels

$$\alpha_{\dot\sigma} = \frac{1}{2}\,\text{atan}\left[\frac{2\Delta\tau_{z\theta}}{(\Delta\sigma_z - \Delta\sigma_\theta)}\right] \qquad (12)$$

The undrained values of G_{oct} have maxima at $\alpha_{\dot\sigma} = 0°$, 45° and (at a lower level) 90°; minima develop between these angles as a result of the cross-coupling between $(\varepsilon_z - \varepsilon_\theta)$ and $\tau_{z\theta}$. Note from Fig. 23 that the data points from tests D0 and E0 fall below the trend of the undrained tests: G_{oct} is lower in drained cases where significant changes develop in p', because of the further cross-coupling effects between $(\varepsilon_z - \varepsilon_\theta)$ and p'.

Bulk stiffness characteristics

Triaxial experiments reported by Atkinson *et al.* (1990) and Smith *et al.* (1992) have shown that non-linear bulk stiffness characteristics also depend on the recent stress history and current $t-p'$ stress path direction. While most of the tests outlined in Table 2 were undrained, volume changes were monitored during the K_0 consolidation and swelling stages, and in tests D0 and E0. Bulk stiffnesses (K') computed from these experiments¶ are plotted against ε_{vol} in Fig. 24, showing again highly anisotropic behaviour. At the right hand limit is marked the tangent value (33 MPa) applying to continuing normal consolidation at $p' =$ 200 kPa (point A in Fig. 2). The secant and tangent K' curves developed during swelling from points A to B (after ageing at A) fall relatively gently with strain, remaining far above the tangent compression value for point A. Test D0 gave a completely different trend; as noted earlier, the silt's dilatant response to axial compression leads to minimal volume strains, and therefore very high initial K' values; the curve is also steeply non-linear. In contrast, the unexpected vertical compression seen in test E0 led to a far softer and less variable K' characteristic.

As described below, considerable volume changes were observed during the *Mb*stage and *Rα*stage tests performed at OCR = 1. These experiments involved no change in p' (or t) and their volumetric response cannot be described in terms of isotropic bulk stiffness. It is clear that the silt's anisotropy leads to ε_{vol} being cross-coupled to changes in t, b and α, as well as being related to p'.

¶ Average values are reported for the consolidation and swelling data. Individual tests showed close agreement with K' values falling within ±10% at $\varepsilon_{vol} = 0{\cdot}002\%$.

Experiments performed at OCR = 1

Table 2 lists the full suite of experiments. Those on normally consolidated samples include: three 'active' undrained experiments (TC1·0, M0 and $M0_{b=0\cdot3}$ all with $\alpha = 0$); two passive undrained tests (TE1·0, M90); and drained experiments *Mb*stage and *Rα*stage in which t and p' were held constant while b and α were varied independently.

The effective stress paths followed by all the tests are plotted on Fig. 25; comparison with Figs 6 and 7 shows that similar 'large strain' (Y_4 yielding) anisotropies apply to samples at OCRs 1·0 and 1·3.

Figure 26 presents the octahedral shear stiffness curves from the undrained experiments, while Fig. 27 shows those from the drained experiments, in addition to TC1·0 and TE1·0. The first points to note are:

(*a*) The curves are steeply non-linear with no visible linear portion.

(*b*) The response to 'passive' undrained axial extension is far stiffer (even at very small strains) than that in axial compression, showing the opposite trend to that for OCR = 1·3 (see Fig. 22).

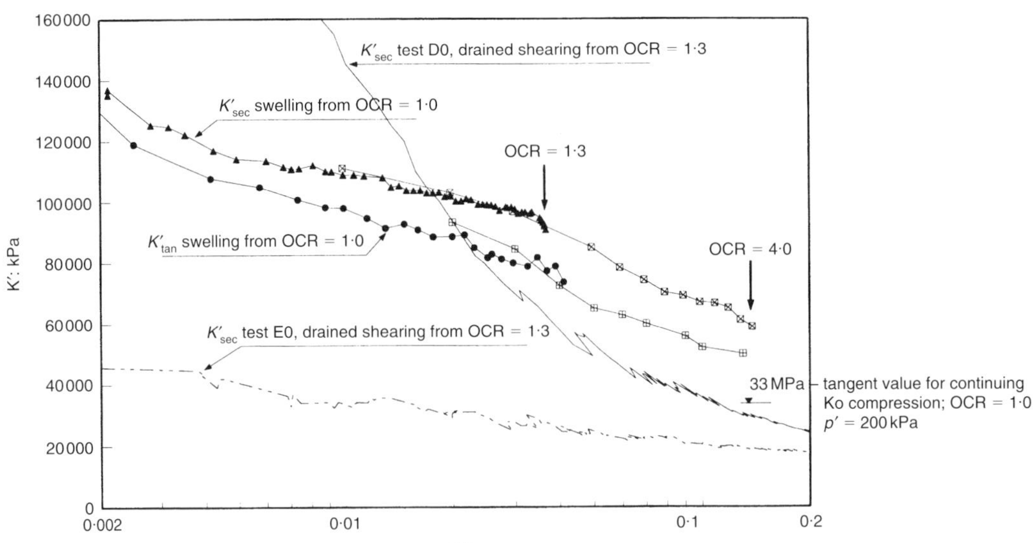

Fig. 24. Bulk moduli during K_0 swelling and during shearing in tests D0 and E0

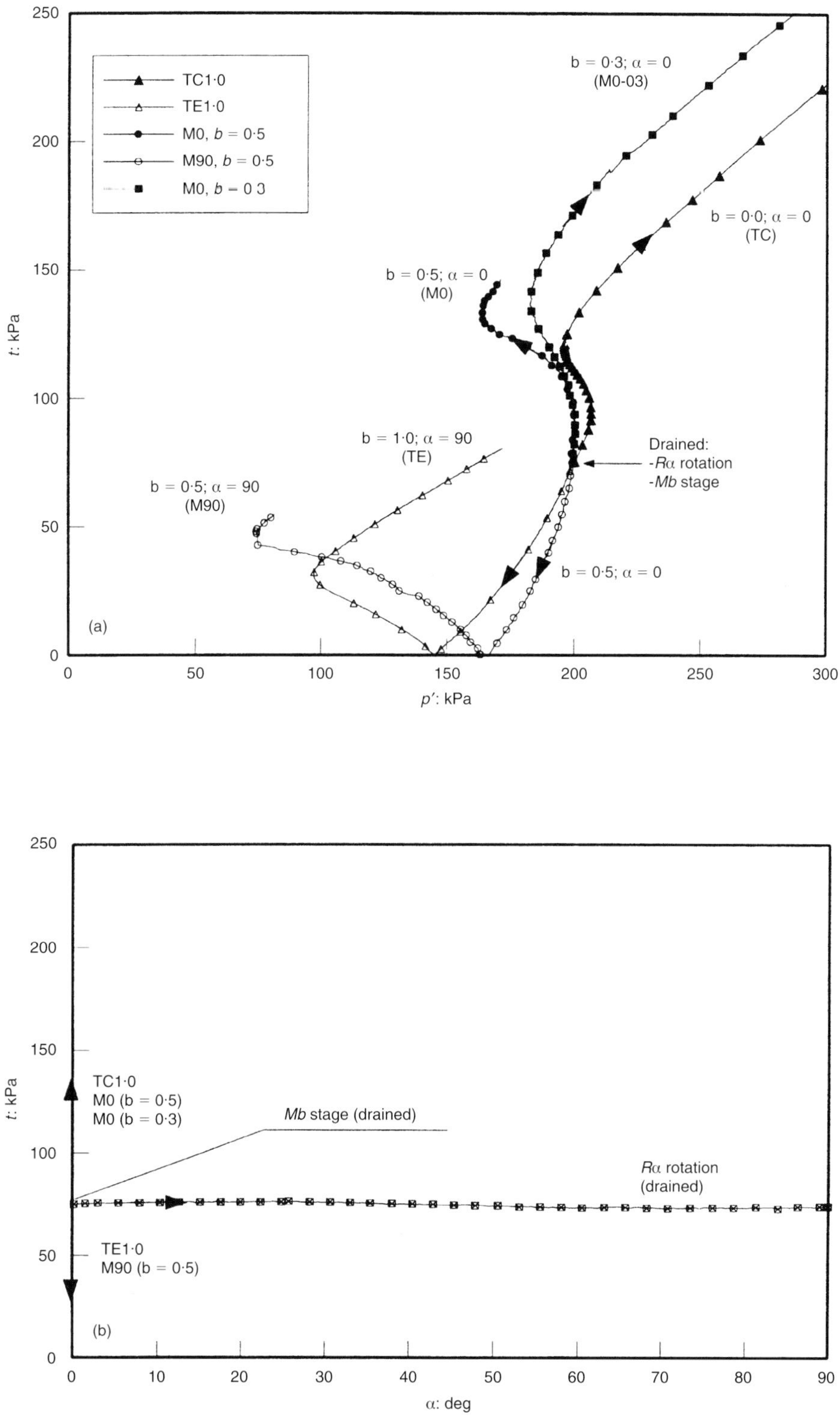

Fig. 25. Effective stress paths from OCR = 1·0: (a) in $t-p'$ plane and (b) in $t-\alpha$ plane

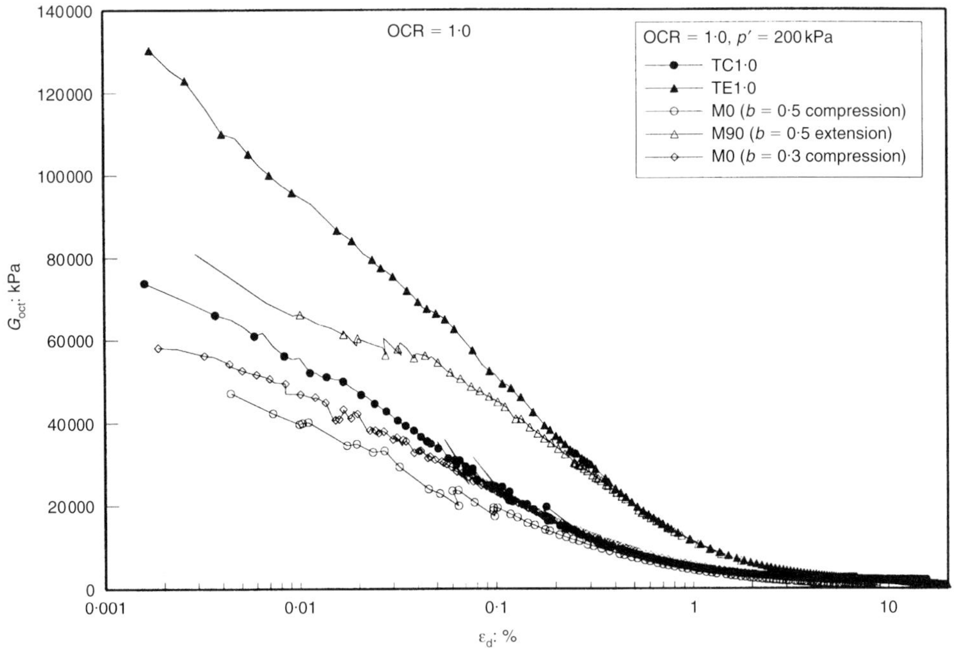

Fig. 26. Octahedral shear stiffness at OCR = 1·0

These two features are typical of K_0 consolidated soils, Jardine (1995). Further points include:

(*c*) The softest octahedral response was that of the drained *Mb*stage experiments, in which σ'_r was loaded while unloading the other two normal effective stresses, see Fig. 4.

(*d*) The early stages of *Rα*stage mainly involved drained increases in $\tau_{z\theta}$ (see Fig. 5) so the stiff octahedral response shown in Fig. 27 indicates particularly high initial values of $G_{z\theta}$.

(*e*) The shearing of tests M0, $M0_{b=0\cdot3}$ and M90 led to lower shear stiffnesses than in the equivalent triaxial tests. G_{oct} reduced by 15% at $\varepsilon_d = 0\cdot01\%$ in compression when b changed from zero to 0·3; increasing b to 0·5 doubled the reduction in G_{oct}.

Further analysis of Mbstage tests

The strain components developed in five typical *Mb*stages are shown in Fig. 28. Recalling from Fig. 4 that the imposed stress changes amounted to: $d\sigma'_r/db = p'/2$, and $d\sigma'_z/db = d\sigma'_\theta/db = -p'/4$, we note:

(*a*) Volume strains of $\approx 0\cdot06\%$ developed even though p' was constant, with subsequent creep increasing ε_{vol} to $\approx 0\cdot16\%$.

(*b*) The ratio $d\varepsilon_z/d\varepsilon_\theta \approx 0\cdot5$, while the stresses changed according to $d\sigma'_z/d\sigma'_\theta = 1$, implying an anisotropic response.

(*c*) The strains developed in the horizontal plane gave $d\varepsilon_r/d\varepsilon_\theta \approx -2$, which matched the ratio of the stress changes ($d\sigma'_r/d\sigma'_\theta = -2$). Recalling that the soil was initially transversely isotropic, the observation shows that (at OCR = 1) similar stiffnesses (E'_r) apply to horizontal loading and unloading.

The response may be investigated further by substituting the *Mb*stage stress conditions into the right-hand side of equation (7) and rearranging to link the stiffness parameters expected at small strains to the developed strain ratios, leading to:

$$\frac{d\varepsilon_r}{d\varepsilon_z} = \frac{-\nu'_{zr} - (2 + \nu'_{\theta r})\dfrac{E'_z}{E'_r}}{1 + \nu'_{rz}\dfrac{E'_z}{E'_r}} \tag{13}$$

$$\frac{d\varepsilon_\theta}{d\varepsilon_z} = \frac{-\nu'_{zr} + (2\nu'_{\theta r} + 1)\dfrac{E'_z}{E'_r}}{1 + \nu'_{rz}\dfrac{E'_z}{E'_r}} \tag{14}$$

If E'_z/E'_r is treated as a simple quantity and is equivalent, because of applied stress changes, to $E'^{unload}_z/E'^{load}_r$, then to solve these two equations requires two assumptions; we note also that stiffnesses may depend on the sign of loading. Adopt-

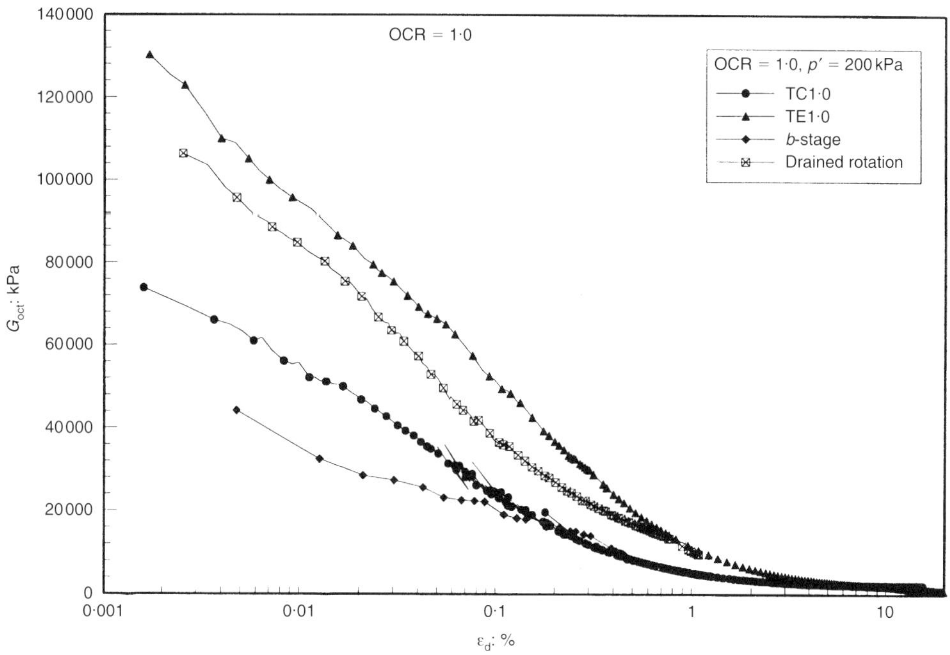

Fig. 27. Octahedral shear stiffnesses for K_0 consolidated silt at OCR = 1·0

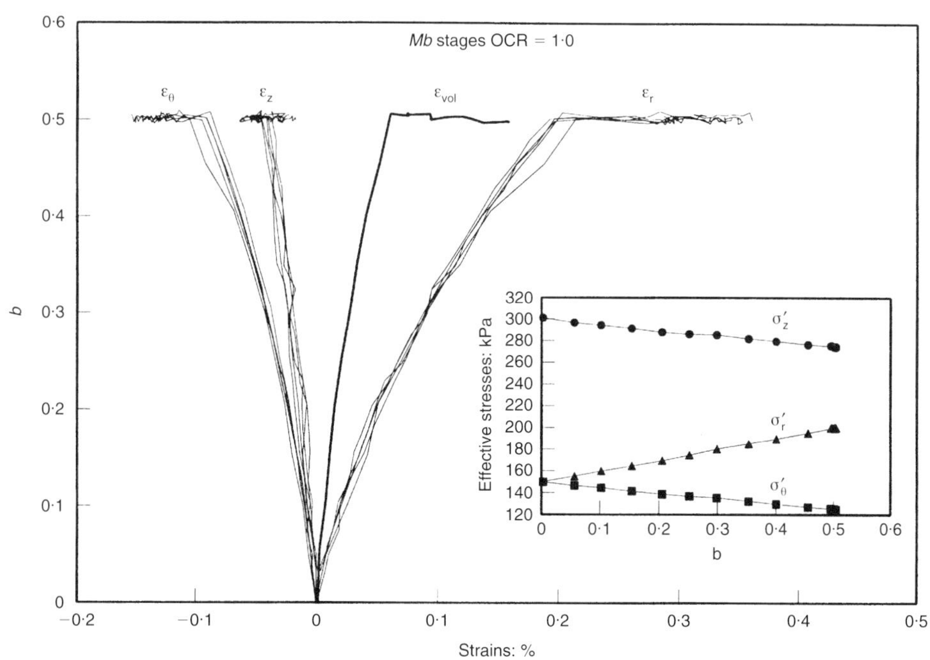

Fig. 28. Strain components developed during drained '*b* stages' in M-series, at OCR = 1·0

ing the values measured at relatively small strains for $\nu'_{\theta r} = 0{\cdot}4$ and $\nu'_{rz} = -0{\cdot}30$ from test E0, leads to predictions (for small strains) of $\nu'_{zr} = 0{\cdot}42$ and $E'^{\text{unloading}}_{z}/E'^{\text{loading}}_{r} = 0{\cdot}95$. Alternatively, if $\nu'_{\theta r}$ remains 0·4 and we take $\nu'_{zr} = 0{\cdot}46$ from test D0, equations (13) and (14) predict $\nu'_{rz} = -0{\cdot}22$, $E'^{\text{unloading}}_{z}/E'^{\text{loading}}_{r} = 1{\cdot}03$. In either case broadly similar sets of stiffness ratios and 'Poisson's ratios' are obtained, providing further information for the compliance coefficients applying at OCR = 1·0.

Further analysis of Rαstage tests

It is also interesting to explore the *Rα*stage tests, where the stress changes shown in Fig. 5 were applied to keep t, p' and b constant while α rotated under drained conditions. Zdravkovic performed five such tests, which terminated at different α values, but strained in the same way initially. Points of interest include the following.

(*a*) As with the 'b stage' tests, significant volumetric creep strains (up to 0·15%) developed during α rotation and aging.

(*b*) The individual strain components, as averaged from the five tests, are plotted in Fig. 29. As with QA0 – which followed the same total stress path from OCR = 1·3, but with undrained conditions (see Fig. 19 and Table 4), all samples tended to compress vertically, even though σ'_z was unloading, with only minor changes developing in ε_r and ε_θ initially. The strain increment directions changed once $\varepsilon_d > 0{\cdot}02\%$: ε_θ became more significant (but still smaller than ε_z) and tensile radial strains developed. These changes were interpreted as 'Y_2' yielding.

(*c*) The variations of $G_{z\theta}$ with $\gamma_{z\theta}$ may be plotted if it is assumed (as in all other tests), that normal stress changes had no effect on $\gamma_{z\theta}$. Data from five tests are given in Fig. 30, showing both very high initial values and fully repeatable trends.

The initial strain response may be analysed by rearranging equation (7). If we assume again that there is no cross-coupling between $\gamma_{z\theta}$ and the normal stresses we find:

$$\frac{d\varepsilon_r}{d\varepsilon_z} = \frac{-\nu'_{zr} + \nu'_{\theta r}\dfrac{E'_z}{E'_r}}{1 + \nu'_{rz}\dfrac{E'_z}{E'_r}} \tag{15}$$

$$\frac{d\varepsilon_\theta}{d\varepsilon_z} = \frac{-\nu'_{zr} - \dfrac{E'_z}{E'_r}}{1 + \nu'_{rz}\dfrac{E'_z}{E'_r}} \tag{16}$$

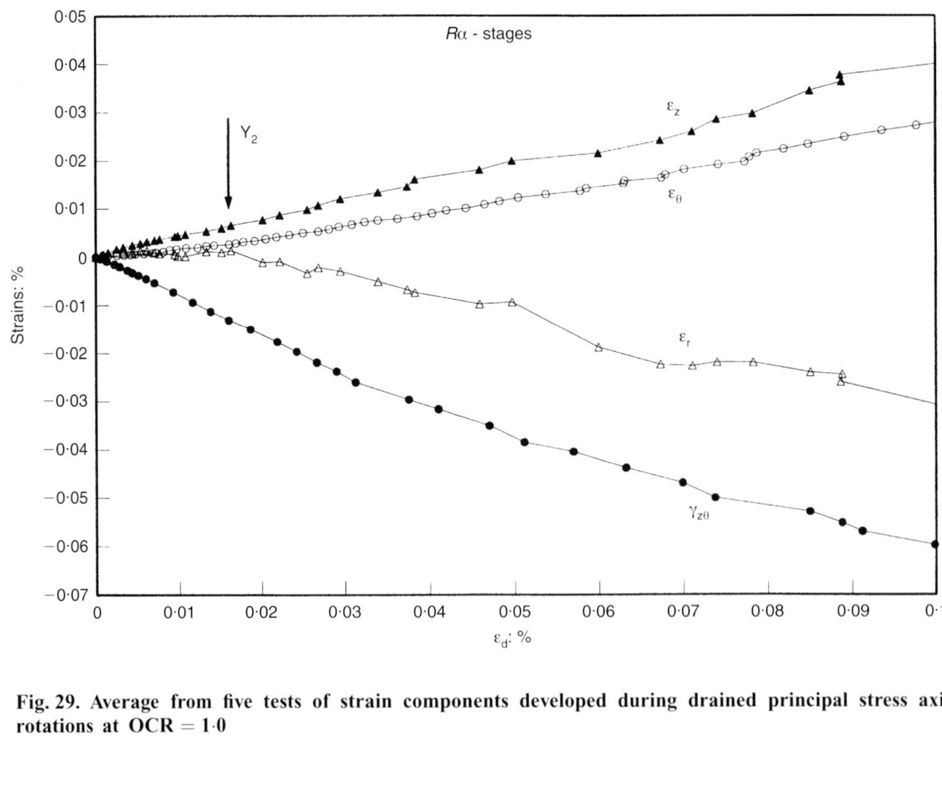

Fig. 29. Average from five tests of strain components developed during drained principal stress axis rotations at OCR = 1·0

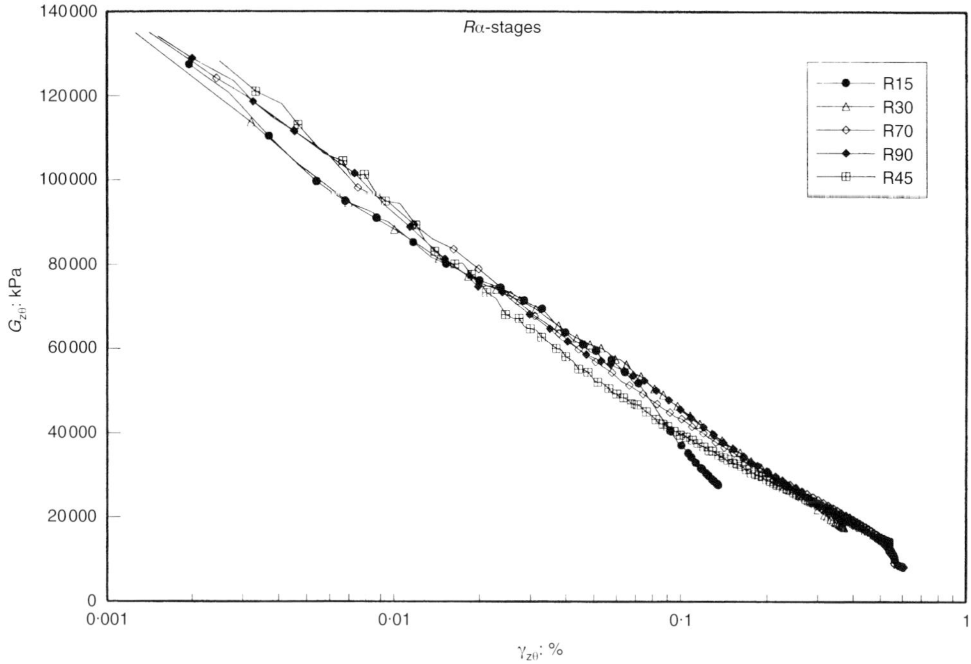

Fig. 30. Torsional shear stiffness during drained principal stress axis rotation at OCR = 1·0

Again we note that the stiffness terms may depend on sign of loading and that, as with other polyaxial tests, additional assumptions are required to obtain solutions. Substituting an average $E_z'^{\text{unloading}}/E_r'^{\text{loading}} \sim 1{\cdot}0$ from the *Mb*stage analysis above, taking $\nu'_{\theta r} = 0{\cdot}4$ from tests D0 and E0 (see Fig. 16), and noting that the measured radial strains were close to zero prior to Y_2 yielding leads to an estimate for initial $\nu'_{zr} \approx 0{\cdot}40$. This result is compatible with the interpretation of the *Mb*stage data at OCR = 1, and test D0 at OCR = 1·3.

However, substituting the observed ratio of $d\varepsilon_\theta/d\varepsilon_z$ into equation (16) leads to apparent ν'_{rz} values between $-3{\cdot}30 \pm 0{\cdot}5$, which have the same sign as those seen in test E0, or interpreted above for the *Mb*stage data, but they are far more negative. The silt's tendency at low OCRs to collapse vertically when loaded on only one horizontal axis, or by $\tau_{z\theta}$ alone (see G0, Fig. 13), is accentuated when the two forms of loading are applied simultaneously.

Further analysis of undrained tests TC1·0 and TE1·0

To complete the interpretation of the stiffness parameters developed at OCR = 1 requires only an analysis of the triaxial tests TC1·0 and TE1·0. Equations (10) and (11) link, for an anisotropic elastic porous material, the drained and undrained stiffnesses developed in compression and extension triaxial tests. Experiments TC1·0 and TE1·0 shared an initial value of $A \approx 0{\cdot}12$; substituting this and $\nu'_{rz} = -0{\cdot}30$ (from test E0, see Fig. 16) leads to predictions of $E_z'^{\text{loading}}/E_u^{\text{loading}} = 0{\cdot}81$ and $E_z'^{\text{unloading}}/E_u^{\text{unloading}} = 0{\cdot}82$. Dividing one ratio by the other and rearranging gives, for small strains:

$$\frac{E_z'^{\text{loading}}}{E_z'^{\text{unloading}}} = \frac{0{\cdot}81}{0{\cdot}82}\frac{E_u^{\text{loading}}}{E_u^{\text{unloading}}} \approx 0{\cdot}5$$

At OCR = 1, loading σ'_z directs the stress path towards the nearby large scale (Y_3) yield surface, leading to a far softer response (even at small strains and with aged samples) than that seen during undrained axial unloading. We note however, that at infinitesimally small strains the ratio might tend to unity.

Table 5 provides a non-dimensional summary of the (pre-Y_2 yield) stiffness parameters interpreted from the tests on OCR = 1 samples. As with the OCR = 1·3, data given in Table 3, the compliance matrix for normally-consolidated silt is generally non-symmetrical, non-linear and non-unique. At both OCRs the silt tends to: (i) compress vertically

Table 5. Summary of stiffness anisotropy at OCR = 1·0 for $\varepsilon_d \geqslant 0{\cdot}003\%$

Parameter	Value	Prediction, source or measurement
$E_z'^{\,load}/E_u^{load}$	~0·8	Above analysis of undrained triaxial tests TC1·0 and TE1·0
$E_z'^{\,unload}/E_u^{unload}$	~0·8	
$E_z'^{\,load}/E_z'^{\,unload}$	~0·5	
$E_z'^{\,unload}/E_r'^{\,load}$	~1·0	Deduced from analysis of strain ratios in *Mb*stage tests
$E_r'^{\,load}/E_r'^{\,unload}$	~1·0	
ν'_{zr}	0·46	Measured in test D0
	~0·4	Deduced from analysis of strain ratios in *Mb*stage and *Ra*stage tests
ν'_{rz}	−0·3	Measured in test E0
	−0·2 to −3·5	Deduced from analysis of strain ratios in *Mb*stage and *Ra*stage tests
$\nu'_{r\theta}$	0·40	Measured in test E0

when loaded by a horizontal normal effective, or shear, stress; and (ii) have normal stiffnesses that depend on the sign of the applied stress change, even at relatively small strains.

The most pronounced effects on stiffness anisotropy of overconsolidation relate to the different ratios applying to vertical or radial loading and unloading (both drained and undrained). Summarizing from Tables 3 and 5:

$$(E_u^{load}/E_u^{unload})_{OCR=1\cdot3} = 1{\cdot}24$$

$$(E_u^{load}/E_u^{unload})_{OCR=1\cdot0} = 0{\cdot}5$$

$$(E_z'^{\,load}/E_z'^{\,unload})_{OCR=1\cdot3} = 1{\cdot}25$$

$$(E_z'^{\,load}/E_z'^{\,unload})_{OCR=1\cdot0} = 0{\cdot}5$$

$$(E_r'^{\,load}/E_r'^{\,unload})_{OCR=1\cdot3} = 0{\cdot}68$$

$$(E_r'^{\,load}/E_r'^{\,unload})_{OCR=1\cdot0} = 1{\cdot}0$$

In both cases torsional shear leads to G_{oct} values that are as high as those for any other mode of shearing. A further difference between the OCR = 1 and OCR = 1·3 tests was their initial effective stress level. Zdravkovic (1996) made a limited check which showed that, with identical OCRs and probing stress conditions, the octahedral shear stiffness depends on p' raised to a fractional power which increases from ≈ 0·5 to 1·0 with increasing strain level. Similar results were found for sand by Porovic & Jardine (1995).

CONCLUSIONS

(*a*) HPF4 silt only exhibits linear (pre-Y_1 yield) stiffness behaviour at very small strains ($\varepsilon_d < 0{\cdot}002\%$).

(*b*) The silt's non-linear stiffness characteristics are strongly dependent on the stress path direction and major principal stress axis inclination.

(*c*) K_0 consolidated samples show a particularly stiff response to torsional shear straining.

(*d*) The silt's compliance matrix is non-symmetrical and non-unique, even at relatively small strains ($\varepsilon_d = 0{\cdot}003\%$). The differences between loading and unloading stiffness terms are particularly marked and provide a major contribution to the soil's stiffness anisotropy.

(*e*) The cross-coupling terms of the compliance matrix are different from those expected conventionally. In particular $\gamma_{z\theta}$ strains can affect ε_z, ε_r and ε_θ strongly and the soil tends to compress vertically when subjected to either torsional shear straining or uniaxial lateral loading. These features are mutually reinforcing when combined (as in the *Ra*stage and QA0 tests).

(*f*) The observed patterns of stiffness anisotropy change with overconsolidation and strain level, and may differ from those invoked at strains much smaller than 0·002% in static or geophysical tests.

(*g*) A second form of yielding, termed Y_2, was evident at relatively small strains in most tests. This occurred after the soil had ceased to be linear or (almost certainly) elastic and was associated with a change of stress increment direction or, in some undrained tests, rate of pore pressure development. Two further forms of yielding, termed Y_3 and Y_4, developed at larger strains.

ACKNOWLEDGEMENTS

The research described was funded by AMOCO (UK) Exploration Company, Health and Safety

Executive (HSE), ORS and PROGRES (YU). The Authors gratefully acknowledge this support and the help and comments of their colleagues of Imperial College, particularly Mr S. K. Ackerley, Professor D. W. Hight and Professor D. M. Potts.

NOTATION

ε_z, ε_r, ε_θ	direct strains in vertical (axial), radial and circumferential direction
$\gamma_{z\theta}$, $\gamma_{\theta z}$	complementary shear strains in $z-\theta$ plane
$\varepsilon_1, \varepsilon_3 = (\varepsilon_z + \varepsilon_\theta)/2 \pm [(\varepsilon_z - \varepsilon_\theta)^2/4 + \gamma_{z\theta}{}^2/4]^{0.5}$	major and minor principal strains
$\varepsilon_2 = \varepsilon_r$	intermediate principal strain
σ_z, σ_r, σ_θ	direct stresses in vertical, radial and circumferential direction
$\tau_{z\theta}$, $\tau_{\theta z}$	complementary shear stresses in $z-\theta$ plane
$\sigma_1, \sigma_3 = (\sigma_z + \sigma_\theta)/2 \pm [(\sigma_z - \sigma_\theta)^2/4 + \tau_{z\theta}{}^2/4]^{0.5}$	major and minor principal stresses
$\sigma_2 = \sigma_r$	intermediate principal stress
$\Delta J = (1/\sqrt{6})[(\Delta\sigma_1 - \Delta\sigma_2)^2 + (\Delta\sigma_2 - \Delta\sigma_3)^2 + (\Delta\sigma_1 - \Delta\sigma_3)^2]^{(1/2)}$	increment of deviatoric stress invariant in terms of principal stresses
$\Delta\varepsilon_d = (2/\sqrt{6})[(\Delta\varepsilon_1 - \Delta\varepsilon_2)^2 + (\Delta\varepsilon_2 - \Delta\varepsilon_3)^2 + (\Delta\varepsilon_1 - \Delta\varepsilon_3)^2]^{(1/2)}$	increment of deviatoric strain invariant in terms of principal strains
$t = (\sigma_1 - \sigma_3)/2$	deviator stress
$p' = (\sigma'_1 + \sigma'_2 + \sigma'_3)/3$	mean effective stress
$b = (\sigma_2 - \sigma_3)/(\sigma_1 - \sigma_3)$	intermediate principal stress parameter
α	the angle between the σ_1 axis and the vertical
$\theta = \mathrm{atan}\,(2\mathrm{d}t/\mathrm{d}p')$	inclination of an effective stress path in $t-p'$ plane
$\alpha_{\dot\sigma}$	the angle between the $\Delta\sigma_1$ axis and the vertical
G_{oct}	octahedral shear stiffness
K'	drained bulk modulus
E_u	undrained Young's modulus
E'_z, E'_v	drained Young's modulus in vertical direction
E'_h, E'_x, E'_r, E'_θ	drained Young's modulus in horizontal direction
G_{vh}, $G_{z\theta}$, G_{zx}	torsional shear modulus in the vertical plane
ν'_{vh}, ν'_{zx}, ν'_{zr}, $\nu'_{z\theta}$	drained Poisson's ratio for horizontal strains due to vertical direct stress changes
ν'_{hv}, ν'_{xz}, ν'_{rz}, $\nu'_{\theta z}$	drained Poisson's ratio for vertical strains due to horizontal direct stress changes
ν'_{hh}, ν'_{xy}, $\nu'_{r\theta}$, $\nu'_{\theta r}$	drained Poisson's ratio for horizontal strains due to a horizontal direct stress change at 90°

APPENDIX 1

Strain and stiffness measurements. Axial, radial, circumferential and torsional strains were measured locally over the central part of the sample in the HCA tests, far from the restraining end platens. ε_z and $\gamma_{\theta z}$ are measured with electrolytic transducers, while the radial and circumferential strains were measured with non-contacting proximity transducers, placed at three locations around both the external and internal circumferences.

Zdravkovic (1996) undertook a series of steps to improve the stability and resolution of the electrolytic type transducers, which were first developed for use in the HCA by Burland & Symes (1982), and then improved by Shibuya (1985). The reference length between the pads attached to the sample is ~ 120 mm and the output signal sensitivity is typically ~ 0·030 V/min.

The main steps were: (i) using a 16 bit A/D converter with particularly high noise rejection, (ii) optimizing the system wiring, (iii) reducing the excitation voltage to 0·5 V and (iv) real time numerical averaging of transducer readings, with each recording representing the mean of 10 individual measurements.

Typical stress–strain curves from an HCA test are presented in Fig. 31. Before starting a new test stage, multiple readings are taken to establish the strain origin.

The recorded stress–strain data show scatter with an almost constant magnitude (±0·00055%), which has an important effect on the early stages of tests. Consequently, for changes in strain < 0·1%, a further manual averaging process is implemented, and a best fitting trend line is drawn through the data points.

The accuracy of individual readings is given by the maximum deviation of the data points about the average curve and a variation of $\Delta\varepsilon \approx \pm 0{\cdot}00055\%$ is typical for the axial and torsional shear results presented in this paper. It is estimated that the interpreted trend curve is reliable to within 1/5 of the scatter, i.e. 0·00011% strain. If so, stiffness may be calculated from the trend line with 11% accuracy at 0·001% strain, and 1·1% accuracy at 0·01% strain. In cases where multiple tests were performed, good repeatability was found at these strain levels: see the *Mb* stages and *Ra* stages data shown in Figs 28 and 30 for example.

The individual radial and circumferential stress–strain curves, measured by proximity transducers, were interpreted in the same way. The interpreted circumferential strain curves have a similar reliability to the axial data (i.e. within ±0·0001% strain), while radial strains show more scatter and the trend curve is thought to be reliable to within ±0·0005% strain.

The Bishop and Wesley triaxial cell was equipped with pendulum devices for local axial strain measurement. Numerical averaging was not used with these and their scatter was typically ±0·0015%. The reliability of their interpreted trend line is also estimated to be 1/5 of the scatter, i.e. ±0·0003% strain. Therefore, the stiffness at 0·001% strain is thought to be estimated with 30% accuracy, while at 0·01% strain the accuracy may be ±3%.

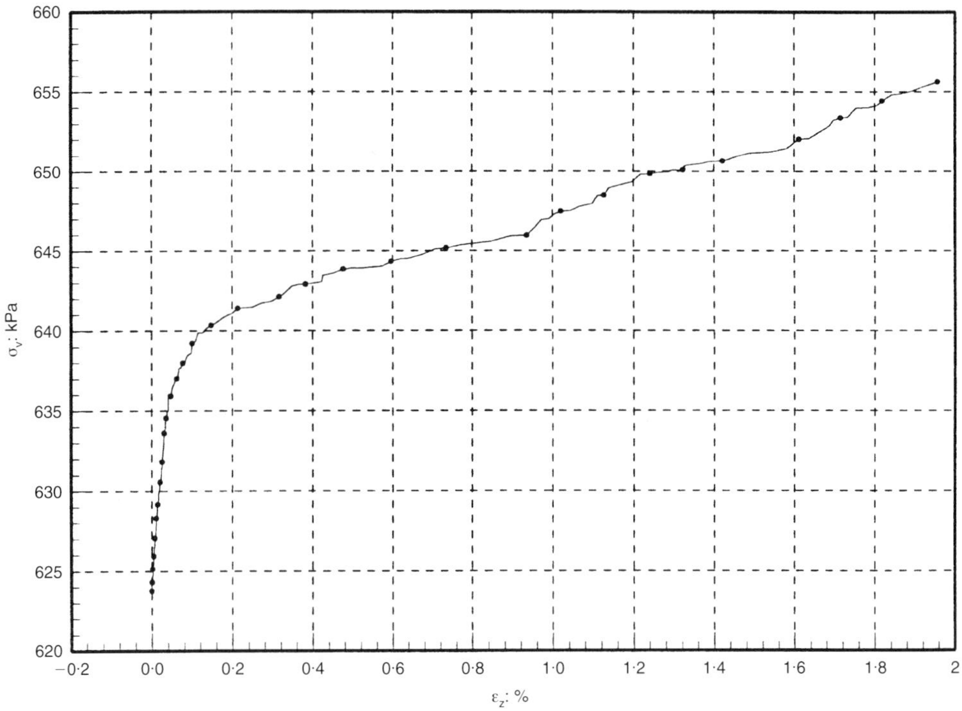

Fig. 31(a). Typical scatter of stress–strain curve for electrolytic type transducers

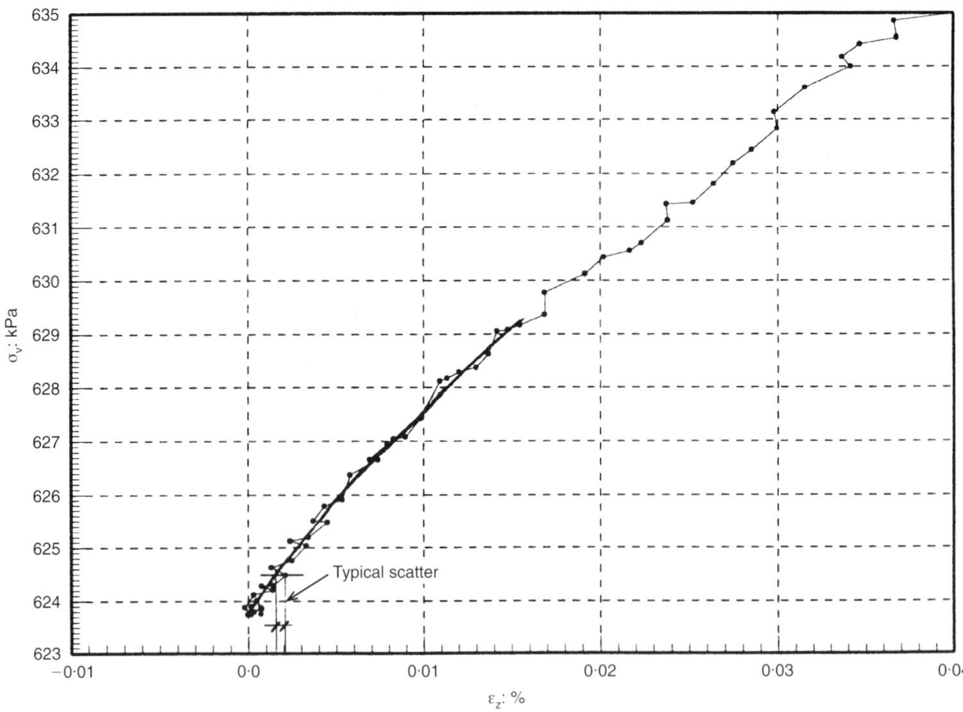

Fig. 31(b). Typical scatter of stress–strain curve for electrolytic type transducers, enlarged small strain view

REFERENCES

Al-Tabbaa, A. & Wood, D. M. (1989). An experimentally based bubble model for clay. *Numerical methods in geomechanics NUMOG III*. Elsevier Applied Science, pp. 91–99.

Atkinson, J. H., Richardson, D. & Stallebrass, S. E. (1990). Effects of recent stress history on the stiffness of overconsolidated soil. *Géotechnique*, **40**, No. 4, 531–540.

Belloti, R., Jamiolkowski, M., Lo Presti, D. C. F. & O'Neill, D. A. (1996). Anisotropy of small strain stiffness in Ticino sand. *Géotechnique*, **46**, No. 1, 115–131.

Burland, J. B. & Symes, M. (1982). A simple axial displacement guage for use in the triaxial apparatus. *Géotechnique*, **32**, No. 1, 62–65.

Connolly, T. (1996). *Laboratory testing on natural Pentre clay silt*. Internal Report, Dept. of Civil Engineering, Imperial College.

Hardin, B. O. (1978). The nature of stress-strain behaviour of soils. *Earthquake engineering and soil dynamics*. Pasadena, Vol. 1, pp. 3–90.

Hardin, B. O. & Blandford, G. (1989). Elasticity of particulate materials. *Journal of Geotech. Eng., Am. Soc. Civ. Engrs* **115**, No. 6, 788–804.

Hight, D. W., Gens, A. & Symes, M. J. (1983). The development of a new hollow cylinder apparatus for investigating the effects of principal stress rotation in soils. *Géotechnique*, **33**, No. 4, 355–384.

Hight, D. W. & Higgins, K. G. (1995). An approach to the prediction of ground movements in engineering practice: background and application. *Keynote Lecture, Pre-failure Deformation of Geomaterials* (eds Shibuya, Mitachi & Miura), Balkema, Vol. 2, pp. 909–945.

Hight, D. W., Bennell, J. D., Chana, B., Davis, P. D., Jardine, R. J. & Porovic, E. (1977). Wave velocity and stiffness measurements of the Crag and Lower London Tertiaries at Sizewell. *Géotechnique Symposium in Print*.

Jardine, R. J., Symes, M. J. & Burland, J. B. (1984). The measurement of soil stiffness in the triaxial apparatus. *Géotechnique*, **34**, No. 3, 323–340.

Jardine, R. J. (1985). *Investigation of pile–soil behaviour, with special reference to the foundations of offshore structure*. PhD thesis, Imperial College, University of London.

Jardine, R. J., Potts, D. M., St. John, H. D. & Hight, D. W. (1991). Some practical applications of a non-linear ground model. *Xth ECSMFE*, Florence, pp. 223–228.

Jardine, R. J. (1992). Some observations on the kinematic nature of soil stiffness. *Soils and Foundations*. **32**, No. 2, 111–124.

Jardine, R. J. (1995). One perspective of the pre-failure deformation characteristics of some geomaterials. *Keynote Lecture, Pre-failure Deformation of Geomaterials* (eds Shibuya, Mitachi & Miura). Balkema, Vol. 2, pp. 855–885.

King, M. S. (1970). Static and dynamic elastic moduli of rocks under pressure. *Rock Mechanics – Theory and Practice, 11th Symposium on Rock Mech., AIME, New York*. pp. 329–351.

Kuwano, R. I. (1996). *Laboratory testing on Dunkirk sand*. Internal Report, Dept. of Civil Engineering, Imperial College.

Lo Presti, D. C. F. (1995). Measurement of shear deformation of geomaterials in the laboratory. *General Report, Pre-failure Deformation of Geomaterials* (eds Shibuya, Mitachi & Miura). Balkema, Vol. 2, pp. 1067–1088.

Love, A. E. H. (1927). *A treatise on the mathematical theory of elasticity*, Chapters III and IV, Edt. Dover Publications, New York.

Pickering, D. J. (1970). Anisotropic elastic parameters for soil. *Géotechnique*, **20**, No. 3, 271–276.

Porovic, E. (1995). *Investigation of soil behaviour using a resonant-column torsional-shear hollow cylinder apparatus*. PhD thesis, Imperial College, University of London.

Porovic, E. & Jardine, R. J. (1995). Some observations on the static and dynamic shear stiffness of Ham River Sand. *Pre-failure Deformation of Geomaterials* (eds Shibuya, Mitachi & Miura). Balkema, Vol. 1, pp. 25–30.

Shibuya, S. (1985). Undrained behaviour of granular materials under principal stress rotation. PhD thesis, Inperial College, University of London.

Simpson, B. (1992). Retaining structures – displacement and design. *32nd Rankine Lecture, Géotechnique*, **42**, No. 4, 539–576.

Smith, P. R. (1992). *The properties of high compressibility clay with special reference to construction on soft ground*. PhD thesis, Imperial College, University of London.

Smith, P. R., Jardine, R. J. & Hight, D. W. (1992). The yielding of Bothkennar clay. *Géotechnique*, **42**, No. 2, 257–274.

Tatsuoka, F. & Shibuya, S. (1991). Deformation characteristics of soils and rocks from field and laboratory tests. *Keynote lecture, IX ACSMFE, Bangkok*, Vol. 2, pp. 101–170.

Zdravkovic, L. (1996). *The stress–strain-strength anisotropy of a granular medium under general stress conditions*. PhD thesis, Imperial College, University of London.

Zdravkovic, L. & Potts, D. M. (1997). *Three-dimensional finite element analyses of the stress and strain nonuniformities in the HCA, using Lade's and Modified Cam Clay constitutive models*. Internal Report, Department of Civil Engineering, Imperial College.

Hayano, K., Sato, T. & Tatsuoka, F. (1997). *Géotechnique* **47**, No. 3, 439–449

Deformation characteristics of a sedimentary soft mudstone from triaxial compression tests using rectangular prism specimens

K. HAYANO,* T. SATO* and F. TATSUOKA*

A new triaxial testing system has been developed for accurately evaluating the deformation characteristics of soft rock. Axial and lateral strains were measured locally along the surfaces of a rectangular prismatic specimen by using, in total, four pairs of local deformation transducers. Sedimentary mudstone samples retrieved by block sampling at Sagamihara Test Site were first compressed isotropically to the field overburden pressure level, followed by drained triaxial compression at different axial strain rates. A series of small unload/reload cycles was applied to define elastic and plastic deformation properties at intermediate loading stages. Very large bedding errors were observed at both the top and bottom ends and at the lateral surfaces of specimens during isotropic compression, and at the ends during triaxial compression. Local strains showed approximately isotropic deformation during isotropic compression. From the results of the drained triaxial compression tests, the Young's modulus and Poisson's ratio, defined for elastic and total strain increments, and their dependences on shear stress level and strain rate were quantified.

KEYWORDS: local strain; Poisson's ratio; sedimentary soft rock; strain rate effect; triaxial compression test; Young's modulus.

Un nouveau système d'essais triaxiaux a été mis au point pour l'évaluation précise des caractéristiques de déformation des roches tendres. Les contraintes axiales et latérales ont été mesurées localement sur les surfaces d'une éprouvette rectangulaire prismatique à l'aide de quatre paires de capteurs de déformation locale. Des échantillons d'argilite sédimentaire prélevés par échantillonnage par blocs sur le terrain d'essai de Sagamihara ont d'abord été comprimés de façon isotrope jusqu'à la pression des morts-terrains avant de subir des essais de compression triaxiale drainés à différents taux de contrainte axiale. On a eu recours à une série de petits cycles décharge/recharge pour définir les propriétés de déformation élastique et plastique aux stades de chargement intermédiaires. On a observé une grande erreur de litage aux extrémités supérieure et inférieure et aux surfaces latérales de l'éprouvette pendant la compression isotrope, ainsi qu'à l'extrémité supérieure pendant la compression triaxiale. Les contraintes locales ont produit une déformation approximativement isotrope pendant la compression isotrope. A partir des résultats des essais de compression triaxiale drainée, on a quantifié le module de Young et le coefficient de Poisson, définis pour les écarts de contrainte élastique et totale, et la façon dont ils dépendent du cisaillement et de l'état de contrainte.

INTRODUCTION

To evaluate accurately the long-term deformation of a sedimentary soft rock deposit and displacements of structures constructed in and on the deposit, both Young's modulus E and Poisson's ratio ν under drained conditions for the soft rock are required for a wide range of strain, from less than about 0·001% to that at peak. The distinction between elastic and plastic deformation properties is also important. It is difficult, however, to evaluate accurately these quantities by conventional triaxial compression tests. So far, the Young's moduli E of several types of sedimentary soft rock have been evaluated by triaxial compression tests measuring axial strains locally (e.g. Jardine *et al.*, 1985; Matthews & Clayton, 1993; Thompson *et al.*, 1993; Clayton *et al.*, 1994; Tatsuoka & Kohata, 1995). On the other hand, the Poisson's ratios ν and the strain rate effects on E and ν for such a wide strain range as cited above have remained unknown.

In view of the above, a new triaxial testing system using a rectangular prismatic specimen was developed. In unconfined and triaxial compression tests on a stiff geomaterial, such as the sedimentary soft mudstone used in the present study, externally measured axial strains may involve large bedding error at the top and bottom ends of the specimen (e.g. Kim *et al.*, 1994). For this reason, measuring

Manuscript received 1 April 1996; revised manuscript accepted 14 February 1997.
Discussion on this paper closes 1 November 1997.
* University of Tokyo.

changes in the distance between two points located on the same axial line on the specimen's lateral surface is now a standard method of local axial strain measurement. In the same way, externally measured lateral strains, such as those obtained from changes in the diameter of a cylindrical specimen, may include large effects of bedding error at the specimen lateral surfaces when the effective lateral stress σ'_h changes. In the present study, rectangular prismatic specimens were used to obtain lateral strains precisely, by measuring changes in the distance between two points located at the same level line on the specimen lateral surface. Local, sensitive measurements of both axial and lateral strains were found to be essential to evaluate Poisson's ratios reliably for small total and elastic strains in triaxial compression tests at constant σ'_h.

TRIAXIAL TESTING SYSTEM

The automated triaxial testing system (Fig. 1) consists of a triaxial cell, a load cell and local deformation transducers (LDTs) inside the triaxial cell, an external axial strain gauge outside the triaxial cell (which was used only for monitoring), two differential pressure transducers (DPTs) measuring the effective lateral stress σ'_h and the amount of water transferred into or expelled from a specimen, a cell pressure system using an air servo with a pressure capacity of 3 MPa, a data acquisition system, and a hydraulic axial pressure loading system. Axial loading is controlled in such a way that the vertical location of the loading ram is always equal to that of a small rod (R in Fig. 1) to an accuracy of less than 0·05 μm as detected by an LVDT with a stroke of ±0·1 mm, which is in turn controlled by a precise gear system. The gear system is free from backlash when reversing the loading direction (Tatsuoka *et al.*, 1994). Using this set-up, the loading system has a large stroke, equal to ±25 mm in this case, with a fixed resolution equal to 0·05 μm.

Local axial and lateral strains were measured by using a pair of vertical LDTs (Goto *et al.*, 1991)

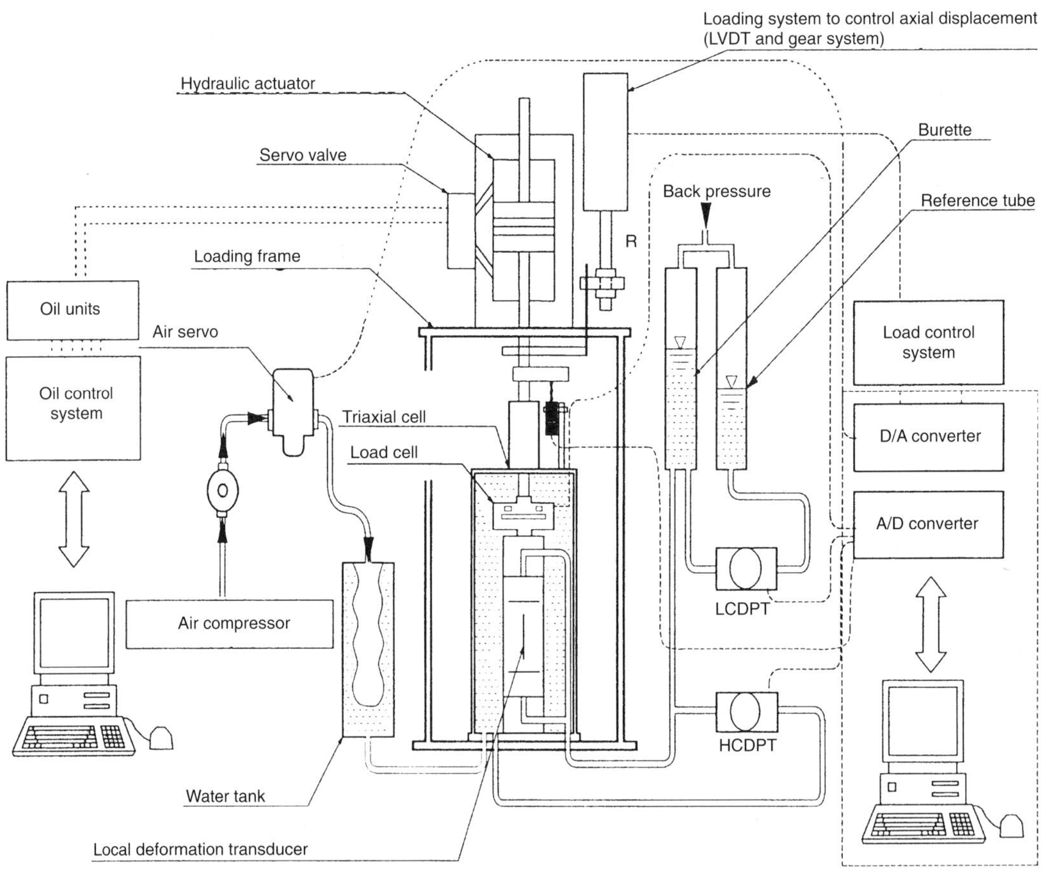

Fig. 1. Triaxial testing system (LCDPT, low-capacity differential pressure transducer; HCDPT, high-capacity differential pressure transducer)

and three other pairs of lateral LDTs as shown in Fig. 2. The LDT is a clip gauge, which detects changes in the distance between the two ends of a bent phosphor bronze strip 0·25 mm thick with electrical resistance strain gauges (e.r.s.) attached at the centre of the strip. Strain is obtained from

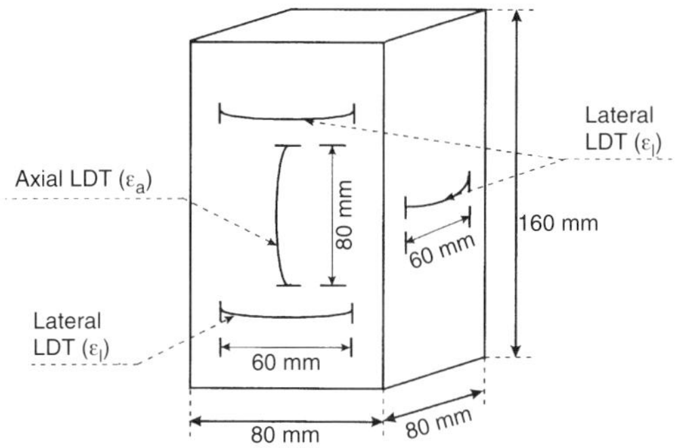

Fig. 2. Rectangular prismatic specimen with local strain measurements

the movement of a set of two targets (hinges) attached to the surface of a membrane. Each rectangular prismatic specimen was 160 mm high 80 mm × 80 mm in cross-section, surrounded by a latex membrane. As the lateral LDTs had a gauge length of 60 mm, which is the shortest that has ever been used, the following two new measures were employed to keep their sensitivity, stability and repeatability high. The same measures were also used for the axial LDTs. First, as shown in Fig. 3, only two active e.r.s. gauges were used because of the width of the phosphor bronze strip, which was reduced to about 2·5 mm to avoid large elastic force. Second, a softer material was used as the lead wire to prevent additional lateral force being applied to each LDT.

The specimen was placed on a pedestal supported by a number of ball bearings as shown in Fig. 4, to allow free lateral sliding while applying no lateral force to the load cell when a shear band was going to appear in the specimen. In fact, this arrangement allowed a single shear band to form smoothly in the specimen.

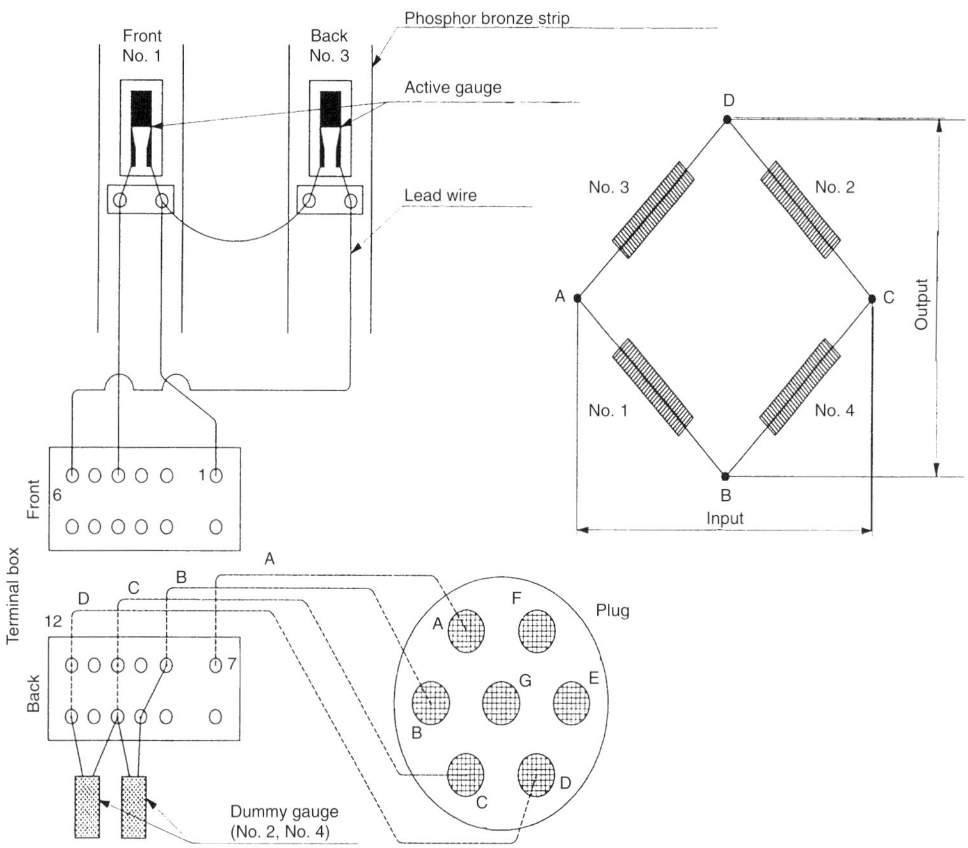

Fig. 3. Electrical connection of short-length LDTs

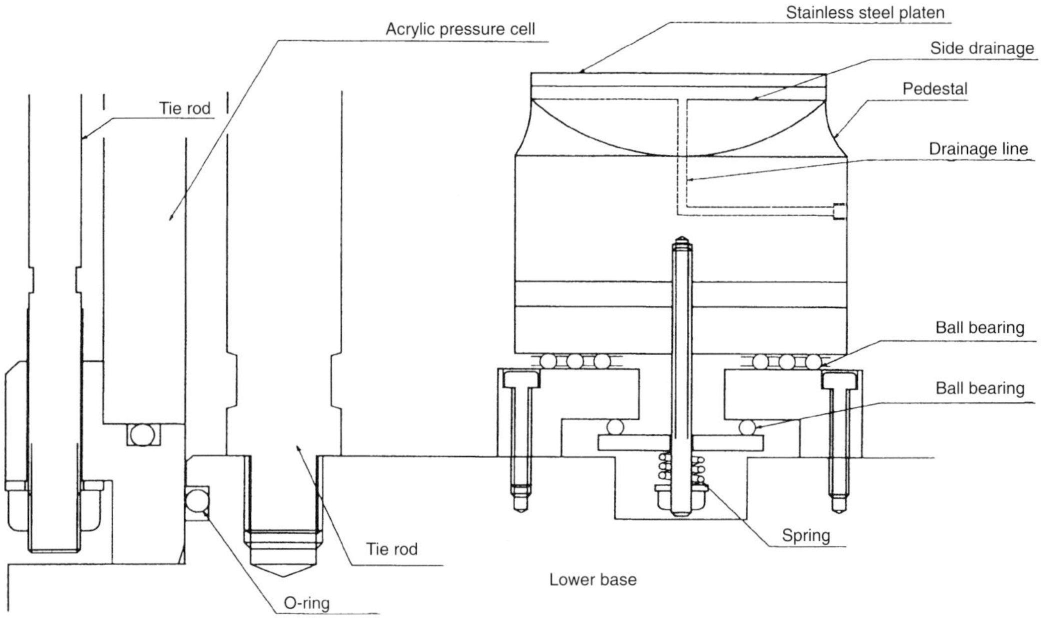

Fig. 4. Sliding mechanism at bottom of pedestal

SAMPLES

A large block sample (about 1·3 m × 1·0 m × 0·5 m) was obtained at a depth of about 50 m from a 1·5 million year old sedimentary soft mudstone at the Sagamihara Test Site (Kim *et al.*, 1994; Tatsuoka *et al.*, 1997). The Sagamihara sedimentary soft mudstones are thought to have been deposited in the late Pliocene to the early Pleistocene epochs. They are well cemented, with a mean diameter D_{50} equal to about 0·006–0·02 mm. Most of them are unweathered and continuous, while some zones have been slightly disturbed by tectonic forces. The reason why they have such high cementation in spite of their young age is not well understood. The block samples were obtained by using an electric saw and occasionally by careful hard carving inside a shaft and tunnel. They were subsequently cored into rectangular prismatic samples using a diamond cutter in the laboratory, in such a way that the axial direction of the samples was the same as the in situ vertical direction. Each specimen was fully saturated by means of a vacuum method, followed by back-pressurizing to 196 kPa in a triaxial cell which was filled with de-aired water (Ampadu & Tatsuoka, 1993). The water content of the tested specimens, measured after the tests, ranged from 22 to 28%, which corresponds to dry unit densities of 1·56–1·68 g/cm^3 and void ratios of 0·98–1·26 (see table inset in Fig. 7(a)). Systematic effects of the scatter in these values on test results were not observed.

Both ends of the specimen were capped with gypsum to ensure a better contact with the end platens; this was considered necessary to ensure more uniform boundary stress distributions. To facilitate drainage of the specimen, side drains of filter paper were used, connected to the side drain of the cap and pedestal (Fig. 4).

The hydraulic conductivity k_v of the mudstone was measured by permeability tests during oedometer tests, and was found to be of the order of 10^{-9} cm/s. Because of the high stiffness of the mudstone, however, the coefficient of consolidation C_v is relatively high; by using secant values of Young's modulus E_{50} and Poisson's ratio ν_{50} at half of the peak deviator stress, about 1 GPa and 0·25 respectively as shown later, we obtain C_v equal to about $1{\cdot}2 \times 10^{-2}$ cm^2/s at pressure levels used in the present study. This value was similar to those evaluated by oedometer tests. According to Bishop & Henkel (1957), the time to failure in a drained triaxial compression test on a cylindrical specimen with a height of $2h$ and a diameter of h, required to ensure 95% of excess pore water pressure is dissipated, is obtained as

$$t_f = \frac{20h^2}{\eta C} \tag{1}$$

where η is a parameter depending on the drainage conditions at the specimen boundaries, and is 32 when side drains are used with capped ends. C is the coefficient of consolidation, which is C_h in this case. Although this equation cannot be directly applied to the present study, where the specimen was a rectangular prism with the sample width h, an approximate value was obtained by assuming $C_h = C_v$ and using the E_{50} and ν_{50} values and $h = 80$ mm. The value of t_f thus obtained is 50 min. As the axial strain at failure is of the order of 0·5%, as shown later, the axial strain rate should be set lower than about 0·01%/min to ensure essentially drained conditions in triaxial compression tests. The axial strain rates employed in the present tests were lower than this limit value.

ISOTROPIC COMPRESSION

Isotropic compression tests were performed from the initial confining pressure of 29·4 kPa up to the field overburden pressure of 470·4 kPa. The axial strain rate was very low, $1{\cdot}25–1{\cdot}69 \times 10^{-5}$/min as measured with LDTs, which is much lower than the limit value discussed above. Volumetric strains were obtained by two methods: $(\varepsilon_{vol})_{burette}$ from the amount of water expelled from the specimen, Δv, measured with a low-capacity differential pressure transducer (LCDPT), and $(\varepsilon_{vol})_{LDT}$ from the local axial and lateral strains $(\varepsilon_a)_{LDT}$ and $(\varepsilon_l)_{LDT}$ measured with LDTs:

$$(\varepsilon_{vol})_{burette} = \Delta v/(\text{initial specimen volume } V) \quad (2)$$

$$(\varepsilon_{vol})_{LDT} = (\varepsilon_a)_{LDT} + 2(\varepsilon_l)_{LDT} \quad (3)$$

Figure 5 shows a typical result. Here, local axial and lateral strains are presented by averaging two and six LDT readings respectively. It is seen that despite the use of end capping, the externally measured axial strain $(\varepsilon_a)_{ext}$ is much larger than the locally measured one $(\varepsilon_a)_{LDT}$. This is also the case with triaxial compression tests as shown later. This error is mainly due to extra compression of a thin disturbed zone formed at the specimen ends during trimming (Kim *et al.*, 1994). This fact suggests that lateral strains obtained by measuring changes in the diameter or width of the specimen while σ'_h is changing may include large errors, supporting the local lateral strain measurement method using lateral LDTs employed in the present study.

It may be seen from Fig. 5 that the difference between external and local volumetric strains is larger than the difference between external and local axial strains. It was confirmed that most of this extremely large difference can be attributed to water drained from the filter paper strips, which had a total area of about 15 000 mm^2 and an unstressed thickness of 0·26 mm. It can be concluded, therefore, that, particularly when side drains of filter paper are used, the volumetric strain of such a stiff material obtained from measurements of the drained water will be totally unreliable. It is considered that $(\varepsilon_{vol})_{burette}$ is also unreliable even in tests not using filter papers, considering that $(\varepsilon_a)_{ext} > (\varepsilon_a)_{LDT}$. This was also the case with triaxial compression tests.

Figure 6 compares axial and lateral strains $(\varepsilon_a)_{LDT}$ and $(\varepsilon_l)_{LDT}$. A nearly isotropic deformation may be seen. Similar results were obtained in the other two tests. The average of $(\varepsilon_a)_{LDT}/(\varepsilon_l)_{LDT}$ for the three tests was 0·015%/0·0138% = 1·09.

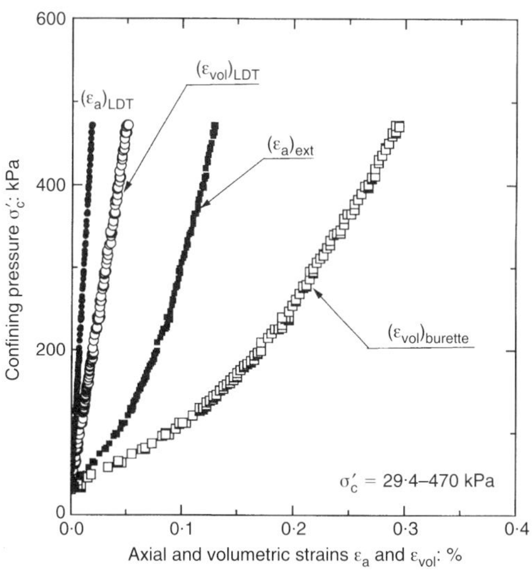

Fig. 5. Volumetric strain during isotropic compression measured by LDT readings compared with that measured by drained water

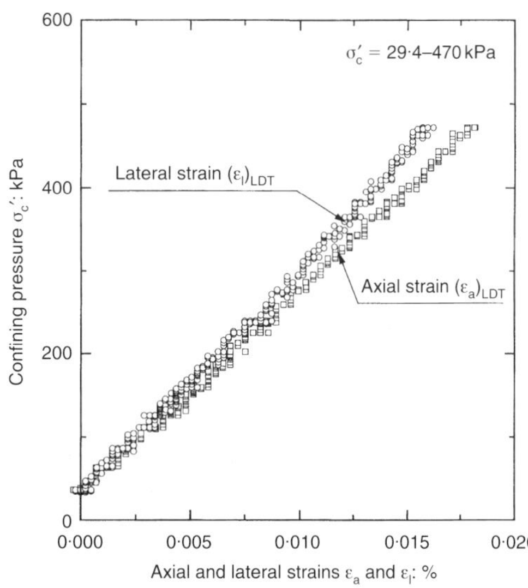

Fig. 6. Axial and lateral strains measured with LDTs during isotropic compression

TRIAXIAL COMPRESSION TESTS

Three drained triaxial compression (TC) tests on rectangular prismatic specimens were performed, one at an externally measured axial strain rate $(\dot{\varepsilon}_a)_{ext}$ of 0·0001%/min and two at 0·001%/min. Fig. 7(a) shows the overall relationships between the deviator stress $q = \sigma'_v - \sigma'_h$ and $(\varepsilon_a)_{LDT}$ for the three rectangular prismatic specimens and one

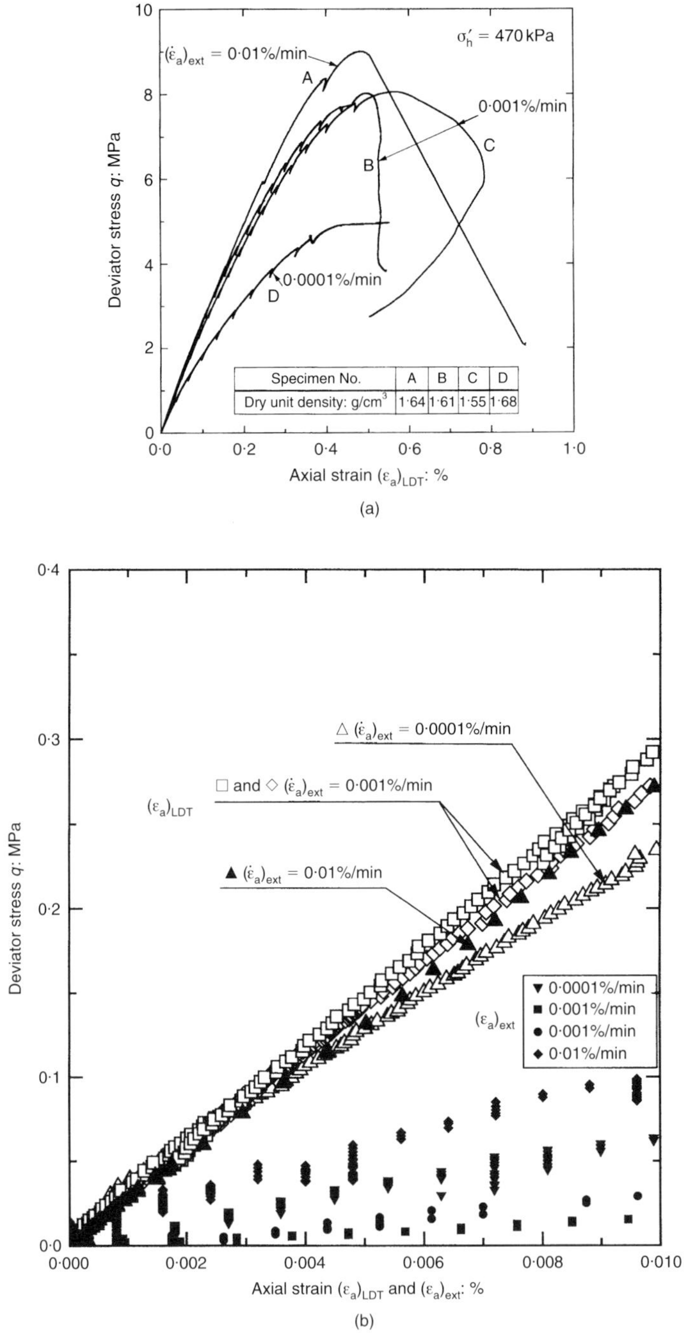

Fig. 7. (a) Summary of deviator stress q and local axial strain $(\varepsilon_a)_{LDT}$ relations for different axial strain rates; (b) enlargement for range of $(\varepsilon_a)_{LDT}$ = 0 to 0·01%

cylindrical specimen. Here, σ'_v and σ'_h are the effective axial and lateral stresses. The TC test on a cylindrical specimen ($d = 50$ mm and $h = 150$ mm, obtained also by block sampling), with end capping and side drains, was performed at an axial strain rate of 0·01%/min, measuring axial strain locally with a pair of LDTs (Wang, 1996). The effects of strain rate are not noticeable on the strain at peak stress state, but the maximum deviator stress q_{max} decreases significantly with the strain rate. Fig. 7(b) shows the relationships at small strains. The effect of bedding error was very large in all the TC tests. Therefore, the rate of locally measured axial strain $(\dot{\varepsilon}_a)_{LDT}$ was much lower than $(\dot{\varepsilon}_a)_{ext}$, owing mainly to the large effects of bedding error and partly to system compliance. The ratio $(\dot{\varepsilon}_a)_{LDT}/(\dot{\varepsilon}_a)_{ext}$ was about 0·23, 0·35 and 0·63 at $(\varepsilon_a)_{LDT} = 0{\cdot}005$, 0·04 and 0·32% respectively.

It is seen that the effect of strain rate on the tangent Young's modulus E_{tan} is very small at strains smaller than about 0·003% (Fig. 7(b)), but it increases with strain (Fig. 7(a)). Figs 8(a) and (b) show the overall relationships between q and lateral strain $(\varepsilon_l)_{LDT}$ and those at small strains for the three prismatic specimens. It may be seen that the lateral strain is more strain-rate dependent and scatters more. Fig. 9(a) shows typical relationships between q and individual LDT readings of $(\varepsilon_l)_{LDT}$. After the peak state, the deformation becomes considerably more non-uniform because of shear banding, and the LDT readings (LDT3 and LDT4), which include shear band deformation, exhibit large rates of increase. On the other hand, the LDT readings (LDT1 and LDT2) decrease after the peak state, owing to elastic rebound in the zones not including shear bands. Fig. 9(b) shows the average lateral strain $(\varepsilon_l)_{LDTA}$, and $(\varepsilon_l)_{LDTB}$ for two orthogonal horizontal directions, and the total average lateral strain $(\varepsilon_l)_{LDTave}$, which were calculated from Fig. 9(a). The pre-peak deformation is not completely axisymmetric. The average of the ratio of $(LDT)_A$ and $(LDT)_B$ at $q/q_{max} = 0{\cdot}5$ was 1·36 for the three specimens. The reasons for this anisotropic behaviour are not known.

During each TC test, to separate elastic and plastic strain components, a number of small unload/reload cycles were applied (Fig. 10). From recoverable strain increments $d\varepsilon_a^e$ and $d\varepsilon_l^e$ in each cycle, the equivalent Young's modulus E_{eq} and Poisson's ratio ν_{eq} were obtained. From total strain increments $d\varepsilon_a^t$ and $d\varepsilon_l^t$ of the stress–strain relations for primary loading, the tangent Young's modulus E_{tan} and tangent Poisson's ratio ν_{tan} were obtained.

Figure 11(a) shows a summary of the values of E_{eq} and E_{tan} plotted against q. It may be seen that the values of E_{eq} are nearly constant with q in each test. Some strain rate effects on E_{eq} are noticeable, but quantitative conclusions cannot be

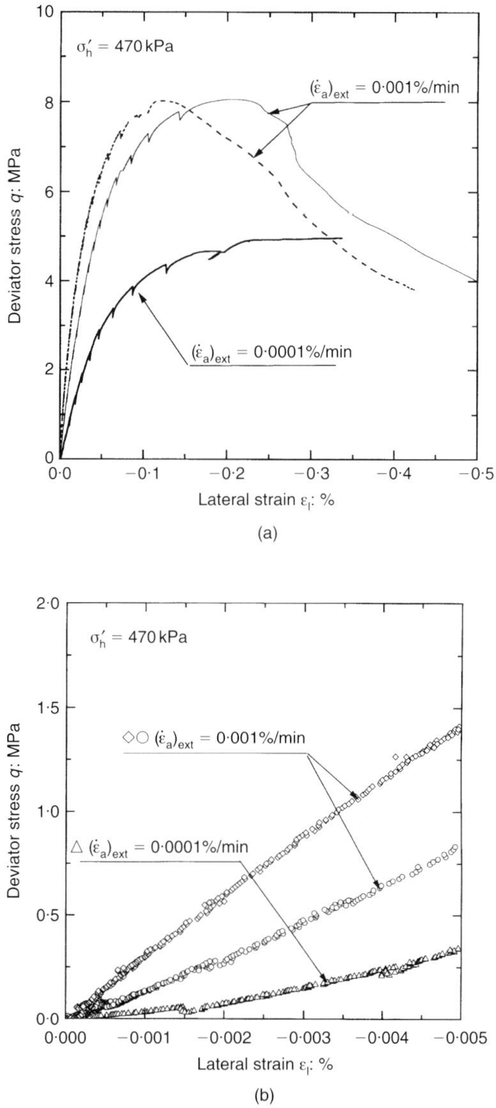

Fig. 8. (a) Summary of deviator stress q and local lateral strain $(\varepsilon_l)_{LDT}$ relations for different axial strain rates; (b) enlargement for range of $(\varepsilon_l)_{LDT} = 0$ to $-0{\cdot}005\%$

obtained owing to the scatter of the data. On the other hand, the strain effects on E_{tan} are more systematic and larger; in particular the E_{tan} values at $(\dot{\varepsilon}_a)_{ext} = 0{\cdot}0001\%/\text{min}$ are noticeably smaller than the others, while the strain rate effects on E_{tan} decrease with decrease in q; the effects on the initial values E_0 are similar to those on E_{eq}. It was found that these strain rate effects on E_{tan} could explain most of a noticeable discrepancy between the stiffness values at strains larger than 0·1%, back-calculated from the field full-scale behaviour

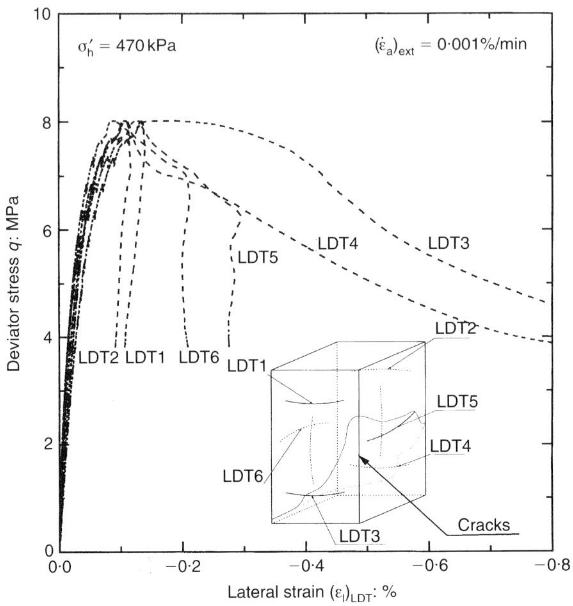

(a)

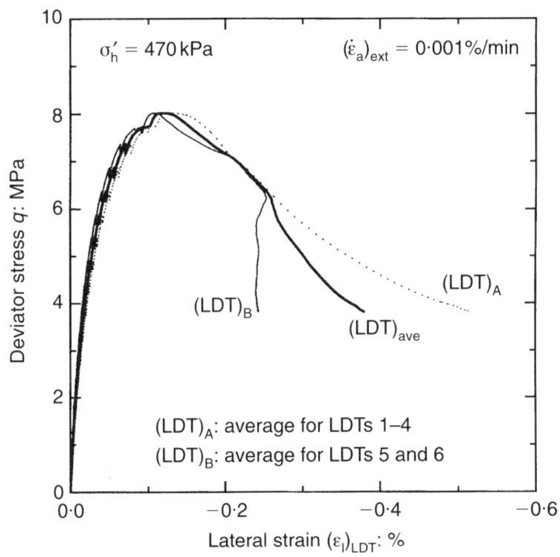

(b)

Fig. 9. Relationships between q and each lateral LDT reading of $(\varepsilon_l)_{LDT}$ during drained TC; (b) relationships between q, $(\varepsilon_l)_{LDTA}$ and $(\varepsilon_l)_{LDTB}$ for two orthogonal horizontal directions and the total average lateral strain $(\varepsilon_l)_{LDTave}$

of the soft rock and the corresponding values from TC tests performed at $(\dot{\varepsilon}_a)_{ext} = 0{\cdot}01\%/\text{min}$, reported by Tatsuoka & Kohata (1995) and Tatsuoka *et al.* (1997).

Figure 11(b) shows the normalized relations obtained by using the respective initial Young's modulus E_0 and peak strength q_{max}. The scatter in the relationships for different strain rates is much smaller than that seen in Fig. 11(a). The plot shown in Fig. 11(b) could be referred to for non-linear numerical analysis.

Figure 12 shows a summary of the values of ν_{eq} and ν_{tan} plotted against the shear stress level q/q_{max}. It may be seen that ν_{eq} is about 0·15 to

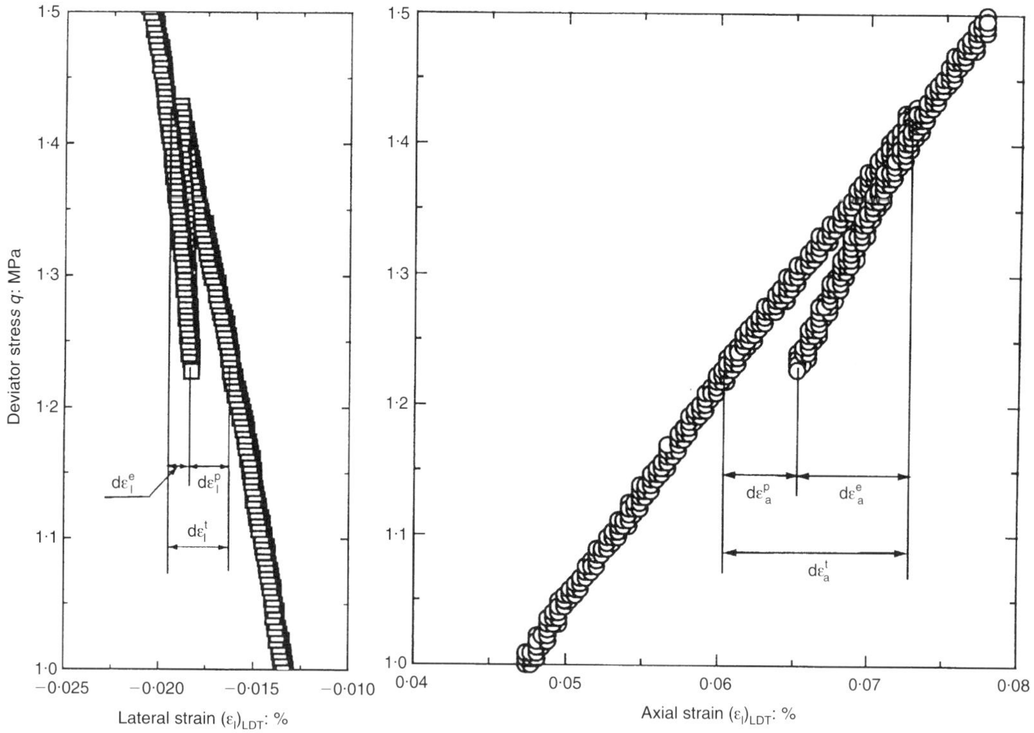

Fig. 10. Relationships between q, axial strain $(\varepsilon_a)_{LDT}$ and lateral strain $(\varepsilon_l)_{LDT}$ for one typical unload/reload cycle

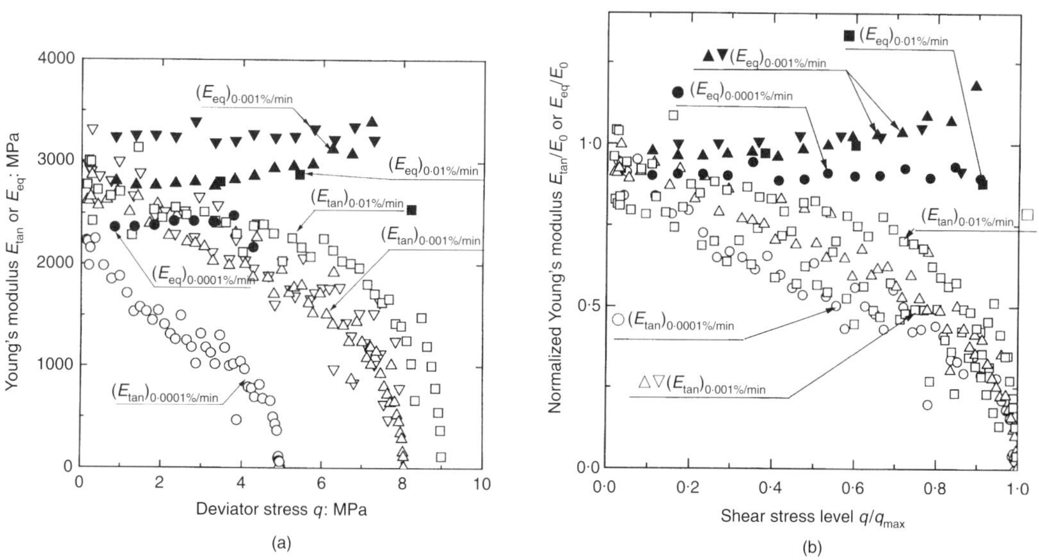

Fig. 11. (a) Sumary of E_{eq} and E_{tan} plotted against deviator stress q; (b) summary of the ratios E_{eq}/E_0 and E_{tan}/E_0 plotted against shear stress level q/q_{max}

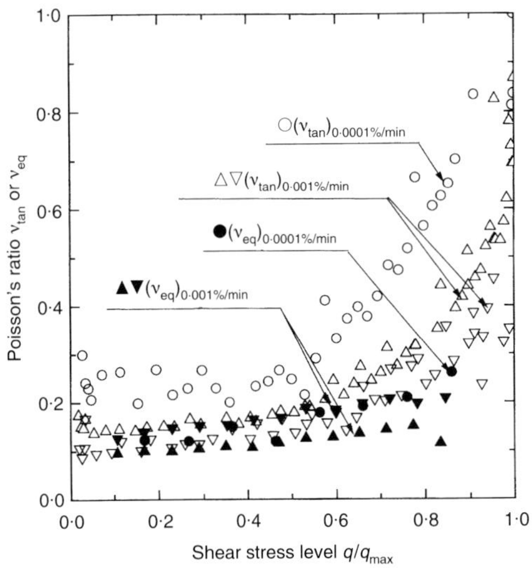

Fig. 12. Summary of ν_{eq} and ν_{tan} plotted against shear stress level q/q_{max}

0·2 and not much affected by strain rate, but gradually increases with q/q_{max}. Fig. 13 shows the relationship between $(\varepsilon_l)_{LDT}$ and $(\varepsilon_a)_{LDT}$ at small strains; recoverable behaviour in both ε_a and ε_l may be seen. The value of ν_{tan} is close to ν_{eq} up to around $q/q_{max} = 0·5$, beyond which ν_{tan} increases rapidly and reaches 0·5 or more at the peak stress state. The behaviour could be explained by the formation of a larger amount of near-vertically oriented micro-cracks on yielding. However, direct evidence was not obtained in these tests. In addition, the values of ν_{tan} increase slightly as the strain rate decreases.

The viscous effects on the stiffness and strength of sedimentary mudstone seen in the data presented in this paper are qualitatively consistent with those obtained by other workers (Nishi *et al.*, 1983). However, more data is required to obtain any general conclusions. Viscous effects on the deformation characteristics at small strains, as observed in laboratory stress–strain tests performed on sedimentary mudstone and other geomaterials (stiff clay, sand and gravel) using a single specimen at a wide range of strain rates were discussed by Di Benedetto & Tatsuoka (1997).

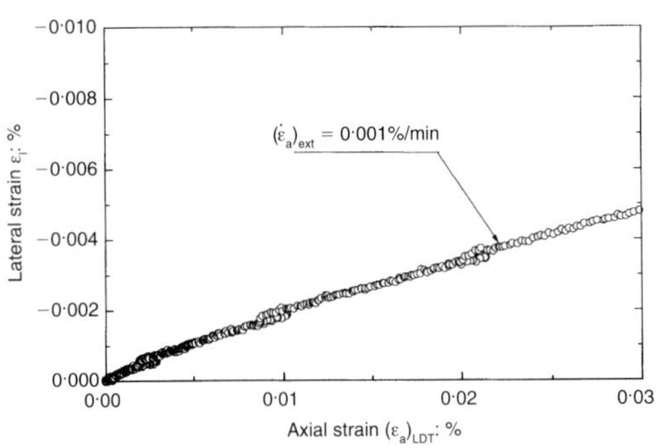

Fig. 13. Relationships between axial and lateral strains for range of $(\varepsilon_a)_{LDT}$ = 0 to 0·03%

CONCLUSIONS

Drained triaxial compression tests on a sedimentary soft rock mudstone were successfully performed, measuring both axial and lateral strains locally for a strain range from less than 0·001% up to around 0·5%. The following conclusions can be derived from the test results.

(*a*) Volumetric strains obtained from the amount of water expelled from the specimen during isotropic and triaxial compression tests were totally unreliable.
(*b*) The deformation during isotropic compression is nearly isotropic.
(*c*) Some effects of strain rate on Young's modulus E_{eq}, defined for elastic increments, was observed. The effects on the tangent Young's modulus E_{tan} were more noticeable, the effects increasing with the shear stress level. The result suggests that such viscous effects should be taken into account when analysing long-term field full-scale behaviour.
(*d*) The drained Poisson's ratio ν_{eq} for elastic strain increments was about 0·15 to 0·2 and was rather independent of strain rate for a range of shear stress level q/q_{max} between 0·0 and 0·8.
(*e*) The drained tangent Poisson's ratio ν_{tan} for total strain increments was close to ν_{eq} for a range of q/q_{max} between 0·0 and 0·5, and increased with q/q_{max} as it exceeded 0·5, reaching 0·5 or more when approaching the peak stress state. The value of ν_{tan} increased slightly as the strain rate decreased.

REFERENCES

Ampadu, S. K. & Tatsuoka, F. (1993). Effects of setting method on the behaviour of clays in triaxial compression from saturation to undrained shear. *Soils Fdns* **33**, No. 2, 14–34.
Di Benedetto, H. & Tatsuoka, F. (1997). Small strain behavior of geomaterials: modelling of strain rate effects. Submitted to *Soils Fdns*.
Bishop, A. W. & Henkel, D. J. (1957). *The measurement of soil properties in the triaxial test*, pp. 124–127. London: Arnold.
Clayton, C. R. I., Gordon, M. A. & Matthews, M. C. (1994). Measurement of stiffness of soils and weak rocks using small strain laboratory testing and field

geophysics. *Proc. 1st Int. Symp. on Pre-failure Deformation Characteristics of Geomaterials, Sapporo, Japan*, Vol. 1, pp. 229–234.

Goto, S., Tatsuoka, F., Shibuya, S., Kim, Y. S. & Sato, T. (1991). A simple gauge for local small strain measurements in the laboratory. *Soils Fdns* **31**, No. 1, 169–180.

Jardine, R. J., Brookes, N. J. & Smith, P. R. (1985). The use of electrical transducers for strain measurements in triaxial tests on weak rock. Technical note, *Int. J. Rock Mech. Min. Sci. and Geomech. Abstr.* **22**, No. 5, 331–337.

Kim, Y. S., Tatsuoka, F. & Ochi, K. (1994). Deformation characteristics at small strains of sedimentary soft rocks by triaxial compression test. *Géotechnique* **44**, No. 3, 461–478.

Matthews, M. C. & Clayton, C. R. I. (1993). Influence of intact porosity on the engineering properties of a weak rock. *Proc. Int. Symp. under the auspices of the ISSMFE, IAEG and ISRM, Athens*, Vol. 1, pp. 693–701.

Nishi, K., Okamoto, T. & Esashi, Y. (1983). Strength–deformation characteristics of mudstone under some kinds of loading conditions and its unificative interpretation. *Proc. Japan Soc. Civ. Engrs* **318**, Oct., 149–158 (in Japanese).

Tatsuoka, F., Sato, T., Park, C. S., Kim, Y. S., Mukabi, J. N. & Kohata, Y. (1994). Measurements of elastic properties of geomaterials in laboratory compression tests. *Geotech. Testing J.* **17**, No. 1, 80–94.

Tatsuoka, F. & Kohata, Y. (1995). Stiffness of hard soils and soft rocks in engineering applications. *Proc. 1st Int. Symp. on Pre-failure Deformation Characteristics of Geomaterials, Sapporo, Japan*, Vol. 2, pp. 947–1063.

Tatsuoka, F., Ochi, K., Tsubouchi, T., Kohata, Y. & Wang, L. (1997). Sagamihara experimental underground excavations in sedimentary soft rock. *Geotech. Engng*, accepted for publication

Thompson, R. P., Newman, R. L. & Davis, P. D. (1993). Advances in the in situ testing of weak mudstone. *Proc. 26th Annual Conf. Engng Group of the Geological Society, Leeds, UK*, pp. 303–310.

Wang, L. (1996). *Study on field deformation characteristics of sedimentary soft rocks by triaxial tests.* Doctoral thesis, University of Tokyo (in Japanese).

Hight, D. W., Bennell, J. D., Chana, B., Davis, P. D., Jardine, R. J. & Porovic, E. (1997). *Géotechnique* **47**, No. 3, 451–474

Wave velocity and stiffness measurements of the Crag and Lower London Tertiaries at Sizewell

D. W. HIGHT,* J. D. BENNELL,† B. CHANA,‡ P. D. DAVIS,§ R. J. JARDINE‖ and E. POROVIC¶

This paper describes an investigation into the body wave velocity and stiffness characteristics of the foundation soils at the site formerly considered for Sizewell C Nuclear Power Station. Results from both in situ and laboratory tests are presented. In situ measurements comprised shear and compression wave velocities using cross-hole and down-hole techniques, including the velocity of horizontally propagating, horizontally polarized shear waves. Laboratory tests included torsional shear tests on hollow cylindrical samples, triaxial compression and extension tests with local measurements of strain, and resonant column tests. Sample quality and the effects of destructuring are evaluated. The anisotropy of the wave velocity and of the stiffness characteristics is highlighted.

KEYWORDS: anisotropy; clays; dynamic measurements; site investigation; stiffness; wave velocities

L'article décrit des recherches sur la vitesse de propagation des ondes de volume et les caractéristiques de rigidité des sols d'assise de l'emplacement proposé de la centrale nucléaire Sizewell C. Il présente les résultats des essais sur place et en laboratoire. Les mesures sur place ont compris la vitesse de propagation des ondes de cisaillement et de compression à l'aide de techniques de forage transversal et descendant, y compris la vitesse de propagation d'ondes de cisaillement horizontales à polarisation horizontale. Les essais en laboratoire ont compris des essais de cisaillement par torsion sur des éprouvettes cylindriques creuses, des essais de compression triaxiale et des essais d'extension, avec la mesure locale des contraintes et des essais de compression de colonne résonante. L'article examine la qualité des échantillons et les effets de la destructuration et met en relief l'anisotropie de la vitesse de propagation des ondes et des caractéristiques de rigidité.

INTRODUCTION

A full dynamic analysis of the interaction of a nuclear power station with its foundation during an earthquake requires as complete a description as possible of the stiffness and body wave velocity characteristics of the ground, including the anisotropy of shear (S) and compression (P) wave velocities, modulus decay and damping. A site investigation carried out in 1994 at the site then being considered for Sizewell C Nuclear Power Station was designed to achieve such a description, within the constraints imposed by the available sampling equipment and laboratory and in situ testing equipment. (Subsequently, a decision was taken on commercial grounds not to proceed with the development, and the planning application was therefore withdrawn.) This paper presents selected results from that investigation, which focus on the seismic and pre-failure deformation properties of the natural soils. Some of the difficulties which were encountered in the investigation are described.

Sizewell is situated on the Suffolk coast. The geology of the site comprises a sequence of Plio-Pleistocene sand (Crag) and Tertiary clays with subordinate sands (i.e. the London Clay, the Woolwich and Reading Beds and the Thanet Beds—the Lower London Tertiaries) overlying the chalk. The strata are essentially flat over the site, with little dip. A simplified stratigraphy of the site is shown in Fig. 1; ground level is at approximately +3 m OD.

The Crag is a very dense, often shelly, shallow marine sand. It is locally silty and contains frequent lenses or layers of silt or clay. Regional geological evidence suggests that the Crag at Sizewell has not been subject to significant ice

Manuscript received 4 November 1996; revised manuscript accepted 14 February 1997.
Discussion on this paper closes 1 November 1997.
* Geotechnical Consulting Group, London.
† University of Wales, Bangor.
‡ Soil Mechanics Ltd, Wokingham.
§ Nuclear Electric Ltd, Barnwood.
‖ Imperial College, London.
¶ Ove Arup and Partners, London; formerly Imperial College.

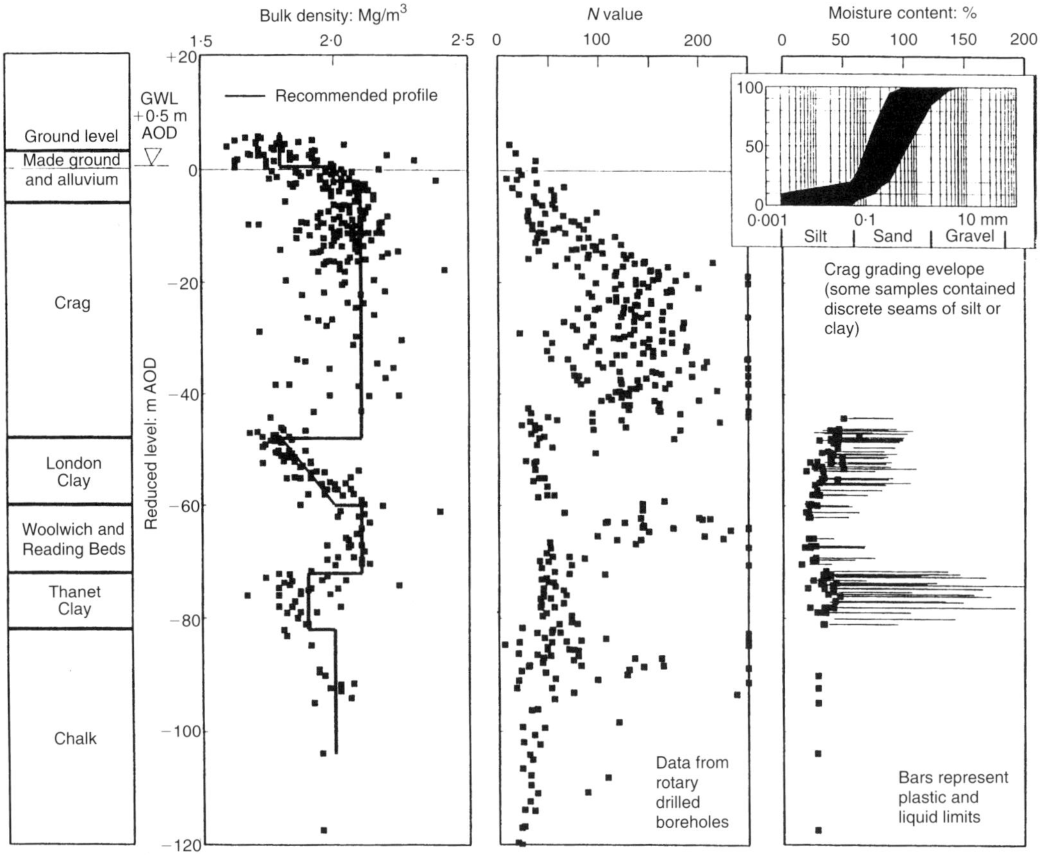

Fig. 1. Simplified stratigraphy of the Sizewell site

loading or removal of overlying sediments. However, it may have been overconsolidated to some extent as a result of natural groundwater lowering during periods of glaciation. The age of the Crag is estimated to be between 0·75 and 2 million years. The London Clay is a very stiff to hard fissured grey clay with silt layers and lenses and occasional ash bands. The stratum is estimated to have had an original thickness of 200–300 m, before being eroded to its present thickness in the Miocene and Pliocene and being overlain by the Crag. The Woolwich and Reading Beds (WRB) comprise fluvio-lacustrine deposits of grey–brown, slightly clayey silty fine sands and very stiff, grey–green, fissured silty clays, often thinly laminated with silt layers. The Thanet Clay is a highly plastic, hard, grey–brown, fissured marine clay which contains localized glauconite and silt pockets. High-pressure CRS oedometer tests on samples of London Clay and Thanet Clay from the site indicate yield stress ratios of 2·3–2·5 at depths of burial of 56 m and 79 m respectively (Cola & Jardine, 1995).

As part of the 1994 site investigation, rotary-cored boreholes were put down to obtain high-quality samples for laboratory testing and to provide access for in situ seismic tests. Cross-hole, down-hole and up-hole seismic tests were carried out in boreholes YC1, YC2 and YC3, while boreholes YA1 to YA5 were used to check the structural geology of the site. Each of these boreholes was geophysically logged. (Surface refraction methods had been used as part of previous investigations at Sizewell, as described by Davis *et al.* (1996).)

The laboratory tests comprised resonant column (RC) tests, carried out at the University of Wales, Bangor (UWB); resonant column–torsional shear (RC–TS) tests, carried out at Imperial College (IC); and undrained triaxial compression (TC) and triaxial extension (TE) tests, carried out by Soil Mechanics Ltd. Details of the samples tested and of the in situ measurements made close to each sampling depth are summarized in Table 1. Values of shear wave velocity V_{hv} and small-strain stiffness $G_{hv}(=\rho V_{hv}^2)$ are given. (The first subscript denotes the direction of propagation and the

Stratum	Borehole No.	Depth: m	Elevation: m OD	Test type	Bulk density of lab specimen Mg/m³		Moisture content of lab specimen		LL: %	PL: %	Assessed V_{hv}: m/s from in situ velocity measurements*			Assessed saturated density ρ: Mg/m³	G_{hv}: MPa		
					Initial	Final	Initial	Final			Min.	Best estimate	Max.		Min.	Best estimate	Max.
Crag	YA4	18·80–19·15	−16·7	TE	2·01	2·01	21·8	20·8	—	—	300	450	560	2·10	189	425	659
	YA1	22·00–22·35	−19·3	TC	2·04	2·06	22·1	21·4	—	—	460	520	560	2·10	444	568	659
	YA1	23·25–23·60	−20·6	RC (UWB)	2·06	—	22·5	—	—	—	480	520	560	2·10	484	568	659
	YC3	36·86–37·12	−34·2	TC	2·13	2·17	19·4	18·8	—	—	550	580	610	2·10	635	706	781
	YC2	37·00–37·35	−34·3	RC (UWB)	2·06	—	20·2	—	—	—	550	580	610	2·10	635	706	781
	YA1	38·10–38·45	−35·4	TE	2·08	2·08	22·0	19·7	—	—	480	570	620	2·10	484	682	807
	YC2	40·85–41·20	−38·2	TC	2·06	2·10	20·3	21·0	—	—	480	570	630	2·10	484	682	833
	YC3	43·16–43·51	−40·5	RC (UWB)	2·07	—	19·3	—	—	—	480	580	630	2·10	484	706	833
	YC1	43·65–44·00	−41·0	TE	2·14	2·19	20·0	19·4	—	—	480	580	630	2·10	484	706	833
London Clay	YA2	50·00–50·40	−47·9	TC	1·78	1·81	40·6	40·6	88	42	260	280	340	1·80	122	141	208
	YC2	50·75–50·90	−48·0	RC (UWB)	1·82	—	40·5	—	—	—	260	280	340	1·80	122	141	208
	YA1	51·80–52·15	−49·1	TE	1·84	1·86	36·1	34·8	91	57	260	280	340	1·85	125	145	214
	YC2	52·70–53·05	−50·0	RC–TS (IC)	1·86	—	37·9	—	79·5	30·4	260	280	340	1·85	125	145	214
	YC1	54·35–54·70	−51·7	RC–TS (IC)	1·99	—	31·8	—	60·0	23·6	260	280	340	1·90	128	149	220
	YA2	54·65–55·00	−52·5	RC–TS (IC)	1·94	—	30·6	—	64·5	26·1	260	280	340	1·90	128	149	220
	YC1	57·50–57·85	−54·8	RC–TS (IC)	1·98	—	29·1	—	62·3	26·2	260	300	370?	1·95	132	176	267
WRB Sand	YA1	65·30–65·65	−62·6	TC	2·04	2·07	23·0	22·1	—	—	460	480	530	2·05	434	472	576
	YC1	66·60–66·90	−63·9	RC (UWB)	1·90	—	26·6	—	—	—	460	480	530	2·00	423	461	562
	YA1	66·70–67·05	−64·0	TE	2·03	2·04	24·0	22·7	—	—	460	480	530	2·05	434	472	576
WRB Clay	YA1	60·45–60·80	−57·8	TC	2·12	2·18	18·9	18·0	—	—	300	?	450	2·10	189	?	425
	YA1	68·30–68·65	−65·6	TC	2·04	2·16	25·0	23·0	47	22	300	?	450	2·05	185	?	415
	YA1	72·45–72·80	−69·8	TC	1·94	1·98	28·7	28·5	74	31	300	?	450	1·85	167	?	375
Thanet Clay	YC2	73·51–73·86	−70·9	TE	1·84	1·86	38·4	37·9	131	34	240	255	290	1·80	104	117	151
	YC3	75·10–75·42	−72·5	TC	1·82	1·85	41·0	40·6	154	39	250	270	310	1·80	113	131	173
	YC2	78·50–78·82	−75·8	RC (UWB)	1·77	—	44·7	—	—	—	260	300	340	1·80	122	162	208
	YA1	80·95–81·30	−78·3	RC–TS (IC)	1·97	—	31·7	—	102·3	37·5	280	380?	500?	1·90	149	274?	475?

* In situ measurements were not always made at the precise sampling depth. Values quoted have slight bias towards low side.

Table 2. Triaxial test data

Borehole No.	Depth: m	Shearing mode	p'_i: kPa	Final consolidation stresses		Volume strains during reconsolidation*				Recoverable strain limit ε_{y2}: %	$\frac{(\sigma_a - \sigma_r)_{max}}{2}$: kPa
						Isotropic stage		Anisotropic stages			
				σ'_a: kPa	σ'_r: kPa	ε_a: %	ε_v: %	ε_a: %	ε_v: %		
Crag											
YA4	18·80–19·15	Extension	—	200	200	0·055	0·161	0·487	0·608	$-0{\cdot}01 < \varepsilon_{y2} < -0{\cdot}005$	−50
YA1	22·00–22·35	Compression	—	240	240	0·022	0·310	0·965	1·137	$0{\cdot}025 < \varepsilon_{y2} < 0{\cdot}05$	+241
YC3	36·86–37·12	Compression	—	400	400	0·319	0·993	0·458	1·253	$0{\cdot}025 < \varepsilon_{y2} < 0{\cdot}05$	+490
YA1	38·10–38·45	Extension	—	420	420	0·169	0·902	0·683	0·748	$-0{\cdot}005 < \varepsilon_{y2}$	−117
YC2	40·85–41·20	Compression	—	450	450	0·125	0·408	0·476	0·838	$0{\cdot}015 < \varepsilon_{y2} < 0{\cdot}02$	+736
YC1	43·65–44·00	Extension	—	480	480	0·274	1·136	1·613	1·623	$-0{\cdot}005 < \varepsilon_{y2}$	−100
London Clay											
YA2	50·00–50·40	Compression	~100	510	360	0·693	1·384	0·250	0·215	$0{\cdot}08 < \varepsilon_{y2} < 0{\cdot}10$	+338
YA1	51·80–52·15	Extension	~106	530	370	0·925	2·299	0·510	0·655	$-0{\cdot}015 < \varepsilon_{y2} < -0{\cdot}01$	−177
WRB Sand											
YA1	65·30–65·65	Compression	—	650	450	0·563	1·626	0·187	0·647	$\varepsilon_{y2} < 0{\cdot}01$	+602
YA1	66·70–67·05	Extension	—	620	470	0·267	1·497	0·165	0·222	$-0{\cdot}005 < \varepsilon_{y2}$	−218
WRB Clay											
YA1	60·45–60·80	Compression	~14	590	410	1·843	3·548	0·327	0·303	$0{\cdot}025 < \varepsilon_{y2} < 0{\cdot}05$	+382
YA1	68·30–68·65	Compression	~50	680	480	2·722	6·153	0·895	1·127	$\varepsilon_{y2} < 0{\cdot}02$	+178
YA1	72·45–72·80	Compression	~150	725	510	0·797	1·576	0·382	0·718	$0{\cdot}02 < \varepsilon_{y2} < 0{\cdot}025$	+446
Thanet Clay											
YC2	73·51–73·86	Extension	~400	750	600	0·436	0·979	0·357	0·371	$-0{\cdot}005 < \varepsilon_{y2}$	−238
YC3	75·10–75·42	Compression	~430	760	600	0·267	1·018	0·386	0·866	$0{\cdot}1 < \varepsilon_{y2} < 0{\cdot}2$	+210

* The relative extents of the isotropic and anisotropic consolidation stages varied. This table should be read in conjunction with Fig. 7.

second subscript denotes the direction of wave polarization.) Tables 2, 3 and 4 provide details from the different laboratory tests.

IN SITU MEASUREMENTS

Conventional cross-hole measurements

A Bison hammer, hydraulically clamped to the internal wall of the plastic casing lining the borehole, was used to provide a source of predominantly vertically polarized shear waves. These waves were detected by two OYO three-component geophones, hydraulically clamped at the same horizontal level as the source in two adjacent in-line boreholes (Fig. 2). Each geophone component was separately recorded on an ABEM Terraloc seismograph. Propagation velocities were computed from the measured difference in travel times to the two geophones (to avoid possible problems with trigger delays) and the difference in path lengths was calculated from the borehole verticality surveys and surface coordinates. The frequency of these horizontally propagating, vertically polarized shear (S_{hv}) waves was approximately 500 Hz.

A borehole sparker was used as a source for generating compression waves and omnidirectional hydrophones were used to receive the signals in adjacent in-line boreholes. This survey was performed at identical levels to those used in the cross-hole shear wave survey. The frequency of the horizontally propagating compression (P_h) waves was approximately 2 kHz.

The success of this type of survey relies on the correct selection of the spacing between the boreholes. Too wide a separation and the detected arrivals could be reflected or refracted events which might not relate to the horizontal level under examination. Too narrow a separation will lead to loss of resolution and accuracy as the borehole, its lining and the surrounding grout (not to mention the disturbed zone around the borehole) will make up a significant portion of the travel path of the waves. A nominal borehole separation of 6–7 metres was adopted as the best compromise in this survey. Owing to departure of the boreholes from vertical, the spacing between boreholes YC1 and YC2 varied from 6·0 m near the surface to 3·6 m at the base. The accuracies achieved in these and other in situ wave velocity measurements are discussed in Appendix 1.

Down-hole survey

This was carried out using a surface source, offset a horizontal distance of 5–6 m from the top of the borehole, to achieve a near-vertical propagation path. The signals were recorded on a twin array of hydraulically clamped, three-component OYO geophones, 6·3 m apart, which were progressively lowered down to the base of the borehole (Fig. 3). For compression waves, a vertical hammer impact onto a metal plate laid on the ground surface was used as the source. For shear waves, a horizontal traction method (a weighted plank with impacts made at each end) was used to provide a source of vertically travelling, horizontally polarized shear (S_{vh}) waves. To aid signal identification, the geophones were aligned so as to have one of their horizontal components parallel to the plank, by means of attached orientation rods. Again, interval velocities were calculated from differences in the travel times to each of the receivers and the known difference in path length. The frequencies of the vertically propagating shear (S_{vh}) and compression (P_v) waves were approximately 60 Hz and 200 Hz respectively.

Horizontally polarised shear wave cross-hole measurements

For the 1994 site investigation, a new seismic borehole source, which was capable of generating

Table 3. Resonant column test data

Stratum	Borehole	Depth: m	Isotropic consolidation stress: kPa	Laboratory				In situ G_{hv}: MPa
				Bender		Resonance		
				V_s: m/s	G_{vh}: MPa	V_s: m/s	$G_{\theta z}$: MPa*	
Crag	YC2	37·00–37·35	480 → 400	350	260	362	279	706
	YC3	43·16–43·51	560 → 470	378	311	396	341	706
London Clay	YC2	50·75–50·90	510 → 425	301	165	285	148	141
WRB Sand	YC1	66·60–66·90	650 → 540	301	185	294	177	461
Thanet Clay	YC2	78·50–78·82	810 → 675	292	154	266	128†	162

* $G_{\theta z}$ at small strains, initial value.

† $G_{\theta z}$ of 151 MPa was measured initially but this degraded rapidly to 128 MPa at small strains. The specimen had been tested previously but slip was thought to have occurred at the specimen/platen interface. No significant change in G_{vh} from benders was found between the two tests.

Table 4. Resonant column–torsional shear test data

Stratum	Borehole	Depth: m	p'_i: kPa		Final consolidation stresses		ε_a during consolidation: %	Lab. $G_{\theta z \max}$: MPa	In situ G_{hv}: MPa	$\frac{(\sigma_1 - \sigma_3)_{\max}}{2}$: kPa
			Before specimen preparation	After specimen preparation	σ'_a: kPa	σ'_r: kPa				
London Clay	YC2	52·70–53·05	53	53	522·0	378·5	1·55	100	145	298
	YC1	54·35–54·70	150	197	550·0	390·0	1·8	101	149	225
	YA2	54·65–55·00	245	307	540·0	380·0	1·3	110	149	238
	YC1	57·50–57·85	50	70	757·0	418·0	2·8	94	176	302
Thanet Clay	YA1	80·95–81·30	340	420	750·0	595·0	1·2	148	274?	337

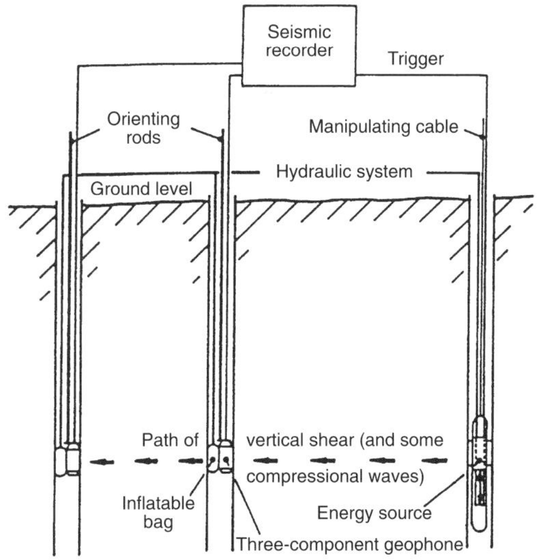

Fig. 2. Set-up for cross-hole measurements

primarily horizontally polarized shear (S_{hh}) waves, was designed and constructed. This used a horizontally oriented solenoid which was energized by passing 24 V DC through the coil for a short time, causing the metal core to be pulled into the coil to impact against a rigid plate. The solenoid was contained in a waterproof housing which was attached to the bottom of a hydraulically clamped, three-component OYO geophone to facilitate coupling the source to the borehole wall, aligning the horizontally polarized shear motion to the required plane, with the aid of the attached orientation rods, and monitoring the exact instant of shear wave generation by recording the signals generated immediately adjacent to the source.

The signals were detected in an adjacent borehole with a twin array of hydraulically clamped, three-component OYO geophones. The array was oriented with respect to the direction of polarization of the generated shear waves by means of the attached orientation rods. The top geophone of the array was placed at the same horizontal level as the source, with the second geophone 6·3 m below this level. Travel times were computed from the differences in arrival times recorded by each geophone and by the geophone attached to the top of the source. This gave very similar compression wave velocities to the conventional cross-hole data. The frequency of the shear waves was approximately 500 Hz.

Initial testing was carried out at −12 m OD in the Crag deposits. Both source and receivers were rotated between shots to confirm that the energy recorded was primarily from horizontally travelling, horizontally polarized shear waves. Once this

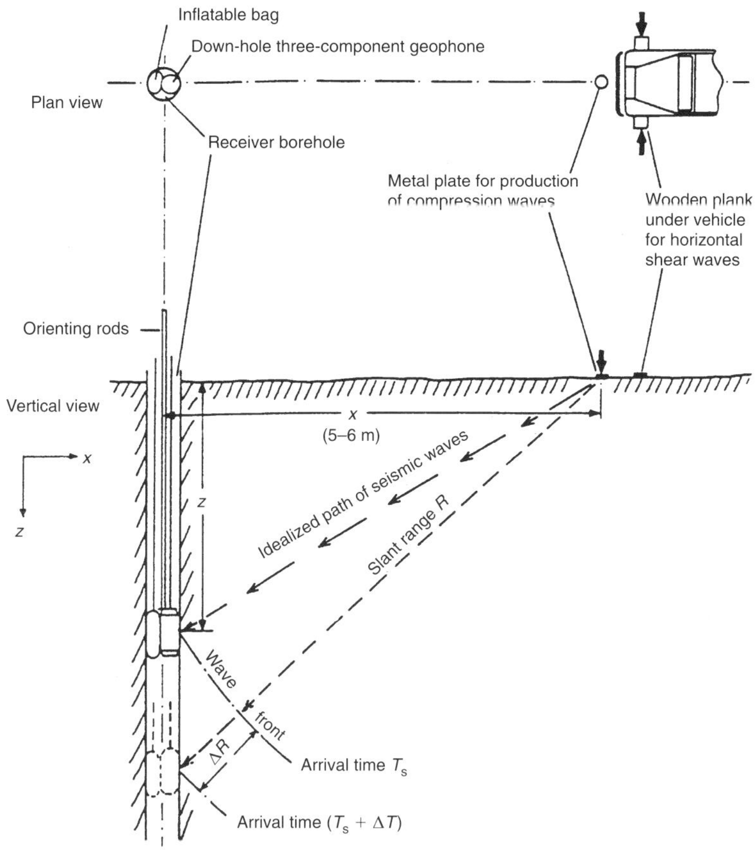

Fig. 3. Set-up for down-hole measurements

was established, the source and receiver array were lowered to an elevation of −47 m OD. Both shear and compression waves were recorded at this depth and at successive 3 m depth intervals, maintaining the source/receiver orientations, until an elevation of −83 m OD was attained.

In situ wave velocity profiles

The general pattern of cross-hole velocities shown in Fig. 4 conforms very well to the lithological boundaries, with slight discrepancies due to refraction effects as the source/receiver array approaches a higher-velocity boundary. (Fig. 4 also shows P wave velocities measured in down-hole tests and these are discussed later.)

The Crag exhibits a steep velocity gradient for V_{hv}, from 130 m/s at −8 m OD to 560 m/s at −20 m OD. V_{hv} then drops to 400 m/s at −25 m OD, followed by a rise to a constant value of 600 m/s from −30 to −43 m OD. V_{ph} in the Crag seems to follow a similar pattern, with a gradient from 1600 to 1950 m/s down to −20 m OD, dropping to 1750 m/s at −25 m OD and then rising to a value close to 2000 m/s from −30 to −43 m OD. The reduced velocity at about −25 m OD appears, from the cores and static cone tests, to be associated with an increase in the frequency and/or thickness of clay layers.

Excluding the zones at the top and bottom of the stratum, where the wave velocities are higher because of refraction through the stiffer adjacent layers, the London Clay is characterized by almost constant velocities of 300 m/s for V_{hv} and 1600 m/s for V_{ph}. The Woolwich and Reading Beds are composed of sands at the location of the boreholes for wave velocity measurements (elsewhere the WRB include some silts and clays). As a result, they exhibit higher velocities than the London Clay, with a V_{hv} of 450 m/s and V_{ph} of 1750 m/s, both showing evidence of a slight increase with depth in the layer. In the Thanet Clay, V_{hv} increases with depth from approximately 250 m/s to 380 m/s, although the higher values near the base of the layer

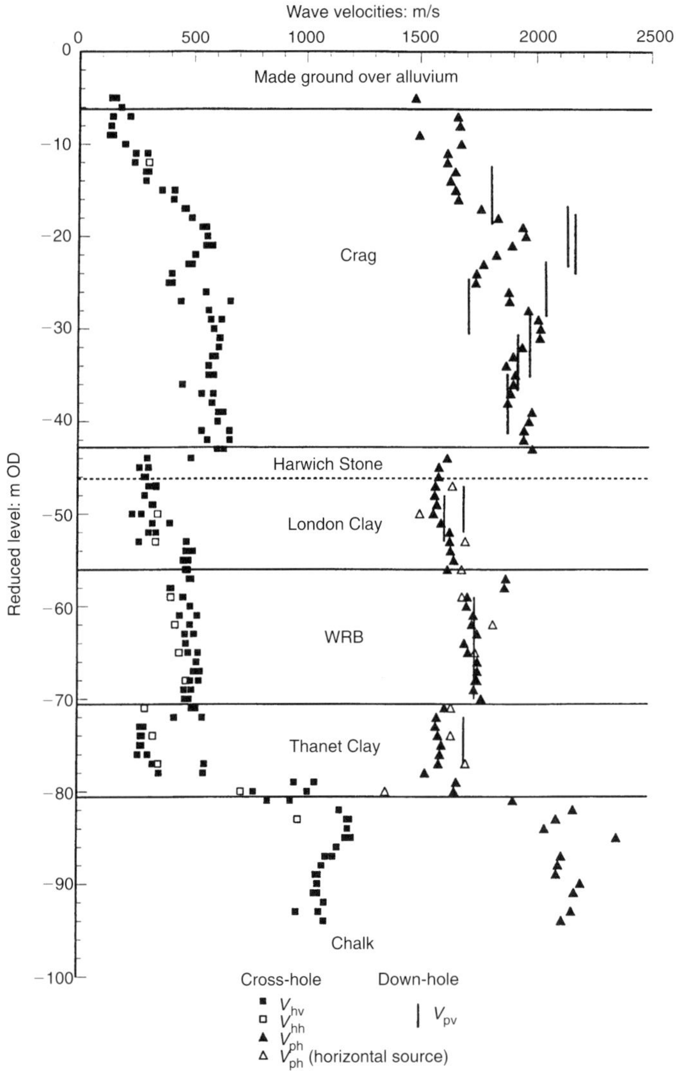

Fig. 4. Profiles of shear and compression wave velocities at Sizewell (all data from 1994 site investigation; see Fig. 16 for more detailed shear wave velocity data)

may have been influenced by refracted arrivals as the top of the Chalk was approached. V_{ph} fluctuates between 1500 and 1650 m/s, but, for the same reasons, increases near the base of the layer.

The Chalk is characterized by a high-velocity zone several metres thick just below its top (V_{hv} = 1200 m/s and V_{ph} = 2350 m/s); this corresponds to the hard ground of the Catton Sponge Bed which is a consistent feature of the Chalk sequence at Sizewell. Below about −85 m OD, however, these velocities reduce to near-constant values of V_{hv} of 1100 m/s and V_{ph} of 2150 m/s.

LABORATORY MEASUREMENTS

Resonant column tests

Resonant column tests were run at UWB on isotropically reconsolidated samples of Crag, London Clay, WRB Sand and Thanet Clay. Cores were carefully trimmed down to a diameter of 70 mm and a length of 140 mm. Testing followed the procedures described by Bennell *et al.* (1984) and Bennell & Taylor Smith (1991). Because of the high stiffness of the samples being tested, special precautions were taken to ensure coupling of the specimen to the apparatus platens. Each specimen was bonded to the top and bottom platens using a plaster-of-Paris cement; this proved to be reasonably successful, even for the limited cement penetration achieved with the clay specimens.

Reconsolidation involved consolidating under an isotropic stress, similar to the in situ vertical effective stress, which was higher than the mean effective stress in the ground, p'_0, and then swelling back to p'_0 before carrying out resonance measurements to the strain limit of the equipment. (The isotropic consolidation stresses are listed in Table 3.) All the resonance measurements were carried out under undrained conditions. Although the specimen was allowed to rest between each set of high-amplitude measurements, no drainage was permitted. Thus, any excess pore water pressures that were generated by the high-amplitude cycling were allowed to build up. Low-strain measurements were made after each high-strain excursion to monitor the specimen recovery and to check for any degradation in small-strain stiffness. Frequencies involved in the resonance tests ranged between 50 and 150 Hz.

Material damping ratios were determined by three methods: magnification factor (steady state), logarithmic decrement (amplitude decay) and width of the amplitude response curve (bandwidth). The methods were found to give remarkably consistent results.

The results of the resonant column tests are presented as follows.

(*a*) Figure 5 shows the variation in secant shear modulus $G_{\theta z}$ with shear strain amplitude. (Note that no corrections have been applied for cyclic degradation due to previous high-shear-strain excursions.) Values for $G_{\theta z \max}$, measured in low-strain-amplitude tests after high-strain excursions, are also shown.

(*b*) Figure 6 shows the variation in shear damping ratio with shear strain amplitude. (The data for damping in the Thanet Clay has been omitted because anomalously high values were recorded—this was believed to be the result of apparatus damping.)

The values of shear wave velocity and of the initial $G_{\theta z \max}$ are summarized in Table 3. The accuracies of these and the other laboratory tests are discussed in Appendix 1.

Bender element measurements

The resonant column apparatus at UWB is equipped with bender elements mounted in the top and bottom platens. Using the techniques described by Bates (1989), measurements of shear wave velocity V_{vh} were made in each sample and these are summarized in Table 3, together with the derived values of G_{vh}. The wave frequencies involved were between 1 and 10 kHz. The path length was taken as the tip-to-tip length of the bender elements, and the arrival time was taken as the first significant jump. In the clays, a slot was pre-cut at each end of the specimen to receive the bender and to reduce the risk of splitting the specimen. The plaster-of-Paris bonding was kept

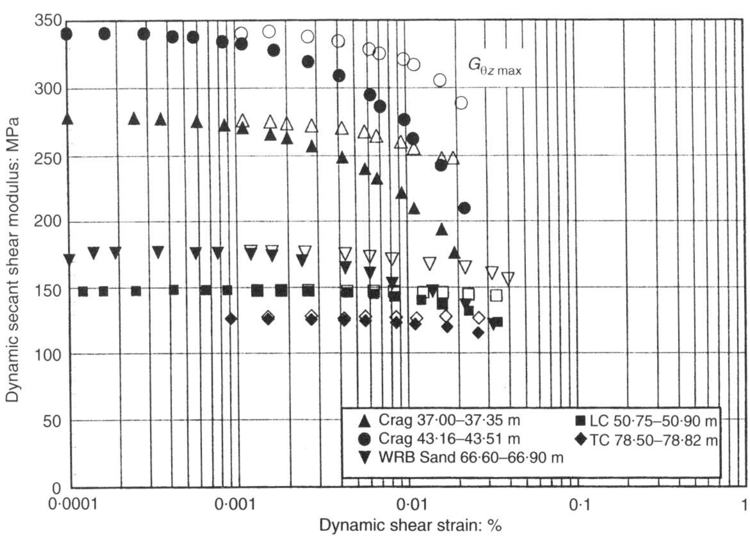

Fig. 5. Dynamic shear modulus from resonant column tests (open symbols denote $G_{\theta z \max}$ values measured after the corresponding high-strain stage represented by the solid symbols)

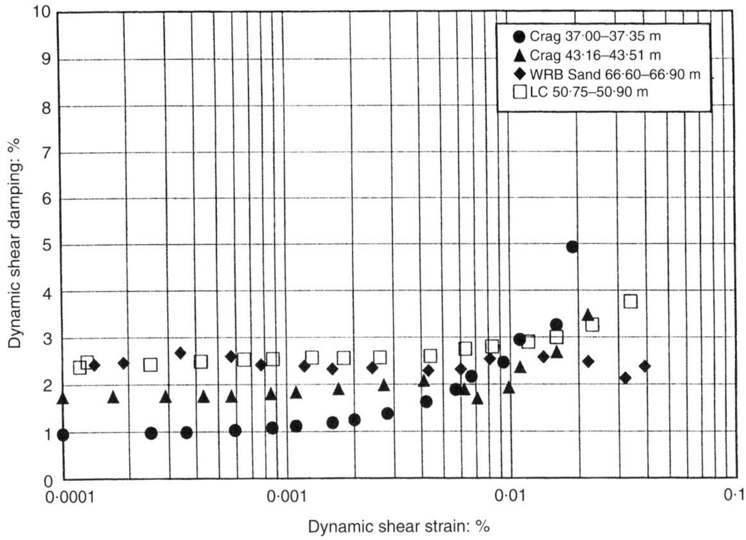

Fig. 6. Dynamic shear damping from resonant column tests

well away from the slot to avoid any influence on the benders.

In the Crag, the G_{vh} values derived from the bender measurements were consistently lower than the $G_{\theta z}$ values from the resonant column (by approximately 10%), but the reverse was true in the London Clay and Thanet Clay. This may be related, at least in part, to differences in frequency and strain amplitude between the two types of measurement, and possibly to additional disturbance around the bender element in the Crag samples.

Undrained triaxial tests

Undrained triaxial compression and extension tests were run on nominally 100 mm diameter × 200 mm high specimens of Crag, London Clay, WRB Sands, WRB Clay and Thanet Clay. Specimens were cut to length from the cores and were set up without further trimming. The specimens carried both axial and radial displacement transducers of the Hall-effect type (Clayton & Khatrush, 1986) and a pore pressure probe, all mounted at mid-height to provide local measurements of axial strain, radial strain and pore pressure. Measurements of the initial effective stress p_i' were made on the clay specimens.

Each specimen was consolidated to its estimated in situ effective stresses, following what was believed to be the most significant event in its recent stress history. Typical consolidation paths that were followed are shown in Fig. 7. For specimens of Crag, K_o was assumed to be 1·0, on the basis of the results of self-boring pressuremeter

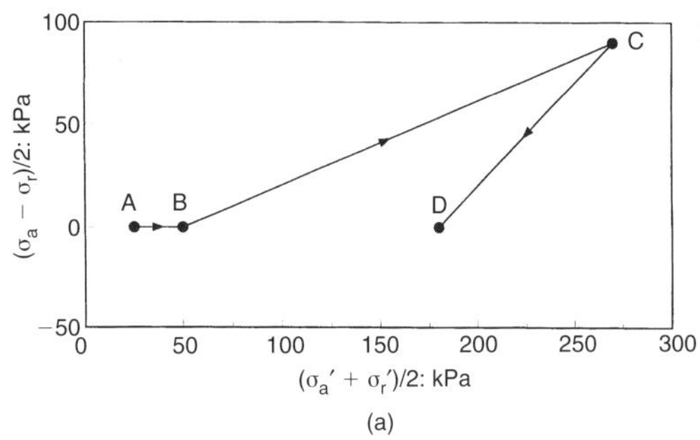

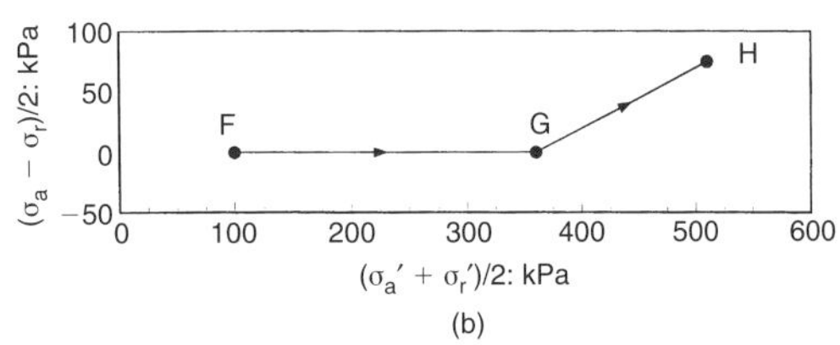

Fig. 7. Reconsolidation paths in triaxial and torsional shear tests: (a) typical path for Crag sample; (b) typical path for Tertiary clay sample (GH = loading by deposition of Crag; location of F varies according to p_i')

tests, and the most recent stress history was considered to be overconsolidation as a result of groundwater lowering during glacial times. For specimens of London Clay, the WRB Sands and Clay, and Thanet Clay, K_o was assumed to be 0·7, 0·7 and 0·8 respectively, and the most significant recent stress history involved loading by deposition of the Crag. The back-pressure during consolidation was 300 kPa.

These tests were run with a series of unload–reload loops of increasing strain amplitude in order to

(*a*) investigate the effects of destructuring on the stiffness characteristics (some destructuring was expected to take place under the stresses imposed by the structures and plant loading)
(*b*) identify recoverable strain limits
(*c*) examine the potential for pore pressure build-up.

Drained creep was permitted at the end of each consolidation stage until the rate of axial strain $\dot{\varepsilon}_a$ was less than 0·002%/h. A 10 min wait was introduced before each unloading and reloading. For undrained shear of the Crag, an axial strain rate of 1%/h was used; for the clays $\dot{\varepsilon}_a$ was 0·2%/h during undrained shear. Triaxial extension was achieved by reducing the axial stress σ_a at constant radial stress σ_r.

The results of the triaxial tests are summarized as follows.

(*a*) Figures 8 and 9 show the effective stress paths observed in the tests on the sands (Crag and WRB) and on the clays (London, WRB and Thanet) respectively.
(*b*) Figures 10–12 show the variations in undrained secant stiffnesses $E_{u\,sec}$ with axial strain, derived from the local measurements; envelopes of the stiffness data from all the unload–reload loops are presented.

Table 2 contains information on p_i', undrained strengths and recoverable strain limits.

The stiffness data for triaxial compression of London Clay falls within the band of data found in the 1988 site investigation, presented by Hight & Jardine (1993). The lack of resolution at very small strains meant that it was not possible to resolve any linear range of behaviour. Problems were also encountered in maintaining a constant deviator stress prior to the unloading and reloading loops.

Resonant column–torsional shear tests

Resonant column–torsional shear (RC–TS) tests were performed at IC on hollow cylindrical specimens of both the London Clay and the Thanet Clay. The RC–TS equipment has been described by Porovic & Jardine (1994); it subjects hollow cylindrical specimens to equal internal and external pressures, an axial force and a torque. The specimens were trimmed to 38 mm ID, 71 mm OD and 191 mm high. The suction in each specimen was measured using the Imperial College suction

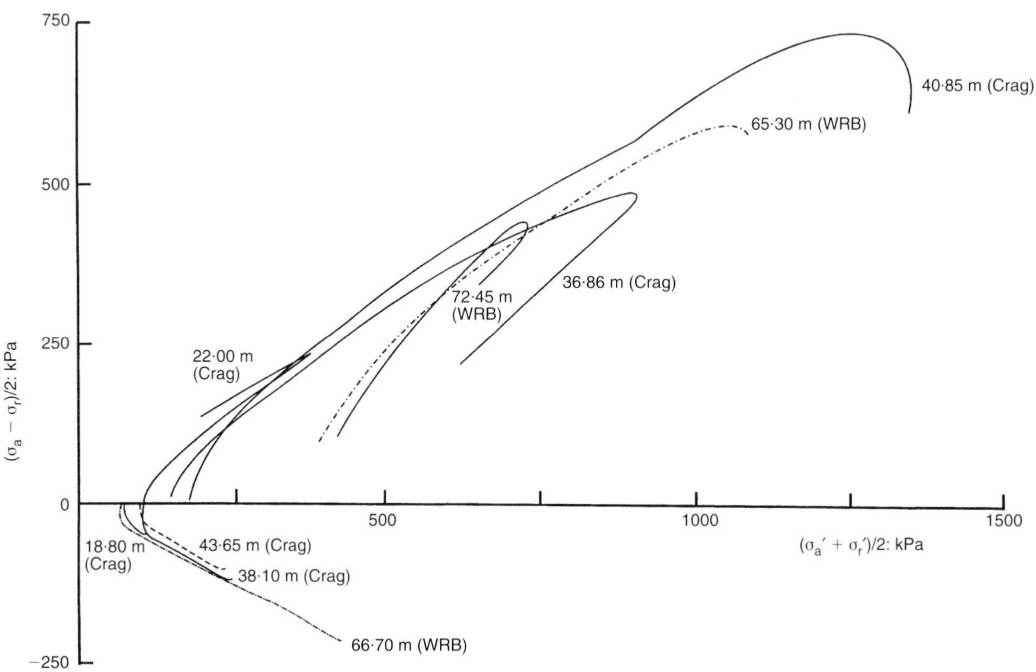

Fig. 8. Effective stress paths during undrained triaxial compression and extension of Crag and WRB sands (migration of effective stress paths during load–unload loops omitted for clarity)

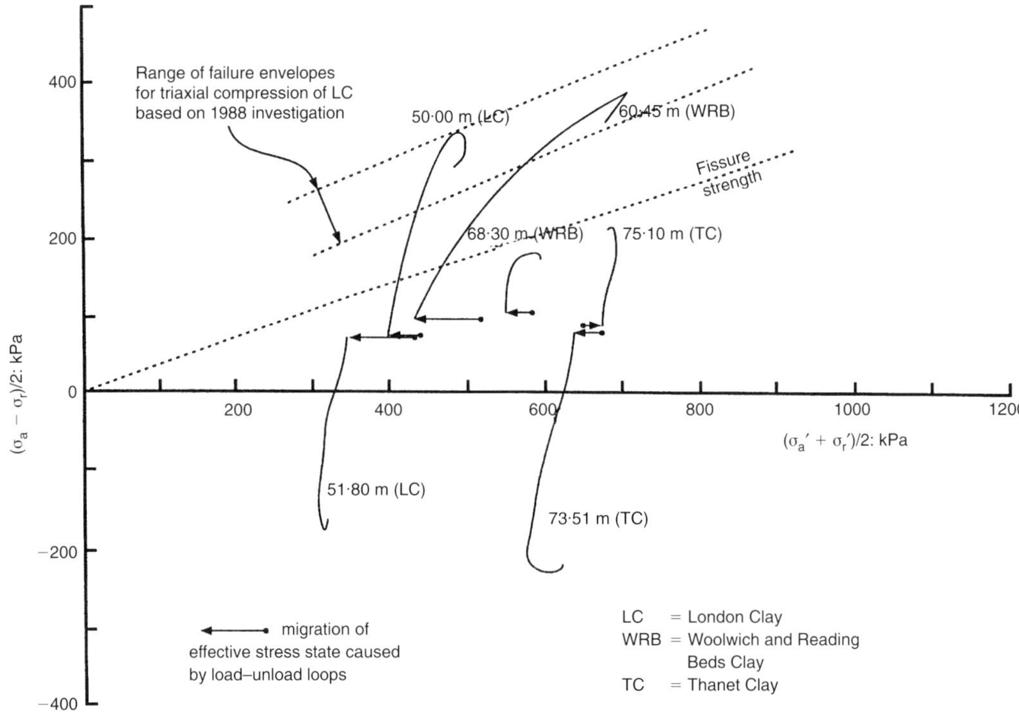

Fig. 9. Effective stress paths during undrained triaxial compression and extension of London Clay, WRB Clay and Thanet Clay

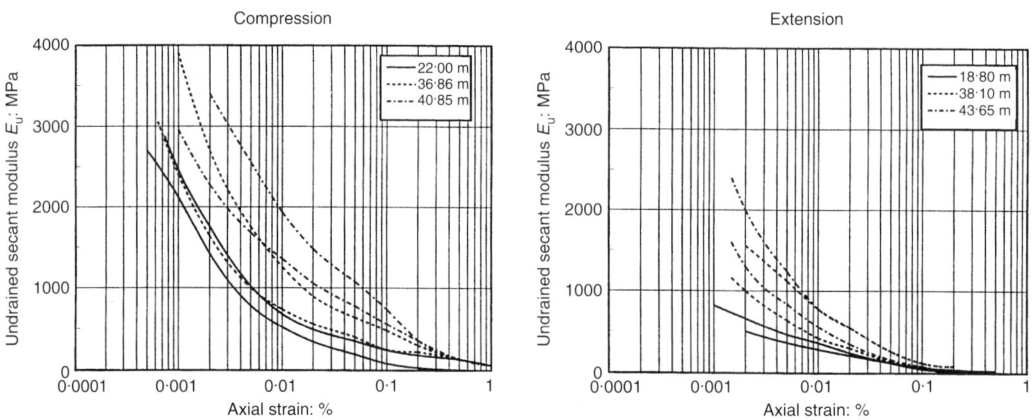

Fig. 10. Envelopes of modulus decay curves for undrained triaxial compression and extension, for Crag

probe (Ridley & Burland, 1993), both before and after trimming, to check on the amount and effects of drying that might occur in the trimming process. The measured suctions are summarized in Table 4.

Four of the specimens were consolidated to their estimated in situ stresses along the stress path shown as path FGH in Fig. 7, using a back-pressure of 300 kPa, except for the deepest sample (from 80·95 m) where air pressure limitations allowed a back-pressure of only 100 kPa. A fifth specimen, YC1-57·5 m, was consolidated beyond its in situ stresses to examine the effects of consolidation under plant loading. Drained creep was allowed at the end of consolidation until the rate of axial strain fell below 0·002%/h. Torsional shear

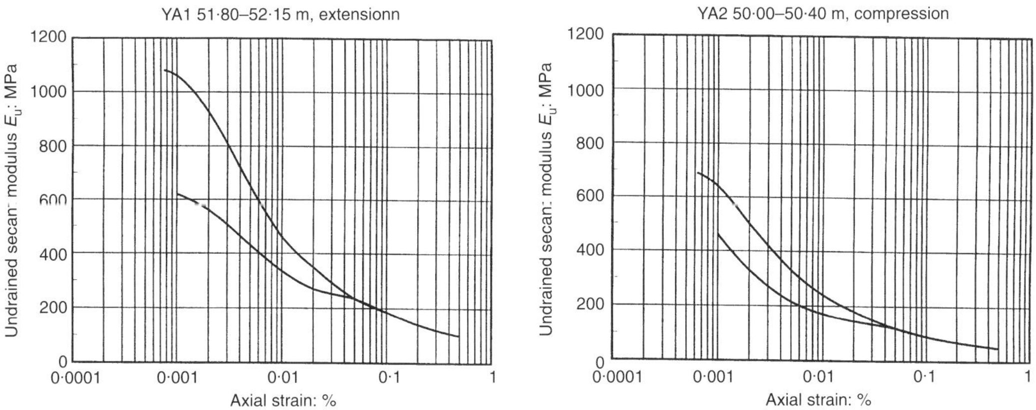

Fig. 11. Envelopes of modulus decay curves for undrained triaxial compression and extension, for London Clay

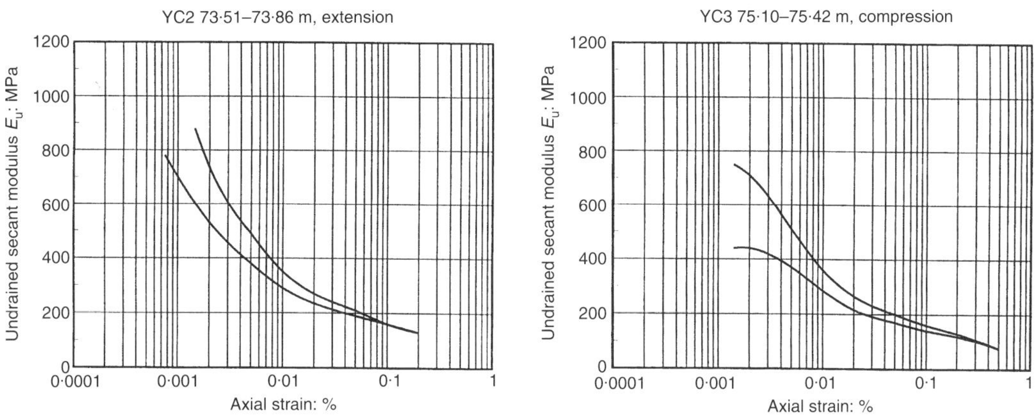

Fig. 12. Envelopes of modulus decay curves for undrained triaxial compression and extension, for Thanet Clay

was applied under undrained conditions (except in test YC2-52·7 m, which was drained) with torsional strain being applied at a rate of 0·2%/h. During torsional shear, the total axial stress and total radial stress were held constant.

Using the resonant column facility in the RC–TS apparatus, measurements of $G_{\theta z\,\max}$ were made at various stages in the consolidation process, immediately prior to undrained torsional shear and at intervals during undrained shear. Frequencies were in the range 120–180 Hz.

The results of the RC–TS tests are presented as follows:

(*a*) applied shear stress $\tau_{\theta z}$ plotted against shear strain $\gamma_{\theta z}$, for all tests (Fig. 13)
(*b*) $G_{\theta z\,\max}$ and the secant shear modulus $G_{\theta z}$ plotted against $\log \gamma_{\theta z}$ for two tests on London Clay (Fig. 14)
(*c*) effective stress paths as $(\sigma_1 - \sigma_3)/2$ plotted against $(\sigma_1' + \sigma_2' + \sigma_3')/3$ (Fig. 15).

This was the first time the RC–TS equipment had been used to test hard clays or to operate at such high stresses, and some difficulties were encountered. After a period of several days air, which had to be retained at the top of the outer cell around the resonator, started to diffuse through the membrane into the drainage lines. To combat this, the testing times, including creep periods, were kept as short as possible and measurement of the initial effective stresses using the suction probe, prior to installation in the cell, helped in this respect.

In a trial test, slip occurred at the top platen/specimen interface when torque was applied after swelling the specimen to low effective stresses. To prevent this happening subsequently, pins were introduced into the top platen for test YC2-52·7 m

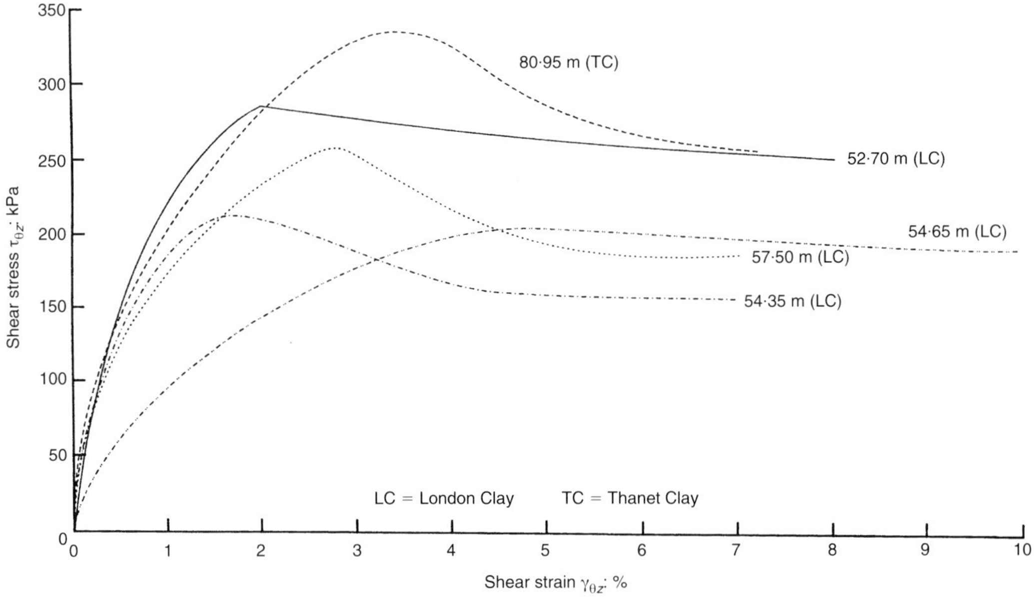

Fig. 13. Shear stress $\tau_{\theta z}$, plotted against shear strain $\gamma_{\theta z}$ in undrained torsional shear tests on London Clay and Thanet Clay

and a cement–sand–silt mix was introduced at the interface in test YA2-54·7 m. For the remaining tests, rougher, porous, sintered bronze platens were used to ensure that failure took place within the specimen. The plaster-of-Paris end caps used in the RC tests at UWB were not employed because of concern that they might affect the strengths developed in torsional shear.

DISCUSSION

Sample quality

The validity of the results of any investigation on natural soil samples depends on the quality of the samples being tested. In an investigation of stiffness characteristics at small strains, it is essential that destructuring, that is, damage to cementing or ageing effects, is avoided. The London Clay and Thanet Clay at Sizewell are known to be structured (Hight & Jardine, 1993). The in situ shear wave velocities in the Crag (Fig. 4 and Table 1) are significantly higher than would be measured in very dense reconstituted sands (see, for example, the data presented for Ticino Sand by Bellotti *et al.* (1996)). This indicates that this material is also structured, which is not surprising in view of its age; cementing is, however, not obvious by inspection. The sampling methods, sample sealing, handling and preparation methods that were employed followed current best commercial practice (Hepton, 1996; Hight, 1996), but this does not guarantee that disturbance would not occur, particularly in materials with a delicate structure, such as the Crag.

One method which has been proposed for assessing whether or not any damage has occurred to cementing or ageing effects is to compare shear wave velocities, measured on the material in situ and on samples in the laboratory (e.g. Hight, 1993). In making this comparison, there should be similarity between the sample in the laboratory and in situ, in terms of stress state, wave propagation and polarization directions, wavelength, etc. Two laboratory measurements are available for this comparison, values of G_{vh} from bender element tests and values of $G_{\theta z\,max}$ from low-strain-amplitude resonant column tests, and these may be compared with in situ measurements of G_{vh} at the sampling depths. Tables 1, 3 and 4 list the estimated in situ values of G_{vh} and Table 5 shows the ratio of G_{vh} to G_{hv} found in situ.

From this comparison, the following is evident.

(*a*) The specimens of Crag and WRB sands tested in the resonant column apparatus had been mechanically disturbed; $G_{\theta z\,max}$ and G_{vh} measured in bender element tests were considerably lower than the in situ values of G_{hv} (and G_{vh}). However, the values of $E_{u\,sec}(= E_{vu})$ derived from the triaxial compression tests on Crag at strains of 0·001% were considerably higher than $3G_{\theta z\,max}$ or even $3G_{hv\,in\,situ}$. ($E_{vu} = 3G$ applies to an isotropic elastic material in which Poisson's ratio is 0·5, i.e. in an isotropic

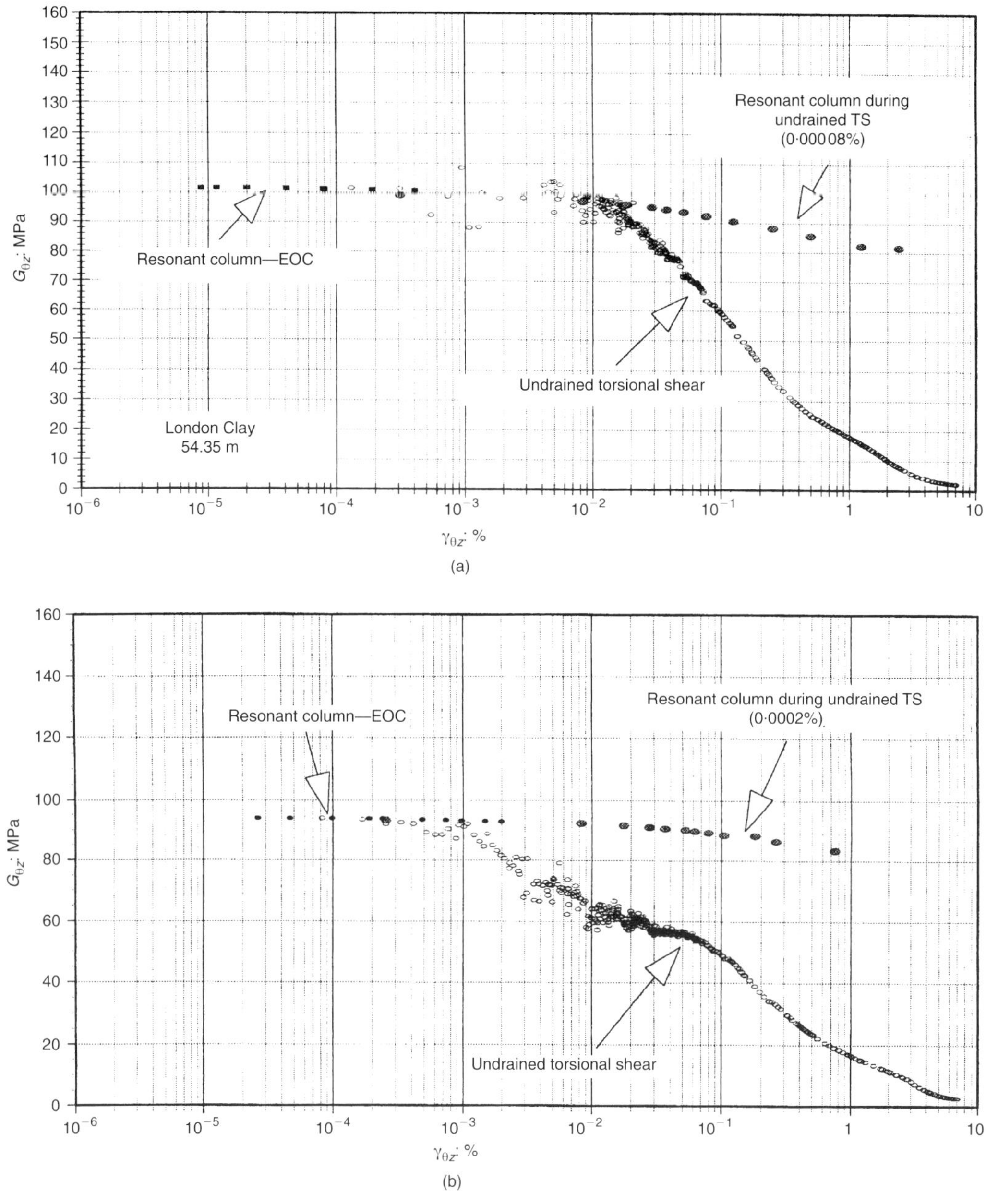

Fig. 14. Shear modulus $G_{\theta z}$ plotted against shear strain $\gamma_{\theta z}$ from undrained torsional shear tests on London Clay: (a) 54·35 m; (b) 57·50 m

saturated soil.) Although there is no link between E_v and G_{hv} in a cross-anisotropic material, the fact that $E_{vu} > 3G_{hv\,in\,situ}$ could indicate that the samples and the triaxial specimens had not been seriously disturbed, whereas the RC specimens were damaged during trimming. (In medium dense dry Ticino Sand, Bellotti *et al.* (1996) found the ratio E_v/G_{vh} to vary with consolidation stress ratio $K(=\sigma'_h/\sigma'_v)$. It ranged between approximately 1·75 at $K = 2$ and 2·75 at $K = 0{\cdot}5$.) Unlike the RC specimens, the triaxial specimens were not trimmed down in diameter. It is our interpretation that the Crag retained its structure after sampling and further evidence to support this view is presented below.

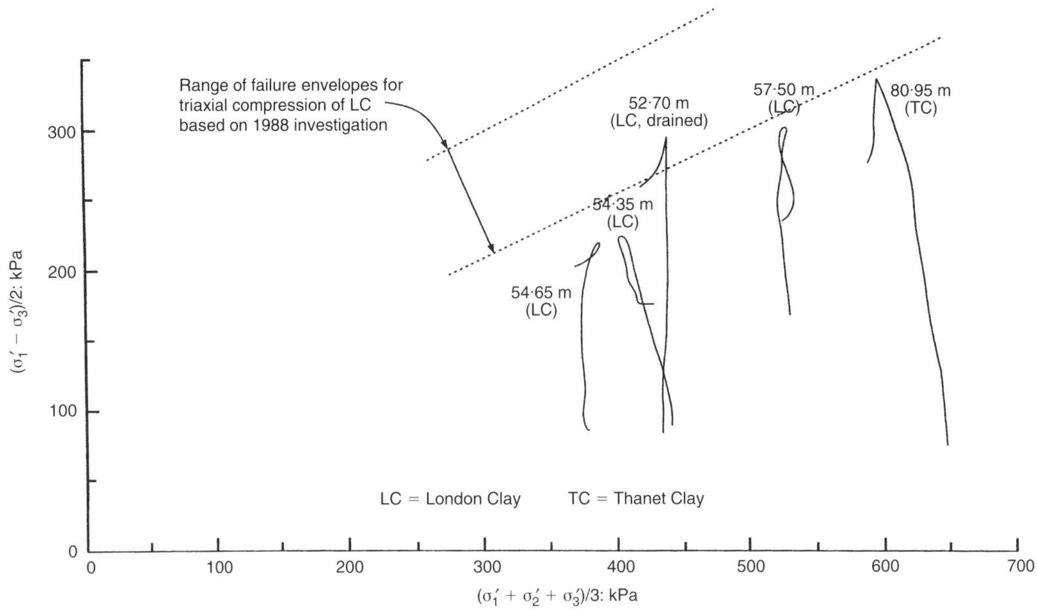

Fig. 15. Effective stress paths during undrained torsional shear of London Clay and Thanet Clay

Table 5. Ratios of shear and compression wave velocities and small-strain stiffnesses

Stratum		V_{vh}/V_{hv}	G_{vh}/G_{hv}†	V_{hh}/V_{hv}	G_{hh}/G_{hv}†	V_{pv}/V_{ph}
Crag	In situ	0·95	0·9	—	—	1·05–1·1
London Clay	In situ	0·9–0·95	0·8–0·9	up to 1·2	up to 1·45	1·05
	Lab.*	0·9–1·3	0·8–1·7	1·4–1·5	2·0–2·3	1·05
WRB	In situ (Sand)	0·95	0·9	0·9	0·8	1·0
	Lab.* (Clay)	0·9	0·8	0·9–1·0	0·8–1·0	1·0
Thanet Clay	In situ	1·0	1·0	1·15	1·3	1·05
	Lab.*	1·4	2·0	1·25–1·5	1·6–2·3	0·95–1·0

Frequencies: Hz

	In situ	Laboratory (unconfined)
V_{vh}	60	1000
V_{hv}	500	1000
V_{hh}	500	1000
V_{pv}	200	250 000
V_{ph}	2000	250 000

* Measurements on unconfined U100 samples.
† Based on shear wave velocities.

(*b*) The samples of London Clay probably survived the rotary coring without significant damage; there was close agreement between the laboratory measurement of $G_{\theta z\,\max}$ on a solid RC specimen (148 MPa) and the in situ value of G_{hv} (141 MPa), and the absence of damage is consistent with previous findings related to rotary cores of London Clay. However, the RC–TS tests on hollow cylindrical specimens gave $G_{\theta z\,\max}$ values that were consistently lower, by approximately 35%, than the corresponding in situ values. The following may have contributed to these differences.

(i) The preparation of the RC–TS specimens involved a relatively complex trimming operation, which was carried out on samples which had, in some cases, swelled to relatively low effective stresses and which dried to some extent during trimming; movement on fissures during trimming was a possibility.

(ii) Creep periods were shortened in the RC–TS tests to minimize problems of air diffusion.

(iii) There may have been imperfect compliance between the specimens and the end platens, recalling that plaster-of-Paris end caps were not used, and some softening local to the ends could have occurred when installing the specimens.

In comparing the values of $G_{\theta z\,max}$ measured in the UWB RC tests and the IC RC–TS tests on London Clay, the differences in consolidation paths should be noted; the former involved isotropic consolidation and swelling to the in situ mean effective stress, while the latter involved consolidation to the in situ anisotropic stress state.

(*c*) In the Thanet Clay sample from 78·5 m, both G_{vh} from the bender test and the initial value of $G_{\theta z\,max}$ from the RC test were similar to the in situ value of G_{vh}, again suggesting the absence of significant disturbance. (Because of the high plasticity of the Thanet Clay, a reasonable tolerance to disturbance would be expected, although displacement on fissures remains a possibility.) Again, the value of $G_{\theta z\,max}$ from the RC–TS test was less than G_{vh} in situ.

Reductions in effective stress when rotary-coring the sands and clays are inevitable. The size of this reduction in the clays can be seen in Tables 2 and 4 by comparing the measurements of p'_i with the estimated in situ stresses. The Thanet Clay sustained high suctions; this is again consistent with its high plasticity and low permeability. Suctions retained in the London Clay and WRB Clay were lower and variable, probably because of the presence of more permeable seams, particularly in the WRB. In the WRB clays, swelling to very low effective stresses occurred in some cases, resulting in gross disturbance which was evident from the large volume changes recorded during reconsolidation. Volume strains during reconsolidation also indicate the possibility of damage to sample YA1-51·8 m from the London Clay, which was subject to triaxial extension, and of additional damage to sample YC1-43·65 m from the Crag.

In summary, we believe the following.

(*a*) The data from triaxial compression and extension on the Crag may be representative of in situ behaviour; the data from resonant column tests requires correction for some destructuring.

(*b*) The data from triaxial compression and resonant column tests on solid specimens of London Clay is representative of in situ behaviour; the data from RC–TS tests may require some upward correction. The same is true for Thanet Clay, except that the data from triaxial extension is also probably reasonable.

(*c*) The data from all tests on the WRB clays and sands has been significantly affected by sample disturbance; this is why this data has not been presented and discussed in detail.

Anisotropy

Shear wave velocities. Detailed comparisons of the different shear wave velocities are presented in Fig. 16. In the Crag, the down-hole shear wave velocities V_{vh} generally form a lower bound to the vertically polarized cross-hole shear wave velocities V_{hv}, and can be said to be approximately 5% lower than the average values of V_{hv}. The trial test with the solenoid source in the Crag, at a depth of -12 m OD, gave a value of V_{hh} in close agreement with the V_{hv} measurements at that depth.

In the London Clay below the Harwich Stone, the down-hole shear wave velocities V_{vh} are 5–10% lower than the conventional cross-hole velocities V_{hv}. In addition, the horizontally polarized cross-hole shear wave velocities V_{hh} are significantly higher than V_{hv}, by up to 20%. The hierarchy of shear wave velocities below the Harwich Stone is, therefore, $V_{vh} < V_{hv} < V_{hh}$. In the WRB Sand, considering data for the middle of the stratum which is unaffected by the clays above and below, this hierarchy is modified to $V_{hh} < V_{vh} < V_{hv}$.

V_{vh} and V_{hv} appear to be similar in the Thanet Clay although the comparison is made difficult by the increase in V_{hv} at the base of the layer and the 6·3 m depth over which V_{vh} is averaged. The V_{hh} values are, nevertheless, consistently higher than V_{hv}, by approximately 15% in the upper part of the layer. Table 5 summarizes the ratios of shear wave velocities in each stratum and the ratios of small-strain shear stiffnesses to which they lead. In considering these ratios it should be borne in mind that V_{hv} will have been influenced by the presence of any stiffer horizontal layers, whereas V_{vh} represents an average for a 6·3 m depth; the ratios should be regarded, therefore, as indicative only.

Shear wave velocities were measured in the laboratory on unconfined U100 samples of the clays, taken in an earlier investigation at the adjacent Sizewell B site (Davis, 1982). Both vertically and horizontally polarized shear waves were transmitted across the samples, to give the velocities V_{hv} and V_{hh}, and horizontally polarized shear waves were transmitted along the axes of the samples, to give V_{vh}, at frequencies of approximately 1 kHz. The values of shear wave velocities that were measured are shown in Table 6. The

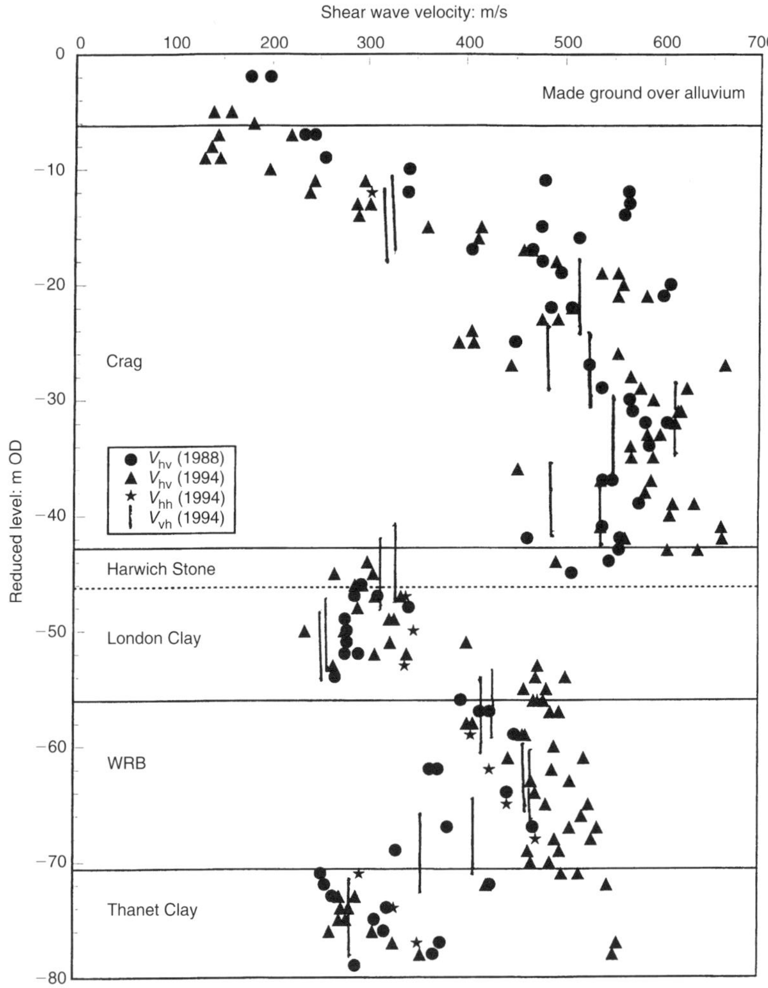

Fig. 16. Comparison of different shear wave velocities (adjustments were made to take account of small differences in stratum levels between sites of 1988 and 1994 measurements)

Table 6. Shear and compression wave velocities in unconfined samples*

Stratum	Reduced level: m OD	Shear wave velocities: m/s			Compression wave velocities: m/s	
		V_{hv}	V_{hh}	V_{vh}	V_{ph}	V_{pv}
London Clay	−48·5†	154	232	143	1404	1488
	−57·2	191	271	254	1581	1647
WRB Clay	−59·1	267	262	236	1725	1724
	−68·2	267	235	238	1757	1726
Thanet Clay	−75·1	206	314	296	1621	1634
	−82·2	282	350	—	1688	1627

* Samples obtained from Sizewell B site where stratum boundary levels are slightly lower than at Sizewell C site.
† Sample not fully saturated.

values of V_{hh} and V_{hv} in the London Clay samples were much lower than the values measured in situ at the Sizewell C site, whereas in the Thanet Clay, similar values of V_{hh} were measured in the laboratory and the field. This may reflect a difference in the response of the two clays to sampling, which can be expected on the basis of their differences in plasticity and fabric; it emphasizes the need to reconsolidate the samples to in situ stresses before making estimates of in situ values of wave velocity.

Furthermore, the anisotropy of shear wave velocity found in the unconfined samples of London Clay and Thanet Clay was quite different from that found in situ; in the laboratory V_{hh} was 40–50% higher than V_{hv} in the London Clay and 25–50% higher in the Thanet Clay. The ratios of velocities are compared with the ratios of in situ values in Table 5. The differences between V_{hv} and V_{vh} in the unconfined samples are initially surprising, but may be related to the effect of thin silt seams within the specimens, which are important with the shorter wavelengths involved in the laboratory tests.

Compression wave velocities. For compression waves, values of V_{pv} are provided in terms of interval velocities from down-hole surveys. In Fig. 4, V_{pv} values have been plotted over depth intervals that were entirely confined to a single lithological unit, that is, an interval that does not cross a geological or velocity boundary. In the Crag, apart from the lower part, values of V_{pv} exceed V_{ph} by between 150 and 300 m/s, that is, by approximately 5–10%. For the London and Thanet Clays, the values of V_{pv} exceed V_{ph} by 50–100 m/s, again indicating the existence of in situ velocity anisotropy. In the sands of the Woolwich and Reading Beds, V_{pv} is very similar to V_{ph}. In a previous site investigation for Sizewell C, weak anisotropy of the P wave velocity was found in the Chalk, with V_{pv} being greater than V_{ph} by 200–300 m/s.

Compression wave velocities were also measured by Davis (1982) on unconfined samples of the clays, for waves travelling along the axis of the sample and across the sample, generated by ultrasonics at approximately 250 kHz. The results are presented in Table 6 and the ratios of velocities are compared with the in situ values in Table 5.

Stiffness and modulus decay. A comparison between modulus decay curves from different types of test on Crag, London Clay and Thanet Clay is presented in Fig. 17. The data from triaxial tests is plotted as $E_{u\,sec}$ against axial strain ε_a. The data from RC and TS tests is plotted onto axes of $G_{\theta z}$ and $\gamma_{\theta z}$, the scales for which have been set such that $G_{\theta z} = E_u/3$ and $\gamma_{\theta z} = \sqrt{3}\varepsilon_a$; the lack of superposition of the curves indicates anisotropy, providing, of course, the specimens are similar in all respects, including levels of sample disturbance.

In making the comparisons between the triaxial compression and triaxial extension tests, it should be noted that undrained shear of the Crag starts from an isotropic stress state; the initial compression loop involves a reversal of stress path direction and the initial extension loop involves a continuation of the previous stress path direction, but with $\sigma_r > \sigma_a$ (Fig. 7). In contrast, undrained shear of the clays starts from a stress state with $\sigma'_a > \sigma'_r$ and, this time, the initial triaxial extension loop involves the reversal of stress path direction and constitutes axial unloading.

The following is evident.

(*a*) There is anisotropy of stiffness over a wide range of strains. For each material, the stiffest response was found in triaxial compression. In the Thanet Clay, the stiffness in triaxial extension was reasonably similar to that in triaxial compression. In the London Clay, the lower stiffness in triaxial extension may be related in part to some disturbance of that particular sample, as discussed previously. In the Crag, the lower stiffness in triaxial extension is partly a reflection of inherent anisotropy. In dry Ticino Sand, Bellotti *et al.* (1996) found that $E_h > E_v$ under isotropic stresses, on the basis of $V_{ph} > V_{pv}$; in contrast, in the Crag, V_{pv} is in the range 1·05–1·1 V_{ph}, probably because of the effects of ageing.

(*b*) Where the strain ranges for the modulus decay curves from resonant column and monotonic torsional shear tests overlap (e.g. in Figs 14(b), 17(c) and 18), the differences in response illustrate the strong dependence of stiffness degradation on the rate and form of loading, in particular whether it is monotonic or cyclic. Rapidly applied reversals of stress direction, as in the resonant column, result in a much reduced rate of degradation of stiffness with strain, compared with monotonically applied shear. This, together with the omnidirectional and rapid reversal of shearing involved in an earthquake, helps to explain why the observed response to shaking is often stiffer than predicted (Electric Power Research Institute, 1988).

(*c*) The differences in the normalized modulus decay curves for the resonant column tests, shown in Fig. 18, are consistent with the usual trend of a steepening decay curve with reducing soil plasticity (e.g. Dobry & Vucetic, 1987). The extent of the initial plateau in the decay curves in the RC tests on the Crag and WRB sands has probably been reduced to some extent by the effects of destructuring in the RC specimens.

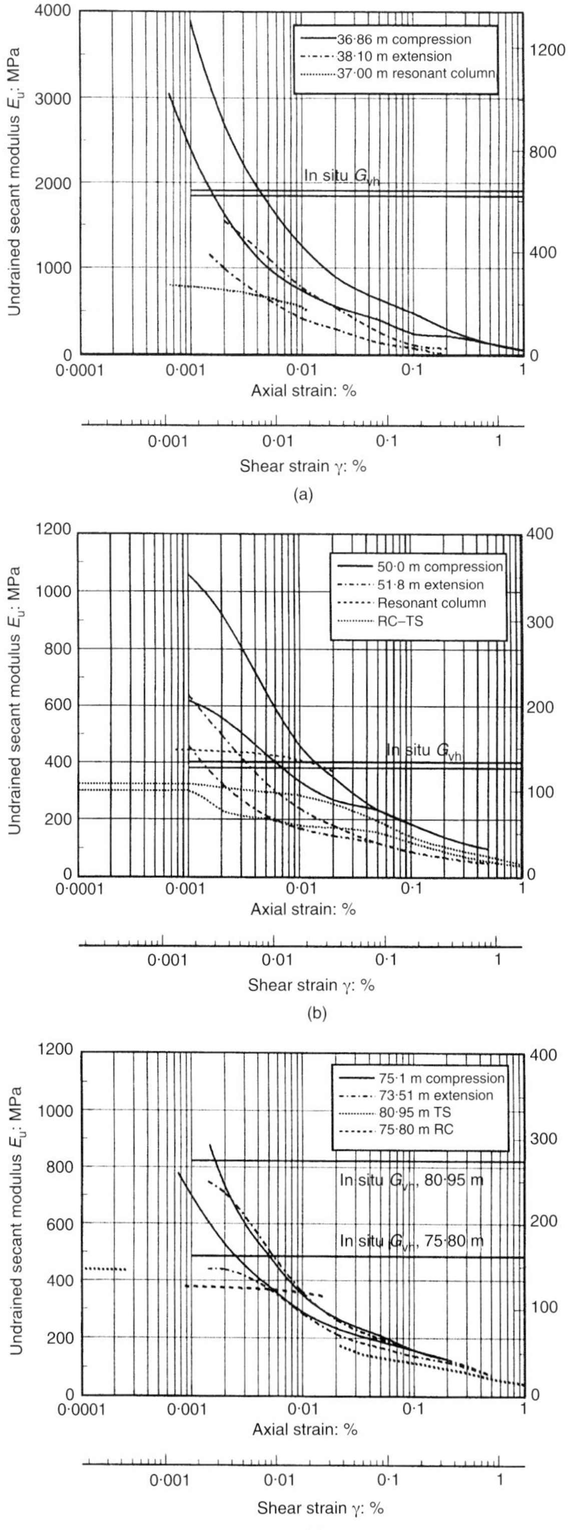

Fig. 17. Modulus decay for different shearing modes: (a) Crag; (b) London Clay; (c) Thanet Clay

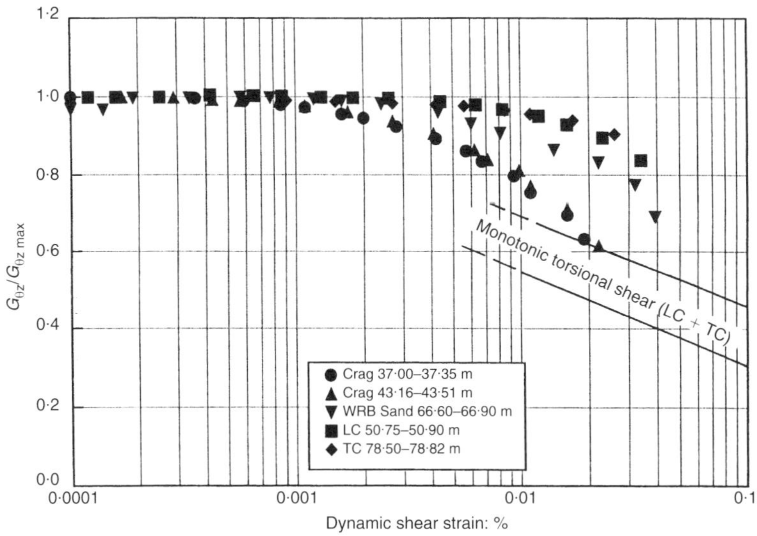

Fig. 18. Normalized modulus decay curves from resonant column and torsional shear tests

Critical strain levels

Three critical strain levels are valuable when assessing stiffness characteristics and the effects on these characteristics of, for example, shearing rate and cyclic loading: ε_{y1}, the strain marking the limit to linear elastic behaviour; ε_{y2}, the strain marking the limit to recoverable behaviour; and ε_{y3}, the strain marking the onset of large scale yielding. These strain levels vary with stress path and principal stress direction (e.g. Jardine, 1992; Hight & Higgins, 1995).

As noted previously, it was not possible to resolve a linear elastic range in the triaxial tests, and any interpretation of ε_{y1} from resonant column tests is influenced by the effect of reversals in direction of straining. Nevertheless, the values of ε_{y1} indicated by the resonant column tests on solid samples were the following: 0·0004% for the disturbed Crag, 0·002% for the disturbed WRB Sand, 0·003% for the London Clay, and approximately 0·004% for the Thanet Clay. Lower values were found for London Clay in the RC–TS tests.

Evidence is available from different sources to identify ε_{y2}: the development of pore pressures in RC and RC–TS tests; the reduction in $G_{\theta z\ max}$ during TS and high-amplitude RC tests, assuming the reduction is the result of destructuring by strains greater than ε_{y2}; and the development of permanent strains after unload–reload loops in triaxial tests.

Positive pore pressures began to develop in TS tests on the London and Thanet Clays at shear strains of between 0·02 and 0·05%; no pore pressure build-up was observed in the RC tests, which reached maximum shear strains of 0·034% and 0·026% in the London and Thanet Clays, respectively. In the RC tests on the Crag, in which the specimens are thought to have been disturbed, pore pressure build-up began at strains as low as 0·005%.

Reductions in $G_{\theta z\ max}$, measured at low strain amplitudes after high-strain excursions in RC tests, occurred at shear strains of between 0·002% and 0·005% in the Crag, at shear strains above 0·005% in the WRB sand, at shear strains above 0·008% in the London Clay and above approximately 0·02% in the Thanet Clay. In the TS tests, reductions in $G_{\theta z\ max}$ occurred at strains of approximately 0·02% in the Thanet Clay and approximately 0·01% in the London Clay.

The recoverable strain limits assessed from unload–reload loops in the triaxial tests are summarized in Table 2. In triaxial compression, ε_{y2} in the London Clay lies between 0·08% and 0·1%, while in the Thanet Clay, ε_{y2} lies between 0·1% and 0·2%. In marked contrast, ε_{y2} for triaxial extension lies between 0 and −0·005% in the Thanet Clay and between −0·01% and −0·015% in the London Clay. In the Crag, ε_{y2} in compression lay between 0·025% and 0·05% in two cases and between 0·015% and 0·02% in another; in triaxial extension, ε_{y2} was generally less than or around −0·005%.

The results confirm that for compression ε_{y2} increases with plasticity, as found by Jardine *et al.* (1991), and highlight the dependence of ε_{y2} on shearing mode and the particular sensitivity of the soils to extension strains.

Destructuring

Destructuring, or damage to structure, is presumed to occur at strains beyond ε_{y2} and to result in a build-up in pore pressure and a reduction in stiffness. It is of relevance when considering the effects of excavation for, and loading by, the power station on the stiffness of its foundation, and when assessing corrections to be made for the effects of sample disturbance.

In the London and Thanet Clays, the effect of increasing strains in unload–reload loops in undrained triaxial tests produced no significant change in modulus decay. The modulus decay curves for each unload–reload loop fitted within the envelopes shown in Figs 11 and 12, with no clear tendency for high-strain loops to fall closer to the lower boundary of the envelope. Similarly, the reductions in $G_{\theta z\,\max}$ in both the RC and the TS tests were relatively small for shear strains up to 1%. It was noticeable, however, that the London Clay sample which was consolidated beyond the in situ stresses before torsional shear (YC1-57·5 m) showed a lower $G_{\theta z\,\max}$, despite the higher stress state, indicating that destructuring had possibly occurred as a result of the additional consolidation.

In contrast to the clays, destructuring of the Crag was evident in triaxial compression and extension at increasing levels of strain in the unload–reload loops. Examples are presented in Fig. 19 which show major changes in stiffness and modulus decay with progressive destructuring. (In these tests, reductions were also taking place in mean effective stress as a result of a build-up in pore pressure.) These effects in the Crag indicate that a significant proportion, if not all, of the structure has been retained in the triaxial samples.

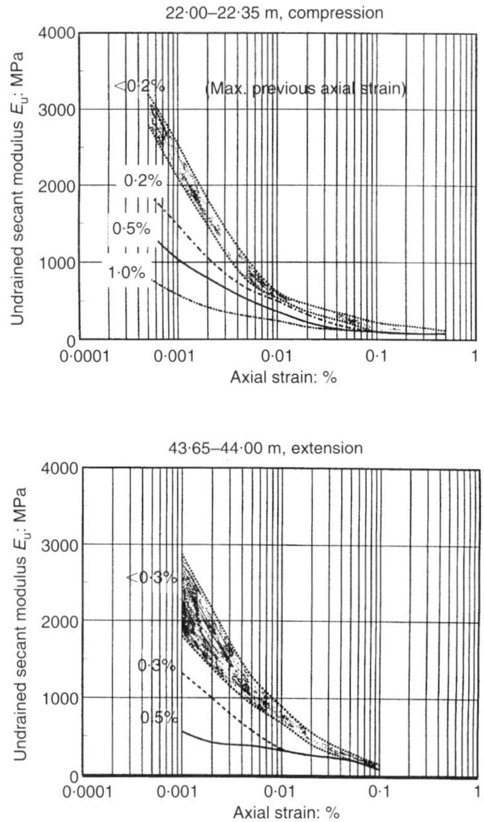

Fig. 19. Effects of destructuring in Crag by unload–reload cycles

CONCLUSIONS

There is anisotropy in the shear and compression wave velocities, and hence in the small-strain stiffnesses, in the Crag and Lower London Tertiaries at Sizewell; this conclusion is based on in situ seismic measurements. In the Crag, $V_{vh} = 0{\cdot}95V_{hv}$; in the London Clay, V_{vh} is in the range $0{\cdot}9$–$0{\cdot}95V_{hv}$ and V_{hh} is up to $1{\cdot}2V_{hv}$; in the WRB Sand, $V_{vh} = 0{\cdot}95V_{hv}$ and $V_{hh} = 0{\cdot}9V_{hv}$; in the Thanet Clay $V_{vh} \approx V_{hv}$ and $V_{hh} = 1{\cdot}15V_{hv}$. (It must be noted, however, that the down-hole measurements of V_{vh} represent an average value over a 6·3 m depth interval.) Generally, V_{pv} is in the range $1{\cdot}05$–$1{\cdot}1V_{ph}$, except in the WRB sands where V_{pv} and V_{ph} are equal.

The anisotropy is both inherent and stress-induced. This is illustrated, for example, by the ratio between V_{hh} and V_{hv}, which was higher in the unconfined samples of London and Thanet Clay than in situ, where vertical stresses are higher than horizontal stresses, inducing a larger increase in V_{hv} than in V_{hh}. The dependence of shear wave velocities on the stresses in the directions of propagation and polarization was first proposed by Roesler (1979).

Butcher & Powell (1996) reported V_{hv} to be larger than V_{vh} at a London Clay site where K_o was greater than 1·0. The fact that, at Sizewell, V_{hv} was again larger than V_{vh} in the London Clay, yet K_o was less than 1, suggests that their relative magnitudes are not only determined by stress state. Anisotropy resulting from layering may be playing an important part in accounting for the differences between V_{vh} and V_{hv}. Certainly, fabric features were significant in the clays and would have influenced the propagation of shear waves of different wavelengths in different ways.

The non-linear stiffness characteristics of the soils were found in the laboratory tests to depend strongly on shearing mode. $G_{\theta z}$ was consistently less than $E_{vu}/3$, where E_{vu} was measured in undrained triaxial compression tests. In reconstituted silts, $G_{\theta z}$ has been found to be higher than $E_{vu}/3$ at an OCR of 1·3 (Zdravkovic & Jardine, 1996). The opposite finding for the Crag is probably related to the effects of ageing. Triaxial compression (axial loading) led to higher stiffnesses at small strains than triaxial extension (axial unloading), particularly in the Crag, although this was influenced by the stress history imposed in the laboratory.

Recoverable strain levels increased with soil plasticity but varied with shearing mode. For example, recoverable behaviour was observed in torsional shear on London Clay and Thanet Clay for shear strains up to 0·02–0·05%, whereas in triaxial compression, recoverable axial strain limits were between 0·08% and 0·1% in London Clay and between 0·1% and 0·2% in Thanet Clay. Each material was easily disturbed by extension strains, the recoverable axial strain limit often being less than −0·005%.

The Crag was easily damaged by shear strains, including those imposed during sample trimming, and this led to a progressive reduction in stiffness. The London Clay and Thanet Clay appeared to be more robust, although the London Clay was damaged, with a consequent reduction in stiffness, by strains associated with consolidation beyond the in situ stresses. Destructuring can seriously affect the investigation of stiffness properties on samples of natural soil.

The form of the modulus decay was affected by shearing mode, rate of shear and whether loading was cyclic or monotonic.

ACKNOWLEDGEMENT

The authors wish to thank Nuclear Electric Ltd for allowing them to present the data included in this paper.

APPENDIX 1. ACCURACY OF MEASUREMENTS AND SOURCES OF ERROR

The following accuracies have been assessed for the various in situ and laboratory measurements.

In situ

Compression wave velocity V_p	±5%
Shear wave velocity V_s	±4%
Derived shear stiffness G_{max}	±10%

Laboratory

Resonant column (UWB and IC).

Frequency, measured to ±0·1 Hz, accurate to	±0·25%
Shear wave velocity V_s	±2%
Derived shear stiffness G_{max}	±4%
Damping	±5%

Bender element transducers.

Shear wave velocity V_s	±2%
Derived shear stiffness G_{max}	±5%

Torsional shear.

$G_{\theta z}$ at $\gamma_{\theta z} = 0{\cdot}001\%$	±20%
$G_{\theta z}$ at $\gamma_{\theta z} = 0{\cdot}01\%$	±5%

Errors in the assessment of in situ shear wave velocities from cross-hole tests can occur when measurements are made close to a high-velocity boundary, formed by a neighbouring stratum or a stiffer inclusion, which gives rise to refraction arrivals. The influence of these boundaries has been highlighted in the paper.

Down-hole measurements represent average values of shear wave velocity over the separation length between the two sets of geophones. They may be influenced by the travel path being non-vertical, but this would be insignificant for the measurements below 30 m reported here.

Errors may occur in the resonance measurements because of imperfect coupling between the specimen and the end platens. This can give rise to an underestimate of G_{max} and an overestimate of the rate of modulus decay with increasing strain, since the problem gets worse as the strain amplitude increases. Particular attention was paid to the coupling in the experiments reported here and we believe that these errors were avoided in the resonant column tests on solid specimens.

At the higher strain amplitudes in the resonant column test, the stiffness may reduce progressively with increasing numbers of vibration cycles. It is difficult to control the number of cycles when establishing resonance and the amount of degradation that is measured may depend to some extent on the time taken to establish resonance.

Errors in bender element measurements may be associated with incorrect path lengths and incorrect interpretation of the arrival time. The arrival of the shear wave was taken as the first major inflection of the signal. The sample length of 150 mm was such that the signals were relatively free from near-field masking effects.

In contrast to the geophysical tests, the static torsional shear and triaxial tests rely on the direct measurement of strain for the determination of stiffness. The accuracy with which the strains can be measured becomes less as the strain level reduces, as shown by the values quoted above for the torsional shear. In the triaxial tests the locally measured axial strains could be resolved to 0·0015%. Individual data points were accurate to ±15% at an axial strain of 0·001%. However, the difficulties encountered in maintaining the stresses constant at the end points in the load–unload cycles led to additional uncertainties in stiffness, encompassed within the envelopes presented here for the modulus decay from the triaxial tests.

REFERENCES

Bates, C. R. (1989). Dynamic soil property measurements during triaxial testing. *Géotechnique* **39**, No. 4, 721–726.

Bellotti, R., Jamiolkowski, M., Lo Presti, D. C. F. & O'Neill, D. A. (1996). Anisotropy of small strain stiffness in Ticino sand. *Géotechnique* **46**, No. 1, 115–131.

Bennell, J. D. & Taylor Smith, D. (1991). A review of laboratory shear wave techniques and attenuation measurements with particular reference to the resonant column. In *Shear waves in marine sediments* (eds J. M. Hovem *et al.*), pp. 83–93. Kluwer.

Bennell, J. D., Taylor Smith, D. & Davis, A. M. (1984). Resonant column testing of marine soils. *Proc. Oceanology Int. Conf., Brighton*, OI 1.9. Society of Underwater Technology.

Butcher, A. P. & Powell, J. J. M. (1996). Practical considerations for field geophysical techniques to assess ground stiffness. *Proc. Int. Conf. on Advances in Site Investigation Practice*, pp. 701–714. London: Thomas Telford.

Clayton, C. R. I. & Khatrush, S. A. (1986). A new device for measuring local axial strain on triaxial specimens. *Géotechnique* **36**, No. 4, 593–597.

Cola, S. & Jardine, R. J. (1995). *Compressibility of natural overconsolidated clays from Sizewell (Suffolk).* Imperial College internal report.

Davis, A. M. (1982). *Static–dynamic properties of sedimentary materials.* PhD thesis, University of Wales.

Davis, P. D., Eldred, P. J. L., Bennell, J. D., Hight, D. W. & King, M. S. (1996). Site investigation for seismically designed structures. *Proc. Int. Conf. on Advances in Site Investigation Practice*, pp. 715–726. London: Thomas Telford.

Dobry, R. & Vucetic, M. (1987). Dynamic properties and seismic response of soft clay deposits. *Proc. Int. Symp. Geotechnical Engineering of Soft Soils, Mexico City*, 2, pp. 51–87.

Electric Power Research Institute (1988). *Soil response to earthquake ground motion.* EPRI Report NP 5747, Palo Alto, California.

Hepton, P. (1996). Deep rotary cored boreholes in soils using wireline drilling. *Proc. Int. Conf. on Advances in Site Investigation Practice*, pp. 269–280. London: Thomas Telford.

Hight, D. W. (1993). A review of sampling effects in clays and sands. *Proc. SUT Conf. on Offshore Site Investigation and Foundation Behaviour*, pp. 115–146.

Hight, D. W. (1996). Moderator's Report to Session 3: Drilling, boring, sampling and description. *Proc. Int. Conf. on Advances in Site Investigation Practice*, pp. 337–360. London: Thomas Telford.

Hight, D. W. & Higgins, K. G. (1995). An approach to the prediction of ground movements in engineering practice: background and application. Keynote lecture. *Proc. 1st Int. Conf. on Pre-failure Deformation Characteristics of Geomaterials, Sapporo, Japan*, Vol. 2, pp. 909–945.

Hight, D. W. & Jardine, R. J. (1993). Small-strain stiffness and strength characteristics of hard London tertiary clays. In *Geotechnical engineering of hard soils—soft rocks* (eds A. Anagnostopoulos, F. Schlosser, N. Kalteziotis & R. Frank), pp. 533–552. Rotterdam: Balkema.

Jardine, R. J. (1992). Some observations on the kinematic nature of soil stiffness. *Soils Fdns* **32**, No. 2, 111–124.

Jardine, R. J., St John, H. D., Hight, D. W. & Potts, D. M. (1991). Some practical applications of a non-linear ground model. *Proc. 10th ECSMFE, Florence,* pp. 223–228.

Porovic, E. & Jardine, R. J. (1994). Some observations on the static and dynamic shear stiffness of Ham river sand. *Proc. 1st Int. Conf. on Pre-failure Deformation Characteristics of Geomaterials, Sapporo, Japan,* Vol. 1, pp. 25–30.

Ridley, A. M. & Burland, J. B. (1993). A new instrument for the measurement of soil moisture suction. *Géotechnique* **43**, No. 2, 321–324.

Roesler, S. K. (1979). Anisotropic shear modulus due to stress anisotropy. *J. Geotech. Engng Div. Am. Soc. Civ. Engrs* **105**, GT7, 871–880.

Zdravkovic, L. & Jardine, R. J. (1996). The stiffness anisotropy of a silt under general stress conditions. *Géotechnique*. Symposium in Print.

Rampello, S., Viggiani, G. M. B. & Amorosi, A. (1997). *Géotechnique* **47**, No. 3, 475–489

Small-strain stiffness of reconstituted clay compressed along constant triaxial effective stress ratio paths

S. RAMPELLO,* G. M. B. VIGGIANI† and A. AMOROSI*

The work described in the paper is an experimental investigation of the effects of anisotropic stress states and histories on the small-strain shear stiffness of reconstituted clays as measured with bender element tests. A brief review is made of previous work dealing with the dependence of small-strain shear modulus on stress state and history in isotropic conditions of confinement, and the rational link that exists between void ratio, *e*, effective stress, σ', and overconsolidation ratio, OCR, is considered. Under isotropic stress conditions, only two out of the three variables *e*, σ' and OCR are required to describe the dependence of the small-strain stiffness of a clay on current state and history. The same approach proved useful in the analysis of the data from tests with anisotropic stress conditions. The values of the small-strain shear modulus of reconstituted samples compressed along stress paths with constant triaxial effective stress ratios were higher than those measured on isotropically compressed samples. The observed differences increased with increasing stress ratio and could be explained only in part in terms of the smaller void ratios of the anisotropically compressed samples at the same mean effective stress. Simple relationships to evaluate the stiffness indexes of clays that have undergone anisotropic stress histories are proposed.

KEYWORDS: clays; laboratory tests; stiffness.

L'article décrit des recherches expérimentales sur les effets de l'état de contrainte anisotrope et des contraintes anisotropes déjà subies sur la rigidité au cisaillement d'argiles reconstituées soumises à de petites contraintes, la rigidité etant measurée a l'aide de languettes piezoelectriques vibrantes (bender elements). Il présente un aperçu de travaux antérieurs sur la variabilité du module de cisaillement, dans le cas de petites contraintes, en fonction de l'état de contrainte et des contraintes déjà subies, dans des conditions de confinement isotropes, et examine le lien rationnel entre l'indice des vides (*e*), la contrainte intergranulaire (σ') et le coefficient de préconsolidation (OCR). En conditions de contrainte isotropes il suffit de connaître deux des trois variables *e*, et σ' OCR pour décrire la variation de la rigidité au cisaillement de l'argile soumise à de petites contraintes en fonction de l'état de contrainte et des contraintes déjà subies. La même méthodologie s'est avérée utile pour analyser les résultats d'essais réalisés en conditions de contrainte anisotropes. Les valeurs du module de cisaillement d'éprouvettes reconstituées, comprimées le long des parcours de contrainte avec des rapports constants de contrainte intergranulaire triaxiale sont supérieures aux valeurs mesurées sur des éprouvettes à compression isotrope. Les écarts observés augmentent avec le rapport de contrainte et s'expliquent en partie par le fait que, pour une même contrainte intergranulaire moyenne, les éprouvettes à compression anisotrope ont un indice des vides moins élevé. L'article propose de simples rapports pour évaluer les indices de rigidité des argiles qui ont subi des contraintes anisotropes.

INTRODUCTION

Any calculation of ground movements around an engineering structure requires a knowledge of soil stiffness. Even at relatively small strains, the stress–strain behaviour of soils is highly non-linear and the stiffness can vary significantly over the range of strains of interest in civil engineering. The assessment of reliable stress–strain relationships for soils, for both static and dynamic deformation problems, requires a correct evaluation of the shear stiffness at small strains, G_0, and of the shape of the stiffness degradation curve (e.g. see Burghignoli *et al.*, 1991).

By assuming that the mechanical behaviour of the soil is linear elastic inside a relatively small

Manuscript received 31 October 1996; revised manuscript accepted 21 March 1997.
Discussion on this paper closes 1 November 1997.

* Università degli Studi di Roma 'La Sapienza'.
† Università degli Studi di Roma 'Tor Vergata'.

kinematic yield surface, within a large-scale yield locus (e.g. Al-Tabbaa & Wood, 1989; Stallebrass, 1990), the elastic behaviour of soils is in principle restricted to overconsolidated states and appropriately small increments of unloading and reloading. For normally consolidated states, an elastic shear modulus cannot be defined in an increment of loading, even if infinitesimal. In dynamic tests, in which body waves propagate within the sample, however, an infinitesimal expansion of the yield locus can be imagined due to the cyclic nature of wave propagation, and an elastic shear modulus can be defined for normally consolidated states.

Most of the available experimental evidence on the elastic shear modulus of clays has been collected by dynamic laboratory tests (resonant column and pulse wave) on reconstituted and undisturbed samples of soils subjected to isotropic conditions of confinement and isotropic stress histories.

Under these conditions, in addition to the dependence of the bulk modulus on mean effective stress, the dependence of the shear modulus on mean effective stress has also been clearly established (e.g. see Wroth & Houlsby, 1985). If the incremental elastic shear and bulk moduli are assumed to be functions of the mean effective stress only, cyclic variations of deviatoric stress q can result in continuous creation of energy by the material (Zytynski *et al.*, 1978). The problem is not particularly relevant for monotonic loading, but can induce significant errors for cyclic loading conditions. With reference to axi-symmetric stress states, the elastic bulk and shear moduli should more properly depend both on mean effective stress, p', and deviatoric stress, q (e.g. see Houlsby, 1985; Lade & Nelson, 1987).

For fine-grained soils, the dependence of G_0 on q was studied by Viggiani & Atkinson (1995a) in triaxial tests, in which reconstituted samples of kaolin underwent successive stress paths at either constant axial effective stress or constant radial effective stress oscillating around isotropic conditions, in both triaxial compression and triaxial extension. Subsequently, Jovičić & Coop (1996) carried out triaxial tests in which reconstituted samples of kaolin followed a series of constant-p' paths after stages of isotropic compression at successively higher confining stress. In both cases, the values of G_0 at selected values of the stress ratio $\eta = q/p'$ were measured using piezoceramic bender elements, and the influence of the deviatoric stress on G_0 was observed to be barely discernible.

In recent years, a number of experimental studies of the anisotropy of the small-strain stiffness induced in fine-grained soils by anisotropic states of stress and anisotropic loading histories, both in the laboratory and in the field, have been carried out (e.g. see Saada *et al.*, 1978; Abbis, 1981; Jamiolkowski *et al.*, 1994; Jovičić & Coop, 1996). Very often, reference has been made to a cross-anisotropic description of the small-strain stiffness with the plane of isotropy corresponding to the horizontal or the depositional plane.

In the work presented in this paper, the small-strain shear stiffness of the soil has been measured using bender elements embedded in the top and base platens of triaxial cells. Under these conditions the particle motion is always horizontal and the direction of propagation of the shear wave is always vertical, so that the only shear stiffness that can be measured is $G_{0(\mathrm{vh})}$ $(= G_{0(\mathrm{hv})})$, where the subscripts v and h refer to the vertical and horizontal directions. Unless samples are trimmed horizontally, the technique adopted in the present work cannot measure the anisotropy of the stiffness of the soil. The objective of the experimental work, therefore, is restricted to assessing the effects of anisotropic stress states and loading histories on the small-strain shear stiffness of the soil, $G_{0(\mathrm{vh})}$, in the vertical plane.

BACKGROUND

Hardin (1978) suggested that the small-strain shear modulus of soils under isotropic stress conditions depended on stress state and history in the following way

$$\frac{G_0}{p_\mathrm{a}} = Sf(e)\,\mathrm{OCR}^k\left(\frac{p'}{p_\mathrm{a}}\right)^n \tag{1}$$

where $f(e)$ is a decreasing function of the void ratio, the mean effective stress, p', accounts for the dependence of the small-strain stiffness on the state of stress, and the overconsolidation ratio, OCR, accounts for the overall stress history of the soil. In equation (1) the introduction of the atmospheric pressure, p_a, as a reference stress makes the expression independent of the choice of the units, and S, n and k become dimensionless material parameters which account for the nature of the soil. Similar relationships originally proposed by Hardin & Richart (1963) for sands were written in terms of the initial void ratio, e_0. When dealing with clays, the current value of the void ratio, e, must be considered in evaluating the function $f(e)$.

The increase in the small-strain stiffness of soils as porosity reduces is experimentally very well established. Much effort has been devoted in the past to proposing alternative expressions for the function $f(e)$ which links observed values of G_0 to changes in void ratio (Hardin & Black, 1969; Ishihara, 1982; Jamiolkowski *et al.*, 1994). The numerical value of the stiffness multiplier S in equation (1) depends on the choice of the void ratio function, $f(e)$, and of the reference stress (p_a in equation (1)).

In equation (1) the existence of a compressibility relationship between void ratio and effective stress is ignored, although clays experience major changes of void ratio in connection to changes of effective stress.

Several experimental and theoretical studies (e.g. see Weiler, 1988; Houlsby & Wroth, 1991) have underlined that, for a given clay under isotropic stress conditions, the current values of mean effective stress and overconsolidation ratio are sufficient to describe the state of the soil, and that one of the three variables e, p' and OCR in equation (1) is redundant. To account for small changes of void ratio at constant effective stress, such as those due, for instance, to creep, the overconsolidation ratio needs to be defined using the unloading–reloading line through the current state point, which varies as the void ratio of the soil changes at constant mean effective stress.

For fine-grained soils and isotropic stress conditions, Viggiani (1992) proposed the following expression for G_0

$$\frac{G_0}{p_r} = S^* \left(\frac{p'}{p_r}\right)^{n^*} R^m \tag{2}$$

where the symbols S^*, n^* and m have similar meanings to but different values from S, n and k in equation (1), p_r is a reference stress introduced to satisfy dimensions, and the overconsolidation ratio, R, is defined as p_y'/p'. The mean effective yield stress p_y' was determined at the intersection of a straight line of slope κ through the current state of the soil and the virgin isotropic compression line; κ was taken as the average slope of the unloading–reloading cycle in the $\ln(p')$–v plane. Although Viggiani (1992) adopted a normalizing reference stress $p_r = 1$ kPa, in the following a value of $p_r = 100$ kPa has been used, which is close to atmospheric pressure, commonly used as reference pressure (e.g. see equation (1)).

In equation (2) the contribution of the void ratio vanishes since e is related to p' for normally consolidated states, and accounted for by R and p' if the clay is overconsolidated, making the function of void ratio $f(e)$ unnecessary. According to equation (2), the small-strain shear modulus of a clay under isotropic conditions depends on the current stress state and on a state variable, the mean effective yield stress.

An alternative method of locating the state of the soil relative to the virgin compression line is that of projecting the state of the soil in the p'–v plane horizontally to obtain the equivalent consolidation pressure

$$p'_e = p_r \exp\left(\frac{N - v}{\lambda}\right) \tag{3}$$

where N is the specific volume ($v = 1 + e$) at a reference mean effective stress p_r, and λ is the slope of the virgin compression line in the $\ln(p')$ $-v$ plane (see Fig. 1).

The relationship between R and p'_e is

$$R = \left(\frac{p'_e}{p'}\right)^{1/\Lambda} \tag{4}$$

where $\Lambda = (\lambda - \kappa)/\lambda$.

Using equation (4), Rampello *et al.* (1994a, b) rewrote equation (2) as

$$\frac{G_0}{p_r} = S^* \left(\frac{p'}{p_r}\right)^{n^*} \left(\frac{p'_e}{p'}\right)^{c} \tag{5}$$

where $c = m/\Lambda$.

Equation (5) can be easily manipulated so that the dependence of G_0 on the current value of void ratio is explicit

$$\frac{G_0}{p_r} = S^* f(e) \left(\frac{p'}{p_r}\right)^{(n^* - c)} \tag{6}$$

where

$$f(e) = \frac{A}{\exp(Be)} \tag{7}$$

and $A = \exp[B(N - 1)]$ and $B = c/\lambda$ are constants that depend on the compressibility parameters N and λ and the stiffness index c.

When comparing equations (2), (5) or (6) with expressions of the type given in equation (1), care must be taken to consider consistent quantities. Thus, for instance, the stress level exponent n in equation (1) must be compared with the difference

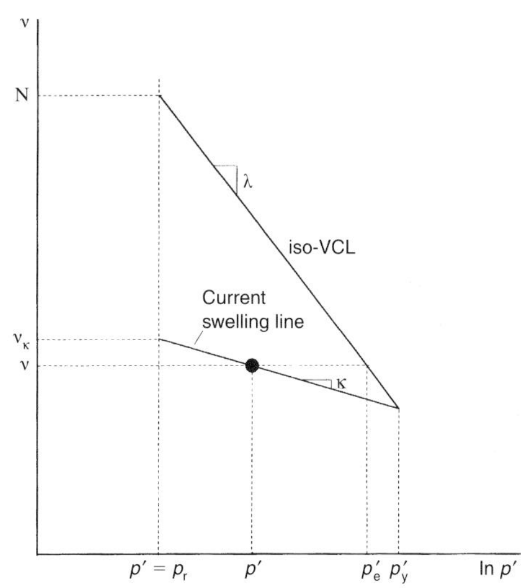

Fig. 1. Equivalent consolidation pressure

$(n^* - c)$ in equation (6) and not with n^* in equations (2) and (5).

As expected, in equation (6) the analytical contribution of the overconsolidation ratio, R, vanishes since each combination of current void ratio and mean effective stress can only correspond to one value of R; for a clay under isotropic conditions, if the state of the soil is expressed as a combination of e and p', further effects of overconsolidation, other than determining the current value of e, are not expected. Consistent with this statement, Jamiolkowski *et al.* (1994) showed that, with an empirical void ratio function in the form $f(e) = e^{-x}$, the influence of OCR in equation (1) was negligible ($k \cong 0$). It would be misleading, however, to argue from equation (6) that in clays overconsolidation has little effect on G_0, at least as incorrect as to argue from equation (2) that the influence of void ratio on G_0 is negligible; equations (2) and (6) represent alternative but equivalent approaches when the relationship between e, p' and R is accounted for.

An expression that is often used for sands to take into account the effects of non-isotropic conditions of confinement (e.g. see Ni, 1987; Hardin & Blandford, 1989) is

$$\frac{G_0}{p_a} = S_{ij} f(e)\,\mathrm{OCR}^k \frac{(\sigma'_i \sigma'_j)^{n/2}}{p_a{}^n} \tag{8}$$

in which σ'_i and σ'_j are the principal effective stresses in the plane in which G_0 is measured. In the case of body wave propagation within the soil, directions i and j correspond to the directions of wave propagation and particle motion, respectively. As briefly discussed in the introduction, if G_0 is measured in a triaxial apparatus fitted with bender elements, the direction of wave propagation is along the axis of the sample so that σ'_i corresponds to the axial effective stress, σ'_a, and σ'_j corresponds to the radial effective stress, σ'_r. Equation (8) can be rewritten in terms of the stress invariants for axi-symmetric conditions $q = (\sigma'_a - \sigma'_r)$, $p' = (\sigma'_a + 2\sigma'_r)/3$

$$\frac{G_0}{p_r} = S_{ij} f(e)\,\mathrm{OCR}^k \left(\frac{p'}{p_r}\right)^n \beta^{n/2} \tag{9}$$

where $\beta = [1 + (\eta/3) - (2\eta^2/9)]$ is a function of the stress ratio $\eta = q/p'$. According to equation (9), G_0 depends directly on the porosity of the soil through the function $f(e)$, on the mean effective stress raised to a power n and on the overconsolidation ratio raised to a power k. Parameter β embodies the effect of the stress ratio, η, or of the value of the deviatoric stress, q, at a given value of p'.

The same comments as above, concerning the redundancy of one of the three variables e, p' and OCR used to describe the current state of a clay apply to equation (9). Equation (9) predicts higher values of G_0 for anisotropically compressed samples than for isotropically compressed samples, at a given mean effective stress. This is both because anisotropically compressed samples have smaller void ratios than isotropically compressed samples at the same mean effective stress, and because for anisotropically compressed samples $\beta > 1$, while for isotropically compressed samples $\beta = 1$. This idea will be explored further in the analysis of the experimental results.

APPARATUS AND TEST PROCEDURES

The tests discussed in this paper were carried out on 38 mm diameter, 76 mm high specimens in computer-controlled stress path triaxial cells.

The small-strain shear modulus of the soil was measured using bender elements of the type developed at the Norwegian Geotechnical Institute by Dyvik & Madshus (1985), fixed into the top and bottom platens of the triaxial cells. Bender element tests are becoming a standard technique to measure the shear modulus of soils at very small strains. Details of the way bender elements can be incorporated in a triaxial apparatus are given by Viggiani (1992) and Jovičić (1995), while the methods used to interpret the results of bender element tests are discussed, among others, by Viggiani & Atkinson (1995b) and Jovičić *et al.* (1996).

A square input wave, with a frequency of 50 Hz and an amplitude of 10 V (20 V peak to peak), was used to drive the transmitter element. The wavelength of the received shear wave was estimated from the ratio V_s/f, where V_s is the velocity of the shear wave through the sample and f is the average frequency of the received signal. The wavelength of the shear wave was calculated to be generally in the range of $0{\cdot}3L$ to $0{\cdot}5L$, where L is the effective length of the wave travel path through the sample. Under these conditions, due to near-field effects, an analysis based on the time of first deflection of the received signal would lead to a gross underestimate of the travel time of the shear wave (Mancuso & Vinale, 1988). In this work the arrival of the shear wave at the other end of the sample was identified by the first sharp reversal of polarity of the received signal, as shown in Fig. 2. Using this procedure, the arrival time of the shear wave may have been underestimated by only about 7% (Viggiani & Atkinson, 1995b; Jovičić *et al.*, 1996). However, it must be emphasized that the choice of the arrival time was consistent throughout the analysis of the data, and this error does not apply to the comparison between shear moduli. The accuracy of the time measurements on the oscilloscope was about 0·002 ms. The minimum travel times corresponding to the highest values of

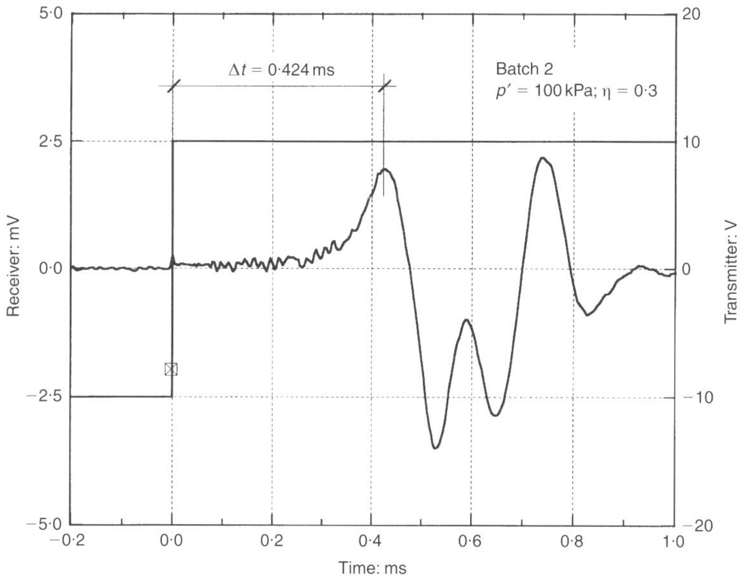

Fig. 2. Bender element trace for a square input wave in reconstituted Vallericca Clay

confining stress were about 0·2 ms, which is equivalent to a maximum error of less than 2% on the calculated values of G_0.

MATERIAL TESTED AND EXPERIMENTAL PROGRAMME

The tests were carried out on reconstituted samples of Vallericca Clay of Plio-Pleistocene age and marine origin.

The small-strain stiffness of natural and reconstituted Vallericca Clay was studied by Rampello & Silvestri (1993), Rampello *et al.* (1994a) and D'Onofrio (1996), while the compressibility and the shear strength of the clay were investigated by Rampello *et al.* (1993), Burland *et al.* (1996) and Amorosi (1996).

Two batches of reconstituted material were formed by mixing the natural clay from two different block samples into a slurry at a water content of about 1·5 times its liquid limit. The slurries were one-dimensionally compressed by incremental loading in a large oedometer up to a nominal vertical effective stress of 100 kPa.

The average index properties and initial water contents of the two batches of reconstituted material are listed in Table 1. The soil is a medium plasticity and activity clay with a calcium carbonate content of about 30%. The small differences in the index properties resulted from using two different blocks to prepare the reconstituted samples.

Table 1. Index properties of the tested material

Batch	G_s	CF: %	W_L: %	I_P: %	w_0: %
1	2·704	48	50·7	28·4	42·1
2	2·714	48	49·6	26·6	39·8

In the triaxial cell the reconstituted samples were first consolidated under isotropic stress conditions at $p' = 50$ kPa, and then brought to the desired value of stress ratio $\eta = q/p'$ along drained constant-p' paths. Bender element measurements were then made at points along radial effective stress paths with constant values of η (Fig. 3). Three tests were carried out for each batch: values of stress ratio $\eta = 0{\cdot}0$, 0·5 and 0·7 and $\eta = 0{\cdot}0$, 0·3 and 0·7 were used for tests on samples from batch 1 and batch 2, respectively. For each test, loading–unloading–reloading stages were carried out in order to describe a regular grid of mean effective stress p' and overconsolidation ratios R. A typical sequence of the compression states attained in a test is shown in Fig. 4 together with lines of constant mean effective stress and overconsolidation ratio. The samples were brought to states with $p' = 50$–400 kPa and $R = 1$–8. Due to the hysteresis of the unloading–reloading loops, nominal values of R are indicated in Fig. 4.

Bender element tests were carried out at the intersections of constant-p' and constant-R lines.

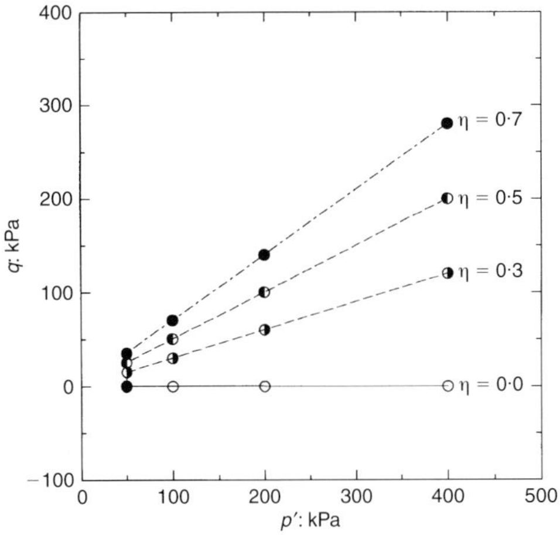

Fig. 3. Constant-η stress paths applied to the reconstituted samples

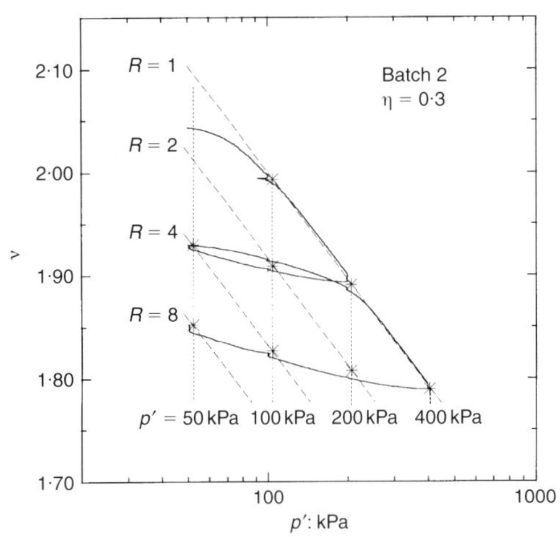

Fig. 4. Typical sequence of compression states in the tests

This sequence of loading was found to be a convenient method of assessing the influence of overconsolidation ratio at a given mean effective stress, and the influence of mean effective stress at a given overconsolidation ratio.

Samples tested under anisotropic stress conditions were compressed or swelled under continuous drained loading or unloading; the rate of variation of mean effective stress was $\dot{p}' = 2$ kPa/h for virgin compression, and $\dot{p}' = 4$ kPa/h for unloading or reloading. Only during the isotropic tests was the stress applied in discrete increments followed by consolidation. At the end of each stage of continuous loading or unloading, measurements of G_0 were taken after an interval of time sufficient to ensure dissipation of small excess pore water pressure. This was checked by periodically closing the drainage lines and verifying that the increments in pore water pressure were less than 1–2% of the mean effective stress. For the isotropic test on batch 1 and the constant $\eta = 0{\cdot}7$ test on batch 2 no measurements of G_0 at overconsolidated states were taken due to malfunctioning of the bender elements.

COMPRESSIBILITY CHARACTERISTICS

In Figs 5(a) and 5(b) the compression and swelling curves observed in the tests are shown in the $\log(p')$–v plane. Fig. 5(a) shows the results from tests carried out at different values of η on samples obtained from batch 1, while Fig. 5(b) refers to samples from batch 2.

The vertical paths at the end of the stages of

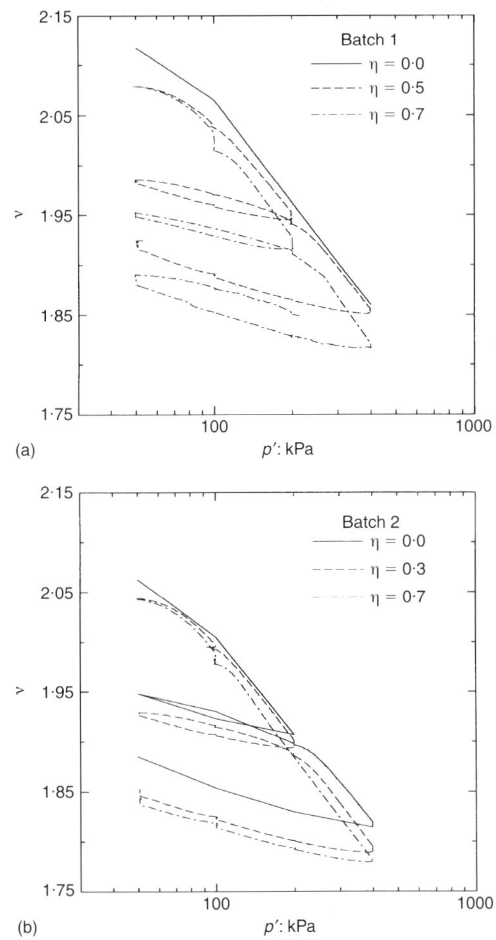

Fig. 5. Results of constant-η compression tests on reconstituted samples: (a) batch 1; (b) batch 2

continuous drained compression or swelling represent small additional changes in specific volume corresponding to the attainment of equilibrium conditions. These variations of v are related to creep and/or dissipation of small residual excess pore water pressure at the end of the stages. The slopes of the compression and swelling curves were evaluated using the equilibrium states.

The anisotropic virgin compression curves are approximately parallel to the corresponding isotropic lines, with decreasing values of specific volume N_η at a reference stress $p_r = 100$ kPa as the value of η increases. A single value of λ was adopted to interpret each set of data, and a single value of κ was found to be in fair agreement with the average slope of the unloading–reloading cycles of all tests, irrespective of the value of η. The compression parameters listed in Table 2 refer to the $\ln(p')-v$ plane, although the data are plotted on a $\log(p')$ scale. The lower values of N_η and λ for batch 2 can be attributed to the difference in the index properties listed in Table 1.

The axial strains ε_a and the volumetric strains ε_v measured during consolidation carried out under different values of η are shown in Fig. 6. In

Table 2. Compression characteristics

Batch	η	N_η	λ	κ	Λ
1	0·0	2·065	0·148	0·03	0·797
	0·5	2·050			
	0·7	2·023			
2	0·0	2·008	0·140	0·03	0·786
	0·3	1·992			
	0·7	1·974			

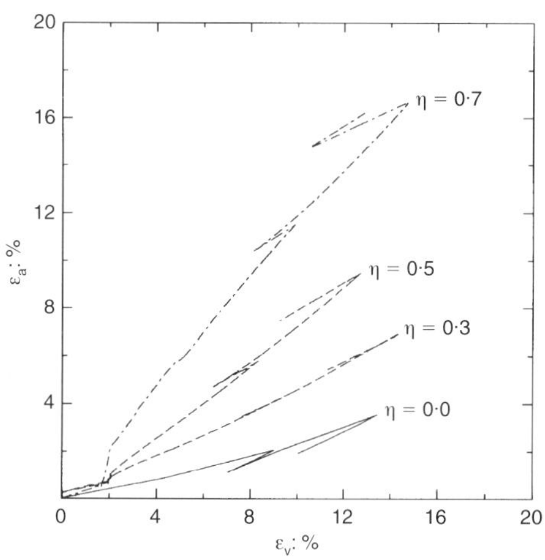

Fig. 6. Drained strain paths during compression under constant η

isotropic compression, the ratio $\Delta\varepsilon_a/\Delta\varepsilon_v$ is about 0·33 for the last increment of mean effective pressure $p' = 200-400$ kPa, which is consistent with isotropic soil behaviour, while it is about 0·25 for values of p' in the range of 50–200 kPa. For values of p' less than about 200 kPa, the soil behaviour may still be affected by the one-dimensional stress history experienced by the reconstituted material in the large oedometer. During virgin compression the ratio of the axial to the volumetric strains is nearly constant in a test and increases for tests with greater stress ratios. Also, the ratio between the cumulative strains measured during constant-η and isotropic virgin compression is constant and independent of p'. The angle between the strain paths traced in the unloading–reloading cycles and those followed during virgin compression increases with increasing stress ratio.

SHEAR STIFFNESS AT SMALL STRAINS

In Fig. 7(a), for the test carried out at $\eta = 0{\cdot}3$, the values of the small-strain shear modulus G_0, divided by the reference pressure $p_r = 100$ kPa, are plotted against the ratio p'/p_r on a bilogarithmic scale; in Fig. 7(b) the non-dimensional shear modulus G_0/p_r is plotted against the specific volume v on a semilogarithmic scale. The thin dashed and dotted lines in Fig. 7 represent lines of constant nominal overconsolidation ratio and mean effective stress, respectively. The dependence of G_0 on p' and v, shown in Fig. 7, is representative of the behaviour observed in the other tests for both isotropic and anisotropic compression histories; G_0 increases with increasing mean effective stress and decreasing specific volume and, at the same mean effective stress, it increases with increasing overconsolidation ratio.

In the next section, the comparison between test results obtained at different values of stress ratio is first limited to normally consolidated states. Also, due to the slight differences in the index properties of the two batches, data from the two sets of samples are analysed separately in the following sections.

Normally consolidated states

For normally consolidated states, the values of $\log(G_{0(nc)}/p_r)$ are plotted in Figs 8(a) and 8(b) against $\log(p'/p_r)$. The open circles refer to samples tested under isotropic stress conditions, the full circles to samples tested under a constant value of stress ratio $\eta = 0{\cdot}7$, while the other two symbols refer to values of $\eta = 0{\cdot}5$ (batch 1) and $\eta = 0{\cdot}3$ (batch 2). For both sets of samples, data points fall close to parallel lines irrespective of the stress ratio experienced by the samples during compression, and increasing values of small-strain

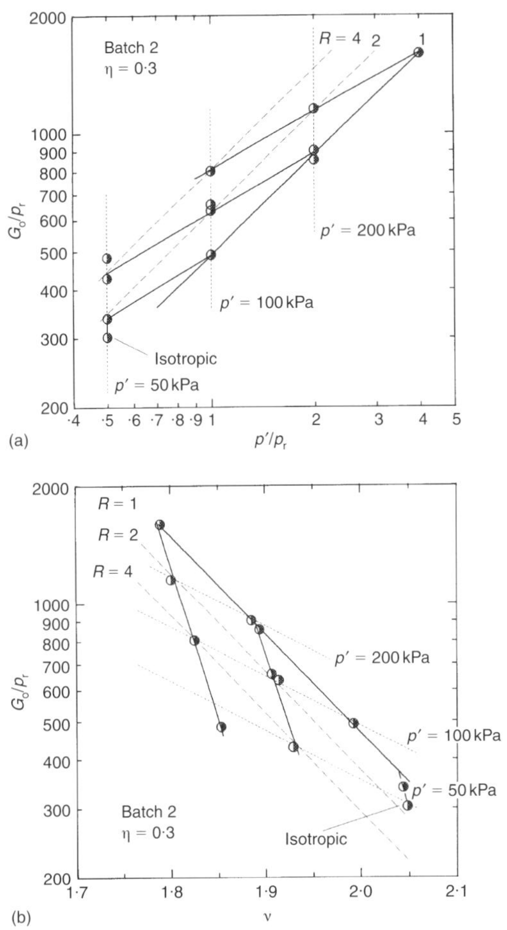

Fig. 7. Small-strain shear modulus against: (a) mean effective stress; (b) specific volume

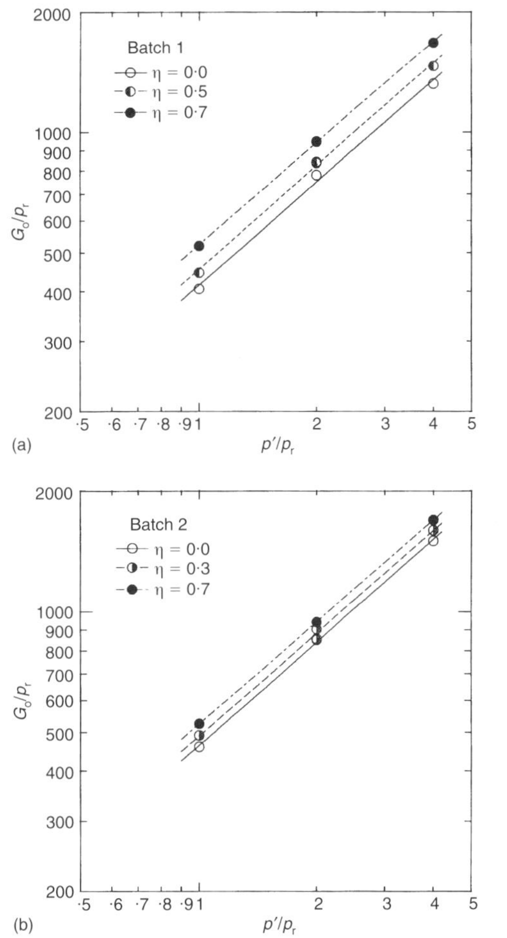

Fig. 8. Small-strain shear modulus against mean effective stress for normally consolidated states under different values of η: (a) batch 1; (b) batch 2

shear modulus are observed with increasing stress ratio at the same mean effective stress.

Applying the same relationship proposed for isotropic stress conditions in equation (2) to anisotropic compressed samples, $G_{0(\mathrm{nc})}$ can be expressed as

$$\frac{G_{0(\mathrm{nc})}}{p_\mathrm{r}} = S_\eta \left(\frac{p'}{p_\mathrm{r}}\right)^n \tag{10}$$

so that the increase of $G_{0(\mathrm{nc})}$ with p' is governed by a single index n for both isotropic and anisotropic compression paths, while increasing values of S_η, which represent the value of $G_{0(\mathrm{nc})}$ at the reference stress, are obtained for increasing values of η. For simplicity, both in equation (10) and in the following sections, the asterisks have been omitted from the stiffness parameters. Table 3 lists the values of S_η and n as obtained from the least squares best-fit lines of data points from tests on normally consolidated samples. The ratio of S_η to the corresponding value S obtained under isotropic conditions ranges from about 1·06 to 1·26 for η increasing from 0·3 to 0·7. The values of n are in the upper bound of the range found in the literature for comparable values of the plasticity index (e.g. see Viggiani & Atkinson, 1995a).

A diagram that proved useful in the interpretation of the data for normally consolidated states is given in Fig. 9. In Fig. 9(a) the specific volumes attained by two samples in isotropic and anisotropic (η = constant) virgin compression are plotted against $\ln(p')$; the two compression lines are characterized by the same value of λ and different values of N and N_η. In Fig. 9(b) the values of $\ln(G_{0(\mathrm{nc})})$ obtained at the states in Fig. 9(a) are

Table 3. Stiffness characteristics

Batch	η	S_η (nc states)	S_η (nc + oc states)	n	c
1	0·0 0·5 0·7	415 455 523	410 464 521	0·851	0·367
2	0·0 0·3 0·7	463 489 524	449 503 536	0·851	0·358

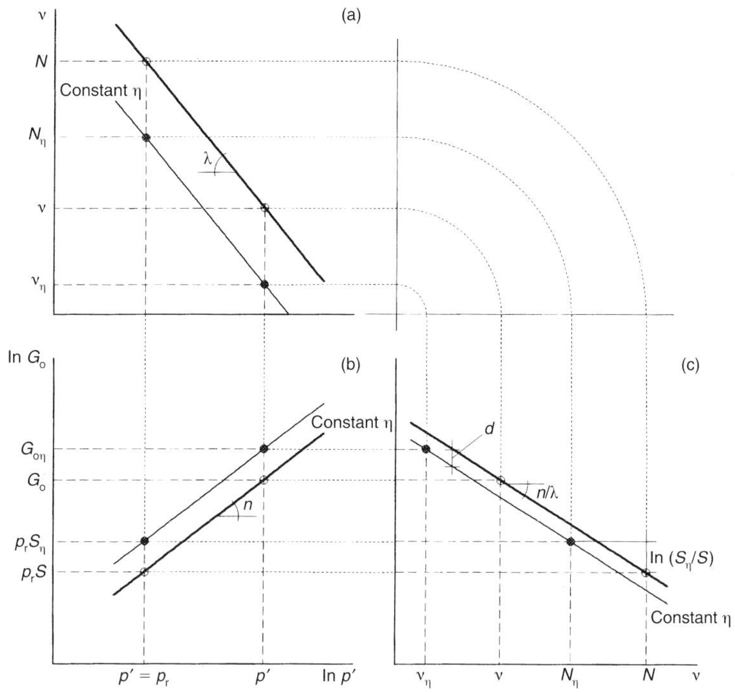

Fig. 9. Diagram for the dependence of G_0 on p' and v for nc samples compressed under isotropic and anisotropic stress conditions

plotted against $\ln(p')$. Consistent with the observations, for both samples, the increase of $G_{0(\text{nc})}$ with p' is governed by the same value of n, the line relative to the anisotropically compressed sample plotting parallel and above the one corresponding to the isotropic stress history. From Figs 9(a) and 9(b) it can be seen that a plot of $\ln(G_{0(\text{nc})})$ against v (Fig. 9(c)) would yield two parallel lines at a slope n/λ and separated by a vertical distance d, where d is given by the following equation

$$d = \ln\left(\frac{G_{0\eta(\text{nc})}}{G_{0(\text{nc})}}\right) = \ln\left(\frac{S_\eta}{S}\right) - \frac{n}{\lambda}(N - N_\eta) \quad (11)$$

However, data points for both sets of samples shown in Figs 10(a) and 10(b) plot close to single lines in the plane v–$\log(G_0/p_r)$, irrespective of the stress ratio experienced by the samples during virgin compression, suggesting that the distance d in Fig. 9(c) is essentially zero. The slopes of the best-fit lines in Fig. 10 are equal to the ratio n/λ to within 3 and 5% for batches 1 and 2, respectively. If d is taken to be zero, from equation (11) it follows that

$$\frac{S_\eta}{S} = \exp\left(\frac{n}{\lambda}(N - N_\eta)\right) \quad (12)$$

Overconsolidated states

Test results for overconsolidated states can be analysed using equation (5), proposed for isotropically compressed samples, written in the form

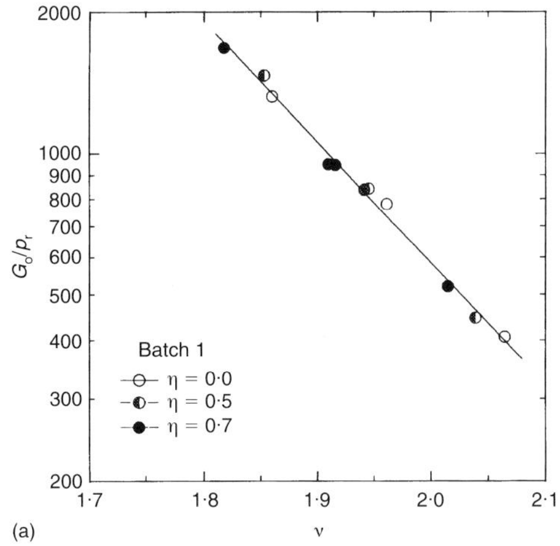

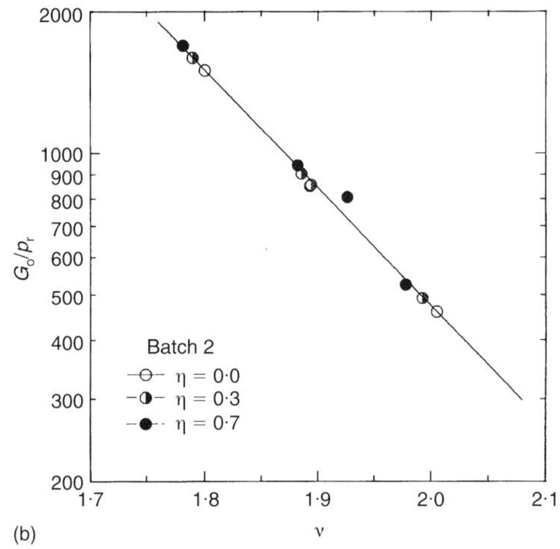

Fig. 10. Small-strain shear modulus against specific volume for normally consolidated states under different values of η: (a) batch 1; (b) batch 2

$$\frac{G_0}{p_r} = S_\eta\left(\frac{p'}{p_r}\right)^n\left(\frac{p'_{e\eta}}{p'}\right)^c = \frac{G_{0(nc)}}{p_r}\left(\frac{p'_{e\eta}}{p'}\right)^c \tag{13}$$

in which the equivalent pressure, $p'_{e\eta}$, is determined on the constant-η virgin compression line relevant to each sample

$$p'_{e\eta} = p_r \exp\left(\frac{N_\eta - \nu}{\lambda}\right) \tag{14}$$

Index c can be evaluated by plotting $\log(G_0/G_{0(nc)})$ against $\log(p'_{e\eta}/p')$. Figs 11(a) and 11(b) show that for both sets of samples, irrespective of the value of η during compression or swelling and despite some scatter, data points can be fitted by a single line of slope c. Values of c listed in Table 3 are in the range of those observed for comparable

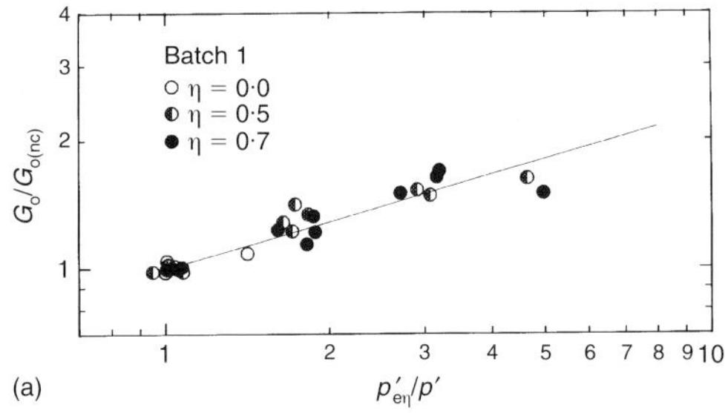

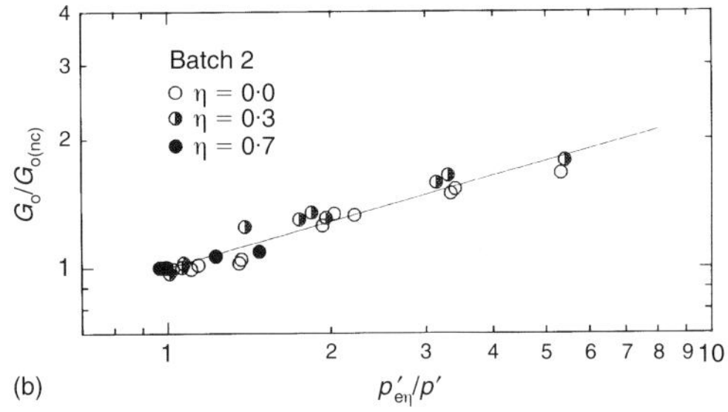

Fig. 11. Evaluation of stiffness index c: (a) batch 1; (b) batch 2

values of the plasticity index (Rampello *et al.*, 1994a).

If data from each test are normalized by dividing by the appropriate equivalent pressure $p'_{e\eta}$, test results for normally consolidated states and for overconsolidated states are rendered compatible since the differences in the void ratios are accounted for in the normalization procedure. This normalization procedure is similar to that commonly used to analyse shear strength data (Hvorslev, 1937) with the difference that, because the dependence of G_0 on p' is non-linear ($n \neq 1$), the equivalent pressure raised to the power n must be used for normalization.

By dividing equation (13) by $(p'_{e\eta})^n$, it is possible to write

$$\left(\frac{G_0}{p_r}\right)\left(\frac{p_r}{p'_{e\eta}}\right)^n = S_\eta\left(\frac{p'}{p'_{e\eta}}\right)^{(n-c)} \tag{15}$$

Data normalized according to equation (15) are plotted in Figs 12(a) and 12(b) for samples from batch 1 and 2, respectively. After normalization, data from each test, both for normally consolidated and overconsolidated states, plot close to best-fit lines obtained using a slope of $(n - c)$. The values of the stiffness multipliers S_η, previously only estimated using the data from normally consolidated states (see Fig. 8), can now be better evaluated by considering the whole set of data from each test in Fig. 12. These values of S_η and the corresponding ratios S_η/S are listed in Tables 3

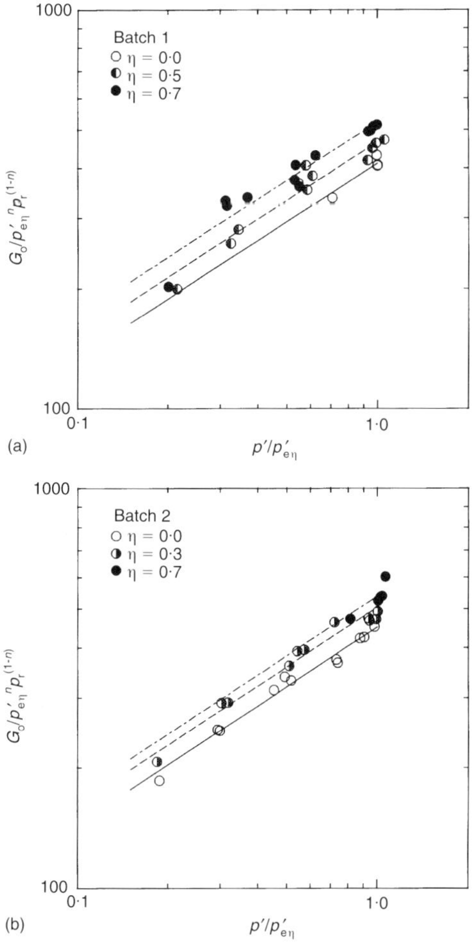

Fig. 12. Small-strain shear modulus normalized by the equivalent pressure on the virgin compression line of each sample: (a) batch 1; (b) batch 2

and 4, respectively. For samples tested under a constant value of $\eta = 0{\cdot}3$ or $0{\cdot}5$, S_η/S is about 1·13, while for tests carried out at $\eta = 0{\cdot}7$ $S_\eta/S = 1{\cdot}27$ for batch 1 and $S_\eta/S = 1{\cdot}20$ for batch 2.

DISCUSSION

In Fig. 9(c) the vertical distance d between the $\ln(G_0)-v$ curves for isotropic and anisotropic stress conditions was not specified, but data from normally consolidated states appeared to plot around single lines, irrespective of the value of η, that is, d tended to zero (see Fig. 10). If the small-strain shear modulus depends solely on the current state, as expressed by v and p', the value of d cannot be zero. In Fig. 13 the compression paths of two samples, one normally consolidated under a constant stress ratio and the other one overconsolidated under isotropic conditions, are shown in the $\ln(p')-v$ plane as O′–B and O–A–B, respectively. If G_0 depends only on v and p', the values of G_0 at point B should be the same for the two samples. The small-strain shear moduli at B, where the current value of p' is equal to $p'_{e\eta}$, can be calculated as

$$\frac{G_0}{p_r} = S\left(\frac{p'_{e\eta}}{p_r}\right)^n \left(\frac{p'_e}{p'_{e\eta}}\right)^c \tag{16}$$

for the isotropic overconsolidated sample and

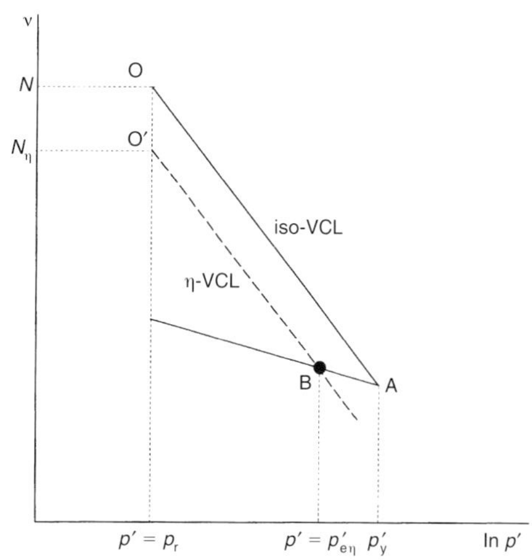

Fig. 13. Compression paths for isotropic (O–A–B) and anisotropic (O′–B) stress conditions

Table 4. Observed and predicted stiffness after anisotropic and isotropic compression

Batch	η	$N - N_\eta$	S_η/S: observed	S_η/S: equation (12)	S_η/S: equation (18)	S_η/S: equation (20)	d
1	0·5	0·015	1·13	1·09	1·04	1·09	−0·05
	0·7	0·042	1·27	1·27	1·11	1·17	−0·14
2	0·3	0·016	1·12	1·10	1·04	1·08	−0·06
	0·7	0·034	1·20	1·23	1·09	1·15	−0·12

$$\frac{G_0}{p_r} = S_\eta \left(\frac{p'_{e\eta}}{p_r}\right)^n \tag{17}$$

for the anisotropic normally consolidated sample. Eliminating G_0/p_r

$$\frac{S_\eta}{S} = \left(\frac{p'_e}{p'_{e\eta}}\right)^c = \exp\left(\frac{c}{\lambda}(N - N_\eta)\right) \tag{18}$$

By combining equations (11) and (18), the following expression for d is obtained

$$d = \frac{(c-n)}{\lambda}(N - N_\eta) \tag{19}$$

i.e. $d \neq 0$.

Table 4 lists the values of d predicted by equation (19). Negative values of d imply that, if constant-η normally consolidated states had been equivalent to isotropically overconsolidated states lying on a constant-R line, data for anisotropically compressed normally consolidated samples should have plotted in Fig. 10 slightly below data corresponding to samples normally consolidated under isotropic stress conditions. The values of S_η/S predicted by equation (18) from the hypothesis that G_0 depends solely on ν and p' are also listed in Table 4; they are in the range of 1·04–1·11 for values of η ranging from 0·3 to 0·7, and underestimate the observed values of S_η/S by 8–13%.

An alternative way to show that anisotropic normally consolidated states cannot be considered equivalent to isotropic overconsolidated states lying on a constant-R line is that of normalizing all data by the equivalent pressure on the isotropic virgin compression line (iso-VCL) raised to the power n, $(p'_e)^n$; in this case p'_e is used in equation (15) rather than $p'_{e\eta}$. In this manner, again the implicit assumption is made that the small-strain shear modulus depends solely on mean effective stress and void ratio. Thus, after normalization, in a plot of $\log[(G_0/p_r)(p_r/p'_e)^n]$ against $\log(p'/p'_e)$, data points should collapse around a single line of slope $(n - c)$, irrespective of the value of η.

Figures 14(a) and 14(b) show that, when normalized using $(p'_e)^n$, data from samples tested under constant values of η ($\neq 0$) plot closer to those obtained under isotropic stress conditions than in Fig. 12. In Fig. 14, however, the distance between the least squares best-fit lines of given slope $(n - c)$ is not eliminated, which confirms that other factors, in addition to the current values of mean effective stress and void ratio, must be considered to explain the differences in the values of G_0.

In Table 4, a better agreement with the observed values of S_η/S is given in equation (12), which, based on the experimental results in Fig. 10, was obtained by assuming $d = 0$ in equation (11).

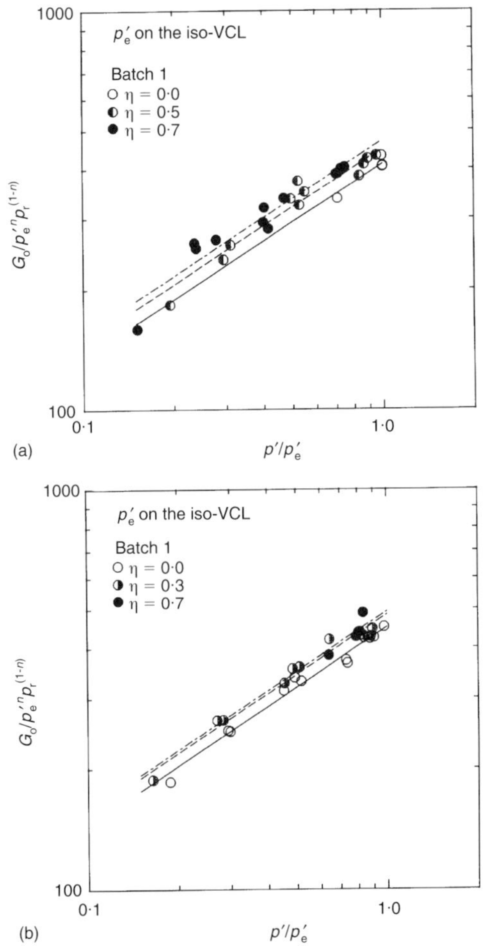

Fig. 14. Small-strain shear modulus normalized by the equivalent pressure on the isotropic virgin compression lines: (a) batch 1; (b) batch 2

Equation (12), however, does not allow the factors affecting the differences observed in the values of G_0 to be distinguished. Equation (18), on the other hand, was derived by assuming that the small-strain shear modulus is only a function of ν and p', in other words, neglecting the influence on the small-strain stiffness of the different strain histories experienced by the samples during isotropic and anisotropic compression. These are likely to produce different particle arrangements at the same current state (ν, p'), thus contributing to different values of G_0.

Following the approach proposed in equation (9), an attempt was made to simulate the effects of different strain histories on G_0 by multiplying the values of S_η/S predicted by equation (18) by $\beta^{n/2}$, such that

$$\frac{S_\eta}{S} = \exp\left(\frac{c}{\lambda}(N - N_\eta)\right)\beta^{n/2} \tag{20}$$

The values of S_η/S predicted by equation (20) are also included in Table 4. Equation (20) yields values of S_η/S that are in better agreement with the experimental results than those predicted by equation (18), although, for both batches, S_η/S is still slightly underestimated by 4–8% for values of η in the range of 0·3–0·7.

CONCLUSIONS

The research described in the paper was an experimental investigation of the stiffness of fine-grained soils at very small strains in triaxial conditions.

The values of G_0 measured using bender elements on reconstituted samples of Vallericca Clay compressed along constant effective stress ratio paths were observed to be higher than the corresponding values measured at the same p' on isotropically consolidated reconstituted samples.

The test results were interpreted using empirical relationships similar in form to those proposed for isotropically consolidated samples.

For normally consolidated states the variation of G_0 with mean effective stress was observed to be governed by a single exponent n, irrespective of the value of η. For overconsolidated states, the exponent c, which governs the variation of G_0 with the ratio $p'_{e\eta}/p'$, was also found to be independent of the value of η. The observation that n and c are independent of η implies that both stiffness parameters can be simply determined by bender element tests on isotropically compressed samples.

The difference in the values of the small-strain shear moduli measured after anisotropic constant-η and isotropic consolidation was quantified using the ratio S_η/S, where S_η and S represent the values of G_0 at the reference pressure $p' = p_r = 100$ kPa after constant-η and isotropic compression, respectively. The observed ratios S_η/S were about 1·10 to 1·20–1·27 for values of η in the range of 0·3–0·5 to 0·7.

Based on the experimental observation that data points from normally consolidated states plot close to a single line in the ν–$\log(G_0/p_r)$ plane, irrespective of the value of η, an empirical relationship (equation (12)) was obtained which closely predicts the values observed for the ratios S_η/S. According to equation (12) a value for the ratio S_η/S can be conveniently obtained once the compression parameters N, N_η and λ are known.

The difference in the values of G_0 for samples which underwent different compression histories is likely to depend on both the different void ratios developed at the same mean effective stress and the different strain histories. The influence of the different void ratios at the same mean effective stress can be eliminated by normalizing test results using the equivalent pressure on the isotropic virgin compression line (equation (18)). In this way the influence of different strain histories was isolated and recognized to be discernible.

Equation (20), derived from previous expressions proposed for anisotropic stress states (e.g. see Hardin & Blandford, 1989), makes it possible to distinguish between the different factors that affect the small-strain shear stiffness. As for the relationships proposed for isotropic stress conditions, the rational link that exists between void ratio, state of stress and overconsolidation ratio is taken into account by equation (20).

The experimental data presented in the paper confirm that parameter β is capable of accounting for the different strain histories experienced by the samples during anisotropic constant-η and isotropic compression paths.

NOTATION

A, B	constants used to define $f(e)$
c	exponent of p'_e/p' and $p'_{e\eta}/p'$
CF	clay fraction (%)
d	vertical distance of isotropic and anisotropic lines in ν–$\ln(G_0)$ plane
e	void ratio
$f(e)$	void ratio function
G_0	small-strain shear modulus
$G_{0(nc)}$	small-strain shear modulus for normally consolidated states
G_s	specific gravity
I_P	plasticity index
iso-VCL	isotropic virgin compression line
k	exponent of OCR
L	effective length of the travel path for the shear wave
m	exponent of R
n, n^*	exponent of mean effective stress
OCR	overconsolidation ratio
p_r	reference stress
p'	mean effective stress
$\dot{p}'$	rate of mean effective stress
p'_e	equivalent pressure
p'_y	yield mean effective stress
q	deviatoric stress
R	mean effective stress yield ratio
S, S^*	stiffness multiplier for isotropic stress history
S_{ij}	stiffness multiplier for anisotropic conditions
S_η	stiffness multiplier for anisotropic stress history (η = constant)
t, Δt	travel time of the shear wave
V_s	shear wave velocity
W_L	liquid limit
w_0	initial water content
β	function of stress ratio η
ε_a	axial strain
ε_v	volumetric strain
η	stress ratio
η-VCL	constant-η virgin compression line

κ swelling index in natural log scale
λ compression index in natural log scale
Λ plastic volumetric strain ratio $(\lambda - \kappa)/\lambda$
v specific volume
N specific volume at the reference stress for isotropic virgin compression
N_η specific volume at the reference stress for constant-η virgin compression
ρ total mass density
σ'_a axial effective stress
σ'_i principal effective stress in the direction of wave propagation
σ'_j principal effective stress in the direction of particle motion
σ'_r radial effective stress

REFERENCES

Abbis, C. P. (1981). Shear wave measurements of the elasticity of the ground. *Géotechnique* **31**, No. 1, 94–104.

Al-Tabbaa, A. & Wood, D. M. (1989). An experimental based "bubble" model for clay. *Proceedings of the 3rd international conference on numerical models geomechanics*, Niagara Falls, pp. 91–99.

Amorosi, A. (1996). *Il comportamento meccanico di una argilla naturale consistente*. University of Rome 'La Sapienza', doctoral thesis.

Burghignoli, A., Pane, V. & Cavalera, L. (1991). Modelling stress–strain–time behaviour of natural soils—monotonic loading. *Proc. 10th Eur. Conf. Soil Mech., Florence* **3**, 961–979.

Burland, J. B., Rampello, S., Georgiannou, V. N. & Calabresi, G. (1996). A laboratory study of the strength of four stiff clays. *Géotechnique* **46**, No. 3, 491–514.

D'Onofrio, A. (1996). *Comportamento meccanico dell'argilla di Vallericca in condizioni lontane dalla rottura*. University of Naples 'Federico II', doctoral thesis.

Dyvik, R. & Madshus, C. (1985). Laboratory measurements of G_{max} using bender elements. *Proceedings of the ASCE Annual Convention: advances in the art of testing soils under cyclic conditions*, Detroit, pp. 186–196.

Hardin, B. O. (1978). The nature of stress–strain behaviour for soils. State-of-the-art-report. *Proceedings of a special conference on earthquake engineering and soil dynamics*, pp. 3–90. Pasadena: American Society of Civil Engineers.

Hardin, B. O. & Black, W. L. (1969). Closure to 'Vibration modulus of normally consolidated clay'. *J. Soil Mech. Fdn Div., Am. Soc. Civ. Engrs* **95**, SM6, 1531–1537.

Hardin, B. O. & Blandford, G. E. (1989). Elasticity of particulate materials. *J. Geotech. Engng Div., Am. Soc. Civ. Engrs* **115**, No. 6, 788–805.

Hardin, B. O. & Richart, F. E. (1963). Elastic wave velocities in granular soils. *J. Soil Mech. Fdn Div., Am. Soc. Civ. Engrs* **89**, SM1, 33–65.

Houlsby, G. T. (1985). The use of a variable shear modulus in elastic-plastic models for clays. *Comput. Geotech.* **1**, 3–13.

Houlsby, G. T. & Wroth, C. P. (1991). The variation of shear modulus of a clay with pressure and overconsolidation ratio. *Soils Fdns* **18**, No. 3, 138–143.

Hvorslev, M. J. (1937). Über die Festigkeitseigenschaften Gestörter Bindinger Böden. Danmarks Naturvidenskabelige Samfund. *Ingeniorvidensk. Skr. A*, No. 45.

Ishihara, K. (1982). Evaluation of soil properties for use in earthquakes response analysis. *Proceedings of the international symposium on numerical models in geomechanics*, Zurich, pp. 237–259.

Jamiolkowski, M., Lancellotta, R. & Lo Presti, D. C. F. (1994). Remarks on the stiffness at small strains of six Italian clays. *Proc. 1st Int. Conf. 'Pre-failure Deform. Character. Geomater.', Sapporo* **2**, 817–836.

Jovičić, V. (1995). *Small strain shear modulus of soils and soft rocks*. Geotechnical Engineering Research Centre, City University, London, Report GE/5/95.

Jovičić, V. & Coop, M. R. (1996). The measurement of stiffness anisotropy in clays with bender element tests in the triaxial apparatus. *Geotech. Testing J.*, submitted.

Jovičić, V., Coop, M. R. & Simić, M. (1996). Objective criteria for determining G_{max} from bender element tests. *Géotechnique* **46**, No. 2, 357–362.

Lade, P. V. & Nelson, R. B. (1987). Modelling the elastic behaviour of granular materials. *Int. J. Num. Anal. Methods Geomech.* **11**, 521–542.

Mancuso, C. & Vinale, F. (1988). Propagazione delle onde sismiche: teoria e misura in sito. *Atti del convegno CNR su deformazioni dei terreni ed interazione terreno-struttura in condizioni di esercizio*, Monselice, pp. 115–138.

Ni, S. H. (1987). *Dynamic properties of sand under true triaxial stress states from resonant column and torsion shear tests*. The University of Texas, Austin, PhD thesis.

Rampello, S. & Silvestri, F. (1993). The stress–strain behaviour of natural and reconstituted samples of two overconsolidated clays. *Proc. Int. Symp. Geotech. Engng Hard Soils – Soft Rocks, Athens* **1**, 769–778.

Rampello, S., Georgiannou, V. N. & Viggiani G. (1993). Strength and dilatancy of natural and reconstituted Vallericca clay. *Proc. Int. Symp. Geotech. Engng Hard Soils – Soft Rocks, Athens* **1**, 761–768.

Rampello, S., Silvestri, F. & Viggiani G. (1994a). The dependence of small strain stiffness on stress state and history for fine grained soils: the example of Vallericca clay. *Proc. 1st Int. Conf. Pre-failure Deform. Character. Geomater., Sapporo* **1**, 273–278.

Rampello, S., Silvestri, F. & Viggiani G. (1994b). The dependence of G_0 on stress state and history in cohesive soils. *Proc. 1st Int. Conf. Pre-failure Deform. Character. Geomater., Sapporo* **2**, 1155–1160.

Saada, A. S., Bianchini, G. F. & Palmer Shook, L. (1978). The dynamic response of anisotropic clay. *Proceedings of a special conference on earthquake engineering and soil dynamics*, pp. 777–801. Pasadena, American Society of Civil Engineers.

Stallebrass, S. E. (1990). *Modelling the effect of recent stress history on the deformation of overconsolidated soils*. City University, London, PhD thesis.

Viggiani, G. (1992). *Small strain stiffness of fine grained soils*. City University, London, PhD thesis.

Viggiani, G. & Atkinson, J. H. (1995a). Stiffness of fine-grained soil at very small strains. *Géotechnique* **45**, No. 2, 249–265.

Viggiani, G. & Atkinson, J. H. (1995b). The interpretation of the bender element tests. *Géotechnique* **45**, No. 1, 149–155.

Weiler, W. A. (1988). Small strain shear modulus of clay. *Proceedings of a special conference on earthquake engineering and soil dynamics*, pp. 331–345. Park City: American Society of Civil Engineers.

Wroth, C. P. & Houlsby, G. T. (1985). Soil mechanics—property characterisation and analysis procedures. *Proc. 11th Int. Conf. Soil Mech., San Francisco* **1**, 1–55.

Zytynski, M., Randolph, M. F., Nova, R. & Wroth, C. P. (1978). On modelling the unloading–reloading behaviour of soils. *Int. J. Numer. Anal. Methods Geomech.* **2**, 87–93.

Cuccovillo, T. & Coop, M. R. (1997). *Géotechnique* **47**, No. 3, 491–508

Yielding and pre-failure deformation of structured sands

T. CUCCOVILLO* and M. R. COOP†

Many natural sands have a structure that is imparted to them during their geological life and that significantly modifies their pre-failure behaviour when compared with that of the reconstituted material. Unlike clays, for which emphasis has been given to both the bonding and fabric components of structure, for sands structure has generally been simply identified with bonding, whereas the effects of fabric have often been overlooked. An understanding of the way in which these components of structure influence the yielding and stiffness of these materials is required to formulate constitutive models that can adequately predict their response. The paper examines the pre-failure behaviour of two structured sands in triaxial compression performed over a wide range of pressures. Since yielding occurred at small strains, laboratory techniques were developed to produce reliable and accurate determinations of stiffness ranging from 1 microstrain to failure. The influence of structure on the shear stiffness was investigated by comparing the behaviour of intact and reconstituted soils after accounting for differences in state. To explain how bond degradation developed, the study focused in particular on the shear behaviour between yielding and failure. The bond degradation resulted in a progressive transformation of the structured soil into a frictional material, giving rise to changes in the yield stress and shear stiffness that contrasted with the strain-hardening behaviour of the unbonded sands. As bonding degraded, the variation of the shear stiffness with state was seen to depend on which structural feature was predominant.

KEYWORDS: elasticity; fabric/structure of soils; laboratory tests; sands; soft rocks; stiffness.

De nombreux sables naturels ont une structure qui leur est impartie pendant leur vie géologique et qui modifie considérablement leur comportement avant rupture par comparaison à celui de matériaux reconstitués. Alors qu'on accorde autant d'importance à l'adhésion qu'à la contexture des argiles, on a tendance à assimiler la structure des sables à la simple notion d'adhésion, en ignorant souvent les effets de la texture. Il faut pouvoir comprendre l'influence que les éléments de la structure exercent sur l'élasticité et la rigidité de ces matériaux pour pouvoir prédire leur réaction de façon fiable. L'article examine le comportement avant rupture de deux sables structurés dans des essais de compression triaxiale effectués à une grande variété de pressions. Vu que la limite élastique est atteinte à de petites contraintes, on a mis au point des techniques de laboratoire qui permettent de déterminer de façon fiable et précise la rigidité, depuis la microcontrainte jusqu'à la rupture. On a examiné l'influence de la structure sur la rigidité au cisaillement en comparant le comportement de sols non remaniés à celui de sols reconstitués, après avoir tenu compte des différences d'état. Pour expliquer la dégradation progressive de l'adhésion, l'étude s'est concentrée plus particulièrement sur le comportement de cisaillement entre la limite élastique et la rupture. La dégradation de l'adhésion entraîne la transformation progressive du sol structuré en un matériau sans cohésion, donnant lieu à une modification de la limite d'élasticité et de la rigidité au cisaillement, par contraste au rhéo-durcissement des sables non adhérents. A mesure que l'adhésion se détériore, on constate que la variation de la rigidité au cisaillement en fonction de l'état dépend de la caractéristique structurale prédominante.

INTRODUCTION

In the last decade much of the research on the fundamental behaviour of soils has focused on the study of natural soils that possess a structure deriving from processes taking place in their geological life. The term 'structure' was first introduced for clays (Mitchell, 1976), and has been described by Burland (1990) as being the combination of bonding and fabric. For a coarse-grained soil, bonding can only take the form of interparticle cementing since interparticle forces are insignificant. Most previous work on natural sands has been focused

Manuscript received 17 Ocotber 1996; revised manuscript accepted 27 March 1997.
Discussion on this paper closes 1 November 1997.
* South Bank University, formerly City University, London.
† City University, London.

on this component of structure. Much less emphasis has been placed on the fabric of these soils (i.e. the arrangement and distribution of the particles comprising the soil). Using the above definition of structure, the behaviour of two coarse-grained soft rocks will be examined. The two soft rocks will be regarded as naturally structured sands, and their behaviour will be interpreted highlighting and contrasting the influence of each of the two structural components.

The critical state theory (Roscoe *et al.*, 1958) considers the shear behaviour of reconstituted soils as being solely controlled by the state of the soil defined as the combination of deviator stress (q), mean effective stress (p') and specific volume (v). Strain-softening and strain-hardening modes of shear behaviour are associated with volumetric compression and dilation, respectively, and are therefore controlled by the volumetric changes that the soil experiences in moving from its initial state to the critical state. For frictional materials such as reconstituted soils, there is considerable evidence that the stress–strain behaviour is largely non-linear (e.g. see Jardine *et al.*, 1991) and that the shear stiffness at any strain depends on the current state of the soil (e.g. see Viggiani & Atkinson, 1995).

As a result of their structure, natural soils may show patterns of behaviour both prior to and during shearing that do not conform to the behavioural framework of reconstituted soils and that cannot be accounted for by state alone. These patterns of behaviour can be seen more clearly through comparisons with the behaviour of the corresponding reconstituted soil (e.g. see Leroueil & Vaughan, 1990). Despite the large amount of research on natural soils, there have been fewer systematic experimental studies of the mechanics of natural sands, and these have been mainly focused on the behaviour of bonded sands at relatively large strains. The principal limitation to the study of their behaviour at relatively small strains has been the inappropriateness of many existing testing techniques to determine the very high values of stiffness that characterize these materials. The difficulty of sampling natural sands in the absence of bonding and, when present, the frequent variability of the bonding led many researchers to focus initially on artificially cemented sands (Clough *et al.*, 1981; Coop & Atkinson, 1993; Huang & Airey, 1993).

The outcome of these studies both on naturally and artificially bonded soils resulted in the formulation of a number of mathematical models for these materials (e.g. see Gens & Nova, 1993; Lagioia & Nova, 1995). Because of the limitations of the experimental data available, their quantitative agreement could be verified only at strain levels which for many natural sands are too large to detect those essential aspects of behaviour related to structure. The initial elastic behaviour assumed by these models has been confirmed by laboratory results recently obtained using small strain measuring techniques (e.g. see Tatsuoka & Shibuya, 1992). Some of the factors that influence the elastic threshold and the stiffness of these materials are, however, still unclear. As yielding is accompanied by bond degradation and the development of relatively large strains, an understanding of these factors, as well as the accurate identification of the onset of yielding, is important in the prediction of the performance of engineering structures in these types of material (Vaughan, 1988).

This paper analyses and contrasts the shear behaviour of two substantially different types of naturally structured sands as they were sheared in triaxial compression from very small strains to failure, examining particularly the influence of the two components of structure on the shear stiffness and yielding. New techniques were implemented for the accurate determination of small strain stiffness and yielding. The testing programme consisted of drained and undrained shearing performed over a wide range of pressures, the undrained shearing including loading–unloading cycles performed to study the influence of bond degradation on the yielding of the materials. The use of high pressure tests allowed general patterns of behaviour to be identified that are not evident if tests are conducted over the 'standard' soil mechanics pressure range, which generally does not exceed around 800 kPa.

The investigation of the natural soils was supported by an experimental study of the reconstituted soils that were obtained by mechanically breaking the bonding of the intact soils. The comparison was used to highlight the influence of structure and bond degradation on the shear stiffness and yielding of the intact materials.

SOIL DESCRIPTIONS

The materials considered in the experimental work were a calcarenite from the site of the offshore gas platform of North Rankin in Western Australia and a silica sandstone from the Lower Greensand series of Kent in England. The two materials were deliberately chosen to compare and contrast the yielding and shear stiffness of two naturally cemented sands having contrasting geological histories and representing extreme cases in terms of soil constituents.

The calcarenite consisted of a medium carbonate sand of biogenic origin bonded by calcium carbonate whereas the silica sandstone was a medium quartz sand bonded by iron oxide. The calcarenite therefore represented a material with weak grains bonded by a comparatively strong cement, in contrast with the silica sandstone which represented a material with strong grains bonded by a relatively weak cement. The calcarenite is of Pleistocene age and is currently under its maximum overburden pressure whereas the silica sandstone is older, being of Cretaceous age, and has previously been subjected to higher stresses than those acting at present.

In the calcarenite the cement precipitated from the saturated sea water during or soon after deposition (Fookes, 1988), preventing large strains from developing in the soil under the increasing overburden stress. This resulted in the soil having low densities and an open fabric. Apthorpe *et al.* (1988) found that the distribution of the degree of bonding, amount of cement and particle packing at Rankin were controlled by the environmental conditions during deposition rather than by the depth of burial. The soils had been deposited at a variety of depths within a tropical coastal environment, from middle shelf to intertidal, the stronger cementation being associated with the shallower water depths. The thin section of the calcarenite in Fig. 1 shows the open fabric of the material. Despite the fact that the presence of cement has preserved the depositional fabric, there appears to have been little effect of the gravitational field on this fabric, which is fairly random. The cement appears as a thin white fringe of crystals growing outwards from the grain surfaces and partly infilling the voids that under cross-polarized light appear as black areas.

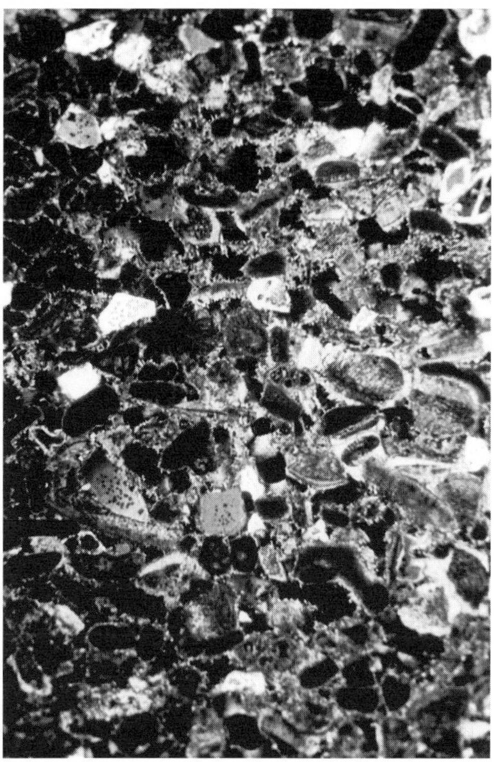

Fig. 1. Thin section of a calcarenite sample under cross-polarized light

The silica sandstone was deposited in a shallow water marine environment (Casey, 1961), the cement in this case being formed in the later stages of the diagenesis when the environmental conditions favoured its precipitation from the groundwater flowing through the sediments (Warren, 1995). Prior to the cement deposition, high overburden pressures had caused a substantial increase in density leading to the development of a well-defined fabric that can be seen in the scanning electron micrographs in Fig. 2. This fabric is characterized by the large area of the intergranular contacts. The iron oxide appears white and, for the grains in contact, the cement coating is likely to weld the grains together. Dapples (1972) observed that iron oxide in silica sands provides only weak bonding because of its poor adherence to the quartz grains. Because of the low cement content and the absence of matrix material, the very low value of its specific volume (about 1·45) is entirely the result of the dense packing of the grains.

The calcarenite samples were obtained from 83 mm cores retrieved at various depths as indicated in Table 1. Triple-tube wireline core barrels were used for sampling with surface-set diamond coring bits (Nyland, 1988). As there are no in situ measurements of the very small strain stiffness for comparison with the laboratory data, the sample disturbance cannot be established precisely. It is likely, however, that the high degree of interparticle bonding would minimize the disturbance. Using similar rotary core samples, the agreement between field and laboratory values of the maximum shear modulus found by Tatsuoka *et al.* (1996) is better for arenaceous weak rocks than argillaceous, and indicates a degree of sampling disturbance that is small in these materials. They also identified that cracks in weak rocks induced by sampling could significantly reduce the initial stiffness at the start of shearing, followed by an increase in stiffness as the cracks close. This feature was not observed in the tests conducted here on the calcarenite core samples. The potential effects of sampling disturbance were also minimized by trimming the cores down to a 50 mm diameter for the laboratory tests.

As Table 1 shows, the samples had a relatively large variation of specific volume, which provided the opportunity of studying the influence of specific volume as well as mean effective stress and bonding on the yielding of this material. The variation of specific volume does not correlate with the depth of burial, confirming the geological interpretation of Apthorpe *et al.* (1988).

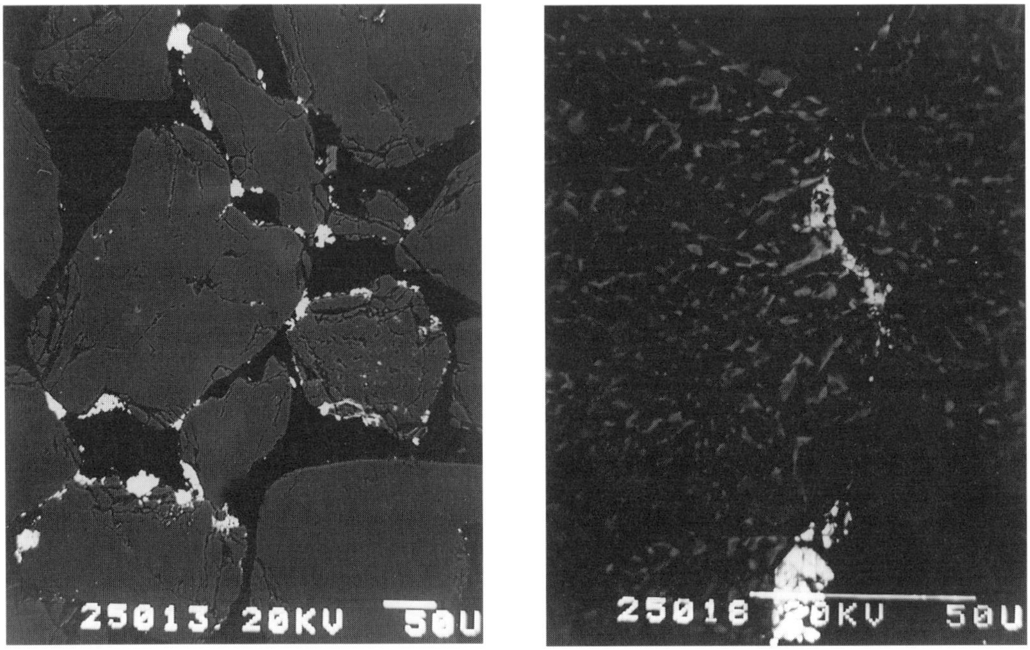

(a) (b)

Fig. 2. Scanning electron micrographs of the silica sandstone with different degrees of magnification: (a) sample with polished surface; (b) sample with unpolished surface

Table 1. Depth of retrieval and initial specific volumes of calcarenite samples

Sample No. denomination	Depth below the sea bed: m	v
N3	96·6	1·779
N2	96·8	1·840
C1	127·4	1·738
N7	128·3	1·770
C3	129·9	1·850
RAN6	133·7	1·854
RAN4	134·5	1·850
RAN3	134·8	1·863
RAN5	135·5	1·871
RAN1	137·1	1·914
NF1	137·6	2·030
N5	141·4	1·675
N4	141·7	1·980

The silica sandstone samples were obtained from a block retrieved at a depth of 5 m below ground surface from a recently excavated face in a quarry located near Maidstone in Kent. It is estimated that the maximum past total overburden stress was about 9 MPa.

Unconfined compression tests were not carried out, but from an extrapolation of the drained triaxial test data at low confining stresses, the unconfined compression strength of the silica sandstone was estimated to be about 600 kPa. For the calcarenite a range of unconfined compressive strengths (500–2000 kPa) was estimated. Just as the specific volume varied greatly between samples, so did the strength.

The reconstituted soils were prepared by breaking the cement bonds while being careful not to damage the sand particles. The intact samples were submerged in water, and the bonds were broken either by rubbing the intact soil between the fingers, in the case of the calcarenite, or using a rubber pestle for the silica sandstone. These methods resulted in a reconstituted soil of single grains rather than cemented clumps. Observation of thin sections of the reconstituted soils did not reveal any significant breakage of the grains by the reconstitution process. The particle size distributions for the reconstituted soils are shown in Fig. 3. These are considered to be representative of those of the intact materials. For the silica sand the small amount of fines confirms that the reconstitution process had left the cement attached to the larger particles rather than adding to the fines fraction. The larger proportion of fines in the reconstituted calcarenite is consistent with the slightly better graded nature of this soil that can be seen in the thin section.

A variety of different methods was tried to estimate the maximum and minimum specific volumes of the reconstituted calcarenite and silica sandstone. The minimum specific volume that

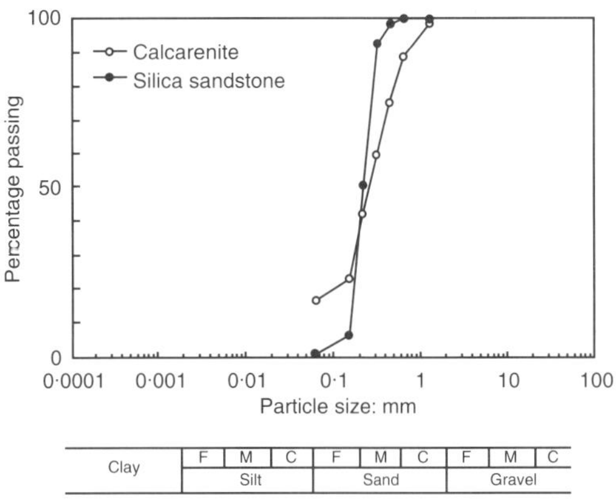

Fig. 3. Particle size distribution

could be achieved for the reconstituted silica sandstone was 1·66. This was obtained using heavy compaction under water. A similar specific volume was also achieved by pouring dry sand through a sieve into a mould, from a height of approximately 1 m. Compaction could not be used for the reconstituted calcarenite, again because of the difficulty of particle breakage. The minimum specific volume of 1·72 for this soil was achieved by vibration. Maximum specific volumes of 1·92 and 1·78 for the reconstituted calcarenite and silica sandstone, respectively, were both produced by preparing frozen samples. Dry sand was pluviated from a height of approximately 5 cm above the water surface in a sample mould. The sample was frozen and then transferred to the pedestal, where it was left to melt, a procedure which avoided any disturbance and consequent change of density during setting up.

LABORATORY EQUIPMENT AND PROCEDURES

The experimental work consisted of triaxial compression tests designed to study the yielding of the two structured sands and to determine the variation of the shear stiffness as the samples approached failure. The tests were performed in two high pressure triaxial cells (Taylor & Coop, 1990), able to reach confining pressures of 10 and 70 MPa, respectively. The two cells were computer controlled, with the high pressures supplied by hydraulic pumps.

The shear stiffness was determined by measuring axial strains internally using a pair of miniature linear variable differential transformers (LVDTs), which measured strains over the central length of the sample (Fig. 4). For these relatively coarse-grained materials characterized by high values of stiffness at small strains, axial strains measured by a transducer mounted externally are highly unreliable as they are affected by bedding errors. Preliminary tests showed that the stiff volumetric response of these materials during isotropic compression resulted in the persistence of bedding errors and non-uniformity of strains even at high values of confining pressure up to 10 MPa. When shearing was performed at lower confining pressures, bedding errors and non-uniformity of strains were observed even as far as the peak state, as the strains were still too small (about 0·1%) to eliminate this effect. External axial strain measurements were also affected by compliance errors, mainly due to the deflection of the internal load cell, which were of the same magnitude as the internally measured strains. Further details of the LVDT system of internal strain measurement and of the data reduction procedures are given by Cuccovillo & Coop (1997).

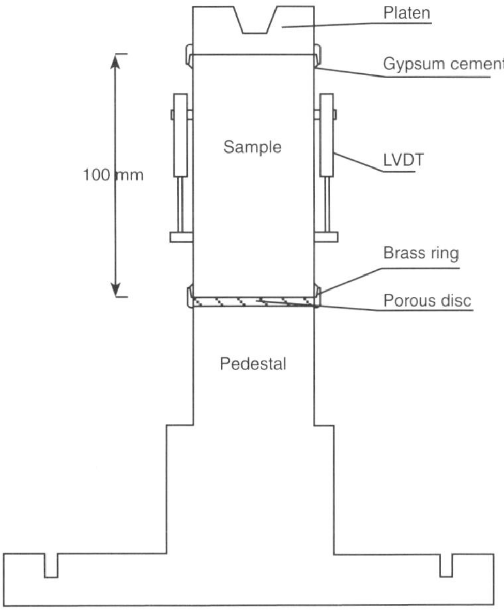

Fig. 4. Set-up of sample in the apparatus with internal LVDTs

Preliminary tests also showed that in order to determine the yield points confidently, it was necessary to resolve axial strains down to 0·0001%. This would guarantee that the stress–strain behaviour could be defined well below the elastic threshold over the whole stress range enclosed by the yield surface. The noise in the output signal of the LVDTs was identified as the major cause of low accuracy at small strains, and this was reduced within the above limit through modifications to both the hardware and the software of the transducer data-logging system. Similar measures were used to reduce the noise of the output signal from the internal load cell to within $\pm 2N$.

Using conventional apparatus and sample preparation procedures, stress–strain data with highly non-uniform strains were obtained as shown in Fig. 5(a). When axial strains in the sample are non-uniform, shear modulus determinations based on the average axial strain measurements from two diametrically opposed internal gauges are correct only as long as the material behaves elastically. When the strains are non-uniform, yielding does not occur simultaneously throughout the sample, so that the yield point and shear stiffness determined on the basis of the average axial strain would be incorrect. It is therefore crucial that uniform strains are guaranteed during the test, and this can be achieved only if the sample is loaded vertically and concentrically.

For the installation of the intact samples in the cell, new devices were designed to ensure the accurate coaxiality of the sample and apparatus and the perpendicularity of the longitudinal axis to the horizontal surfaces (Cuccovillo, 1995). In addition to these checks, it proved necessary to use smaller tolerances in the preparation of the intact samples than those specified by the International Society of Rock Mechanics (ISRM, 1981). These were hand trimmed to a diameter of $50 \pm 0{\cdot}02$ mm and a length of $100 \pm 0{\cdot}01$ mm. After the geometric adjustments, the sample was fixed in its correct position by using gypsum cement around the sample at the top and bottom platens (see Fig. 4).

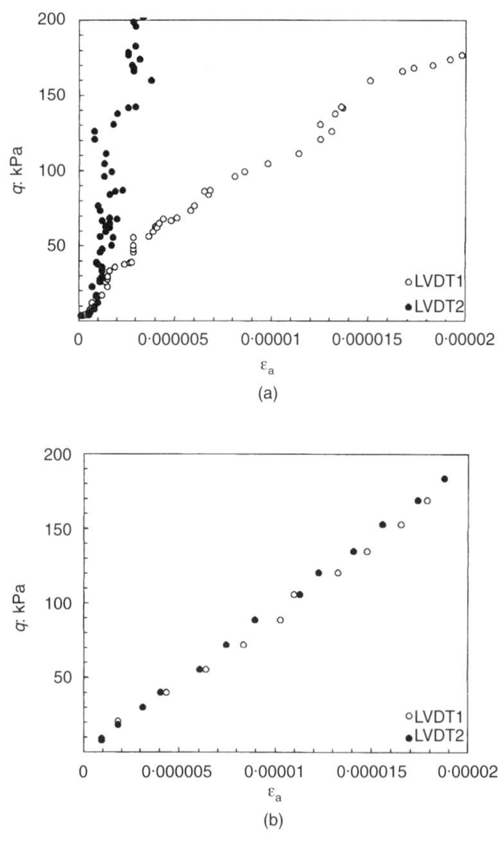

Fig. 5. Typical stress–strain curves from the two local axial gauges: (a) before modifications of testing techniques; (b) after modifications

Despite the temperature-controlled environment, relatively high displacement rates were required to minimize the effects of temperature drift on the measurements, since these effects could be significant during the initial part of shearing because of the very small strains being resolved. When a rigid connection is used between a stiff sample and the axial loading system, these displacement rates can result in an increase in the deviatoric load during shearing that can be so rapid that it becomes difficult to acquire data sufficiently quickly to detect the initial part of the stress–strain curve. A new connection between the sample and the internal load cell was therefore designed. This was based on a modified form of suction cap so that the sample connection to the axial loading system could be made before isotropic loading, avoiding any disturbance at the start of shearing. The new connection had a flexibility which was sufficiently low to prevent sudden unloading of the sample during strain-softening, but high enough to enable the use of both a high displacement rate of the motorized loading frame and a low initial rate of loading on the sample. The flexibility of this connection also attenuated the effects of any minor eccentricity of the axial load that was still present after the sample had been assembled, and Fig. 5(b) shows the quality of the stress–strain data after the testing procedures had been modified. Further details of the new connection are given by Cuccovillo (1995).

STIFFNESS AND YIELDING OF THE SILICA SANDSTONE

The study of the stiffness and yielding of the silica sandstone involved both drained shearing and undrained shearing. In the drained tests the samples were monotonically loaded to failure following constant-p' stress paths. The data from the drained shearing were analysed in terms of a tangent stiffness ($\delta q/3\delta\varepsilon_s$) plotted against the logarithm of either the deviator stress or the shear strain (ε_s), the former being preferred for the identification of the stresses at yield. Since the strains were small, $\delta\varepsilon_s$ was determined as

$$\delta\varepsilon_s = \delta\varepsilon_{al} - \frac{\delta\varepsilon_v}{3} \tag{1}$$

where $\delta\varepsilon_{al}$ was the average local axial strain and $\delta\varepsilon_v$ was the volumetric strain measured by the volume gauge, both being calculated as linear strains. From the pore pressure response during undrained loading, Cuccovillo (1995) concluded that the silica sandstone had isotropic properties so that the stiffness calculated represented the tangent shear modulus G. At the start of shearing, the volumetric strains measured by an Imperial College-type volume gauge were affected by bedding errors so that there would have been no point in trying to improve the lower accuracy of the volumetric strain measurements (about 0·001%). Therefore, given that the material was isotropic and a constant-p' stress path was followed, the readings up to yield were neglected. The tangent stiffness was determined using a linear regression over a set of five data points in the $q-\varepsilon_s$ curve, the whole stress–strain curve being made up of about 1000 points. A smaller scatter of the stiffness data can be achieved by extending the number of data points in the regression, but this would smooth the change of stiffness when yielding occurs, preventing the accurate determination of the yield point.

Figures 6 and 7 show the variation of the shear modulus from two tests performed at values of p' of 1800 and 5600 kPa, respectively. The stiffness data define a plateau followed by a very rapid decrease of stiffness as failure is approached, the stiffness reducing by more than an order of magnitude for an order of magnitude increase in strain. For both tests the constant value of the modulus is equal to about 3050 MPa. At the confining stress of 1800 kPa the strain at yield was about 0·00015 (i.e. 0·015%). This is an order of magnitude greater than the highest threshold yield strain of 0·0015% that Lo Presti (1995) suggested was appropriate for uncemented sands, but, as will be discussed later, this is not a unique value. The increased elastic range is one consequence of the interparticle bonding.

For shearing performed at higher mean effective stresses, the data in Figs 8 and 9 show a smooth variation of the shear modulus with no well-defined yield. In this case the samples were tested in the 70 MPa apparatus, where the stiffness could be determined only for deviator stresses higher than 300 kPa. This was because no type of suction cap could be used with the very high pressures, so that the connection of the axial loading ram to the sample at the start of shearing prevented the definition of the very small strain behaviour. The comparison between Figs 8 and 9 shows that at any value of q the stiffness increases with p'.

To assess the reversibility of the pre-yield stress–strain response and to study the mechanism of bond degradation and its effects on stiffness and yielding, strain-controlled undrained shearing cycles were performed (Table 2). These were conducted at a nominally constant rate of axial strain of about 0·003%/min. The maximum extent of each probe was defined by a limiting deviator stress, as shown in the table, and the strains imposed by each probe were in the approximate range

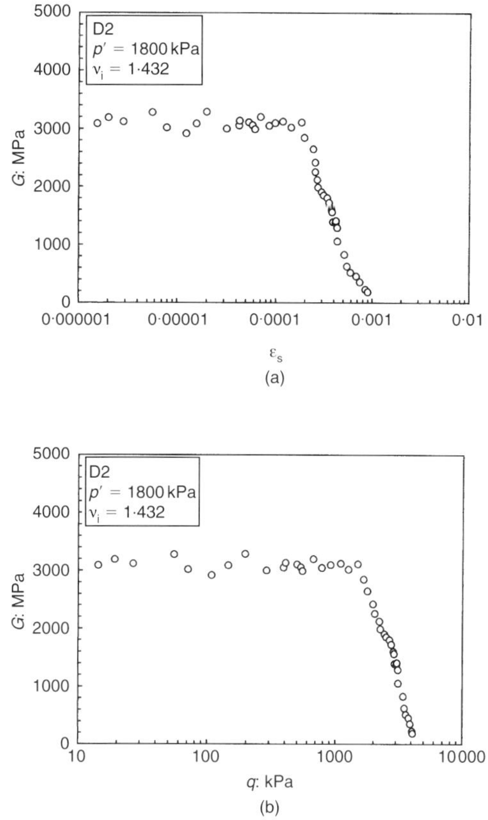

Fig. 6. Variation of G with (a) shear strain and (b) deviator stress for the silica sandstone at relatively low confining pressures

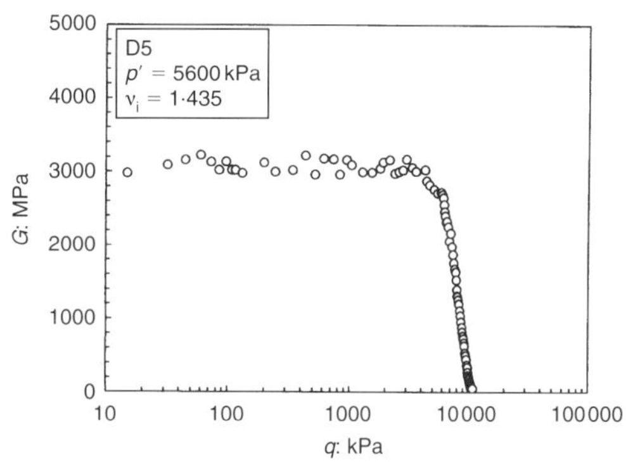

Fig. 7. Variation of G with deviator stress for the silica sandstone at intermediate confining pressures

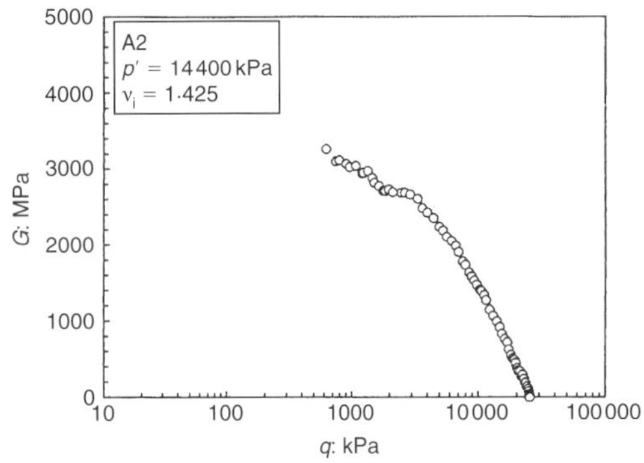

Fig. 8. Variation of *G* with deviator stress for the silica sandstone at high confining pressures

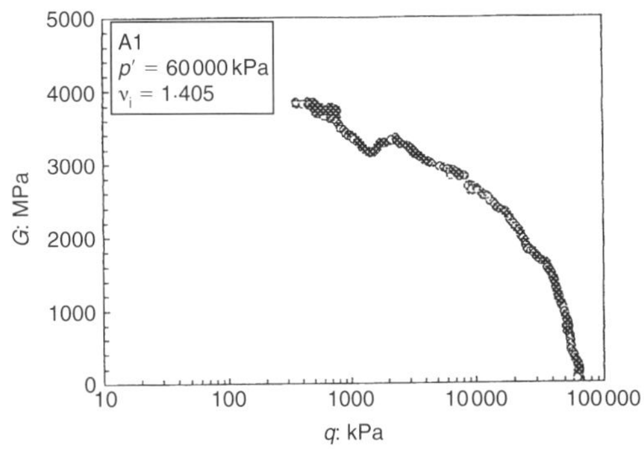

Fig. 9. Variation of *G* with deviator stress for the silica sandstone at very high confining pressures

0·005–0·1%. The test consisted of a series of loading–unloading shearing probes of increasing magnitude from the same isotropic state at an initial value of mean effective stress (p_i') of 1760 kPa. For each value of the maximum deviator stress (q_{max}), the cycles were repeated from three to five times to assess any influence on the values of stiffness and yield point. At the end of each cycle, any excess pore pressure was allowed to dissipate in order to re-establish the same p_i'. The pore pressure dissipation resulted in negligible volumetric strains so that, although all the probes were conducted on the same sample, its specific volume was kept reasonably constant during the test, as can be seen in Table 2. For each loading path the tangent shear modulus G was determined as $\delta q/3\delta\varepsilon_{al}$ using a linear regression.

Figure 10 shows the stress–strain data from three cycles performed up to a deviator stress of 900 kPa (series B probes). For all three cycles the stress–strain relationships are linear and substantially reversible. In Fig. 11(a) it can be seen that the values of the shear modulus lie within a scatter of ±6% of the average value of 2750 MPa. The difference between the maximum shear modulus (G_{max}) determined in the drained and undrained tests was found to arise from having neglected the elastic volumetric strains in the undrained shearing that, since the pore fluid had a compressibility which was similar to that of the soil skeleton, were not negligible. The shear modulus was constant also throughout the subsequent three cycles performed up to a deviator stress of 2100 kPa (series C probes), as shown in Fig. 11(b). However, in the subsequent set of probes (Fig. 12) performed up to a deviator stress of 4200 kPa (series D probes), the stress–strain curves showed hysteresis and irreversible strains that were interpreted as being the result of bond degradation starting to develop in the material after first yielding. Fig. 12(b) shows that during each loading probe the shear modulus

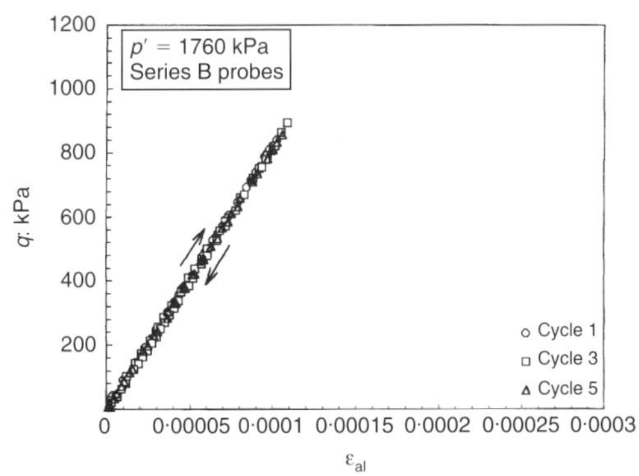

Fig. 10. Pre-yield stress–strain relationships from undrained shearing of the silica sandstone

Table 2. Testing programme for the undrained cycles performed on the silica sandstone

Series	No. of cycles	p_i': kPa	q_{max}: kPa	v
A	5	1760	450	1·445
B	5	1760	900	1·445
C	5	1760	2100	1·445
D	3	1760	4200	1·444
E	3	1760	6000	1·443

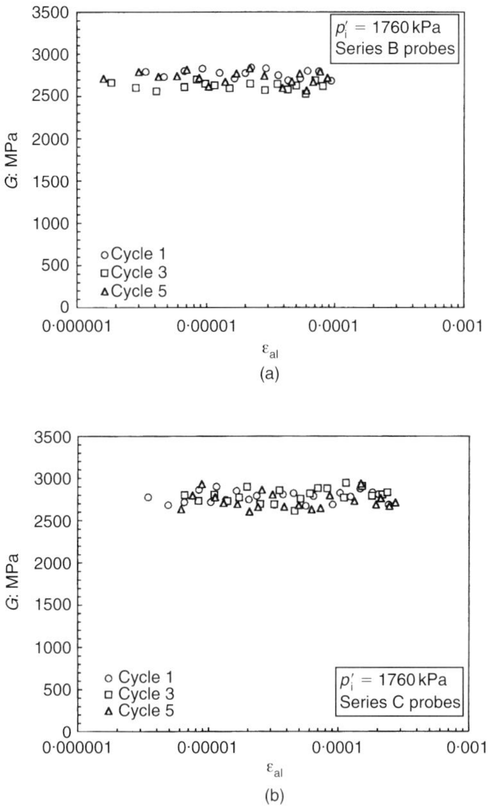

Fig. 11. Values of G_{max} for the silica sandstone from undrained cycles performed prior to yielding

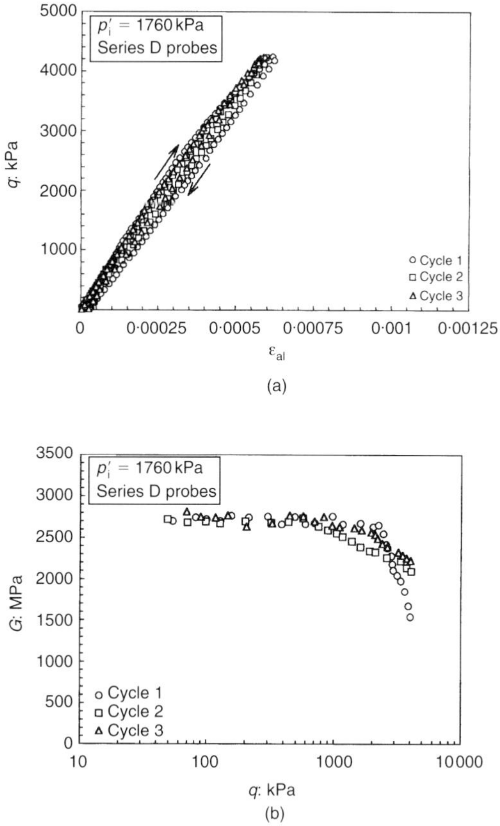

Fig. 12. Undrained shearing of the silica sandstone: (a) postyield stress–strain relationships and (b) variation of *G*

defines initially a plateau, followed by a decrease of stiffness. As the sample is repeatedly loaded, the range of linear behaviour reduces, and the yield point occurs at lower deviator stresses, followed by a smoother decrease of the shear modulus.

The effective stress paths for the series D probes (Fig. 13) are illustrative of the pore water pressure response during undrained cycling. On the first probe, a higher pore pressure of about 300 kPa was generated, which was very much larger than that on subsequent probes and which was also irreversible. Because this was not repeated for subsequent probes, it was assumed to result from the influence of the bedding errors on the pore pressure response and was believed not to be solely the result of the yielding that occurred during this cycle. The direction of the undrained stress path is not vertical even within the elastic range shown by the stiffness data. Cuccovillo (1995) demonstrated this to be the result of the compressibility of the pore water. When this is accounted for, the pore pressure response of the sandstone gave a vertical effective stress path. Since the stress–strain behaviour was

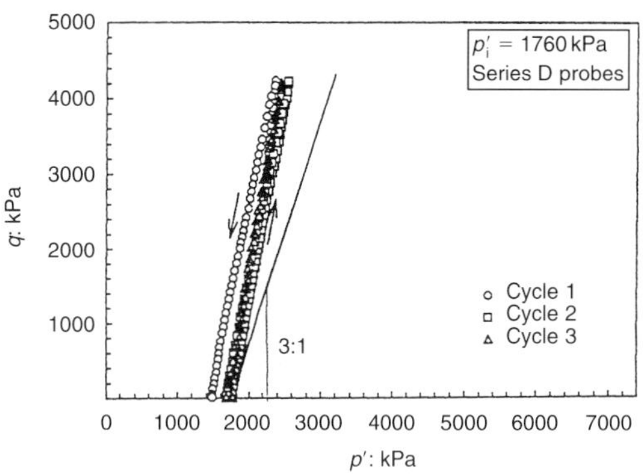

Fig. 13. Effective stress paths and pore water pressure responses from undrained shearing of the silica sandstone

found to be substantially elastic, this is consistent with the behaviour of an isotropic material for which the shear and volumetric responses would be decoupled. A similar type of pore pressure response was also seen for the calcarenite. Yield may be detected from the pore pressure response for the second and following probes as the deviation from an initially straight path but, as illustrated by the data for the silica sandstone, this provides a much less clear means of identification than the stiffness data.

For the final three probes performed up to a q_{max} of 6000 kPa (series E probes), the behaviour during the first probe will be influenced by the plastic straining that the sample had undergone during the previous set of cycles to a q_{max} of 4200 kPa. Nevertheless, the pattern of a continued decrease in the deviator stress at yield is clear, as shown in Fig. 14. The non-linear behaviour of the material was now associated with larger plastic strains than those observed for the smaller cycles. The last reloading of this series of probes caused a reduction of G_{max} of almost 30%, reflecting the extent to which the degradation of bonding had progressed in the material. Therefore the process of bond degradation that began at the first yield developed progressively during the test, causing first a decrease of the yield stress and finally a decrease of the maximum shear modulus. Since the yield stress changes with the number of cycles and the stress level at which they were carried out, it is clear that there is not a unique, or critical, threshold strain corresponding to the onset of yield which occurred in the approximate range of 0·005–0·05% shear strain.

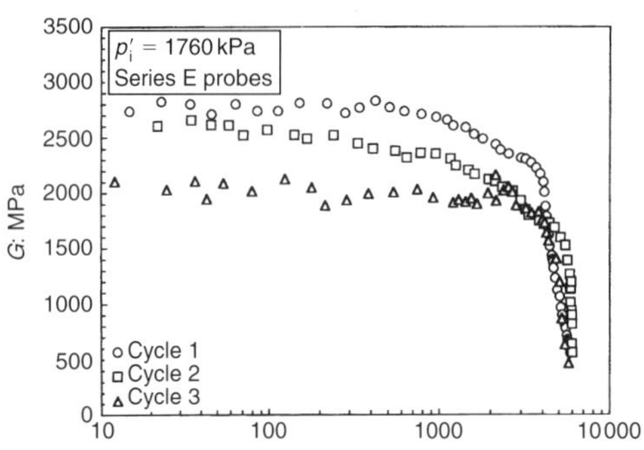

Fig. 14. Variation of *G* with the number of undrained cycles for the silica sandstone

STIFFNESS AND YIELDING OF THE CALCARENITE

For the calcarenite, undrained shearing probes were again performed to investigate the reversibility of the pre-yield stress–strain response and the mechanism of bond degradation. The undrained probes were also used, in this case, to study the variation of the maximum shear modulus with mean effective stress. The sample (C3) was brought to the required value of p_i' and then subjected to the series of cycles shown in Table 3. At any value of p_i' the maximum deviator stress in the probes was chosen to keep the excess pore pressure relatively small (typically below about 20% of p_i'), so that its dissipation at the end of the cycles produced only negligible variations of the specific volume. At the lower values of p_i', the maximum deviator stress was kept well below the peak state to prevent localization of strains in the sample. Conversely, at values of p_i' where the sample strain-hardened, the maximum deviator stress was made high enough to produce yielding and subsequent bond degradation throughout the sample.

At the lowest values of p_i' (series A and B probes), where the probes were performed at states well below the isotropic yield stress, the stress–strain relationship was found to be practically linear, with only a small amount of hysteresis and substantially recoverable strains as shown in Fig. 15. Fig. 16 shows that for a given cycle the shear modulus has a scatter of about 5% around the average over the whole range of strains. For both series A and B probes, there was no significant variation of the average modulus with the number of cycles or with the mean effective stress. As for the silica sandstone, after accounting for the compressibility of the pore fluid, the effective stress paths followed during undrained shearing were consistent with the material having isotropic properties. Larger pore pressures were again seen for the first cycle of each series as a result of bedding errors.

As the initial mean effective stress was increased but still kept below the isotropic yield

Table 3. Testing programme for the undrained cycles performed on the calcarenite (sample C3)

Series	No. of cycles	p_i': kPa	q_{max}: kPa	v
A	3	250	1000	1·850
B	3	800	1000	1·840
C	3	1550	680–920	1·835
D	3	2000	600–780	1·825
E	3	8400	1000	1·540

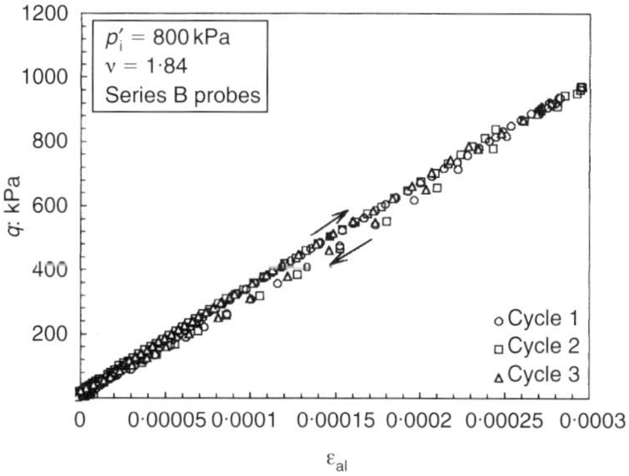

Fig. 15. Pre-yield stress–strain relationships from undrained shearing of the calcarenite

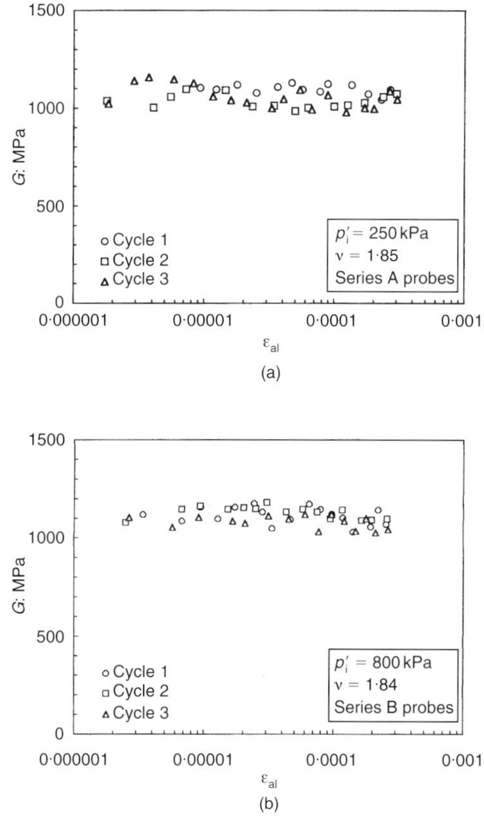

Fig. 16. Values of G_{max} for the calcarenite from undrained cycles performed at low values of p' and prior to yielding

stress, yielding occurred during shearing with the development of large plastic strains, as shown by the stress–strain curves of the series C probes in Fig. 17. First yield was interpreted as the onset of bond degradation. For the first loading of the series

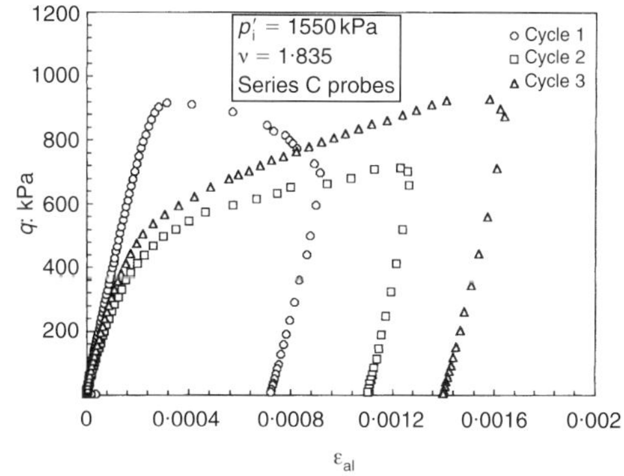

Fig. 17. Post-yield stress–strain relationships from undrained shearing of the calcarenite

C probes, Fig. 18(a) shows that the shear modulus remains initially constant and decreases rapidly after a well-defined yield. This first yield at a deviator stress of about 550 kPa again corresponds to a strain of about 0·00015 (i.e. 0·015%) although, as for the silica sandstone, this was not found to

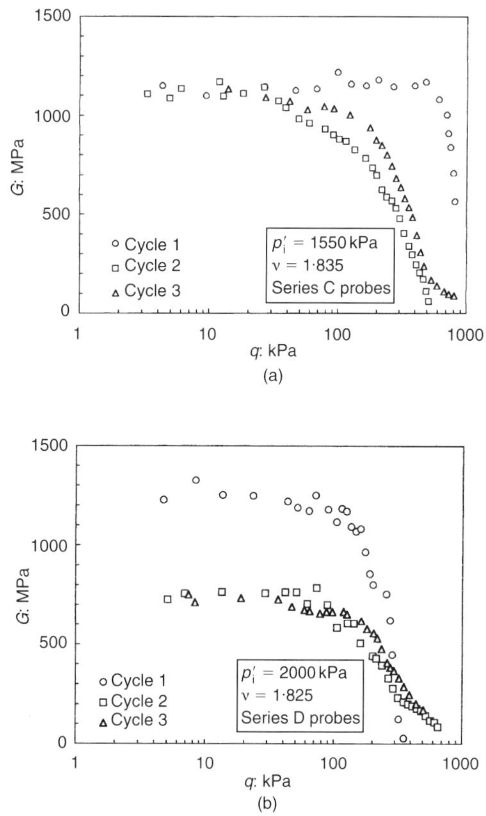

Fig. 18. Variation of G with the number of undrained cycles for the calcarenite at intermediate values of p'

be a constant. As was also seen for the silica sandstone, the value of G_{max} on reloading does not vary but the range of linearity decreases as the yield point moves back to a lower deviator stress and lower strain (0·0005% on cycle 3). The second and third cycles also show a more gentle decrease of stiffness after yielding. For the first probe performed in the following series (series D probes), the value of G_{max} increased slightly in comparison with that measured for the series C probes, as can be seen in Fig. 18(b). This indicated that there was by now some influence of the change in state (p' and v) on the stiffness. At this point, subsequent reloading caused a significant decrease in the value of G_{max}, indicating that a substantial deterioration of bonding took place throughout the sample. The deterioration now affected not only the location of the yield point but also the value of stiffness.

In the probes performed at the highest value of p_i' (series E probes), shown in Fig. 19, the value of G_{max} is higher than those observed in the last two probes performed at a p_i' of 2000 kPa. This reflected the influence of both the increase of the mean effective stress and the decrease of the specific volume that the sample experienced at this stage during the isotropic compression. The influence of state on the values of stiffness indicates that the bonded soil transformed progressively towards a frictional material. In the second and third loading cycles, the value of G_{max} did not change as the initial state of the soil did not vary but the linear range extended to higher deviator stresses, so that the soil now had a strain-hardening mode of behaviour.

COMPARISON BETWEEN INTACT AND RECONSTITUTED SOILS

The shear stiffness of the intact soils was compared to that of the corresponding reconstituted soil to highlight the influence of the two components of structure. For reconstituted clays, the influence of state and overall stress history on stiffness are both accounted for once the values of v and p' are fixed. For sands, on the other hand, a distinction needs to be made between compacted and overconsolidated samples, since these may exhibit different stiffnesses even when v and p' have the same values (Jovičić & Coop, 1997).

For the reconstituted silica sandstone, the lowest specific volume that could be obtained by compaction was 1·66. A specific volume comparable to that of the intact soil could only be achieved by overconsolidation as in the case of sample R9 that was isotropically compressed to a maximum p' of 70 MPa on the normal compression line (NCL) and then unloaded (Fig. 20). The low specific volume of the intact soil was unlikely to be the result of overconsolidation since the estimated value of the maximum past overburden pressure was an order of magnitude less than the maximum pressure required to bring the reconstituted soil to a similar state. The large reduction in density experienced by the intact soils under the overburden pressure was more likely to have been due to pressure solution of quartz occurring at the grain surfaces where the contact stresses were significantly higher than the stresses due to the overburden. This interpretation also explains the well-defined fabric observed in the intact soil characterized by a large grain contact area, and is supported by the studies performed by Maxwell (1964) and Dusseault & Morgenstern (1979) on geologically old sands similar to this silica sandstone. The particle arrangement and contacts once disrupted as in the reconstituted soil could not be reproduced through simple compaction, resulting in much lower densities.

A qualitative comparison can, however, still be made between the stiffness of intact and reconsti-

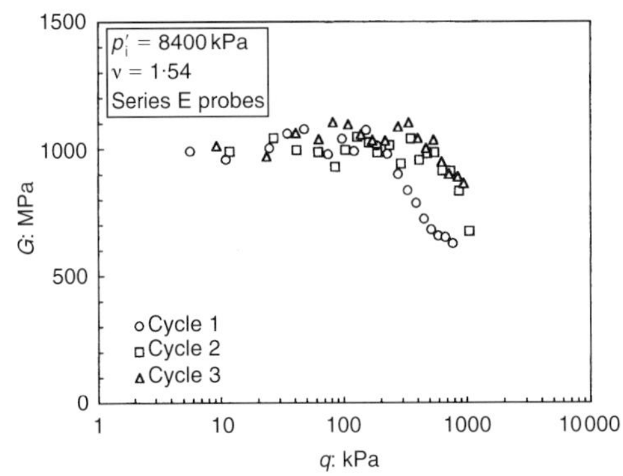

Fig. 19. Variation of G with the number of undrained cycles for the calcarenite at high values of p'

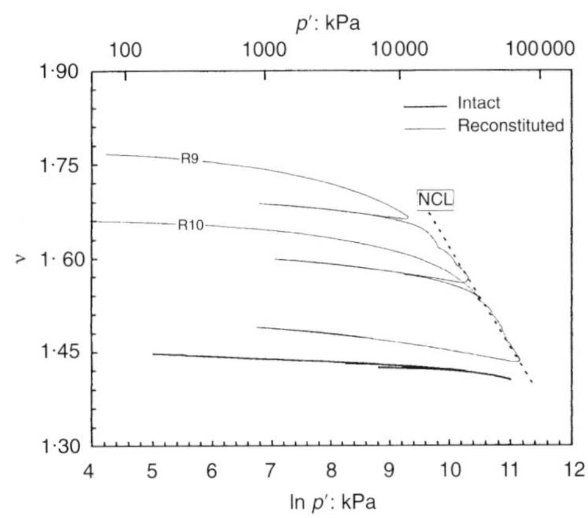

Fig. 20. Isotropic compression of intact and reconstituted silica sandstone

tuted soils for the silica sandstone, as in Fig. 21, which shows the results obtained for samples R9 and A2, both sheared from similar states following constant-p' stress paths. It should be noted that the state of the intact sample was beyond the isotropic yield point, which occurred at a p' of 11 MPa, as clearly shown in Fig. 22 by the variation of the bulk modulus (K') with mean effective stress. The data shown in the figure have been computed considering the volumetric strain equal to three times the average local axial strain. Good agreement was obtained with the volumetric strain data from the volume gauge at confining stresses over 5 MPa, once the bedding errors had been eradicated. This confirmed the validity of the assumption of isotropy. After isotropic yielding, the cementing of the initially intact sample was likely to have been disrupted, whereas much higher stresses would be required to strain this sample sufficiently to disrupt its fabric. Therefore, the influence of fabric appears to be the main factor explaining the large difference of stiffness between the initially intact and reconstituted soils.

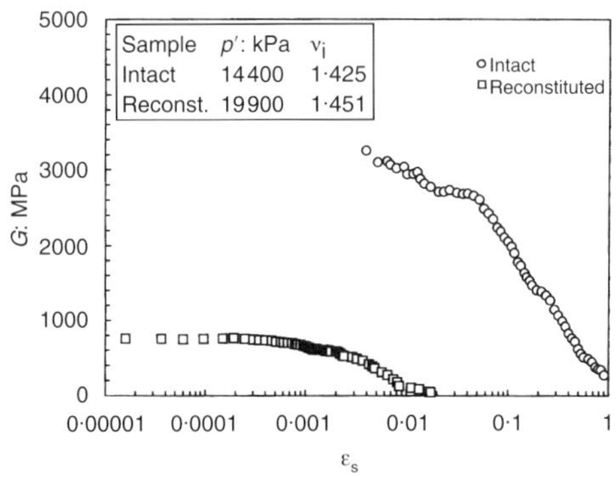

Fig. 21. Comparison between the shear moduli of intact and reconstituted silica sandstone

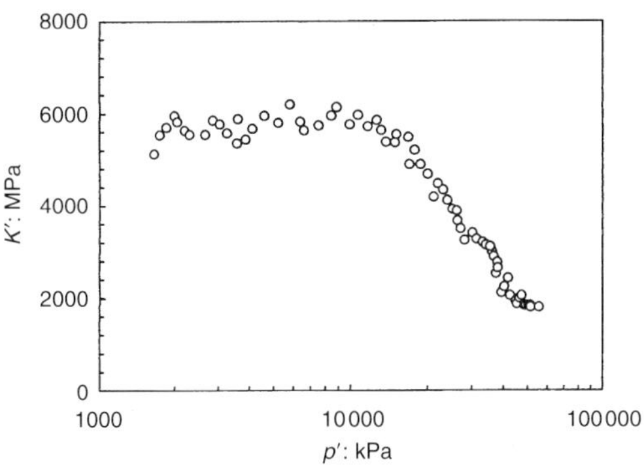

Fig. 22. Variation of the bulk modulus with mean effective stress for the silica sandstone

Unlike the silica sandstone, the intact calcarenite had densities which could be achieved by compacting the reconstituted soil. In this case the early precipitation of the cement during deposition prevented large strains developing in the soil during compression under the overburden stress. This, together with the fact that no pressure solution process had occurred of the type seen in the silica sandstone, meant that the soil had not developed a pronounced fabric either in terms of evident particle orientation or interlocking. At a given state, any difference in response between the intact and reconstituted soils of the calcarenite can therefore be thought of as being the result of bonding alone. To highlight the influence of bonding on the stiffness and yielding of the intact soil, undrained loading–unloading shearing probes were performed on a reconstituted sample at similar initial mean effective stresses as for the intact soil (Table 4). For a given value of p_i' the maximum deviator stress reached in each probe was chosen to keep the specific volume substantially unchanged.

Figures 23(a)–23(d) show the variation of the shear modulus during the probes performed at each value of the initial mean effective stress. At any value of p_i', the stiffness plateau defined during the first loading path extended only over a small range of deviator stresses with shear strains at yield in the range 0·0005–0·005%, and the response was largely non-linear. Successive probes had the effect of reducing the non-linearity of the soil, increasing the values of the deviator stress and strain at yielding. This typical strain-hardening response contrasts with the decrease of deviator stress at yielding seen for the intact soil in Fig. 18 that was produced by the degradation of bonding, initiated after first yielding. A strain-hardening response was only observed for the intact soil at the highest confining stress when the bonding had been largely degraded so that the behaviour of the soil tended to be frictional.

As can be seen from Figs 16, 18, 19 and 23 and the summary plots of Fig. 24, comparisons between the stiffness graphs of the intact and reconstituted soils at similar values of p_i' show that the presence of bonding in the intact material resulted in higher values of G_{max} and a larger elastic range than the reconstituted soil, despite the slightly higher specific volume of the intact sample.

INFLUENCE OF STATE AND BONDING ON G_{max}

For the silica sandstone the influence of the volume:stress state and bonding on G_{max} can be assessed by comparing the variation of G_{max} in the drained tests. The samples sheared before isotropic

Table 4. Testing programme for the undrained cycles performed on the reconstituted calcarenite

Series	No. of cycles	p_i': kPa	q_{max}: kPa	v
A	3	300	180	1·805
B	2	800	300–750	1·715
C	3	2000	1500	1·635
D	3	8400	2000	1·505

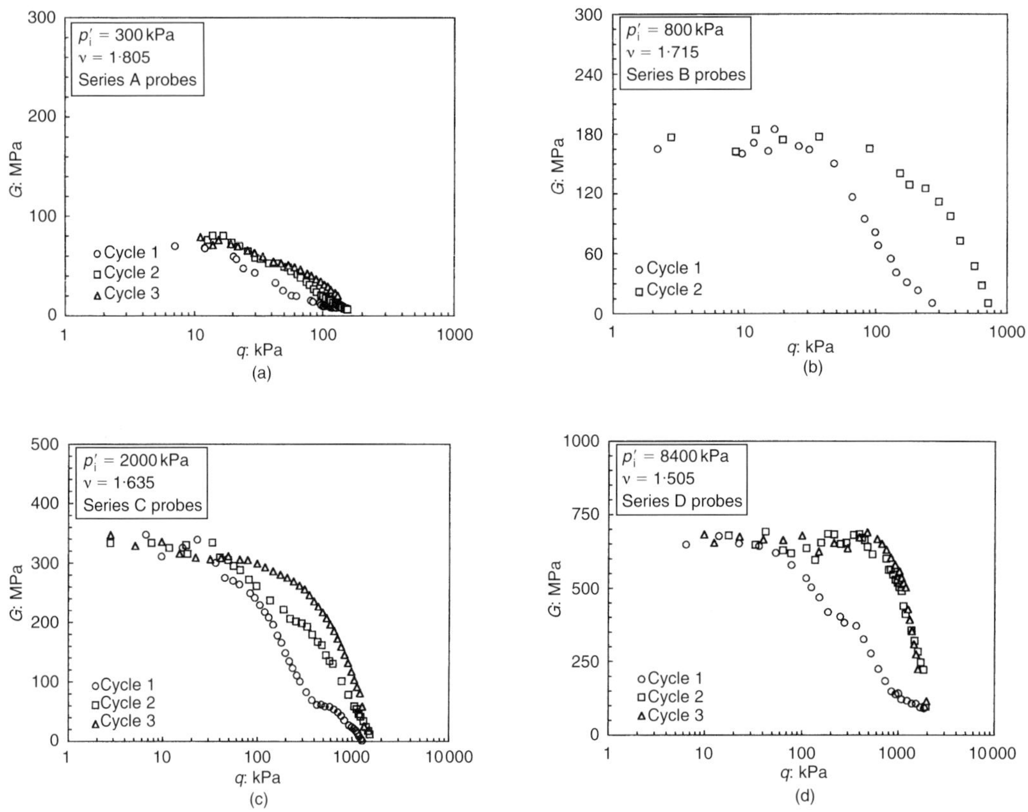

Fig. 23. Variation of G with the number of undrained cycles for the reconstituted calcarenite at increasing values of p'

yielding all had similar specific volumes at the end of the isotropic compression due to the high value of the bulk modulus. Differences in state were therefore only caused by the different mean effective stresses during shearing, which were found, however, not to influence the values of G_{max} (see Figs 6 and 7). For the samples sheared after isotropic yielding, as in Figs 8 and 9, the extrapolation of the stiffness graphs to the initial stages of shearing indicates an increase of the values of G_{max} with mean effective stress. This suggests that once the bonding had degraded the shear modulus of the soil, which was likely still to be influenced by the fabric of the soil, was also influenced by state.

For the calcarenite, the influence of state and bonding on G_{max} can be analysed by comparing the maximum shear moduli of the intact and reconstituted soils from the various undrained probes (Fig. 24). For the reconstituted soil, the value of G_{max} for each series of probes was constant, so that a single data point is shown at any particular value of mean effective stress or specific volume. For the intact soil, the data points labelled as A1, B1, C1, D1 and E1 refer to the first loading probes of the A, B, C, D and E series, whereas point D3

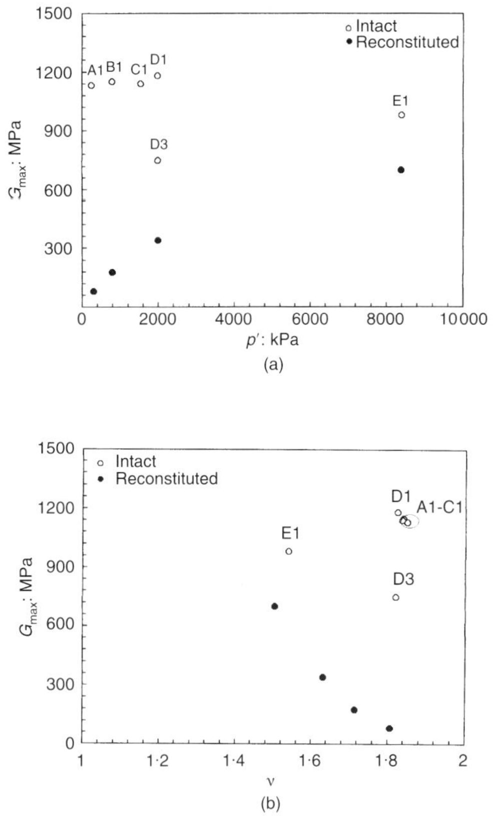

Fig. 24. Values of G_{max} for intact and reconstituted calcarenite as a function of (a) mean effective stress and (b) specific volume

shows the reduced value of G_{max} observed in the third loading cycle of the series D probes.

It can be seen that for the reconstituted soil a variation in state caused a significant variation of G_{max}. Because of this dependency of stiffness on the state (i.e. the combination of both specific volume and mean effective stress), comparison cannot be easily made with G_{max} data for other carbonate sands. Nevertheless, the values given here for the reconstituted soil are broadly consistent with those measured for other carbonate sands by Fioravante *et al.* (1995) and Jovičić & Coop (1997) using dynamic methods.

For the intact soil, the variation in state did not influence the values of G_{max} (points A1 to D1), and the influence of bonding resulted in values of stiffness that were higher than those of the reconstituted soil at similar states. From point D1 onwards, sufficient deterioration of bonding occurred in the sample to result in a significant reduction of stiffness during repeated cycling (points D1 and D3). A further increase in p', with a consequent reduction in v, resulted then in an increase of G_{max} (points D3 and E1).

The variation of G_{max} with state was further analysed by referring the state of both intact and reconstituted soils to their respective isotropic boundaries. Fig. 25 shows results from the isotropic compression of intact and reconstituted soils and the isotropic boundaries defined by both soils. For high values of v, the isotropic boundary of the intact soil (intact IB) lay to the right of the boundary of the reconstituted soil identified by the NCL. In Fig. 26(a) the values of G_{max} for both soils are plotted against the normalized variable p'/p'_{IB}, where p'_{IB} is an equivalent pressure taken on the NCL for the reconstituted soil and on the intact IB for the intact soil, respectively determined as

$$p'_{IB} = \exp\frac{2{\cdot}695 - v}{0{\cdot}132} \tag{2}$$

for the reconstituted soil and

$$p'_{IB} = \exp\frac{3{\cdot}658 - v}{0{\cdot}235} \tag{3}$$

for the intact soil. The data from the reconstituted soil show that for states during first loading up to the NCL, G_{max} increases approximately linearly, so that

$$\frac{G_{max}}{p_r} = 990\,000\frac{p'}{p'_{IB}} - 282\,000 \tag{4}$$

where p_r is a reference pressure equal to 1 kPa, used to make equation (4) non-dimensional. This equation is only valid for the shear moduli measured during first compression of a sample of one particular density. It is not valid for states located on the normal compression line, for which G_{max} will continue to increase while p'/p'_{IB} remains

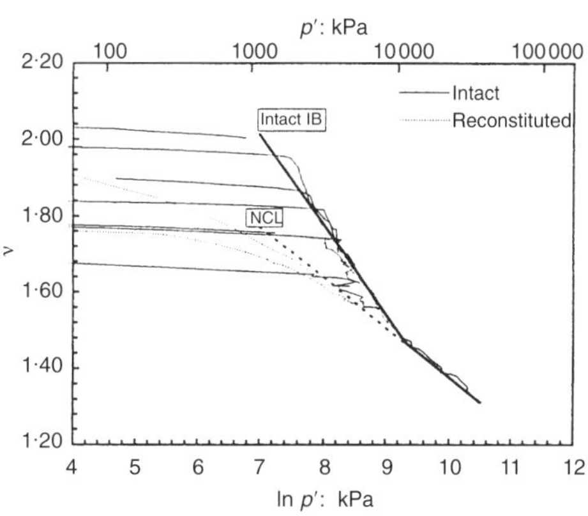

Fig. 25. Isotropic compression data and state boundaries for intact and reconstituted calcarenite

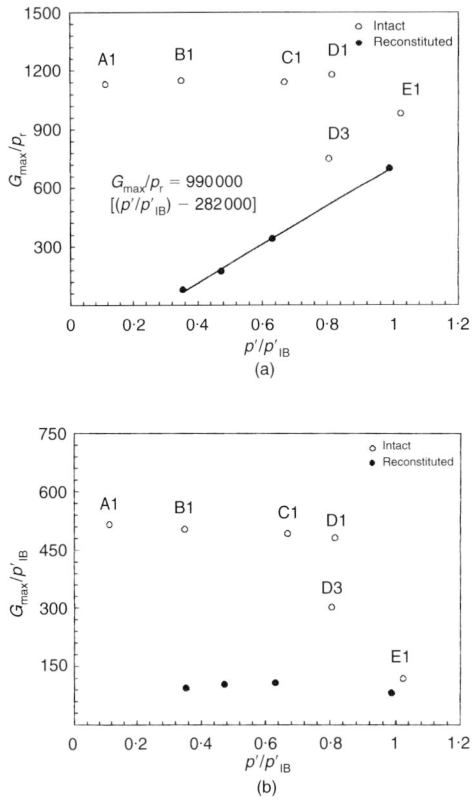

Fig. 26. Comparison between values of (a) G_{max} and (b) normalized G_{max} for intact and reconstituted calcarenite at equivalent states

constant at 1. Fig. 26(a) allows comparison between the variation of G_{max} with state for the intact soil after yield (points D3 and E1) and that of the reconstituted soil. For a given change of the normalized variable, both soils experienced a similar variation of stiffness.

Because of the influence that state has on the stiffness, a more correct comparison of the stiffnesses of the intact and reconstituted soils was made by normalizing the values of G_{max} of the intact and reconstituted soils, again using their respective values of p'_{IB}. This normalization is consistent with the work of Hardin (1978), who recognized that the stiffness of uncemented soils would be dependent on the mean effective stress, void ratio and overconsolidation ratio. The normalizing parameter p'_{IB} is solely a function of the specific volume for each soil, as shown by equations (2) and (3), and the principal departure from previous work is the necessity of having two different functions, one for the natural and one for the reconstituted soil. The normalization also accounts for the fact that for coarse-grained soils, Jovičić & Coop (1997) found that it is the distance of the current state from the boundary surface which is of key importance, rather than the overconsolidation ratio as had been used by Hardin. Values of G_{max}/p'_{IB} were then plotted against p'/p'_{IB} in Fig. 26(b). The graph shows that as the states of the intact and reconstituted soils move towards their corresponding isotropic boundaries, their normalized stiffnesses tend to converge. This suggests that for a value of p'/p'_{IB} equal to 1, the different absolute values of stiffness of the intact and reconstituted soils resulted from differences in state, and that the contribution of bonding to the stiffness must have been substantially removed after cycle D1. Higher stresses could not be reached because of the limitation of the confining stress achievable in the apparatus. It is possible, however, that even if the state of the intact soil had been brought on the NCL of the reconstituted soil, some differences would have continued to exist between the values of G_{max} of the two soils, indicating a small but persistent influence of bonding in the initial stages of shearing. This structural influence was, however, much less than that seen for the silica sandstone after isotropic yield.

CONCLUSIONS

Structured sand samples have been found to have higher shear stiffness than equivalent reconstituted samples at comparable states. The contribution of the component of structure arising from bonding to the shear stiffness can be relied upon only as long as yielding is prevented. In this case the shear stiffness was found to be independent of state and could be described within an elastic theory.

Yielding in structured sands is marked by a rapid decrease of stiffness and by a progressive deterioration of bonding with the development of plastic strains. Cycles of reloading after first yielding and prior to failure were found first to reduce the threshold of the elastic stress–strain behaviour and eventually to decrease the maximum shear modulus. This mode of behaviour was interpreted as being the result of the progressive transformation of the bonded soil into a frictional material, and contrasted with the typical strain-hardening response of the unbonded reconstituted soils.

As bonding deteriorated after first yielding, the shear stiffness was shown to become dependent on state. For the calcarenite it was found that due to bond degradation the contribution of structure to stiffness reduced. As a consequence, as the state of the intact soil approached the normal compression line, the value of G_{max} tended to decrease and converge towards the value of G_{max} of the reconstituted soil. Conversely for the silica sandstone, despite bond degradation, the influence of structure on the shear stiffness still persisted so that the values of the stiffness were very much higher than those of the reconstituted soil at comparable states. As the state of the intact soil moved towards the normal compression line, the shear stiffness was found not to reduce after isotropic yielding but to increase. A decrease of the shear stiffness was only seen for the silica sandstone under cycles of shear stress at constant state. This could be interpreted as the effect of the loss of bonding which must precede any disruption to the particle arrangement.

The different ways in which state influenced the shear stiffness after the initiation of bond degradation reflected the different structural features of the two materials. In sands where the influence of structure arises predominantly from bonding, as in the calcarenite, the values of the shear stiffness after first yielding decrease as the bonding degrades, even if the mean effective stress and density increase. Conversely, in sands where the influence of structure predominantly arises from an interlocking fabric as in the silica sandstone, the values of the shear stiffness remain high despite bond degradation, and even increase when the mean effective stress and density increase. In the formulation of constitutive models there may be a need to account for the effects of bond degradation on the maximum shear modulus as well as on the yield stress of cemented sands and to differentiate the way in which state influences the shear stiffness, depending on the predominant structural feature present in the material.

ACKNOWLEDGEMENTS

The research was entirely funded by the EPSRC, to whom the authors are grateful. The calcarenite samples were kindly provided by BP International. The authors are also grateful to Dr A. Mochizuki for his advice on apparatus development.

REFERENCES

Apthorpe, M., Garstone, J. & Turner, G. J. (1988). Depositional setting and regional geology of North Rankin A foundation sediments. *Proc. Int. Conf. Calcareous Sediments, Perth* **2**, 357–366.

Burland, J. B. (1990). On the compressibility and shear strength of natural clays. *Géotechnique* **40**, No. 3, 329–378.

Casey, R. (1961). The stratigraphical paleontology of the Lower Greensand. *Paleontology* **3**, Part 4, 487–621.

Clough, G. W., Sitar, N. & Bachus, R. C. (1981). Cemented sands under static loading. *J. Geotech. Engng Div., Am. Soc. Civ. Engng* **107**, GT6, 799–818.

Coop, M. R. & Atkinson, J. H. (1993). The mechanics of cemented carbonate sands. *Géotechnique* **43**, No. 1, 53–67.

Cuccovillo, T. (1995). *Shear behaviour and stiffness of naturally cemented sands*. City University, London, PhD thesis.

Cuccovillo, T. & Coop, M. R. (1997). The measurement of local axial strains in triaxial tests using LVDTs. *Géotechnique* **47**, No. 1, 167–172.

Dapples, E. C. (1972). Some concepts of cementation and lithification of sandstones. *Bull. Am. Ass. Petroleum Geol.* **56**, No. 1, 3–25.

Dusseault, M. B. & Morgenstern, N. R. (1979). Locked sands. *J. Engng Geol.* **12**, 117–131.

Fioravante, V., Capoferri, R., Hameury, O. & Jamiolkowski, M. (1995). Deformation characteristics of uncemented Quiou sand. *Proc. Int. Symp. Pre-failure Deform. Character. Geomater., IS-Hokkaido* **1**, 55–61.

Fookes, P. G. (1988). The geology of carbonate soils and rocks and their engineering characterisation and description. *Proc. Int. Conf. Calcareous Sediments, Perth* **2**, 787–806.

Gens, A. & Nova, R. (1993). Conceptual bases for a constitutive model for bonded soils and weak rocks. *Proc. Int. Symp. Geotech. Engng Hard Soils – Soft Rocks, Athens* **1**, 485–494.

Hardin, B. O. (1978). The nature of the stress-strain behaviour of soils. *Proc. ASCE Geotech. Engng Div., Specialty Conf.* **1**, 3–90.

Huang, J. T. & Airey, D. W. (1993). The effects of cement and density on an artificially cemented sand. *Proc. Int. Symp. Geotech. Engng Hard Soils – Soft Rocks, Athens* **1**, 553–560.

International Society of Rock Mechanics (1981). Rock characterisation testing and monitoring. In *ISRM suggested methods* (ed. E. T. Brown). Oxford: Pergamon Press.

Jardine, R. J., Potts, D. M., St. John, H. D. & Hight, D. W. (1991). Some practical applications of a non-linear ground model. *Proc. 10th Eur. Conf. Soil Mech., Florence* **1**, 223–228.

Jovičić, V. & Coop, M. R. (1997). Stiffness of coarse-grained soils at small strains. *Géotech. Symp.*, In Print, **47**, No. 3, 545–567.

Lagioia, R. & Nova, R. (1995). An experimental and theoretical study of the behaviour of a calcarenite in triaxial compression. *Géotechnique* **45**, No. 4, 633–648.

Leroueil, S. & Vaughan, P. R. (1990). The general and congruent effects of structure in natural soils and weak rocks. *Géotechnique* **40**, No. 3, 467–488.

Lo Presti, D. C. F. (1995). General Report: Measurement of shear deformation of geomaterials in the laboratory. *Proc. Int. Symp. Pre-failure Deform. Character. Geomater., IS-Hokkaido* **1**, 1067–1088.

Maxwell, J. C. (1964). Influence of depth, temperature and geologic age on porosity of quartzose sandstone. *Bull. Am. Ass. Petroleum Geol.* **48**, No. 5, 697–709.

Mitchell, J. K. (1976). *Fundamentals of soil behaviour*. New York: Wiley.

Nyland, G. (1988). Detailed engineering geological investigation of North Rankin “A” Platform. *Proc. Int. Conf. Calcareous Sediments, Perth* **2**, 503–512.

Roscoe, K. H., Schofield, A. N. & Wroth, C. P. (1958). On the yielding of soils. *Géotechnique* **8**, No. 1, 22–53.

Tatsuoka, F. & Shibuya, S. (1992). Deformation characteristics of soils and rocks from field and laboratory tests. *Proc. 9th Asian Regional Conf. SMFE, Bangkok* **2**, 101–170.

Tatsuoka, F., Kohata, Y., Tsubouchi, T., Murata, K., Ochi,

K. & Wang, L. (1996). Sample disturbance in rotary core tube sampling of soft rock. *Proc. ICE Conf. on Advances in Site Investigation Practice*, London, 281–292.

Taylor, R. N. & Coop, M. R. (1990). Stress path testing of Boom Clay from Mol, Belgium. *Procs. Conf. on the Engineering Geology of Weak Rock*, Leeds, 89–98.

Vaughan, P. R. (1988). Characterising the mechanical properties of in-situ residual soils. *Procs. 2nd Int. Conf. on Geomechanics in Tropical Soils*, Singapore, 469–487.

Viggiani, G. & Atkinson, J. H. (1995). Stiffness of fine-grained soil at very small strains. *Géotechnique* **45**, No. 2, 249–265.

Warren, C. (1995). Personal communication. Sir William Halcrow and Partners.

Jiang, G.-L., Tatsuoka, F., Flora, A. and Koseki, J. (1997). *Géotechnique* **47**, No. 3, 509–521

Inherent and stress-state-induced anisotropy in very small strain stiffness of a sandy gravel

GUAN-LU JIANG,* FUMIO TATSUOKA,† ALESSANDRO FLORA‡ and JUNICHI KOSEKI§

Deformation characteristics at strains less than about 0·001% in compacted well-graded sandy gravel were evaluated by measuring vertical and horizontal strains locally in very small amplitude cyclic tests on square prismatic specimens. The very small strain Young's modulus E^0 exhibits strong inherent anisotropy and stress-state-induced anisotropy. The very small strain Poisson's ratio ν^0_{vh} is a function of stress ratio. Effects of ageing on the values of E^0_v and ν^0_{vh} are discussed.

KEYWORDS: anisotropy; deformation; elasticity; gravels; laboratory equipment; laboratory tests.

On a évalué les caractéristiques de déformation d'un gravier sableux compacté et à granulométrie continue à des contraintes inférieures à 0,001 p. 100 environ, en mesurant les contraintes verticales et horizontales locales dans des essais cycliques à très faible amplitude réalisés sur des éprouvettes prismatiques carrées. Le module de Young pour très faibles contraintes E^0 met en évidence une forte anisotropie inhérente et une anisotropie provoquée par l'état de contrainte. Le coefficient de Poisson pour très faibles contraintes $\nu^0{}_{vh}$ est une fonction du rapport de contrainte. L'article examine les effets du vieillissement sur les valeurs de $E^0{}_v$ et $\nu^0{}_{vh}$.

INTRODUCTION

To evaluate movements of ground and structure and soil–structure interaction at working loads, the deformation characteristics of soil at very small strains and at various stress states are required. Recent studies have revealed that deformations of most granular materials at strains less than about 0·001% are essentially recoverable and strain-rate insensitive (i.e. elastic) (e.g. Tatsuoka, Sato, Park, Kim, Mukabi & Kohata, 1994a; Tatsuoka, Teachavorasinskun, Dong, Kohata & Sato, 1994b). The Young's modulus and Poisson's ratio at strains less than 0·001% will herein be denoted as E^0 and ν^0. The very small strain stiffness is usually anisotropic (Stokoe, Lee & Lee, 1991; Bellotti, Jamiolkowski, Lo Presti & O'Neill, 1996). Despite the importance of inherent and stress-state-induced anisotropy in the very small strain stiffness, most conventional laboratory testing methods are not capable of accurately and sufficiently detecting such behaviour. In the present investigation, very small amplitudes of cyclic axial stress or horizontal stress were applied to square prismatic triaxial specimens of a sandy gravel to evaluate its very small strain deformation parameters.

Manuscript received 9 October 1996; revised manuscript accepted 17 April 1997.
Discussion on this paper closes 1 November 1997.
* Japan Railway Technical Research Institute.
† University of Tokyo.
‡ University of Napoli.
§ Institute of Industrial Science, University of Tokyo.

TESTING PROGRAMME

Figure 1 shows a large square prismatic specimen of a well-graded crushed sandstone (called Chiba gravel), which was prepared in 14 layers. Each layer was compacted by vertical vibration for 15 min using a 19·5 kgf weight vibrator, followed by scraping a thin top layer before placing the next layer. To simulate the field conditions to which the results of the present study were to be applied, the specimens were kept slightly wet when prepared (see Table 1) and throughout each test. The initial dry densities of the specimens are listed in Table 1. It seems that some scatter in the densities was due to the fixed vibration time with slight variations in the material properties and water contents among the specimens. Figure 2 shows the grain size distribution curves before and after Test TACA, which will be described later. The two curves are practically the same, showing negligible grain breakage. Axial deformations of each specimen were measured (1) outside the triaxial cell with an LVDT (linear variable displacement transformer), (2) on the top cap of the specimen with a pair of gap sensors (proximeters), and (3) on its

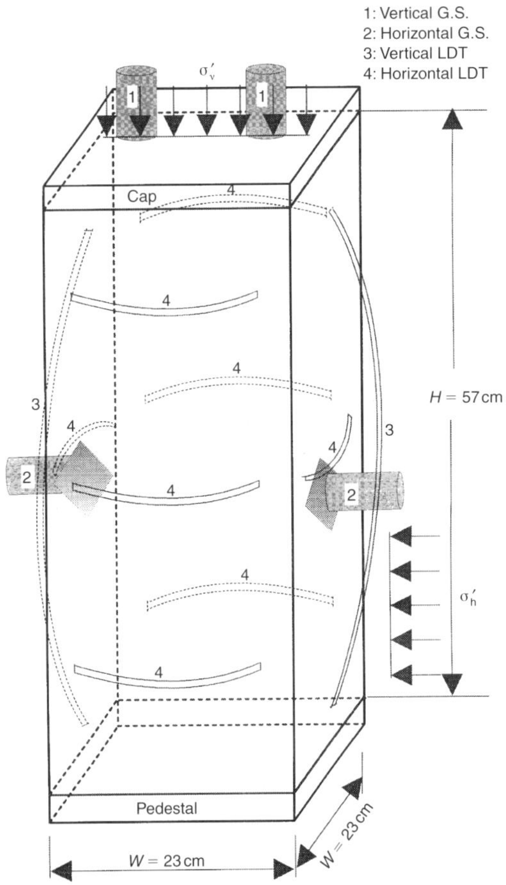

Fig. 1. Square prismatic specimen

lateral surface with a pair of LDTs (local deformation transducers) (Goto, Tatsuoka, Shibuya & Sato, 1991). Lateral deformations of the specimen were measured locally by using four pairs of laterally-placed LDTs at three levels of the specimen, and a pair of gap sensors at the mid-height of the specimen. More details of the apparatus are described in Tatsuoka *et al.* (1994a, b) and Hoque, Sato & Tatsuoka, (1996).

The testing programme is described in Table 1. In Test TACA, the specimen was isotropically compressed to $\sigma'_v = \sigma'_h = 490$ kPa and unloaded to 49 kPa, followed by a series of anisotropic compression stages at different stress ratios $R = \sigma'_v/\sigma'_h$, following the sequence indicated in Fig. 3. At each small circular point shown in Fig. 3, the specimen was aged for about 30 min or more, followed by a series of very small amplitude cyclic tests to measure the very small strain deformation properties of the specimen. The cyclic tests consisted of five very small unload–reload cycles of axial stress σ'_v at a frequency f of 0·05 Hz while keeping the lateral stress σ'_h constant (vertical cyclic tests) and then five very small unload–reload cycles of σ'_h at $f = 0{\cdot}005$ Hz while keeping σ'_v constant (lateral cyclic tests). This small f (= 0·005 Hz) was necessary to perform accurate and stable lateral cyclic tests (Hoque *et al.*, 1996). The effects of using these different f values were evaluated in some special tests, and very small strain Young's moduli at $f = 0{\cdot}005$ Hz and 0·05 Hz were found to be nearly the same (see Fig. 13a of Di Benedetto & Tatsuoka, 1997). Note that the two principal stresses in the lateral directions were always the same as each other. Lastly, the specimen was brought to failure in triaxial compression (TC) at $\sigma'_h = 49$ kPa while a single very small unload–reload cycle of σ'_v was applied at a number of intermediate loading stages. Four cyclic triaxial tests with increasing amplitudes of σ'_v were performed at $\sigma'_h = 49$, 117·6, 176·4, 215·6 kPa (Tests 10, CY1, CY2 and CY3). All the anisotropic and isotropic compression and triaxial tests were performed at an axial strain rate of about 0·09% per min.

Figures 4(a), (b) and (c) show typical results from vertical and lateral cyclic loading tests at an isotropic stress state. Values of ε_v and ε_h were

Table 1. List of tests

Test name	γ_d: kN/m^{3*}	e_0*	ω: %†	Stress path	σ'_h: kPa	σ'_v: kPa
TACA	18·0	0·49	3·35	Isotropic and anisotropic compression	49 ~ 490	49 ~ 970·2
				Triaxial compression	49	49 ~ 529·2‡
10	18·13	0·47	3·34	Cyclic triaxial ($\Delta\sigma'_h = 0$)	49	49 ~ 303·8‡
CY1	18·62	0·43	3·35		117·6	117·6 ~ 597·8‡
CY2	18·64	0·41	3·35		176·4	176·4 ~ 1097·6‡
CY3	18·62	0·43	3·35		215·6	215·6 ~ 1195·6‡

* Initial void ratio measured at $\sigma'_v = \sigma'_h = 19{\cdot}6$ kPa.
† Water content at sample preparation.
‡ Peak value at failure condition.

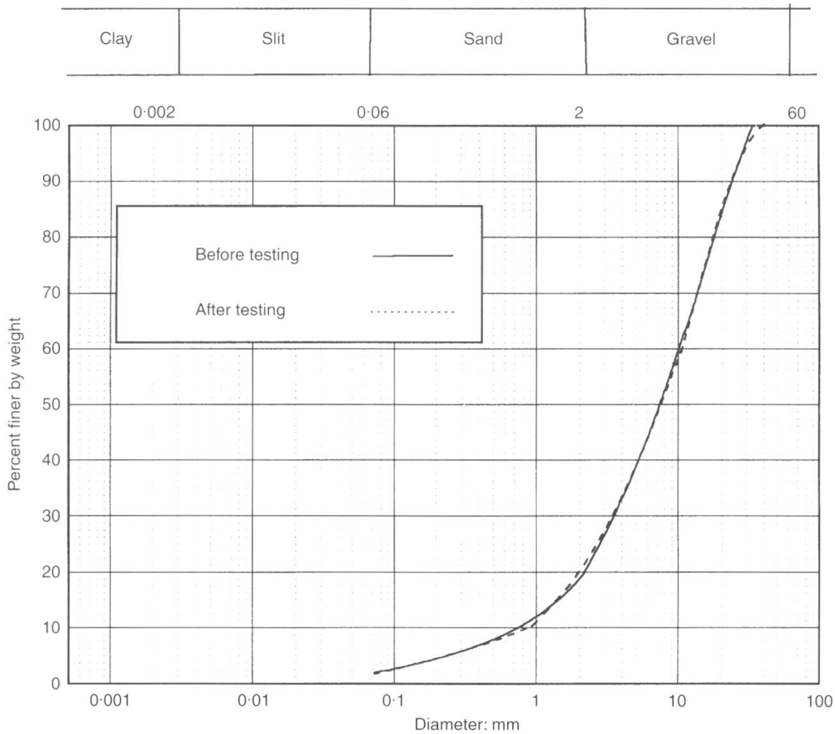

Fig. 2. Grain size distribution curves of the test material (Test TACA)

obtained by averaging readings of, respectively, the vertical and lateral LDTs. Hysteresis loops were almost closed when the single amplitude strain was smaller than about 0·001% (Fig. 4(c)). The Young's modulus $E_\nu = (d\sigma'_\nu/d\varepsilon_\nu)_{d\sigma'_h=0}$ and the Poisson's ratio $\nu_{\nu h} = -(d\varepsilon_h/d\varepsilon_\nu)_{d\sigma'_h=0}$ were obtained from linear regression of data as shown in Figs 4(a), (b) and (c); the average values obtained in this way from five cyclic tests are quoted. The equation for lateral strain increment $d\varepsilon_h$ in cross-anisotropic materials with the vertical axis of symmetry is

$$d\varepsilon_h = (1/E_h)\,d\sigma'_h - (\nu_{hh}/E_h)\,d\sigma'_h - (\nu_{\nu h}/E_\nu)\,d\sigma'_\nu \quad (1)$$

From equation (1), together with the condition that $d\sigma'_\nu = 0$ in the lateral cyclic loading tests, we obtain:

$$E_h = F_h(1 - \nu_{hh}); \quad F_h = (d\sigma'_h/d\varepsilon_h)_{d\sigma'_\nu=0} \quad (2)$$

where ν_{hh} is the Poisson's ratio in the horizontal planes. Each F_h value was obtained as shown in Fig. 4(b). The value of $\mu_{h\nu} = -(d\varepsilon_\nu/d\varepsilon_h)_{d\sigma'_\nu=0}$ shown in Fig. 4(b) is equal to $2\nu_{h\nu}/(1 - \nu_{hh})$. Furthermore, the measured values of $\mu_{h\nu}$ from most of the other cyclic tests were not as good as the one shown in this figure due to the fact that strain amplitudes $d\varepsilon_\nu$ were generally very small (Hoque *et al.*, 1996). For these two reasons, the Poisson's ratio $\nu_{h\nu}$ will not be discussed in this paper. To evaluate the values of E_h and $\nu_{h\nu}$ more confidently and to obtain the values of ν_{hh}, the horizontal stress in one direction σ'_{h1} is needed to be changed independently of the other two principal stresses σ'_{h2} and σ'_ν.

TEST RESULTS

The results from the isotropic loading and unloading stages in Test TACA are shown in Fig. 5. The difference between the LVDT and gap sensor measurements of axial strain reflects the effects of the compliance of the axial loading system, while the difference between the gap sensor and LDT measurements of axial strain is due to bedding errors at the top and bottom ends of the specimen. Also, the difference between the gap sensor and LDT measurements of lateral strain is due to the effects of membrane penetration on the lateral surfaces of the specimen. Similar errors in ε_ν and ε_h values were also observed in the very small amplitude vertical and lateral cyclic tests. These results show the necessity of using a prismatic specimen to evaluate lateral strains free from membrane penetration effects by using, for example, the method shown in Fig. 1. The difference between

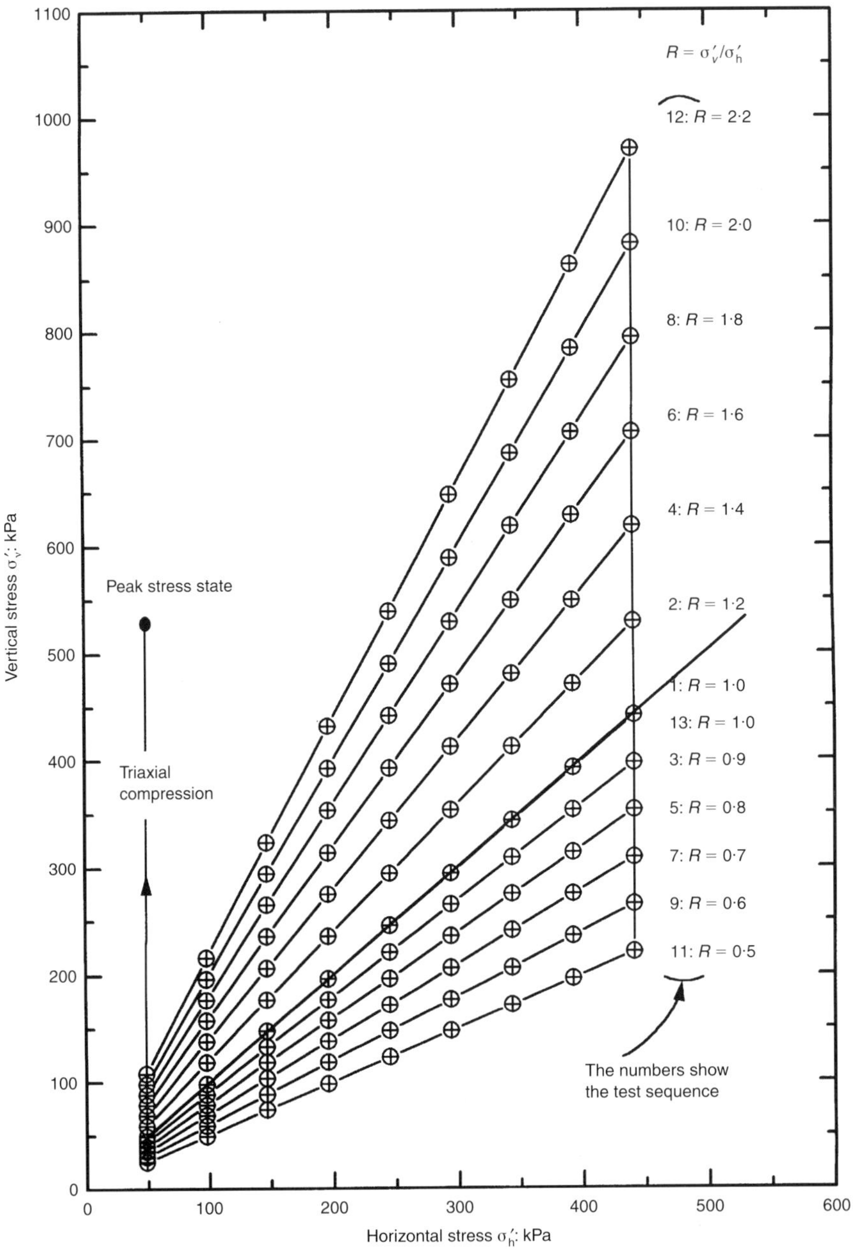

Fig. 3. Stress states of very small cyclic loading tests

the axial strain ε_v (LDT) and the lateral strain ε_h (LDT) shows inherent anisotropy in the deformation for a large strain range. The yield stress of about $\sigma'_v = \sigma'_h = 120$ kPa seen in the relations perhaps reflects the level of compaction used in preparing the specimen.

When the single amplitude axial strain $d\varepsilon_{v-sa}$ was larger than 0·001% as shown in Figs 4(a) and (b), measured values of $E_v(d\varepsilon_{v-sa})$, $F_h(d\varepsilon_{h-sa})$ and $\nu_{vh}(d\varepsilon_{v-sa})$ were corrected to the corresponding values E_v^0, F_h^0 and ν_{vh}^0, defined for $d\varepsilon_{v-sa}$ or $d\varepsilon_{h-sa} = 0{\cdot}001\%$. The correction was made by using the average relationships between $A = E_v(d\varepsilon_{v-sa})/E_v^0$ and $B = \nu_{vh}(d\varepsilon_{v-sa})/\nu_{vh}^0$ and $d\varepsilon_{v-sa}$

shown in Fig. 6 such that:

$$E_\nu^0 = E_\nu(\mathrm{d}\varepsilon_{\nu-\mathrm{sa}})/A, \quad F_\mathrm{h}^0 = F_\mathrm{h}(\mathrm{d}\varepsilon_{\mathrm{h-sa}})/A \qquad (3)$$

$$\nu_{\nu\mathrm{h}}^0 = \nu_{\nu\mathrm{h}}(\mathrm{d}\varepsilon_{\nu-\mathrm{sa}})/B \qquad (4)$$

The data presented in Fig. 6 were obtained from another cyclic triaxial test using the same triaxial apparatus on the same sandy gravel isotropically consolidated to 49 kPa. The same relation A was used for the correction of F_h to F_h^0 while substituting each amplitude of horizontal strain $\mathrm{d}\varepsilon_{\mathrm{h-sa}}$ into the horizontal axis in Fig. 6. For all the data, the

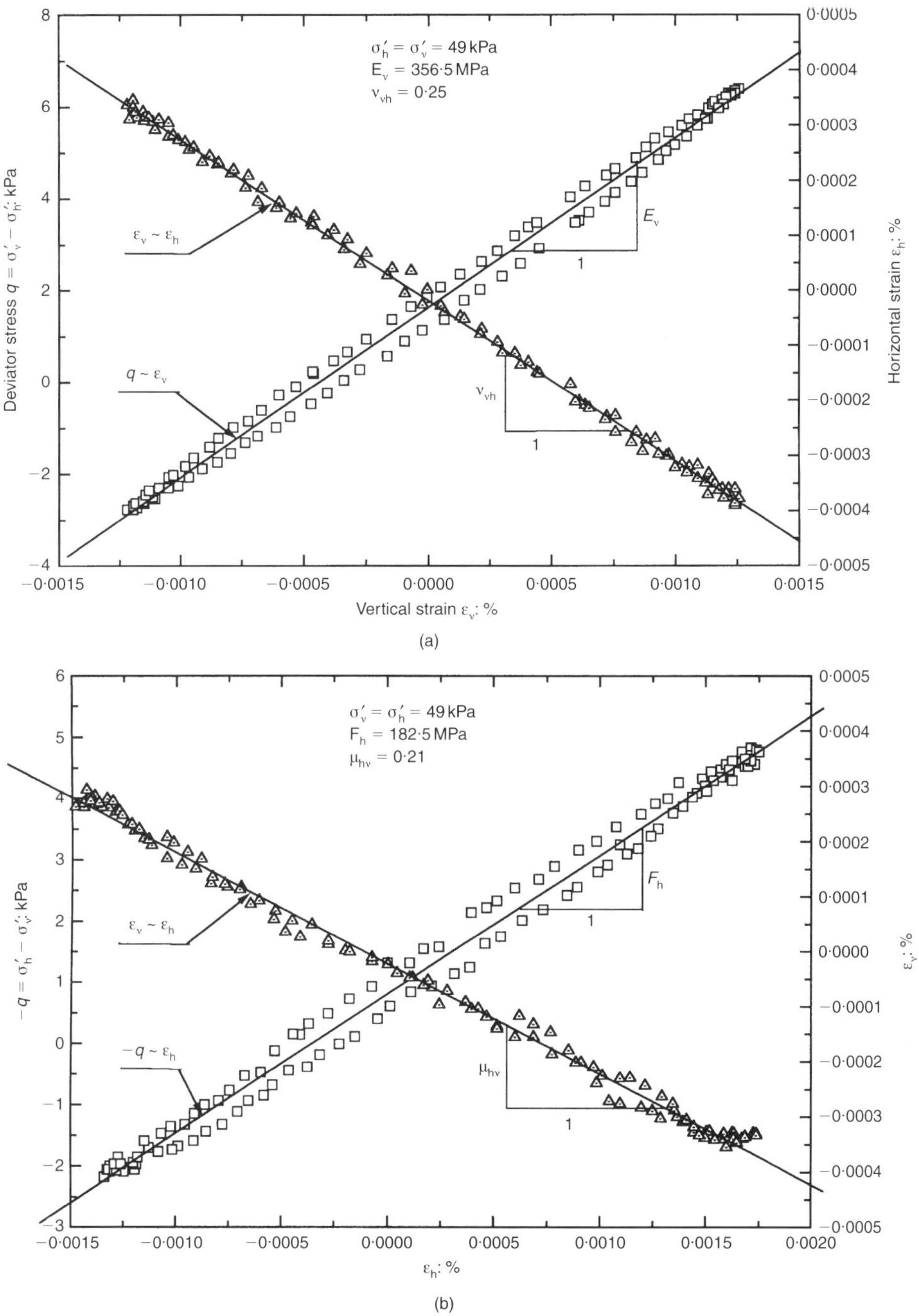

Fig. 4(a) and (b). **continued overleaf**

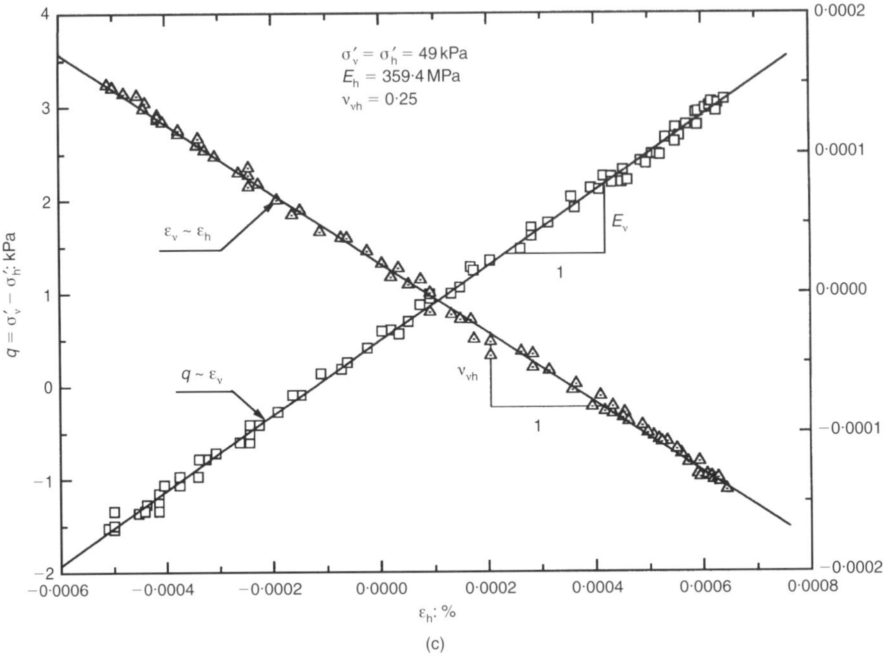

Fig. 4. Typical results from (a) vertical cyclic tests and (b) horizontal cyclic tests with relatively large strain amplitudes and (c) vertical cyclic tests with relatively small strain amplitude

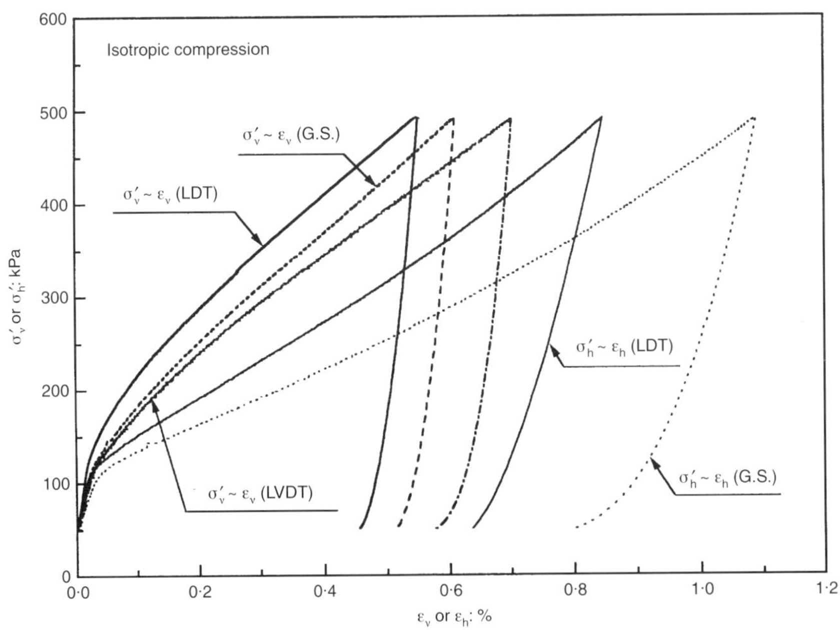

Fig. 5. Results of the isotropic compression test

corrections did not exceed 10% of the measured values.

Figure 7 shows the relationships among $E^0_\nu/f(e)$, $F^0_{\rm h}/f(e)$, $\nu^0_{\nu{\rm h}}$ and $\sigma'_\nu = \sigma'_{\rm h}$ from Test TACA. Fitted linear relations are also presented. These data were obtained at isotropic stress states before and after very small cyclic tests were performed at anisotropic stress states shown in Fig. 3. The

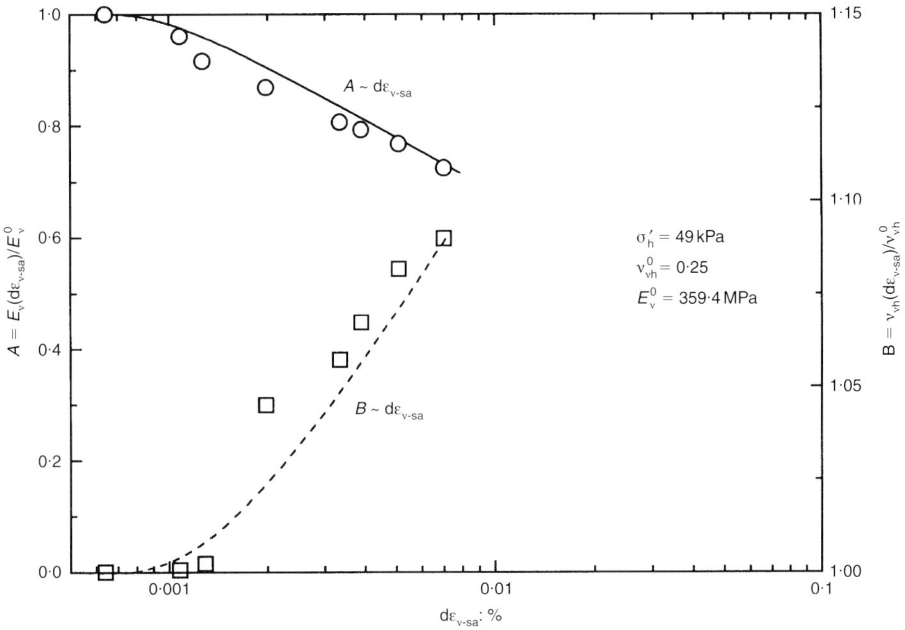

Fig. 6. Correction curves used to obtain the elastic deformation moduli and Poisson's ratio

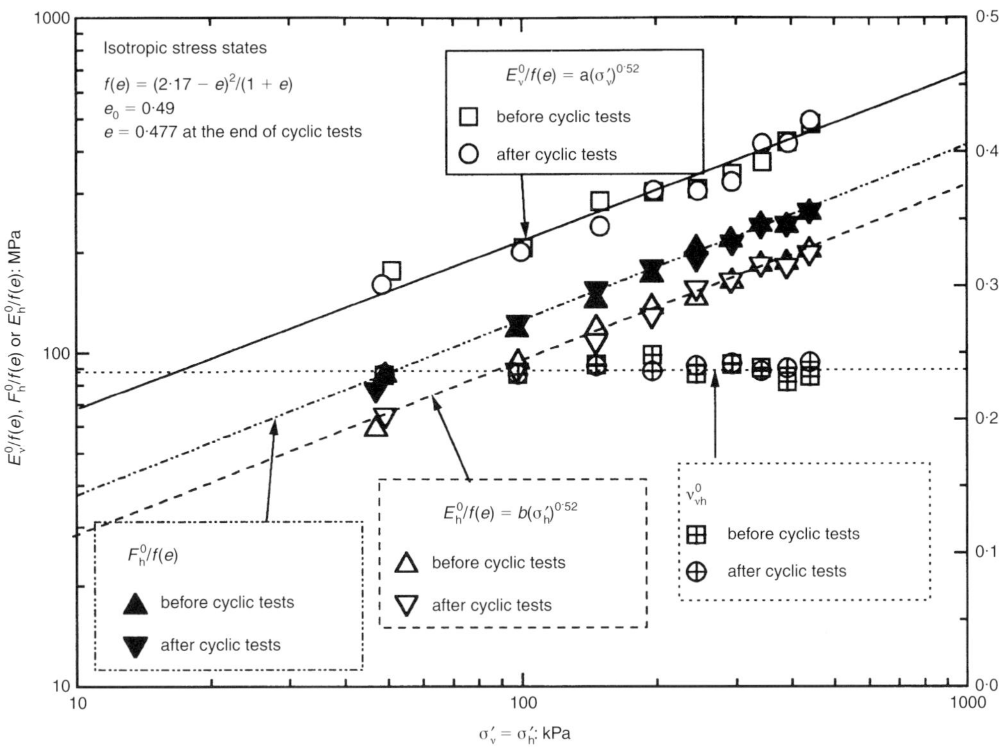

Fig. 7. Young's moduli E_v^0, F_h^0, E_h^0 and Poisson's ratios ν_{vh}^0 at isotropic stress states; $\nu_{hh}^0 = 0{\cdot}24$ was used to obtain E_h^0

void ratios 0·49 and 0·477 indicated in Fig. 7 (and Fig. 8) are, respectively, the initial values listed in Table 1 and that measured at the end of cyclic tests (i.e. at the start of triaxial compression to failure). To exclude the influence of small variations in void ratios at various stress states (and among the different specimens), the calculated values of E_ν^0 and F_h^0 were normalized by using the empirical equation $f(e)$ presented in Fig. 7 (Hardin & Richart, 1963). This equation has been found to be valid for this type of gravel as for many other granular materials (e.g. Iwassaki, Tatsuoka & Takagi, 1978). According to this equation, the difference in E_ν^0 and F_h^0 between the two void ratios shown in Fig. 7 is 2·4%.

The E_h^0 values obtained as $F_h^0/(1-\nu_{hh}^0)$ by assuming the ν_{hh}^0 value were also plotted against $\sigma_\nu' = \sigma_h'$ in Fig 7. The values of ν_{hh} and ν_{hh}^0 could not be obtained from the present study, as the two principal stresses were always the same. As shown later, the $\nu_{\nu h}^0$ values are a function of the stress ratio σ_ν'/σ_h'. Furthermore, the two lateral principal stresses σ_{h1}' and σ_{h2}' are the same. For these two reasons, at a constant stress ratio $\sigma_{h1}'/\sigma_{h2}'$ equal to 1·0, ν_{hh}^0 was assumed to be independent of σ_ν'/σ_h'. At this stage, it was also assumed that ν_{hh}^0 would be the same with $\nu_{\nu h}^0$ at isotropic stress state, equal to 0·24 as shown below. At isotropic stress states, ν_{hh}^0 and $\nu_{\nu h}^0$ would not be exactly the same as each other. It is likely, however, that at isotropic stress states on the horizontal planes, the values of ν_{hh}^0 are much less than 0·5, as suggested by Bellotti *et al.* (1996). In that case, the effects of possible errors in this assumption on '$1-\nu_{hh}^0$', thus on 'E_h^0 values' would be relatively insignificant. This point will be discussed again later. It may be seen from Fig. 7 that both the E_ν^0 and E_h^0 values are a function of $(\sigma_\nu')^{m_0} = (\sigma_h')^{m_0}$ with a power $m_0 =$ 0·52, while the $\nu_{\nu h}^0$ value is almost independent of isotropic pressure level, and that all these values are independent of the stress history. On the other hand, the ratio of E_ν^0/E_h^0 is about 2·2 within the stress range examined, which indicates large inherent anisotropy in very small strain stiffness. This result is consistent with a much stiffer response in the vertical direction during loading and unloading of isotropic compression shown in Fig. 5.

In Fig. 8, the values of $E_\nu^0/f(e)$ obtained at the various isotropic and anisotropic stress states and those during the triaxial compression test (Test TACA) are plotted against the value of σ_ν', while the values of $E_h^0/f(e)$ are plotted against σ_h'. The average relations for $E_\nu^0/f(e)$ and $E_h^0/f(e)$ at

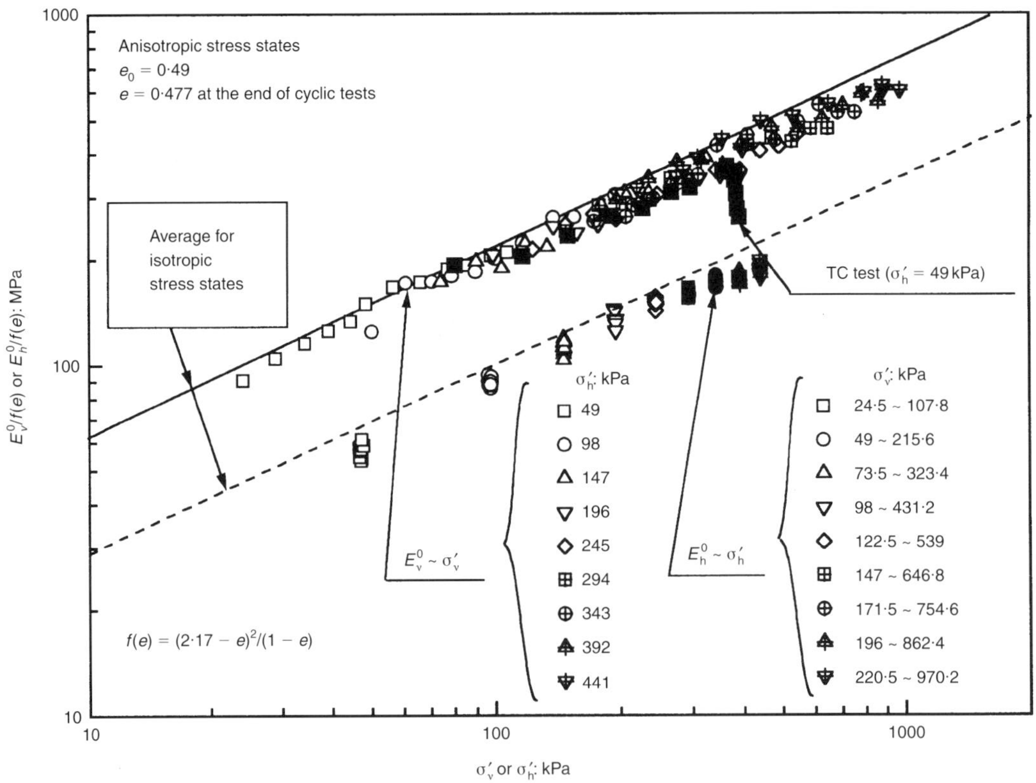

Fig. 8. E_ν^0 versus σ_ν' and E_h^0 versus σ_h'; ν_{hh}^0 = 0.24 was used to obtain E_h^0

isotropic stress states are also shown by solid and broken lines, respectively. From Fig. 8, the following trends of behaviour can be noted. First, within the wide range of anisotropic stress states, the E_v^0 values measured after ageing at each stress point are a function of $(\sigma'_v)^m$ with m being about 0·5, while they are independent of σ'_h. Similarly, the E_h^0 values are a function of $(\sigma'_h)^m$ with m being about 0·5, while they are independent of σ'_v. These facts suggest that the ratio E_v^0/E_h^0 increases in proportion to $(\sigma'_v/\sigma'_h)^m$ (Fig. 9). The data points shown in Fig. 9 are for the values of E_h^0 obtained by using $\nu_{hh}^0 = 0{\cdot}24$. It may be seen that they follow approximately the relation cited above with $m = 0{\cdot}52$. More rigorously, however, the average relation of the data is not linear. A solid curve shown in this figure is the average relation obtained by square regression analysis (the broken curve will be discussed later). This property can be called stress-state-induced anisotropy, which is consistent with the results obtained from the similar tests on clean sands (Hoque *et al.*, 1996) and body wave velocity measurements in Ticino sand (Bellotti *et al.*, 1996).

In Fig. 10, the Poisson's ratio ν_{vh}^0 obtained at the various isotropic and anisotropic stress states and those obtained during the triaxial compression test (Test TACA) are plotted against σ'_v/σ'_h. The scatter in the ν_{vh}^0 values is smallest at and near $\sigma'_v/\sigma'_h = 1{\cdot}0$, since the hysteresis loops are very stable, showing negligible drifting in strains with cyclic loading. The general trend is that the ν_{vh}^0 value increases in proportion to $(\sigma'_v/\sigma'_h)^n$. The power n obtained by regression analysis of these data within a stress ratio range of $\sigma'_v/\sigma'_h = 0{\cdot}5 \sim 2{\cdot}2$ (Fig. 3) is equal to 0·30. This value is close to a half of m_0 (i.e. $m_0/2 = 0{\cdot}26$).

The compliance matrix becomes symmetric when $\nu_{vh}^0/E_h^0 = \nu_{vh}^0/E_v^0$ as for ideal cross-anisotropic elastic media, which is achieved when the following conditions are satisfied:

$$\nu_{vh}^0 = (\nu_0\sqrt{a})(\sigma'_v/\sigma'_h)^n \quad (5a)$$

$$\nu_{hv}^0 = (\nu_0\sqrt{a})(\sigma'_h/\sigma'_v)^n \quad (5b)$$

where $n = m_0/2$, a is equal to E_v^0/E_h^0 at $\sigma'_v/\sigma'_h = 1{\cdot}0$, and ν_0 is the Poisson's ratios ν_{vh}^0 or ν_{hv}^0 when $\sigma'_v/\sigma'_h = (1/a)^{1/2n}$, where $E_v^0 = E_h^0$. When based on this postulate, equation (5), one possible assumption as to the ν_{hh}^0 value, other than $\nu_{hh}^0 = 0{\cdot}24$, would be that ν_{hh}^0 is equal to ν_0, considering that this isotropic condition prevails in all

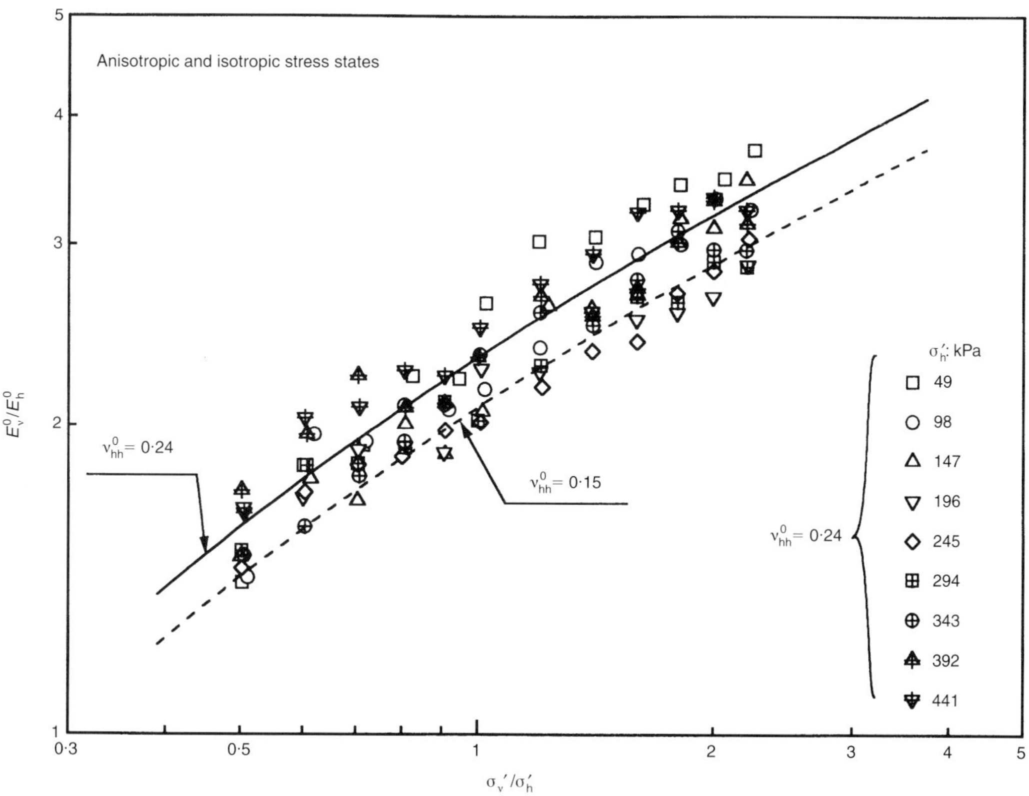

Fig. 9. Ratio E_v^0/E_h^0 versus stress ratio σ'_v/σ'_h

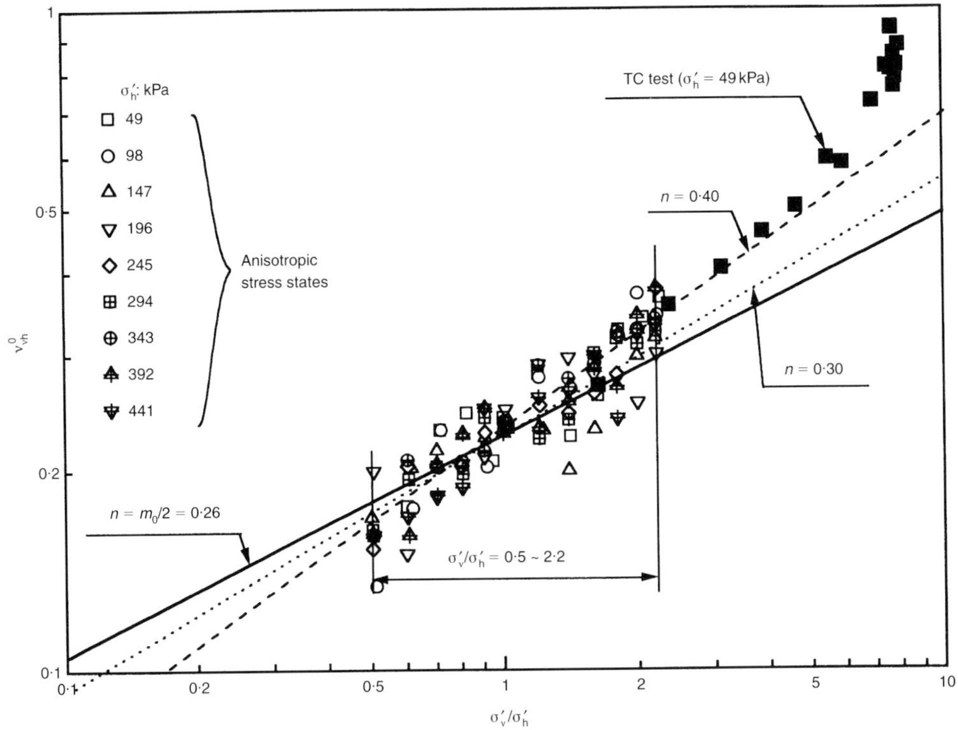

Fig. 10. Poisson's ratio ν^0_{vh} versus σ'_v/σ'_h

the directions within the specimen. This ν_0 value was obtained by trial and error. That is, a stress ratio σ'_v/σ'_h equal to 0·3 was found to satisfy the condition that $E^0_v/\{F^0_h(1-\nu^0_{vh}\}=1{\cdot}0$ with the ν^0_{vh} value satisfying equation (5a) (with $n = 0{\cdot}26$) and giving $\nu_0 = 0{\cdot}15$ (see Figs 9 and 10). The average relation for $\nu^0_{hh} = \nu_0 = 0{\cdot}15$ shown in Fig. 9 was obtained by vertically shifting the relation for $\nu^0_{hh} = 0{\cdot}24$. Then, for $\nu^0_{hh} = 0{\cdot}15$, the value of E^0_v/E^0_h when $\sigma'_v/\sigma'_h = 1{\cdot}0$ is equal to about 2·0, compared with 2·2 when $\nu^0_{hh} = 0{\cdot}24$ is used. That is, the ratio E^0_v/E^0_h is similar for $\nu^0_{hh} = 0{\cdot}24$ and 0·15.

It may be seen from Figs 8 and 10 that the values of E^0_v and ν^0_{vh} evaluated for each very small unload–reload cycle of σ'_v applied when σ'_v/σ'_h was less than about 2 during the triaxial compression test are similar to the respective corresponding values measured after ageing (i.e. sufficient consolidation) at each stress state shown in Fig. 3. However, as the stress state is approaching failure in the triaxial compression test, the values of E^0_v gradually become smaller than the respective corresponding value evaluated after ageing at the same σ'_v, while the ν^0_{vh} values gradually become larger than those obtained by extrapolating the relation with $n = 0{\cdot}30$ (see Fig. 10).

To confirm the trend mentioned above, the same analysis was made for the results from the four cyclic tests with large amplitudes of σ'_v at a constant σ'_h. Figure 11 shows the relationships between $\sin\phi_{mob}$ and shear strain $\gamma = \varepsilon_v - \varepsilon_h$. For convenience, positive and negative values of $\sin\phi_{mob}$ were defined for σ'_v, respectively, larger and smaller than σ'_h. In each test, a single very small unload–reload cycle of σ'_v was applied at each intermediate loading state (i.e. without ageing). In Fig. 12, the values of $E^0_v/\{f(e){\cdot}(\sigma'_v/p_a)^{0{\cdot}52}\}$ from these cyclic loading tests are shown together with those evaluated at the isotropic and anisotropic stress states shown in Fig. 3. This normalization, using the atmospheric pressure p_a, was to remove the effects of variations in σ'_v on E^0_v values. A similar comparison is made for ν^0_{vh} in Fig. 13. From Figs 12 and 13, it may be seen that the values of E^0_v and ν^0_{vh} measured during the large amplitude cyclic triaxial tests are consistently, respectively, smaller and larger than the corresponding values evaluated at the same σ'_v after ageing at each stress state. The deviation became larger with the increase in accumulated strains. The reduction in E^0_v is particularly large for triaxial extension stress states (i.e. when $\sin\phi_{mob}$ is negative). This behaviour has been observed in a simi-

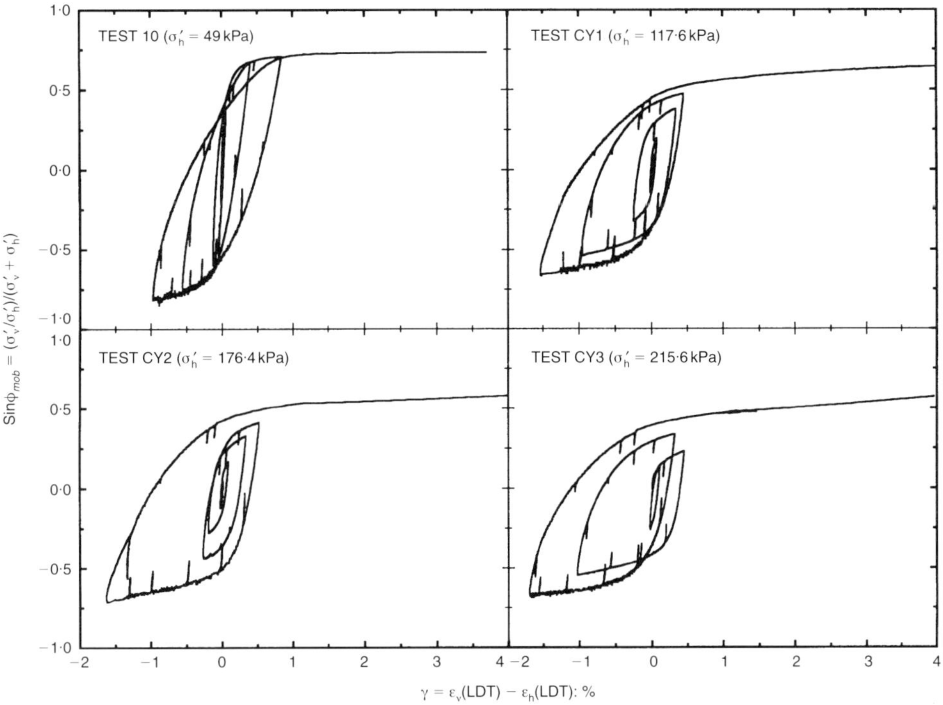

Fig. 11. Overall stress–strain relations from the cyclic triaxial tests

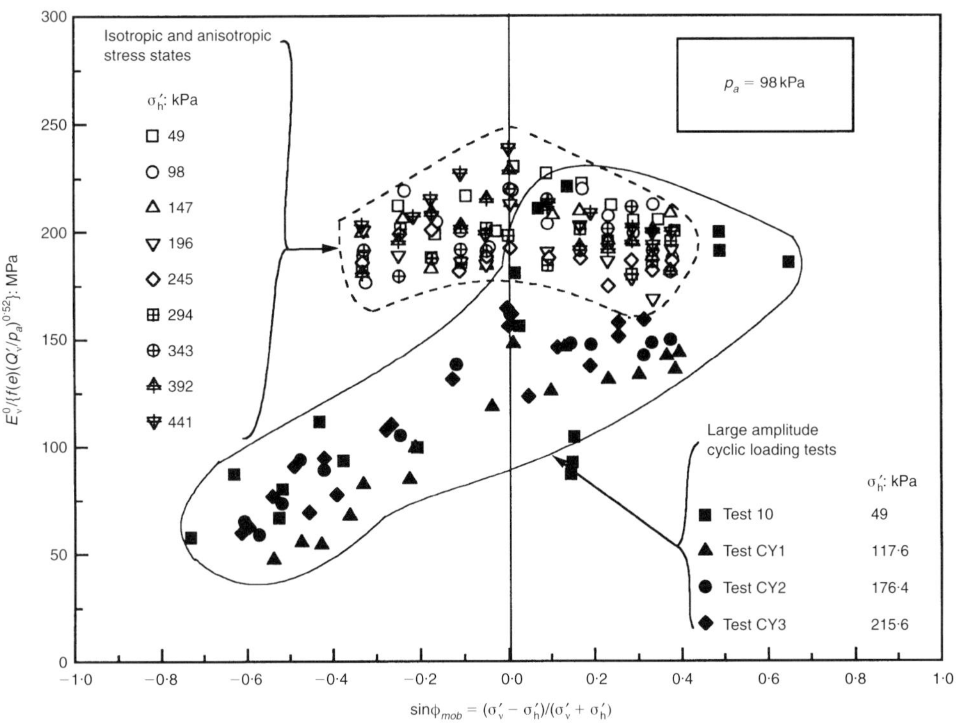

Fig. 12. $E^0_v/\{f(e)\cdot(\sigma'_v/p_a)^{0.52}\}$ **versus** $\sin\phi_{mob}$

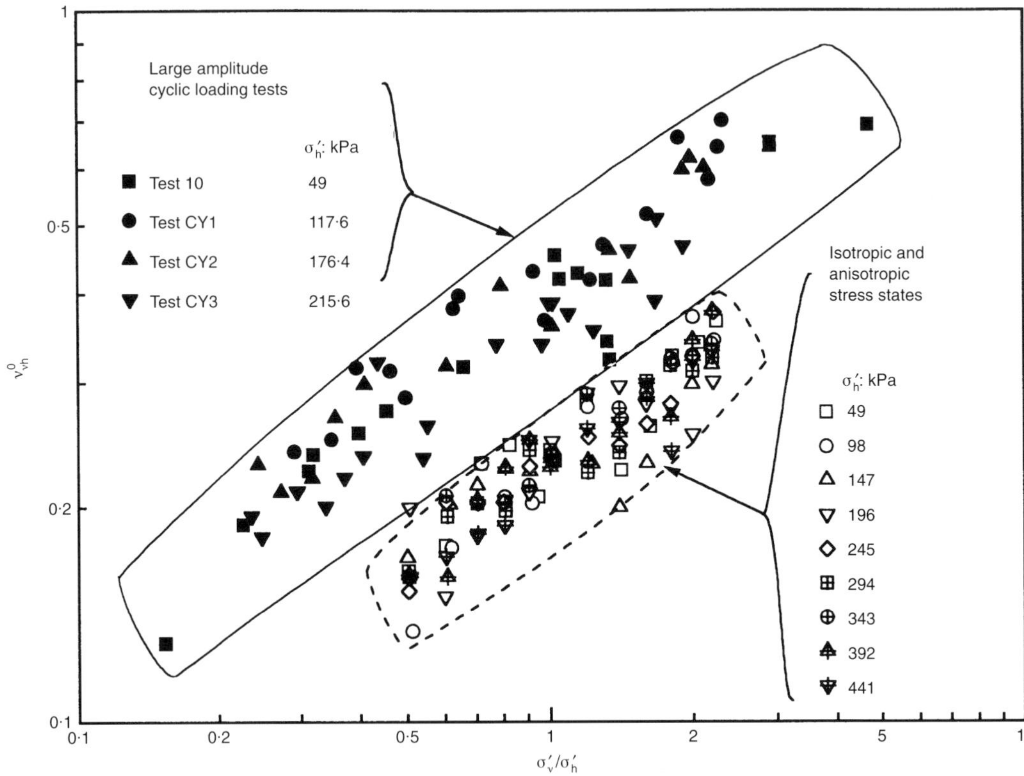

Fig. 13. ν^0_{vh} versus σ'_v/σ'_h

lar test (Flora, Jiang, Kohada & Tatsuoka, 1994), and it may be explained by the effects of damage to the fabric by shear deformation and decrease in the damage effects by ageing.

CONCLUSIONS

The following conclusions can be drawn from the test results presented above.

(*a*) To evaluate accurately the axial and lateral strains of a coarse-grained material by triaxial tests with variable axial and lateral stresses, local axial and lateral strain measurements free from effects of bedding error and membrane penetration, as achieved in the present study, are essential.

(*b*) Deformation characteristics during loading and unloading in isotropic compression with a large range of stress amplitude were highly anisotropic, with a stiffer response in the vertical direction, in which the specimen was compacted.

(*c*) At isotropic stress states, the Young's moduli, E^0_v and E^0_h, defined for major principal strain increments less than 0·001% occurring, respectively, in the vertical and horizontal directions, increased in proportion to approximately $(\sigma'_c)^{0\cdot52}$ (σ'_c is the confining pressure). The ratio E^0_v/E^0_h was about 2·0. This fact together with the above (*b*) indicates that the specimen had similar inherent anisotropy in both elastic and plastic deformation.

(*d*) For a wide range of anisotropic stress states, E^0_v measured after ageing at each stress point was essentially a function of $(\sigma'_v)^m$ with m equal to about 0·5, while E^0_h was essentially a function of $(\sigma'_h)^m$. Therefore, the ratio E^0_v/E^0_h increased approximately in proportion to $(\sigma'_v/\sigma'_h)^m$ (i.e. stress-state-induced anisotropy). The Poisson's ratio ν^0_{vh} increased in proportion $(\sigma'_v/\sigma'_h)^n$ with n similar to $m/2$.

(*e*) The values of E^0_v and ν^0_{vh} observed in each very small unload–reload cycle of σ'_v applied without ageing at stress ratios σ'_v/σ'_h less than about 2 during monotonic triaxial compression loading were similar to those measured after ageing at each stress state. However, as the stress state approached the failure state, the values of E^0_v and ν^0_{vh} measured during triaxial compression became consistently, respectively, smaller and larger than the corresponding values measured after ageing at each stress state. This trend was observed also in the values of E_v^0 and ν_{vh}^0 measured during large amplitude cyclic triaxial tests. This behaviour may be explained by the effects of damage to the fabric by shear deformation and recovery from the damage by ageing.

REFERENCES

Bellotti, R., Jamiolkowski, M., Lo Presti, D. C. F. & O'Neill, D. A. (1996). Anisotropy of small strain stiffness in Ticino sand. *Géotechnique* **46**, No. 1, 115–131.

Di Benedetto, H. & Tatsuoka, F. (1997). Small strain behavior of geomaterials: modelling of strain rate effects. *Soils and Foundations*, **37** (to appear).

Flora, A., Jiang, G. L., Kohada, Y. & Tatsuoka, F. (1994). Small strain behavior of a gravel along some triaxial stress paths, *Proc. Int. Symp. on Prefailure Deformation Characteristics of Geomaterials, IS-Hokkaido '94*, Balkema, **I**, 279–285.

Goto, S., Tatsuoka, F., Shibuya, Y. & Sato, T. (1991). A simple gauge for local strain measurements in the laboratory. *Soils and Foundations* **31**, No. 1, 169–180.

Hardin, B. O. & Richart, F. E. (1963). Elastic wave velocities in granular soils. *J. Soil Mech. Found. Div., Am. Soc. Civ. Engrs* **89**, No. SM1, 33–65.

Hoque, E., Sato, T. & Tatsuoka, F. (1996). Performance of LDTs for the use in triaxial tests. *Geotechnical Testing Journal, ASTM*, **19**, No. 4, 411–420.

Iwassaki, T., Tatsuoka, F. & Takagi, Y. (1978). Shear moduli of sands under cyclic torsional shear loading. *Soils and Foundations* **18**, No. 1, 39–56.

Stokoe, K. H., Lee, J. N. K. & Lee, H. H. (1991). Characterization of soil in calibration chambers with seismic waves. *Proc. Int. Symp. on Calibration Chamber Testing*, Clarkson Univ., Postdam, NY (ed. Huang), Elsevier. 363–376.

Tatsuoka, F., Sato, T., Park, C. K., Kim, Y. S., Mukabi, J. N. & Kohata, Y. (1994a). Measurements of elastic properties of geomaterials in laboratory compression tests. *Geotechnical Testing Journal, ASTM* **17**, No. 1, 80–94.

Tatsuoka, F., Teachavorasinskun, S., Dong, J., Kohata, Y. & Sato, T. (1994b). Importance of measuring local strains in cyclic trixial tests on granular materials. *Dynamic Geotechnical Testing II*, ASTM STP 1213, 288–302.

Cotecchia, F. & Chandler, R. J. (1997). *Géotechnique* **47**, No. 3, 523–544

The influence of structure on the pre-failure behaviour of a natural clay

F. COTECCHIA* and R. J. CHANDLER†

Results are presented of an extensive laboratory and field investigation of a Pleistocene stiff clay in its natural and reconstituted states, including constant rate of strain oedometer and triaxial tests over a wide stress range, scanning electron microscopy and chemical micro-analysis. The relationships between the structure of the clay, its geological history and its mechanical response are demonstrated. The structure of the clay is strongly influenced by bonding, apparently largely due to a film of amorphous calcite which developed *in situ* at the end of normal consolidation. The effects of structure on the stiffness of the clay are identified, with sensitivity as the parameter quantifying the mechanical effects of structure. A frictional flow rule is found generally to be applicable over a wide strain range, including yield, even when strong bonding is present, although its form is influenced by structure. A new normalization shows that the gross yield surface is isotropically hardening in terms of both volume and structural change.

KEYWORDS: constitutive relations; fabric/structure of soils; in situ testing; microscopy; shear strength; stiffness.

L'article présente les résultats de recherches approfondies, en laboratoire et sur le terrain, sur une argile raide du Pléistocène à l'état naturel et reconstituée, y compris des essais oedométriques et triaxiaux à taux de contrainte constant, sur un large éventail de contraintes, ainsi que les résultats de travaux de microscopie électronique à balayage et de micro-analyse chimique. Les auteurs montrent les rapports entre la structure de l'argile, son histoire géologique et sa réaction mécanique. La structure de l'argile est fortement influencée par l'adhérence, qui semble en grande partie due à une pellicule de calcite amorphe qui s'est formée sur place à la fin de la consolidation normale. Les auteurs identifient les effets de la structure sur la rigidité de l'argile, la sensibilité étant le paramètre qui quantifie les effets mécaniques de la structure. On constate qu'un écoulement de frottement s'applique généralement à un large éventail de contraintes, y compris à la limite d'élasticité, même quand l'adhérence est forte, bien que sa forme soit influencée par la structure. Une nouvelle normalisation montre que la surface brute d'écoulement subit un durcissement isotrope tant en volume qu'en changement structural.

INTRODUCTION

In a soil mechanics context, the terms 'structure' and 'bonding' have generally been used to attempt to identify elements that may or may not be physical components of the soil. Here, following, for example, Lambe & Whitman (1969), the term 'structure' is used to combine 'fabric', the arrangement of the component particles, and 'bonding', defined as those interparticle forces which are not of a purely frictional nature. Bonding is not necessarily (although it may be) a solid link, particularly with clays, but is the consequence of such phenomena as electrostatic, electromagnetic or other forces acting to connect the particles which have developed during the geological life of the soil. Bonding is thus the combined result of such factors as the minerology, water chemistry during deposition, pressure, temperature and organic content, and it depends on mechanical conditions such as deposition and consolidation rates and strain paths (Mitchell & Houston, 1969; Sides & Barden, 1970; Sfondrini 1975; Mitchell, 1976). The structure of a natural soil thus depends on the physical and chemical conditions that applied during deposition, consolidation, ageing and unloading.

The basic elements of structure exist in all clays, and thus structure should be seen as the source of similarities in the behaviour of different clays. It follows that, in comparing the mechanical behaviour of natural and 'reconstituted' (Burland,

Manuscript received 23rd October 1996; revised manuscript accepted 21 March 1997.
Discussion on this paper closes 1 November 1997.

* Formerly Imperial College of Science, Technology and Medicine, London; now Politecnico di Bari.
† Imperial College of Science, Technology and Medicine, London.

1990) clays with a view to contributing to a better understanding of an overall clay behaviour, one should first look to the similarities in behaviour. Of course, there will be differences, of which sensitivity (Terzaghi, 1944; Skempton & Northey, 1952), which often characterizes the behaviour of natural clay, though not reconstituted clay, is perhaps the most important.

Following this approach, this study endeavours to make a contribution to the understanding of pre-failure soil behaviour by comparing the mechanical behaviour of a strongly bonded natural clay with the behaviour of the same clay when reconstituted, the two clays having different structure and sensitivity. The pre-failure consequences of the structure of the clay that are examined here are the *in situ* stress state of the natural clay, and the oedometer and triaxial compression behaviour. Full details of sampling, testing and interpretation procedures, as well as other background material, are given in Cotecchia (1996); for reasons of space much of this information is presented here only in outline.

THE GEOLOGICAL HISTORY AND INDEX PROPERTIES OF PAPPADAI CLAY

The natural clay that is the subject of this study is the Pappadai Clay, a Pleistocene clay from the Montemesola Basin, near Taranto, Italy, the basic engineering geology of which is described by Cotecchia & Chandler (1995). In summary, this clay was deposited about 1.3 million years ago, in a quiet sea where reducing conditions applied. There was subsequent erosion of some 120 m at the sampling location, resulting in the clay being overconsolidated.

Block samples were taken from a depth of 25·4 m during the construction of a reservoir draw-off shaft. Fig. 1 shows the present geological and hydrogeological situation. The current vertical effective stress at the sampling depth is 415 kPa, while the preconsolidation stress deduced from geological evidence is about 1300 kPa.

Typical index and other properties of the clay, together with the mineralogical composition, are given in Table 1. The high proportions of illite, chlorite and smectite, sedimented in a saline marine environment, will have resulted in flocculation, giving rise to a bookhouse-type of fabric (Fig. 2). The decrease in the proportion of smectites with depth, along with an increase in the proportions of non-swelling minerals and a decrease in activity, all suggest that diagenesis (i.e. chemical and/or other modifications to both soil particles and bonding during the period of burial) has occurred in the Pappadai Clay. It is likely, therefore, that there will have been substantial changes to the structure of the clay, which will have occurred at the stress levels required to generate the mineralogical

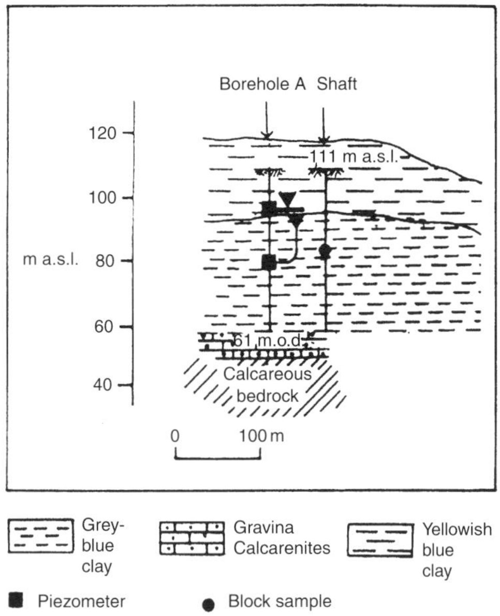

Fig. 1. Geological and hydrogeological conditions at the sampling site

Table 1. Properties and minerology of natural Pappadai Clay

Properties	Specific gravity	2·69
	Liquid limit	65%
	Plastic limit	30%
	Natural water content	≈31%
	Organic content	2·4%
	Carbonate content	28%
Mineralogy	Quartz	3%
	Feldspar	1%
	Carbonate	22%
	Dolomite	6%
	Kaolinite	12%
	Chlorite	14%
	Illite	20%
	Smectite	12%
	Intergrades	10%
	Total	100%

changes, that is, at the stresses that applied towards the end of normal consolidation.

In an attempt to explain the observed present-day depth–water content profile, the geological history of the Pappadai Clay was modelled (Cotecchia & Chandler, 1995), assuming a sequence of stages of (i) normal deposition, (ii) erosion, (iii) deep drying, and (iv) weathering and re-establishment of the ground water to the present level (Fig. 3(a)). The normal consolidation of the natural clay (the

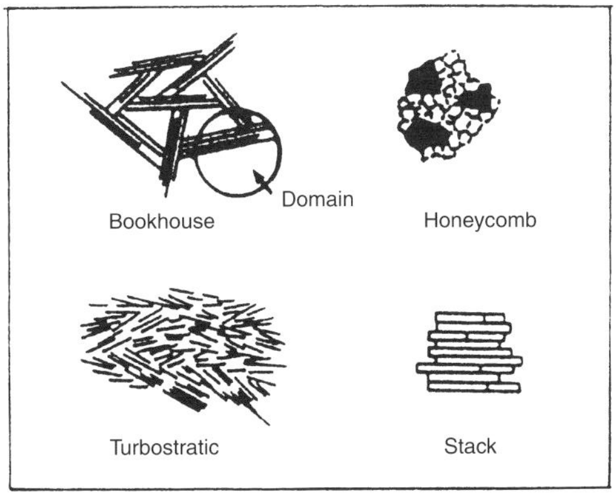

Fig. 2. Idealized clay fabrics. (After Sides & Barden, 1970)

sedimentation compression curve, SCC; Terzaghi, 1941) was assumed to be that of a clay of low to medium sensitivity and thus is assumed to lie in the middle of the range of natural clay compaction behaviour reported by Skempton (1970) (Fig. 3(b)). The model makes reasonable predictions of the present water content profile (stage (iv); Cotecchia & Chandler, 1995), the non-linear shape of which (dotted line, Fig. 3(a)) is the consequence of deep desiccation followed by weathering at shallow depths. The predicted history of the block samples tested, from original deposition (O) to the present state (B), is given by the path O–P–N–M–B. The observed present state, at $\bar{B}$, lies marginally below B, consistent with the predictions of the model. The intrinsic compression line (ICL; Burland, 1990), obtained from a disaggregated sample of the Pappadai Clay prepared at a water content of 1·6 × liquid limit, consolidated in an oedometer to a vertical stress (σ_v') of 200 kPa, is also shown in Fig. 3(a). The distance between the ICL and the SCC reflects the difference in structural strength between the natural and the reconstituted clay that developed during normal consolidation.

THE STRUCTURE OF NATURAL AND RECONSTITUTED PAPPADAI CLAY

The microstructure of the Pappadai Clay from the block samples was compared with samples of the clay reconstituted to point A in Fig. 3(a). A Cambridge 500 scanning electron microscope, an apparatus well documented in the literature (e.g. see Smart & Tovey, 1982), was used to examine the microfabric of the Pappadai Clay. The samples were prepared by freeze-drying, and were then broken to form approximately vertical and horizontal surfaces, which were gold coated. The resulting electron microscope images were then examined visually.

Natural clay

At medium magnification the fabric of the natural clay seen on vertical fractures is densely packed (Fig. 4). The clay particles are generally aggregated in 'domains', that is, groups of parallel particles (see Fig. 2), which originated as flocs during deposition (Sides & Barden, 1970; Cotecchia *et al.*, 1982). Groups of subhorizontally oriented domains can be seen in Fig. 4, suggesting that the fabric is well oriented. However, some areas are more densely packed than others, such as regions A and B, and both 'stacks' of domains in face-to-face contact (see Fig. 2), and irregularly intersecting domains can be seen, forming voids of up to 10–20 µm in size (Fig. 5). Stacks also occur interbedded with zones of randomly flocculated fabric (Fig. 6), and individual particles in edge-to-face contact are seen in Fig. 7.

These observations appear to confirm that the natural Pappadai Clay at deposition had a 'bookhouse' fabric, subsequently compressed so that there are areas of oriented clay, intermixed with remnants of the original fabric. Fabric of this type, a 'compacted bookhouse fabric' (Sides & Barden, 1970), is very similar to that anticipated by Sfondrini (1975) for one-dimensionally normal consolidated flocculated clay (Fig. 8). Uniform orientation is not observed; rather, non-uniform strains occurred during compression, so that an open honeycomb fabric is interspersed with densely packed, well-oriented stacks of domains.

In spite of the relatively high carbonate content (Table 1) there is little evidence of calcite crystals, which may in part be explained by the relatively high incidence of carbonate fossils.

Reconstituted clay

Figure 9 shows the fabric on a vertically fractured surface of reconstituted clay. The fabric appears more open than the natural clay, reflecting the considerable difference in the preconsolidation pressures of the two clays. This fabric has, none the less, much in common with that of the natural clay. Oriented areas, with stacks of domains, can be recognized, together with randomly oriented flocculated areas. Fig. 10 shows an example of open honeycomb fabric. Thus, as with the natural clay, a non-uniform reorientation of the clay particles has occurred during compression. The natural clay is, however, far more complex, with less regular alternations of oriented and flocculated fabrics, probably the consequence of diagenesis.

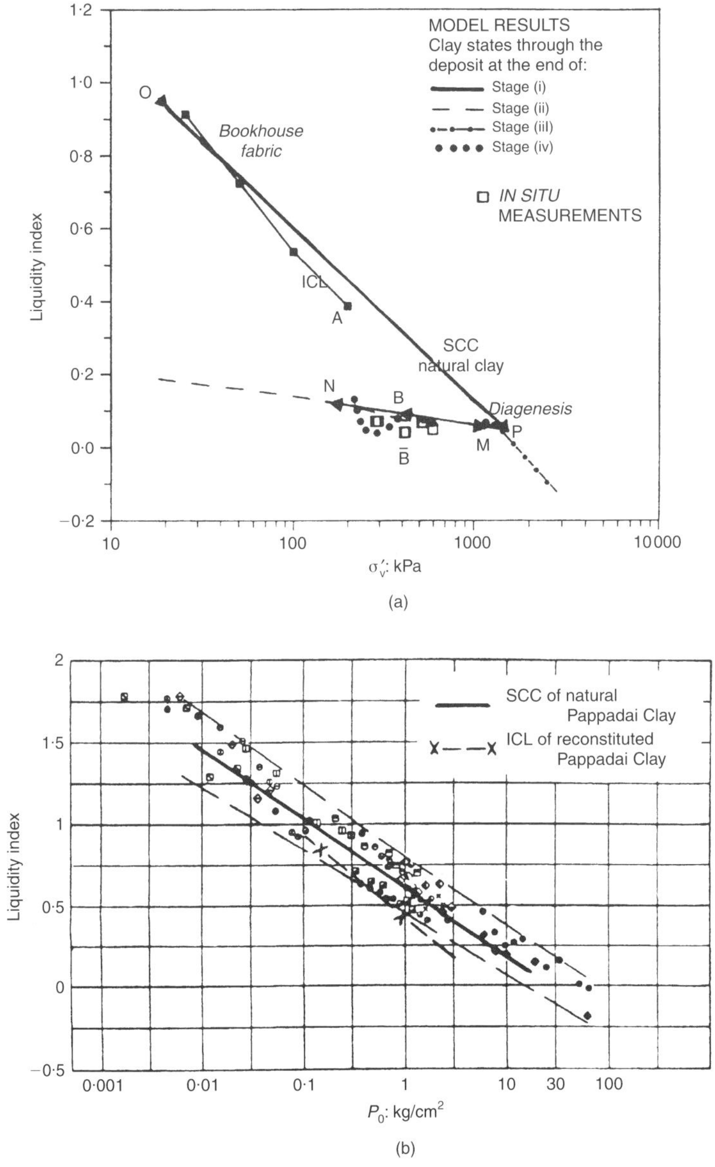

Fig. 3. State history of the natural and reconstituted Pappadai Clay: (a) results of modelling geological stages (i)–(iv), and (b) SCC assumed to apply to natural Pappadai Clay (from Skempton, 1970).

Chemical micro-analysis: methods and results

A microprobe analysis was used in the electron microscope to examine the chemical composition of selected areas of samples of the natural clay. Carbon-coated freeze-dried samples were used, and locations as small as 1 μm were investigated, the results being presented as diffractograms indicating the chemical elements present.

The technique was 'calibrated' by scanning such fabric elements as framboid pyrite crystals and fossil fragments, and in two such cases the size of the sulphur and iron peaks (Fig. 11(a)) and the calcium peak (Fig. 11(b)) are as expected. The diffractogram of Fig. 11(c), typical of many, relates to a clay particle and shows the expected mica components (Si, Al and K), but also shows a high proportion of calcium, which seems to be coating the clay particle. The diffractogram of Fig. 11(d),

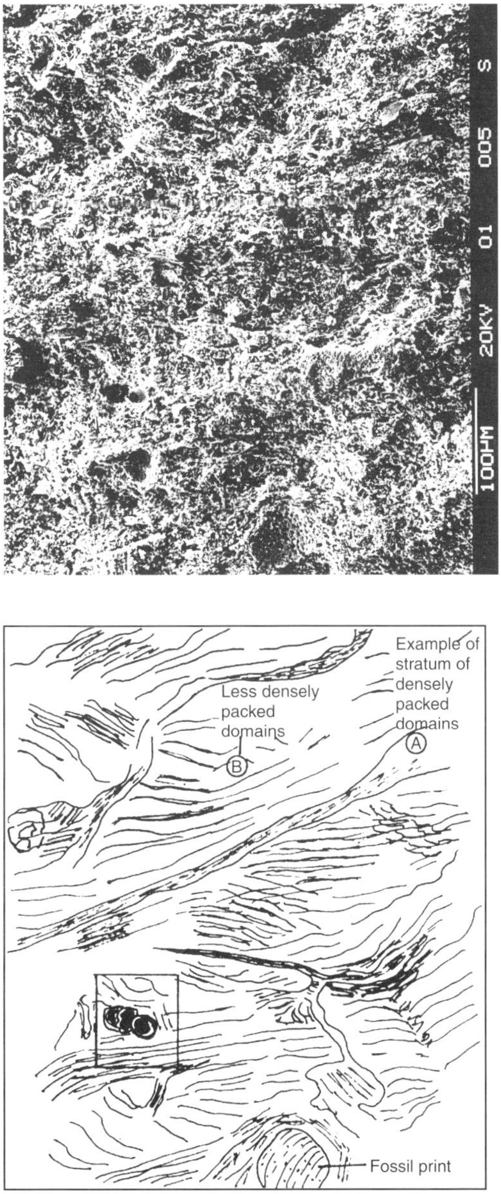

Fig. 4. Undisturbed Pappadai Clay: overall view of the fabric on a vertical fracture

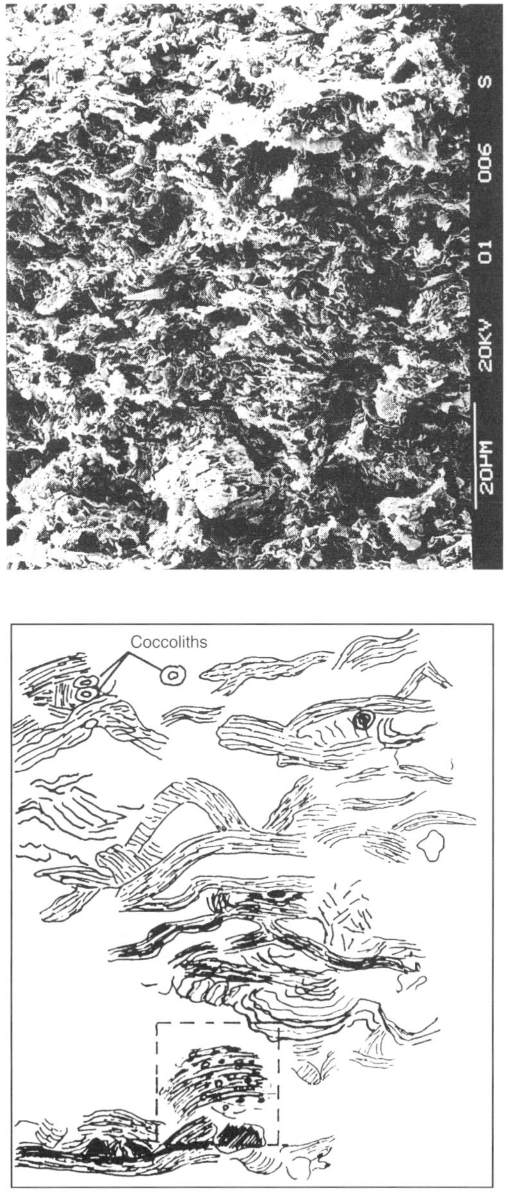

Fig. 5. Undisturbed Pappadai Clay: fabric on a vertical fracture. Details of parallel and intersecting domains; calcite crystals occur within the dashed line

again typical, relates to a 'bridge' between two domains. As before, the basic mica components appear, together with a considerable proportion of calcium.

These results show that calcium is widespread throughout the fabric of the clay, the calcium content being higher at interparticle contacts. As previously noted, calcium is not generally present in crystalline form, so it is presumably amorphous, and since it is ubiquitous it doubtless coats and binds the particles.

The deposition of the carbonate coating is unlikely to have occurred during deposition since precipitation of calcite from solution produces crystalline, not amorphous, calcite. Moreover, the model of the geological history of the clay (see Fig. 3(a)) would not explain the present water content profile unless it were controlled by normal

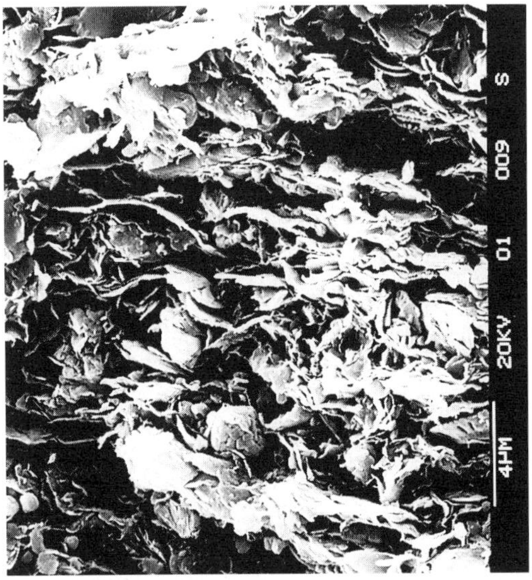

Fig. 6. Undisturbed Pappadai Clay: fabric on a vertical fracture. Stacks and random flocculated fabric

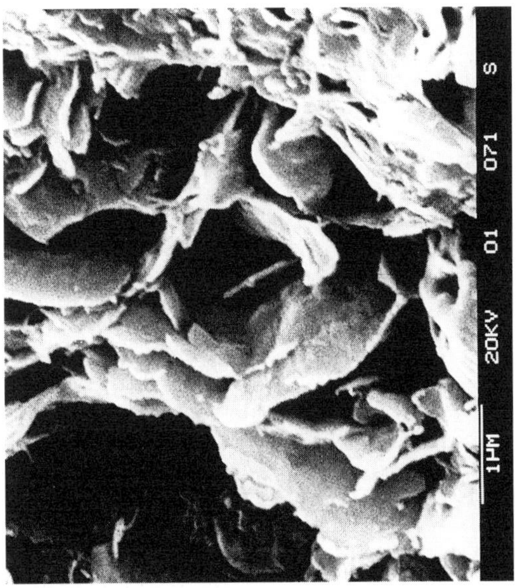

Fig. 7. Undisturbed Pappadai Clay: fabric on a vertical fracture. Edge to edge and edge to face particle contacts

consolidation, so the diagenetic calcite film was probably deposited after normal consolidation as a result of precipitation from supersaturated pore fluid. Similar formation of authigenic calcite has been observed elsewhere in soils rich in carbonate fossils (e.g. see Sfondrini, 1975; O'Brien & Slatt, 1990).

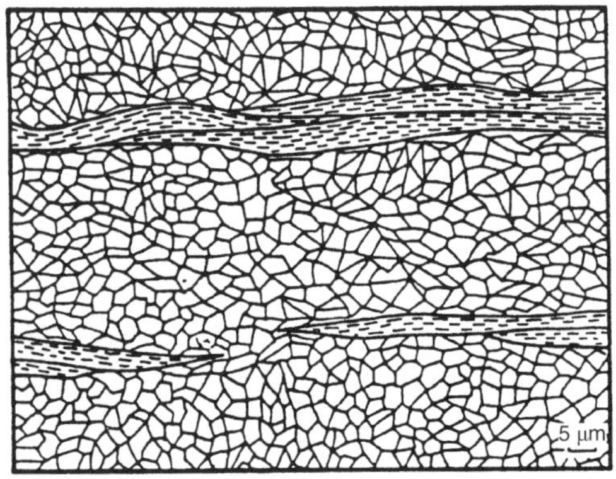

Fig. 8. Scheme of flocculated fabric compressed one-dimensionally. (After Sfondrini, 1975)

THE CURRENT STATE OF PAPPADAI CLAY *IN SITU* AND AFTER SAMPLING

A series of self-boring pressuremeter tests was carried out at depths to 45 m in a boring 80 m from the shaft where the block samples were obtained (borehole A, Fig. 1). The pressuremeter was of the type designed for soft rock, with a capacity of 20 MPa (Clarke & Allan, 1989). Since the ground surface level and depths correspond in boring and shaft, it is assumed that the *in situ* stresses for the block samples are those for the pressuremeter tests at the corresponding depth.

The horizontal effective stresses were determined noting the piezometric measurements made in borehole A. Zero pore pressures were taken to apply down to a depth of 18·4 m, below which either hydrostatic pressures with respect to this depth (assumption 1), or pressures lower than hydrostatic to allow for downward seepage into the underlying bedrock (assumption 2; Fig. 1), were assumed. Fig. 12 shows that this range of values of pore pressure does not have much effect on the value of K_0, which at the sample depth (25·4 m) is about 1·6, corresponding to a mean effective stress of about 575 kPa. This compares quite well with the suctions in the block samples, which were measured using various techniques to be about 650 kPa.

Effect of diagenesis on the values of K_0

The K_0 – depth profiles shown in Fig. 12 for the four geological stages were calculated using

$$K_0 = (1 - \sin\phi') \tag{1}$$

for stage (i),

$$K_0 = (K_0)_{nc}\mathrm{OCR}^{\sin\phi'} \tag{2}$$

for stages (ii) and (iv), and

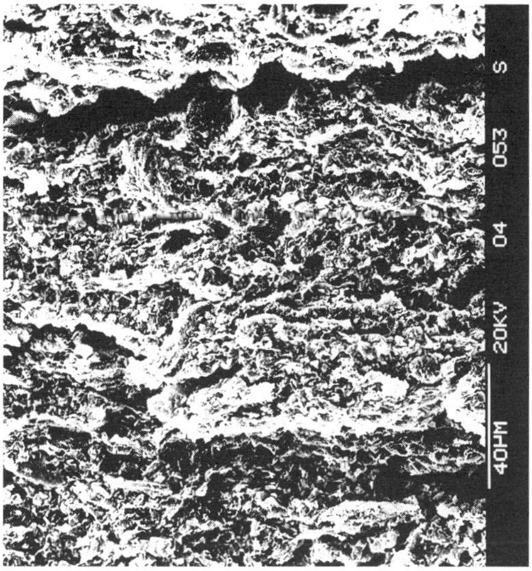

Fig. 10. Reconstituted Pappadai Clay: fabric on a vertical fracture. Detail of honeycomb fabric

Fig. 9. Reconstituted Pappadai Clay: fabric on a vertical fracture

$$K_0 = (K_0)_{nc}[\mathrm{OCR}/\mathrm{OCR}_{max} + \tfrac{3}{4}(1 - \mathrm{OCR}/\mathrm{OCR}_{max})] \quad (3)$$

for stage (iii) (Mayne & Kulhawy, 1982). The value of ϕ' used is 23°, which, as is seen later, is consistent with the value at critical state. The measured values of K_0 are higher than calculated using equations (1)–(3), perhaps not surprising since these empirical expressions for K_0 were derived from data for soft and reconstituted clays and thus do not apply to clays such as the Pappadai Clay that have undergone diagenesis.

To illustrate the differences between the observed stress state and the Mayne & Kulhawy predictions, the relationships between q and p' for the sampling depth have been calculated for geological stages (i)–(iv) (Figs 3(a) and 12), and are plotted in Fig. 13, where they are compared with the measured *in situ* stress state at about the depth of sampling. It is seen that the natural clay is subject to a higher mean stress and lower deviatoric stress than would be expected if overconsolidation resulted from simple unloading, as implicitly assumed by equations (1)–(3).

Although the Cam Clay model (Schofield & Wroth, 1968) simply assumes non-linear elastic behaviour in unloading and does not model the hysteresis of the unload–reload loops, it is none the less of value in helping to identify the various factors that contribute to the observed values of K_0. Using the Cam Clay model, the increase in K_0 on unloading depends on the value of Poisson's ratio, which in turn can be shown to vary with the overconsolidation ratio and shear modulus (Cotecchia, 1996). Thus the increase in K_0 with the overconsolidation ratio is greater with higher values of shear modulus. Thus if, after normal consolidation, the natural clay became stiffer as a result of diagenesis, the observed higher values of K_0 would be expected. This seems to be the case

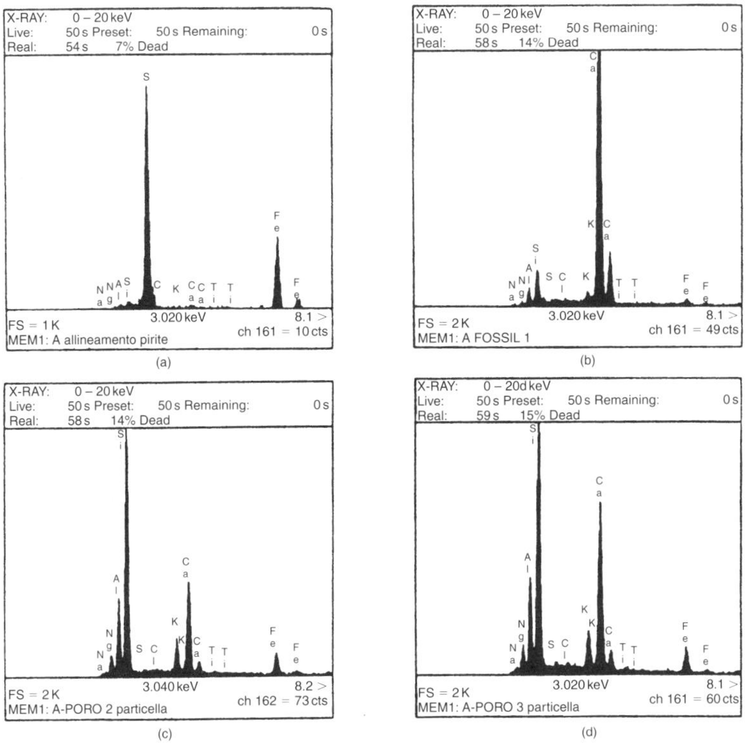

Fig. 11. Diffractograms resulting from the micro-chemical analysis

for Pappadai Clay, whose bonding was strengthened by diagenesis prior to unloading.

OEDOMETER TESTING

Oedometer tests with incremental loading (test series OED) were carried out, to a maximum stress of about 7 MPa in a conventional oedometer, and also with constant rate of strain compression (test series CRS) in a high-pressure oedometer to a vertical stress about 25 MPa, with drainage to the upper boundary of the sample and pore pressures measured at the base. In the following figures, both the OED and CRS tests on the reconstituted clay are distinguished from those on the natural clay by the suffix 'R'. The constant rate of strain tests were interpreted using a modified version of the theory developed by Wissa *et al.* (1971). In all tests, corrections were applied for machine compliance. More details of the apparatus and testing will be published elsewhere, but some of the data which highlight the effect of the structure of the clay are discussed briefly.

Representative compression and swelling curves for natural and reconstituted clay are shown in Fig. 14. For clarity, some compression stages prior to swelling are omitted; only the swelling stage prior to compression is shown in the case of OED7. In the case of the reconstituted clay the average yield stress is about 200 kPa, consistent with the maximum stress attained in the consolidometer. Beyond yield, the CRS reconstituted sample defines the ICL for the Pappadai Clay to 22 MPa.

The initial swelling curve of OED7 shows the very stiff response of the natural clay. On compression, the clay is similarly stiff, with 'gross yield', the threshold of major plastic strains (subsequently referred to as 'yield') occurring beyond the ICL line. The yield stress (σ_y') of the natural clay from the OED tests is about 2600 kPa, much higher than the equivalent vertical stress ($\sigma_e^* = 750$ kPa) required to bring the reconstituted clay to the same void ratio. The stress ratio $\sigma_y'/\sigma_e^* = 3{\cdot}5$ is a measure of the higher strength of the structure of the natural clay compared to the reconstituted clay, and is presumably related to the sensitivity of the clay. Since the yield stress of the natural clay exceeds the geological preconsolidation stress, σ_p', it lies to the right of the SCC. Thus the ratio σ_y'/σ_e^*, already greater than unity during normal consolidation (see Fig. 3(a)), has increased further to 3·5 as a result of diagenetic bonding. Conse-

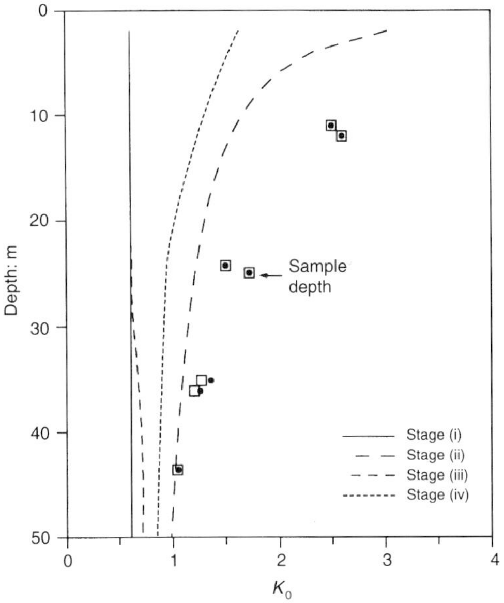

Fig. 12. Values of K_0 measured *in situ* (squares and circles) compared with the variation of K_0 with depth obtained by modelling stages (i)–(iv) of the geological history of the clay. The squares correspond to piezomatic assumption 1 (see text); the circles to assumption 2

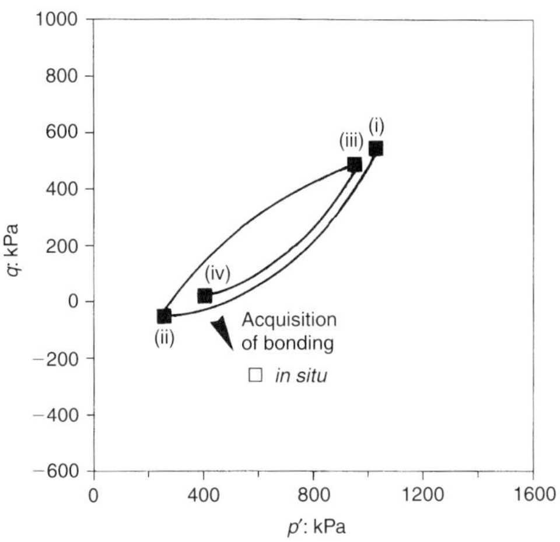

Fig. 13. Modelled q–p' states at the sampling depth at end of each geological stage (solid squares), compared with the measured *in situ* stress state of the clay (open squares)

quently the yield stress ratio, YSR, σ_y'/σ_v' ($= 6{\cdot}0$), of the natural clay is greater than the geological overconsolidation ratio, which is 3·1.

The swell sensitivity (defined as the ratio of the swelling index of the reconstituted clay to that of the natural clay, C_s^*/C_s; Schmertmann, 1969; Burland, 1990) is about 2·5 preyield. This ratio reduces rapidly beyond yield, becoming unity at vertical stresses about 5000 kPa. As bonding will restrain swelling, this reduction in swell sensitivity illustrates the loss of some bonding with straining beyond yield, which is not observed with the stable, insensitive reconstituted clay. None the less, the convergence of the compression curves of the two clays shown in Fig. 14 is slow, the natural clay still carrying higher stresses than the reconstituted clay even at a void ratio as low as 0·3. Thus the natural clay retains a structure stronger than the reconstituted substantially beyond yield, and perhaps at all stages of loading.

Despite their different structures, the two clays follow similar patterns of behaviour, with, in qualitative terms, similar trends of variation with stress level of such properties as compression index, stiffness and permeability, and with a similar magnitude of strain-rate dependency (Cotecchia, 1996).

As an example, one aspect of the similarity of the clays that is of particular relevance to pre-failure behaviour, the variation with YSR of stiffness in K_0 compression, $M(= 1/m_v)$, will be discussed. Fig. 15 shows the typical variation of the values of M with vertical stress observed in the tests on the natural and the reconstituted clay. The M_{nc}–σ_v' relationship for the natural clay is similar to that observed for the reconstituted clay, but with different values of the coefficient and exponent. Similar linear relationships to those shown in Fig. 15 are obtained for all tests on both natural and reconstituted clay specimens.

Since the mechanical behaviour of reconstituted clays can generally be normalized with respect to volume, as with the critical state model (Schofield & Wroth, 1968), it is to be anticipated that the stiffness, M, should be capable of similar normalization. Such a result may be obtained by normalizing the current value of M along each oedometer loading curve by the value of the stiffness at yield, M_{nc} (Fig. 16). This is equivalent to normalization by volume, since, ignoring strain-rate effects, for a given value of σ_v' the void ratio is related to a single yield stress (Fig. 16), which in turn is related uniquely to M_{nc}.

The normalized stiffnesses of the reconstituted clay are shown in Fig. 17 for different loading paths. In tests CRSR2 and CRSR4 the clay was continuously compressed, whereas in test OEDRb the clay was first compressed beyond yield (OEDRb(1)) and then unloaded and reloaded to successively higher consolidation pressures (OEDRb(2) and OEDRb(3)). (Only the reloading stiffness is plotted for OEDRb(2) and OEDRb(3).)

If a similar normalization is applied to the stiffness of the natural clay both during first loading and in reloading beyond yield, the normalized stiffness of the clay postyield is less than that

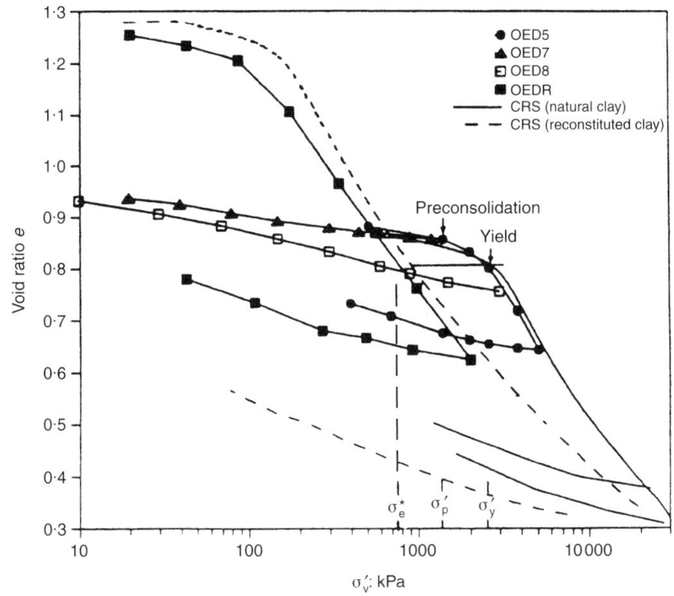

Fig. 14. Compression behaviour of the natural and reconstituted clay

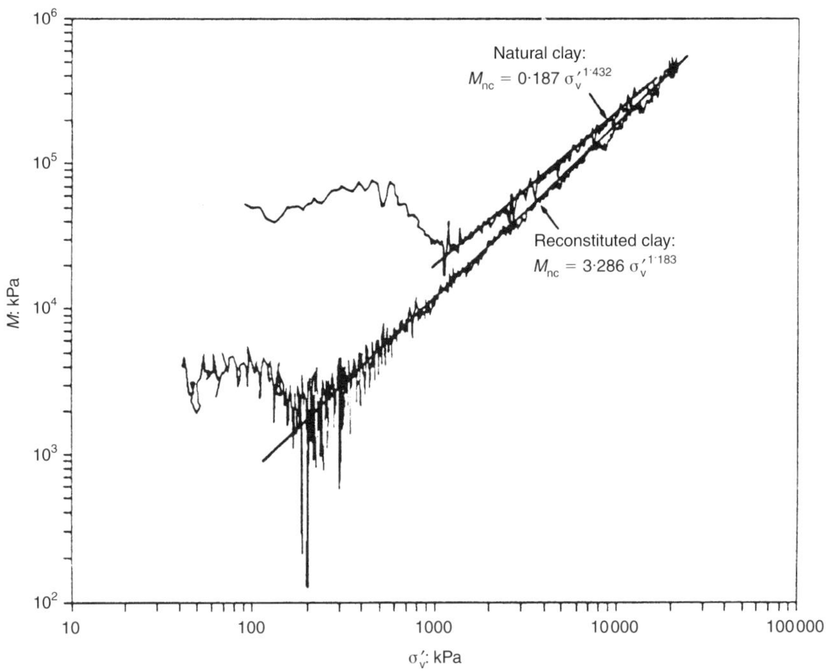

Fig. 15. M–σ_v' relationships for the natural and the reconstituted clay

exhibited by the clay during first loading, the consequence of the degradation of structure post-yield. However, this structural effect may be eliminated by further normalizing M/M_{nc} by the stress ratio σ_y'/σ_e^*. This is shown in Fig. 18, where the data points OED10(2) and OED10(3) refer to reloading beyond yield. A near unique relation with the YSR is obtained, similar to that obtained

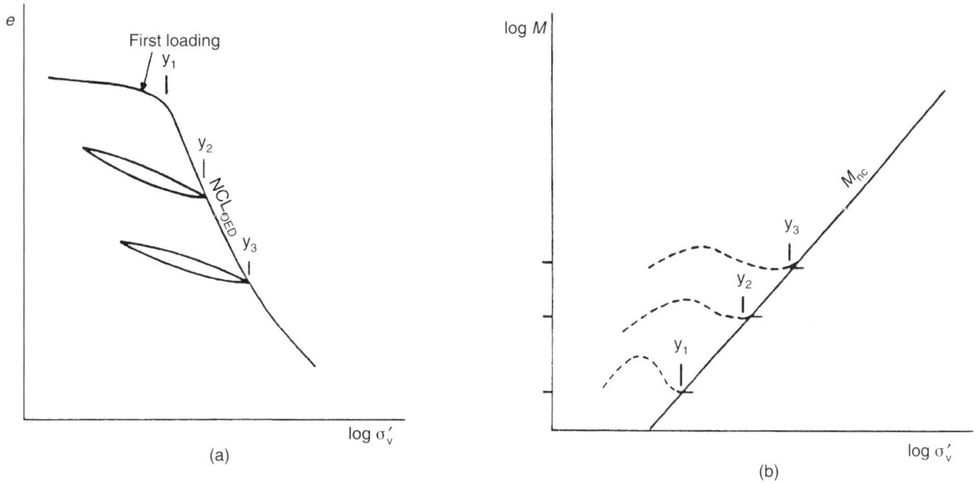

Fig. 16. Idealized (a) e–σ_v' and (b) M–σ_v' behaviour

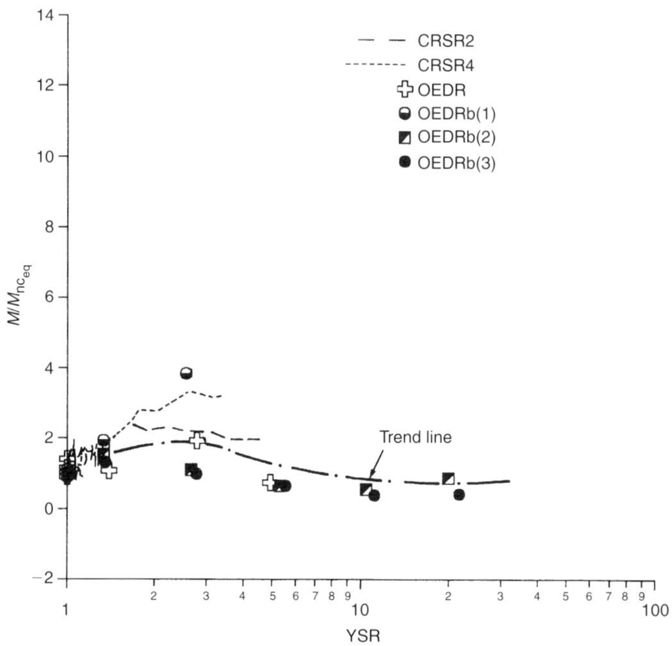

Fig. 17. Stiffness of the reconstituted clay in K_0 compression normalized with respect to volume

for the reconstituted clay, strongly suggesting that σ_y'/σ_e^* is an important parameter for the quantification of the effects of soil structure.

TRIAXIAL TESTING

Triaxial tests were carried out to investigate the elastoplastic constitutive laws of the natural and reconstituted clay. Drained and undrained tests were carried out at values of p' well below the *in situ* value ($\approx$ 600 kPa), as well as up to 3800 kPa. Strains in the small to intermediate range were measured both with local strain transducers and externally, the latter being corrected for machine compliance.

The discussion here is limited to pre-failure behaviour. However, the development of the flow rule requires knowledge of the ultimate state,

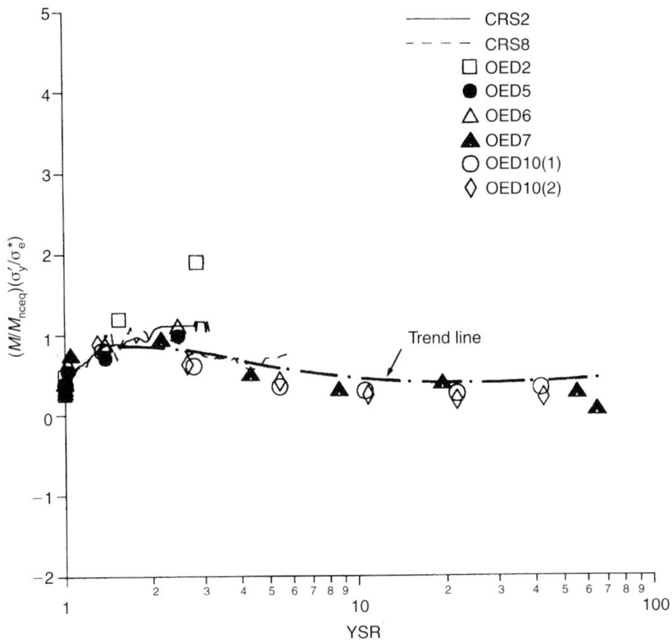

Fig. 18. Stiffness of the natural clay in K_0 compression normalized with respect to both volume and structural changes

which is discussed briefly. In this context, it is worth recording that with tests on the natural clay, shear surfaces formed during strain softening, and the stresses at that stage were computed using the techniques suggested by Chandler (1966, 1968).

Triaxial compression and shearing data

The isotropic compression data for the reconstituted clay in the $\nu - p'$ plane are plotted in Fig. 19, where it is seen that the gradient of the isotropic normal compression line (INCL), λ^*, is 0·204. In test TR5, K_0 conditions applied in the stress range 200–260 kPa, yielding a value of K_0 of 0·57. If this value of K_0 is applied to the oedometer test CRSR1 the two sets of data form a continuous linear relation with the common slope λ^*, confirming the numerical value of K_0 and the applicability of the critical state model to the reconstituted Pappadai Clay. The value of λ^* is consistent with the value of the compression index, $C_c{}^*$, measured in the oedometer tests.

In the normal consolidation range of behaviour the natural clay yielded values of λ of 0·254 in both isotropic compression and K_0 compression, as shown in Fig. 20, where, for clarity, only some results are plotted. Thus, despite the very different structures of the two clays they exhibit very similar behaviour in $\nu - p'$ space, differing only in magnitude, with the isotropic yield stress for the natural clay being substantially greater than the equivalent stress on the INCL for the reconstituted clay. Similarly, the value of λ for the natural clay exceeds the value for the reconstituted clay.

Differences as well as similarities in behaviour are also observed. In Fig. 20, the preyield isotropic loading–unloading data plot close to a straight line with gradient $\kappa = 0{\cdot}02$. The value of κ increases substantially postyield, reaching the reconstituted value κ^* at the highest stresses. The soil behaviour during both isotropic and K_0 unloading–reloading can be explained satisfactorily with a non-linear isotropic elastic model in which κ depends on the mean stress reached in isotropic normal consolidation.

In K_0 compression (Fig. 21), the value of K_0 falls preyield, then rises postyield to the constant value of 0·63. This presumably reflects the degradation of some elements of the natural structure; similar behaviour has been noted with other weakly cemented soils (Chandler, 1967; Lerouil & Vaughan, 1990; Coop & Atkinson, 1993).

The stress paths of the tests on both reconstituted and natural clay are plotted in normalized form in Figs 22(a) and 22(b), respectively. The data are normalized by the equivalent pressure on the INCL, $p_e{}^*$. In Fig. 23 the data are plotted in the $\nu - \ln p'$ plane, and it is seen that both the isotropic and K_0 normal consolidated samples of reconstituted clay reach the critical state on a straight line of gradient λ^*, yielding the stress ratio $M^* = 0{\cdot}91$. The sample on the dry side of the

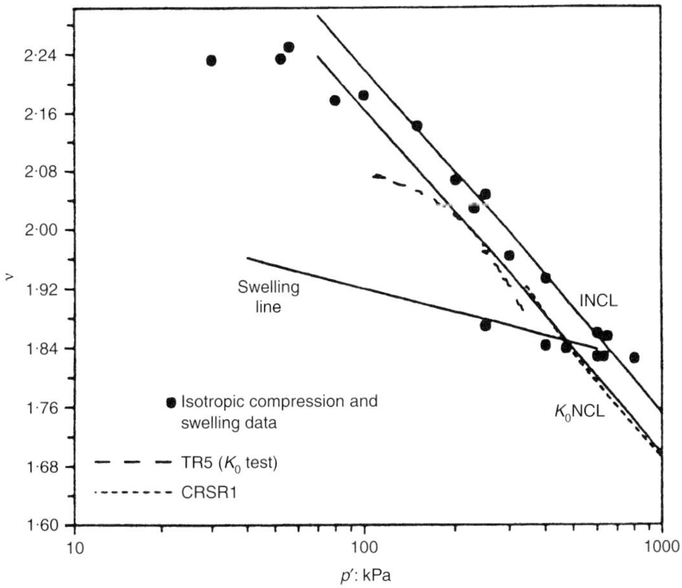

Fig. 19. Reconstituted clay: isotropic and K_0 compression and swelling

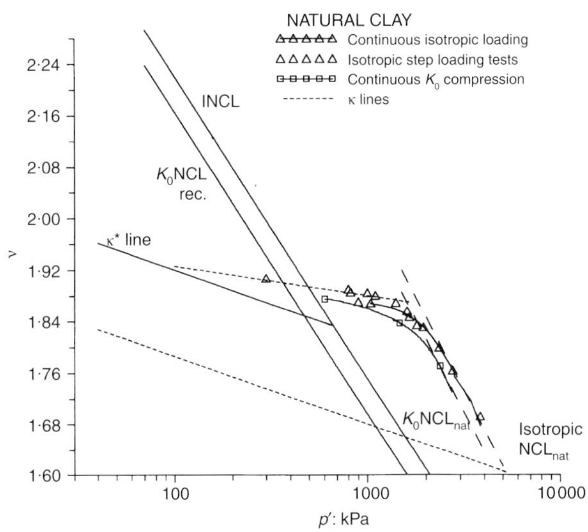

Fig. 20. Natural and reconstituted clay: isotropic and K_0 compression and swelling

critical state line (TR2) exhibited strain localization after reaching the peak value of q/p'; allowing for this (see above), the stress path is found ultimately to reach critical state. In the case of sample TR3, which had an overconsolidation ratio of 1·33, localization occurred close to the critical state.

Taking the stress ratio q/p', rather than q, as representing soil strength, yield during shearing is taken as the point at which there is a significant reduction in the gradient of the $q/p'-\varepsilon_s$ curve. The yield states thus defined are shown for the natural clay in Fig. 22(b). Beyond yield, the natural clay at $p'/p_e^* < 2$ (YSR $>$ 2) strain softens in terms of q/p' while moving up the current state

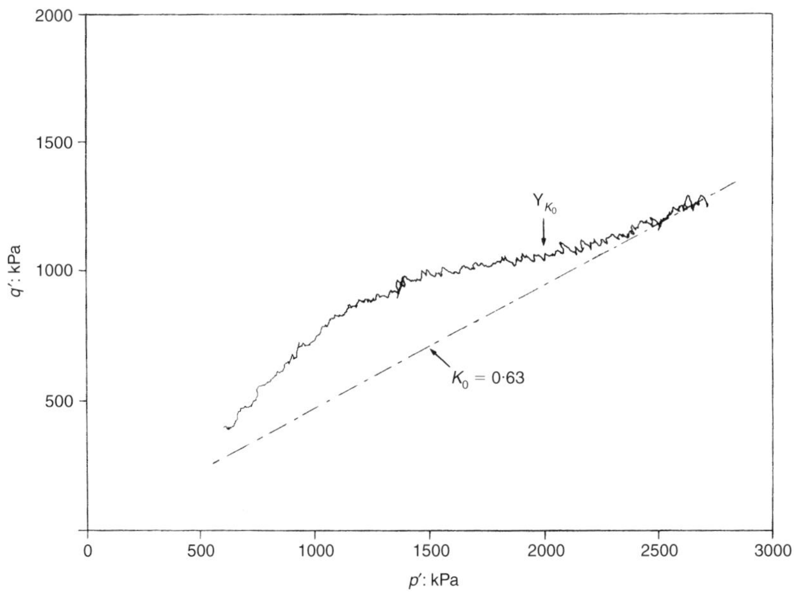

Fig. 21. K_0 compression of the undisturbed clay

boundary surface as far as peak deviatoric stress with dilation occurring, similar to reconstituted clays. At YSR < 2 the behaviour of the natural clay beyond yield differs from that of the reconstituted clay, as the clay strain softens while still contracting.

With appropriate corrections for the changing cross-sectional area when a failure surface occurred, all samples reached a plastic strain increment ratio $d\varepsilon_v^p/d\varepsilon_s^p$ of zero (calculated as defined later) at a common ultimate stress ratio $M = 0{\cdot}85$, that is, the critical state was reached at a stress ratio only slightly lower than M^*. This point is discussed in detail by Cotecchia (1996). As shown in Fig. 23, in the v–$\ln p'$ plane the critical state line of the natural clay is parallel to the isotropic and K_0 compression lines, as with the reconstituted clay, and lies to the right of the INCL.

It appears that structural differences between the natural and reconstituted states remain, even at the critical state. This is not surprising since structural differences have been shown to continue to exist even after compression to the highest stresses (see Fig. 14), and the structures of the natural clay and the reconstituted clay differ at all stages of loading.

Triaxial testing: clay stiffness at small to intermediate strains

In Fig. 24 the undrained triaxial compression stiffnesses of the natural and reconstituted clay samples are compared. The stiffness measured with the self-boring pressuremeter test (over the initial 0·2% of the reloading cycles) at the sampling depth is also shown, and further compared with that of a triaxial sample trimmed horizontally. Various authors (e.g. see Viggiani & Atkinson, 1995) have proposed an expression for the small strain shear modulus, G:

$$G/p_r = A(p'/p_r)^n(R)^m \tag{4}$$

where p_r is a normalizing pressure taken as unity, R is the isotropic overconsolidation ratio, and the coefficients A and n vary with strain level. So as to normalize the stiffness for confining stress p', the values of n (Fig. 24) have been derived from the relationship between n and the plasticity index (Viggiani & Atkinson, 1995). The normalized tangent stiffnesses, $G/(p')^n$, are plotted against either R, or the yield stress ratio in isotropic compression, YSR_{is}. The stiffness measured in the pressuremeter test is normalized by p' *in situ*.

The conclusion drawn from Fig. 24 is that there is little difference between the normalized small to intermediate strain behaviour of the natural and reconstituted clay, in contrast to the large strain behaviour, where, as will be seen, the differences are substantial. This observation supports similar findings by Rampello *et al.* (1994) and by Coop *et al.* (1995) for natural clay samples that were likely to have had a higher degree of sampling disturbance than the Pappadai Clay.

The log–log relation between normalized stiff-

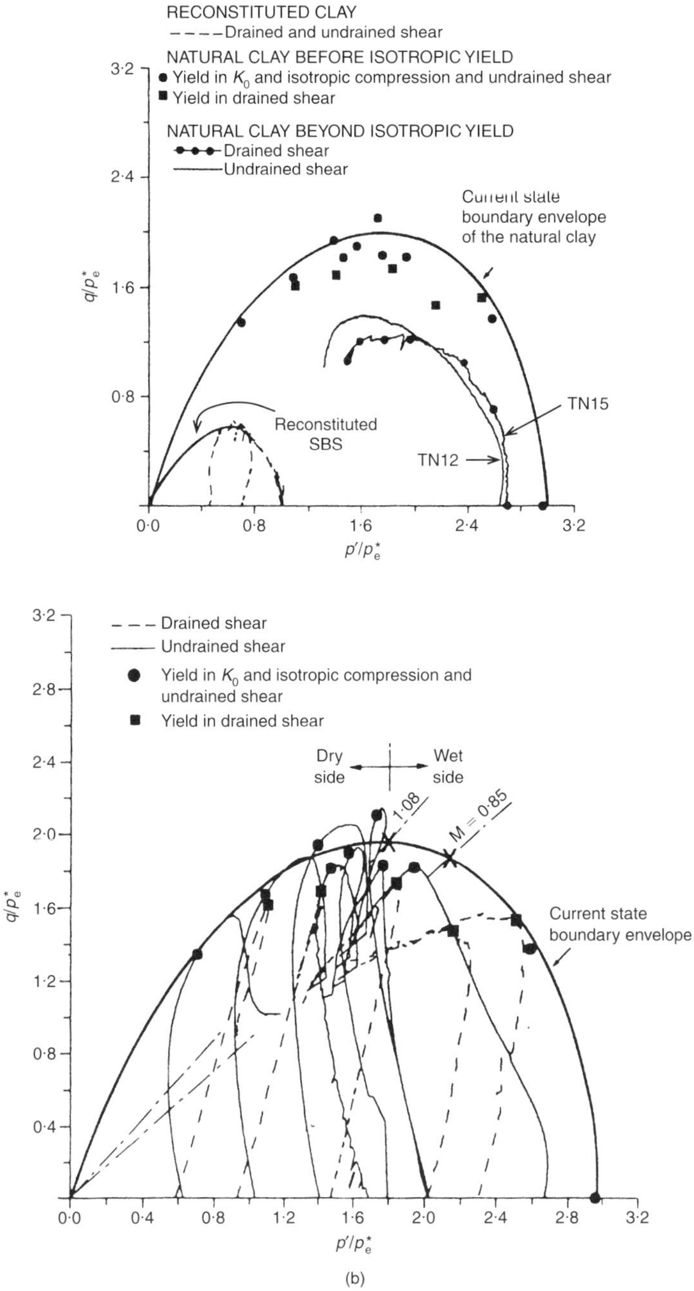

Fig. 22. Normalized stress paths and yield states: (a) current yield states of the natural clay, stress paths of the natural clay compressed beyond isotropic yield, and stress paths of the reconstituted clay; (b) stress paths of the natural clay specimens consolidated to states pre-isotropic yield

ness and clay state in Fig. 24 does not appear to be linear as implied by equation (4), perhaps because the Pappadai data relate to a greater stress range than used in previous studies.

The horizontal sample proved to be much stiffer than the vertical samples, but the result agrees well with the horizontal *in situ* stiffness measured with the self-boring pressuremeter.

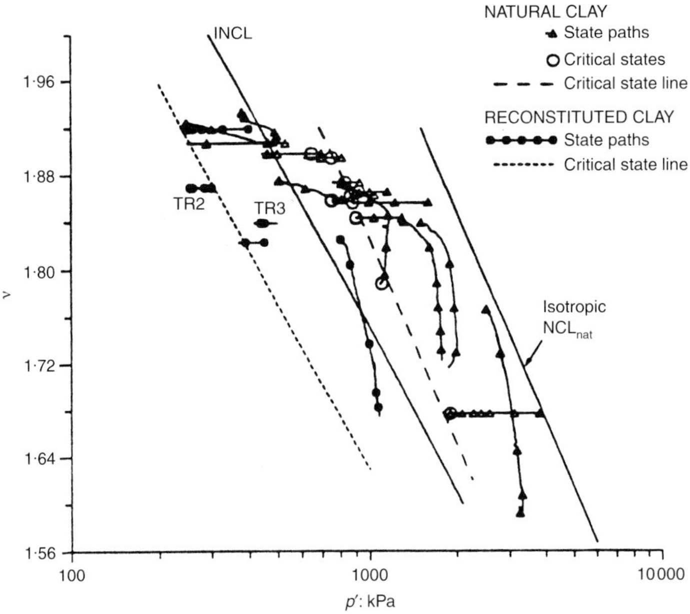

Fig. 23. v–p' paths and critical states of both natural and reconstituted clay specimens

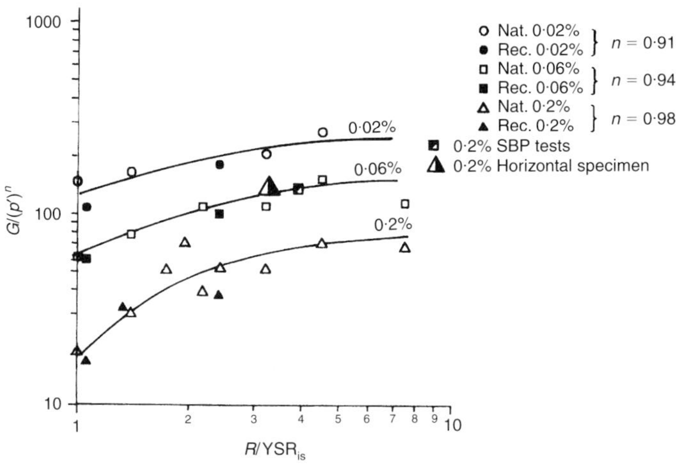

Fig. 24. Normalized shear stiffnesses of both natural (Nat.) and reconstituted (Rec.) clay

Flow rules

At shear strains greater than about 0·02% both the natural and reconstituted clay exhibit irreversible strains, and are thus behaving elastoplastically, differing from the non-linear elastic behaviour observed with both isotropic and K_0 unloading. The plastic strain increment ratios $d\varepsilon_v^p/d\varepsilon_s^p$ are plotted against stress ratio in Fig. 25 for the reconstituted clay, and in Fig. 26 for the natural clay. The plastic volumetric strains were obtained using the elastic constant κ previously defined, while the elastic shear strains were considered to be negligible since shear moduli as high as 100 MPa were measured for the natural clay and as high as 40 MPa for the reconstituted clay. The data for the reconstituted clay all fit closely the linear relation

$$q/p' = Q - A(d\varepsilon_v^p/d\varepsilon_s^p) \quad (5)$$

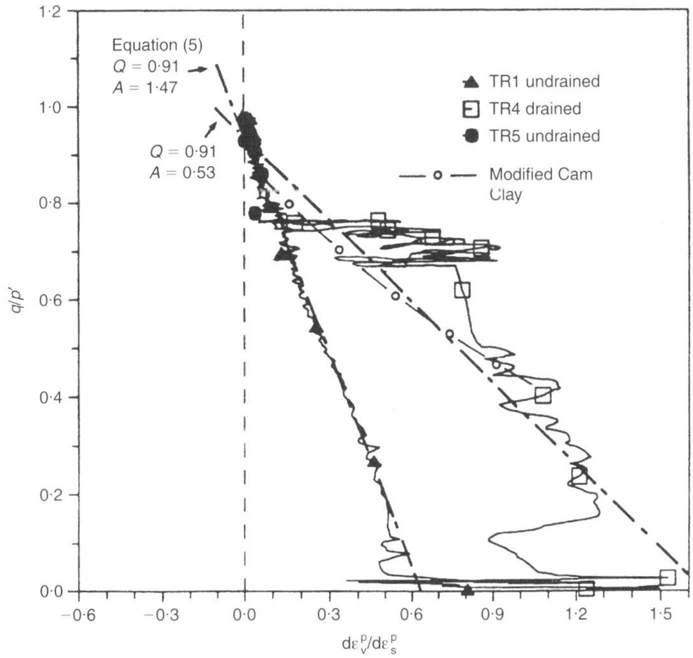

Fig. 25. Reconstituted clay: flow rules during shear ($q/p' = 1/3$) for isotropically (TR1, TR4) and K_0 (TR5) normal consolidated samples

where the intercept Q in this case is numerically equal to M^*, the stress ratio at the critical state (= 0·91), and A is 1·47 in undrained shear and 0·53 in drained shear. This flow rule describes the behaviour of the clay better than either the Cam Clay or modified Cam Clay (Roscoe & Burland, 1968) predictions (see Fig. 25). The similarity with Cam Clay-type behaviour is evident; since the Cam Clay flow rule interprets the behaviour as purely frictional, it follows that the present linear flow rule is also 'basically frictional' in nature, although there are apparently other, non-frictional elements of behaviour present since the coefficient $A \neq 1$. As with the Cam Clay, the intercept Q is a constant representing the frictional properties of the soil (Schofield & Wroth, 1968).

For the natural clay, a similar linear relationship applies from small strains to beyond yield (Fig. 26). Again, this linear behaviour may, for the same reasons as previously, be described as basically frictional. In all tests, Q, which represents the friction parameter for the natural material, here defined as M_{nat}, has a value of 1·08, higher than the stress ratio M at the critical state (0·85). The coefficient A is greater than unity for drained shearing (Fig. 26(a)) and less than unity for undrained shearing (Fig. 26(b)), as with the reconstituted clay. The non-frictional elements of behaviour that result in A differing from unity are thought to be due to structural degradation, the structural changes differing with different stress paths. All the drained stress paths were the same, as were all the undrained paths, so that A differs for the two stress paths that were used, but does not differ with the initial volume–stress state, which varied from test to test.

Thus, it appears that the principal difference between the pre-failure flow rules of the natural and reconstituted clay is the value of the parameter Q, which seems to be influenced by structure, and which imparts to the natural Pappadai Clay the behaviour of a sensitive clay. Referring to Fig. 22(b), yield coincides with the point of maximum q/p' in all cases, so that the natural clay samples which lie on the wet side of the critical state line harden to a yield stress ratio lower than Q, while on the dry side samples harden to a higher stress ratio than Q. Beyond yield, structural changes result in a deviation from the linear flow rule, leading to purely frictional behaviour at the critical state.

Yield surfaces and hardening law

The data in Fig. 22(b) clearly identify the 'current' state boundary surface of the natural clay, that is, the boundary states of the clay that was not compressed to yield before shearing. Indeed, the current yield points lie sufficiently close to the current state boundary surface so that, as with the

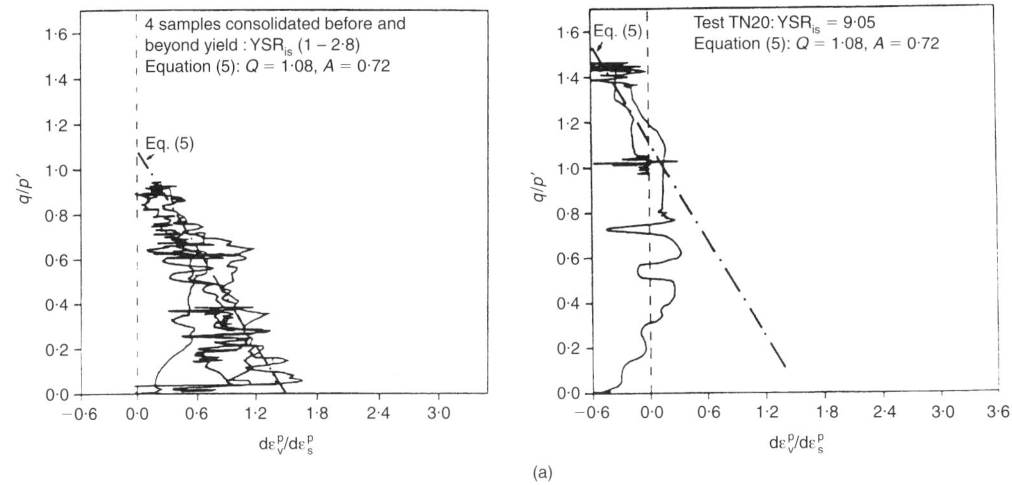

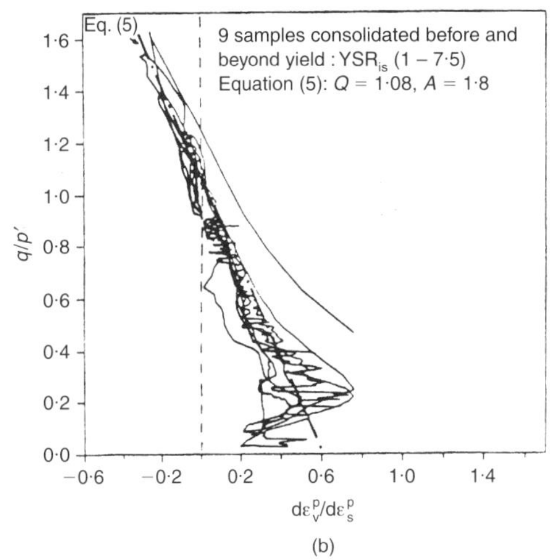

Fig. 26. Natural clay: flow rules in (a) drained shear and (b) undrained shear. Note that in drained shear those samples which follow a vertical path have YSR$_{is}$ > 2

reconstituted clay (Fig. 22(a)), it is difficult to distinguish the yield and the boundary states. It will be noted that the drained stress paths do not quite reach the current boundary surface; this is probably in part a strain-rate effect due to the much slower strain rates used for the drained tests.

The current state boundary surfaces of both clays have a similar 'dome' shape (rather than the modified Cam Clay ellipse) to that observed for other one-dimensionally consolidated clays (Graham *et al.*, 1988), which is assumed to reflect the anisotropic structure of both clays which has not been erased by the subsequent application of isotropic compression.

Since the stress paths of the two natural clay samples consolidated beyond yield (TN12 and TN15) both plot within the current boundary surface (Fig. 22(a)), they demonstrate that, unlike the reconstituted clay, the state boundary surface of the natural clay is not satisfactorily normalized by p_e^*. The reason for this is that p_e^* normalizes with respect to the volume change of the reconstituted clay (controlled by λ^*), whereas the natural clay changes volume with respect to λ, which is numerically greater than λ^*. It is thus necessary to find a normalizing parameter that reflects the volumetric changes of the natural clay, and in turn reflects the changes of the soil structure.

Such a parameter is provided by a stress at the intersections between the postyield isotropic compression curve of the natural clay and the swelling line which corresponds to the current volume of the clay, as for example p_y':

$$p'_y = \exp\left[(N - v - \kappa \ln p')/(\lambda - \kappa)\right] \qquad (6)$$

where N is the specific volume on the isotropic

compression line for a value of p' of unity. Fig. 27 shows the results of this normalization, in which the yield states fit the current yield surface of the clay. The samples sheared post-isotropic yield still lie below the current yield curve, so that a unique yield curve is not obtained by normalizing solely with respect to volume. It therefore appears that the yielding behaviour of the natural clay is affected by the changes of structure which occur with changing volume.

It was shown earlier that the stiffness in K_0 compression, M, follows a unique relationship with YSR when normalized with respect to volume using the stiffness parameter M_{nc}, and with respect to the structural changes occurring in the clay using the stress ratio σ_y'/σ_e^*. Similarly, the triaxial stress paths for the natural clay have been normalized by p_y' and by the ratio p_y'/p_e^*, p_e^* being the mean stress on the INCL corresponding to the volume at isotropic yield of the natural clay, and are shown in Fig. 28. This further normalization appears to result in a unique representation of the yield curve of the natural clay.

These normalizations demonstrate that the yield behaviour of the natural clay is isotropically hardening, not only with respect to volume but also with a factor depending on the structural changes occurring in the clay. These structural changes may be quantified using the parameters σ_y'/σ_e^* or p_y'/p_e^*. However, the behaviour of the clay when sheared beyond yield does not appear to be isotropically hardening with respect to the value of σ_y'/σ_e^* used preyield, since beyond yield the state paths fall below the state boundary surface to reach the critical state (Figs 27 and 28). However, the value of σ_y'/σ_e^* probably changes after yield in shear, which might keep the stress paths close to a simple normalized boundary surface, but this cannot be demonstrated.

CONCLUSIONS

The modelling of the geological history of the strongly bonded Pappadai Clay has established the difference between the actual behaviour of the clay *in situ* during normal consolidation and unloading and its behaviour had it been a reconstituted clay. The sensitivity of the clay during normal consolidation is found to have been lower than in the present state, the result of diagenetic changes *in situ*. These caused strengthening of the clay bonding, the result of a carbonate film at the interparticle contacts. This has increased both the clay stiffness and the yield stress ratio, which have in turn increased the value of K_0 *in situ*.

Despite major differences in structure, the natur-

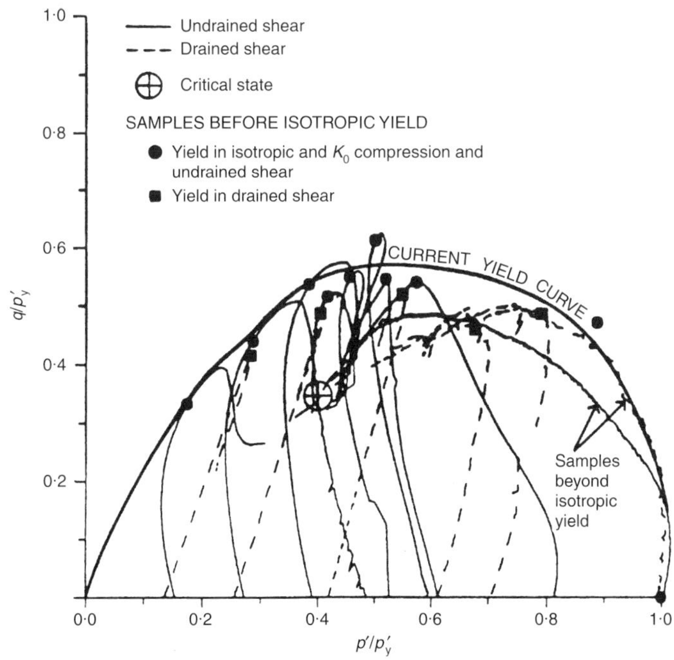

Fig. 27. Natural clay stress paths, consolidated to both pre- and postyield, normalized with respect to volume by the yield pressure p_y'

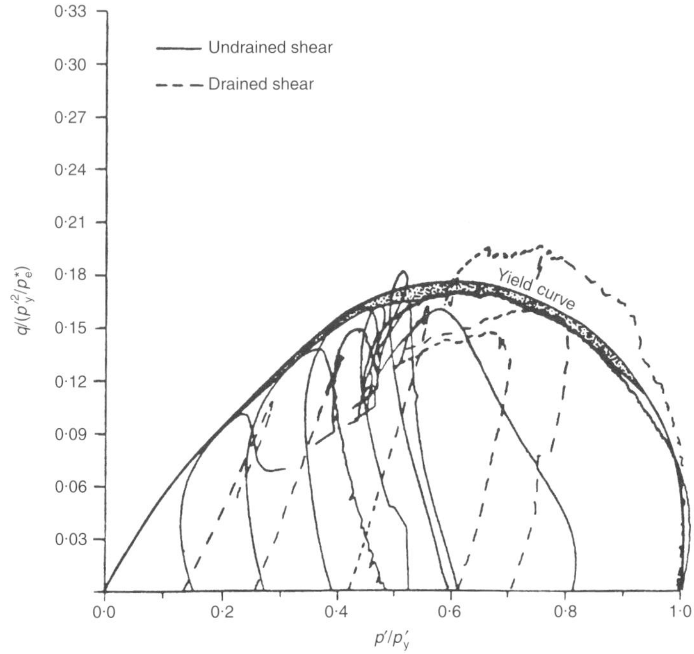

Fig. 28. Natural clay stress paths, consolidated to both pre- and postyield, normalized with respect to volume by the yield pressure p_y' and with respect to structure by the stress ratio p_y'/p_e^*

al and reconstituted clay are shown to follow similar patterns of behaviour. Both have parallel isotropic, K_0 compression and critical state lines in the v–ln p' plane. Their small strain stiffness during shear are similar when normalized, and follow a unique relationship with YSR. At larger strains up to and beyond gross yield, both the clays follow a similar, basically frictional flow rule, although their friction parameters are structure-dependent. The gross yielding behaviour of the two clays is isotropically hardening, although the natural clay not only strain hardens with respect to volume but its hardening law is also influenced by the structural changes that develop with strain. Straining degrades some unstable structural elements of the clay, which thereafter no longer contribute to the strength or the sensitivity of the clay.

The change in sensitivity of the clay, which can be quantified using the stress ratio σ_y'/σ_e^*, therefore quantifies the change in structure. Pre-failure the clay behaviour can be normalized with respect to both volume and the ratio σ_y'/σ_e^*, so that the soil behaviour is shown to be isotropically hardening with respect to volume and σ_y'/σ_e^*. Post-failure these two parameters do not successfully normalize the behaviour of the natural clay.

The comparison between the natural and reconstituted behaviour of the Pappadai Clay has therefore demonstrated that the two different structures give rise to many similarities in behaviour, the differences when they occur being almost all related to sensitivity. Since the two clays have very different structures, the reconstituted clay having a simple structure and the natural clay being far more complex, it is likely that the conclusions drawn from this study may also prove to be applicable to other clays. Moreover, the similarities between the pre-failure mechanics of the two materials, from small strains up to failure, indicate that constitutive models developed for reconstituted clays should not require much modification to enable them to be applied to natural clays.

ACKNOWLEDGEMENTS

We thank Professor F. Veniale (University of Pavia) for the mineralogical analyses and for help with the interpretation of the scanning electron micrographs.

NOTATION

A coefficient of the basically frictional flow rule; coefficient of the G_0-R relationship

C_c compression index
CRS high-pressure constant rate of strain oedometer test
C_s swelling index
$C_s{}^*/C_s$ swell sensitivity
e void ratio
G shear modulus
ICL intrinsic compression line
INCL isotropic normal consolidation curve of the reconstituted clay
K_0 coefficient of earth pressure at rest, $\sigma_h{}'/\sigma_v{}'$
LI liquidity index
M q/p' at the critical state; stiffness in K_0 compression
M_{nat} intercept of the flow rule of the natural clay
M_{nc} stiffness M beyond yield
m_v coefficient of compressibility in K_0 compression
N specific volume on the isotropic normal consolidation line for $p' = 1$ kPa
nc normal consolidation
OCR overconsolidation ratio: $\sigma_p{}'/\sigma_v{}'$
OED conventional oedometer test
p' mean normal effective stress
$p_e{}^*$ (equivalent) pressure on the INCL
$p_y{}'$ mean effective stress at yield in isotropic compression
q deviatoric stress
Q intercept of the basically frictional flow rule
R overconsolidation ratio in terms of p'
SCC sedimentation compression curve
u pore water pressure
YSR yield stress ratio, $\sigma_y{}'/\sigma_v{}'$
YSR_{is} yield stress ratio in isotropic compression, $p_y{}'/p'$
ε_v^p plastic volumetric strain
ε_s^p plastic shear strain
κ gradient of the swelling line
λ gradient of the normal consolidation lines with constant q/p'
v specific volume
$\sigma_e{}^*$ equivalent vertical effective stress on the ICL
$\sigma_p{}'$ vertical (geological) preconsolidation pressure
$\sigma_v{}'$ vertical effective stress
$\sigma_y{}'$ vertical effective stress at yield
ϕ' friction angle
$\phi_{cs}{}'$ friction angle at critical state
* denotes parameters applying to the reconstituted clay

REFERENCES

Burland, J. B. (1990). On the compressibility and shear strength of natural clays. *Géotechnique* **40**, 329–378.

Chandler, R. J. (1966). The measurement of residual strength in triaxial compression. *Géotechnique* **16**, 181–186.

Chandler, R. J. (1967). Discussion. *Proc. Geotechn. Conf. Oslo* **2**, 177–178.

Chandler, R. J. (1968). A note on the measurement of strength in the undrained triaxial test. *Géotechnique* **18**, 261–266.

Clarke, B. G. & Allan, P. G. (1989). A self-boring pressuremeter for testing weak rock. *Proc. 12th ICSMFE, Rio de Janeiro* **1**, 211–214.

Coop, M. R. & Atkinson, J. H. (1993). The mechanics of cemented carbonate sands. *Géotechnique* **43**, 53–67.

Coop, M. R., Atkinson, J. H. & Taylor, R. N. (1995). Strength, yielding and stiffness of structured and unstructured soils. *Proc. 11th ECSMFE, Copenhagen* **8**, 55–62.

Cotecchia, F. (1996). *The effects of structure on the properties of an Italian Pleistocene clay*. University of London, PhD thesis.

Cotecchia, F. & Chandler, R. J. (1995). The geotechnical properties of the Pleistocene clays of the Pappadai valley, Taranto, Italy. *Q. J. Engng Geol.* **28**, 5–22.

Cotecchia, V., Federico, A. & Trizzino, R. (1982). Microtessiture di sedimenti argillosi. Analisi delle tipologie e terminologie, ed applicazione nel campo della geotecnica. *Geol. Appl. Idrogeol.* **17**, 1–34.

Graham, J., Crooks, J. H. A. & Lau, S. L. K. (1988). Yield envelopes: identification and geometric properties, *Géotechnique* **38**, 125–134.

Lambe, T. W. & Whitman, R. V. (1969). *Soil mechanics*. New York: Wiley.

Leroueil, S. & Vaughan, P. R. (1990). The general and congruent effects of structure in natural soils and weak rocks. *Géotechnique* **40**, 467–488.

Mayne, P. W. & Kulhawy, F. H. (1982). K_0–OCR relationships in soils. *Proc. ASCE* **108**, GT6, 851–872.

Mitchell, J. K. (1976). *Fundamentals of soil behaviour*. New York: Wiley.

Mitchell, J. K. & Houston, W. N. (1969). Causes of clay sensitivity. *Proc. ASCE*, SM3, 845–871.

O'Brien, N. R. & Slatt, R. M. (1990). *Atlas of argillaceous rocks*. Berlin: Springer-Verlag.

Rampello, S., Silvestri, F. & Viggiani, G. (1994). The dependence of small strain stiffness state and history for fine grained soils: the example of Vallericca Clay. *Proc. Int. Symp. Pre-failure Deform. Characterist. Geomater. Sapporo* **1**, 273–278.

Roscoe, K. H. & Burland, J. B. (1968). On the generalized stress-strain behaviour of wet clay. In *Engineering plasticity* (eds J. Heyman & F. A. Leckie), pp. 535–609. Cambridge: Cambridge University Press.

Schmertmann, J. H. (1969). Swell sensitivity. *Géotechnique* **19**, 530–533.

Schofield, A. N. & Wroth, C. P. (1968). *Critical state soil mechanics*. London: McGraw-Hill.

Sfondrini, G. (1975). Caratteristiche microtessiturali e microstrutturali di alcuni sedimenti argillosi connesse con la natura ed il tipo delle sollecitazioni subite. *Geol. Appl. Idrogeol.* **10**, 300–320.

Sides, G. & Barden, L. (1970). The microstructure of dispersed and flocculated samples of kaolinite, illite and montmorillonite. *Can. Geotech. J.* **8**, 391–400.

Skempton, A. W. (1970). The consolidation of clays by gravitational compaction. *Q. J. Geol. Soc. London* **125**, 373–412.

Skempton, A. W. & Northey, R. D. (1952). The sensitivity of clays. *Géotechnique* **3**, 30–53.

Smart, P. & Tovey, N. K. (1982). *Electron microscopy of soils and sediments: techniques*, p. 264. Oxford: Oxford University Press.

Terzaghi, K. (1941). Undisturbed clay samples and undisturbed clays. *J. Boston Soc. Civ. Engng* **28**, 211–231.

Terzaghi, K. (1944). Ends and means in Soil Mechanics. *Engng J. (Canada)* **27**, 608.

Viggiani, G. & Atkinson, J. H. (1995). Stiffness of fine-grained soil at very small strains. *Géotechnique* **45**, 249–265.

Wissa, A. E. Z., Christian, J. T., Davis, E. H. & Heiberg, S. (1971). Consolidation at constant rate of strain. *Proc. ASCE* **97**, SM10, 1393–1413.

Jovičić, V. & Coop, M. R. (1997). *Géotechnique* **47**, No. 3, 545–561

Stiffness of coarse-grained soils at small strains

V. JOVIČIĆ* and M. R. COOP†

The accurate predictions of deformations around geotechnical structures depend on the correct definition of the stiffness–strain curve of the soil, and in particular on the elastic stiffness at very small strains, which, being the starting point of the curve, may be regarded as a benchmark parameter for a given soil. Whereas for clays the factors controlling stiffness are well understood, for sands the current analyses tend to be empirical. The paper presents the results of bender element tests examining the elastic stiffness of sands, which is then related to the stiffness at larger strains determined in triaxial tests using a new system of local axial strain measurement. Three sands with very different geological origins were tested over a wide range of stresses, allowing a new general framework for stiffness to be established. By means of correct normalization of the data, unique relationships are derived for each soil which demonstrate that the confining stress and volumetric state relative to the normal compression line are principal controlling factors as they would be for clays. The paper shows, however, that for sands the means of arriving at its initial volume–stress state is also important, in particular whether this is by geological overconsolidation or compaction during the depositional process.

KEYWORDS: calcareous soils; dynamics; elasticity; laboratory tests; sands; stiffness.

La précision des prévisions des déformations autour de structures géotechniques repose sur une définition correcte de la courbe rigidité/déformation du sol, et en particulier sur la rigidité élastique à très faibles contraintes, qui, vu qu'elle constitue le point de départ de la courbe, peut être considérée le paramètre de référence d'un sol donné. On comprend bien les facteurs qui régissent la rigidité des argiles, mais, dans le cas des sables, les analyses actuelles tendent à être empiriques. L'article présente les résultats d'essais de résistance à la flexion réalisés pour examiner la rigidité élastique des sables et la mettre ensuite en corrélation avec la rigidité à des contraintes plus fortes, établie dans des essais triaxiaux à l'aide d'un nouveau système de mesure locale des contraintes axiales. Les essais ont porté sur trois sables d'origines géologiques très différentes, dans un large éventail de contraintes, ce qui a permis de formuler un nouveau schéma général pour la rigidité. Grâce à une normalisation correcte des données, les auteurs ont pu établir pour chaque sol des rapports individuels qui montrent que la contrainte de confinement et la contrainte hydrostatique par rapport à la ligne de compression normale sont les principaux facteurs déterminants, tout comme dans le cas des argiles. L'article montre toutefois que, dans le cas des sables, le processus qui aboutit à l'état initial de contrainte hydrostatique est aussi un facteur important, qu'il s'agisse de préconsolidation géologique ou de tassement pendant la sédimentation.

INTRODUCTION

The behaviour of most soils even at small strains is generally assumed to be non-linear, the typical relationship between strain and stiffness having the form shown in the schematic diagram of Fig. 1. In this diagram, and throughout this paper the shear modulus, G, is calculated from the tangent to the stress–strain curve for a monotonic loading path at any particular strain. Atkinson & Sallfors (1991) suggested that the strain levels dividing the very small, small and larger strain levels would be around 0·001% and 1%, respectively. For stiffer soils the strains around geotechnical structures have been found to be within the small to very small ranges (Burland, 1989) so that the effect of the non-linearity is pronounced and non-linear numerical analyses of such materials have therefore become relatively common (e.g. Jardine *et al.*, 1991; Stallebrass & Taylor, 1997). Within such models the initial elastic stiffness at very small

Manuscript received 17 October 1997; revised manuscript accepted 11 March 1997.
Discussion on this paper closes 1 November 1997.

* Arup Geotechnics, formerly City University.
† Geotechnical Engineering Research Centre, City University.

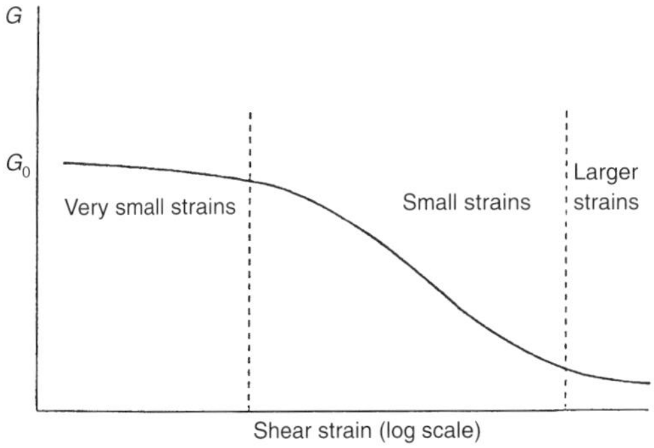

Fig. 1. Idealized stiffness:strain curve (after Atkinson & Sallfors, 1991)

strains (G_{max} or G_0) is relatively important, being the starting point of the stiffness–strain curve for any soil which is not currently yielding. This stiffness has often been measured by dynamic methods either in the laboratory or field, and using bender elements fitted in a triaxial apparatus Viggiani & Atkinson (1995) established a framework for G_0 in fine-grained soils, relating the stiffness to both the state and nature of the soil. The current paper sets out to define the corresponding framework for coarse-grained soils which will be necessary if similar non-linear constitutive models are to be used.

Viggiani & Atkinson also made comparisons between the dynamic measurements of G_0 made with bender elements and the measurements of stiffness calculated from the stress–strain behaviour during shearing probes, thereby extending the general framework of stiffness to larger strain levels. At that time stiffnesses from continuous loading tests could only be measured accurately down to a strain of about 0·01%, leaving a considerable gap between these stiffnesses and those from the dynamic tests for which the strain level was less than 0·001% (Dyvik & Madshus, 1985). Only more recently have local axial strain transducers become sufficiently accurate to enable direct comparisons between dynamic and continuous loading stiffnesses in the triaxial to be made (Goto *et al.*, 1991; Cuccovillo & Coop, 1997a). Nevertheless, the correspondence of dynamic and continuous loading stiffnesses and the general shape of the stiffness–strain curve has been demonstrated for sands by means of comparisons between resonant and torsional shear tests carried out in the resonant column apparatus (Iwasaki *et al.*, 1978; Ni, 1987; Bolton & Wilson, 1989) so that the value of G_0 has been found not to be rate dependent. Georgiannou *et al.* (1991) made similar observations for clays using the same methods with the addition of some continuous loading triaxial data, but again only for larger strain levels. They suggested that the yield strain which defined the end of the small initial linear elastic plateau, i.e. the end of the very small strain region on Fig. 1, was dependent on the plasticity of the soil being of the order of 0·01% for a plastic clay but only about 0·001% for sands. More recently, Lo Presti (1995) has proposed an elastic limit strain of 0·001% for any uncemented soil. Only some soft rocks particularly of the cemented-sand type (Cuccovillo & Coop, 1997a) have an extended elastic range.

The stiffness of sands has long been recognized to be controlled by the confining stresses (e.g. Hardin & Black, 1966), and many researchers have expressed this relationship as a power function relating G to the mean normal effective stress, p' (e.g. Wroth & Houlsby, 1985):

$$\frac{G}{p_r} = A\left(\frac{p'}{p_r}\right)^n, \tag{1}$$

where p_r is a reference pressure of 1 kPa used to make the parameters A and n dimensionless. If equation (1) is used to refer only to the elastic stiffness, G_0, then A and n are material parameters (here they will be referred to as A_0 and n_0). Considering Hertzian theory for a material comprising of ideal elastic spheres Duffy & Mindlin (1957) identified that the value of n would be $1/3$. The same equation can also be used, however, for the tangent stiffness at larger strains, in which case A and n will not be constants but will be dependent on both the strain level and the stress path followed. Experimental data from dynamic and continuous loading tests (Wroth *et al.*, 1979) showed that at small strains the exponent n had a value of 0·5. At large strains while n increased, tending towards a value of 1, A decreased, reaching zero at the critical state when the tangent stiffness would also be zero.

For clay soils the importance of the volumetric state has been well established (e.g. Ladd & Foot, 1974), so that following the work of Wroth & Houlsby (1985) and Houlsby & Wroth (1991), Viggiani & Atkinson showed that the following form of expression for G_0 was generally applicable to reconstituted clays:

$$\frac{G_0}{p_r} = A_0\left(\frac{p'}{p_r}\right)^{n_0} {R_0}^m, \tag{2}$$

where R_0 is the overconsolidation ratio expressed in terms of p' and the parameters A_0, n_0 and m are related to the plasticity of the soil. In contrast, the influence of state on the stiffness of sands has perhaps received less attention, although Hardin & Black (1966) identified that both the

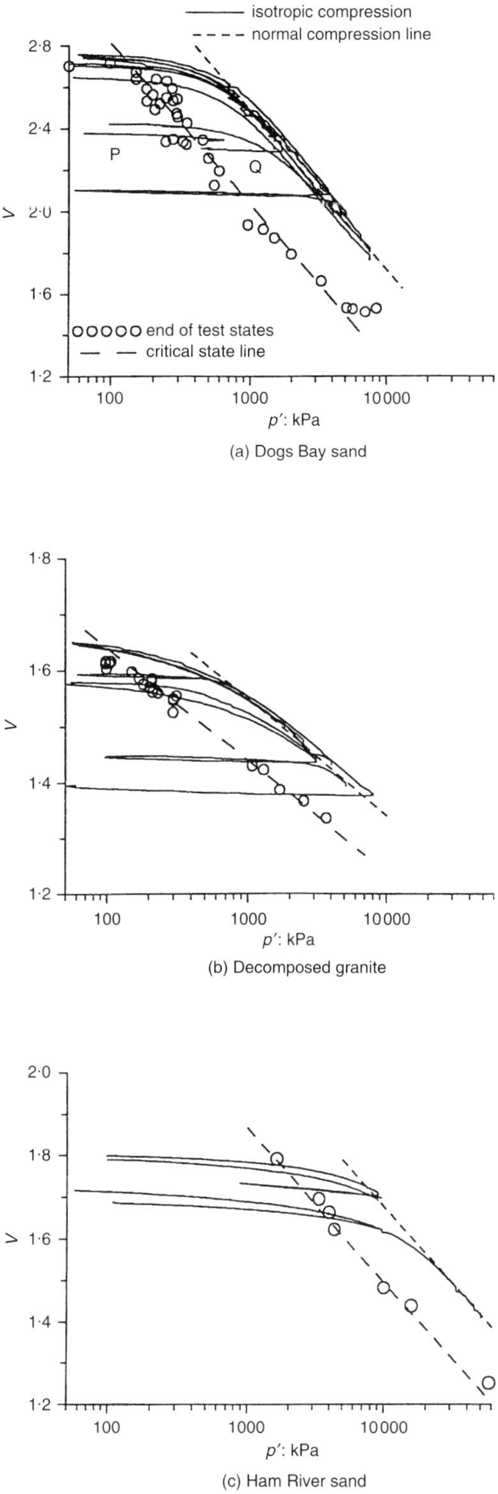

Fig. 2. Normal compression and critical state lines for three sands (adapted from Coop & Lee, 1993)

voids ratio (e) and the stress history were important. In consequence, a general framework for the stiffness behaviour of sands has, so far, not been established.

From the results of triaxial tests conducted up to high stress levels Coop & Lee (1993) identified the general features of the mechanics of coarse-grained soils at large strains. Fig. 2 illustrates how compression to high stresses resulted in each sand reaching a unique isotropic normal compression line (NCL) which was independent of the initial density. The terminology used in their paper, and also in the current work, is different to that used by some other authors, (e.g. Jefferies, 1993) who have used the term 'normal compression line' to refer to the first loading compression curves even before they reach the unique line identified above, so that there are an infinity of NCLs for samples of different initial densities. Coop & Lee used the term NCL to refer only to the unique line towards which all the compression curves converge. This is consistent with the usage by Pestana & Whittle (1995) who have since referred to the NCL as the 'limiting compression curve'.

Providing shearing was continued to large enough strains, Coop & Lee identified critical states that could be found which plotted on a locus, the critical state line (CSL), which was parallel to the NCL. The location of the critical state line was independent of the stress path taken to reach it. The authors showed that particle crushing was associated with the large plastic volumetric strains which occur both during compression down the NCL or during shearing to a critical state. Each point on the NCL or CSL was associated with a unique degree of breakage which was quantified by means of Hardin's (1985) 'relative breakage'. Some authors (e.g. Been *et al.*, 1991) have suggested that the CSL increases gradient with the onset of particle crushing at high stresses. For the three soils shown in Fig. 2 the CSLs were found to be straight on a graph of v plotted against $\ln(p')$, within the range of densities tested and any slight tendency to curvature at the lowest stress levels was believed by the authors to be the result of apparatus restraint inhibiting dilative behaviour.

A key difference between the behaviour of sands and clays was identified by Coop & Lee as the fact that while, within the engineering range of stresses, the loosest possible clay samples would have states that lie on the normal compression line, for sands the initial volumetric state is governed by the method of the sample preparation (or geological deposition). During first loading of sands there is then a gradual yield as the soil state curves slowly towards the NCL. The curvature of the first loading lines results from plastic volumetric straining which occurs even before sample reaches the NCL, as can be seen, for example, for test 8db (Fig. 8) which was unloaded prior to reaching a normally consolidated state. This contrasts with the very rigid, elastic unload–reload behaviour and results from the fact that the particle contacts are not disturbed during unloading and reloading. The particle crushing only restarts when the previous preconsolidation pressure is reached, so giving a very well-defined yield for reloading. A second principal difference between sands and clays is then that for a sand a given stress–volume state can be reached in various different ways.

Figure 3 shows undrained stress paths for two samples (P and Q) which both have the same specific volume (v) and are both sheared from the same initial isotropic stress. One of the samples had reached this initial state by being compressed to the NCL and then being unloaded; this will be referred to as being overconsolidated for consistency with clays. The other sample was compacted mechanically before loading and had arrived at the initial state under first loading only. The shearing behaviour of the overconsolidated sample was then much stiffer than that of the compacted although the critical states were similar. It was believed that this resulted from the greater degree of particle breakage that the overconsolidated sample had undergone during compression and the consequent effect on the distribution of particle contacts or fabric.

SOILS TESTED

For consistency with the previous work at larger strains the same three sands were tested, these having been originally chosen for their diversity of origins and characteristics. Grading curves are given in Fig. 4 and scanning electron micrographs of typical particles in Fig. 5.

The Ham River sand is a poorly-graded quartz sand which has rounded particles as a result of the erosive and transport processes that it has undergone. The Dogs Bay sand is a biogenic carbonate sand which is comprised predominantly of relatively unbroken foraminifera and mollusc shells. It is again poorly-graded and this, together with the open and angular nature of the particles, gives rise to its high voids ratios. Its behaviour was investigated in detail by Coop (1990).

The decomposed granite was tested extensively by Lee & Coop (1995) and is of Korean origin. The present research was aimed at investigating the small strain behaviour of sands in their reconstituted states as Viggiani & Atkinson had done for clays. The behaviour of natural sands may however be influenced by the presence of a natural structure resulting from their geological origin (Cuccovillo & Coop, 1997a) as is now well established for clays (Burland, 1990). Thus, although the decomposed granite had a weathering grade V (Geological Society, 1990) the in situ structure of the soil was removed mechanically before use. Particles greater than 5 mm were also discarded. Being the product of weathering rather than erosion and transport, the soil is well graded with angular to sub-angular particles.

Although the particles appear discrete and robust each is formed of an amalgam of different minerals, some inherited from the parent rock (e.g. quartz and feldspars) and some the products of weathering (e.g. kaolin, mica and smectite). The particles are also criss-crossed with fractures, again produced by weathering.

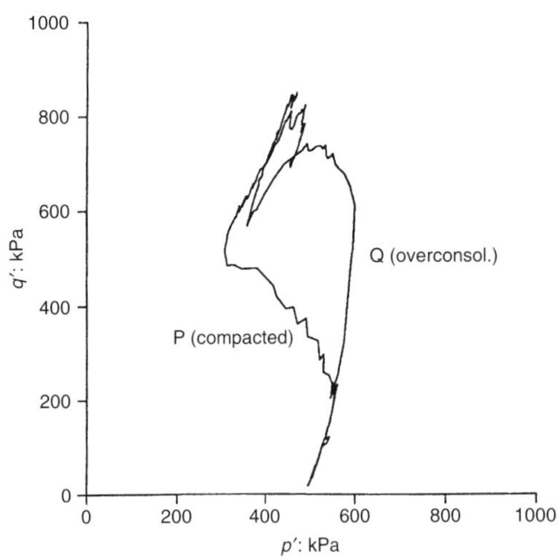

Fig. 3. Undrained stress paths for compacted and overconsolidated samples of Dogs Bay sand (after Coop, 1990)

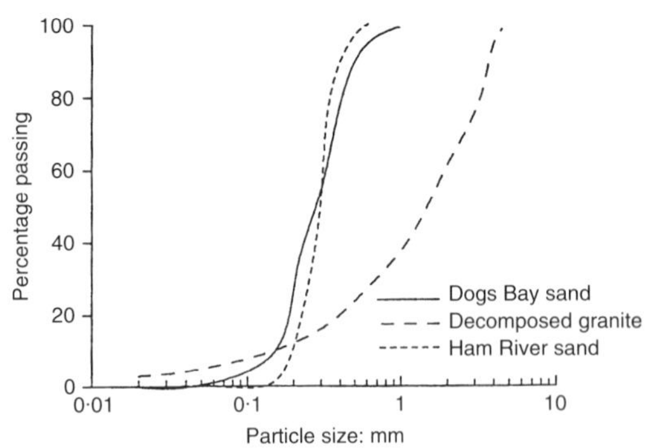

Fig. 4. Grading curves for three sands (after Coop & Lee, 1993)

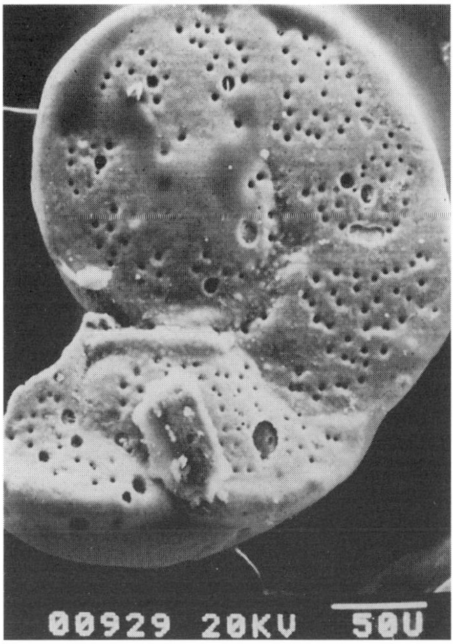

(a)

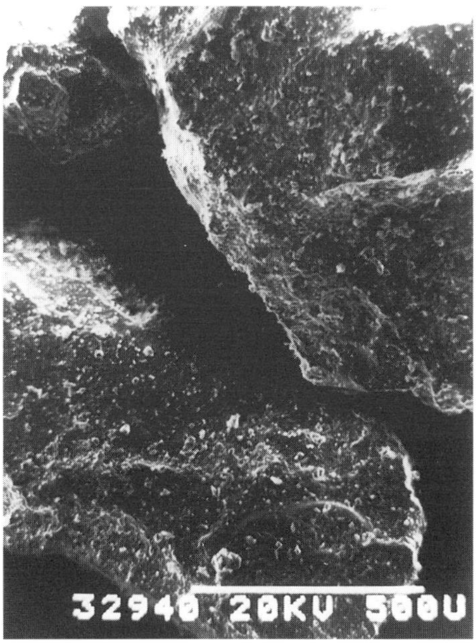

(b)

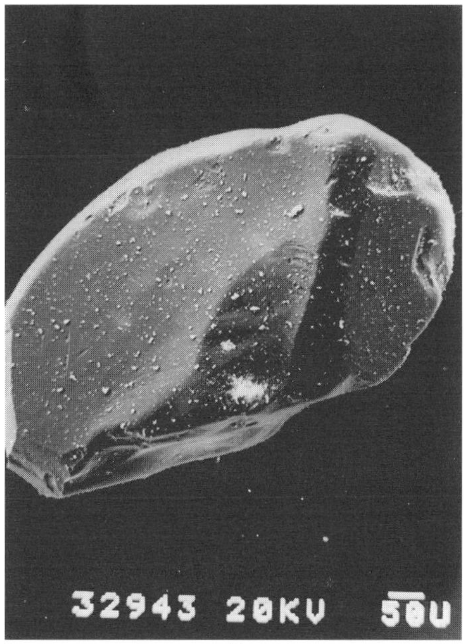

(c)

Fig. 5. Scanning electron micrographs of typical soil particles (a) Dogs Bay sand (b) decomposed granite (c) Ham River sand

APPARATUS AND PROCEDURES

Tests at low to medium pressures were conducted in a computer-controlled hydraulic triaxial apparatus while tests at higher pressures were conducted in two high pressure triaxial apparatus, both of which were again computer-controlled, one with a cell pressure capacity of 5 MPa and the other 70 MPa. A sample diameter of 50 mm was used in the high pressure apparatus and 38 mm in the hydraulic cell. All three apparatus had similar instrumentation and were fitted with bender elements (Shirley & Hampton, 1977). Fig. 6 shows a schematic diagram of the instrumentation of the hydraulic cell. For the shearing probe tests the axial strains were measured with local axial strain transducers attached directly to the sample. In the

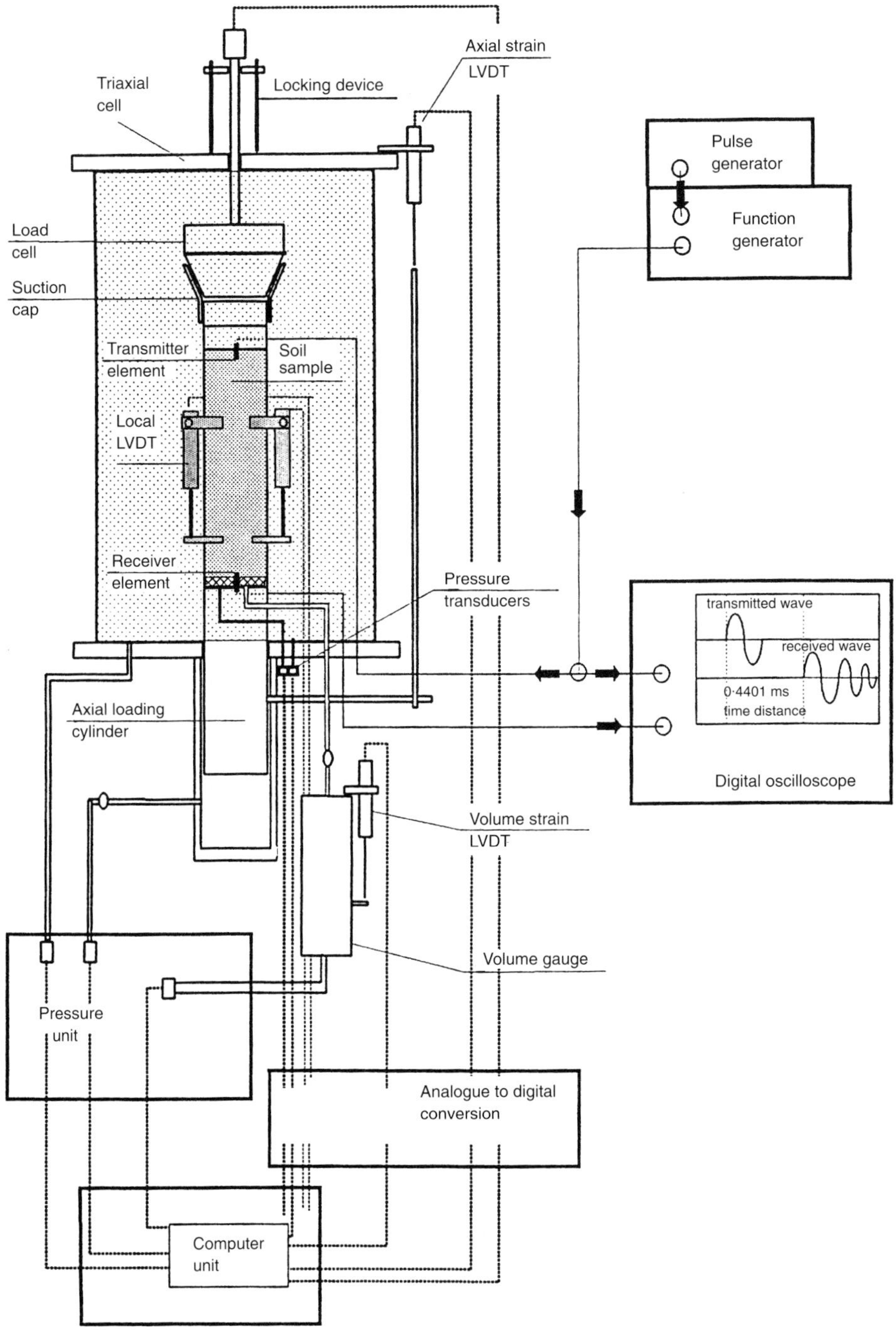

Fig. 6. Schematic layout of the standard pressure triaxial apparatus with bender elements

5 MPa apparatus, Hall effect transducers were used (Clayton & Khathrush, 1986) and in the standard pressure cell miniature submersible low-voltage displacement transducers (LVDTs) (Cuccovillo & Coop, 1997b).

The samples were created by a variety of techniques. For the Dogs Bay and Ham River sands the soil was generally pluviated directly into the membrane held within a mould placed on the platen, tamping the soil in layers if necessary to achieve the desired density. The loosest samples of Dogs Bay sand were created by the wet compaction method in which the soil was gently compacted into the membrane in a partially saturated state. Carbon dioxide was then circulated through the sample prior to flooding it with distilled water to aid saturation when the back pressure was subsequently applied. Pluviation could not be used for the decomposed granite because of the segregation which would result for such a well-graded soil and so all of the samples were created by wet compaction. Fuller details of the preparation and testing techniques are given by Coop (1990), Lee & Coop (1995) and Jovičić (1997).

The samples were subjected to continuous stages of isotropic loading and unloading, bender-element measurements of the stiffness being made while the stages continued. Long periods of rest were avoided as the creep that would occur during these periods resulted in significant changes of stiffness. Fig. 7 shows creep data from tests on Dogs Bay sand which was the soil most affected by creep. In these tests the isotropic stress had been applied quickly before the zero time of the plots and the values of stiffness are normalized with respect to the initial value at zero time. There is an increase of up to 15% over a period of about three days, the effect being proportionately worse at lower stress levels. It is also interesting to note that, while the effect is more severe for samples undergoing first loading, there is also a measurable increase for samples undergoing either unloading or reloading. The increase in stiffness cannot therefore be attributable solely to volumetric creep as no such volumetric changes were seen for the latter samples. It is therefore probably an aging phenomenon and similar observations have been made by, among others, Schmertman (1991). On the basis of these results, rates of loading were chosen that were roughly proportional to the current stress, around 0·5% of the current isotropic stress per hour. This rate would give a change in G_0 of only 2% during the first day of a delay occurring in the test, and resulted in tests lasting typically for a few weeks.

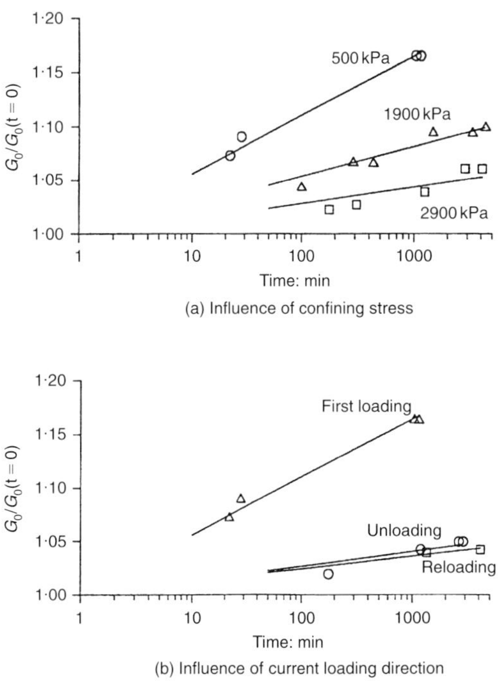

Fig. 7. Creep test data for Dogs Bay sand

Generally, a sinusoidal input wave to the bender element was used with a frequency high enough to avoid the near field effect (Jovičić *et al.*, 1996). The values of stiffness were calculated from the velocity of travel of the wave through the sample (V_s) as follows:

$$G_0 = \rho V_s^2 \tag{3}$$

where ρ is the mass density of the sample. The travel velocity V_s was calculated from the time lag between sending and receiving the wave, which was read directly off the oscilloscope, and the distance between the tips of the two bender elements, which was calculated from the initial dimensions of the sample together with the strains measured up to the point the bender element reading was taken. Gajo (1996) has identified that for sands this method of calculating G_0 may give rise to an error of up to 10–15% as a result of the inertial coupling of the two phases of the material. At the time of writing it is not certain whether a correction to the data is required, nor what its magnitude should be. The majority of the measurements were however taken using frequencies lower than the critical frequency identified by Gajo as being a threshold below which the effect is small. The correction would also be similar for all stress levels and densities of each of the three sands. It would therefore have an absolute rather than relative effect on the stiffnesses so that the conclusions of the current paper would not be affected. It is also interesting to note that in the direct comparisons of dynamic and continuous loading stiffnesses that will be discussed later there is good agreement.

For some tests small undrained shearing probes were conducted at intervals during the test. The shearing was generally continued only to an axial strain of about 0·1% because of the need to have sufficient compression of the sample during the isotropic loading stage between successive probes as to erase the effect of the previous probe. Bender element measurements of stiffness were generally not made during the probes as while large deviatoric stresses can change significantly the value of G_0 measured, the stress changes imposed during the probes were too small to have a measurable effect.

Within this paper a choice has been made to examine the stiffness measured as a function of $p'(=[\sigma'_a+2\sigma'_r]/3)$. Roesler (1979) has, however, demonstrated that for wave propagation dynamic measurements of stiffness the value of the principal stress acting out of the plane of the wave motion has no impact on the stiffness measured, so that this stress may be omitted from the function relating G_0 to the confining stresses. All of the measurements of G_0 presented in this paper have, however, been made under an isotropic stress state so it is unimportant whether or not the out-of-plane stress appears in the stress invariant.

STIFFNESS AT VERY SMALL STRAINS

The loading paths during isotropic compression of the test samples are shown in plots of v against $\ln(p')$ in Fig. 8 and the bender element measurements of G_0 in Figs 9–13. For each test many readings of G_0 were taken (typically about 50) and they are too numerous for the individual data points to be shown on Figs 9 and 11. Instead the lines shown have been constructed by joining individual points with straight lines and there has been no smoothing of the data. On first loading, the values of G_0 for all densities tend towards a

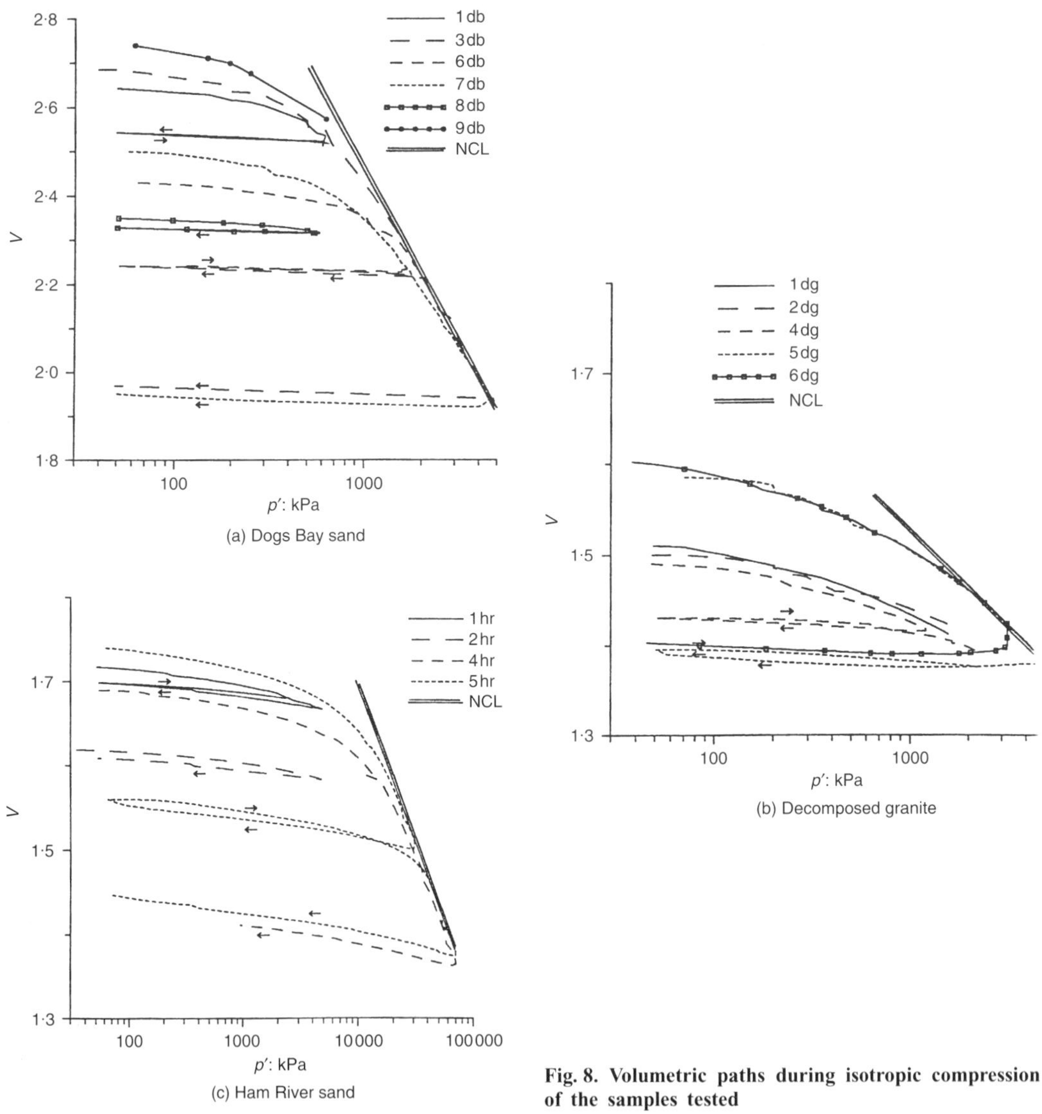

Fig. 8. Volumetric paths during isotropic compression of the samples tested

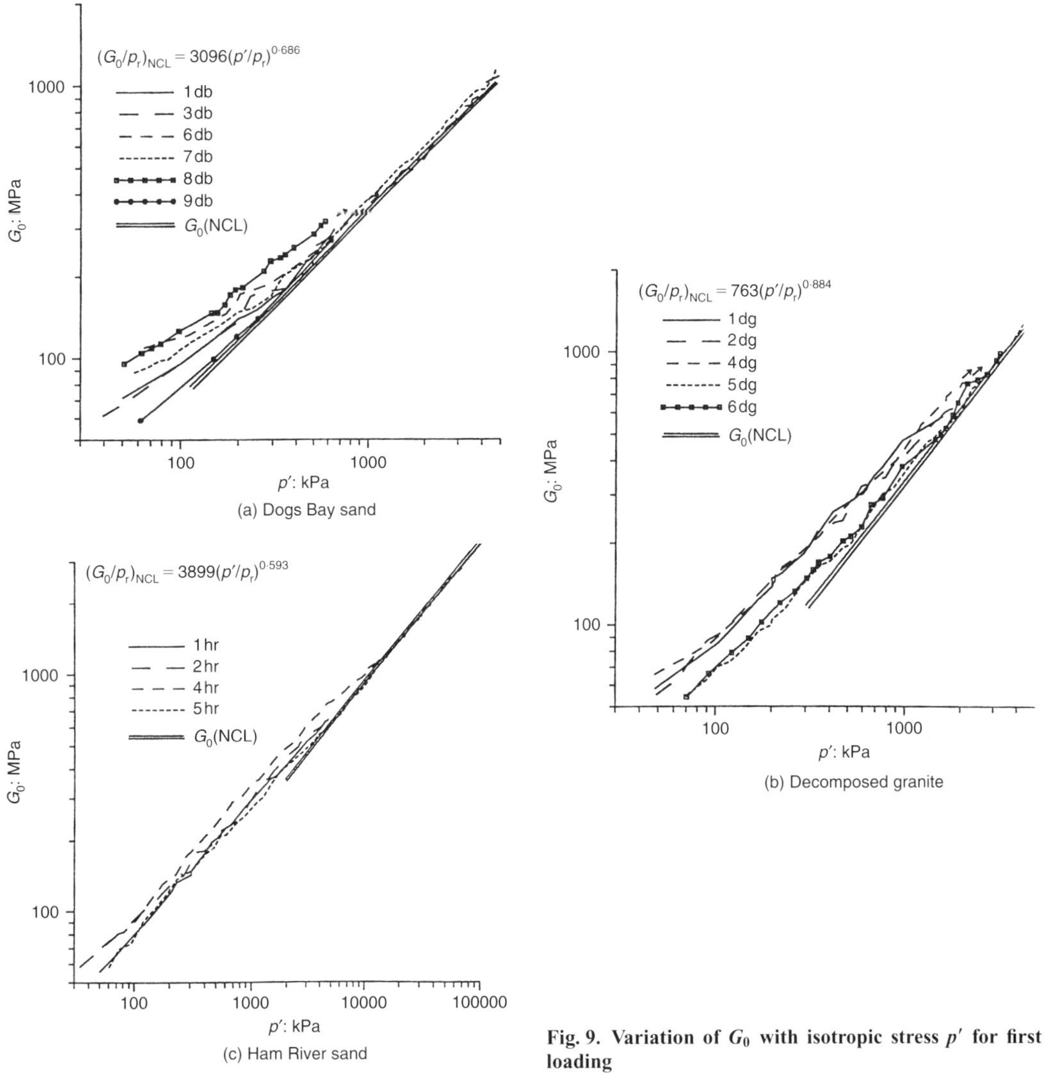

Fig. 9. Variation of G_0 with isotropic stress p' for first loading

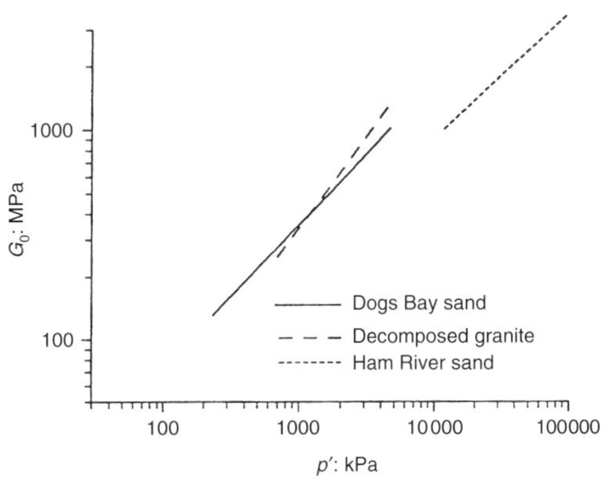

Fig. 10. Comparison of G_0 on the normal compression line for the three sands

unique line which corresponds to the NCL in the volumetric plane and may be characterized by equation (1).

Sample 8db of Dogs Bay sand was extremely dense and since it was tested in the standard pressure apparatus it could not be brought to a normally compressed state. Consequently the G_0 values for this test also stop short of reaching the unique line of stiffness for normally compressed samples. The gradients of these lines (n_0) and the intercepts (A_0) range from 0·593 to 0·884 and from 760 to 3900 respectively, but, as illustrated in Fig. 10, while there is significant variation in these values the general positions of the NCLs in terms of $\log(G_0)$ and $\log(p')$ are fairly similar for these three very different sands. Viggiani & Atkinson related the values of A_0 and n_0 for normally

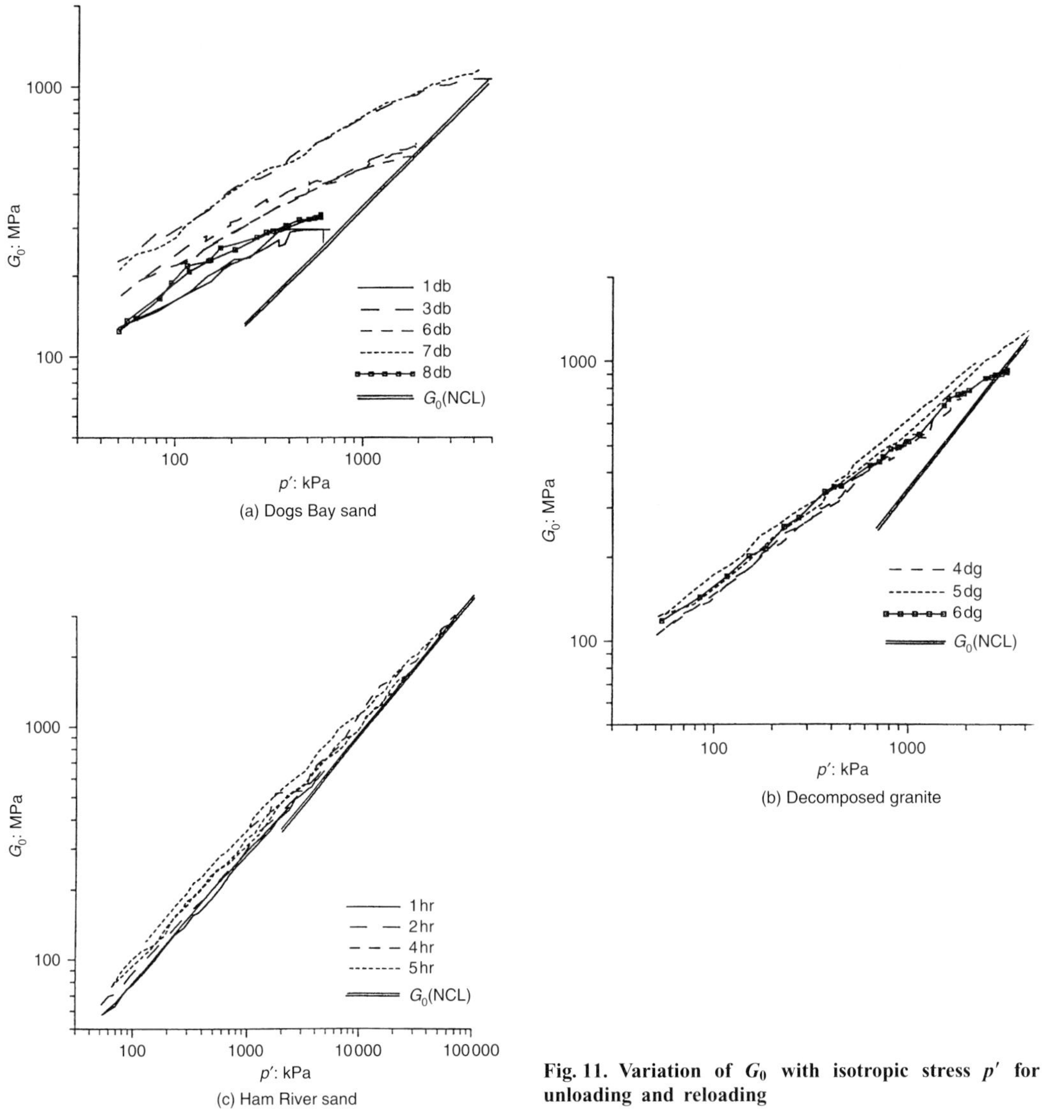

Fig. 11. Variation of G_0 with isotropic stress p' for unloading and reloading

consolidated clays to the plasticity of the soil. For sands it is likely that there is a similar relationship with the nature of the soil but with data for only three materials it is too early to determine what might be the controlling parameters.

The gradients of the first loading stiffness data prior to reaching the NCL are not constant; the approach path to the NCL being slightly curved. The values of n_0 in this range are however approximately in the range 0·5–0·7 and so are closer to values quoted in the literature but still much greater than the value of 1/3 derived from Hertzian theory. Ko & Scott (1967) proposed a theoretical model for sands which accounted for the increased number of contacts between particles during compression, and showed that this would result in a higher volumetric stiffness than that predicted by Hertzian theory. Similarly, by means of numerical analyses of a random system of spheres, Dobry *et al.*, (1989) also highlighted the importance of the number of particle contacts on the soil stiffness. This is likely to be the reason for the higher exponents seen here, particularly since these models did not account for particle breakage which would increase still further the number of contacts per particle. Another important feature of the data is that whereas the stiffness on the NCL is unique for a particular stress, that on the first loading lines before reaching the NCL depends on the soil's initial density. It is interesting to note that there is

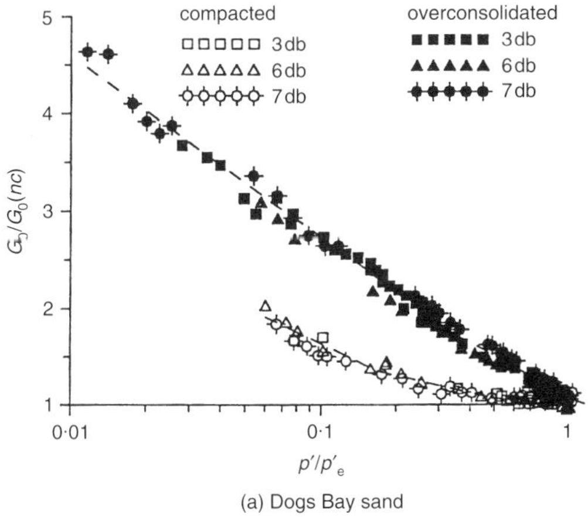

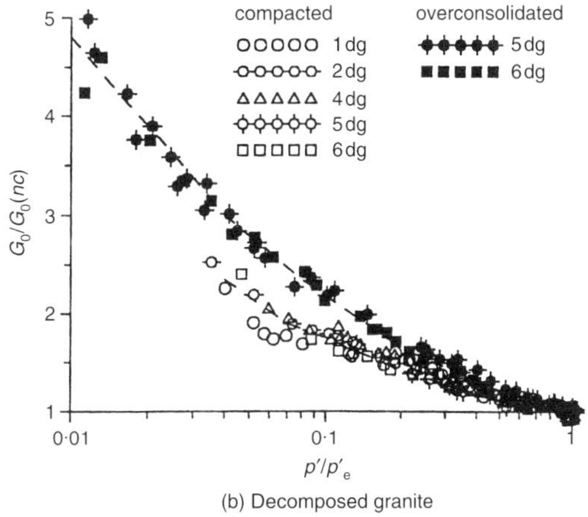

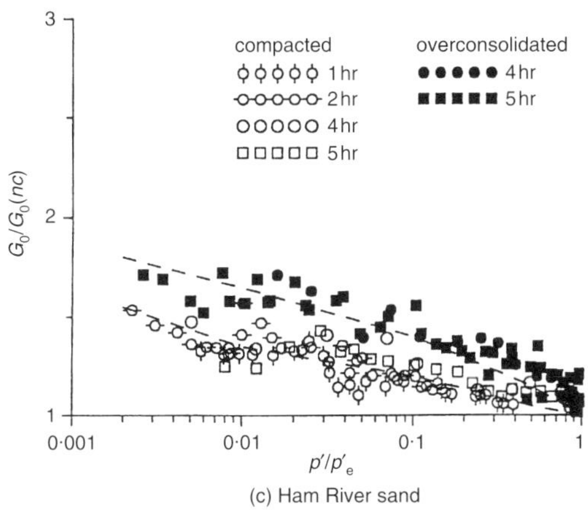

Fig. 12. Variation of G_0 with normalized volumetric state

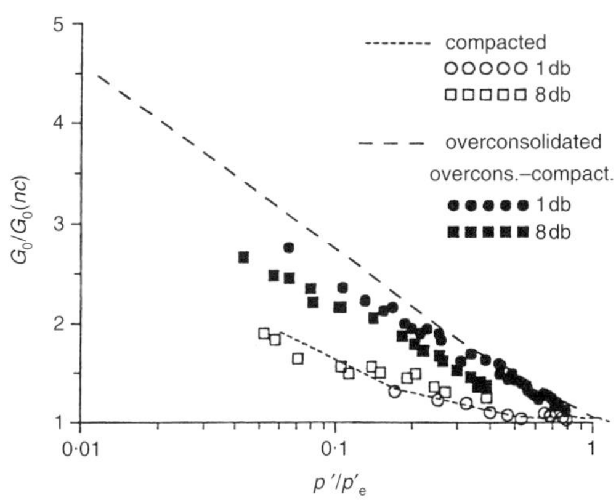

Fig. 13. Influence of G_0 of unloading prior to reaching NCL for Dogs Bay sand

a tendency to reach the NCL in plots of $\log(G)$ against $\log(p')$ before reaching it in terms of the volume. For example the loosest sample created of Dogs Bay sand (test 9db) reaches a unique line in terms of $\log(G)$ plotted against $\log(p')$ at about 200 kPa but does not reach the NCL in volumetric space until about 700 kPa.

The stiffnesses measured during isotropic unload–reload loops are shown in Fig. 11. These data lie on curved lines above the line representing G_0 for the NCL, each line corresponding to a swelling line in the plot of v against $\ln(p')$ so that its location is controlled by the preconsolidation pressure. There is no hysteresis in these lines corresponding to the lack of such a loop in the volumetric data so that the stiffnesses, measured during reloading are indistinguishable from those measured during unloading. At the start of unloading however there is often a small section of the line where the stiffness does not change much. This feature results from the creep of the soil which gave continued volumetric compression during the initial part of unloading.

Given that the stiffnesses of both the overconsolidated samples and those during first loading prior to reaching the NCL are greater than those of normally consolidated samples, the question arises as to whether there is a unique relationship between the volumetric state and stiffness. For convenience the term overconsolidated will be reserved only for those samples which have been unloaded after reaching the NCL and 'compacted' will be defined to mean any sample which is undergoing first loading prior to reaching the NCL no matter how loose its initial state. The volumetric state of the samples has therefore been defined by normalizing the current p' of the sample with respect to an equivalent pressure p_e' taken at the same specific volume on the NCL:

$$p_e' = \exp\left(\frac{N - v}{\lambda}\right) \tag{4}$$

where λ is the gradient of the NCL and N its intercept at a p' of 1 kPa. This is analogous to using the overconsolidation ratio for a clay. Alternatively, the current state parameter of the soil could be used (Jefferies, 1993). Similarly the values of G_0 for the overconsolidated and compacted samples have been normalized with respect to a G_0 for a normally-consolidated sample at the current value of mean effective stress:

$$\left(\frac{G_0}{p_r}\right)_{(nc)} = A_0\left(\frac{p'}{p_r}\right)^{n_0} \tag{5}$$

Using this method of presenting the data all of the stiffnesses of samples lying on the NCL plot as a single point at 1:1 as shown in Fig. 12. As for clays there is a general increase of the normalized stiffness with increasing distance from the NCL (decreasing p'/p_e'), but, whereas all of the compacted samples define one unique line, the overconsolidated samples have higher stiffnesses and define a different line, but one which is also unique for all samples at all volumes. For the overconsolidated samples there is again no evidence of hysteresis, the measurements taken during unloading and reloading defining the same line. Using the semi-logarithmic axes chosen some of the relationships are linear while others are curved, but in each case the most important feature is that they are unique. This means that the value of n does not vary with p'/p_e' so that for a given value of p'/p_e' the $\log(G_0)$ against $\log(p')$ relationship for either compacted or overconsolidated samples would plot as a straight line parallel to and above the line for normally consolidated samples on Fig. 10.

There is a considerable difference between the influence that state plays for the various soils; the steepest relationships being seen for the carbonate sand while for the silica sand the volumetric state is less important. The approach curves of the compacted soils to the NCL on Fig. 12 are also fairly flat, particularly for the carbonate sand, this being the reason why it appears to reach the NCL on the $\log(G)$ against $\log(p')$ graph prior to reaching it in volumetric space. This feature, together with the fact that previous work has often been done on silica sands over a limited range of stresses, may explain why previous research has sometimes concentrated more on the importance of the stress state rather than of the specific volume on the stiffness of sands.

The above framework does not account for the possibility that a sample may be unloaded prior to its reaching the NCL. Such a sample will be defined as being overconsolidated–compacted and Fig. 13 shows how its stiffness varies with its volumetric state. From its initial state, as the sample is isotropically loaded its stiffness moves down the line for the compacted samples towards the point representing the NCL. Before reaching this point, however, the sample is unloaded so that the value of p'/p_e' reduces again. Its normalized stiffness is then greater than it had been during first loading but lower than that of samples which have been unloaded from the NCL, so that the line defined by the stiffness during unloading lies below that for the truly overconsolidated soil. It can now be seen that the two lines for the normalized stiffnesses identified in Fig. 12 define the limits of possible values, so that the space between them is the domain of the possible stiffnesses that might result from different combinations of compaction and overconsolidation. The size of this domain again varies considerably between the three sands, being most prominent for the carbonate sand and least prominent again for the silica sand. This may result from the fact that the former soil is that which is the most prone to particle breakage so that there could be a greater effect of the crushing on the nature and number of particle contacts than would be the case for the solid rounded silica particles.

STIFFNESS AT LARGER STRAINS

Figure 14 shows comparisons between the measurements of G_0 made with the bender elements and the tangent stiffnesses (G_u) calculated from the stress–strain data from the undrained shearing probes. For each probe about 500 data points were logged and Fig. 15 gives typical stress–strain data for one of the probes. The axial strains for these

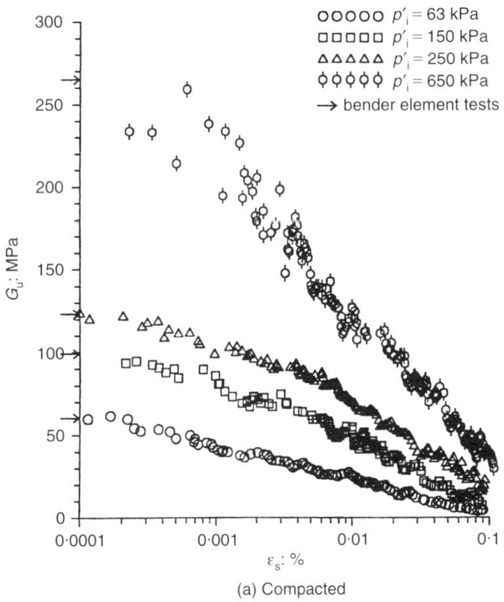

Fig. 14. Variation of G_0 with strain for a set of undrained triaxial compression tests

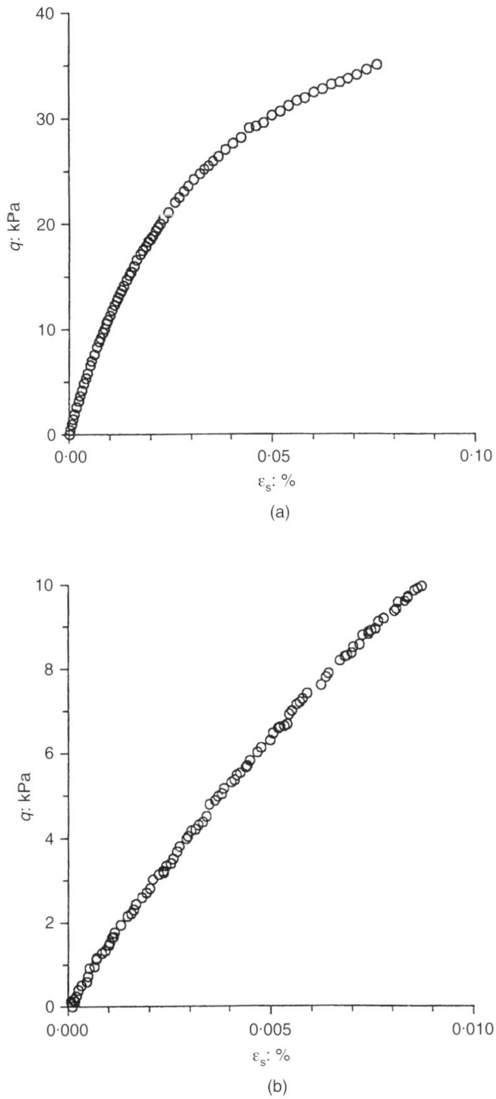

Fig. 15. Stress:strain data for undrained probe at $p_i' = 63$ kPa

data were all measured using a new system of two miniature LVDTs attached directly to the sample (Cuccovillo & Coop, 1997b). An average strain for the two transducers is given in Fig. 15 although the difference between them was typically less than 5% of the current reading. The stiffnesses shown in Fig. 14 have been calculated by means of a linear regression through an odd number of data points on the stress–strain curve (typically 11), the resulting value of tangent stiffness then being plotted against the value of strain at the middle data point of the range. The range of data considered for the linear regression was then incremented by one and the next value of stiffness calculated. There are therefore as many values of stiffness on Fig. 14 as data points on the stress–strain curve, with the exception that at the very smallest strains at the start of the probe (i.e. below 0·0001%) the random noise in the readings dominates, giving rise to unreliable and widely-varying stiffnesses which have been omitted. Problems with stick–slip behaviour also in many cases restricted the minimum strain at which the stiffness could be defined, particularly for the overconsolidated samples. For the compacted samples there is, however, very good agreement between the dynamic measurements and the continuous loading stiffnesses at the strain of 0·0001%. In contrast to previously published data, however, (e.g. Georgiannou *et al.*, 1991) there is little evidence of a plateau of constant stiffness extending to 0·001% and it is not therefore known whether the trend of increasing stiffness would have continued to yet smaller strains. Such comparisons between the dynamic and continuous loading stiffnesses should bear in mind not only the possibility of differences arising from the effect of inertial coupling (Gajo, 1996), as discussed earlier, but also the possibility that there may be some inherent anisotropy in the soil which would also give rise to a difference in the stiffnesses measured by the two means. It is unlikely that the strains within the sample, up to the points when the probes were conducted, would have been sufficient to erase any anisotropy arising from the preparation methods.

For the overconsolidated samples the shapes of the stiffness–strain curves are again fairly similar for each stress level as they are for the compacted samples. There is, however, a noticeable difference between the two, with a more pronounced drop in stiffness for the overconsolidated samples between G_0 and the first continuous loading measurement (G_u) at 0·001%. There is then a more gentle reduction of stiffness with strain and so a less severe non-linearity than is seen for the compacted samples. For the compacted samples the stress path rotation for the undrained shearing probes was by definition +90° (Atkinson *et al.*, 1990) and, for the overconsolidated, both rotations of +90° and −90° were used although there was no noticeable difference in the stiffnesses between the two. Other stress path rotations, while having the same G_0 might, however, give significantly different stiffnesses at larger strain levels because of the recent stress history effect.

The larger strain stiffnesses for the compacted samples under first loading are compared in Fig. 16 to the NCL defined for the G_0 values. A set of sub-parallel lines is defined, converging towards higher stress levels exactly as Viggiani & Atkinson (1995) had seen for clays. As the strain level increases so the intercept A reduces towards zero and the gradient n increases, probably towards a value of 1·0 (Fig. 17), although the probes could not be continued to sufficiently high strains to

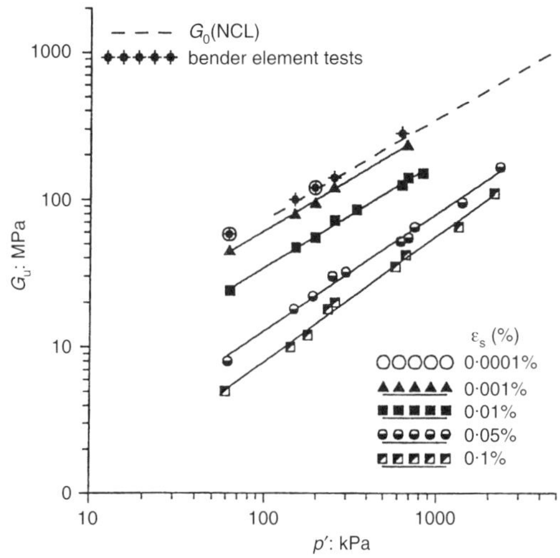

Fig. 16. Variation of G_u with stress and strain for first loading of Dogs Bay sand

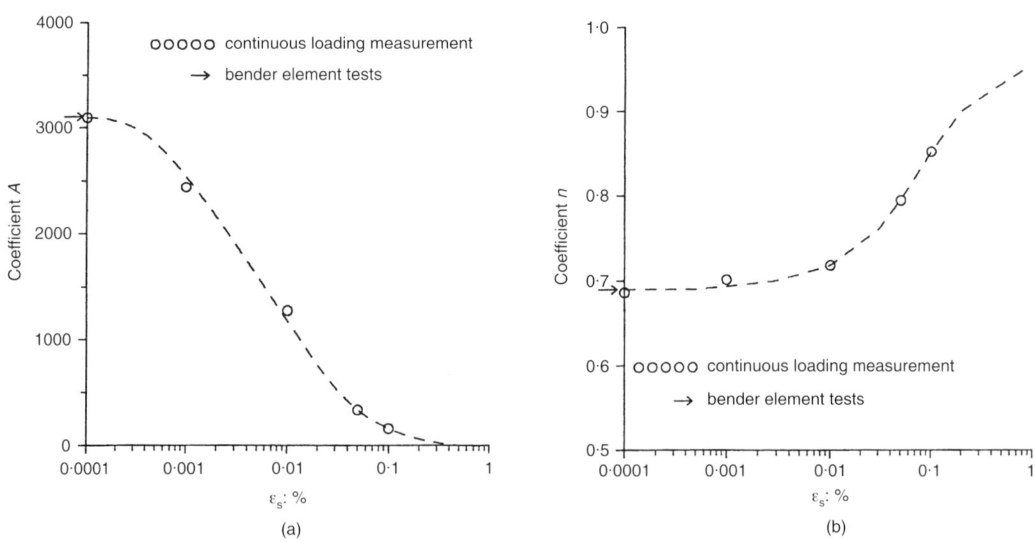

Fig. 17. Variation of the parameters A and n with strain level for Dogs Bay sand

confirm this. It should be emphasized that the shape of the stiffness backbone curve and hence the values of A and n, unlike A_0 and n_0, at strains beyond the elastic range are not fundamental soil properties but are dependent on both the loading path the sample was subjected to and perhaps the recent stress history.

For the overconsolidated samples Fig. 18 shows the change in the normalized stiffness with the volumetric state. The diagram is similar to Fig. 12 except that a logarithmic $G_u/G_0(nc)$ axis has been used for convenience when plotting the larger strain data. The samples for which data are shown were not quite truly overconsolidated as, because of the limited capacity of the apparatus in which the tests were conducted, the samples were unloaded from states slightly short of reaching the NCL, so that, as was seen in Fig. 13, the G_0 values plot slightly below the line for the truly overconsolidated samples. Nevertheless a clear pattern can be seen with unique lines for each strain level for samples of various densities.

INFLUENCE OF PORE WATER

Coop & Lee (1995) found that the large strain behaviour of the decomposed granite was severely affected by the presence or absence of pore water. Fig. 19 illustrates the difference between a completely saturated sample and one which was air dried, both of which had the same initial specific volume. Measurements of the volumetric strain of the dry samples had been made either with an inner chamber (Bishop & Donald, 1961) or by using both internal radial and axial strain transducers. While pore water was not found to influence the critical state friction angle, it was found that a lower

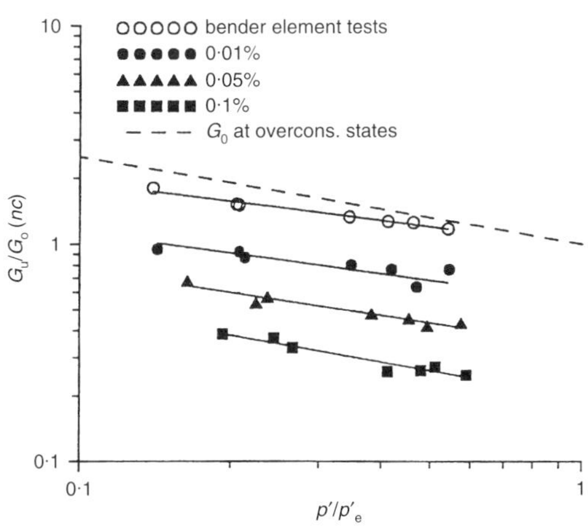

Fig. 18. Variation of G_u with normalized volumetric state and strain for overconsolidated samples of Dogs Bay sand

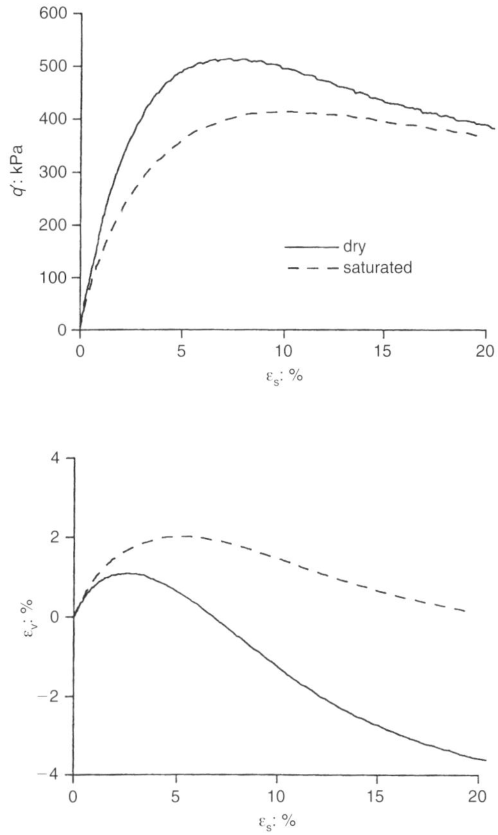

Fig. 19. Comparison of stress:strain data for drained triaxial tests on saturated and dry samples of decomposed granite (adapted from Lee & Coop, 1995)

degree of breakage of the dry samples gave rise to greater dilation or a less pronounced compression during shearing, so that the boundary surface of the dry soil lay at significantly higher volumes than that of the saturated. The greater strength of the soil particles when dry was attributed to suction within the particles, in the micro-fractures and clay minerals. A similar but smaller effect was also seen for the Dogs Bay sand which also has a porous particle fabric, but no effect was seen for the silica sand whose particles are crystalline and so not porous.

The stiffnesses at a strain of 0·1% have been reinterpreted within the new framework (Fig. 20). All of the samples were compacted, with specific volumes ranging from 1·31 to 1·67. Although the data are for drained tests because of the necessity of comparing wet and dry behaviour, the patterns of behaviour observed are the same. When normalized, the stiffnesses of these samples again define a unique line, both below that of the undrained tests on overconsolidated samples at 0·1% and that of the G_0 values of the compacted samples.

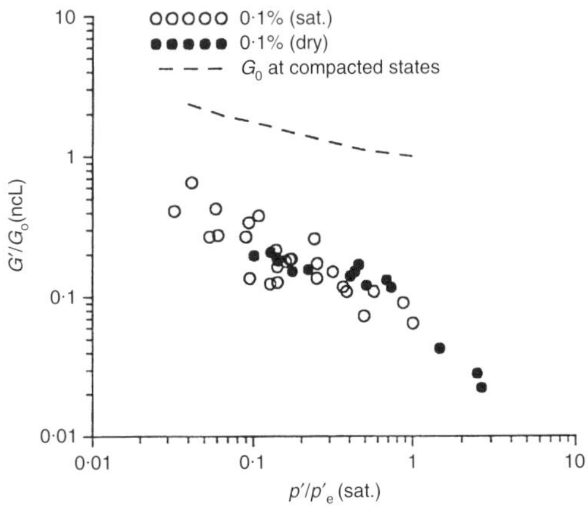

Fig. 20. Comparison of shear stiffness G' measured for drained triaxial compression tests on dry and saturated samples of decomposed granite

Again the values of p'_e have been calculated on the NCL of the saturated soil so that the dry samples starting from states on the dry NCL have initial values of p'/p'_e greater than 1·0. Nevertheless the dry and saturated soil have the same stiffness at small strains in contrast to the large strain behaviour. It can be concluded from this that while the large strain behaviour is dominated by the influence of pore water on particle breakage, that at small strains is not.

CONCLUSIONS

By means of bender element testing and the measurement of stiffnesses at larger strains with local triaxial instrumentation a new framework for the stiffness of reconstituted sands has been proposed. The framework emphasizes not only the influence of the confining stress but also that of the volumetric state, and similarities have been highlighted with that previously found to apply to reconstituted clays. The framework is based on the behaviour of three sands with very different mineralogies and geological origins, so is likely to be generally applicable and should provide the basis for future improvements in the non-linear constitutive modelling of sands. In order to obtain a complete understanding it was necessary to use high pressure testing, but it should be emphasized that the framework is appropriate for all stress levels.

A particular feature of the framework is that truly overconsolidated sands and those which have only undergone first loading have significantly different stiffnesses, so that the geological history of the soil deposition and its subsequent loading history would have an influence on its stiffness, quite apart from the effects of any natural structure which it acquires from diagenetic processes.

At the start of the work creep was identified as having a significant effect on the stiffness of sands. It was beyond the scope of this work to investigate in detail the influence of creep, but while it may affect the absolute values of stiffness it is unlikely to have a great effect on the framework proposed. Creep was found to be more pronounced in compacted samples than overconsolidated and so would tend to reduce the differences in stiffness that have been observed between them.

ACKNOWLEDGEMENTS

The research was entirely funded by the EPSRC to whom the authors are grateful. The authors would also like to thank Mr L. Martyka for his contribution in achieving the high quality of signal conditioning and Mr K. Osborne for his help in performing the high pressure tests.

NOTATION

A_0	non-dimensional factor relating G_0 to p'
A	non-dimensional factor relating G to p'
e	void ratio
G	tangent shear modulus
G_0	tangent shear modulus at very small strains
G'	tangent shear modulus for drained shearing
G_u	tangent shear modulus for undrained shearing
$G_0(nc)$	shear modulus at very small strains for normally consolidated soil
m	exponent relating G_0 to R_0 for clays
n_0	exponent relating G_0 to p'
n	exponent relating G to p'
p'	mean effective stress
p'_e	mean effective stress taken at the same specific volume on the normal compression line
p_r	reference pressure (taken as 1 kPa)
p'_i	initial p' for undrained shearing probe
R_0	overconsolidation ratio of a clay expressed in terms of p'
V_s	shear wave velocity
v	specific volume
ϵ_s	shear strain
λ	gradient of the normal compression line in graph of v plotted against $\ln(p')$
N	intercept of the normal compression line at $p' = 1$ kPa
ρ	soil density

REFERENCES

Atkinson, J. H. & Sallfors, G. (1991). Experimental determination of soil properties. *Proc. 10th Eur. Conf. Soil Mech., Florence*. Vol. 3, pp. 915–956.

Atkinson, J. H., Richardson, D. & Stallebrass, S. E. (1990). Effect of recent stress history on the stiffness of overconsolidated soil. *Géotechnique* **40**, No. 4, 531–540.

Been, K., Jefferies, M. G. & Hachey, J. (1991). The critical state of sands. *Géotechnique* **41**, No. 3, 365–381.

Bishop, A. W. & Donald, I. B. (1961). The experimental study of partly saturated soil in the triaxial apparatus.

Proc. 5th Intl. Conf. Soil Mechs. 7 Found. Eng., Paris. Vol. **1**, pp. 13–21.
Bolton, M. D. & Wilson, J. M. R. (1989). An experimental and theoretical comparison between static and dynamic torsional shear tests. *Géotechnique* **39**, No. 4, 585–599.
Burland, J. B. (1989). The Ninth Bjerrum Mem. Lecture: Small is beautiful – the stiffness of soils at small strains. Can. Geot. J. **26**, 499–516.
Burland, J. B. (1990). On the compressibility and shear strength of natural clays. 30th Rankine Lecture: *Géotechnique* **40**, No. 3, 329–378.
Clayton, C. R. I. & Khathrush, S. A. (1986). A new device for measuring local axial strains on triaxial specimens. *Géotechnique* **36**, No. 4, 593–597.
Coop, M. R. (1990). The mechanics of uncemented carbonate sands. *Géotechnique* **40**, No. 4, 607–626.
Coop, M. R. & Lee, I. K. (1993). The behaviour of granular soils at elevated stresses. *Proc. Wroth Memorial Symposium: Predictive soil mechanics*, pp. 186–198. London: Thomas Telford.
Coop, M. R. & Lee, I. K. (1995). The influence of pore water on the mechanics of granular soils. *Proceedings XI ECSMFE. Copenhagen.* Vol. 1, pp. 63–73.
Cuccovillo, T. & Coop, M. R. (1997a). Yielding and prefailure behaviour of structured sands. *Proc. Géotechnique Symp. in Print*, **47**, No. 3, 491–508.
Cuccovillo, T. & Coop, M. R. (1997b). The measurement of local axial strains in triaxial tests using LVDT's. *Géotechnique* **47**, No. 1, 167–171.
Dobry, R., Ng, T. T. & Petrakis, E. (1989). Deformation characteristics of granular soils in the light of particulate mechanics. *Atti delle conferenze di geotecniche di Torino*, Vol. 3, pp. 1–34. Torino: Politecnico di Torino.
Dyvik, R. & Madshus, C. (1985). Laboratory measurements of G_{max} using bender elements. *Proc. ASCE Annual convention: Advances in the art of testing soils under cyclic conditions*, Detroit, Michigan.
Duffy, J. & Mindlin, R. D. (1957). Stress–strain relations and vibrations of a granular medium. *J. Appl. Mech. Trans. ASME* **24**, 585–593.
Gajo (1996). The effects of inertial coupling in the interpretation of dynamic soil tests. *Géotechnique* **46**, No. 2, 245–259.
Geological Society (1990). Tropical residual soils, Engineering Group working party report. *Q. J. Engng Geol.* **23**, No. 1, 1–101.
Goto, S., Tatsuoka, F., Shibuya, S., Kim, Y. S. & Sato, T. (1991). A simple gauge for local small strain measurements in laboratory. *Soils & Foundations* **31**, No. 1, 169–180.
Georgiannou, V. N., Rampello, S. & Silvestri, F. (1991). Static and dynamic measurement of undrained stiffness of natural overconsolidated clays. *Proc. X ECSMFE, Florence.* Vol. 1, pp. 91–96.
Hardin, B. O. & Black, W. L. (1966). Sand stiffness under various triaxial stresses. *J. Soil Mech. & Found. Div. ASCE 92*, SM2, pp. 27–42.
Hardin, B. O. (1985). Crushing of soil particles. *J. Geotech. Eng.* **111**, No. 10, 1177–1192.
Houlsby, G. T. & Wroth, C. P. (1991). The variation of shear modulus of a clay with pressure and overconsolidation ratio. *Soils & Foundations* **31**, No. 3, 138–143.
Iwasaki, T., Tatsuoka, F. & Takagi, Y. (1978). Shear moduli of sands under cyclic torsional shear loading. *Soils & Foundations* **18**, No. 1, 39–56.
Jardine, R. J., Potts, D. M., St. John, H. D. & Hight, D. W. (1991). Some practical applications of a non-linear ground model. *Proc. X ECSMFE, Florence*, Vol. 1, pp. 223–228.
Jefferies, M. G. (1993). Nor-Sand: a simple critical state model for sand. *Géotechnique* **43**, No. 1, 91–103.
Jovičić, V., Coop, M. R. & Simić, M. (1996). Objective criteria for determining G_{max} from bender element tests. *Géotechnique* **46**, No. 2, 357–362.
Jovičić, V. (1997). The measurement and interpretation of small strain stiffness of soils. City University, London, PhD thesis.
Ladd, C. C. & Foot, R. (1974). New design procedure for stability of soft clays. *Proc. ASCE, J. Geotechnical Eng. Div.* Vol. 100, No. GT7.
Lee, I. K. & Coop, M. R. (1995). The intrinsic behaviour of a decomposed granite soil. *Géotechnique* **45**, No. 1, 117–130.
Ko, H. Y. & Scott, R. F. (1967). Deformation of sand in hydrostatic compression. *J. Soil Mech. & Found. Div. ASCE* Vol. 93, No. SM3, pp. 137–156.
Lo Presti, D. C. F. (1995). *General Report: Measurement of shear deformation of geomaterials in the laboratory. Pre-failure deformation of geomaterials*, pp. 1067–1088. Rotterdam: Balkema.
Ni, S. H. (1987). Dynamic properties of sand under true triaxial stress states from resonant column/torsion shear tests. University of Texas at Austin, PhD thesis.
Pestana, J. M. & Whittle, A. J. (1995). Compression model for cohesionless soils. *Géotechnique* **45**, No. 4, 611–633.
Roesler, S. K. (1979). Anisotropic shear modulus due to stress anisotropy. *J Geotech. Eng. ASCE* **105**, No. 7, 871–880.
Schmertman, J. H. (1991). The mechanical aging of soils. The twenty-fifth Karl Terzaghi Lecture. *J. Geotech. Eng.* **117**, No. 9, 1288–1330.
Shirley, D. J. & Hampton, L. D. (1977). Shear-wave measurements in laboratory sediments. *J. Acous. Soc. America* **63**, No. 2, 607–613.
Stallebrass, S. E. & Taylor, R. N. (1997). The development and evaluation of a constitutive model for the prediction of ground movements in overconsolidated clay. *Géotechnique*, **47**, No. 2, 235–253.
Viggiani, G. & Atkinson, J. H. (1995). Stiffness of fine-grained soil at very small strains. *Géotechnique* **45**, No. 2, 249–265.
Wroth, C. P. & Houlsby, G. T. (1985). Soil mechanics, Property characterisation and analysis procedure. *Proc. XI ICSMFE. San Francisco*, pp. 1–55.
Wroth, C. P., Randolph, M. F., Houlsby, G. T. & Fahey, M. (1979). A review of engineering properties of soils with particular reference to the shear modulus. University of Oxford. OUEL Report 1523/84.

Kohata, Y., Tatsuoka, F., Wang, L., Jiang, G. J., Hoque, E. & Kodaka, T. (1997). *Géotechnique* **47**, No. 3, 563–580

Modelling the non-linear deformation properties of stiff geomaterials

Y. KOHATA,* F. TATSUOKA,† L. WANG,‡ G. L. JIANG,* E. HOQUE§ and T. KODAKA‖

The modelling of non-linear stiffness, based on elastic characteristic moduli for stiff geomaterials, is proposed, on the basis of the results of a large number of triaxial compression tests on sedimentary soft rocks, tuff, weathered granite, cement-treated soils, densely compacted gravels and sands. Axial strains from less than 0·001% to about 1% were measured by using a local gauge, while elastic Young's moduli E^e were obtained by applying a series of small unload/reload cycles of axial stress during each monotonic loading test. Strain increments less than about 0·001% were nearly recoverable and strain-rate-independent. For sedimentary soft rocks and cement-treated soils, the elastic Young's moduli E^e from laboratory tests using high-quality undisturbed samples were very similar to the equivalent values obtained from field seismic surveys. The dependence of E^e on stress state is summarized, and the effects of geomaterial type are analysed. The non-linearity during shear deformation is regarded as being caused by damage to the elastic deformation properties and the occurrence of plastic strains. The effects of cementation and structuration on elastic deformation properties are discussed. A method to estimate the tangent Young's modulus E_{tan} of a given geomaterial at a given stress state from the E^e value evaluated by field seismic surveys and the non-linear tangent stiffness model is proposed.

KEYWORDS: deformation; elasticity; fabric/structure of soils; laboratory tests; stiffness.

L'article présente une modélisation de la rigidité non linéaire basée sur les modules d'élasticité de matériaux rigides, qui repose sur les résultats d'un grand nombre d'essais de compression triaxiale sur des roches tendres sédimentaires, du tuf, du granit désagrégé, des sols traités au ciment, des graviers fortement tassés et des sables. Un extensomètre local a permis de mesurer des contraintes axiales allant de moins de 0,001% à 1% environ, tandis qu'une série de cycles de décharge/recharge de contraintes axiales pendant chaque essai de charge monotone a permis d'obtenir des modules d'élasticité E^e de Young. Les écarts de contrainte de moins de 0,001% étaient presque récupérables et ne dépendaient pas du taux de contrainte. Dans le cas des roches tendres sédimentaires et des sols traités au ciment, les essais de laboratoire sur des échantillons non perturbés de grande qualité ont produit des modules d'élasticité E^e très semblables aux valeurs correspondantes fournies par les relevés sismiques sur le terrain. L'article résume la façon dont l'élasticité E^e dépend de l'état de contrainte et analyse les effets du type de matériau. La modélisation suppose que la non-linéarité du cisaillement est causée par la détérioration de l'élasticité et l'apparition de déformations plastiques. L'article examine les effets de la cimentation et de la structuration sur les propriétés des matériaux. Enfin, il propose une méthode qui permet d'estimer le module tangent de Young E_{tan} d'un matériau donné à un état de contrainte donné, à partir de la valeur E^e évaluée à l'aide des levés sismiques sur le terrain et du modèle de rigidité tangentielle non linéaire.

INTRODUCTION

The deformation characteristics at small strains of stiff geomaterials (e.g. sedimentary soft rocks, cement-treated soils and well-compacted granular materials) are important in many geotechnical engineering problems, for example, precise prediction of the displacements of the foundations of large important structures or the ground movements caused by deep excavation. In such cases, strains in the ground at working loads are smaller than about 0·1%, or at the largest, about 0·5% (e.g. Burland, 1989; Tatsuoka & Kohata, 1995). Therefore, the elastic deformation properties and stiffness at these strains are essential design parameters. However, in many case histories in the past, the stiffnesses of the ground determined by conventional laboratory

Manuscript received 9 October 1996; revised manuscript accepted 14 February 1997.
Discussion on this paper closes 1 November 1997.
* Railway Technical Research Institute, Tokyo.
† University of Tokyo.
‡ Chuo Kaihatsu Corporation, Japan.
§ Bangladesh University of Engineering and Technology.
‖ Institute of Industrial Science, University of Tokyo.

and field testing methods, except for field seismic survey, were found to be too small to explain the full-scale field behaviour.

Results of recent triaxial tests with accurate axial strain measurements using very small strains have revealed the following.

(*a*) Axial strains obtained from the axial displacement of the loading piston or the top cap in conventional triaxial tests are larger than the true values, owing to the effects of bedding error and system compliance.
(*b*) Apparently different stiffnesses obtained from different laboratory and field tests and from full-scale field behaviour can be linked to each other only when the effects of influencing factors including sample disturbance, the dependences of stiffness on strain level and pressure level, and viscous effects are properly considered.
(*c*) For cohesionless soils, the elastic Young's modulus defined for the major elastic principal strain increment occurring in a certain direction is a unique function of the normal stress in that direction (e.g. Jamiolkowski *et al.*, 1991; Tatsuoka & Shibuya, 1992; Kohata *et al.*, 1994; Tatsuoka & Kohata, 1995; Jardine, 1995).

Therefore, the elastic properties of geomaterials become anisotropic under anisotropic stress states (i.e. stress-system-induced anisotropy).

Strain non-linearity is a common feature of a variety of geomaterials. This property has usually been represented by the relationship between the secant Young's modulus E_{sec} and axial strain ε_v (or between secant shear modulus G_{sec} and shear strain γ). However, this relationship is not unique, but depends on the definition of the origin for strain and on the stress path followed. Therefore, the use of the tangent modulus E_{tan} (or G_{tan}) as a function of the current stress state (i.e., as a function of shear stress level and pressure level), instead of E_{sec} (or G_{sec}), can be more appropriate.

This paper describes first the dependence of the elastic Young's modulus E^e on the current stress state for various stiff geomaterials. Then, a method is proposed to obtain the value of E_{tan} of a given geomaterial at a given stress state, from the stress-state-dependent E^e value, while taking into account the non-linearity due to shearing.

TRIAXIAL TESTING METHOD

Test materials

The test materials and the specimen dimensions (height h and diameter ϕ for solid cylindrical specimens and height h and width w for rectangular prismatic specimens) were as follows:

(*a*) air-dried Hostun sand ($D_{50} = 0{\cdot}31$ mm and $U_c = 1{\cdot}94$) (h 150 mm, ϕ 75 mm), prepared by pluviating particles in air
(*b*) an air-dried well-graded crushed sandstone, namely Nerima gravel ($D_{50} = 3{\cdot}9$ mm and $U_c = 14{\cdot}0$) (h 600 mm, ϕ 300 mm and h 570 mm, w 230 mm × 230 mm), well-compacted to a dry density $\rho_d = 1{\cdot}77–1{\cdot}98$ g/cm^3 by vibrator compaction (Jiang *et al.*, 1997); and similar gravels, namely Chiba gravel ($D_{50} = 6{\cdot}6$ mm and $U_c = 12{\cdot}1$) (h 570 mm, w 230 mm × 230 mm), vibration-compacted to $\rho_d = 1{\cdot}80$ g/cm^3, and Nagoya gravel ($D_{50} = 4{\cdot}76$ mm and $U_c = 91{\cdot}8$) (h 600 mm, ϕ 300 mm), vibration-compacted to $\rho_d = 2{\cdot}21–2{\cdot}25$ g/cm^3 (Kohata *et al.*, 1994)
(*c*) saturated, undisturbed samples from a rock-fill dam core constructed by compacting weathered sandstone ($D_{50} = 1{\cdot}5$ mm and $U_c = 676$), sampled 21 years after construction (h 200 mm, ϕ 82 mm) (Tatsuoka & Kohata, 1995)
(*d*) saturated core samples of sedimentary soft silt-to-sandstone of about 5 million years age (Kobe group) (h 100 mm, ϕ 50 mm), obtained by block sampling at the bottom of an excavation taken to a depth of 61 m at the construction site of Anchor 1A for the Akashi Kaikyo (Strait) Suspension bridge, one of the Honshu–Shikoku bridges
(*e*) saturated core samples of sedimentary soft mudstone of about 2 million years age (Kazusa group) (h 150 mm, ϕ 55 mm), obtained by block sampling at the bottom of an excavation taken to a depth of 70 m at the Sagamihara Test Site (Tatsuoka & Kohata, 1995; Tatsuoka *et al.*, 1997)
(*f*) saturated samples of cement-treated sandy soil (CTSS-dry), block-sampled from a test pit for full-scale placement tests (h 630 mm, w 250 mm × 250 mm) (Uchida *et al.*, 1993; Tatsuoka *et al.*, 1997a) (this cement-treated sandy soil, called 'dry type', was used to construct the flat part of the Kisarazu man-made island on the Trans-Tokyo Bay Highway; the material was produced, first, by mixing sand in air at a natural water content (about 10%) with cement containing 7·5% by weight of dry sand and an anti-segregation chemical, and then by pouring into the sea through a large-diameter pipe; the strength of this material after placement beneath the sea is lower than that of another type of CTSS prepared in a slurry state prior to underwater placement.

Triaxial testing systems

Small triaxial apparatuses were used for testing sand, weathered sandstone and sedimentary soft mudstone, and a large triaxial apparatus was used for testing gravels and cement-treated sandy soil. The specimens of sand and gravel were kept air-dried throughout the tests. A very sensitive load cell was installed inside the triaxial cell. Local axial strains were measured accurately by means of a pair of local deformation transducers (LDTs) (Goto *et al.*, 1991) attached to the surface opposite diametrically of the cylindrical specimen or to opposite surfaces of a rectangular prismatic specimen. An external displacement transducer and two proximity transducers were also used to measure the displacements of the loading piston and of the top cap of the specimen respectively. For the small cylindrical specimens of sand, lateral strains were obtained by measuring the change in the specimen diameter with two sets of three proximity transducers, positioned opposite each other around the specimen diameter (Tatsuoka *et al.*, 1994). For the rectangular prismatic specimen of gravel, lateral strains were obtained by averaging the readings from four pairs of lateral LDTs placed horizontally on the lateral surfaces of the specimen (Hoque *et al.*, 1996).

Each specimen was isotropically consolidated (except where stated otherwise), followed by drained triaxial compression (TC) at an axial strain rate of 0·06%/min under a constant confining pressure. A number of small unload/reload cycles were applied during TC.

Three types of Young's moduli were defined as shown in Fig. 1. The initial Young's modulus E_0 was obtained from the initial linear elastic part of the stress–strain curve at axial strains ε_v less than about 0·001%, measured with LDTs. The tangent modulus $E_{tan} = dq/d\varepsilon_v$ was obtained from each primary stress–strain curve. The equivalent Young's modulus E_{eq} was obtained from the peak-to-peak secant modulus for each small unload/reload cycle, for which the strains were essentially elastic. Therefore, for a given deviator stress increment, E_{tan} and E_{eq} reflect the total and elastic strain increments $d\varepsilon_v{}^t$ and $d\varepsilon_v{}^e$ respectively.

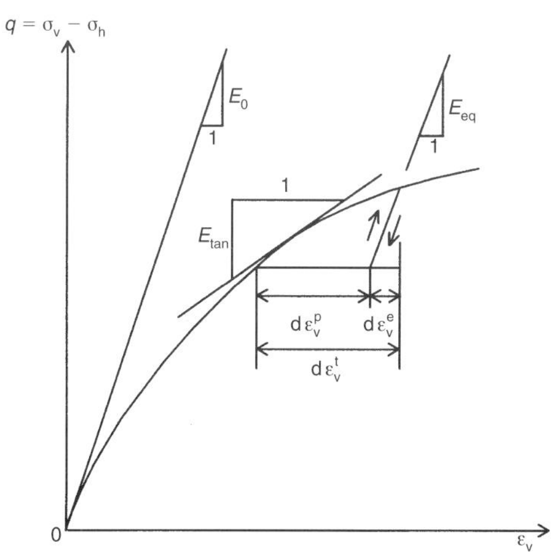

Fig. 1. Definitions of Young's moduli

TYPICAL STRESS–STRAIN RELATIONS OF GEOMATERIALS

Figures 2–7 present typical stress–strain curves for the various materials. Each figure is in three parts, (a), (b) and (c), each with a successively decreasing strain scale. The values of E_{tan} are presented later.

The following two types of stress–strain curve were obtained from the TC tests. In the first type, the value of E_{tan} decreased consistently from the start of loading until the peak stress state (decreasing E_{tan} type). In the second type, as the axial stress increased, first the value of E_{tan} decreased, but soon it started increasing, until as failure was approached, it began decreasing again (increasing E_{tan} type).

Decreasing E_{tan} type

This type of stress–strain curve was observed in most of the TC tests on virgin samples of sands and gravels, which had not been pre-sheared. Typical examples (marked 'virgin') are shown in Fig. 2 (Hostun sand), Fig. 3 (well-graded, crushed sandstone gravel) and Fig. 4 (undisturbed dam-core material). This type was also observed for the well-cemented geomaterials, namely sedimentary soft mudstone from Sagamihara and cement-treated sandy soil (CTSS-dry) (Figs 5 and 6). These specimens were consolidated isotropically to $\sigma_h{}' =$ 0·70 MPa for Sagamihara mudstone and 196 kPa for the CTSS-dry specimen.

The following features of behaviour may also be noted from Figs 2–6. (*a*) Owing to the effects of bedding error and system compliance, the externally measured axial strains are considerably larger than the locally measured ones, and therefore the stress–strain relations based on conventional axial strain measurements are totally unreliable. (*b*) The behaviour at small strains is more linear and more recoverable for the cemented materials than for the uncemented granular materials. It is considered that the decreasing E_{tan} type stress–strain relation results from an increasing rate of development of plastic strains with q, while the effects of the increase in the elastic Young's modulus with an increase in q (as discussed later) are masked.

Increasing E_{tan} type

In addition to the result from the virgin sample (HOSTN6), the results from the other TC tests on the three specimens of Hostun sand, which had

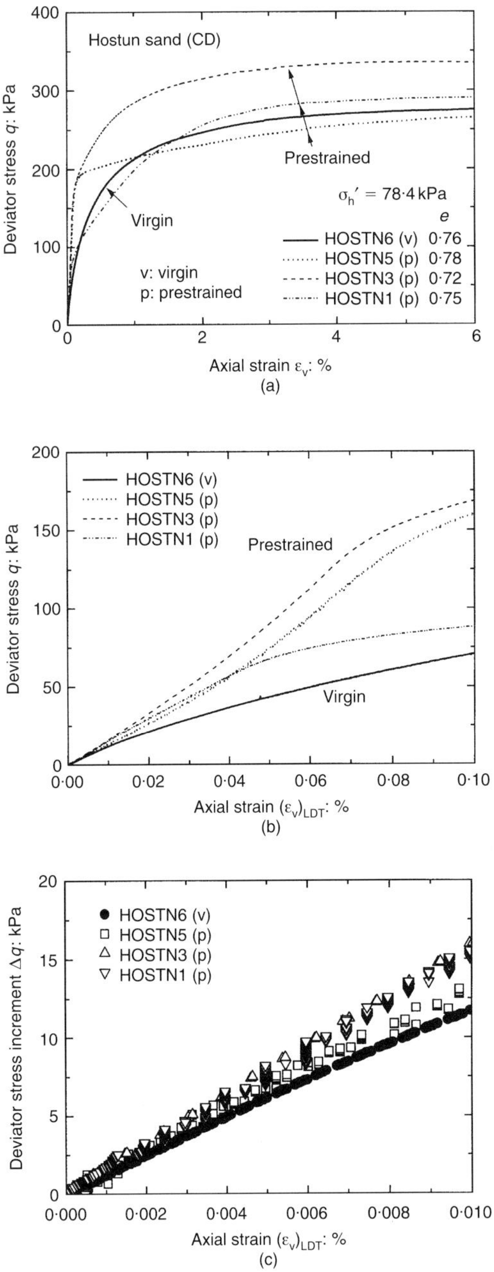

Fig. 2. CD TC test results, Hostun sand

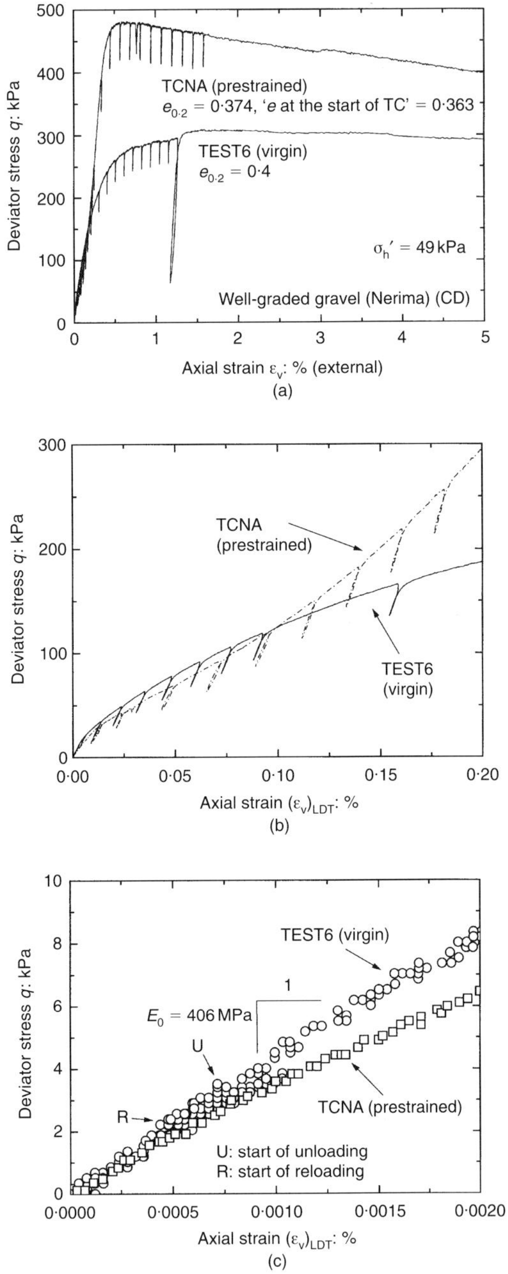

Fig. 3. CD TC test results, well-graded crushed gravel (Nerima)

been cyclically prestrained under drained conditions, are presented in Fig. 2. The cyclic prestraining (CP) was applied symmetrically about an isotropic or anisotropic neutral stress state after isotropic consolidation. Specimens HOSTN5 and HOSTN3, which had been isotropically consolidated to 78·4 kPa, were subjected to 65 000 and 40 000 cycles respectively of axial stress between 78·4 and 196 kPa, with a single-amplitude axial strain at the residual condition $(\varepsilon_v)_{SA}$ equal to about 0·03%, until the residual axial strains $(\varepsilon_v)_{res}$ became 0·98% and 0·1% respectively at the end of CP. Specimen HOSTN1 was subjected to CP at

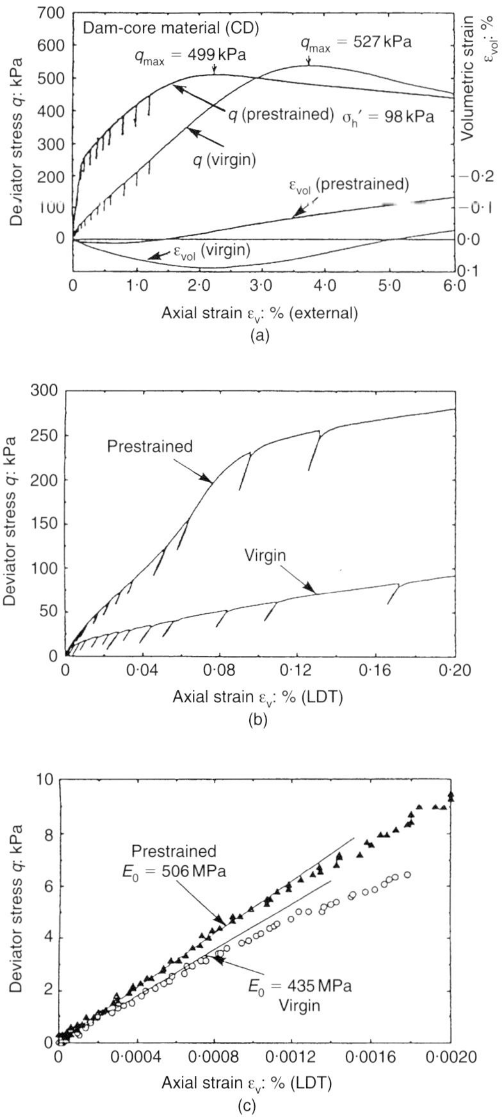

Fig. 4. CD TC test results, undisturbed dam-core material

three stages: CP-1 ($\sigma_v' = 78{\cdot}4-196$ kPa, 25 000 cycles of CP, $(\varepsilon_v)_{SA}$ about 0·03% and $(\varepsilon_v)_{res} = 0{\cdot}23\%$), CP-2 ($\sigma_v' = 39{\cdot}2-196$ kPa, 30 000 cycles of CP, $(\varepsilon_v)_{SA}$ about 0·06% and $(\varepsilon_v)_{res} = -0{\cdot}08\%$) and CP-3 ($\sigma_v' = 24{\cdot}5-132{\cdot}3$ kPa, 120 000 cycles of CP, $(\varepsilon_v)_{SA}$ about 0·05% and $(\varepsilon_v)_{res} = -0{\cdot}01\%$). In Fig. 2(a), axial strains ε_v measured with a proximeter are used after approximately $\varepsilon_v = 1{\cdot}5\%$, where the LDTs stopped functioning. In Fig. 2(c) for comparison purposes, these relationships have been plotted after shifting to a common origin.

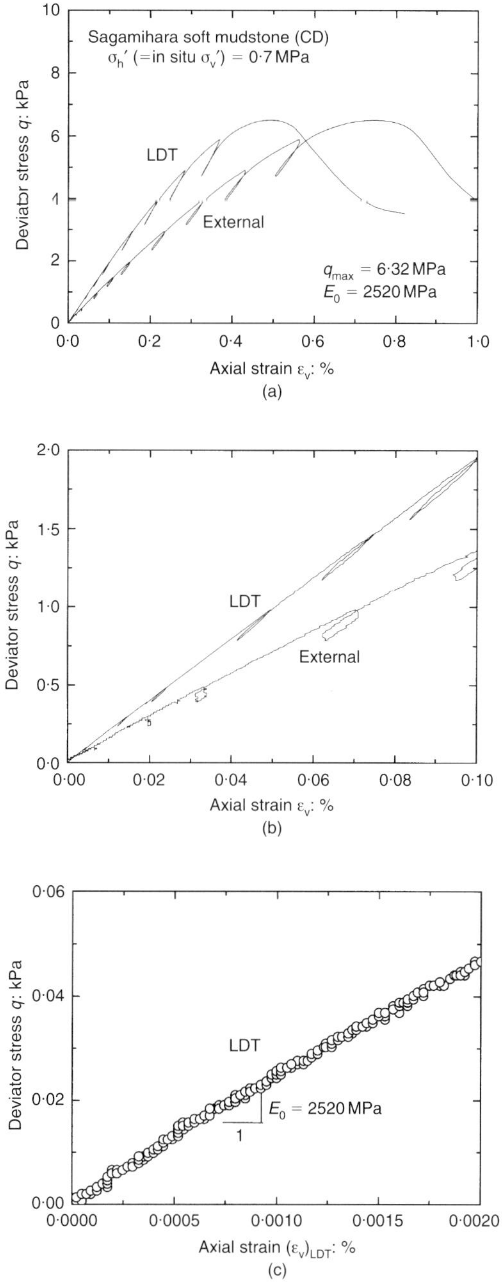

Fig. 5. CD TC test results, Sagamihara soft mudstone

The following features may be noted.

(*a*) For specimens HOSTN3 and HOSTN5, which had experienced CP entirely on the TC side, the stress–strain relations exhibit a noticeable S-shape (Fig. 2(b)).

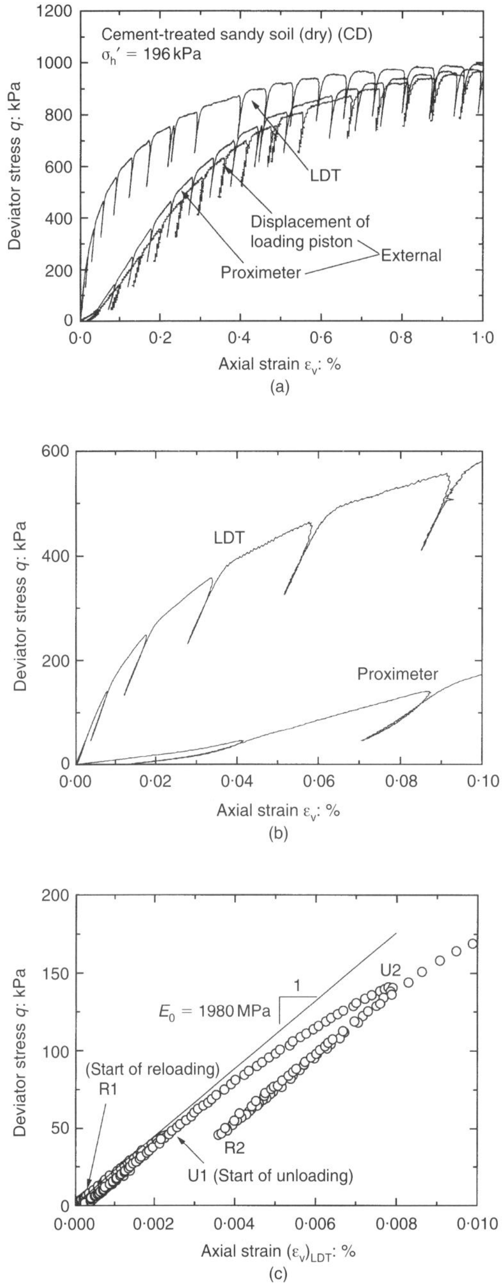

Fig. 6. CD TC test results, cement-treated sandy soil (dry type)

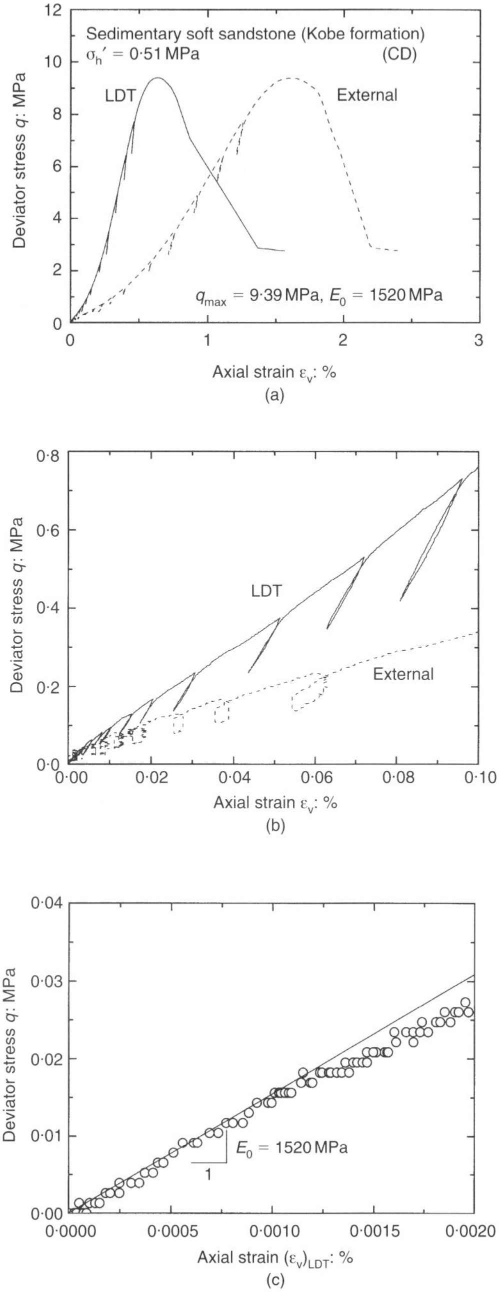

Fig. 7. CD TC test results, sedimentary soft silt-to-sandstone (Kobe formation)

(*b*) Owing to the effects of CP, the linear range in the stress–strain relations increased up to about $\varepsilon_v = 0{\cdot}01\%$ (Fig. 2(c)).

(*c*) Specimen HOSTN1 exhibits a softer response than that of the virgin specimen (HOSTN6) at intermediate strain levels during monotonic loading (Fig. 2(a)). This result is considered to be due to the effect of CP-3, during which the specimen came close to the failure state in triaxial extension.

Figures 3(a)–(c) compare the results from the CD TC tests on the virgin specimen (TEST6) and the prestrained specimen (TCNA) for Nerima gravel. For the virgin specimen, the behaviour is recoverable up to ε_v of about 0·001% (Fig. 3(c)). The cyclically prestrained specimen had been subjected to cyclic loading in the axial and lateral directions alternately at 108 stress points for a range of σ_h' of 49–441 kPa and a range of σ_v'/σ_h' of 0·5–2·2. The single amplitude of axial strain in the 'vertical' cyclic loading tests and that of lateral strain in the 'horizontal' cyclic loading tests was about 0·002%. The numbers of cycles at each stress point were 6 and 3 respectively in the vertical and

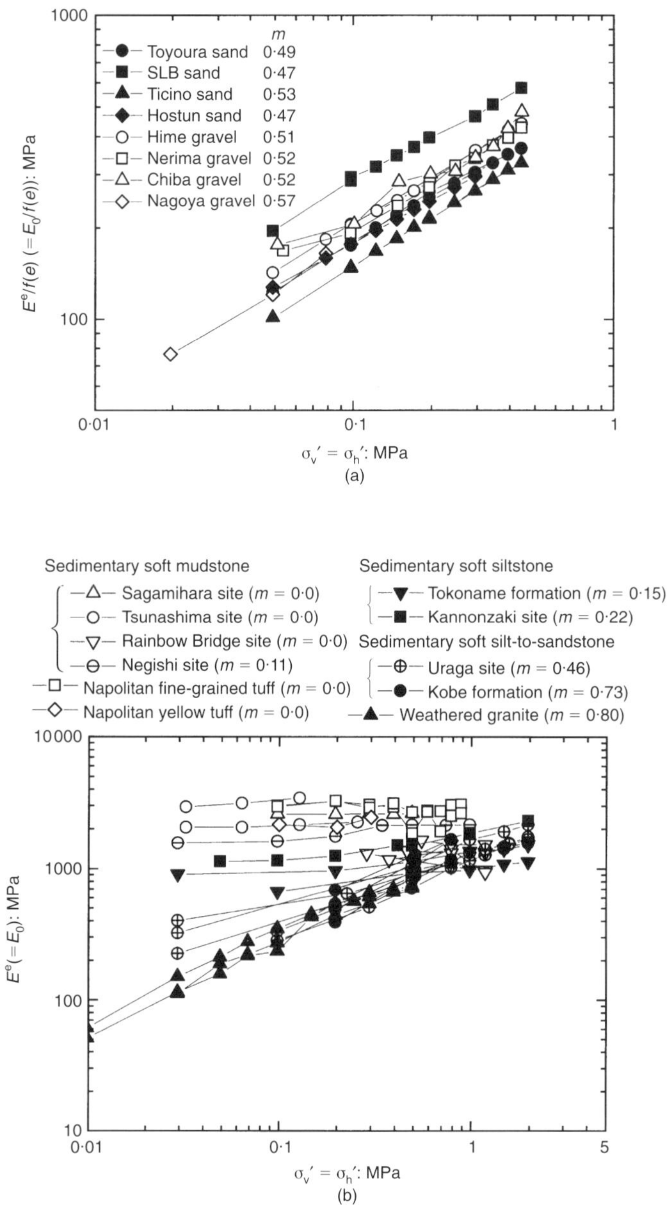

Fig. 8. Summary of the relationships between the elastic Young's modulus E_0 and the confining pressure σ_h': (a) uncemented granular geomaterials; (b) cemented geomaterials (sedimentary soft rocks)

horizontal cyclic loading tests. The total numbers of cycles were 648 in the vertical cyclic loading tests and 324 in the horizontal cyclic tests. It is seen that the stress–strain relation of the CP specimen is S-shaped (Figs 3(a) and (b)) and it has a larger linear initial part (Fig. 3(c)), similar to the case of Hostun sand. The peak strength of the CP specimen is considerably larger than that of the virgin specimen. This difference is caused partly by the difference in void ratio and partly by the effects of the CP and overconsolidation.

The stress–strain relations from the TC tests at $\sigma_h' = 98$ kPa of the undisturbed dam-core samples are shown in Figs 4(a)–(c). The CP specimen had been consolidated anisotropically to a stress state of $\sigma_v' = 196$ kPa and $\sigma_h' = 98$ kPa, followed by 10 000 cycles of drained cyclic loading with $(\varepsilon_v)_{SA}$ about 0·02%. As a result of the relatively large amount of CP, the E_0 value increased slightly, while the linearity increased noticeably and the stress–strain relation became S-shaped.

It has been shown above that the stress–strain relations of sand and gravel can become S-shaped as a result of cyclic prestraining. Similar but more noticeably S-shaped stress–strain relations were obtained from CD TC tests on 'virgin' specimens of sedimentary soft silt-to-sandstone of the Kobe formation (Fig. 7).

The elastic Young's modulus E_{eq} increases with an increase in the axial stress; this is more or less reflected in the E_{tan} value. It is considered that in the tests on sand, gravel and soft rock which exhibited the increasing E_{tan} type of stress–strain relation, this effect of the increase in E_{eq} overwhelmed the effect of increasing plastic strain increments, resulting in an increase in the E_{tan} value with an increase in q (i.e. S-shaped stress–strain relation).

PRESSURE-LEVEL-DEPENDENCE OF ELASTIC YOUNG'S MODULUS

This feature is one of the key components of the model. Fig. 8 summarizes the relationships between the elastic Young's modulus E_0 and the confining pressure σ_h' obtained at different isotropic stress states. This type of data for cement-treated soils is reported in Tatsuoka *et al.* (1995a). In these tests, each E_0 value was obtained by TC loading up to an ε_v of about 0·005% and subsequent unloading, repeated at different isotropic and anisotropic stress states using a single specimen. In Fig. 8(a), the E_0 values have been normalized by using a function $f(e) = (2{\cdot}17 - e)^2/(1 + e)$ (Hardin & Richart, 1963) to account for differences in the void ratio, which was 28% at the largest. This void function has been found to be relevant to cohesionless sands by a number of other workers (e.g. Iwasaki *et al.*, 1978). The values of E_0 for each type of geomaterial can be expressed as

$$E_0 = C\sigma_v'^m \qquad (1)$$

It may be seen from Fig. 8(b) that the stiffnesses of the sedimentary soft mudstones and siltstones are independent of pressure change (i.e. $m = 0$), while those of the sedimentary silt-to-sandstone and the weathered granite are as sensitive to pressure as the granular materials. The values of the power m for the granular materials are similar to each other and nearly equal to 0·5. A power $m = 0{\cdot}5$ for dry Hostun sand was also reported by Hicher (1996). It was also found that for sedimentary soft rocks having similar geological ages, the value of the power m increases with an increase in the mean diameter D_{50}, in contrast to nearly no dependence of m on D_{50} for the uncemented granular materials (Fig. 9; the symbols for the uncemented granular materials are the same as those used in Fig. 8(a)). It is considered that for similar geological ages, cementation becomes weaker as the particle size increases.

STRESS-SYSTEM-INDUCED ANISOTROPY OF ELASTIC YOUNG'S MODULUS

Figure 10 shows the Young's moduli E_v and E_h defined for major elastic principal strain increments occurring in the axial and lateral directions respectively, plotted against the axial stress σ_v' and the horizontal stress σ_h' for a well-graded crushed sandstone gravel (Nerima gravel). Similar results for Chiba gravel have been reported by Jiang *et al.* (1997). The values of E_v, denoted by E_{eq} above, were obtained from vertical cyclic loading tests, applying small cyclic axial stresses σ_v' with a single-amplitude axial strain of about 10^{-5} while

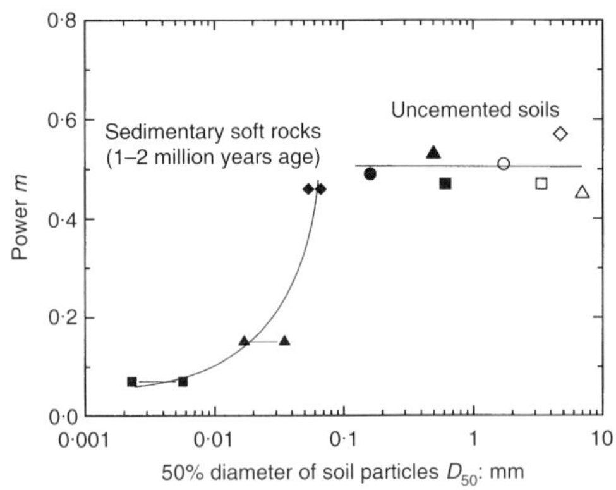

Fig. 9. Dependence of power m on D_{50} for stiff geomaterials

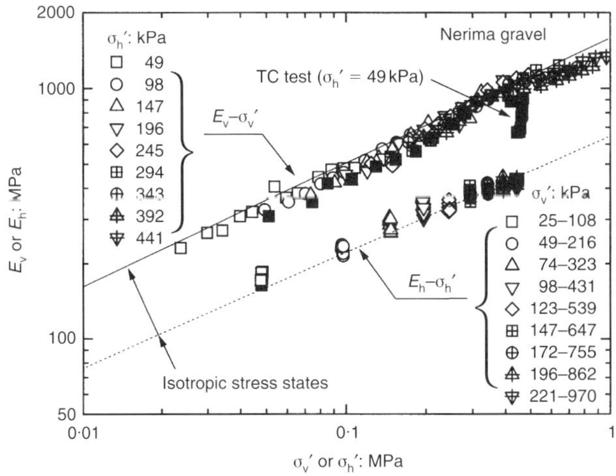

Fig. 10. Relationships between E_v and σ_v' and between E_h and σ_h', for Nerima gravel

keeping the two lateral stresses σ_h' constant. The values of E_h were also obtained from horizontal cyclic loading tests in which the two lateral stresses σ_h' were changed cyclically, with a small-amplitude lateral strain of about 10^{-5} while keeping σ_v' constant. These cyclic loading tests were performed at various isotropic and anisotropic stress states after having consolidated the specimen sufficiently at each stress state. According to the theory of elasticity, the following equations can be derived for the vertical and horizontal cyclic loading tests:

$$E_v = \Delta\sigma_v'/\varepsilon_v, \ \nu_{vh} = -\Delta\varepsilon_h/\Delta\varepsilon_v \tag{2}$$

$$E_h = (1-\nu_{hh})\Delta\sigma_h'/\Delta\varepsilon_h, \ 2\nu_{hv}/(1-\nu_{hh}) = -\Delta\varepsilon_v/\Delta\varepsilon_h \tag{3}$$

The values of ν_{hh} could not be determined from these cyclic tests (Jiang *et al.*, 1997). In the present study, it was assumed that ν_{hh} = 'ν_{vh} at isotropic stress states'. Similar tests were performed for other types of granular materials (Hoque *et al.*, 1996), and the values of 'ν_{vh} at isotropic stress states' were found to be 0·17, 0·10, 0·16, 0·14, 0·24 and 0·24 for Toyoura, SLB and Ticino sands, and Hime, Nerima and Chiba gravels respectively. When the values of ν_{hh} are of the same order as these values, which are much smaller than unity, the effects of possible errors in the values of '$1 - \nu_{hh}$' (i.e. the values of E_h based on equation 3) are not significant. For example, the ratio (1 − 0·25)/(1 − 0·15) is 0·88. In Fig. 10, different symbols are used for E_v values measured at different σ_h' values, while the same symbol is used for E_h values measured at each σ_h' value and the range of σ_v' covered at each σ_h' value is indicated. It may be seen from Fig. 10 that E_v is a unique function of σ_v' and independent of σ_h', while E_h is a unique function of σ_h' and independent of σ_v'. Then, the ratio E_v/E_h becomes a function of σ_v'/σ_h', as summarized in Fig. 11 for the granular materials.

The following trends of behaviour can be noted from Fig. 11.

(*a*) The ratio E_v/E_h at σ_v'/σ_h' = 1·0 reflects the inherent anisotropy, which is noticeably large for compacted Nerima and Chiba gravels and

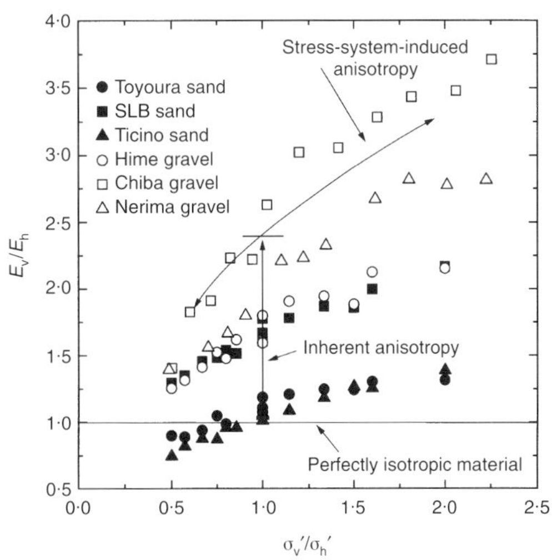

Fig. 11. Stress-system-induced anisotropy for sands and gravels

for subrounded SLB sand and Hime gravel, and small for air-pluviated subangular Toyoura and Ticino sands. These values of E_v/E_h at $\sigma_v'/\sigma_h' = 1{\cdot}0$ are affected by the assumption regarding the ν_{hh} values. To confirm the inherent anisotropy observed in this study, more sophisticated tests, where only one horizontal stress component σ_h' is changed while keeping the other horizontal component σ_h' and the vertical component σ_v' constant (i.e. true triaxial tests) will be required.

(*b*) E_v/E_h increases as σ_v'/σ_h' increases, for all the materials. The stress-system-induced anisotropy of the elastic deformation properties of sand as shown above has been discussed by many other researchers (Jamiolkowski *et al.*, 1991; Lo Presti & O'Neill, 1991; Stokoe *et al.*, 1991; Tatsuoka & Kohata, 1995). For the testing conditions in the present study, cross-anisotropic elasticity is a realistic model for the materials.

The stress-state dependence of the stiffness of sedimentary soft silt-to-sandstone was examined by measuring the values of elastic modulus E_v after sufficient consolidation at each isotropic or anisotropic stress state (only under TC stress conditions), using a single specimen with a confining pressure up to 1960 kPa and an axial stress up to 3136 kPa. It may be seen from Fig. 12 that the values of E_v are controlled by σ_v' to a much larger extent than they are by σ_h', particularly at low pressure levels. These results suggests that the elastic properties of sedimentary soft rock also become anisotropic at anisotropic stress states.

Correspondingly, the elastic modulus E_v of sand and gravel increases as the axial stress σ_v'

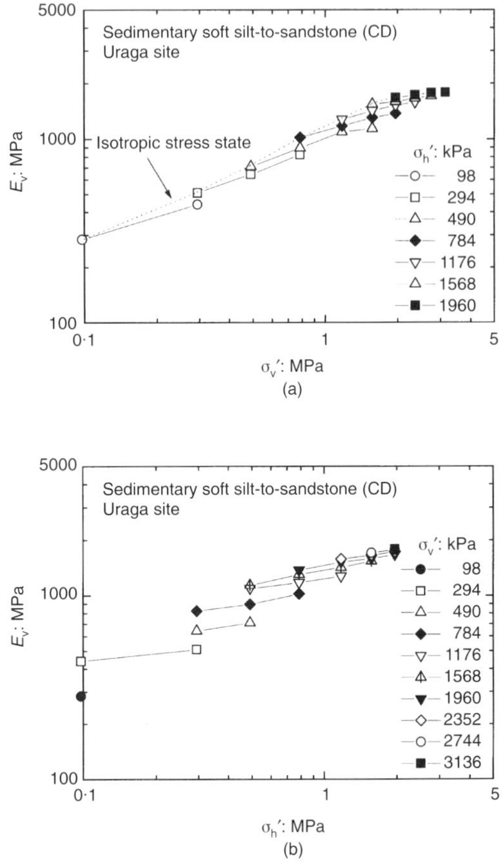

Fig. 12. Stress-system-induced anisotropy for sedimentary soft silt-to-sandstone, Uraga site: (a) E_v–σ_v' relations; (b) E_v–σ_h' relations

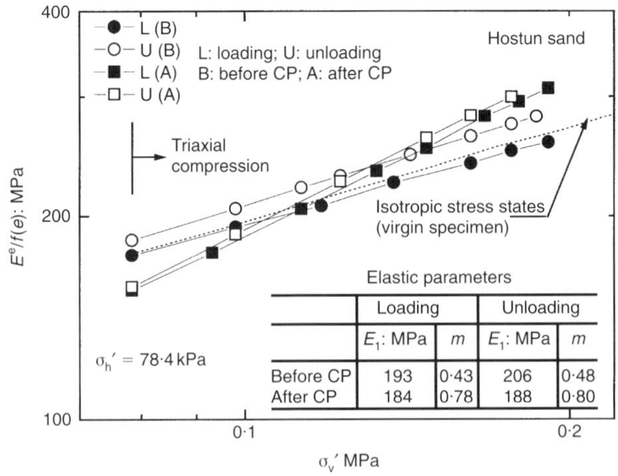

Fig. 13. Relationships between $E^e/f(e)$ and σ_v' along the CP stress path for Hostun sand

increases during triaxial compression (TC) at a constant σ_h' (Hoque *et al.*, 1996; Jiang *et al.*, 1997). It may be seen from Fig. 13 that for Hostun sand (test HOSTN3, Fig. 2), the E_{eq} values measured during the first cycle of loading and unloading in TC at a constant σ_h' (78·4 kPa) change as the axial stress σ_v' changes in the same way as they do at isotropic stress states. This is also the case with the E_v values measured after a large number of cyclic loads with the same axial stress amplitude, but the power m has increased noticeably, owing to the effects of cyclic prestraining.

Figure 14 shows the pressure-level dependence of E_0 and E_v (= E_{eq}) for uncemented soils and cemented soil. Note that $(\sigma_v'/\sigma_h')_c$ is the consolidated stress ratio. The E_0 values were obtained after sufficient consolidation, either at different anisotropic stress states of $(\sigma_v'/\sigma_h')_c = 2{\cdot}0$ (Fig. 14(a)) or at isotropic stress states (Figs 14(b)) and (c)). The E_{eq} values were measured by applying small unload/reload cycles during monotonic TC shear to failure. It is seen from Fig. 14 that the values of E_v (= E_{eq}) of both the dam-core material (Fig. 14(a)) and the sedimentary soft silt-to-sandstone (Fig. 14(b)) increase noticeably with σ_v' during TC, as do the E_0 values. However, the rate of increase of E_{eq} with an increase of σ_v' during TC is smaller than that of E_0. On the other hand, the E_0 value of the artificially cemented material (Fig. 14(c)) does not increase with σ_v' at isotropic stress states, while the E_v value decreases with increasing σ_v' during TC. The observations described above could be considered to be due to both damage to the microstructure by shear straining and creep deformation during TC. That is, a sort of microstructure should have been formed by heavy compaction of Nerima gravel. For the dam-core material, some microstructure should have been formed in the dam core during heavy compaction and by ageing effects during the subsequent long period of time, and also to some extent by cyclic prestraining in the laboratory. Natural and artificial cementation of sedimentary silt-to-sandstone and CTSS-dry material, which do not have very strong cementation, are also damaged by shearing.

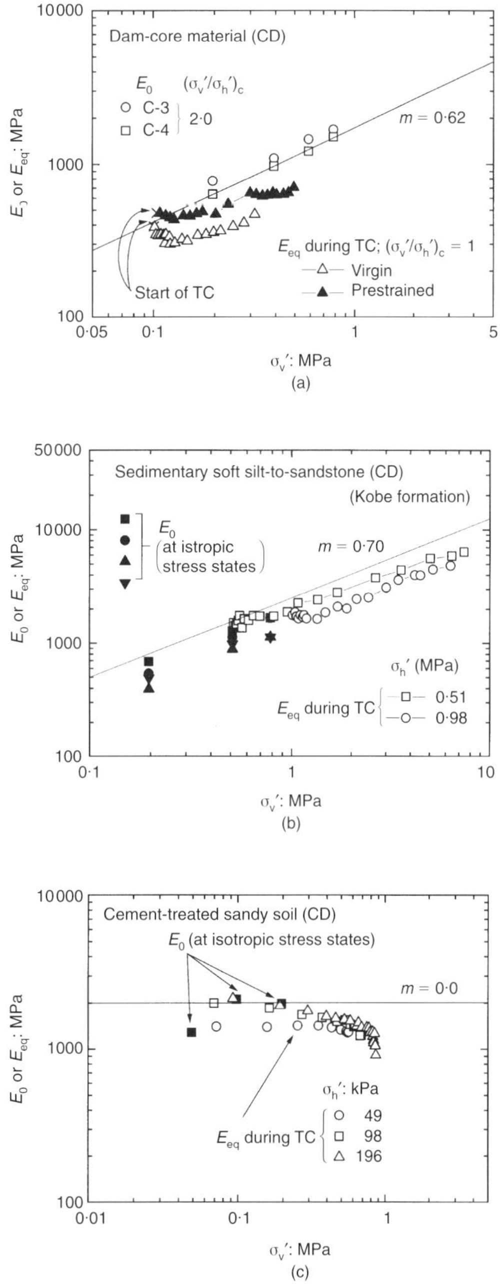

Fig. 14. Pressure level dependence of E_0 at isotropic stress states and of E_{eq} during TC: (a) dam-core material; (b) sedimentary soft silt-to-sandstone; (c) cement-treated sandy soil (dry type)

COMPARISON OF G_0 AND G_f FOR CEMENTED GEOMATERIALS

It is very convenient when the elastic deformation modulus E_0 or $G_0 = E_0/2(1+\nu)$ can be estimated from the field shear wave velocity V_s. It has been shown that the G_0 values of high-quality sand samples measured by TC tests are very similar to the respective field values $G_f = \rho V_s^2$ (Tokimatsu & Hosaka, 1986), and this is also the case with stiff clays (Tatsuoka *et al.*, 1995a, 1995b).

Figure 15 compares the elastic shear moduli G_0 obtained from TC tests with the respective values of G_f from field seismic surveys, for cemented geomaterials and stiff Pleistocene clays. In this comparison, some effects of anisotropy on the G_0

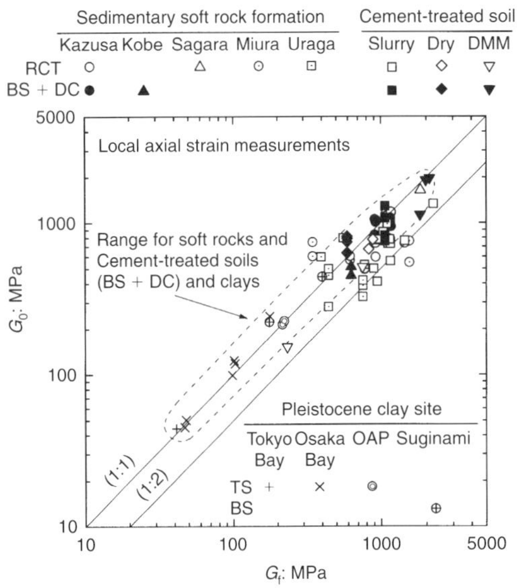

Fig. 15. Summary of relations between G_0 and G_f for cemented geomaterials (data for sedimentary soft rocks of Sagara, Miura and Uraga groups from Tatsuoka *et al.*, 1995a); BS, block sampling; DC, direct coring; DMM, deep mixing method; RCT, rotary coring; TS; fixed-piston thin-wall sampling; OAP; Osaka Amenity Park in Osaka City

and G_f values may be involved. Sedimentary soft mudstone of the Kazusa group at Sagamihara exhibited only slight inherent anisotropy (Tatsuoka *et al.*, 1997), and the effects of stress-system-induced anisotropy would also be very small, considering that the power m is nearly zero (Fig. 8(b)). Because of scatter among the data, however, it is difficult to evaluate the possible effects of anisotropy on the G_0 and G_f values. In Fig. 15, 'DMM' refers to the deep mixing method by which a clay deposit can be improved by being mixed with cement slurry at a large depth. The values of G_0 were evaluated from the respective E_0 values obtained from local axial strains, by using an appropriate value of Poisson's ratio. Each data point represents the average value of several G_0 values obtained from a set of TC tests for a similar condition, plotted against the equivalent value of G_f. For both the cemented materials and the clays, the values of G_0 are generally close to the respective values of G_f. However, the values of G_0 from TC tests using samples of cemented materials obtained by rotary core tube sampling (RCT) showed a large scatter; some of the values are noticeably lower than the respective G_f values. In contrast, the values of G_0 of the least-disturbed samples, obtained by block sampling (BS) and direct sampling (DS), are generally much closer to the respective G_f values. This result also indicates that G_f and E_f represent the elastic shear and Young's moduli of the geomaterial in the field; they can be used in evaluating ground movements and structure displacements when the ground is subjected to static and dynamic loads, taking into account the effects of inherent and stress-system-induced anisotropy, non-linear effects and viscous effects, among others.

MODELLING OF NON-LINEAR DEFORMATION PROPERTIES

The model presented below is based on the following two types of non-linearity: non-linearity due to pressure change, which is modelled through the dependence of the elastic Young's moduli on pressure as described above, and non-linearity due to shear straining. The second part will be discussed first.

Dependence of stiffness on shear strain or shear stress level

A plot of the secant Young's modulus E_{sec} against axial strain ε_v has been the most popular method to represent the strain non-linearity of the stiffness of geomaterials. However, E_{sec} is not unique, but is controlled by the stress state at the start of the loading or the definition of the origin for the axial strain ε_v. As the value of E_{tan} is not affected by the definition of the origin for ε_v, the relationship between E_{tan} and ε_v obtained by TC tests, for Hostun sands and Nerima gravel, is plotted in Figs 16(a) and (b). However, the ε_v value depends on the definition of the origin for ε_v. One of the other, more general, methods of presentation would be the E_{tan} against q/q_{max} plot, as shown in Fig. 17. This relation is much more independent of the definition of the origin of ε_v. In this figure, the values of $E_{eq}(= E_v)$ measured during TC are also shown. Also plotted against q/q_{max} is the Young's modulus E^e defined for elastic axial strain increments measured at undamaged elastic states (i.e. at isotropic stress states), given as

$$E^e = E_0(\sigma_v'/(\sigma_v')_0)^m \tag{4}$$

where σ_v' and $(\sigma_v')_0$ are the axial stresses during and at the start of TC, and E_0 is the value of the undamaged Young's modulus E^e measured at the start of TC. It is known that the structure of geomaterials can be damaged by large plastic strains associated with isotropic compression up to relatively large pressures. It is assumed here that the effect of this damage is insignificant within the range of pressure involved in these tests. It may be noted that the rate of increase of E_{eq} with q/q_{max} is similar to the one that is obtained from equation

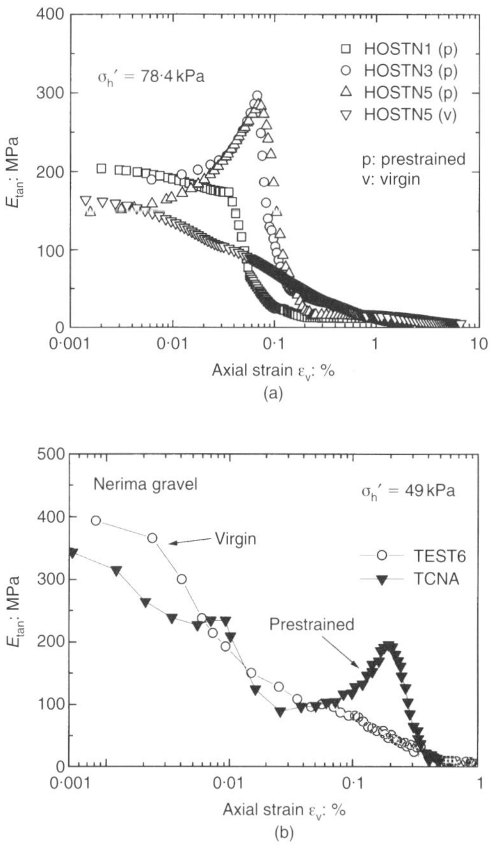

Fig. 16. Relationships between E_{tan} and $\log \varepsilon_v$: (a) Hostun sand; (b) Nerima gravel

(4) for Hostun sand (Fig. 17(a)) and Sagamihara soft mudstone (Fig. 17(d)). However, E_{eq} becomes smaller than E^e as q/q_{max} increases, for Nerima gravel (Fig. 17(b)), the dam-core material (Fig. 17(c)), cement-treated sandy soil (Fig. 17(e)) and Kobe sandstone (Fig. 17(f)). The reason for this may be the effects on the E_{eq} values of damage to the structure by shear straining and creep during TC, as discussed in relation to Fig. 14. On the other hand, as seen from Fig. 17(d), the effects of damage are not noticeable on the E_{eq} values of well-cemented sedimentary soft mudstone from Sagamihara. Note that this mudstone showed a negligible pressure dependence of E_0 at isotropic stress states (i.e. $m = 0{\cdot}0$).

Modelling

The relations shown in Fig. 17 depend greatly on the type of geomaterial. Furthermore, as can be seen from Figs 16 and 17, the non-linearity of E_{tan} in the cyclically prestrained specimen is very peculiar, and it is difficult to model this behaviour in a unified manner together with that of the virgin specimens. An attempt has been made to model the very different patterns of behaviour shown above for a wide variety of stiff geomaterials having different stress–time histories, on the basis of the elastic deformation characteristics, introducing the damage function $f(\mathrm{SL})$, the plasticity function $g(\mathrm{SL})$ and the combined function $h(\mathrm{SL}) = f(\mathrm{SL})g(\mathrm{SL})$ (Fig. 18) (Kohata *et al.*, 1994; Tatsuoka & Kohata, 1995). The function $h(\mathrm{SL})$ represents the total non-linearity of E_{tan} caused by shear deformation.

The relationship between the ratio E_{eq}/E^e and the stress ratio $\mathrm{SL} = q/q_{max}$ could be called the damage function f. The ratio E_{eq}/E^e is equal to 1·0 at SL = 0 and decreases with SL. As the ratio E_{tan}/E_{eq} represents the ratio of elastic axial strain increment $d\varepsilon_v{}^e$ to total axial strain increment $d\varepsilon_v{}^t$, the relationship between E_{tan}/E_{eq} and SL could be called the plasticity function g. The ratio E_{tan}/E_{eq} is always less than unity and decreases continuously with SL, from 1·0 at SL = 0·0. The ratio E_{tan}/E^e is equal to $(E_{tan}/E_{eq})(E_{eq}/E^e)$, thus the function h representing the relationship between E_{tan}/E^e and SL is a combination of the damage function f and the plasticity function g. The relationships between E_{eq}/E^e, E_{tan}/E^e and SL are shown in Fig. 19. The E_{tan}/E^e against SL relations for the cyclically prestrained specimens of Hostun sand, Nerima gravel and the dam-core material are different from those for the respective virgin specimens. Yet, the functions h for the virgin and prestrained specimens have a similar feature in that the ratio E_{tan}/E^e decreases with SL, from 1·0 at SL = 0·0. Therefore, they could be modelled by similar equations, taking into account the effects of strain history. When the function h (i.e. the $E_{tan}/E^e - q/q_{max}$ relation) is known, the relationship between E_{tan} and $\mathrm{SL} = q/q_{max}$ can be obtained by knowing the E^e value. The value of E^e at a certain $\sigma_v{}'$ can be obtained by using the relationship $E^e = E_f(\sigma_v{}'/(\sigma_v{}')_0)^m$, where the value of E_f is the value measured at a certain vertical stress ($\sigma_{v0}{}'$) in the field. The power m could be evaluated by relevant laboratory tests. To obtain the value of m in a first approximation, the relationships shown in Fig. 8 can be referred to.

It may be noted from Fig. 19 that the rate of decrease of E_{tan}/E^e with SL still varies greatly among different types of geomaterial having different degrees of cementation and different strain histories. It may also be noted that, generally, a cemented material which exhibits a larger rate of decrease of E_{eq}/E^e with SL shows a larger rate of decrease of E_{tan}/E^e. Fig. 20(a) shows the relationships between $f(\mathrm{SL})$ and $g(\mathrm{SL})$ when SL is equal to 0·2 and Fig. 20(b) shows the relationships between $f(\mathrm{SL} = 0{\cdot}2)$ and the power m obtained

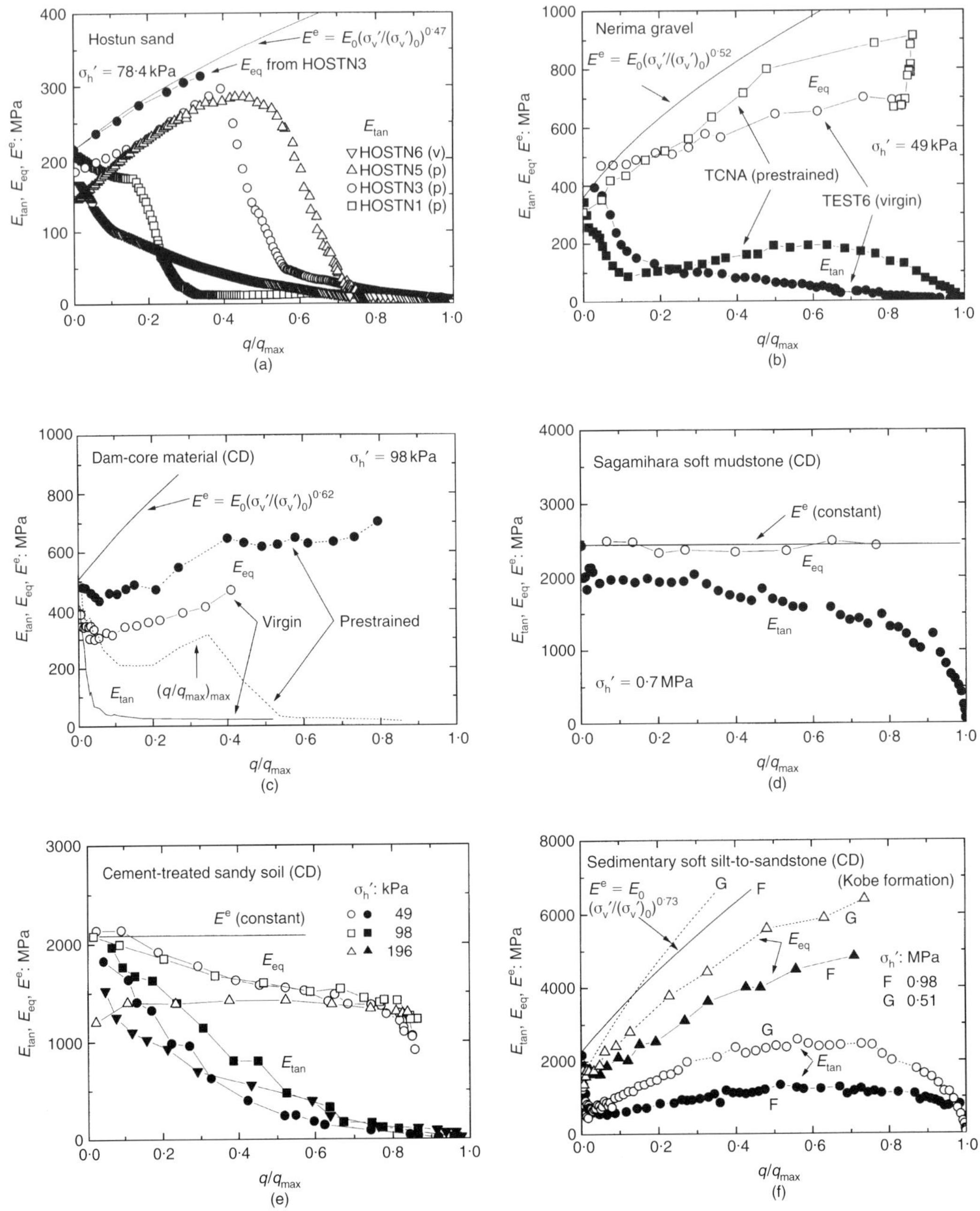

Fig. 17. Relationships between E_{tan}, E_{eq}, E^e and q/q_{max} obtained from CD TC tests: (a) Hostun sand; (b) Nerima gravel; (c) undisturbed dam-core material; (d) Sagamihara soft mudstone; (e) cement-treated sandy soil (dry type); (f) sedimentary soft silt-to-sandstone (Kobe formation)

from the TC tests on the cemented materials, where the confining pressure ranged from 0·1 MPa to 1·96 MPa. The general trend seen from Fig. 20(a) is that a cemented material which is prone to the effect of damage, exhibiting a lower value of $f(\mathrm{SL} = 0{\cdot}2)$, shows more non-linearity by plastic shear straining, with a lower value of $g(\mathrm{SL} = 0{\cdot}2)$. It may be seen from Fig. 20(b) that a material

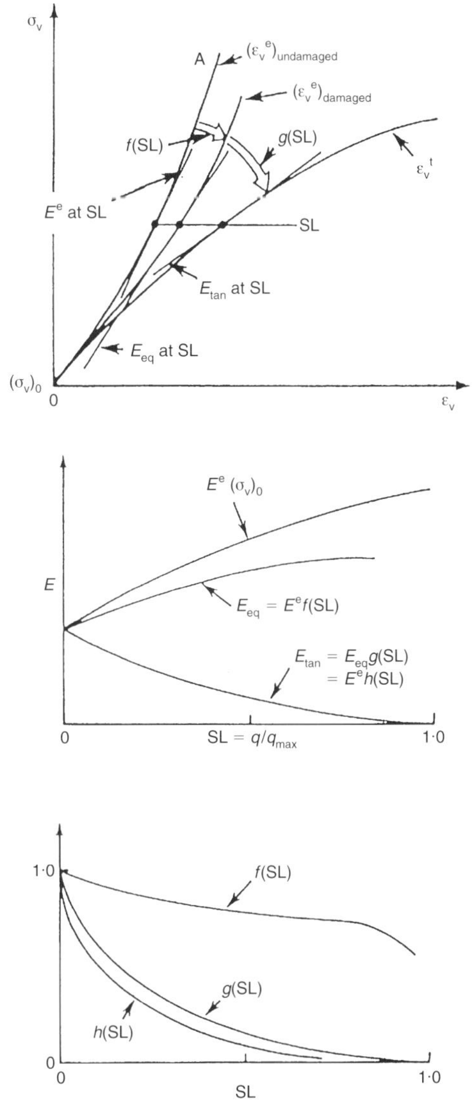

Fig. 18. Schematic illustration of the damage function $f(SL) = E_{eq}/E^e$, the plasticity function $g(SL) = E_{tan}/E_{eq}$ and the combined function $h(SL) = f(SL)g(SL)$

which has a higher value of m exhibits lower values of $f(SL = 0{\cdot}2)$. All these results indicate that for cemented materials, the effects of pressure level on the elastic modulus and non-linearity by shear deformation, which consist of the effect of damage and the occurrence of plastic deformation, become more important as the degree of cementation decreases. On the other hand, when $g = E_{tan}/E_{eq}$ is closer to unity, elastic strain increments are more dominant than plastic strain increments. In such a case, modelling based on elastic deformation characteristics is more relevant.

The values of f and g at SL = 0·2 for the dam-core material are also plotted in Fig. 20(a), and the values of f at SL = 0·2 and m are shown in Fig. 20(b). It may be seen that these data points are located in the same zones as the soft rocks, indicating that the mechanism of non linearity is similar for cemented materials (i.e. soft rocks) and structured uncemented materials (i.e. dam-core material).

On the other hand, for less-structured uncemented materials (sands and gravels), the decrease in $g = E_{tan}/E_{eq}$ with SL is much larger than the decrease in $f = E_{eq}/E^e$ with SL, as shown in Fig. 21. This means that non-linearity by shear deformation, which is reflected in $h = E_{tan}/E^e$, is mainly due to the occurrence of plastic deformation, which is reflected in $g = E_{tan}/E_{eq}$, but not due to destructuration or damage to cementation, which is reflected in $f = E_{eq}/E^e$.

In this paper, Poisson's ratio at small strains has not been discussed because only a limited amount of data for stiff geomaterials is available (e.g. Hoque *et al.*, 1996; Jiang *et al.*, 1997; Hayano *et al.*, 1997).

CONCLUSIONS

The following conclusions were derived from the results of triaxial compression tests on relatively stiff geomaterials.

(*a*) The deformation at strains less than about 0·001% was nearly recoverable. The initial elastic Young's modulus E_0 from laboratory tests using high-quality, undisturbed samples was very similar to the value obtained from field seismic surveys. This result suggests that field seismic surveys are very useful in evaluating the full-scale deformation of ground and the displacement of structures.

(*b*) For the granular materials, the elastic Young's modulus defined for the major elastic strain increments in the axial direction was a unique function of the axial stress, while that defined for strain increments in the lateral direction was a unique function of the lateral stress. That is, the elastic deformation characteristics of granular materials become anisotropic at anisotropic stress states. Because of this feature, for a material having a noticeably stress-dependent elastic Young's modulus, the Young's modulus defined for elastic axial strain increments increased noticeably as the axial stress increased during triaxial compression.

(*c*) For small strains occurring in a given stiff geomaterial, a recoverable strain increment $d\varepsilon_v^e$ is the major component of the total strain increment $d\varepsilon_v{}^t$. In such a case, the tangent Young's modulus E_{tan} defined for $d\varepsilon_v{}^t$ could be linked to the elastic Young's modulus evaluated at the stress state where the E_{tan} value was obtained.

On the basis of the facts described above, a model for the non-linear tangent Young's moduli E_{tan} has been proposed, which consists of functions representing the pressure-level dependence of the elastic Young's modulus, the damage to the elastic

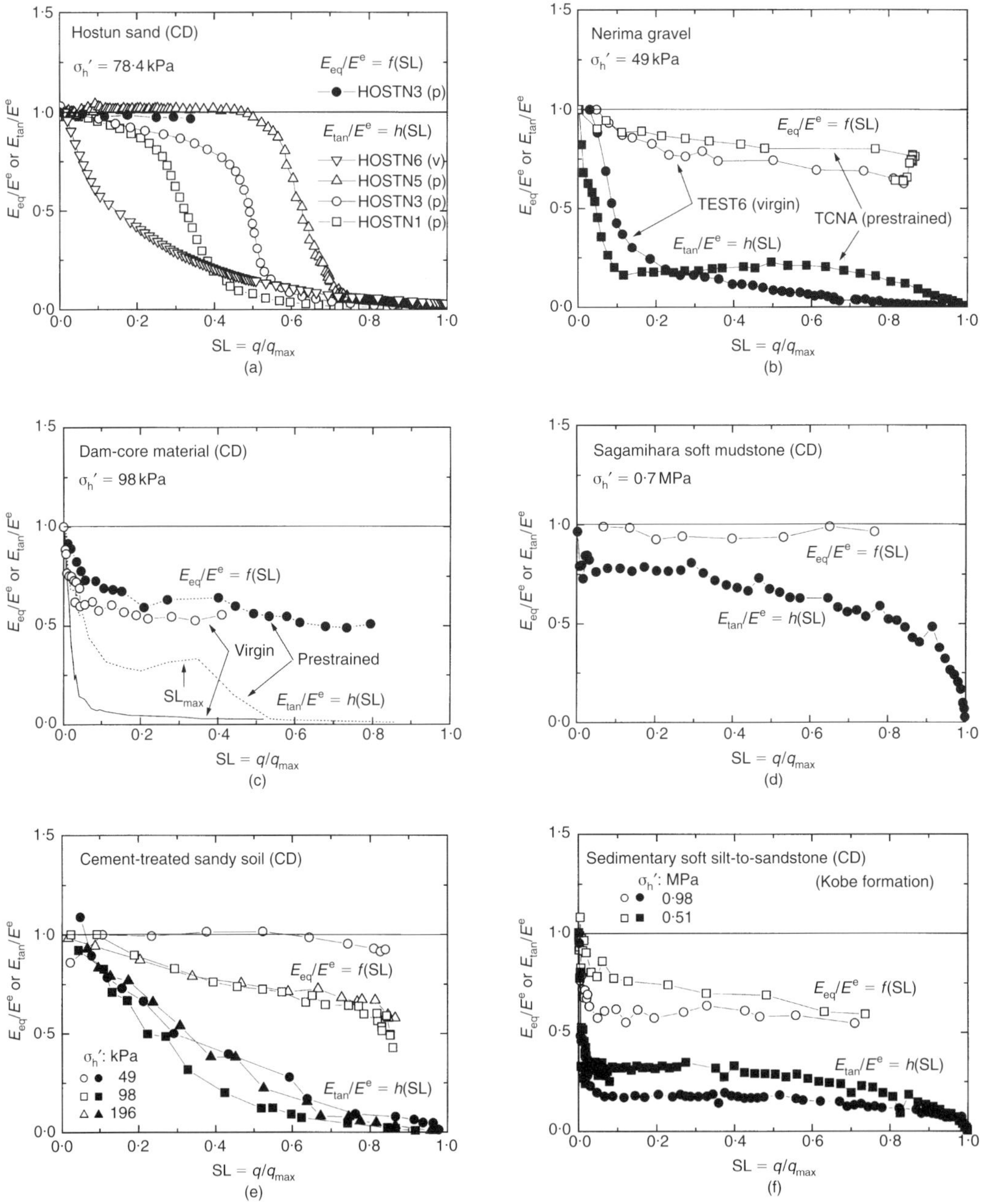

Fig. 19. Relationships between $f(\text{SL})$, $h(\text{SL})$ and shear stress level $\text{SL} = q/q_{max}$: (a) Hostun sand; (b) Nerima gravel; (c) undisturbed dam-core material; (d) Sagamihara soft mudstone; (e) cement-treated sandy soil (dry type); (f) sedimentary soft silt-to-sandstone (Kobe formation)

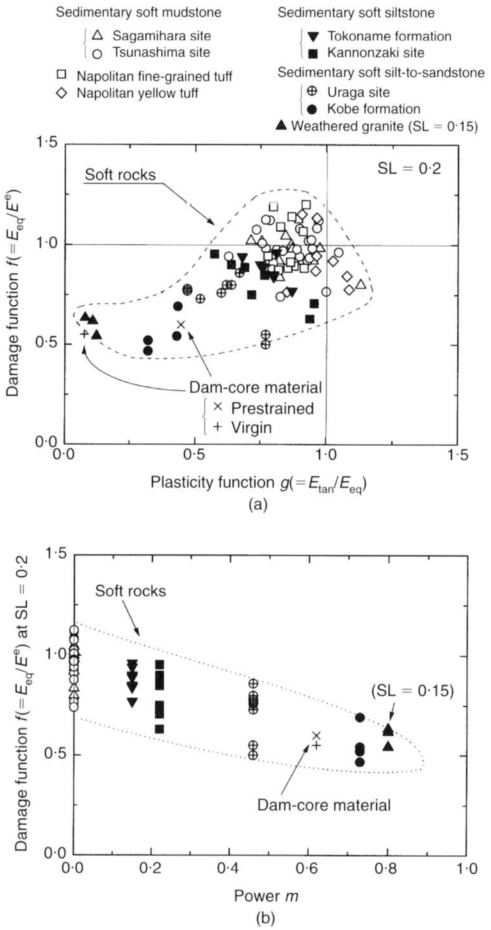

Fig. 20. (a) Relations between f(SL = 0·2) and g(SL = 0·2) for soft rocks and dam-core material; (b) relations between f(SL = 0·2) and the power m

Young's modulus and the ratio of the elastic to the plastic strain increments.

ACKNOWLEDGEMENTS

All the tests presented in this paper were performed at the Institute of Industrial Science, University of Tokyo. Hostun sand was provided by Professor H. Di Benedetto of the Ecole Nationale des Travaux Publics de l'Etat, Lyon, France. The authors are indebted to their colleague Mr T. Sato for his continuous support given to the experimental work. Ms M. Torimitsu is also appreciated for her help in preparing the manuscript.

REFERENCES

Burland, J. B. (1989). Ninth Laurtis Bjerrum Memorial Lecture. Small is beautiful—the stiffness of soils at small strains. *Can. Geotech. J.* **26**, 499–516.

Goto, S., Tatsuoka, F., Shibuya, S., Kim, Y.-S. & Sato, T. (1991) A simple gauge for local small strain measurements in the laboratory. *Soils Fdns* **31**, No. 1, 169–180.

Hardin, B. O. & Richart, F. E. (1963). Elastic wave velocities in granular soils. *J. ASCE* **89**, No. 1, 33–65.

Hayano, K., Sato, T. & Tatsuoka, F. (1997). Deformation characteristics of sedimentary soft rock from triaxial compression tests using rectangular prism specimens. *Proc. Géotechnique Symposium in Print*, **47**, No. 3, 439–449.

Hicher, P. Y. (1996). Elastic properties of soils. *J. Geotech. Engng* **122**, No. 8, 641–648.

Hoque, E., Tatsuoka, F. & Sato, T. (1996). Measuring anisotropic elastic properties of sand using a large triaxial specimen. *Geotech. Testing J., ASTM* **19**, No. 4, 411–420.

Iwasaki, T., Tatsuoka, F. & Takagi, Y. (1978). Shear moduli of sands under cyclic torsional shear loading.

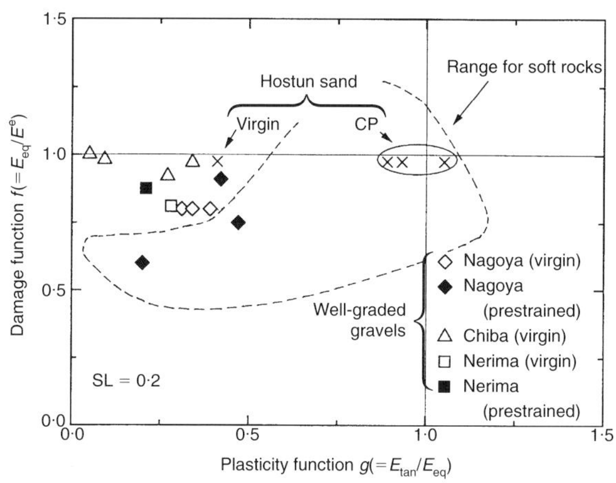

Fig. 21. Relations between f(SL = 0·2) and g(SL = 0·2) for uncemented geomaterials

Soils Fdns, Japan. Geotech. Soc. **18**, No. 1, 39–56.

Jamiolkowski, M., Leroueil, S. & Lo Presti, D. C. F. (1991). Design parameter from theory to practice. *Geo-Coast '91, Yokohama*, pp. 877–917.

Jardine, R. J. (1995). One perspective of the pre-failure deformation characteristics of some geomaterials. In *Pre-failure deformation of geomaterials* (eds S. Shibuya, T. Mitachi and S. Miura), Vol. II, pp. 947–1063. Balkema.

Jiang, G.-L., Tatsuoka, F., Flora, A. & Koseki, J. (1997). Inherent and stress state induced anisotropy in a small strain stiffness of very sandy gravel. *Proc. Géotechnique Symposium in Print*, **47**, No. 3, 509–521.

Kohata, Y., Tatsuoka, F., Dong, J., Teachavorasinskun, S. & Mizumoto, K. (1994). Stress state affecting elastic deformation moduli of geomaterials. In *Pre-failure deformation of geomaterials* (eds S. Shibuya, T. Mitachi and S. Miura), Vol. I, pp. 3–9. Balkema.

Lo Presti, D. C. F. & O'Neill, D. A. (1991). Laboratory investigation of small strain modulus anisotropy in sands. *Proc. ISOCCT1, Clarkson University, Potsdam, NY* (ed. Huang), pp. 213–224. Elsevier.

Stokoe, K. L. II, Lee, J. N.-K. & Lee, S.-H. (1991). Characterization of soil in calibration chambers with seismic waves. In *Calibration chamber testing*, (ed. Huang), pp. 363–376. Elsevier.

Tatsuoka, F. & Shibuya, S. (1992). Deformation characteristics of soils and rocks from field and laboratory tests. *Proc. 9th Asian Regional Conf. on Soil Mechanics Foundation Engineering*, Vol. 2, pp. 101–170.

Tatsuoka, F., Teachavorasinskun, S., Dong, J., Kohata, Y. & Sato, T. (1994). Importance of measuring local strains in cyclic triaxial tests on granular materials. In *Dynamic geotechnical testing II*, ASTM, STP 1213, pp. 288–302.

Tatsuoka, F. & Kohata, Y. (1995). Stiffness of hard soils and soft rocks in engineering applications. In *Pre-failure deformation of geomaterials* (eds S. Shibuya, T. Mitachi and S. Miura), Vol. II, pp. 947–1063. Balkema.

Tatsuoka, F., Kohata, Y., Ochi, K. & Tsubouchi, T. (1995a). Stiffness of soft rocks in Tokyo metropolitan area—from laboratory tests to full-scale behaviour. *Proc. Int. Workshop on Rock Foundation of Large-Scaled Structure*, pp. 3–17. Balkema.

Tatsuoka, F., Lo Presti, D. C. F. & Kohata, Y. (1995b). Deformation characteristics of soils and soft rocks under monotonic and cyclic loads and their relations. *Proc. Int. Conf. on Recent Advances in Geotechnical Earthquake Engineering and Soil Dynamic, St. Louis*, Vol. II, pp. 851–879.

Tatsuoka, F., Ochi, K., Tsubouchi, T., Kohata, Y. & Wang, L. (1997). Sagamihara experimental underground excavations in sedimentary soft rock. *Geotech. Engng.*

Tatsuoka, F., Uchida, K., Imai, K., Ouchi, T. & Kohata, Y. (1997a). Properties of cement-treated soils in Trans-Tokyo Bay Highway project. *Ground Improvement* **1**, No. 1, 37–57.

Tokimatsu, K. & Hosaka, Y. (1986). Effect of sample disturbance on dynamic properties of sand. *Soils Fdns, Japan. Geotech. Soc.* **26**, No. 1, 53–64.

Uchida, K., Shioi, Y., Hirukawa, T. & Tatsuoka, F. (1993). The Trans-Tokyo Highway Project—a huge project currently under construction. *Proc. Int. Conf. on Transportation Facilities through Difficult Terrain* (eds Wu and Barrett), pp. 57–87. Balkema.

Fodil, A., Aloulou, W. & Hicher, P. Y. (1997). *Géotechnique* **47**, No. 3, 581–591

Viscoplastic behaviour of soft clay

A. FODIL,* W. ALOULOU* and P. Y. HICHER†

Based on an extensive experimental programme, this paper presents the stress–strain–time behaviour of a natural soft clay. The clay was part of the foundation soil of Le Flumet dam, whose delayed settlements were very large. The main aim of this study was to explain, to quantify and to anticipate that settlement. Stiffness and strength of clays are both functions of time. Studying the relation (q–ε–t) through conventional laboratory tests showed their dependence upon the loading rate in a stress–strain plane. The effect of time was also observed during creep tests (triaxial and oedometer) or during stress–relaxation tests. A constitutive model based on the critical state concept and viscoplasticity theory was used to predict the viscous behaviour. The model was based on Hujeux's elastoplastic model and it was focused on the plastic flow function which was modified. The model parameters were determined by carrying out different simulations. The model was validated using all the experimental test results along different stress–strain paths.

KEYWORDS:

Se basant sur une étude expérimentale détailée cet article présente l'effet du temps sur le comportement mécanique d'une argile naturelle. L'argile étudiée est prélevée sous les fondations du barrage du Flumet, qui présente un tassement important. Le but principal de cette étude est d'expliquer, de quantifier et de prédire ce tassement. On peut dire que la résistance et la déformation des argiles sont toutes les deux fonction du temps. L'étude de la relation (q–ε–t) à travers les essais conventionnels au laboratoire montre sa dépendance de la vitesse d'essai sur le plan (q–ε). Cependant les effets du temps peuvent aussi être observés soit durant les essais de fluage (triaxial ou oedometrique), soit durant les essais de relaxation des contraintes. Un modèle basé sur le concept d'état critique et la théorie de la viscoplasticité est utilisé pour prédire le comportement visqueux. Ce modèle est issu de modèle élastoplastique de Hujeux, il converge autour de la loi d'écoulement qui a été modifié. Les paramètres sont déterminés en procédant à différents calages sur les résultats expérimentaux suivant différents chemins de contraintes et de déformations.

INTRODUCTION

It is well known that time has an effect on the mechanical properties of geomaterials. This effect is closely connected to an evolution of the geomaterials properties with time which is revealed, for example, by creep and stress–relaxation phenomena, and by the influence of the strain rate. Two main topics will be examined: the fundamental experimental findings concerning the influence of time on the mechanical properties; and mathematical modelling of time-dependent behaviour.

An experimental programme was performed on a soft clay which was taken from the foundations of 'Le Flumet' dam (EDF: Electricité de France) in order to investigate its time-dependent behaviour. The experimental programme involved drained creep tests, stress–relaxation tests, undrained monotonic triaxial tests at different strain rates and oedometer tests. We will first present these tests and analyse them in terms of viscoplastic mechanisms.

TESTING PROGRAMME

Tested clay

The following tests were performed on two clay materials, one was taken from a depth of 9 m and the other from a depth of 12 m. The Atterberg limits were:

Depth	LL	PL	IP
9 m	38%	24%	14
12 m	42%	30·5%	11·5

The unit weight of the clay specimen was 1·9 g/cm^3 and their water content was 30–45%.

Test conditions

Different types of test were conducted in different conditions: triaxial tests – drained stress–relaxation tests, drained creep tests and monotonic

Manuscript received 30 October 1996; revised manuscript accepted 21 April 1997.
Discussion on this paper closes 1 November 1997.
* Ecole Centrale Paris.
† Université de Nantes.

undrained tests with different strain rates; and oedometer tests.

In triaxial tests, the vertical load was measured inside the cell in order to avoid the problem of friction between the piston and the cell, especially during the stress–relaxation tests and creep tests. The vertical displacement was measured outside the cell.

TIME EFFECT

Influence of strain rate

Stress–strain relationship. In Fig. 1, the stress–strain curves of undrained triaxial tests, performed at constant strain rates $3{\cdot}33 \times 10^{-7}$, $3{\cdot}33 \times 10^{-6}$, $3{\cdot}33 \times 10^{-5}$ (per second), are presented. No general trend could be derived from these test results because the natural heterogeneities of the clay specimen were more influencial than the strain rate. The maximum strength was mainly related to the natural water content. To eliminate these natural variations, we used the method proposed by Leinenklugel (1967) which consists of shearing the same specimen while changing the rate of strain in a ratio of 10 during the test. In Fig. 2 we present one of these tests.

The change in strain rate during the test produced a sudden increase or decrease in the deviatoric stress, which could be evaluated to represent about 5% of the stress level at any point of the stress–strain curve. The influence of the strain rate was therefore not very pronounced compared to the values available in the literature (Bjerumm, Simons & Torblaa, 1958; Bjerumm (1973); Hicher, 1985; Flavigny, 1987; Murayama & Shibata, 1964; Singh & Mitchell, 1969; Graham, Crooks & Bell, 1983). A value of 10% increase for a strain rate multiplied by 10 is a more common value. This could probably be related to the low plasticity index of this particular clay. This result could contribute to explain the results in Fig. 1. Because of the small strength variations, other factors can influence more widely the mechanical behaviour of the clay.

Oedometer tests

Secondary compression of clays has been widely investigated. It corresponds to the delayed settlement during oedometric consolidation after the end of the hydrodynamic consolidation. This is attributed to the clay viscosity.

The vertical strain ε_1 or the void ratio e are generally plotted versus the logarithm of time. The slope of these curves leads to the definition of the coefficient of secondary compression $C\alpha$. For clays this slope does not seem to diminish with time unlike in the case of sands. Fig. 3 presents some results obtained on the clay specimen studied. We can notice that the duration of primary consolidation was short (only a few minutes), leading to a mean value of the coefficient of permeability $k = 2 \times 10^{-9}$ m/s. This high permeability can be explained by very small sand layers, present in the clayed materials.

In Fig. 4, we present the variation of C_α with stress $[C\alpha = (1 + eo)\,\mathrm{d}\varepsilon/\mathrm{d}\log(t)]$. C_α increases with the effective vertical stress but above the consolidation pressure, the increase is less significant, as already obtained for other clays by

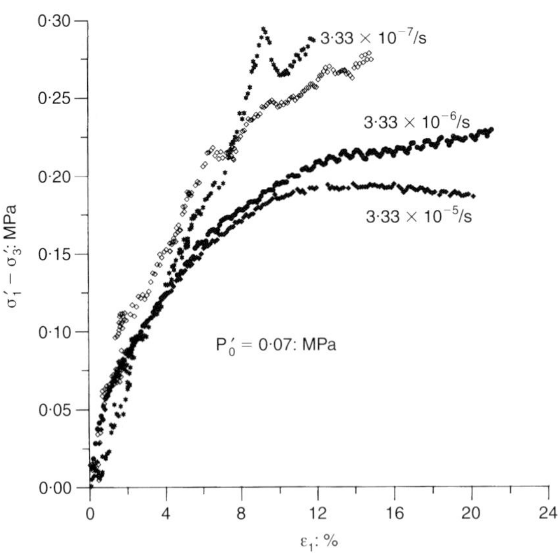

Fig. 1. Influence of strain rate on the CU triaxial tests

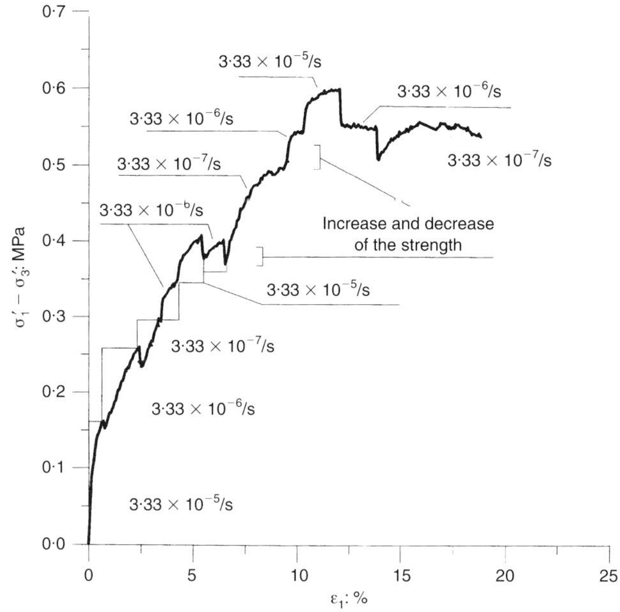

Fig. 2. Influence of strain rate on the CU triaxial tests

Murayama & Shibata, (1964). They observed for an alluvial clay an increase of $d\varepsilon/d\log(t)$ with stress up to 1·5 until the consolidation pressure.

The consolidation curves in conventional axes $(e, \log\sigma'_v)$ lead to the definition of the compressibility index C_c (Fig. 5), we notice that the preconsolidation pressure value is around 0·04 MPa whereas the value of the in situ vertical effective stress is around 63 kPa (9 m sample) and 85 kPa (12 m sample). This is probably due to sample disturbance.

To explain the effect of time on the oedometric behaviour, we present in Fig. 6 the results for different durations of consolidation. The consolidation curves are parallel to each other, so the duration of consolidation does not affect the compressibility index but the preconsolidation pressures obtained increase with the strain rate (Leroueil *et al.*, 1985).

Triaxial creep tests

Creep is defined as the increase in deformation with time under constant effective stress. In this type of tests, the effective axial and lateral stresses are maintained constant and the values of the axial and volumetric strains are measured as a function of time.

EXPERIMENTAL CONDITIONS

The specimens were first consolidated under an isotropic state of stress and then a shear stress was applied. After that, at a given axial strain (around 1% and 3% for each test), the shear stress was held and the axial and volumetric strain were measured as a function of time.

Creep tests require important precautions:

(*a*) The lateral membrane should avoid osmosis between the cell water and the specimen which could happen during long duration testing. For this, we used a neoprene membrane.
(*b*) The measurement of the axial load was taken with a transducer inside the cell.
(*c*) It was necessary to load the specimen at a constant strain rate small enough ($3{\cdot}33 \times 10^{-6}$/s) to ensure a fully drained test up to the creep level and then to follow the delayed deformations.
(*d*) Samples were drained only through the bottom end.

Deviatoric creep tests

In such a test it is common to distinguish three phases: primary creep characterized by a decreasing axial strain rate; secondary creep during which the strain rate remains constant (stationary creep); and tertiary creep with increasing strain rate (accelerating creep).

Primary creep is always present, tertiary creep is observed only for stress levels close to failure stresses, whereas secondary creep is seldom observed (Flavigny, 1987; Hicher, 1985).

In the performed tests, we observed only prim-

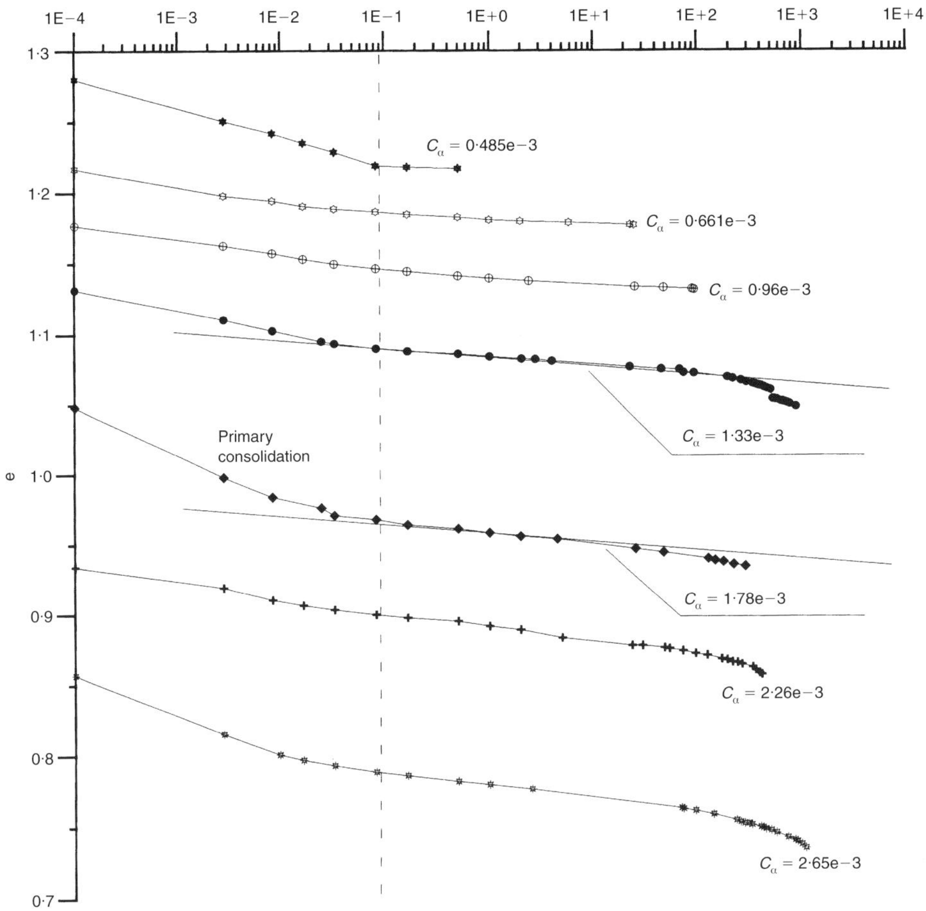

Fig. 3. Oedometer tests on 9 m depth specimen

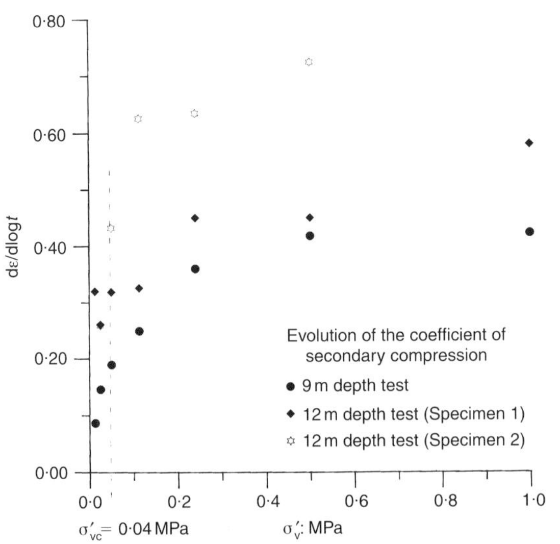

Fig. 4. Evolution of the secondary compression coefficient

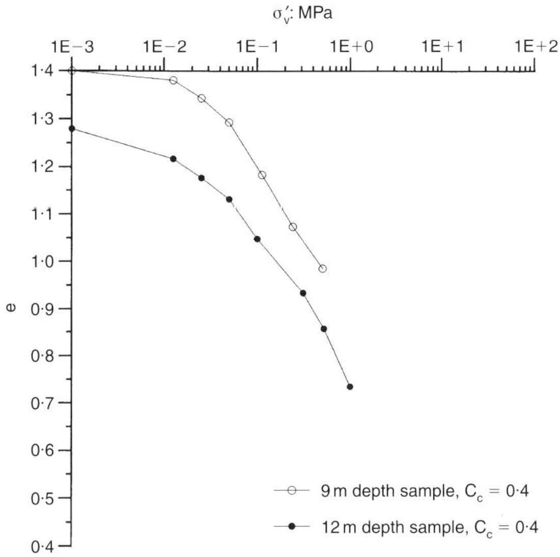

Fig. 5. Oedometer tests on clay samples

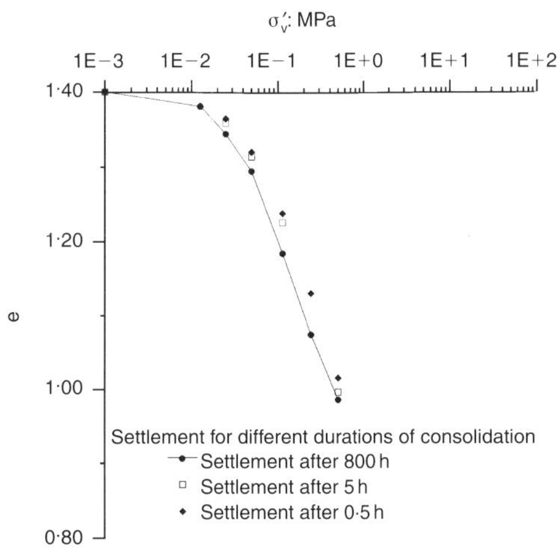

Fig. 6. Oedometer test with different durations of consolidation

ary creep (Fig. 7(a), 7(b)). That was foreseeable because the tests were performed in drained conditions and the stress levels were far from the maximum stress level. The initial time represents the beginning of the creep phenomenon. The same behaviour was observed for the two clay materials.

Plots of $\log(\mathrm{d}\varepsilon/\mathrm{d}t)$ versus $\log(t)$ indicated that the slope m was around 0·7. The slope m represents the capability in soil to be subjected to creep deformations (Singh & Mitchell, 1969; Tavenas *et al.*, 1978).

Singh & Mitchell (1969) used this slope in a phenomenological relationship in order to represent creep in soils:

$$\mathrm{d}\varepsilon/\mathrm{d}t = Ae^{\alpha q}(t_1/t)^m \qquad (1)$$

where α and A are soil parameters; t_1 is the time reference (1 min, 1 h, ...); and m was found to be independent of the stress level.

Stress–relaxation tests

Stress–relaxation is defined as the decrease in deviatoric stress with time when the deformation is stopped. Theoretically, when the deformation is stopped, the deviatoric stress decreases immediately and continues to decrease at lower rate (Fig. 8).

Several tests were performed on the two clay materials in drained conditions with different initial strain rates: $3{\cdot}33 \times 10^{-8}$, $3{\cdot}33 \times 10^{-7}$ and $3{\cdot}33 \times 10^{-6}$ (/s) (Fig. 9). We notice that the stress–relaxation levels do not affect the subsequent stress–strain behaviour.

The stress–relaxation tests were performed for each test (on the same sample) at about 1%, 3% and 6% of axial strain and lasted for at least 24 h. The effect of the strain rate is presented in Fig. 10 at the same strain level (3%) on three different tests. In the same conditions, the decrease of stress was higher for the tests performed with a higher strain rate.

We observed that the stress did not decrease immediately but began to decrease only after 10–30 s in most of the tests. Fig. 11 represents an example of this observation.

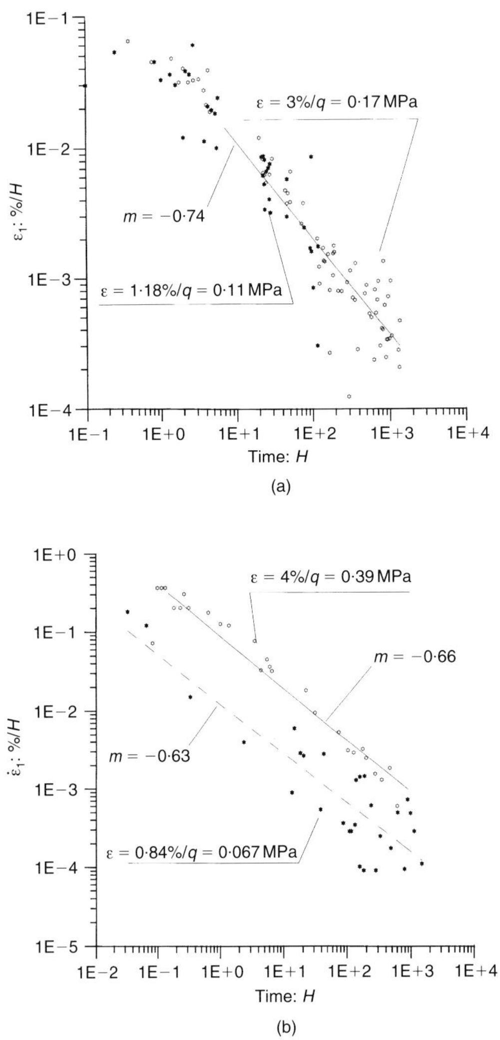

Fig. 7. Creep test on (a) 9 m and (b) 12 m depth specimen

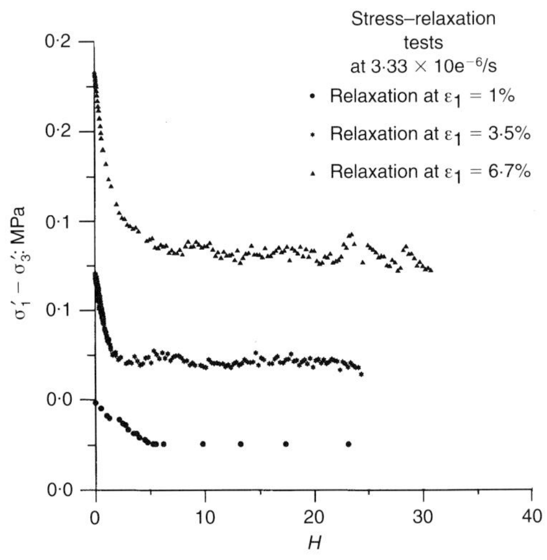

Fig. 8. Stress–relaxation test on 9 m depth specimens

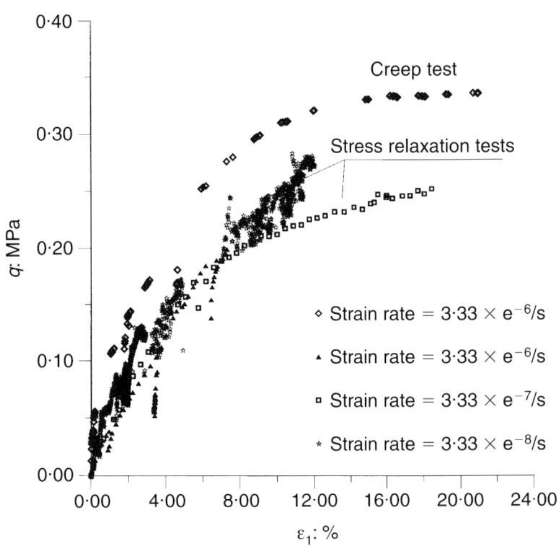

Fig. 9. Stress–strain behaviour during stress–relaxation test on 9 m depth specimens at different strain rates

Conclusion

The different tests (creep, stress–relaxation and oedometer tests) performed on this natural soft clay revealed a common behaviour concerning the evolution of the strain rate and of the deviatoric stress. In fact, it was observed that the effect of time was similar on the strain rate and on the deviatoric stress evolutions. The log (strain rate) in creep tests decreased linearly with log (time) following a slope m. A similar evolution was observed in oedometer tests. Plotting $\log(de/dt)$ versus $\log(t)$ indicated a linear relationship with a slope between 0·79 and 0·9 (Fig. 12).

Concerning the stress–relaxation tests, the evolution of the log of the deviatoric stress rate with the log of time showed a linear relationship with a slope of around 0·65.

The values of the different slopes mentioned above fall in the same range, although they appear to be dependent on the stress and strain paths. The difference in the values, however, could also be related to the natural variations in the soil samples and to the experimental errors.

Consequently, the slope m seems to be an important parameter that reflects the viscous clay behaviour. For a given sample and a given mech-

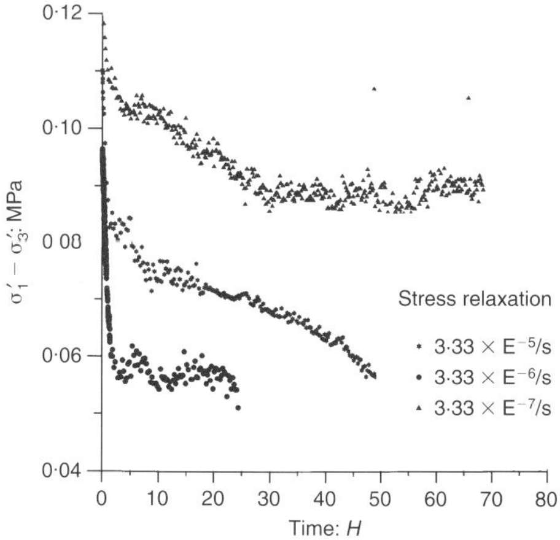

Fig. 10. Stress–relaxation test on 9 m depth specimens at different strain rates

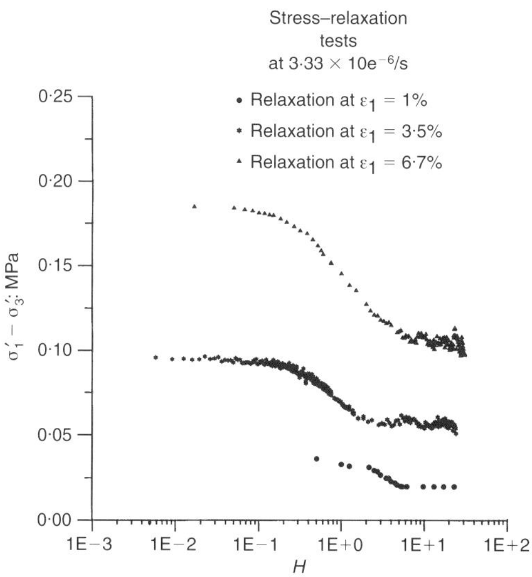

Fig. 11. Stress–relaxation test on 9 m depth specimen

anical loading, the value of m appears to be independent of the initial strain amplitude, which indicates that the viscous behaviour has the same influences on the stress–strain response regardless of the strain amplitude.

MODELLING THE VISCOPLASTIC BEHAVIOUR OF SOFT CLAY

Description of the viscoplastic model

The viscoplastic model is a multimechanism model with three deviatoric mechanisms and an isotropic one. It was developed based on Hujeux's elastoplastic model (Hujeux, 1985; Aubry, Kodaissi & Meimon, 1985).

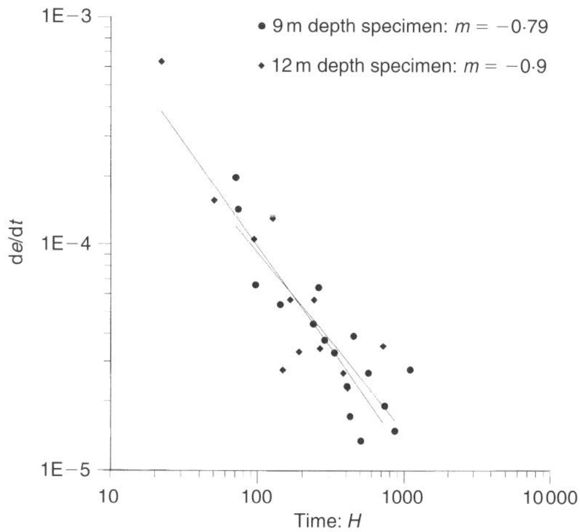

Fig. 12. Evolution of void ratio with time in oedometer tests

We placed ourselves in the scope of Perzyna's theory (Perzyna, 1963). The plastic flow rate has the following form:

$$\overline{\dot{\varepsilon}^{\mathrm{vp}}} = \langle \Phi(F) \rangle \overline{\frac{\partial g}{\partial \sigma}} \tag{2}$$

$\langle \mathrm{A} \rangle = \mathrm{A}$ if $\mathrm{A} > 0$ and $\langle \mathrm{A} \rangle = 0$ if $\mathrm{A} \leqslant 0$.

$\overline{\partial g / \partial \sigma}$ represents the direction of the strain increment normal to the potential surface g, whereas $\langle \Phi(F) \rangle$ controls the amplitude of the plastic strain rate. $\Phi(F)$ is called the plastic flow function.

Many investigators proposed a variety of functions for $\Phi(F)$. In the case of our model, it initially took the following form:

$$\Phi(F) = \frac{\left(\dfrac{F}{F_0}\right)^n}{\mu} \tag{3}$$

where n and μ are two parameters that take into account the viscous properties of the material, F is the value of the yield function at the actual stress state and F_0 is a reference value to insure the homogeneity of the relation.

Deviatoric mechanism. The plastic flow for each mechanism k is presented by the deviatoric plastic strain $(e^{\mathrm{p}})_{\mathrm{k}}$ and the volumetric plastic strain $(\varepsilon_v{}^{\mathrm{p}})_{\mathrm{k}}$

$$\partial_t e^{\mathrm{p}}{}_{\mathrm{k}} = \langle (f_{\mathrm{k}}{}^{\mathrm{M}}/f_0)^{\mathrm{n}}/\mu^{\mathrm{D}} \rangle \psi_{\mathrm{k}}{}^{\mathrm{D}} \tag{4}$$

$$\partial_t \varepsilon_v{}^{\mathrm{p}}\mathrm{k} = \langle (f_k{}^{\mathrm{M}}/f_0)^{\mathrm{n}}/\mu^{\mathrm{D}} \rangle \psi_{\mathrm{k}}{}^{\mathrm{v}} \tag{5}$$

The components of the functions ψ_{k} are defined by:

$$(\psi_k{}^D)_{ii} = (\sigma_{ii} - \sigma_{jj})/2q_k \quad (6)$$

$$(\psi_k{}^D)_{jj} = -(\sigma_{ii} - \sigma_{jj})/2q_k \quad (7)$$

$$(\psi_k{}^D)_{ij} = \sigma_{ij}/2q_k \quad (8)$$

$$(\psi_k{}^D)_{kl} = 0 \text{ for any value of 1 if } p^* = p_k \quad (9)$$

$$\psi_k{}^V = \sin\phi - (s_k \cdot \psi_k{}^D)/p_k \quad (10)$$

where $q_k = (1/4(\sigma_{ii} - \sigma_{jj})^2 + \sigma_{ij}{}^2)^{1/2}$; ϕ is the friction angle at critical state; s_k is a vector whose norm is the radius of Mohr's circle.

In the deviatoric plane of each mechanism, the yield surface is considered to be a circle. This circle is normalized by the critical friction factor $F_k(p_k, p_c)$ and thus the yield surface is a circle with radius $r_k{}^M$:

$$f_k{}^M(p_k, s_k, \varepsilon_v{}^p, r_k{}^M) = |s_k/F_{k1}| - r_k{}^M \quad (11)$$

$F_k(p_k, \varepsilon_v{}^p)$ is defined by:

$$F_k(p_k, \varepsilon_v{}^p) = \{1 - b\log(p^*/p_c(\varepsilon_v{}^p)\}\, p_k \sin\phi \quad (12)$$

with: $p_k = 1/2(\sigma_{ii} + \sigma_{jj})$; $p^* = p$ or p_k; $s_k = (1/2(\sigma_{ii} - \sigma_{jj}), \sigma_{ij})$; $p_c = p_{co}\exp(\beta\varepsilon_v{}^p)$; $\varepsilon_v{}^p = \sum_{k=1,4}(\varepsilon_v{}^p)^k$.

The parameters σ, β, p_{co}, and p_c are, respectively, the critical state friction angle, the plastic compressibility modulus, the initial critical pressure corresponding to the initial void ratio and the critical pressure corresponding to the actual void ratio.

The parameter b can have a value between zero and one. It influences, in the Mohr's plane, the role of volumetric hardening on the deviatoric yield function. If p is higher than p_c, the resistance capacity of the material to friction is lower than that given by the Mohr-Coulomb. If p is lower than p_c, the resistance capacity of the material to friction is higher than that given by the Mohr-Coulomb. If p is equal to p_c, both criteria coincide at the critical state. The parameter b, in fact, allows to regulate this influence. If b is equal to one, we have the Cam-clay yield function. If b is null, we meet the Mohr-Coulomb criterion.

The internal mobilized shear friction in the mechanism k is defined by:

$$\partial_t r_k{}^M = \langle (f_k{}^M/f_0)^n/\mu^D \rangle (1 - r_k{}^M)^2/a^M \quad (13)$$

where a^M is a numerical parameter of the model.

Isotropic mechanism. It is asssumed that this mechanism (fourth mechanism) produces purely volumetric deformations:

$$\partial_t \varepsilon_4{}^p = \langle (f_4{}^M/f_0)^n/\mu^v \rangle (p/3d\,p_c(\varepsilon_v{}^p)) \quad (14)$$

In primary loading, the yield surface has the following form:

$$f_4{}^M(p, \varepsilon_v{}^p, r_4{}^M) = |p/d\,p_c(\varepsilon_v{}^p)| - r_4{}^M \quad (15)$$

The evolution of r_4 is given by:

$$\partial_t r_4 = \langle (f_4{}^M/f_0)^n/\mu^V \rangle (1 - r_4)^2 p_{ref}/(\varepsilon_v{}^p) \quad (16)$$

where c is a numerical parameter of the model and d is the distance between the virgin consolidation curve and the critical state curve in the $(e, \mathrm{Log}(p))$ plane. The formulations concerning the case of cyclic loading will not be presented here.

Modification of the flow function

A power type relation does not very well reflect the experimental results. The stress evolution for a given deformation is rather proportional to $\log\dot{\varepsilon}$. Therefore, we retained a different formulation proposed by Laigle (1992). This formulation is based on experimental results, such as the linear relation of $(\log(d\varepsilon/dt), q)$ and its slope variation, which is a function of ε.

The new relation is as follows:

$$\Phi(F) = \mu\left[\exp\left(n\left(\frac{F}{F0}\right)\right) - 1\right] \quad (17)$$

From a numerical point of view, the new formulation allows us to limit the increase of q with the strain rate and therefore enlarges the domain of application of the viscoplastic model.

This modified version of the viscoplastic model was applied to predict the behaviour of the soft clay. The objective was to be able to determine a set of parameters which could lead to a good simulation of the clay behaviour for the different stress–strain paths corresponding to the previous testing.

Determination of model parameters

The viscoplastic model parameters can be divided into the following three groups.

(*a*) A group of parameters directly measured from experimental curves, such as the elastic parameters E (Young's modulus), ν (Poisson's ratio), the critical state parameters: ϕ, p_{co}, d, the compressibility modulus β. The parameters were initially determined from the test results on our clay samples.

(*b*) A set of numerical parameters which could be called the elastoplastic parameters, appearing in the different equations of the model. These parameters describe the evolution of the hardening variables and they can only be determined by comparison between numerical and experimental curves, by means of an optimization procedure. The value ranges of these parameters for different kinds of materials are given by Michalski & Rahma (1989).

(*c*) The viscous parameters n and μ, which are the key of our investigation, can be determined from creep tests, stress–relaxation tests and oedometer tests in the secondary compression range. We will be interested in: the rate of change of the deviatoric stress q as the deformation is held constant, in the case of a stress–relaxation test; the creep rate as the deviatoric stress is held constant, in the case of a drained creep test; and the rate of change of the void ratio e during an oedometer test in secondary compression.

We will concentrate our presentation on the viscous parameters.

The influence of the viscoplastic parameters

First, the elastoplastic parameters were estimated using the procedures 1 and 2 mentioned in the previous sections. To study the influence of the viscoplastic parameters, only the viscous parameters μ and n were changed and each one at a time.

In a stress–relaxation test, the rate of change of the deviatoric stress q, as the deformation is held constant, is related directly to the viscous properties of the material. In addition, the other parameters of the model have a small influence on the decrease of the deviatoric stress. Therefore, the stress–relaxation tests provide a direct measure between the loading point and the yield surface which represents a purely viscous datum in the viscoplastic behaviour theory. Fig. 13 shows the experimental and the numerical curves giving the variation of q with time for different values of μ and n. Concerning the influence of each viscoplastic parameter, we can make the following remarks: the parameter μ affects the resistance of the material and the deviator stress is increased as the value of the parameter μ is decreased; the parameter n affects both the resistance of the material, the rate of change of the deviator stress and its total change. As the value of n is increased, the resistance is decreased, the rate of change of q (the curvature) is increased and the total change of q is decreased.

For a creep test, the results are usually presented in the form of curves giving the axial strain rate versus time on logarithmic scales. Fig. 14 indicates that the creep rate is increased as μ is decreased or as n is increased and that the influence of n is greater than that of μ.

In the case of the oedometer test, to study the influence of the viscoplastic parameters, it is necessary to compare the coefficients of the secondary compression C_α in the plane e–$\log(t)$ obtained from the numerical simulation with those of the experimental curves. In general, the slopes of the curves C_α are higher when we decrease the values of μ or n.

Model parameters

Our objective is to find a couple of parameters (μ, n) which can give us simulation curves that can best fit the experimental curves for the stress–relaxation, creep, and oedometer tests (curves in the time-related planes) at the same time.

By studying the influence of the viscous parameters on these tests, we conclude that the best values of the viscous parameters for this clay are

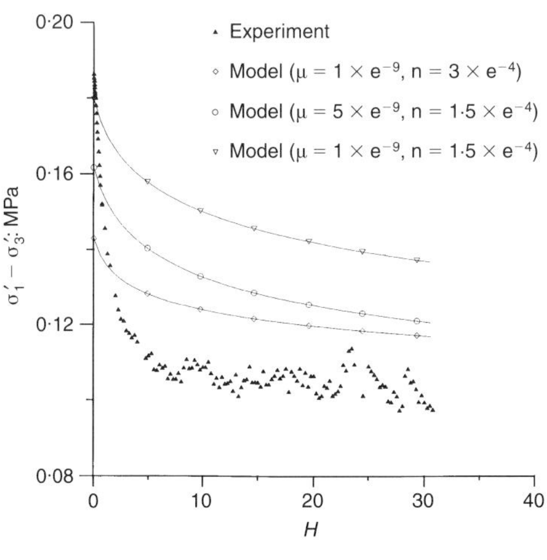

Fig. 13. Effect of parameters on the stress–relaxation modelling

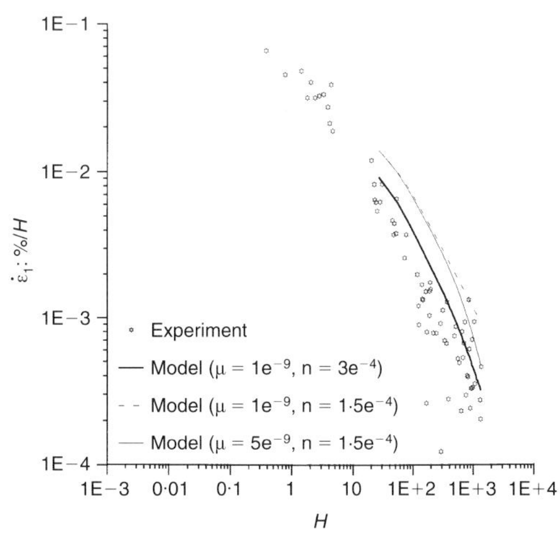

Fig. 14. Effect of parameters on the creep modelling

$\mu = 10^{-9}$ and $n = 3 \times 10^{-4}$. The results are shown in Fig. 15 for the creep tests (presented in Fig. 7(a)), Fig. 16 for the stress–relaxation tests under drained conditions, and Fig. 17 and 18 for the oedometer tests. For clarity, three stress levels only are plotted in Fig. 17.

The stress–relaxation test in Fig. 16 was done at a strain rate of $3{\cdot}33 \times 10^{-6}$/s.

Simulation of creep triaxial and oedometric secondary consolidations gave a good agreement with experimental data. This was not the case for stress–relaxation tests. The phenomenon was qualitatively well simulated but the computed decreases of the axial stress was smaller than the experimental ones. The explanation lies in the fact that we could not avoid a stress–relaxation of the loading system which created a small unloading of the clay specimen. This unloading was added to the stress–relaxation of the specimen, which was therefore overestimated by the experimental data.

After determining the viscoplastic parameters, the set of numerical elastoplastic parameters were adjusted by comparing the numerical and the experimental curves in the other planes (example: the

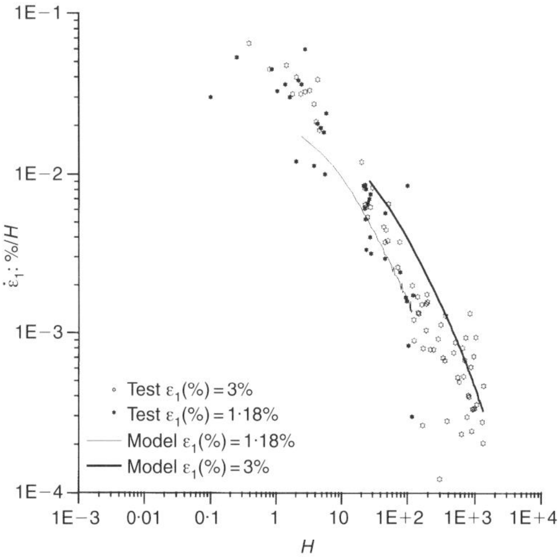

Fig. 15. Predicted and measured results for creep test

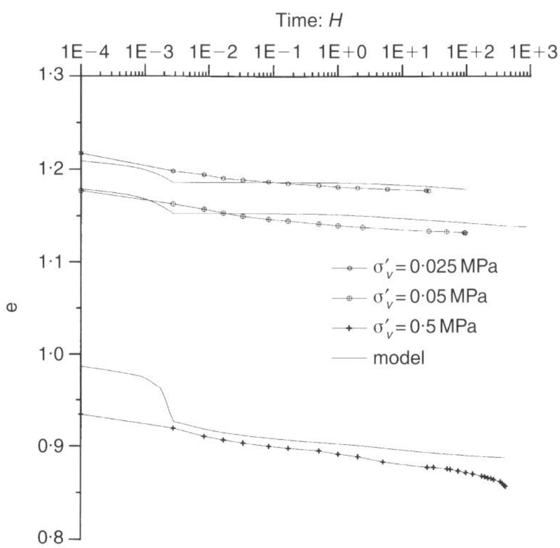

Fig. 17. Predicted and measured results for oedometer test

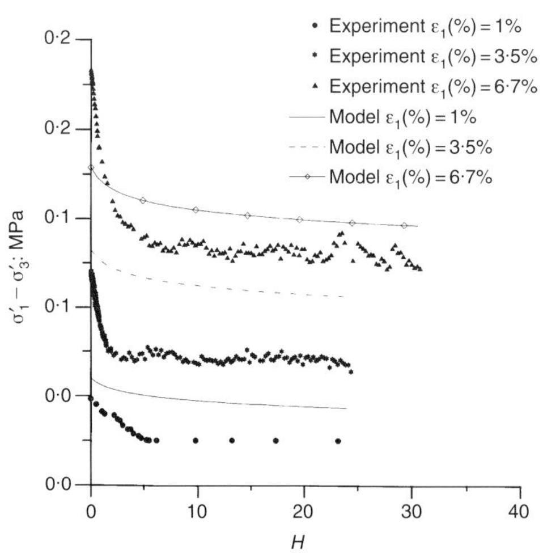

Fig. 16. Predicted and measured results for stress–relaxation tests

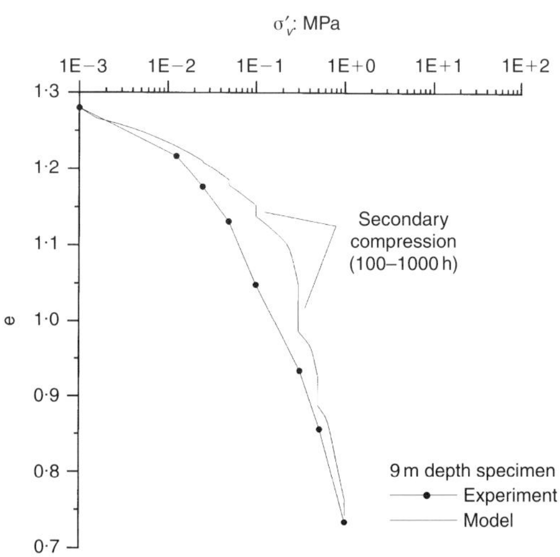

Fig. 18. Predicted and measured results for oedometer test

$q - \varepsilon_1$ and the $\varepsilon_v - \varepsilon_1$ planes for the stress–relaxation test, etc ...) for the different tests. Standard values for clays were obtained.

CONCLUSION

Stress–strain–time behaviour has been investigated by means of laboratory tests of a soft clay obtained from the foundations of Le Flumet dam (EDF).

The existence of the viscosity aspect of the material was confirmed and characterized by different tests: stress–relaxation, creep and oedometric tests. It was found that the parameter m (in equation (1)) seems to quantify well the viscous behaviour on different stress and strain paths. In our case m was around 0·75.

A viscoplastic constitutive model developed to predict this viscous clay behaviour was presented. The constitutive equations were based on Hujeux's elastoplastic model and Perzyna's theory of viscoplasticity. A set of numerical parameters appropriate for the natural clay was determined by carrying out different simulations and comparing the results with those of the various experimental tests performed. The simulations were done only for tests carried out on 9 m depth specimens and for drained conditions. The simulations for undrained condition tests and for 12 m depth specimens were not presented in this paper.

In fact, the model was not capable of predicting secondary compression, creep and stress–relaxation with the same precision; this can be due to the natural dispersion of the clay properties. However, overall, the results of the simulations were satisfactory.

The future work is now, having introduced the constitutive model in the finite element code GEFDYN, to calculate the delayed settlement of Le Flumet dam and to compare our results with the in situ instrumentation results.

ACKNOWLEDGEMENT

This work is part of a research contract between the French Ministry of Research (MENESR) and Electricité de France (EDF/SEPTEN).

REFERENCES

Aubry, D., Kodaissi, E. & Meimon, Y. (1985). A viscoplastic constitutive equation for clays including a damage law. *Fifth International Conference on Numerical Methods in Geomechanics*. Nagoya, Japan. **I**, 421–428.

Bjerumm, L. (1973). Problems of soil mechanics and construction on soft clays. *Proc. VIII ICSMFE*, Moscow.

Bjerumm, L., Simons, N. & Torblaa, I. (1958). The effect of time on the shear strength of a soft marine clay. *Proc. Conf. on Earth Pressure Problems*. Bruxelles. **I**.

Flavigny, E. (1987). Propriétés visqueuses des géomatériaux. *Manuel de rhéologie des géomatériaux*. Presse de l'Ecole Nationale des Ponts et Chaussées, Paris.

Graham, J., Crooks, J. H. A. & Bell, A. L. (1983). Time effects on the stress–strain behaviour of soft natural clays. *Géotechnique* **33**, No. 3, 327–340.

Hicher, P. Y. (1985). *Comportement mécanique des argiles saturées sur divers chemins de sollicitations monotones et cycliques. Application à une modélisation élastoplastique et viscoplastique*. Thèse de doctorat d'état, Université, Paris VI.

Hujeux, J. C. (1985). Une loi de comportement pour le chargement cyclique des sols. *Génie Parasismique*. Presses des Ponts et Chaussées, Paris.

Laigle, F. (1992). *Modèle de comportement viscoplastique de l'Ecole Centrale de Paris*. Rapport EDF-CNEH Technolac, le Bourget du Lac.

Leroueil, S. (1985). Stress–strain–strain-rate relation for the compressibility of natural clays. *Géotechnique*, **35**, No. 2, 159–180.

Leinenklugel (1967). *Deformations und Festigkeitsverhalten bindinger Ersstoffe*. Thèse Karlsruhe, Heft 66.

Michalski, E. & Rahma, A. (1989). *Modélisation du comportement des sols en élastoplasticité*. Rapport de BRGM.

Murayama, S. & Shibata, T. (1964). Flow and stress-relaxation of clays. Rheology and soil mechanics, *Symposium IUTAM*, Grenoble.

Perzyna, P. (1963). The constitutive equations for work hardening and rate sensitive plastic materials. *Proc. Vibrational Problem*, **4**, No. 3, 281–290.

Singh, A. & Mitchell, J. K. (1969). Creep potential and rupture of soils. *Proceedings of the Seventh International Conference on Soil Mechanics and Foundation Engineering*, Mexico. **I**. 379–384.

Tavenas, F. (1978). Creep behavior of an undisturbed lightly overconsolidated clay. *Canadian Geotechnical Journal* **15**.

Shibuya, S., Hwang, S. C. & Mitachi, T. (1997). *Géotechnique* **47**, No. 3, 593–601

Elastic shear modulus of soft clays from shear wave velocity measurement

S. SHIBUYA,* S. C. HWANG* and T. MITACHI*

Case records comprising the results of downhole seismic surveys collected at nine sites worldwide, together with comparative results of laboratory bender element tests on reconstituted clay samples, were examined in an attempt to quantify the shear modulus of normally consolidated clays at very small strains of the order of 0·0001%. The shear modulus G_{max} under the current state of stresses is given in a formula which includes a newly proposed void ratio function. An empirical expression incorporating the new void ratio function is also proposed for practical use in estimating G_{max} profiles with depth in natural soil deposits from routinely available borehole data. In addition, effects of soil structure and ageing on G_{max} are discussed on the basis of a case record.

KEYWORDS: clays; compressibility; elasticity; in situ testing; laboratory tests; stiffness.

On a examiné les dossiers de cas comprenant les résultats de relevés sismiques à foration descendante, obtenus dans dix points du globe, ainsi que les résultats comparatifs d'essais d'éléments de flexion en laboratoire sur des échantillons d'argile reconstitués, en vue d'essayer de quantifier le module de cisaillement d'argiles normalement consolidées à de très petites contraintes de l'ordre de 0,0001 p. 100. Le module de cisaillement G_{max}, à l'état de contrainte existant est donné dans une formule qui comprend une nouvelle fonction proposée d'indice des vides. Une expression empirique comprenant la nouvelle fonction d'indice des vides est également proposée pour l'application pratique à l'estimation des profils G_{max} en fonction de la profondeur dans le dépôts de sol naturel à partir de données de foration descendante couramment disponibles. De plus, les effets de la structure du sol et du vieillissement sur G_{max} sont examinés dans le contexte d'un cas d'espèce.

INTRODUCTION

Abundant experimental data are now available to support the existence of the pseudo-elastic shear modulus of geomaterials at shear strains of the order of about 0·0001%. The shear modulus for this very small strain region represents the maximum value that will occur at the current density and stress state when the material is subjected to either monotonic or cyclic shear. The G_{max} value is hardly influenced by rate of shear straining or the type of loading (i.e. monotonic or cyclic shear). However, the value of hysteretic damping associated with G_{max} is not exactly zero, so that the terminology 'pseudo-elastic' is employed for describing G_{max} (Tatsuoka & Shibuya, 1992; Shibuya *et al.*, 1992, 1995; Lo Presti *et al.*, 1993). An attempt is made in this paper to quantify G_{max} of normally consolidated soils having no true cohesion or cementation, by considering the link with their effective overburden stress state and void ratio, both of which can be obtained through ordinary site investigation practice.

PREVIOUS WORK ON QUANTIFICATION OF SMALL-STRAIN SHEAR MODULUS OF SOILS

Theory on shear wave propagation in an elastic body tells us that the G_{max} value of soils from in situ and laboratory shear wave velocity (V_s) measurements is given by

$$G_{max} = \rho V_s^2 \tag{1}$$

where ρ is the soil density. According to Hardin (1978), the G_{max} value of soils subjected to isotropic consolidation may be expressed as

$$G_{max} = Cf(e)\sigma'^{n}_{c}\sigma_r^{(1-n)}(\mathrm{OCR})^k \tag{2}$$

where C is an empirical constant, $f(e)$ is a void ratio function, σ'_c is the current effective confining

Manuscript received 28 October 1997; revised manuscript accepted 24 April 1997.
Discussion on this paper closes 1 November 1997.
* Hokkaido University.

stress, σ_r is the reference stress, OCR is the overconsolidation ratio and n and k are exponents. Note that C, n and k will take different values for different soils. The value of C also depends on the choice of the reference stress σ_r.

Laboratory tests on various kinds of geomaterials show that mechanical overconsolidation has a negligible influence on G_{max} (i.e. $k = 0$ in equation (2)) when the results are interpreted with an appropriate form of $f(e)$ (e.g. Tatsuoka & Shibuya, 1992; Shibuya *et al.*, 1992; Jamiolkowski *et al.*, 1994). It is also well known that G_{max} is independent of the magnitude of normal stress on the plane through which the shear body wave propagates (Roesler, 1979; Yu & Richart, 1984; Stokoe *et al.*, 1994). The results suggest the use of $s' = (\sigma'_1 + \sigma'_3)/2$ instead of σ'_c in equation (2) when the soil is subjected to an anisotropic stress state.

This means that where the shear wave propagates in the vertical direction, with the particle motion in a horizontal direction, G_{max} may be expressed by the following relationship (e.g. Jamiolkowski *et al.*, 1994):

$$G_{max} = S_{vh} f(e) \sigma_r^{(1-nv-nh)} \sigma'^{\,nv}_{v} \sigma'^{\,nh}_{h} \qquad (3)$$

where S_{vh} is a material constant reflecting the current soil structure, σ'_v is the effective vertical stress, σ'_h is the effective horizontal stress and nv and nh are exponents. Equation (3) can be used to interpret values of G_{max} measured in conventional down-hole seismic surveys on level ground or in laboratory tests using vertically cut specimens. The effects of the two principal stresses on G_{max} are more or less equal to each other, implying $nv = nh$ in equation (3). Therefore, equation (3) becomes

$$G_{max} = A f(e) \sigma_r^{(1-2nv)} \sigma'^{\,2nv}_{v} \qquad (4)$$

in which the non-dimensional parameter A is equal to $S_{vh} K_0^{\,nv}$, where K_0 is the coefficient of earth pressure at rest, σ'_h/σ'_v.

IN SITU SEISMIC SURVEYS AND LABORATORY BENDER ELEMENT TESTS

A database comprising the results of down-hole seismic surveys performed at nine sites worldwide is examined in this paper. Soil profiles, as well as the details of the survey, are described for each site in the relevant reference (Table 1). The investigation at the Higashi-Ohgishima site was carried out in the upper sand deposit as well as in the lower clay deposit, whereas the rest dealt solely with cohesive soils. The subsoils at these sites are essentially normally consolidated or lightly overconsolidated clays with any overconsolidation due mainly to the effects of ageing.

Figure 1 shows a consolidometer equipped with a set of bender elements to measure the shear wave velocity through the specimen (Dyvik & Madshus, 1985). In this apparatus, developed at Hokkaido University, the circular specimen of nominal dimensions 60 mm dia. and 20 mm high undergoes one-dimensional consolidation. The vertical stress on the specimen is measured by means of a load cell mounted on the base, the measurement of which is thus unaffected by any mechanical friction on the loading piston. The horizontal stress can also be measured by monitoring the hoop strain in the thin section of the consolidation ring.

A total of eight reconstituted samples (one for each of the clays 10 to 17 in Table 1) were prepared by consolidating a slurry of clay with an initial water content twice the liquid limit. The slurries of Ariake, Drammen, Higashi-Ohgishima, Kurihama and Hachirougata clays were each prepared by using intact samples retrieved at the exact site where the in situ seismic survey was carried out. The rest, MC-clay and NSF-clay, were powder clays supplied by commercial firms in Japan (for details, see Shibuya *et al.*, 1995, 1996). Each specimen was trimmed out of the one-dimensionally consolidated block of clay, for which the σ'_v direction during consolidation coincided with the σ'_v direction at the preconsolidation stage. The preconsolidation pressure was 100 kPa for all the samples. In a similar way, two intact specimens of Ariake clay were trimmed out of the sample retrieved by a thin-wall sampler.

On placing the clay specimen into the consolidation ring, a back pressure of 200 kPa was applied to the inner water-filled chamber in order to improve the degree of saturation (Fig. 1). The consolidation stress was then increased incrementally to a level which was five times the preconsolidation stress for the reconstituted samples, and ten times the in situ effective overburden pressure for the intact specimens of Ariake clay (Fig. 2). In each step of loading, the principal stresses, together with shear wave velocity, were measured at an instant when the excess pore water pressure at the bottom of the specimen was fully dissipated.

The bender elements protruded into the soil sample by 4 mm on each side. The transmitter mounted on the top plate was excited using a driving square wave with a frequency of 50 Hz. The travel distance of the shear wave was taken as the tip-to-tip distance between the bender elements. Furthermore, the arrival time of the shear wave was determined at the first clear peak as measured by the receiver (see Viggiani & Atkinson, 1995).

Table 1. Results of in situ seismic surveys and laboratory bender element tests

			ρ_s: g/cm^3	W_L: %	I_p	m	C'	K_0 (nc)	A (for $\sigma_r = 98{\cdot}1$ kPa in equation (9))	In equation (3) with $f(e) = (1+e)^{-2{\cdot}4}$		Reference
										nv ($= nh$)	S_{vh} (for $\sigma_r = 1$ kPa)	
In situ seismic survey	1	Ariake	2·60–2·66	71–127	28–116	1·27	0·33	—	30 800	—	—	Shibuya & Tanaka (1996)
	2	Bothkennar	2·86 (assumed)	53–87	28–59	1·15	0·25	—	27 100	—	—	Hight *et al.* (1992)
	3	Drammen	2·78–2·81	34–48	20–30	1·25	0·17	—	19 900	—	—	Lunne *et al.* (1976)
	4	Fucino	2·65 (assumed)	93–120	52–77	0·97	0·16	—	27 400	—	—	A.G.I. (1991)
	5	Hachirougata	2·46–2·67	112–239	61–152	1·22	0·43	—	27 500	—	—	
	6	Higashi-Ohgishima (upper sand)	2·66–2·68	Not determined	Not determined	0·76	0·003	—	25 900	—	—	Shibuya & Tanaka (1996) (rows 5–8)
	7	Izumo	2·69–2·91	62–152	19–107	2·76	0·90	—	22 700	—	—	
	8	Kurihama	2·62–2·75	73–118	56–75	0·75	0·12	—	23 100	—	—	
	9	St Alban	2·70 (assumed)	18–63	10–30	2·13	0·70	—	28 700	—	—	Lefebvre *et al.* (1994)
Bender element test on reconstituted clay samples	10	Ariake	2·77	82	37	0·93	0·10	0·48	29 000	0·34	2800	
	11	Drammen	2·78	96	52	0·76	0·05	0·50	10 700	0·32	2620	
	12	Higashi-Ohgishima (lower clay)	2·72	67	25	0·84	0·09	0·50	18 900	0·31	2510	
	13	Kiyohoro	2·74	41	19	0·68	0·06	0·45	15 200	0·27	2670	This study (rows 10–17)
	14	Kurihama	2·70	98	59	0·84	0·12	0·52	21 400	0·27	2960	
	15	MC-clay	2·78	87	49	0·64	0·12	0·75	19 900	0·20	3490	
	16	NSF-clay	2·98	55	26	0·75	0·08	0·58	19 400	0·30	2500	
	17	Hachirougata	2·53	120	54	0·94	0·11	0·55	24 300	0·34	2350	

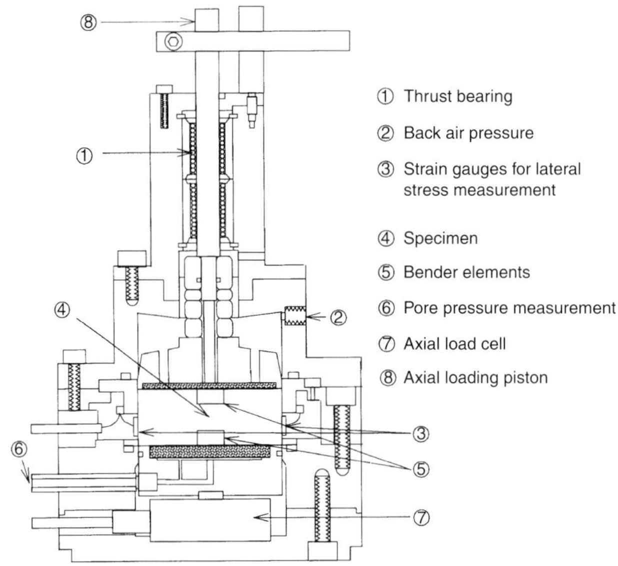

Fig. 1. Consolidometer equipped with bender elements

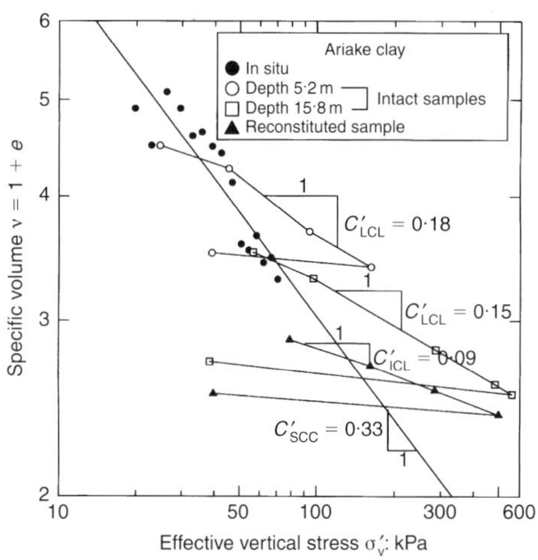

Fig. 2. Relationships between specific volume and effective overburden stress for Ariake clay

PROPOSAL OF A NEW VOID RATIO FUNCTION

A relationship between G_{max} and void ratio expressed in terms of effective stress state, for naturally sedimented or reconstituted soils, is formulated here on the following grounds.

Assumption 1

For a given soil in a state close to normal consolidation, K_0 is a constant and hence A is a constant in equation (4).

Assumption 2

For a given soil the compression law for the soil is characterized by a linear relationship between the specific volume $v = 1 + e$ and σ'_v in their double logarithmic representation (Butterfield, 1979):

$$v/v_r = (\sigma'_v/\sigma'_{vr})^{-C'} \quad (5)$$

where v_r is the value of specific volume at a reference stress σ'_{vr} and C' represents the compression index. The relationship between v and σ'_v for the natural soil deposits studied in this paper is shown in Fig. 3(a), in which the slope refers to the compression index associated with the sedimentary compression curve (SCC) (Burland, 1990). Similar relationships for the reconstituted samples are shown in Fig. 3(b), in which the slope refers to the compression index of the intrinsic compression line (ICL) (Burland, 1990). The values of C'_{SCC} and C'_{ICL} were determined by linear regression (Table 1). It is interesting that the slopes for the SCC cover a range of C' between 0·12 and 0·9 for eight out of the nine clays, which is much larger than the corresponding range of slopes for the ICL (0·05 to 0·12). It may be an effect of the fact that the SCCs are determined at relatively small depths, with the effective overburden stress less than 200 kPa, whereas the ICLs are determined at vertical effective stress values that are much larger.

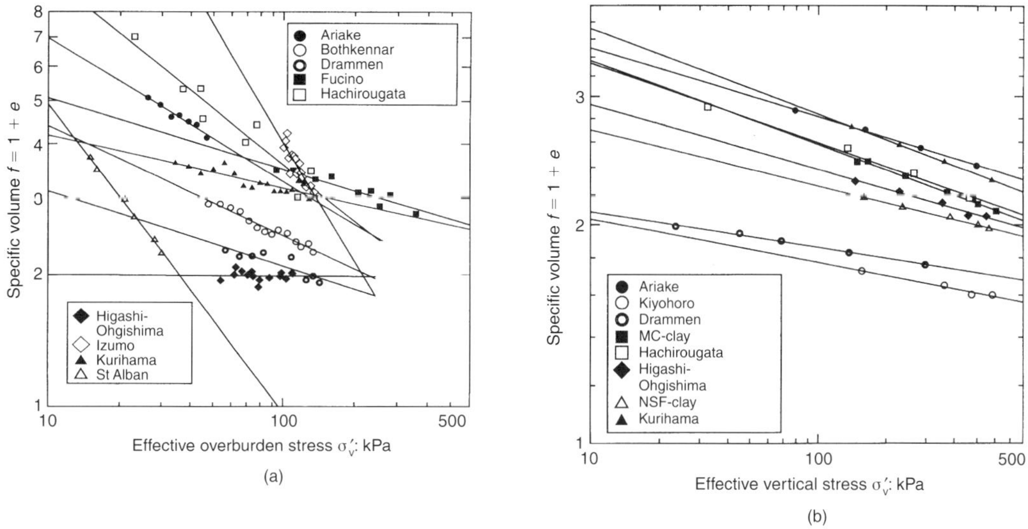

Fig. 3. Compression data: (a) natural soil deposits; (b) reconstituted clays; associated with compression indices C'_{SCC} and C'_{ICL} respectively

Assumption 3

For a given soil, the G_{max} value increases with σ'_v in an exponential form (Shibuya & Mitachi, 1994; Shibuya & Tanaka, 1996), that is,

$$G_{max}/G_{max\,r} = (\sigma'_v/\sigma'_{vr})^m \tag{6}$$

where $G_{max\,r}$ is the value of G_{max} at the reference stress σ'_{vr}. The results are shown in Fig. 4 and the values of m are presented in Table 1.

For mathematical convenience, a void ratio function having the form

$$f(e) = (1+e)^\alpha = v^\alpha \tag{7}$$

is considered in equation (4). Since equations (5), (6) and (7) are all expressed in exponential forms, comparison with equation (4) shows

$$m = 2nv - \alpha C' \tag{8}$$

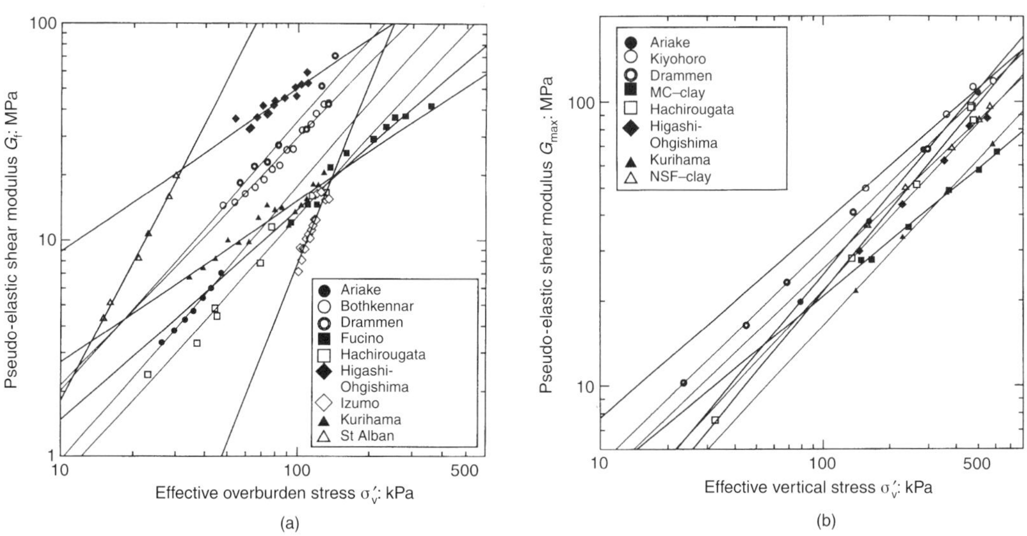

Fig. 4. Dependence of small-strain shear modulus on the current stress: (a) natural soil deposits; (b) reconstituted clays

Accordingly, if α and nv are constants for all soils, α can be determined as the averaged negative inclination of data in a plot of m against C' for different soils.

Figure 5 shows the relationship from the authors' database as listed in Table 1. The solid line with a slope of 2·4 was tentatively obtained as a result of linear regression applied to all the data points, bearing in mind that the exponent m is approximately 0·5 for C' close to zero. This limit at $C' = 0$ is based on previous work which shows that for dense, virtually incompressible sands with C' close to zero, G_{max} varies with σ'_v raised to a power of 0·5 (e.g. Hardin & Richart, 1963; Iwasaki *et al.*, 1978).

On the basis of the results shown in Fig. 5, the following expression is proposed for practical use in estimating G_{max} of soils having no cementation:

$$G_{max} = A(1+e)^{-2 \cdot 4}\sigma_r^{0 \cdot 5}\sigma'^{0 \cdot 5}_v \tag{9}$$

It should be stressed that equations (5) and (6) apply to a specific soil, with the exponents C' and m taking different values for different soils, whereas the proposal is that the exponents in equation (9) ($\alpha = -2{\cdot}4$ and $2nv = 0{\cdot}5$) are meant to be applicable to all soils, with possible variation of the parameter A for different soils. It should also be pointed out that the estimate using equation (9) needs information on e and σ'_v, both of which can be obtained from ordinary borehole or oedometer test data. The values of the empirical parameter A in equation (9), for a reference stress σ_r of 98·1 kPa, are given in Table 1. The A value for each clay was determined by means of linear regression applied to the measured data of $G_{max}/(1+e)^{-2 \cdot 4}$ against $\sigma'^{0 \cdot 5}_v$. The A value spanned mostly over a range between 18 000 and 30 000, with the average about 24 000. A plot of A against plasticity index does not show any particular trend.

DISCUSSION

A comparison of the term $Af(e)$ in equation (4) is made in Fig. 6 for the new empirical expression of equation (9) with alternative expressions proposed in the literature. The results were obtained by applying the reference stress of 1 kPa to these seven expressions, most of which have the exponent $2nv$ of 0·5 in common. The new expression given in equation (9), as well as the previous relationship $G_{max} = 5000e^{-1 \cdot 5}\sigma_r^{0 \cdot 5}\sigma'^{0 \cdot 5}_v$ by Shibuya & Tanaka (1996), are both promising in respect of their apparent capability of estimating G_{max} of not only clays but also sandy soils. This is evident from the fact that for a range of void ratios between 0·5 and 1, typical of sandy soils, the relationships between $Af(e)$ and e for these expressions lie close to that of the well-established relationship for sandy soils proposed by Hardin & Richart (1963).

The applicability of equation (9), together with the alternative expression of Shibuya & Tanaka (1996), to normally consolidated clays is examined in Fig. 7, in which the G_{max} profile from in situ shear wave velocity measurement is compared with the estimated profiles with depth for two sites. In these comparisons, the following expression in use for practical estimate of G_{max} is employed

$$G_{max} = A(1+e)^{-2 \cdot 4}\sigma'^{0 \cdot 5}_v \tag{10}$$

where the units of G_{max} are kPa.

The A parameter in equation (10) refers to the value at $\sigma_r = 98{\cdot}1$ kPa (see equation (9) and Table

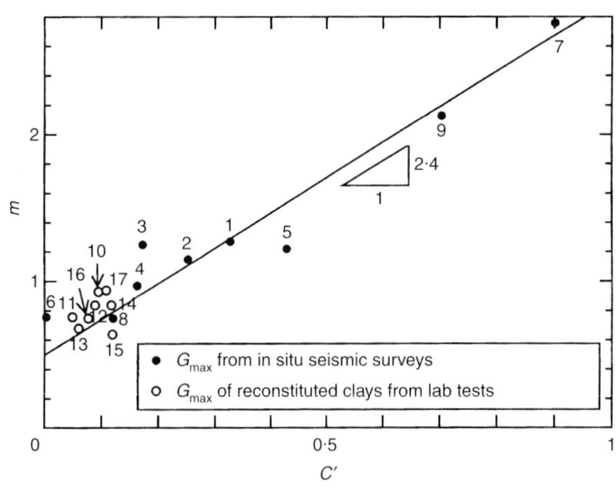

Fig. 5. Determination of exponent α in equation (7) using the authors' database (for details see Table 1)

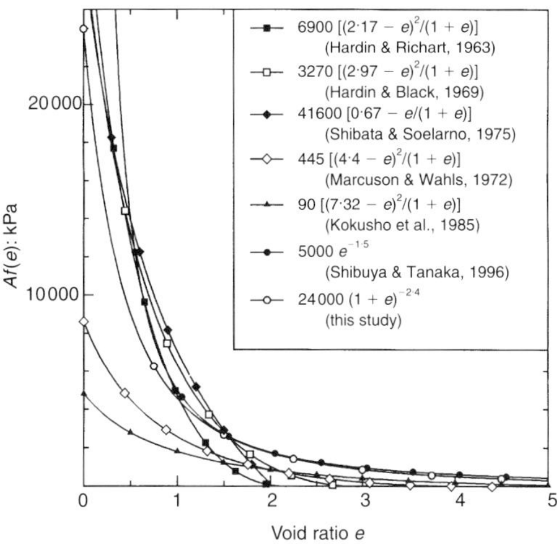

Fig. 6. Comparison of the term $Af(e)$ in equation (2) with empirical expressions previously proposed

1) and σ'_v is given by the in situ effective overburden pressure expressed in kPa. It is demonstrated in these examples that the estimated profiles agree well with the measured G_{max} profile with depth. Similar results are obtained for the other sites listed in Table 1.

Figure 2 shows compression data of Ariake clay, a sensitive soft clay in Japan. Two intact samples retrieved at depths of 5·2 m ($\sigma'_{v\,(in\ situ)} = 25$ kPa) and 15·8 m ($\sigma'_{v\,(in\ situ)} = 52$ kPa) were subjected to recompression for $\sigma'_v < \sigma'_{v\,(in\ situ)}$, virgin compression for $\sigma'_v > \sigma'_{v\,(in\ situ)}$ and unloading. A third sample, reconstituted in the laboratory, was also subjected to a similar consolidation history. For the virgin compression branch, the relationship between v and σ'_v for each intact specimen can be characterized using a straight line in a double logarithmic plot. The values of compression index related to this laboratory compression line (LCL), C'_{LCL} (Shibuya & Mitachi, 1994) were 0·15 and 0·18 for the samples taken from the two different depths. These values were about half of the relevant C'_{SCC} of 0·35 and slightly larger than the C'_{ICL} of 0·09.

The results of measurement of G_{max} in these tests are shown in Fig. 8, in which the examination is made in a plot of $G_{max}/(1+e)^{-2\cdot4}$ against $\sigma'_h\sigma'_v$. The following are to be noted.

(*a*) The relationship can reasonably be approximated by a straight line in spite of the large variety of OCR values in each set of tests. This validates the use of $f(e) = (1+e)^{-2\cdot4}$, by which mechanical overconsolidation has a negligible influence on G_{max} (i.e. $k = 0$ in

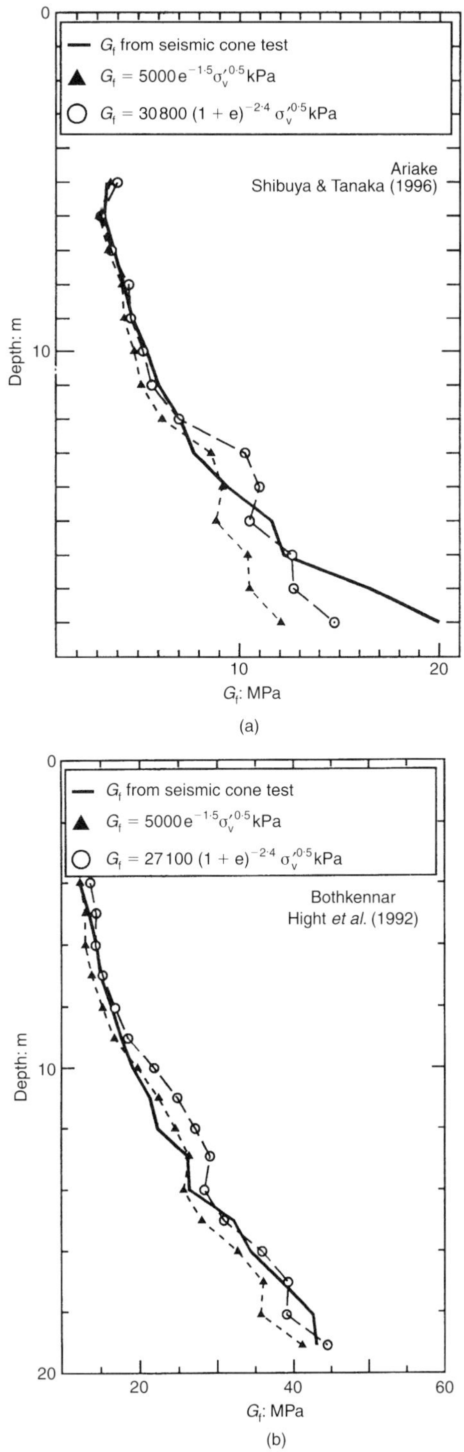

Fig. 7. (a), (b) Two examples to demonstrate the performance of the newly proposed expression given in equation (9) for estimating the G_{max} profile with depth

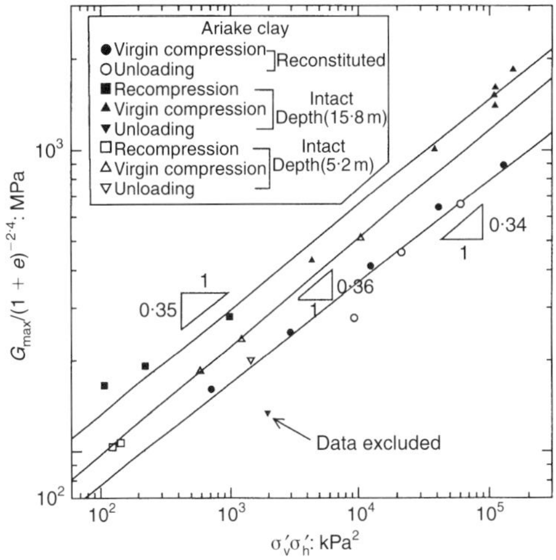

Fig. 8. Variations of G_{max} with σ'_v observed in the consolidometer using intact and reconstituted samples of Ariake clay

equation (2)). The trend was similar in tests on the other reconstituted clays.

(*b*) The slopes of the lines in Fig. 8 were similar for the intact and reconstituted samples, but they differed from the value of $nv = 0{\cdot}25$, and hence $2nv = 0{\cdot}5$, suggested by equation (9). This suggests that for a given soil, an improved fit to the measured data can be achieved by using equation (4) with $f(e) = (1+e)^{-2{\cdot}4}$, together with the fitted nv value for the soil in question.

(*c*) The S_{vh} value given in equation (3) with $f(e) = (1+e)^{-2{\cdot}4}$ and $\sigma_r = 1$ kPa was increasing in the following order of the reconstituted sample ($S_{vh} = 2800$), the intact sample at $d = 5{\cdot}2$ m ($S_{vh} = 2920$) and the intact sample at $d = 15{\cdot}8$ m ($S_{vh} = 3280$).

Point (*b*) above suggests that the exponent nv given by equations (3) and (4) with $f(e) = (1+e)^{-2{\cdot}4}$ is not influenced by the current soil structure or the degree of 'ageing' for a given soil. Conversely, point (*c*) indicates that the value of S_{vh} in equation (3), and hence the A parameter in equation (4), increase as the clay gets more 'aged' and 'structured'.

The measured values of nv, together with S_{vh} at $\sigma_r = 1$ kPa in similar tests on the eight reconstituted clays, are given in Table 1. Note that, except for MC-clay, the S_{vh} values fall in a narrow range between 2300 and 3000. Conversely, the value of $2nv$ showed a variation from 0·40 to 0·68 rather than the constant value of 0·5 assumed in equation (9). It is now clear that the occurrence of a wide range of the A parameter in equation (9) is partially a consequence of artificially forcing $2nv$ to take the value of 0·5 in this equation.

CONCLUSIONS

On the basis of examination of characteristics of the small-strain shear modulus of soft clays, from in situ and laboratory measurements of shear wave velocity, it was found that G_{max} increased with effective overburden stress σ'_v in an exponential manner. Similarly, the compression law for the soils in a state of normal consolidation was successfully characterized by a linear relationship between the specific volume $(1+e)$ and σ'_v in their double logarithmic representation.

Considering these characteristic aspects, a void ratio function $f(e) = (1+e)^{-2{\cdot}4}$ is proposed for relating G_{max} to the current effective stress state and void ratio of a given soil. An empirical expression, $G_{max} = A(1+e)^{-2{\cdot}4}\sigma'^{\,0{\cdot}5}_v$ (in kPa) (equation (10)), is also proposed for practical use in estimating the G_{max} profile with depth in soil deposits, for which an averaged A value of about 24 000 was obtained for the normally consolidated clays examined. The applicability of this expression was satisfactory as examined for all the case records listed in Table 1.

Despite the fact that the exponent of σ'_v in equation (9) has an everaged value of 0·5 for many soils, an improved fit for an individual soil can be achieved by relaxing this requirement, to give $G_{max} = A(1+e)^{-2{\cdot}4}\sigma_r^{1-2nv}\sigma'^{\,2nv}_v$ (equation (4)).

Results of laboratory bender element tests on intact and reconstituted samples from the same origin indicated that soil structure and ageing effects had little influence on the exponent nv in equation (4), with $f(e) = (1+e)^{-2{\cdot}4}$. The material constant S_{vh} reflecting the soil structure was, however, larger for intact specimens than for reconstituted samples. Moreover, for intact samples, the value of S_{vh} increased with soil depth. The results suggest that the A parameter in equations (9) and (4) is affected by soil ageing, and hence G_{max} increases as the clay ages.

Obviously, further study is needed to take account of possible variation of the S_{vh} values, and hence the A parameter, with depth when quantifying the G_{max} profile with depth in 'aged' or 'structured' natural soil deposits.

REFERENCES

Associazione Geotechnica Italiana (1991). Geotechnical characterization of Fucino clay. Italian contribution to the conference theme. *Proc. 10th European Regional Conf. on SMFE, Florence*, 1–14.

Burland, J. B. (1990). On the compressibility and shear strength of natural clays. 30th Rankine lecture. *Géotechnique* **40**, No. 3, 329–378.

Butterfield, R. (1979). A natural compression law for soils (an advance on $e-\log p'$). *Géotechnique* **29**, 469–480.

Dyvik, R. & Madshus, C. (1985). Laboratory measurements of G_{max} using bender elements. *Proc. Am. Soc. Civ. Engrs Conv., Detroit*, pp. 186–196.

Hardin, B. O. & Richart, F. E., Jr. (1963). Elastic wave velocities in granular soils. *J. Soil Mech. Fdns Div., ASCE* **89**, No. SM1, 33–65.

Hardin (1978). The nature of stress-strain behaviour of soils. *Proc. Geotech. Eng. Div. Speciality Conf. on Earthquake Eng, and Soil Dynamics,* **1**, ASCE, Pasadena, 3–90.

Hardin, B. O. and Black, W. L. (1969). Vibration modulus of normally consolidated clay. *Journal of the SMF Div. Proc., ASCE*, **95**, No, SM 6, 1531–1537.

Hight, D. W., Bond, A. J. & Legge, J. D. (1992). Characterisation of Bothkennar clay: an overview. *Géotechnique* **42**, 303–347.

Iwasaki, T., Tatsuoka, F. & Takagi, Y. (1978). Shear moduli of sands under cyclic torsional shear loading. *Soils Fdns* **18**, No. 1, 39–56.

Jamiolkowski, H., Lancellotta, R. H. & Lo Presti, D. C. F. (1994). Remarks on the stiffness at small strains of six Italian clays. In *Pre-failure deformation of geomaterials* (eds S. Shibuya, T. Mitachi & S. Miura), Vol. 2, pp. 817–836. Rotterdam: Balkema.

Kokusho, T., Yoshida, Y. & Esashi, Y. (1982). Dynamic properties of soft clay for wide strain range. *Soils Fdns* **22**, No. 4, 1–18.

Lefebvre, G., Leboeuf, D., Rahhal, M. B., LaCroix, A., Warde, J. & Stokoe, K. H. (1994). Laboratory and field determinations of small-strain shear modulus for a structured Champlain clay. *Can. Geotech. J.* **31**, No. 1, 61–70.

Lo Presti, D. C. F., Pallara, O., Lancellotta, R., Armandi, M. & Maniscalco, R. (1993). Monotonic and cyclic loading behaviour of two sands at small strains. *Geotech. Testing J.* **16**, No. 4, 409–424.

Lunne, T. O., Eide, J. & de Ruiter, J. (1976). Correlations between cone resistance and vane shear strength in some Scandinavian soft to medium stiff clays. *Can. Geotech. J.* No. 13-4, 430–441.

Marcuson, W. G. & Wahls, H. E. (1972). Time effects on dynamic shear modulus of clays. *Proc. Am. Soc. Civ. Engrs* **97**, No. SM1, 1359–1373.

Roesler, S. K. (1979). Anisotropic shear modulus due to stress-anisotropy. *J. Geotech. Engng Div. ASCE* **105**, No. GT 7, 871–880.

Shibata, T. & Soelarno, D. S. (1978). Stress-strain characteristics of clays under cyclic loading. *J. Japan. Soc. Civ. Engrs*, No. 276, 101–110 (in Japanese).

Shibuya, S., Tatsuoka, F., Teachavorasinskun, S., Kong, X. J., Abe, F., Kim, Y. S. & Park, C. S. (1992). Elastic deformation properties of geomaterials. *Soils Fdns* **32**, No. 3, 26–46.

Shibuya, S. & Mitachi, T. (1994). Small strain stiffness of clay sedimentation in a state of normal consolidation. *Soils Fdns* **34**, No. 4, 67–77.

Shibuya, S. & Tanaka, H. (1996). Estimate of elastic shear modulus in Holocene soil deposits. *Soils Fdns* **36**, No. 4, 45–55.

Shibuya, S., Mitachi, T., Fukuda, F. & Degoshi, T. (1995). Strain rate effects on shear modulus and damping of normally consolidated clay. *Geotech. Testing J.*, **8**, No. 3, 365–375.

Shibuya, S., Mitachi, T., Hosomi, A. & Hwang, S. C. (1996). Strain rate effects on stress strain behaviour of clay as observed in monotonic and cyclic triaxial tests. In *Measuring and modeling time dependent soil behavior* (eds T. C. Sheahan & V. N. Kaliakin), ASCE GSP No. 61, RR. 214–227.

Stokoe, K. H., II, Hwang, S. K. & Lee, J. N.-K. (1994). Effects of various parameters on the stiffness and damping of soils at small to medium strains. In *Pre-failure deformation of geomaterials* (eds S. Shibuya, T. Mitachi & S. Miura), Vol. 2, pp. 785–816, Rotterdam: Balkema.

Tatsuoka, F. & Shibuya, S. (1992). Deformation characteristics of soils and rocks from field and laboratory tests. *Proc. 9th Asian Regional Conf. on SMFE, Bangkok* **2**, 101–170.

Viggiani, G. & Atkinson, J. H. (1995). Interpretation of bender element tests. *Géotechnique* **45**, No. 1, 149–154.

Yu, P. & Richart, F. E., Jr. (1984). Stress ratio effects on shear modulus of dry sands. *J. Geotech. Engng, ASCE* **110**, No. 3, 331–345.

Lo Presti, D. C. F., Jamiolkowski, M., Pallara, O., Cavallaro, A. & Pedroni, S. (1997). *Géotechnique* **47**, No. 3, 603–617

Shear modulus and damping of soils

D. C. F. LO PRESTI,* M. JAMIOLKOWSKI,* O. PALLARA,* A. CAVALLARO† and S. PEDRONI‡

The shear modulus (*G*) and damping ratio (*D*) of two natural clays and both siliceous and carbonatic pluvially-deposited dry sands have been measured at the authors' laboratory by means of a Resonant Column/Torsional Shear apparatus for a strain range of 0·0001% up to 1%. Different kinds of tests were performed on hollow cylindrical specimens off all the: (a) monotonic loading torsional shear tests at constant stress rate, (b) cyclic loading torsional shear tests at constant strain rate or at constant frequency and (c) Resonant Column tests. The results of the above tests enabled the authors to analyse the influence of strain level (γ), loading rate ($\dot{\gamma}$), number of loading cycles (*N*) and type of loading (monotonic or cyclic) on the measured values of *G* and *D*.

KEYWORDS: clays; dynamics; laboratory tests; sands; stiffness.

Les auteurs ont mesuré le module de cisaillement (*G*) et le rapport d'amortissement (*D*) de deux argiles naturelles et de dépôts pluviaux de sables secs silicieux et carbonatitiques dans leur laboratoire à l'aide d'essais à colonne résonante et d'essais de cisaillement par torsion dans une plage de contraintes de 0,0001% à 1%. Différents types d'essais ont été effectués avant tout sur des éprouvettes cylindriques creuses: (a) des essais de cisaillement par torsion à charge monotone et à taux de contrainte constant, (b) des essais de cisaillement par torsion à charge cyclique à taux de contrainte constant ou à fréquence constante, et (c) des essais à colonne résonante. Les résultats de ces essais ont permis aux auteurs d'analyser l'influence du niveau de deformation, du taux de charge, du nombre de cycles de charge (*N*) et du type de charge (monotone ou cyclique) sur les valeurs mesurées de *G* et *D*.

INTRODUCTION

The shear modulus (*G*) and damping ratio (*D*) of soils are usually determined by means of cyclic static tests (i.e. torsional or triaxial cyclic tests) or by means of dynamic tests (i.e. Resonant Column tests). The testing procedures and the data reduction for these kinds of tests are described in many papers (see for example: Drnevich *et al.*, 1978; Alarcon-Guzman *et al.*, 1986; Isenhower *et al.*, 1987; Toki *et al.*, 1995). The main differences between static and dynamic tests concern the loading rate (i.e. $\dot{\gamma}$ shear strain rate) and number of loading cycles (*N*). For a static test, after the specimen has been consolidated at given pressures, the cyclic loading stage is applied stepwise under strain or stress control. To perform such a task, a certain number of loading cycles are applied to the specimen using sinusoidal or triangular variation of torque at constant frequency, generally in the range 0·1–1 Hz. Resonant Column tests (RCTs) are also performed stepwise with a sine wave excitation, but in this case it is not possible to control $\dot{\gamma}$ and *N*. The resonance frequency generally falls into the range 30–200 Hz and the number of applied loading cycles is usually of the order of thousands. By assuming a linear variation of deformation with time during cyclic loading, it is possible to compute the current strain rate with the following formula:

$$\dot{\gamma} = 4\gamma f [\%/\mathrm{s}], \qquad (1)$$

where $\dot{\gamma}$ = shear strain rate;
γ = single amplitude shear strain (%);
f = loading frequency (Hz) for static tests;
$f = f_n$ resonance frequency (Hz) for RCTs.

In all other cases, equation (1) provides an equivalent strain rate. RCTs, therefore, adopt much higher strain rates and impose a greater number of loading cycles on the specimen in comparison to static tests.

Alarcon-Guzman *et al.* (1989) showed the great influence of *N* on the shear modulus *G* of dry sands for shear strains ranging from between 0·01

Manuscript received 11 November 1996; revised manuscript accepted 6 March 1997.
Discussion on this paper closes 1 November 1997.

* Department of Structural Engineering, Politecnico di Torino, Corso Duca degli Abruzzi 24, 10129 Torino, Italy.
† University of Catania.
‡ ENEL CRIS of Milan.

and 0·1%; moreover, the number of loading cycles to obtain a stable value of G ranged between 30 and 50. These findings contradict the results obtained by other researchers (Hardin & Drnevich, 1972a, b; Sherif & Ishibashi, 1976; Iwasaki *et al.*, 1978; Kokusho, 1980; Tatsuoka *et al.*, 1991) who showed a moderate influence of N on G during the first 5–10 cycles. The results shown in the present paper try to explain the above-mentioned difference in results.

An increase of the stiffness for an increase of the strain rate is generally recognized (Dobry & Vucetic, 1987; Tatsuoka & Shibuya, 1992). Tatsuoka & Shibuya (1992) consider that, at small strains, the soil behaviour is essentially elastic and hence rate independent. Therefore, the same elastic initial tangent modulus (or small strain stiffness) is expected from dynamic and static (cyclic or monotonic) tests. At larger strains, when the influence of strain rate on the stress–strain curve becomes more and more important, rate effects can mask the soil non-linearity moving up the elastic limit. Therefore, the stress–strain relationship, in a typical quasi-static test performed at constant frequency, can be distorted, as experimentally observed by Isenhower & Stokoe (1981).

This paper provides some information on the rate effect on the shear modulus. In particular the increase of the elastic limit with strain rate is experimentally verified. The stiffness increase with strain rate is evaluated in terms of the coefficient of strain rate:

$$\alpha(\gamma) = \frac{\Delta G(\gamma)}{\Delta(\log\dot{\gamma})G(\gamma, \dot{\gamma}_{\mathrm{REF}})} \qquad (2)$$

i.e. the increase in shear modulus for one logarithmic cycle of strain rate normalized with respect to the shear modulus obtained with a reference strain rate.

Recent works have shown that the damping ratio is much more sensitive than shear modulus to N and $\dot{\gamma}$ (Shibuya *et al.*, 1995; Tatsuoka *et al.*, 1995; Stokoe *et al.*, 1995; Tatsuoka & Kohata, 1995). This is a consequence of the concomitant effects of soil non-linearity (which increases with the strain level), viscosity of soil skeleton (which is predominant at low strain rates) and viscosity of the pore fluid (which is predominant at very large strain rate). The influence of N and $\dot{\gamma}$ on D is also given in the paper.

The G–γ and D–γ curves are used in equivalent-linear analysis. In the case of non-linear analysis the stress–strain relations for virgin or first loading (backbone or skeleton curve) and that for subsequent unloading and reloading should be determined. In principle, the skeleton curve should be determined from monotonic loading tests. The possibility of obtaining the skeleton curve from cyclic tests should be verified by comparing monotonic against cyclic test results. Such a comparison is provided in the present paper for both cohesive and cohesionless soils. Similar comparisons are available in the works by Alarcon-Guzman *et al.* (1989), Tatsuoka & Shibuya (1992) and Lo Presti *et al.* (1993).

TESTING APPARATUS AND PROCEDURES

The Torsional Shear/Resonant Column apparatus (Fig. 1) used for this research is the same as that described by Lo Presti *et al.* (1993). The main characteristics of the apparatus are summarized below.

The apparatus is equipped with an electric motor (SBEL, Arizona) which has eight coils and four magnets. It is capable of a maximum torque of about 1 Nm and can operate under stress control conditions. The motor is driven by means of an arbitrary function generator and a power amplifier. Both the arbitary function generator and the data acquisition system are computer controlled.

The apparatus can house solid and hollow specimens. The latter can be anisotropically consolidated with consolidation stress ratio $K_c = \sigma'_{hc}/\sigma'_{vc} \leq 1$. However, test results presented in this paper mainly concern isotropically-consolidated specimens.

A bath of silicon oil is provided all around the specimen to avoid air migration through the membranes.

An accelerometer is used to determine resonance during RCTs. The shear strain during monotonic loading torsional shear tests (MLTSTs) or cyclic loading torsional shear tests (CLTSTs) is determined by means of a pair of proximity transducers with a resolution of 0·1 μm. The transducer targets are fixed to the drive system at a radial distance of about 30 mm, therefore they rotate with the sample. The proximity transducers are blocked on the frame supporting the coils and monitor the overall rotational movement. The relative distance between the proximity transducers and their targets can be set or reset during a test by moving the transducers along the horizontal direction using a system which enables operation from outside the cell.

There is limited information in the literature on the comparison between overall (or external) and local rotational shear strain measurements (Drnevich, 1978; Hight *et al.*, 1983; Tatsuoka & Shibuya, 1992; Porovic & Jardine, 1994). Generally a good agreement has been found between local and overall rotational shear strains, for the strain interval considered in this research, therefore indicating a moderate influence of the system compliance on the experimental measurements.

The shear strains reported in this paper have been computed by means of the following formula:

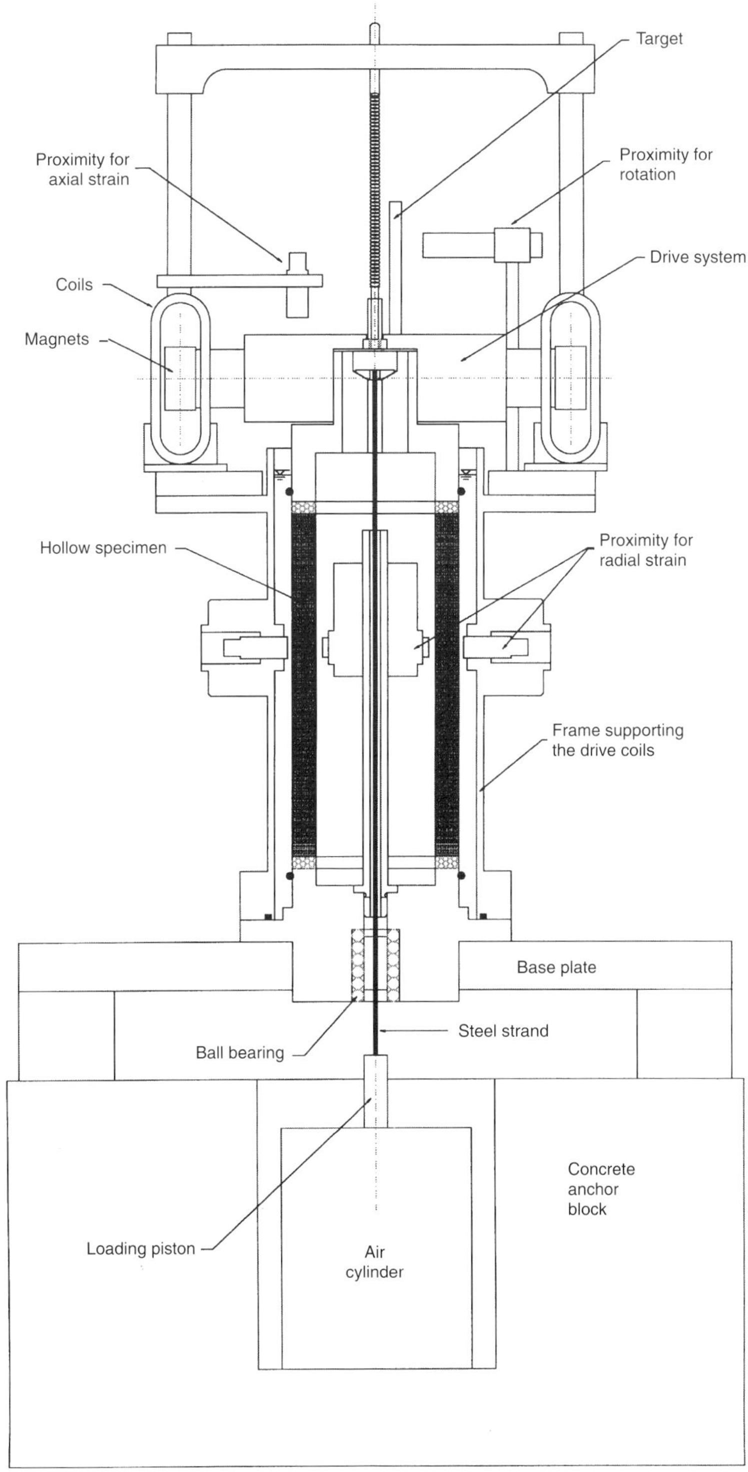

Fig. 1. Internal arrangement of the Torsional Shear/Resonant Column apparatus (Lo Presti *et al.*, 1993)

$$\gamma = \frac{2}{3}\frac{\theta}{h}\frac{R_0^3 - R_i^3}{R_0^2 - R_i^2} \quad (3)$$

where θ is the rotation, h is the specimen height, R_0 is the outer radius of the specimen and R_i is the inner radius of the specimen. For solid specimen $R_i = 0$.

A glass window is available for visual inspection inside the pressure cell. The applied torque is computed by means of a torque–voltage calibration curve (Lo Presti *et al.*, 1993).

Acquisition of the applied voltage and of the output of the two proximity transducers used for rotation measurement was carried out by means of three simultaneously-triggered digital multimeters. Thus, no delay exists between torque and rotation measurements. According to Toki *et al.* (1995) a delay between the torque and the rotation acquisition could introduce a certain error in damping determination.

In the case of RCTs, the damping ratio was determined during forced vibrations by means of the so-called half-power method. No correction was made to account for equipment-generated damping (D_{app}). Stokoe *et al.* (1995) suggest that such a correction should be done after appropriate calibration of each apparatus. In the case of cyclic torsional shear tests, it was decided that a minimum acquisition rate of 30 data points per cycle was necessary. This limit was established by Pallara (1995) in order to accurately define the area enclosed by the loop and hence to obtain stable and accurate damping measurements. The suggestion given by Toki *et al.* (1995) of more than 40 data points per cycle, is more severe.

TEST SOILS AND PROGRAMME

A total of 52 tests were performed on dry, reconstituted hollow cylindrical specimens of three different kinds of sands (Toyoura (TS), Quiou (QS) and Catania (CS)) with an external diameter of 70 mm, an internal diameter of 50 mm and a height of 140 mm.

Table 1 summarizes the main characteristics of the tested sands. Three kinds of tests were performed on these sands

(*a*) MLTSTs at constant stress rate
(*b*) CLTSTs performed under stress control using sinusoidal and triangular torque at constant frequency with three different values of frequency being used ($f = 0{\cdot}01$, $0{\cdot}1$ and $1{\cdot}00$ Hz)
(*c*) RCTs which, in the case of QS and TS, were performed on the same specimens, previously subjected to cyclic loading. For CS, the RCTs were performed on the same specimens, previously subjected to CLTST or MLTST.

Tables 2(a), 2(b) and 2(c) summarize test conditions for static (monotonic and cyclic) and dynamic tests on the sand specimens.

A total of six specimens of two natural Italian clays (Pisa and Augusta) were tested in this research. The clay specimens were reconsolidated to $\sigma'_{vc} = \sigma'_{vo}$ (i.e. the best estimate of the in situ vertical stress). The majority of the tests were isotropically consolidated. Only a few specimens were reconsolidated with $K_c = \sigma'_{hc}/\sigma'_{vc} = K_0$. The K_0 coefficient of earth pressure at rest for the Pisa clay specimens considered is about 0·65. In the case of Augusta clay $K_0 \cong 1$. After reconsolidation the specimens were subjected to three different kinds of tests

(*a*) MLTSTs at constant stress rate
(*b*) CLTSTs performed under stress control. In order to have a constant equivalent strain rate, the loading frequency was decreased with an increase of the shear strain level, according to equation (1). Average strain rates of about 0·04%/min and 0·4%/min were used
(*c*) RCTs.

The tests were performed with closed or opened drainage, as specified in Table 3. In some cases, the same specimen was first subjected to monotonic loading, then to cyclic static loading, after a rest period of 24 h with opened drainage, and eventually, after a further 24 h of rest with opened

Table 1. Main characteristics of the tested sands

Sand	*FC* %	G_s	D_{50} mm	U_c	e_{max}	e_{min}	Mineralogy
Toyoura (TS)	–	2·650	0·22	1·35	0·985	0·611	Quartz
Quiou (QS)	2	2·716	0·75	4·40	1·281	0·831	Carbonate
Catania (CS)	–	2·683	0·24	2·20	0·850	0·592	NA

FC fine content.
G_s specific gravity.
D_{50} mean grain size.
U_c coefficient of uniformity.
e_{max}, e_{min} maximum and minimum void ratio.

Table 2(a). Test conditions for Toyoura sand specimens

Test N	e	$\sigma'_{vc} = \sigma'_{hc}$ kPa	Frequency Hz	Waveform	G_o (static) MPa	G_o (dynamic) MPa
01	0·918	100	0·01	Triangular	90	84
02	0·927	100	0·1	Triangular	85	81
03	0·897	100	1	Triangular	83	78
04	0·931	300	0·01	Triangular	139	132
05	0·920	294	0·1	Triangular	121	119
06	0·902	295	1	Triangular	140	124
07	0·846	108	0·01	Triangular	100	94
08	0·842	101	0·1	Triangular	94	83
09	0·819	100	1	Triangular	84	86
13	0·820	295	Variable	Triangular	160	150
11	0·838	296	0·1	Triangular	154	145
12	0·814	296	1	Triangular	160	142
14	0·980	104	0·01	Triangular		75
15	0·852	98	MLTST		85	
16	0·950	99	MLTST		70	
17	0·896	101	MLTST		78	

Table 2(b). Test conditions for Quiou sand specimens

Test N	e	$\sigma'_{vc} = \sigma'_{hc}$ kPa	Frequency Hz	Waveform	G_o (static) MPa	G_o (dynamic) MPa
01	0·943	95	0·01	Triangular		70
02	0·970	100	0·1	Triangular	60	63
03	0·962	97	1	Triangular	80	70
04	0·900	293	0·01	Triangular	160	153
05	0·892	295	0·1	Triangular		161
06	1·178	98	0·1	Triangular		65
07	1·092	294	0·1	Triangular	122	124
08	1·014	298	1	Triangular	150	143
09	1·152	100	0·01	Triangular	63	59
10	0·957	100	0·1	Sinusoidal	75	67
11	1·020	99	1	Sinusoidal	70	64
12	0·930	296	0·01	Sinusoidal	142	139
13	0·922	296	0·1	Sinusoidal	160	149
15	0·935	101	0·01	Triangular	77	72
16	0·844	100	0·01	Triangular		80
17	0·901	99	0·1	Triangular	89	78
18	0·977	99	1	Triangular	84	70
19	0·877	295	1	Triangular	160	161
20	0·928	297	0·01	Triangular	130	140
21	0·875	298	0·1	Triangular	155	145
22	1·000	98	MLTST		64	
23	1·055	97	MLTST		54	
24	0·863	97	MLTST		69	
25	1·163	97	MLTST		42	
26	1·074	98	MLTST		54	
27	1·107	98	MLTST		57	
28	1·074	98	MLTST		69	

drainage, to an RCT. In two cases, the same specimen experienced two static cyclic loading stages. A list of the tests performed on Pisa and Augusta clays is given in Table 3. The size and shape of the specimens are also indicated in the same table.

EXPERIMENTAL RESULTS

Cohensionless soils

Due to the accuracy in the measurement of very small strains, it was possible to assess the small strain shear modulus (G_o) from MLTSTs, CLTSTs and RCTs. The small strain shear modulus (G_o)

Table 2(c). Test conditions for Catania sand specimens (Cascone, 1997)

Test N	e	$\sigma'_{vc} = \sigma'_{hc}$ kPa	Frequency Hz	Waveform	G_o (static) MPa	G_o (dynamic) MPa
01	0·727	38	1	Triangular	70	71
02	0·705	57		MLTST	90	84
03	0·691	105		MLTST	120	113
04	0·799	39		MLTST	70	62
05	0·770	55	1	Triangular	73	
06	0·778	102		MLTST	95	82
07	0·727	38		MLTST	66	

Table 3. Test conditions for clay specimens

Site test No.	σ'_{vc} kPa	K_c	e	PI	MLTST	CLTST	RCT	G_o (1) MPa	G_o (2) MPa	G_o (3) MPa	Specimen
Pisa 1	112	0·65	1·606	55	D		U	20		23	H
Pisa 2	112	1	1·572	47	D		U	25		28	H
Pisa 3	138	1	1·023	21		U	U		48	48	S
Pisa 4	114	1	0·901	22	U			66			S
Pisa 5	104	1	0·896	22		Two stages (U)			76		S
Augusta 6	182	1	0·684	29		Two stages (U)	U	80	84	85	S

D Drained, U Undrained.
G_o (1) from MLTSTs, G_o (2) from CLTSTs, G_o (3) from RCTs.
H Hollow cylindrical specimen ($R_o = 25$ mm $R_i = 15$ mm h = 100 mm).
S Solid cylindrical specimen ($R = 25$ mm $h = 100$ mm).

values were determined at $\gamma \leqslant 0{\cdot}001\%$. These values are reported in Tables 2(a), (b) and (c) for TS, QS and CS, respectively. It is possible to observe that, irrespective of the strain rate (or frequency), type of loading (cyclic or monotonic) and number of loading cycles, the same values of G_o were obtained from static (monotonic and cyclic) and dynamic tests. However, in the case of QS the G_o values, obtained from MLTSTs, are about 20% lower than those obtained from CLTSTs and RCTs. It is supposed that this difference is due to the very high sand crushability (Lo Presti *et al.*, 1993; Fioravante *et al.*, 1994). In particular, the authors believe that grain crushability is responsible for firstly, the modification of the soil fabric (i.e. increase of the contact area and/or number of contacts), and secondly, a stress–strain response of non-linear type even at $\gamma \leqslant 0{\cdot}001\%$, in the case of MLTST.

The G_o values of TS and QS are plotted against the void ratio in Fig. 2. Two different groups of data can be seen in the figure, depending on the level of consolidation pressure (100 kPa or 300 kPa). The G_o values of CS for specimens consolidated at about 100 kPa are also plotted in Fig. 2. The following analytical expression was used to fit the data of TS and QS (data from MLTST on QS specimens were not considered in the regression analysis):

$$G_o = CF(e)p_a^{1-n} \cdot (p')^n \tag{4}$$

where C is an experimentally determined material constant, $F(e) = e^{-x}$ is the void ratio function (Lo Presti, 1989), p_a is the atmospheric pressure expressed consistent to the unit system adopted and p' is the isotropic effective confinement pressure.

The values of C, n and x obtained for the two sands by means of the regression analysis are reported in the figure. The larger value of the modulus exponent for QS is probably due to the different mineralogy and grain angularity as well as the different coefficient of uniformity.

The experimentally determined G/G_o values of TS are plotted against γ in Fig. 3. The data points refer to the 17 tests listed in Table 2(a). Different symbols have been used to distinguish between CLTST, MLTST and RCT and, for the same kind of test different symbols have been used depending on the confining pressure level ($p' = 100$ or 300 kPa). As far as CLTSTs and RCTs are concerned, it is possible to see that p' has a moderate

	MLTSTs	CLTSTs	RCTs	C	n	x
Toyoura	▬	□	■	724	0·45	1·3
Quiou	X	✧	◆	708	0·62	1·3
Catania	○		●			

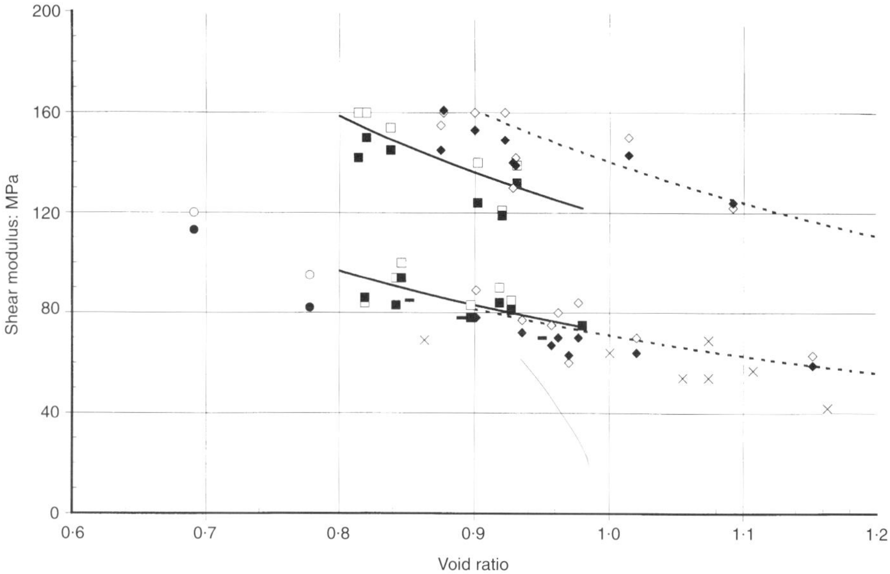

Fig. 2. Small strain shear modulus of tested sands

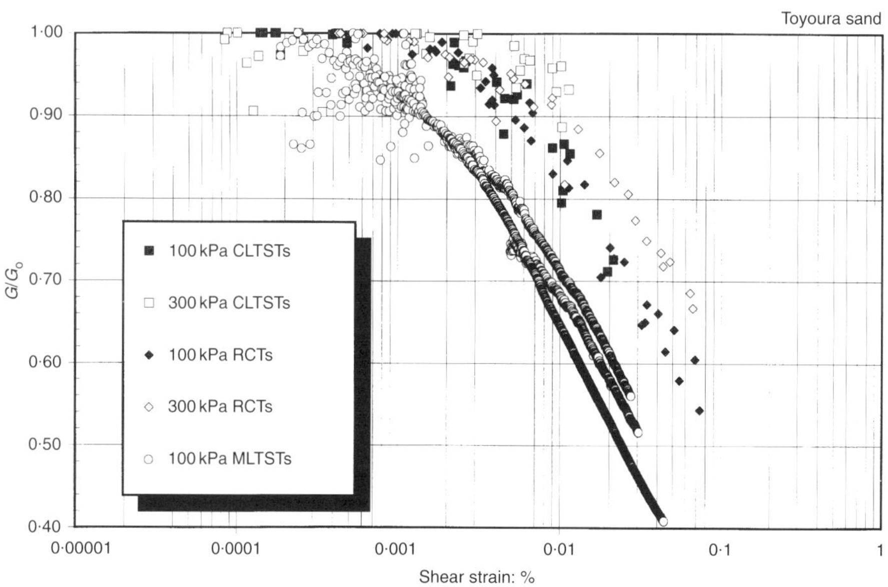

Fig. 3. Normalized shear modulus plotted against shear strain of Toyoura sand from MLTSTs, CLTSTs and RCTs

influence on the G/G_o–γ curves while the frequency (i.e. strain rate) does not have any influence. Typical strain rates for RCTs, CLTSTs and MLTSTs are reported in Fig. 4. As a consequence of the fact that strain rate does not influence the G/G_o–γ curves in a measurable way, it is possible to conclude that the stress–strain relationships of cohesionless soils cannot be distorted in a conventional cyclic test. These results are in good agreement with those obtained by other researchers (Hardin & Drnevich, 1972a, b; Iwasaki *et al.*, 1978; Alarcon-Guzman *et al.*, 1989; Bolton & Wilson, 1989; Tatsuoka & Shibuya, 1992; Porovic & Jardine, 1994).

On the other hand, the trend for monotonic tests is different and exhibits lower values of G for a given shear strain level. This is confirmed by the data shown in Fig. 5. This figure shows the stress–strain curve of a QS specimen (test No. 1) subjected to a cyclic shear stress of $\tau_c = \pm 14{\cdot}8$ (kPa) without any prior stress–strain history. It is possible to see that the stress–strain response during the first quarter of the first cycle is softer than that obtained during the subsequent cyclic loading, unlike what is usually observed during a conventional stepwise test (i.e. for a specimen with a prior stress–strain history). It is reasonable to believe that, in a stepwise test, during each loading step, plastic shear strains take place. Therefore, a given cyclic shear stress τ_c induces larger plastic shear strains in a virgin specimen than in a specimen subjected to a prior stress–strain history, during which a certain amount of the plastic shear strain has already occurred. Moreover, the secant shear modulus related to the reverse point of the first quarter of the first cycle (virgin specimen) matches the results obtained from MLTSTs, while after a certain number of loading cycles it approaches those obtained from CLTSTs and RCTs (Fig. 6).

Shear modulus and damping at very small strains are hardly influenced by N. The influence of N on G and D becomes more and more relevant as the shear strain level increases. The increase of G and the decrease of D with an increase of N is shown in Fig. 7, in the case of a virgin specimen of QS. Results of a stepwise test on TS are shown in Fig. 8. In both cases an average single amplitude shear strain of about 0·03% was imposed to the specimens. It is possible to see that the increase of G with N for the virgin specimen is very large, which is in good agreement with data by Alarcon-Guzman *et al.* (1989), while for the stepwise test the increase of G is less than 20% in agreement with other available data (Hardin & Drnevich, 1972a, b; Sherif & Ishibashi, 1976; Iwasaki *et al.*, 1978; Kokusho, 1980; Tatsuoka *et al.*, 1991). However, a stable value of G was reached after 5 or 10 cycles in both tests. Unlike the shear modulus, D of dry sands is strongly influenced by N even in the case of a stepwise test (Figs 7 and 8). Moreover, even after hundreds of cycles D still decreases. The D value, obtained for the same shear strain level from the RCT performed after the CLTST, is also plotted in Fig. 7. It is possible to see that this value matches the trend of the CLTST results.

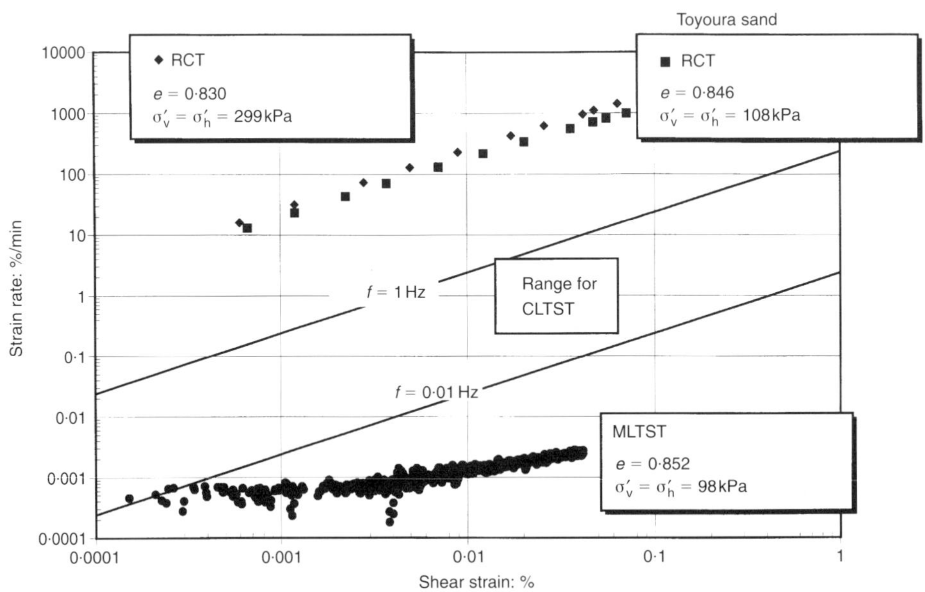

Fig. 4. Typical strain rates during CLTSTs, MLTSTs and RCTs

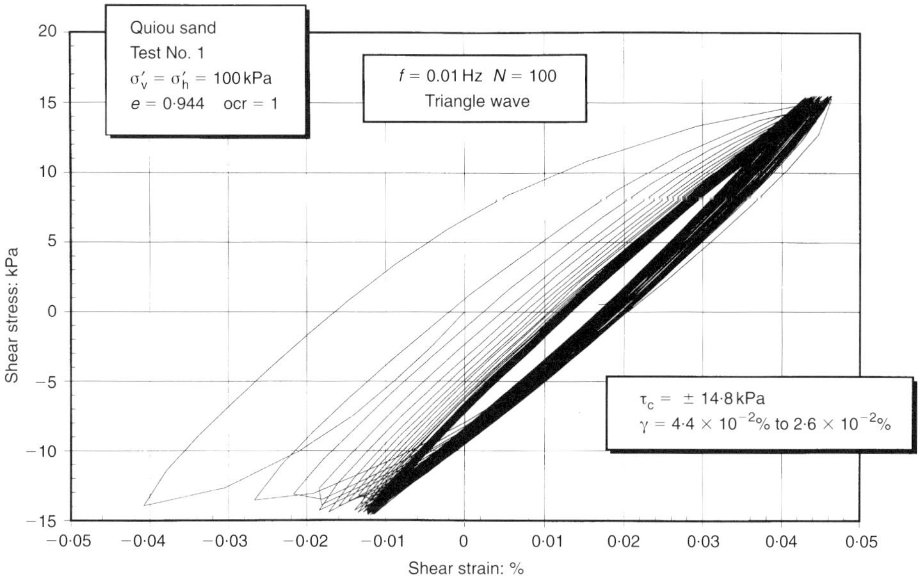

Fig. 5. Stress–strain loops for Quiou sand (virgin specimen)

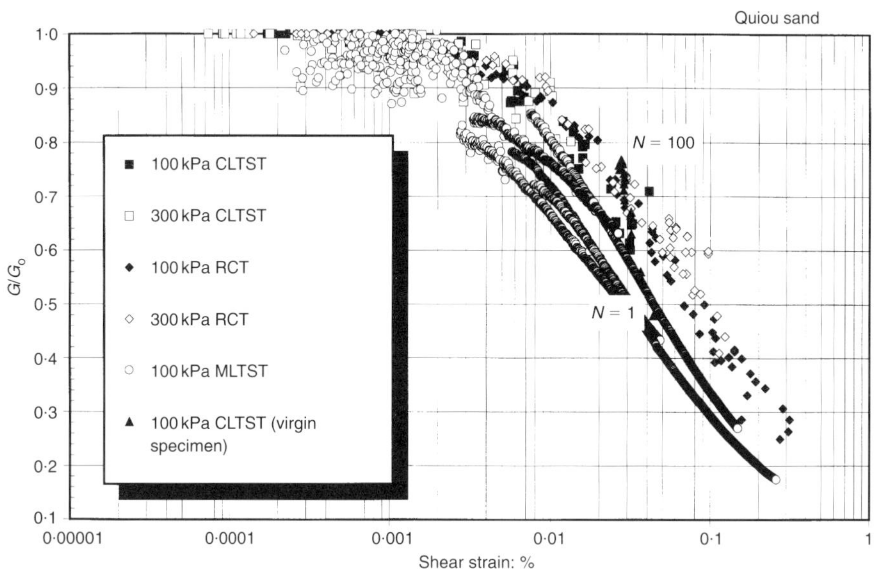

Fig. 6. Normalized shear modulus plotted against shear strain of Quiou sand from MLTSTs, CLTSTs and RCTs

The damping values obtained from RCTs and CLTSTs are compared in Fig. 9 for TS. As far as CLTSTs are concerned, the damping ratio values plotted in Fig. 9 are those obtained after 20 or 30 cycles when the variation of D with N becomes relatively important. It is possible to see that, for a given strain level, the damping ratio from RCT is greater than that from CLTST in the small strain range, whilst, at larger strains, the opposite is true. Similar results were obtained for QS and CS. Data shown by Papa *et al.* (1988) and by Stokoe *et al.* (1995) are in good agreement with those of Fig. 9. At large strains, the effect of N on D can explain the differences between the results of CLTSTs and

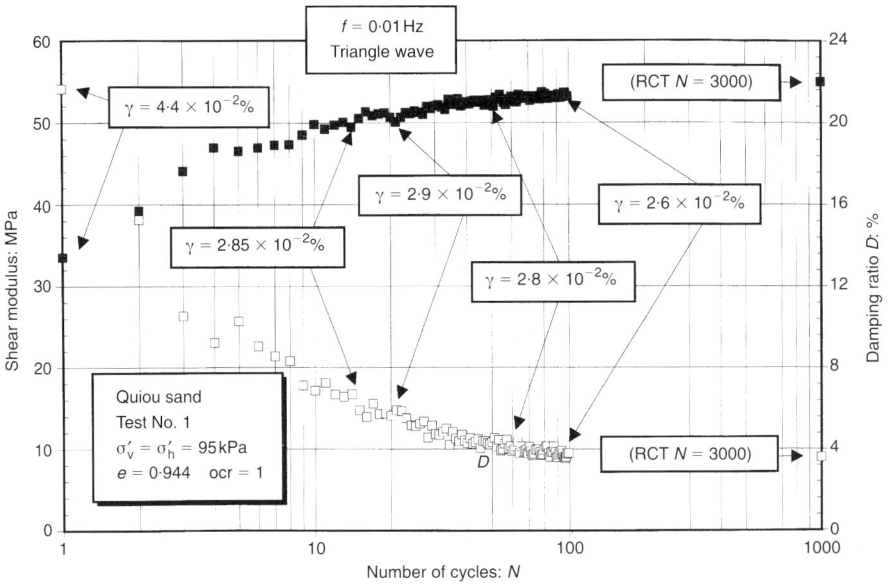

Fig. 7. Shear modulus and damping ratio plotted against number of loading cycles (Quiou sand, virgin specimen)

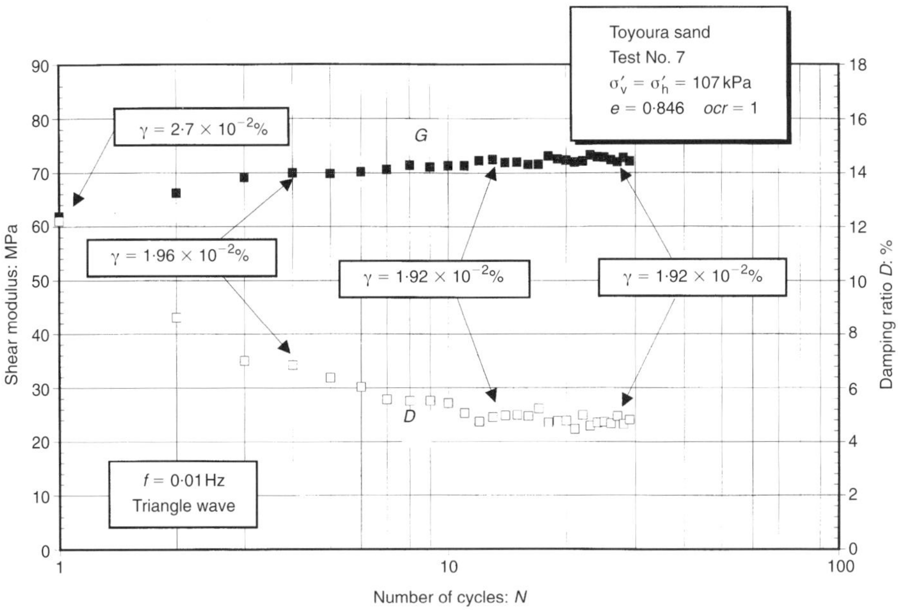

Fig. 8. Shear modulus and damping ratio plotted against number of loading cycles (Toyoura sand, stepwise CLTST)

those of RCTs. Considering that at very small strains the effect of N is negligible, there are two possible explanations for the above-mentioned differences, in the small strain range

(*a*) equipment-generated damping in RCT (Stokoe *et al.*, 1995)

(*b*) rate effect due to the viscosity of the pore fluid (not necessarily water) (Kokusho, 1987).

Cohesive soils

The small strain shear modulus of cohesive soils obtained from MLTSTs, CLTSTs and RCTs is

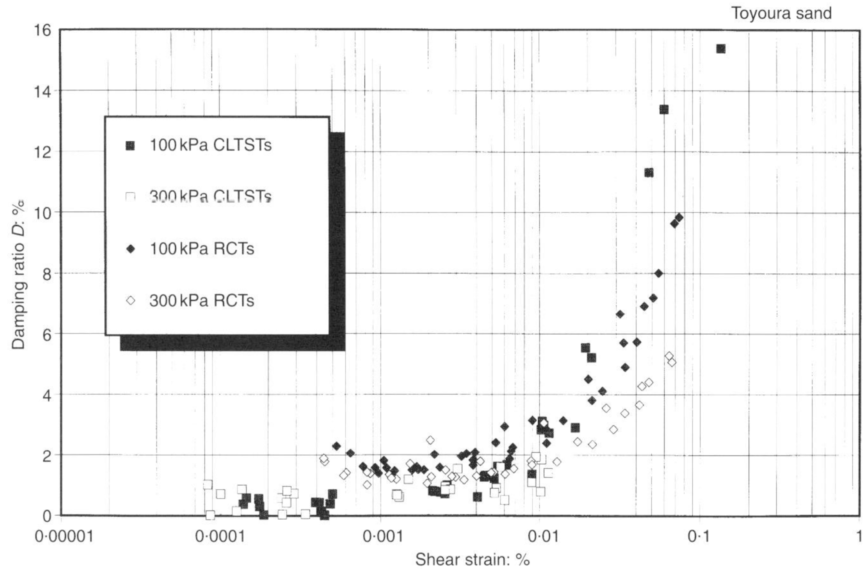

Fig. 9. Damping ratio of Toyoura sand plotted against shear strain from CLTSTs and RCTs

reported in Table 3. Unlike granular soils, a measurable influence of strain rate and type of loading was observed in this case even at very small strains. The rate effect becomes more and more relevant with an increase of the shear strain level, as can be seen in Figs 10 and 11. The G–γ curves obtained from CLTSTs and RCTs, in the case of Augusta and Pisa clays, are compared in Figs 10 and 11 respectively. It is possible to notice that the G–γ curves obtained from RCTs show a greater elastic limit (γ_t^e). Moreover, for a given strain level, G increases with an increase of the strain rate. The method used to compute the coefficient of strain rate $\alpha(\gamma)$ is also shown in Fig. 10.

The experimentally determined $\alpha(\gamma)$ values are summarized in Fig. 12. Other data, available in the literature are also shown in the same figure. These data were obtained by means of undrained compression tests performed on both remoulded and undisturbed clays. The data, on the whole, clearly

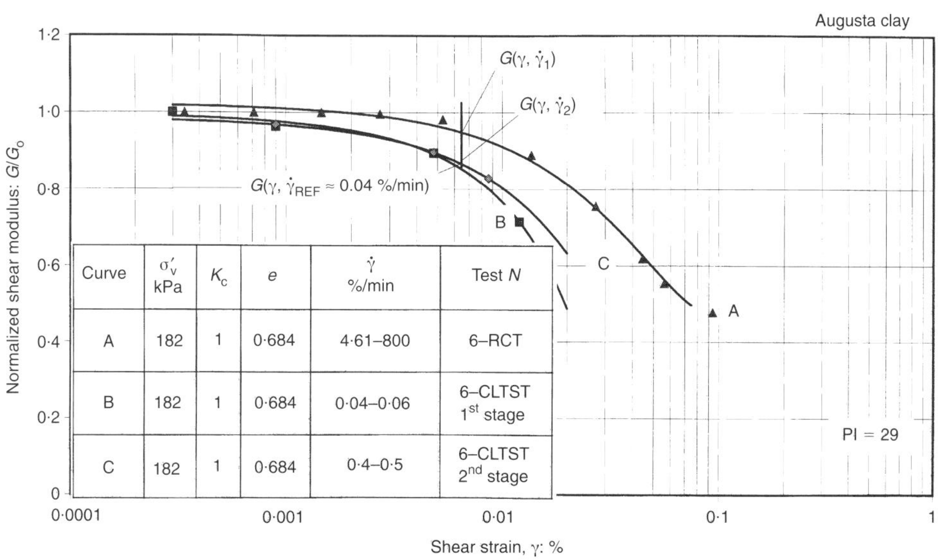

Curve	σ'_v kPa	K_c	e	$\dot{\gamma}$ %/min	Test N
A	182	1	0·684	4·61–800	6–RCT
B	182	1	0·684	0·04–0·06	6–CLTST 1st stage
C	182	1	0·684	0·4–0·5	6–CLTST 2nd stage

Fig. 10. Shear modulus of Augusta clay from CLTST and RCT tests (Lo Presti *et al.*, 1996)

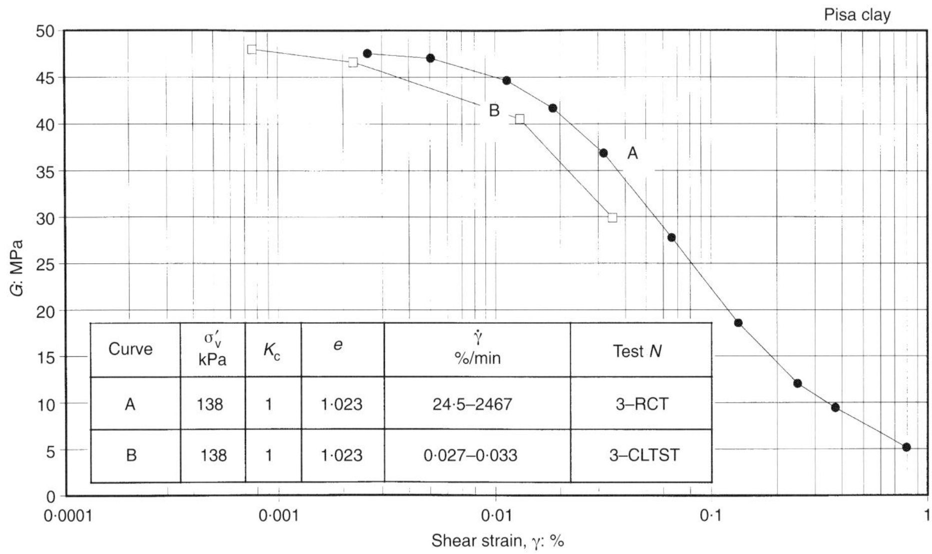

Fig. 11. Shear modulus of Pisa clay from CLTST and RCT tests (Lo Presti *et al.*, 1996)

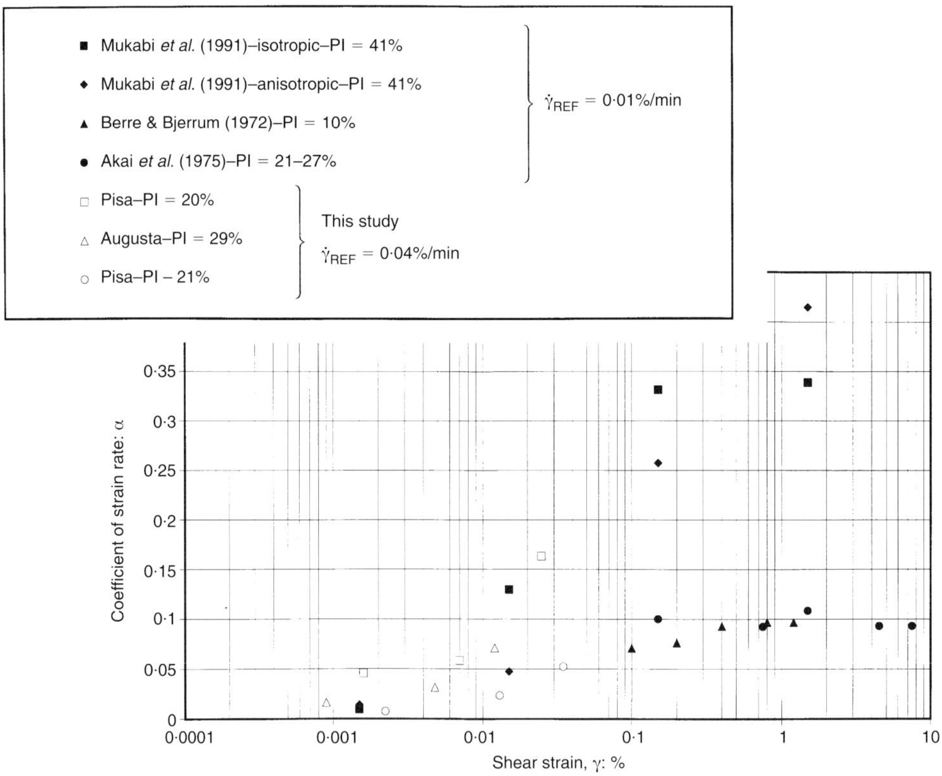

Fig. 12. Coefficient of strain rate plotted against shear strain (Lo Presti *et al.*, 1996)

show that $\alpha(\gamma)$ increases with the plasticity index and, for a given soil, with the shear strain level.

Unlike granular soils, cohesive materials are influenced in a negligible way by the number of loading cycles, at least in the strain interval considered. The shear modulus of Pisa clay, subjected to a single amplitude cyclic shear strain of about 0·03%, exhibits a 10% decrease and the damping ratio exhibits a 15% decrease after 29 cycles. Similar results were obtained for Augusta clay.

A moderate increase of the pore pressure (Δu) was observed during undrained tests, performed on cohesive saturated specimens. The accumulated Δu, for a given shear strain, and the threshold strain level (γ_t), above which the pore pressure build-up was observed, were both dependent on loading rate and the type of loading (monotonic or cyclic) (Lo Presti *et al.*, 1996). In particular γ_t increased whilst Δu decreased with an increase of the strain rate in cyclic tests. Greater γ_t and smaller Δu were observed during MLTST in comparison to cyclic tests. The pore pressure increase was measured at the base of the specimen and the maximum accumulated Δu was of less than 9 kPa, therefore the possible effect on the measured stiffness was considered negligible.

A comparison between the damping ratio values obtained from RCTs and those obtained from CLTSTs is shown in Fig. 13 for Augusta clay. Greater values of D are obtained from RCTs for the whole strain interval investigated. According to Stokoe *et al.* (1995) the D values from RCTs should be corrected by subtracting the equipment-generated damping (D_{app}), which is a frequency dependent parameter. The $D_{app}-f$ calibration curve is, of course, different for each apparatus (Stokoe *et al.*, 1995). However, considering that the apparatus available at the authors' laboratory is quite similar to that used by Stokoe *et al.* (1995), the use of the $D_{app}-f$ calibration curve proposed by Stokoe *et al.* (1995) to correct the experimental measurements of D is not enough to explain the differences seen in Fig. 13 concerning the damping measurements in different test conditions. It is supposed that a relevant part of these differences is due to the rate effect, in agreement with data shown by Shibuya *et al.* (1995) and Tatsuoka *et al.* (1995).

The stress–strain response to monotonic and cyclic loading of two Pisa clay specimens retrieved from the same depth, with the same void ratio and consolidation stresses and subjected to very similar strain rates, are compared in Fig. 14. This kind of comparison clearly shows that the stress–strain response to monotonic loading is softer than that to cyclic loading, as in the case of cohesionless soils. The two curves, which are quite close at very small strains, tend to diverge for increasing shear strain level.

CONCLUSIONS

The experimental programme carried out enabled the comparison of static against dynamic test results and monotonic against cyclic test results for three reconstituted sands and two undisturbed clays. The effect of shear strain, number of loading cycles, loading rate and type of loading on the G

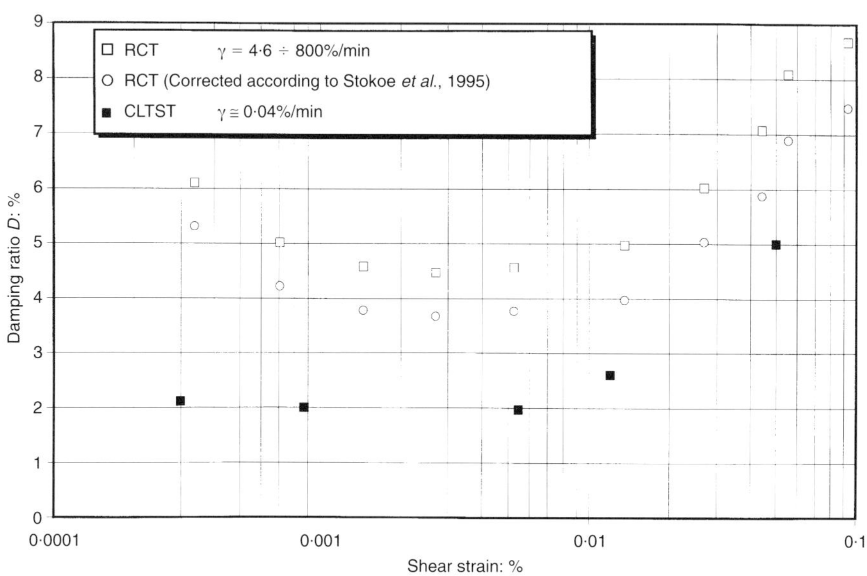

Fig. 13. Damping ratio of Augusta clay from CLTST and RCT tests

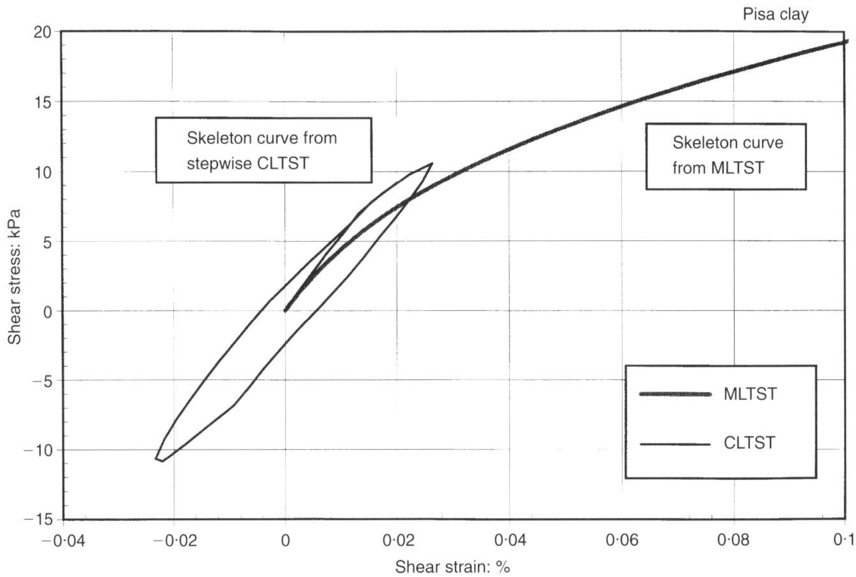

Fig. 14. Stress–strain relations from MLTST and CLTST tests for Pisa clay

and D of the tested soils was investigated. On the basis of the experimental results shown, it is possible to draw the following conclusions.

(*a*) At small strains, the influence of N on the shear modulus and damping ratio of dry sands is not measurable but it becomes more and more important with an increase of the shear strain level. The influence of N is more pronounced in the case of virgin specimens. A stable value of G is approached after the application of less than 10 cycles, while damping ratio is affected by N even after the application of several hundreds of loading cycles.

(*b*) G and D of dry sands are almost insensitive to loading frequency.

(*c*) In the case of clays, G and D are hardly affected by N, at least in the strain interval considered, but are very sensitive to strain rate even at very small strains. However, the rate effect on the shear modulus increases with increasing shear strain level. The strain rate coefficient depends on PI and, for a given soil, on γ.

(*d*) As a consequence of the rate effect, the $G/G_0-\gamma$ curves obtained from conventional static tests performed at constant frequency and RCTs, are rate dependent. Moreover, very high strain rates are experienced during RCTs which makes this kind of test not very suitable for the determination of the $G/G_0-\gamma$ curves.

(*e*) The stress–strain response of a soil element to monotonic loading is softer than that to cyclic loading for both cohesive and cohesionless soils.

ACKNOWLEDGEMENTS

The authors would like to thank Professor Maugeri of the University of Catania, who provided the Augusta samples and Catania sand, Mr R. Maniscalco, the laboratory technician for the help given in setting up the apparatus, the graduate student Mr E. Cascone for the help given in performing the tests on Catania sand and Mrs M. Jones who reviewed the literary form of the text.

Cyclic tests on Quiou and Toyoura sands were performed in the framework of a joint research between ENEL CRIS of Milan and the Politecnico di Torino.

REFERENCES

Akai, K., Adachi, T. & Ando, N. (1975). Existence of a unique stress–strain–time relation of clays. *Soils & Foundations*, **15**, No. 1, 1–16.

Alarcon-Guzman, A., Chameau, J. L. & Leonards, G. A. (1986). A new apparatus for investigating the stress–strain characteristics of sands. *Geotechnical Testing Journal*. GTJODJ **9**, No. 4, 204–212.

Alarcon-Guzman, A., Chameau, J. L., Leonards, G. A. & Frost, J. D. (1989). Shear modulus and cyclic undrained behaviour of sands. *Soils & Foundations* **29**, No. 4, 105–119.

Berre, T. & Bjerrum, L. (1972). Shear strength of normally consolidated clays. *Proc. ICSMFE, Moscow*, paper No. 1/6, pp. 39–49.

Bolton, M. D. & Wilson, J. M. R. (1989). An experimental and theoretical comparison between static and dynamic torsional soil tests. *Géotechnique* **39**, No. 4, 585–599.

Cascone, E. (1997). Analisi sperimentale e modellazione del comportamento dinamico dei muri di sostegno. University of Catania, PhD thesis.

Dobry, R. & Vucetic, M. (1987). Dynamic properties and response of soft clay deposits. State of the art report. *Proc. Int. Symp. Geotech. Eng. Soft Soils, Mexico City* **2**, 51–87.

Drnevich, V. P., Hardin, B. O. & Shippy, D. J. (1978). Modulus and damping of soils by the resonant column method. *Dynamic Geotechnical Testing, ASTM STP 654*, pp. 91–125.

Drnevich, V. P. (1978). Resonant column testing – problems and solutions. *Dynamic Geotechnical Testing. ASTM STP 654*, ASTM, pp. 384–398.

Fioravante, V., Jamiolkowski, M. & Lo Presti, D. C. F. (1994). Stiffness of carbonatic Quiou sand. *Proc. XIII ICSMFE, New Delhi, India* **1**, 163–167.

Hardin, B. O. & Drnevich, V. P. (1972a). Shear modulus and damping in soils: measurement and parameter effects. *J. Soil Mech. Found. Div.* **98**, No. SM 6, 603–624.

Hardin, B. O. & Drnevich, V. P. (1972b). Shear modulus and damping in soils: design equations and curves. *J. Soil Mech. Found. Div.* **98**, No. SM 7, 667–693.

Hight, D. W., Gens, A. & Symes, M. J. (1983). The development of a new hollow cylinder apparatus for investigating the effects of principal stress rotation in soils. *Géotechnique* **33**, No. 4, 355–383.

Isenhower, W. M. & Stokoe, K. H. II, (1981). Strain rate dependent shear modulus of San Francisco Bay mud. *Proc. Int. Conf. Recent Adv. Geotech. Earthquake Eng. & Soil Dyn., St. Louis, Missouri.*

Isenhower, W. M., Stokoe, K. H. II & Allen, J. C. (1987). Instrumentation for torsional shear/resonant column measurements under anisotropic stresses. *Geotechnical Testing Journal* **10**, No. 4, 183–191.

Iwasaki, T., Tatsuoka, F. & Takagi, Y. (1978). Shear moduli of sands under cyclic torsional shear loading. *Soils & Foundations* **18**, No. 1, 39–50.

Kokusho, T. (1989). Cyclic triaxial test on dynamic soil properties for wide strain range. *Soils & Foundations* **20**, No. 2, 45–60.

Kokusho, T. (1987). In-situ dynamic soil properties and their evaluation. *Proc. 8th Asian Regional Conf. on SMFE*, Vol. 2, pp. 215–340.

Lo Presti, D. C. F. (1989). Proprietà dinamiche dei terreni. *XIV Conferenza Geotecnica di Torino*, Dept Struct. Eng., Politecnico di Torino (in Italian).

Lo Presti, D. C. F., Pallara, O., Lancellotta, R., Armandi, M. & Maniscalco, R. (1993). Monotonic and cyclic loading behaviour of two sands at small strains. *Geotechnical Testing Journal* **16**, No. 4, 409–424.

Lo Presti, D. C. F., Jamiolkowski, M., Pallara, O. & Cavallaro, A. (1996). Rate and creep effect on the stiffness of soils. *ASCE Convention, Washington*, 10–14 Nov. 1996, *Geotechnical Special Publication*. No. 61, pp. 166–180.

Mukabi, J. N., Tatsuoka, F. & Hirose, K. (1991). Effect of strain rate on small strain stiffness of Kaolin In: *CU Triaxial Compression, Proc. 26th Japan National Conf. on SMFE, Nagano*, pp. 659–662.

Pallara, O. (1995). Comportamento sforzi-deformazioni di due sabbie soggette a sollecitazioni monotone e cicliche. Dept Struct. Eng., Politecnico di Torino, (in Italian). PhD thesis.

Papa, V., Silvestri, F. & Vinale, F. (1988). Analisi delle proprietà di un tipico terreno piroclastico mediante prove di taglio semplice. *Proc. Gruppo Nazionale di Coordinamento per gli Studi di Ingegneria Geotecnica. Monselice, Italy*, Vol. 1, pp. 265–286.

Porovic, E. & Jardine, R. J. (1994). Some observations on the static and dynamic shear stiffness of Ham River sand. *Proc. Int. Symp. on Pre Failure Deformation of Geomaterials, Sapporo, Japan*, Vol. 1, pp. 25–30.

Sherif, M. A. & Ishibashi, I. (1976). Dynamic shear modulus for dry sands. *J. Geotech. Eng. Div.* **102**, No. GT11, 1171–1184.

Shibuya, S., Mitachi, T., Fukuda, F. & Degoshi, T. (1995). Strain rate effect on shear modulus and damping of normally consolidated clay. *Geotechnical Testing Journal* **18**, No. 3, 365–375.

Stokoe, K. H. II, Hwang, S. K., Lee, J. N.-K. & Andrus, R. (1995). Effects of various parameters on the stiffness and damping of soils at small to medium strains. Keynote Lecture 2, *IS Hokkaido 1994* **2**, 785–816.

Tatsuoka, F., Shibuya, S. & Teachavorasinskun, S. (1991). Discussion: 'Shear modulus and cyclic undrained behaviour of sands', Alarcon-Guzman *et al.*, *Soils and Foundations* **31**, No. 2, 202–209.

Tatsuoka, F. & Shibuya, S. (1992). Deformation characteristics of soil and rocks from field and laboratory tests. Keynote Lecture, *Proc. IX Asian Conference on SMFE, Bangkok*, 1991 **2**, 101–190.

Tatsuoka, F. & Kohata, Y. (1995). Stiffness of hard soils and soft rocks in engineering applications. Keynote Lecture 8, *IS Hokkaido 1994* **2**, 947–1066.

Tatsuoka, F., Lo Presti, D. C. F. & Kohata, Y. (1995). Deformation characteristics of soils and soft rocks under monotonic and cyclic loads and their relations. *3rd Int. Conf. Recent Adv. Geotech. Earthquake Eng. & Soil Dyn., State of the Art 1*, **2**, 851–879.

Toki, S., Shibuya, S. & Yamashita, S. (1995). Standardization of laboratory test methods to determine the cyclic deformation properties of geomaterials in Japan. Keynote Lecture 1, *IS Hokkaido 1994* **2**, 741–784.

Izumi, K., Ogihara, M. & Kameya, H. (1997). *Géotechnique* **47**, No. 3, 619–632

Displacements of bridge foundations on sedimentary soft rock: A case study on small-strain stiffness

K. IZUMI,* M. OGIHARA* and H. KAMEYA†

Full-scale behaviour of the foundations for an 800 m long suspension bridge, Rainbow Bridge, constructed on sedimentary mudstone about 1·5 million years old and completed in 1991 is described. The settlement of the foundation with strain levels in the soft rock of less than 0·1% was overestimated several times by finite-element method simulations based on the stiffness from conventional oedometer tests and plate loading and pressuremeter tests. Field seismic surveys and triaxial compression tests measuring locally axial strains performed after the construction provided very similar elastic stiffness to each other. Based on the elastic modulus while taking into account the strain non-linearity obtained from the triaxial compression tests, the full-scale behaviour of the foundations could be simulated very well.

KEYWORDS: case history; footings/foundations; in-situ testing; laboratory tests; monitoring; soft rocks

L'article décrit le comportement en vraie grandeur des fondations du Rainbow Bridge, pont suspendu de 800 m, construit sur de l'argilite sédimentaire tendre vieille de 2 millions d'années, et terminé en 1991. Le tassement des fondations, avec de niveaux de contrainte dans la roche tendre de moins de 0,1 p. 100 environ, a été largement surestimé dans les modélisations par éléments finis basées sur les valeurs de rigidité fournies par des essais oedométriques classiques, des essais à la table et des essais pressiométriques. Les relevés sismiques et les essais de compression triaxiale réalisés après la construction pour mesurer les contraintes axiales locales ont donné des valeurs de rigidité très semblables. A partir du module d'élasticité, on a pu simuler de très près le comportement en vraie grandeur des fondations, en tenant compte de la non-linéarité des contraintes indiquée dans les essais triaxiaux.

INTRODUCTION

The Rainbow Bridge (Fig. 1(a), (b)) is a huge suspension bridge constructed on sedimentary soft-rocks comprising mostly mudstone from the Neogene to Early Pleistocene (about 1·5 million years in age). As shown in Fig. 2, the surface Holocene deposit consists mostly of soft clayey soil, while a sandy gravel layer 1 m thick is found at the bottom. The Holocene deposit is underlain by sedimentary softrock consisting mostly of mudstones (Table 1). During the construction, the behaviour of the foundation and the supporting ground was carefully monitored, and the observations were used for construction control and adjustment. The observations showed that the strain level in the ground was about 0·01% or less, which was much smaller than those predicted in advance. At the same time, the comparison between the predicted and observed behaviours revealed that the stiffness had been largely underestimated at the design stage.

In order to find reasons for this discrepancy, another series of geotechnical investigations was performed after completion of the bridge by state-of-the-art technology, which included field seismic surveys, undisturbed sampling, triaxial compression tests measuring local axial strains from less than 0·001% to 1% and three-dimensional finite-element method (3D FEM) analysis taking into account the strain non-linearity of the stiffness of the mudstone.

This paper describes the deformation characteristics at relatively small strains of the softrock obtained from the two series of investigation performed before and after construction and also from the observed full-scale behaviour.

Manuscript received 10 October 1996; revised manuscript accepted 21 March 1997.
Discussion on this paper closes 1 November 1997.
* Metropolitan Expressway Public Corporation, Tokyo.
† Oyo Corporation, Tokyo.

DEFORMATION CHARACTERISTICS OF MUDSTONE

Investigation at the design stage

Plate loading tests (PLTs) were performed by using a rigid plate with a diameter of 30 cm at a depth of 40 m on the excavated ground surface in each of the two anchorages. Other PLTs were per-

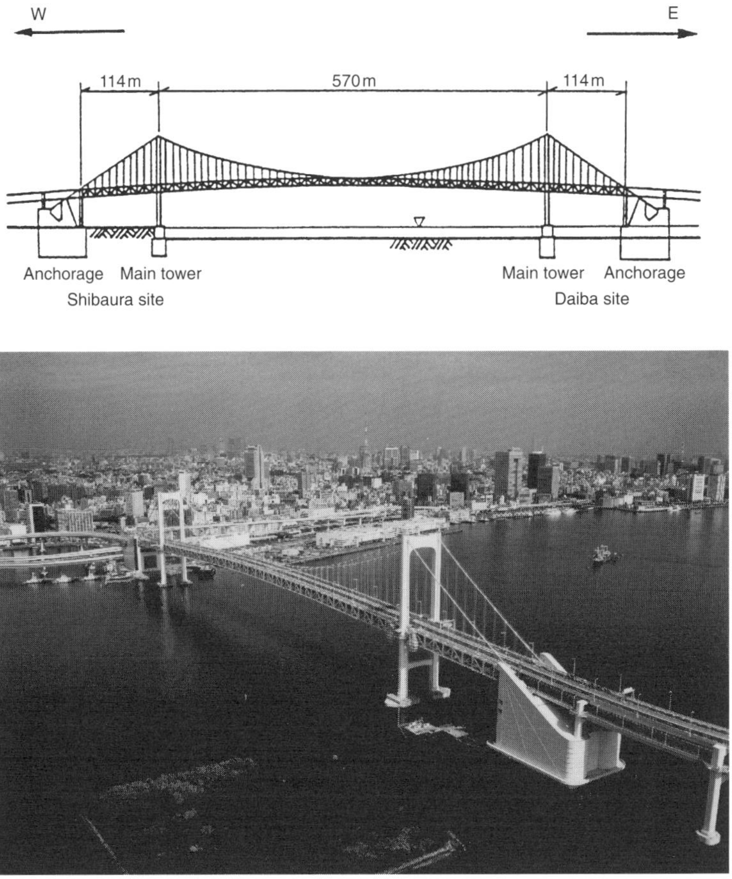

Fig. 1. Side view of the Rainbow Bridge

formed by using a rigid plate with a diameter of 25 cm in a 32 cm diameter borehole. Pressuremeter tests (PMTs) were also performed, employing a monocell probe with a latex membrane installed in a predrilled borehole. Only primary loading was applied in the PMTs.

The Young's moduli of the mudstone along the depth, obtained by conventional linear theory, are shown in Fig. 3. Some differences can be seen in the E values from the PLTs and the PMTs. This difference is probably due to the larger strain levels being involved in the PMTs than in the PLTs, rather than to inherent anisotropy as discussed below.

The deformation and strength characteristics of this soft mudstone formation at Sagamihara have been found to be nearly isotropic (Tatsuoka & Kohata, 1995). This was also confirmed in this investigation by the results of consolidated-undrained CU triaxial compression tests on core samples (5 cm in diameter and 10 cm in height) which had different angles θ of the specimen axial direction relative to the field vertical direction. The samples were cored from 20 cm diameter samples obtained by rotary core tube sampling from a depth of 40 m at the Daiba site. The specimens were isotropically reconsolidated to three different confining pressures σ'_c (Fig. 4). The initial Young's moduli $E_{initial}$ were obtained from the apparently linear portion of each stress–strain curve (as shown in Fig. 6). As the axial strains were obtained by external measurements, the $E_{initial}$ values are much smaller than the true elastic Young's modulus as shown later. It is seen that the $E_{initial}$ values exhibit nearly isotropic behaviour.

Considering that the deformation of the mudstone layer below the bridge foundations is rather one-dimensional, and that tests are relatively simple to perform, a series of oedometer tests were performed on mudstone samples 6 cm in diameter

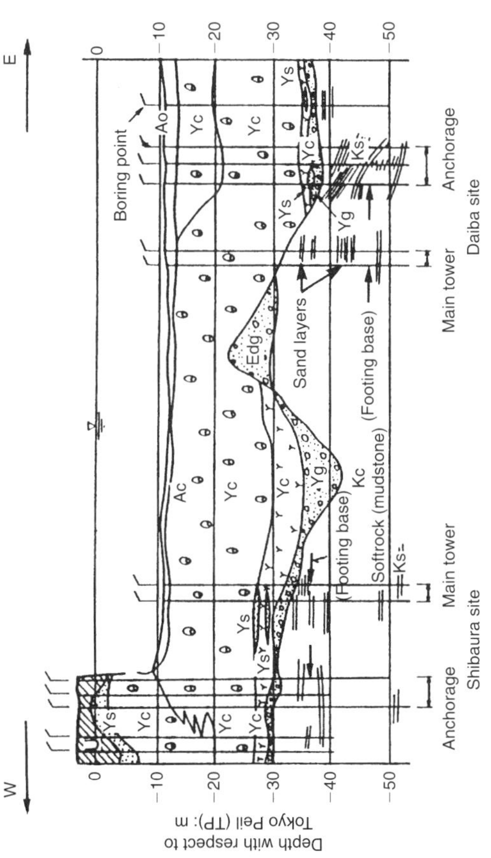

Geological age	Strata	Soil Types	Symbols
Holocene	Recent	Artificial Soil	U
		Floating Mud	Ao
	Yurakucho formation	Silt–clay	Yc
		Sand	Ys
		Gravel	Yg
Pleistocene	Edogawa formation	Gravel	Edg
Pleistocene–Tertiary	Kazusa group	Softrock (mudstone)	Kc
		Uncemented fine sand	Ks

Fig. 2. Geological profile of the ground along the axis of the Rainbow Bridge

Table 1. Properties of the mudstone

Texture*	Sand/gravel content: 6·1%	Silt content: 44·7%	Clay content: 49·2%
Consistency*	$w_L = 60·1\%$,	$w_P = 26·9\%$,	PI = 33·2
Natural moisture content, w_n		30·7%	
Wet density, ρ_t		1·92 g/cm^3	
Void ratio, e		0·86	
Dry density, ρ_d		1·47 g/cm^3	
Consolidation yield stress, p_y		5–7 MPa	
Overconsolidation ratio		7–19 MPa	
Unconfined compressive strength, q_u		2·5–4 MPa	

* Properties of thorough crushed samples of the mudstone.

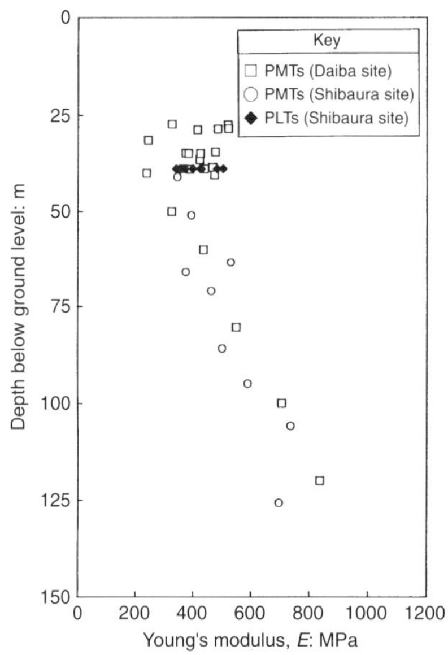

Fig. 3. Young's moduli from field tests performed at the design stage

and 2 cm in height. The samples were obtained by rotary core tube sampling from various depths at the Shibaura site. A typical result is shown in Fig. 5(a). All the samples were loaded to p_y (consolidation yield stress), followed by unloading to the lowest pressure level p_0 which would be realized upon ground excavation, and then followed by reloading. The preloading to the pressure level which far exceeds the original in situ stress level was considered to be necessary to decrease the effects of sample disturbance and bedding and seating errors.

The relationships between the inverse of the tangential coefficient of volume compressibility $(1/m_v)$ and the average consolidation pressure p during reloading are summarized in Fig. 5(b). For the mudstone, since the drained Poisson's ratio ν' is about 0·12 as obtained from consolidated-drained CD triaxial compression tests, the value of $1/m_v = E'(1-\nu')/[(1+\nu')(1-2\nu')]$ is very close to the drained Young's modulus E'. According to Fig. 5(b), the $1/m_v$ values increase with the increase in the depth of sample, particularly immediately after the start of reloading. As a whole, the $1/m_v$ values during reloading are close to the values of E obtained by the PMTs presented in Fig. 3. As shown later, those $1/m_v$ values are too small to explain the observed full-scale behaviour of the foundations. Note that the $1/m_v$ values from the primary loading tests were 50–200 MPa, which are even lower. These lower values of $1/m_v$ are considered primarily due to the effects of bedding and seating errors at the top and bottom surfaces of a specimen.

Triaxial compression tests performed after the construction of the bridge with local deformation transducers

As explained later, the stiffness values of the softrock obtained at the design stage was found to be too small in the light of the full-scale behaviour of the ground observed upon construction of the bridge. It was considered that this was due to conventional field and laboratory testing methods failing to measure stiffness at strains less than those operating in the ground (i.e. less than about 0·1%) (Tatsuoka *et al.*, 1996).

In view of the above, after construction of the bridge, a series of CU and CD triaxial compression tests were performed by using local deformation transducers (LDTs; Goto *et al.*, 1991) for local measurements of axial strains. The specimens were

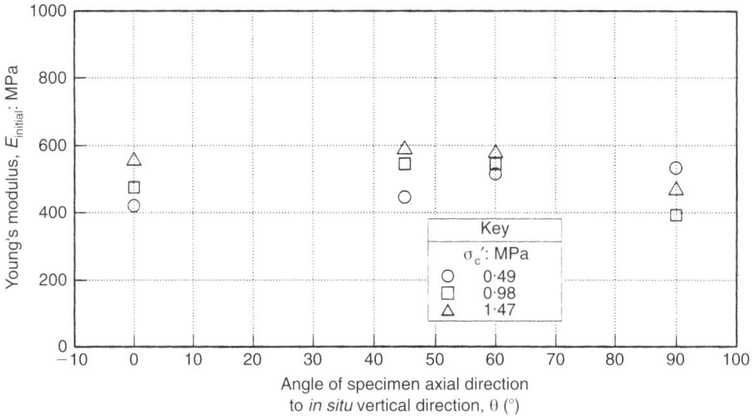

Fig. 4. Young's moduli from triaxial compression tests of mudstone samples with axes in different *in situ* directions

isotropically reconsolidated to the field vertical effective stress. The axial strain rate was 0·002%/min. A typical test result is shown in Fig. 6.

From this and other test results, the initial Young's moduli E_0 were obtained at a strain level of 5×10^{-6} based on locally measured axial strains, and the Young's moduli $E_{initial}$ were obtained from the apparently linear portions of the stress–strain relations based on externally measured axial strains. In this case, the value of E_0 is around twice as large as that of $E_{initial}$. This discrepancy has two origins. The first is extra compression of the drainage layer and bedding errors at the top and bottom ends of specimens. As the errors are considerable for the triaxial test specimens of 10 cm in height, they might have become extremely large for the oedometer test specimens of 2 cm in height, as discussed by Tatsuoka *et al.* (1995a). The second cause is larger strain levels used for defining $E_{initial}$.

The relationships between E_0 and the consolidation confining pressure σ_c' obtained from the CU triaxial compression tests are shown in Fig. 7(a). It is seen that the effect of σ_c' is small at all depths. The values of E_0 are much larger than those presented in Fig. 3.

To examine the effects of drainage conditions on E_0, three specimens were isotropically reconsolidated to the respective in situ overburden pressure. They were subjected to cyclic shear deformation up to a strain level of 0·01% under undrained conditions followed by drained conditions. The test results are shown in Fig. 7(b). The shear moduli G_0 shown in the figure were obtained as $G_0 = E_0/2(1+\nu)$ by assuming the drained and undrained Poisson's ratios to be, respectively, 0·12 and 0·5. It is seen that E_0 (CU) is greater than E_0 (CD), indicating that the Young's modulus is influenced by drainage conditions. The values of G_0 (CU) and G_0 (CD), on the other hand, are almost the same. These findings suggest that drainage conditions should be taken into account when evaluating the deformation of softrock in the field.

The relationships between E_0 and the compressive strength q_{max} of the mudstone at the site and other places (Kim *et al.*, 1994) are shown in Fig. 8. The ratios E_0/q_{max} for softrock of the same formation (Kazusa) obtained at this site and at the Sagamihara and Yokohama sites are very similar to each other. This ratio is around 500 in CD triaxial compression tests, but is around 500–1000 under CU conditions. This is because E_0 is smaller and q_{max} is slightly larger under CD conditions than under CU conditions.

FULL-SCALE BEHAVIOUR AND DISPLACEMENT ANALYSIS

Field observation

Each anchorage is composed of two sections, a caisson 70 m wide by 45 m deep, and an anchorage body which supports cables. The caissons were sunk by the pneumatic caisson method. Deep wells were installed to reduce the necessary compressed air pressure. The base of the foundation was located after an 8 m depth of mudstone was excavated.

The weight of the anchorage (Fig. 9) upon completion of concrete filling of the caisson was approximately 280 000 tf, with a further 140 000 tf for the weight of the anchorage section. The erection of the cables and stiffening girders also introduced cable tension acting on the anchorage.

Figure 10 shows the vertical and horizontal dis-

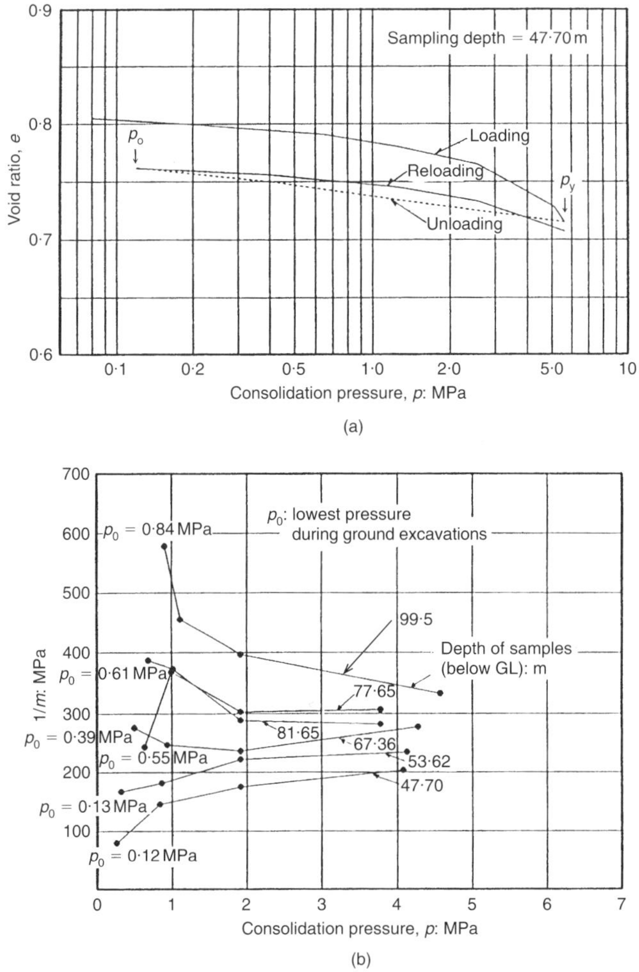

Fig. 5. Results of the oedometer tests: (a) typical p–e relations; (b) $1/m_v$ during reloading

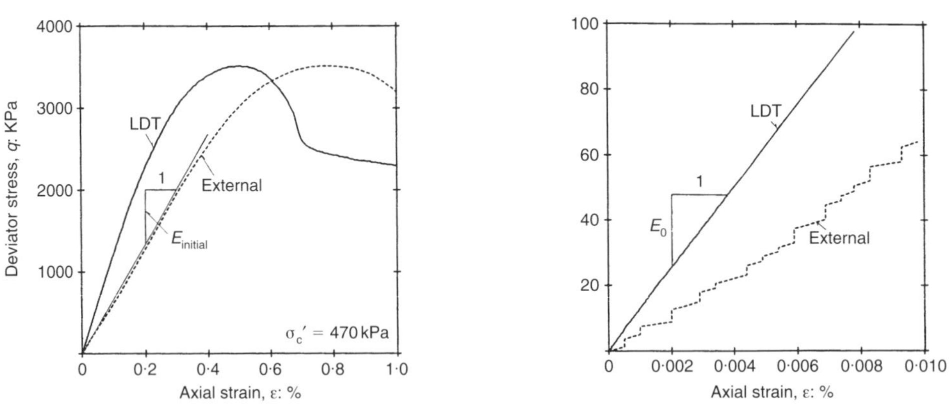

Fig. 6. Typical stress–strain relations using different scales from a CD triaxial compression test

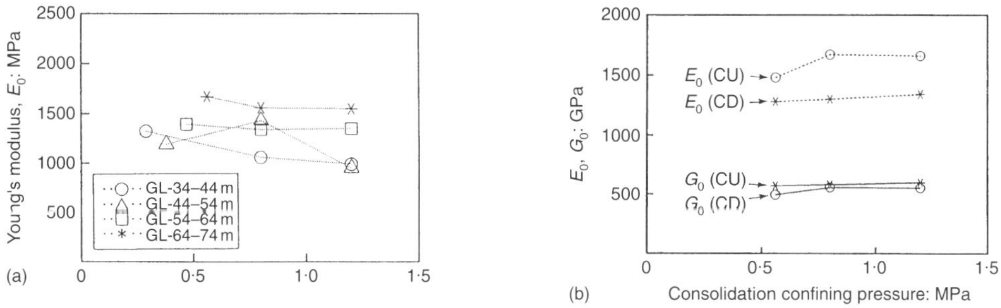

Fig. 7. Dependency of stiffness on: (a) confining pressure (from CD triaxial compression tests); (b) drainage condition

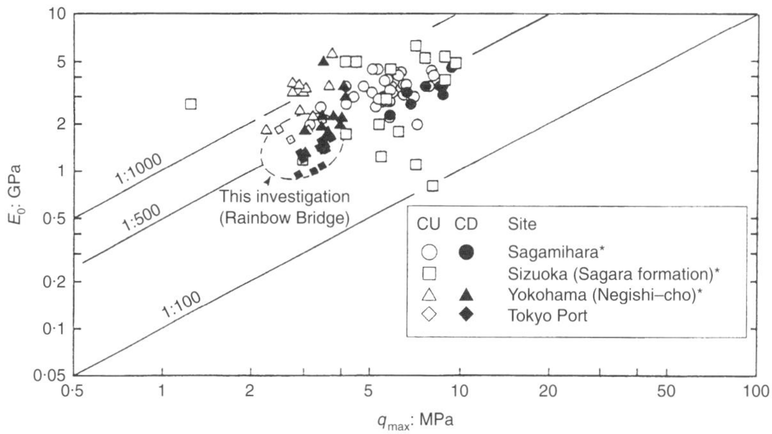

Fig. 8. Relationship between E_0 and q_{max} from triaxial compression tests. *****Data from Kim *et al.* (1994)**

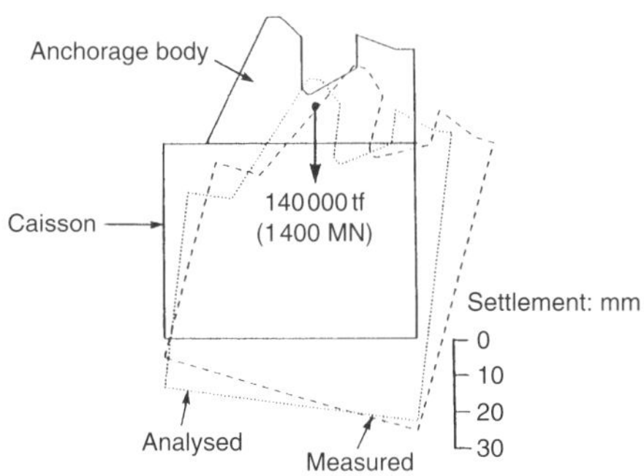

Fig. 9. Comparison between measured and analysed settlements of the anchorage

placements of the two anchorages (see Fig. 1) observed in the boreholes at the Daiba site which occurred at different construction stages of the anchorage body. The boreholes were located 3·5 m away from both the front and rear sides of the anchorage, as shown by the small vertical downward-directed arrows in Fig. 10. The maximum vertical displacement was observed at a depth slightly deeper than the caisson bottom. Horizontal displacement occurred towards the centre of the caisson near the caisson bottom. The reason for this is that the caisson was founded 8 m below the mudstone surface and, during the caisson sinking, gaps were formed between the mudstone and the lateral caisson surface due to the provision of friction cutters at the tip of the caisson. The lateral ground near the observed boreholes displaced towards the centre of the caisson to fill those gaps. This tendency is also confirmed by the numerical analysis described later (see Fig. 17).

Figure 11 shows the settlement at the front and rear edges of the two anchorages, and the averages, caused by the construction of the anchorage body. It is interesting to note that the rate of the average settlement increased as the weight of the anchorage

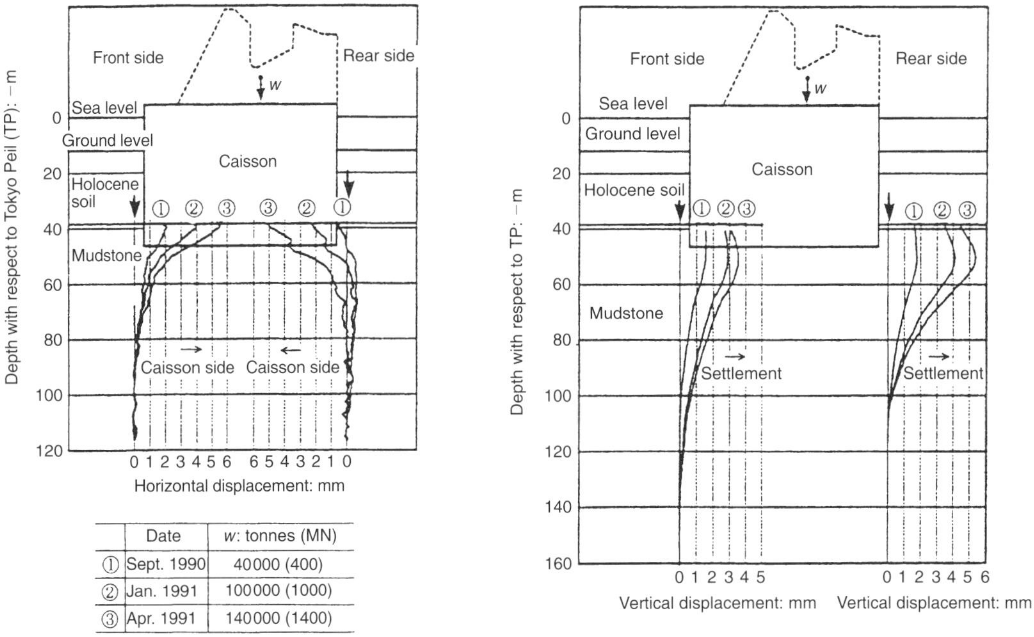

	Date	w: tonnes (MN)
①	Sept. 1990	40000 (400)
②	Jan. 1991	100000 (1000)
③	Apr. 1991	140000 (1400)

Fig. 10. Distribution of vertical and horizontal displacements during the construction of the anchorage body from May 1990. *w*, weight of anchorage body

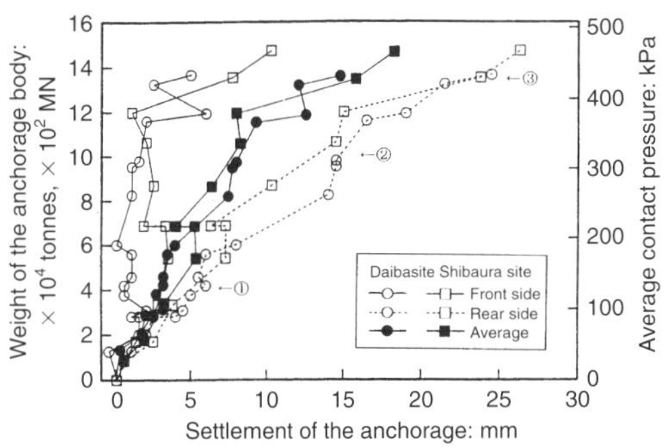

Fig. 11. Settlement of the anchorage during construction of the anchorage body. ①, Sept. 1990; ②, Jan. 1991; ③, Apr. 1991

increased. It is very likely that this behaviour reflects strain non-linearity of the stiffness of the mudstone and a very low dependency of the stiffness on pressure level.

Deformation characteristics of softrock from the full-scale behaviour and their comparison with those from in situ and laboratory tests

The coefficient of vertical subgrade reaction K_v is considered to be proportional to B^n. For the uniform elastic ground, the power n is equal to $-1{\cdot}0$. For the real ground, Yoshida & Yoshinaka (1972), for example, obtained $n = -3/4$ from PLTs conducted on deposits of cemented volcanic ash clay (Kanto loam) using plates of various sizes.

As for the mudstone beneath the bridge, Fig. 12 shows the K_v values obtained from PLTs using plates of variable sizes ($\varnothing = 15$, 30, 60 and 120 cm) performed at a nearby location (Kariba) on the same mudstone formation at the same level with the caisson base, together with those from PLTs in a borehole ($\varnothing = 25$ cm) and PLTs at the bottom of the caisson for the anchorage at Shibaura ($\varnothing = 30$ cm). These values were obtained from the secant gradient of the first unloading–reloading cycle. The values obtained from the load–settlement relation during the anchorage body construction are also plotted in the figure. From these test data points, $n = -0{\cdot}81$ is obtained, which is slightly different from $-3/4$. Although the difference in the n values appears small, the consequence in predicting the settlement of a huge foundation is noticeable, as seen from Fig. 12. The deviation of the n value from $-1{\cdot}0$ may be mainly due to the following two factors. First, the stiffness of the softrock increases with the depth (see Fig. 3). Therefore, PLTs with a smaller B measure smaller stiffness at shallower depths. Second, the softrock exhibits strain non-linearity with small pressure level dependency, and larger strains are

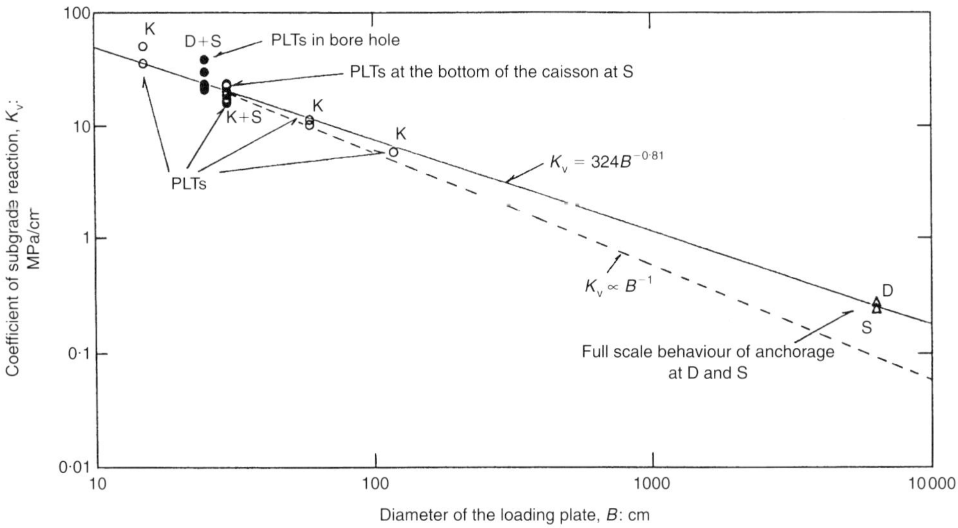

Fig. 12. Size effect on the coefficient of subgrade reaction obtained from plate loading tests and the load–settlement relation during anchorage construction. Sites: D, Daiba; K, Kariba; S, Shibaura

involved in PLTs with a rather fixed settlement of plate used to define K_v. Therefore, PLTs measure smaller stiffness at larger strains.

The values of G_f from suspension PS logging, G_{obs} from the full-scale behaviour, the secant shear modulus G_{sec} obtained from the CD triaxial compression tests using LDTs, and the values of G_p obtained from PMTs are shown along with the corresponding strain levels in Fig. 13. The values of G_{obs} were obtained as

$$G_{obs} = E_{obs}/2(1 + \nu_{drained}) \quad E_{obs} = \Delta\sigma_v'/(\Delta L/L) \tag{1}$$

where $\nu_{drained}$ is the drained Poisson's ratio (0·12), $\Delta\sigma_v'$ is the increase in the effective overburden pressure at the depth of interest along the centre line of the caisson caused by the weight of the anchorage section, calculated by the Boussinesq equation, and L and ΔL are the distance and its change between two vertically adjacent points where the settlement was measured. To obtain the G_{sec} values, a constant Poisson's ratio equal to 0·12 was used. The strain level of G_f in suspension PS logging is around 1×10^{-9} to 5×10^{-6}. The respective loading times are less than a second for G_f, several hours for G_{sec} and G_p, and about 1 year for G_{obs}. The relatively large scatter seen in the data is partly due to the variation in measurement depth. It may be seen, however, that these shear moduli are consistent with each other at the respective strain levels.

The distributions of the values of G_0 from the triaxial compression tests, G_f, and G_{obs} are shown in Fig. 14. It may be seen that despite the differences in testing place (in situ or laboratory), loading methods and loading duration, these shear moduli are very close to each other, while much larger (about five times) than those obtained from the Young's moduli presented in Fig. 3. The values of G_0 are slightly smaller than the values of G_f. This may be attributed to the effect of sample disturbance (Tatsuoka *et al.*, 1995b). In addition, the values of G_{obs} are only slightly smaller than the values of G_f. The difference may be attributed to differences in strain level and loading duration.

FEM analysis of the full-scale behaviour

The analysis was performed for the load increment by the anchorage body construction. The load increment and the observed full-scale behaviour at this stage are summarized as follows:

(*a*) The anchorage body (≈ 140 000 tonnes) was constructed over a period of around 1 year. The centre of gravity of the body was eccentric towards the rear side of the anchorage.
(*b*) The pore water pressure was kept almost constant and the same as the static water pressure. The excess pore water pressure generated during the construction was thought to have been dissipated with the progress of construction.
(*c*) The rebound of the mudstone layer at the preceding stages (caisson construction and caisson filling) was very small. The effect on the stiffness of stress release caused by the excavation seems to be small.

The three methods outlined in Table 2 were

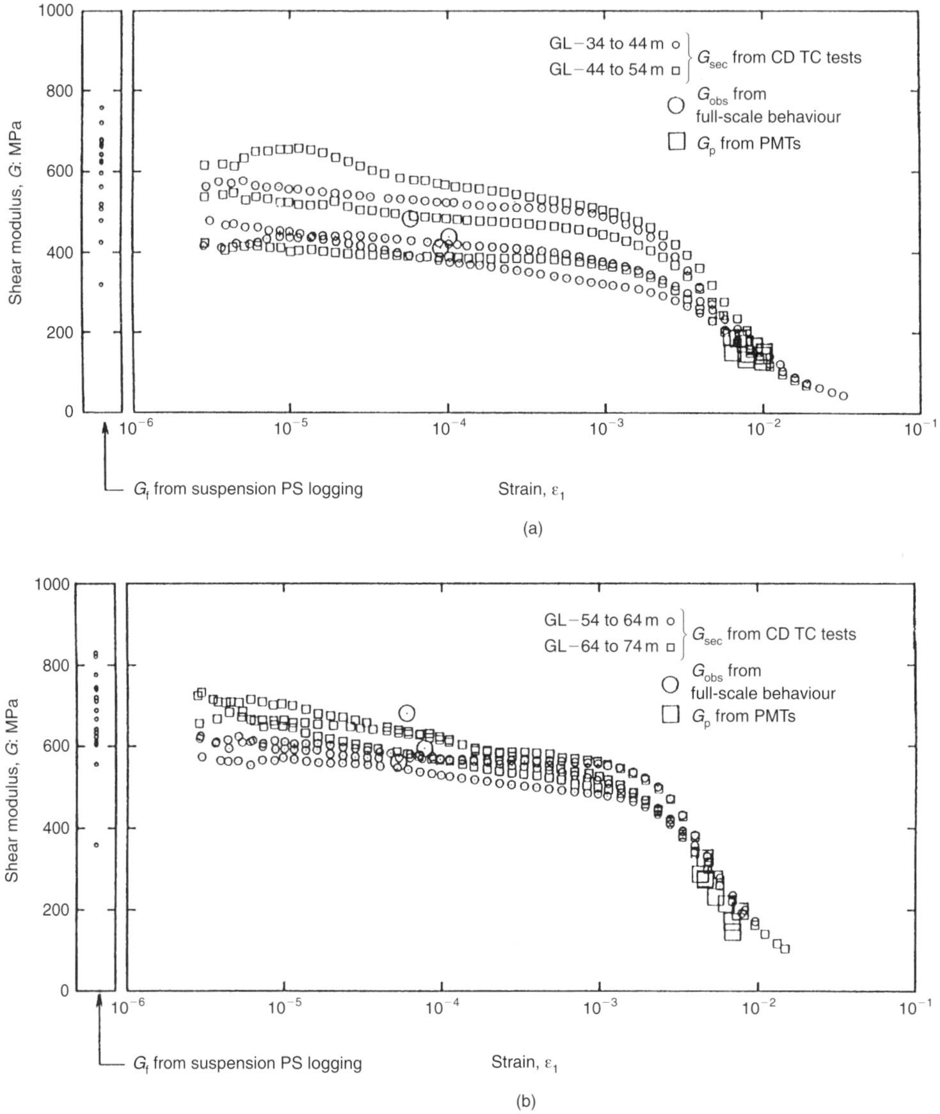

Fig. 13. Relationships between the shear moduli and strain levels: (a) upper mudstone; (b) lower mudstone. TC, triaxial compression; CD, consolidated drained; PS, velocity of primary and secondary wave

employed. In method 1, the linear analysis was carried out using the reloading Young's modulus obtained from the oedometer tests. Method 2 was linear elastic analysis using the Young's moduli determined from the PLTs using a 30 cm diameter plate and PMTs (see Fig. 3). Method 3 was non-linear elastic analysis, where the initial shear modulus G_0 obtained from triaxial compression tests were used (Fig. 14). A drained Poisson's ratio equal to 0·12 was used. This procedure is nearly equivalent to the use of the shear modulus G_f from field shear wave velocity while taking into the drainage conditions. The strain non-linearity curve obtained by averaging the measured relations from the CD triaxial compression tests (Fig. 13) was used.

A 3D FEM model, shown in Fig. 15, was used for methods 2 and 3, while a plane strain model was used for method 1. The results of method 1 were converted into the equivalent values for the

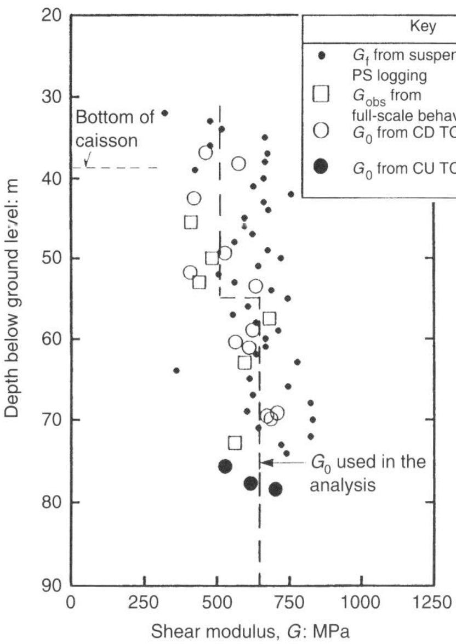

Fig. 14. Comparison of shear moduli from various tests and measurements. TC, triaxial compression; PS, velocity of primary and secondary wave

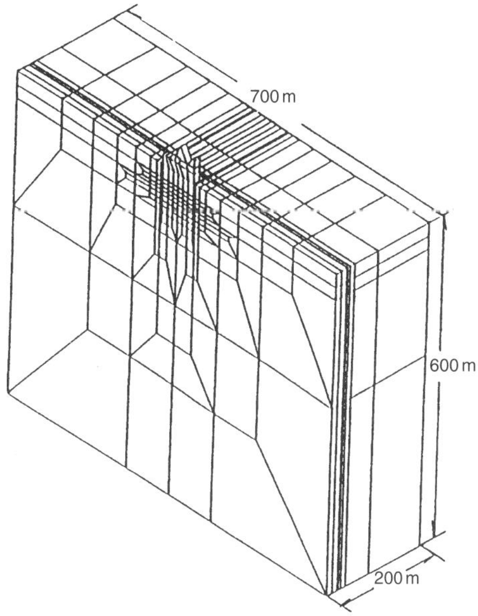

Fig. 15. The FEM model used for methods 2 and 3

3D model using the modification coefficients derived from the comparison between the results for the 3D and plane strain models obtained by method 2.

The settlements of the anchorages obtained from the three analysis methods are compared with the observed values in Fig. 16. For comparison, the settlement obtained from the conventional consolidation calculation using deformation characteristics at primary loading from the oedometer tests are also shown, which are much larger than the observed values. This large overestimation is mainly due to too large values of the compressibility m_v (see Fig. 5(b)).

The settlements obtained by the FEM analysis do not include creep effects. The creep deformation was estimated to be 20% of the instantaneous settlements as simulated, based on the creep tests for 90 days, which are similar to the construction period concerned. It is seen from Fig. 16 that methods 1 and 2 overestimate the observed values by about three and two times, respectively. On the other hand, method 3 simulated very well the observed behaviour; Fig. 9 compares the displacements of the Daiba Anchorage that were measured and simulated by method 3, while Fig. 17 compares the displacements along the boreholes adjacent to the anchorage caisson. These results also show that method 3 is the most relevant.

The overestimation by method 1 is due to a large underestimation of the stiffness by oedometer tests caused by the bedding errors at the top and bottom faces of a specimen. The overestimation by method 2 is due to underestimation of the stiffness by the PLTs and PMTs, perhaps due to strain levels involved that are larger than those operating in the ground (see Fig. 13). Note that the use of Young's modulus obtained by PLTs in FEM analysis is equivalent to the use of a K_v value with

Table 2. Three methods of analysis

Method	Type of analysis	Stiffness	Test type from which the Young's modulus was determined
1	Linear analysis based on effective stress	Young's modulus as a function of effective stress	Oedometer tests
2	Linear analysis	Young's modulus from *in situ* tests	Plate loading tests and pressuremeter tests
3	Non-linear analysis	Young's modulus as a function of strain level	Triaxial compression test using LDT

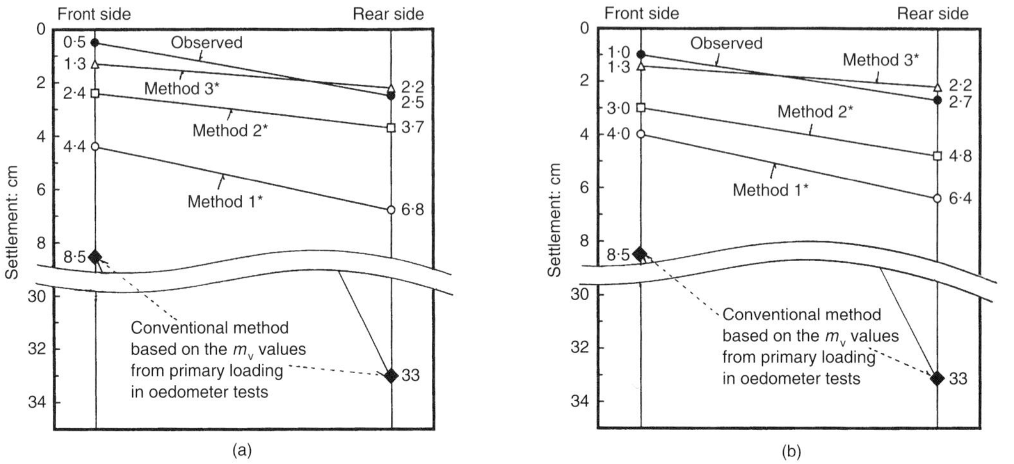

Fig. 16. Comparison between measured and analysed settlements during the construction of the anchorage body: (a) Diaba site; (b) Shibaura site. *Creep settlements were added to analysed values

$n = -1{\cdot}0$. This procedure is not consistent with the fact that $n = -0{\cdot}81$.

As for the long-term behaviour actually measured over the last several years since completion, no noticeable settlement of the anchorages has been reported so far.

CONCLUSIONS

The comparison of full-scale behaviour of the foundations for Rainbow Bridge with simulations based on some conventional methods and the state-of-the-art method revealed the following:

(*a*) FEM simulations using the stiffness values from reloading tests in conventional oedometer tests and those from conventional plate loading and pressuremeter tests overestimate by about three and two times, respectively, the observed settlement of the foundations. Possible reasons are the effects of bedding errors in the oedometer tests and strain levels involved in the field loading tests that were larger than those actually operating in the ground (about 0·01%).

(*b*) To estimate the full-scale settlement based on the coefficient of vertical subgrade reaction K_v from plate loading tests, K_v should be considered to be proportional to (the scale of structure)$^{-0{\cdot}81}$. The use of a power equal to $-1{\cdot}0$ is equivalent to the use of the Young's modulus obtained from plate loading tests in FEM analysis. This factor may be one of the reasons for underestimation of the ground stiffness for large structures.

(*c*) The stiffness values at strains of less than about 0·001% obtained from triaxial compression tests measuring axial strains locally were very close to those from field shear wave velocity.

(*d*) Non-linear 3D FEM analysis based on the following could simulate very well the observed behaviour:

  (i) The initial stiffnesses at strains of less than about 0·001% at different depths were determined by triaxial compression tests using high-quality undisturbed samples while measuring axial strains locally. These values were confirmed by field shear wave velocity measurements.

  (ii) Strain non-linearity of the stiffness which was evaluated by the triaxial compression tests was taken into account.

  (iii) Based on triaxial test results, the effects of pressure change on the stiffness were not taken into account, but those of drainage conditions and creep deformation were taken into account.

(*e*) Based on the experience with these foundations, it can be concluded that similar caissons or raft foundations can be constructed without using pile foundations on this and similar sedimentary softrock without exhibiting intolerable instantaneous and long-term deformation of the ground.

ACKNOWLEDGEMENTS

Our sincere thanks are due to Professor Fumio Tatsuoka of University of Tokyo. His insight and

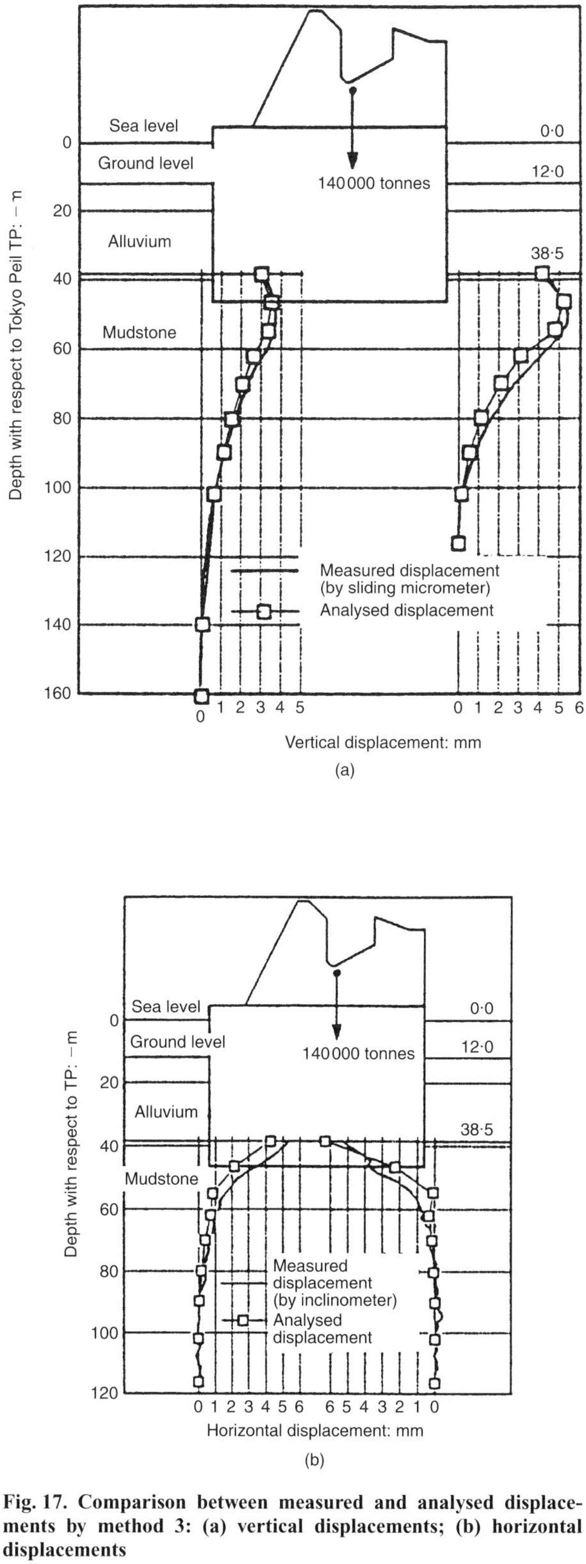

Fig. 17. Comparison between measured and analysed displacements by method 3: (a) vertical displacements; (b) horizontal displacements

enthusiasm have been the inspiration for our interest in this study.

REFERENCES

Goto, S., Tatsuoka, F., Shibuya, S. Kim, Y.-S. & Sato, T. (1991). A simple gage for local small strain measurements in the laboratory. *Soils Fdns* **31**, No. 1, 169–180.

Kim, Y.-S., Tatsuoka, F. & Ochi, K. (1994). Deformation characteristics at small strains of sedimentary soft rocks by triaxial compression tests. *Géotechnique* **44**, No. 3, 461–478.

Tatsuoka, F. & Kohata, Y. (1995). Stiffness of hard soils and soft rocks in engineering applications. Keynote lecture. *Proc. 1st Int. Symp. Prefailure Deform. Character. Geomatera.*, Balkema **2**, 101–170.

Tatsuoka, F., Lo Presti, D. & Kohata, Y. (1995a). Deformation characteristics of soils and soft rocks under monotonic and cyclic loads and their relationships. *Proc. 3rd Int. Conf. Recent Adv. Geotech. Earthquake Engng Soil Dynamics, St Louis* **2**, 85–879.

Tatsuoka, F., Kohata, Y., Ochi, K. & Tsubouchi, T. (1995b). Stiffness of softrocks in Tokyo metropolitan area—from laboratory tests to full-scale behaviour. Keynote lecture. *International symposium on rock foundation of large-scaled structures*. Tokyo, Balkema, pp. 3–18.

Yoshida, I. & Yoshinaka, R. (1972). A method to estimate modulus of horizontal subgrade reaction for a pile. *Soils Fdns* **12**, No. 3, 1–18.

Viana da Fonseca, A., Matos Fernandes, M. & Silva Cardoso, A. (1997). *Géotechnique* **47**, No. 3, 633–651

Interpretation of a footing load test on a saprolitic soil from granite

A. VIANA DA FONSECA,* M. MATOS FERNANDES* and A. SILVA CARDOSO*

In the north-western region of Portugal residual soils from granite are dominant. Their porous cemented structure leads to a complex geotechnical behaviour, quite distinct from that of transported soils with a similar grain size distribution. The paper describes some of the studies carried out at an experimental site on which a fairly homogeneous saprolitic soil 6 m thick was encountered. These studies included: (i) a full-scale load test of a circular concrete footing (1·20 m diameter), under carefully controlled conditions and with extensive monitoring; (ii) in situ testing techniques (SPT, CPT, SBPT, CH) and (iii) laboratory testing on high-quality block samples, including triaxial tests with local strain measurement. The discussion is concentrated on the interpretation of the pressure–settlement curve from the footing load test, for a range of pressures typical of working conditions, by means of linear and non-linear elastic analyses based on stiffnesses from the laboratory and in situ tests. Some practical conclusions are extracted regarding the use of in situ and laboratory test results in analytical and empirical or semi-empirical methods for settlement estimation on these residual soils.

KEYWORDS: residual soils; deformation; footing; full-scale tests; in situ testing; laboratory tests.

Les sols résiduels du granit dominent le littoral de la région nord du Portugal. Les fabrique et structure de ces sols et particullièrement leur matrice cémentée et poreuse conduissent à des comportements géotechniques complexes, très différents des sols transportés avec granulométrie similaire. L'article décrit une partie des études réalisés sur un terrain expérimental où apparaissent des niveaux superficiels typiques des sols saprolitiques du granit de la région. Les études réalisés comprennent: (i) un essai de charge d'une semelle circulaire de béton armé (avec 1·20 m de diamètre), soigneuse et extensivement instrumentée; (ii) diverses essais in situ (SPT, CPT, SBPT, CH); (iii) essais en laboratoire sur des échantillons de haute qualité retirés des blocks taillés en puits, comprennant des essais triaxiaux avec instrumentation interne des déformations. La discussion est concentrée dans l'interprétation de la courbe de chargement-tassement obtenu de l'essai de charge de la semelle en utilisant des résultats des essais in situ et en laboratoire, considérant les niveaux de charge typiques des conditions de service. On extrait des conclusions sur l'application des résultats des essais in situ et en laboratoire aux méthodes empiriques et semi-empiriques pour l'évaluation des tassements des fondations en sols résiduels du granit.

INTRODUCTION

In the north-western region of Portugal residual soils from granite are dominant. The thickness of these regional saprolitic soils may sometimes attain 20 m, with more common values of 5–10 m. Due to their specific genesis such soils present complex characteristics, which are a consequence, on the one hand, of their overall variability and inhomogeneity and, on the other hand, of the spatial arrangement and distribution of the particles and pore spaces. Coarse and resistant quartz grains are bonded by fragile clayey plagioclase bridges, resulting in a fabric with medium to high porosity. The feldspars are subjected to intense weathering processes, typical of high average annual precipitation and well-drained ground profiles. The mechanical and hydraulic behaviour of these soils is often quite distinct from that of transported soils with similar densities and grain size distributions.

Most of the current buildings and structures in the region are founded on footings on these soils. The behaviour of shallow foundations under typical working conditions, in which stress and strain levels induced in the ground are low to moderate, seems to be considerably dependent on the cemented structure of the saprolitic soil. This has been one of the main issues addressed in a research programme at the University of Porto on the geotechnical characterization and modelling of regional saprolitic soils.

A major part of the experimental work was carried out at a given site on which a fairly homogeneous saprolitic soil 6 m thick was encountered. The central element of the experiments was

Manuscript received 15 October 1996; revised manuscript accepted 18 April 1996.
Discussion on this paper closes 1 November 1997.
* Faculdade de Engenharia da Universidade do Porto.

the execution of a load test on a 1·20 m diameter rigid reinforced concrete footing. In association with this test an extensive geological and geotechnical characterization was undertaken, including in situ testing [SPT, CPT, SBPT, cross-hole (CH)] and laboratory testing for chemical, mineralogical and physical identification, as well as oedometer and triaxial testing on high quality block samples.

In this paper the results of the experimental studies are presented and discussed, mainly in regard to the pre-failure deformation behaviour of the saprolitic soil. The pressure–settlement curve from the footing load test, for typical working load levels, is compared with those obtained by using simplified linear and non-linear elastic analyses based on deformation moduli taken from the laboratory and in situ tests.

GENERAL CHARACTERIZATION OF THE EXPERIMENTAL SITE

Figure 1 shows the experimental arrangement adopted for the footing test and the location of in situ tests in the vicinity.

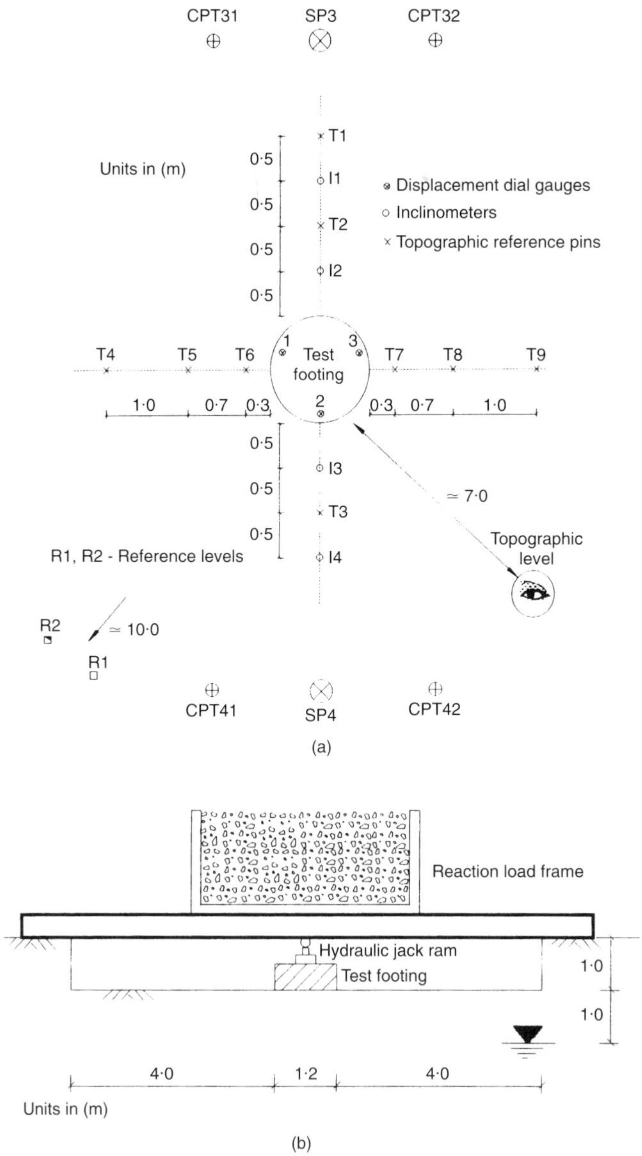

Fig. 1. Plan (a) and detailed cross-section (b) of the experimental site

The results of SPT and CPT tests and the values of the maximum shear modulus, G_0, obtained from cross-hole tests are shown in Fig. 2. It is observed that the CPT cone resistance, q_c, increases linearly with the at-rest effective vertical stress, σ'_{VO} whilst G_0 reveals a very low variation.

The variation of grain size distribution and Atterberg limits with depth is summarized in Table 1. This table contains the results obtained with specimens taken from the SPT sampler and from blocks trimmed in ditches 1–2 m deep. On the whole, the results reveal a fairly homogeneous ground profile which can be classified as SM (silty sand) or SM-SC (silty clayey sand) according to the Classification for Engineering Purposes (ASTM D2487, 1989).

Table 2 includes some natural physical properties of the soil at the experimental site and the corresponding typical range in the region (da Fonseca, Matos Fernandes, Cardoso & Martins, 1994). The degree of saturation, S_r, is very often in the range 90–95% and the corresponding suction levels are very low.

Table 3 presents some state indices commonly used for residual soils. In the Collins (1985) classification the values of X_d correspond to the metastable zone, which characterizes highly porous (flocculated) fabrics with an interparticle cemented matrix.

Figure 3 illustrates the plotting of the CPT test results on the Robertson (1990) classification chart. It can be observed that: (i) almost all values are situated in typical zones associated with cemented, aged or very stiff natural soils; (ii) with respect to grain size distribution the values fall mainly in the zones for sands or silt/sand mixtures.

INITIAL STRESS STATE

Regional experience suggests that values of the coefficient of earth pressure at rest, K_0, are relatively low. This is in agreement with the conclu-

Table 1. Grain size distribution and Atterberg limits

	Depth: m	Clay: %	Silt: %	Sand: %	w_L: %	w_P: %
	1·2	8	25	62	28	25
	1·7	4	29	65	29	24
	2·1	6	26	63	27	19
	2·7	8	29	58	30	21
Borehole	3·5	3	23	68	31	28
SP3	4·0	5	25	63	32	21
	4·8	5	29	59	NP	
	5·2	5	22	67	NP	
	5·6	5	21	67	NP	
	6·2	5	22	67	32	28
	0·8	7	19	64	NP	
	1·2	7	31	55	32	25
	1·7	6	24	65	30	22
	2·1	6	22	63	NP	
	2·5	6	24	59	28	24
Borehole	3·0	5	22	60	28	25
SP4	3·6	5	20	64	28	22
	4·0	4	21	65	28	21
	4·6	5	21	64	NP	
	5·0	5	24	58	31	25
	5·3	4	22	62	NP	
	6·0	4	19	66	28	26
Blocks	1·0–1·5	5–7	14–23	64–67	28–40	19–26

Table 2. Natural physical properties

	γ_s: kN/m^3	e	S_r: %	w: %	k: m/s
Exp. site	25·7–26·5	0·60–0·85	70–100	15–27	10^{-6}–10^{-5}
Regional	25·5–26·7	0·40–0·85	60–100	10–30	10^{-6}–10^{-5}

γ_s: unit weight of solid particles; e: void ratio; S_r: degree of saturation; w: water content; k: coefficient of permeability.

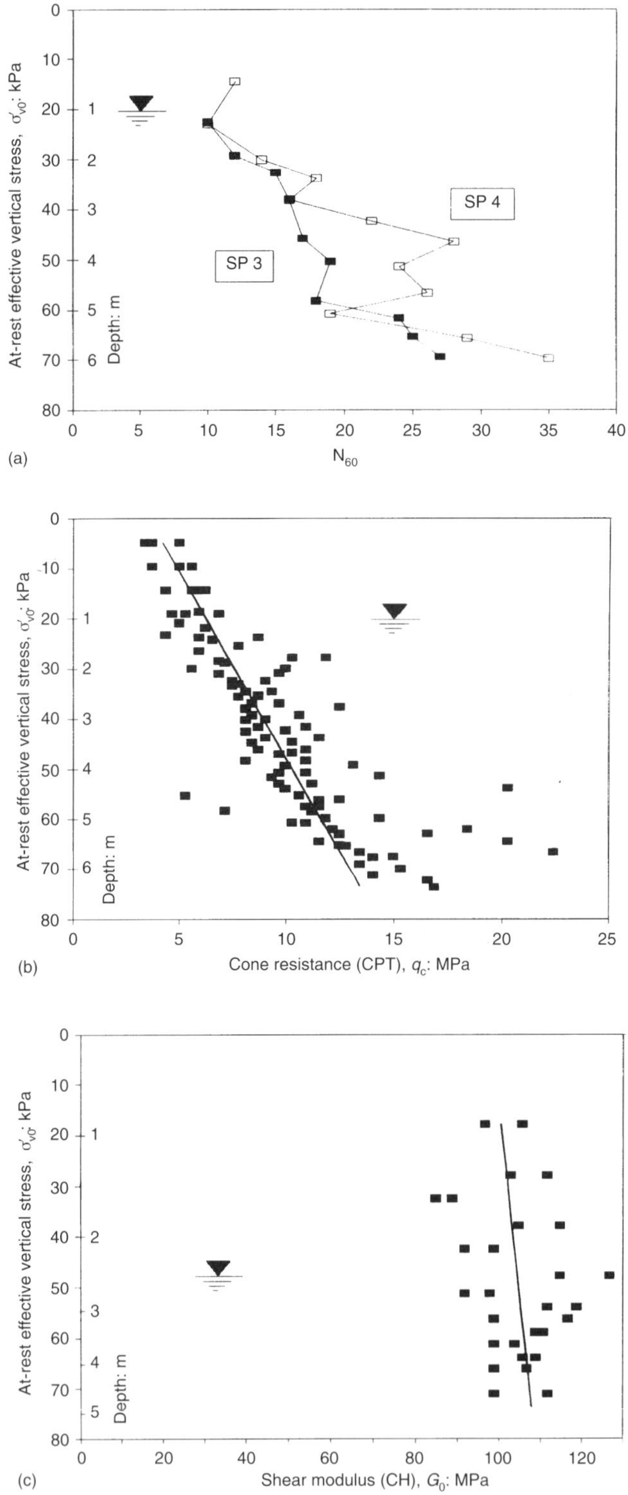

Fig. 2. In situ test results versus the at-rest effective vertical stress: (a) SPT; (b) CPT; (c) CH

Table 3. Weathering and void ratio indices

Index (symbol)	Reference	Values
Lixiviation index (β)*	Rocha Filho *et al.* (1985)	0·40–0·50
Degree of decomposition (X_d)†	Lumb (1962) Collins (1985)	0·59–0·63
Relative void ratio (e_r)‡	Vaughan (1988)	0·72

* $\beta = \text{ba}_1$ (weathered rock)/ba_1 (sound rock), where $\text{ba}_1 = (K_2O + Na_2O)/Al_2O_3$.
† $X_d = (1 - W_f/W_{f0})/(1 + W_f/W_q)$, where W_q is the weight of quartz, unaltered by weathering, and W_f and W_{f0} are the weight of feldspar in the soil and in the parent rock, respectively.
‡ $e_r = (e - e_{opt})/(e_L - e_{opt})$, where e_L and e_{opt} are the void ratio at the liquid limit and at the optimum dry density.

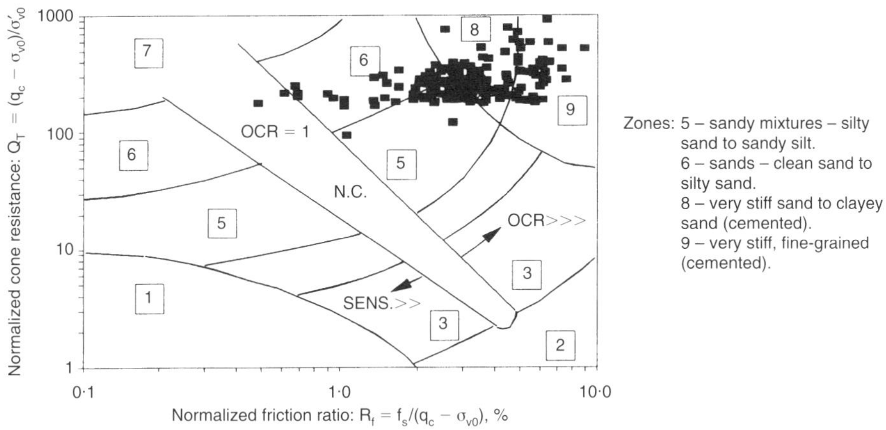

Fig. 3. Soil classification by the Robertson (1990) chart

sions of Vaughan & Kwan (1984) for cases where the weathering processes do not generate expansible minerals.

The four tests with the Cambridge self-boring pressuremeter (SBPT) included in the experimental studies lead to an average value of K_0 equal to 0·38. This value has been globally corroborated by the results of incremental triaxial tests with null radial strain (da Fonseca, 1996).

FOOTING LOAD TEST

General

The instrumentation used for the footing load test consisted of (see Fig. 1): three displacement dial gauges with 50 mm travel and 0·01 mm divisions connected to the footing to record its settlement; nine reference pins to survey the deflection of the ground surface around the footing by means of topographic electronic equipment; and four vertical inclinometer tubes sealed at 6 m depth to measure horizontal displacements in the ground; the inclinometer tubes were also used to observe the position of the water table.

The loading process consisted of 35 load steps, including four unload–reload cycles. Values of the stabilization time fulfilled a criterion commonly considered as appropriate to saprolitic soils (GOST 20276, 1985 – national Soviet standard): the settlement shall not exceed 0·1 mm in a time interval of 4 h. The whole duration of the test was approximately 15 days.

During the test the water level remained practically constant at 1 m depth.

Bearing in mind the number of loading stages, the piezometric readings and the adopted stabilization criterion, one can consider that the test was performed under drained conditions.

The loading process was concluded when the observed settlement reached around 10% of the footing diameter (ASTM D1194, 1989). The load at the last stage corresponded to a pressure of about 1000 kPa.

Pressure–settlement curve

The complete pressure–settlement curve from the footing load test is shown in Fig. 4(a). As can be seen in the enlargement of Fig. 4(b), there is a clear increase of the settlement rate with load for pressure values exceeding around 125 kPa. It should be noted that the time for settlement stabili-

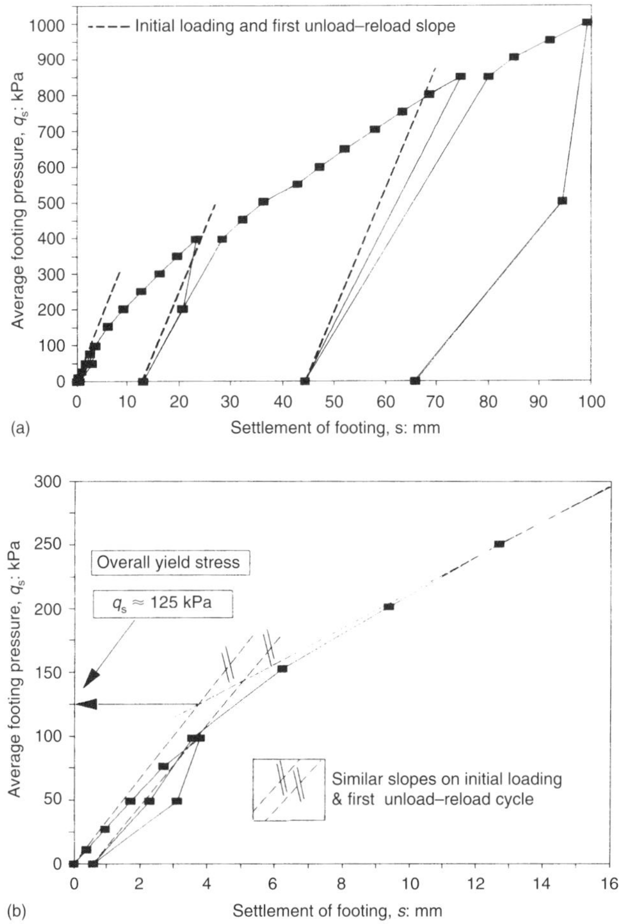

Fig. 4. Pressure–settlement curve of the footing load test: (a) general picture; (b) enlargement for $q_s \leqslant 300$ kPa

zation in each load step significantly increased above this pressure.

Although it is not precisely defined, this pressure seems to represent a transition from an essentially elastic behaviour to a phase in which the cemented structure of the soil is substantially damaged. This seems to be confirmed by the small difference between the inclination of the first loading curve and that of the first unload–reload cycle (Fig. 4(b)). In the subsequent cycles the unload–reload modulus is lower (Fig. 4(a)).

Figure 5 shows the vertical displacement of six of the reference pins on the ground surface around the footing. For the first load stages settlements increase at an approximately constant rate after which a stabilization occurs.

At the end of the test the rotation of the footing was measured with the help of a high precision clinometer. The measurements were performed on a smooth steel plate which had been previously fixed to the top of the footing and carefully levelled. The rotation angles were 1°20′, in the plane of the settlement points T_4 and T_9, and 30′, in the perpendicular plane. This rotation was also appreciated from the settlement measurements of the three dial gauges resting on the footing.

This rotation explains the fact that the settlements of points T_7 and T_8 become clearly smaller than the corresponding ones, T_6 and T_5 respectively, with the approach of the end of the loading process. Consistent differences of settlements between points T_4 and T_9 are not detected due to the large distance to the footing.

In relation to the horizontal displacements at the surface and in the ground in the vicinity of the footing, the recorded values are generally very small, most of them of the order of magnitude of the inclinometer equipment precision. The displacements observed at the two inclinometer tubes closest to the footing are illustrated in Fig. 6 for

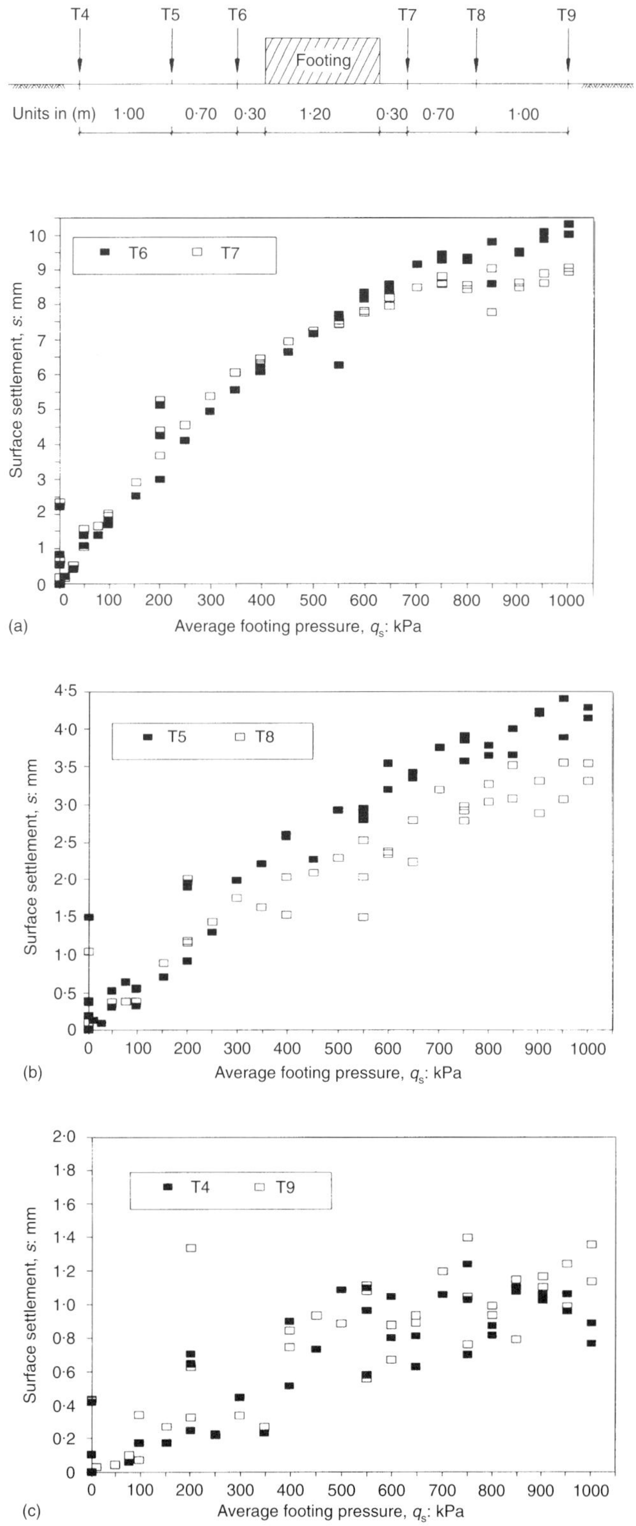

Fig. 5. Settlement of the ground surface around the footing

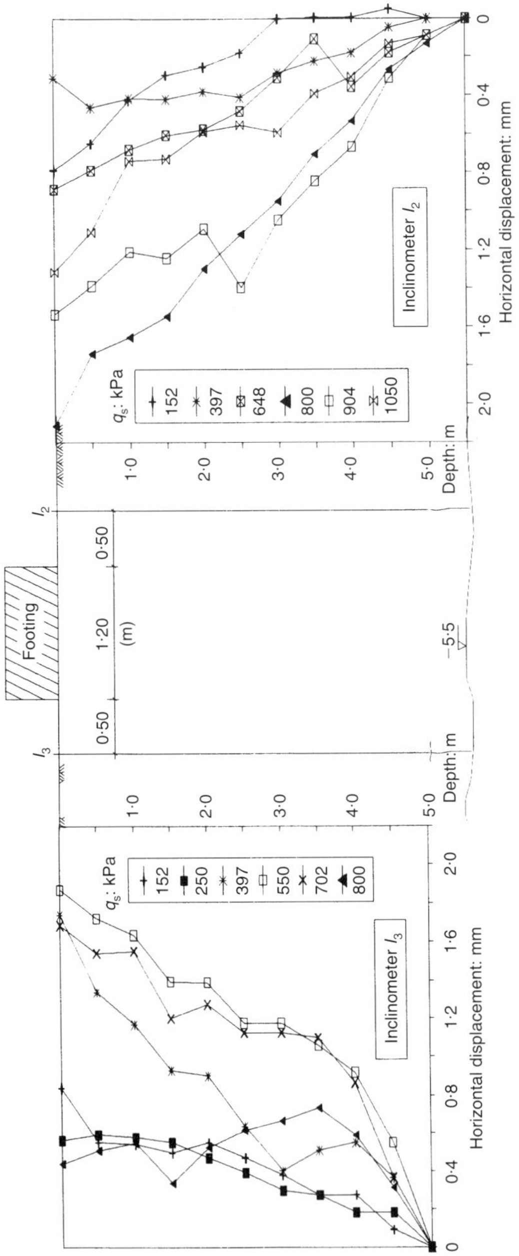

Fig. 6. Horizontal displacement in the ground 0·5 m away from the footing edge

some stages of the test. In spite of some scatter, a general movement towards the loaded area can be observed during the initial loading steps followed by a clear reversal of the displacement direction. Nevertheless, final displacements are still directed towards the footing.

Note on the type of failure

The shape of the pressure–settlement curve is typical of a punching shear failure mode. It should also be noted that the end of the test corresponded to a 100 mm settlement of the footing base while the ground surface pins situated at 0·3 m from its border did not reach values of settlement greater than 10 mm (see Fig. 5(a)). The interpretation of the curve with the purpose of estimating the bearing capacity of the footing is beyond the scope of this paper.

Allowable pressure evaluation

In order to establish the serviceability limit state pressure the criterion proposed by Décourt (1992) was adopted as a reference. This stipulates that the allowable pressure on a shallow foundation on residual soil should correspond in its effect to a settlement of 6·0 mm for a 0·8 m diameter plate load test. This represents a settlement of 0·75% of the diameter of the loading surface, which gives a settlement of 9 mm in the present case.

For the footing test curve (Fig. 4) a 9 mm settlement corresponds to an applied pressure of about 195 kPa.

INDUCED STRAINS IN THE GROUND

Figure 7(a) illustrates the maximum principal (vertical) strain distribution under the centre of the footing obtained from a simplified non-linear elastic analysis, to be presented below, whose results closely fit the first part of the observed pressure–settlement curve. Results are shown for three values of the applied pressure: 100, 200 and 400 kPa, corresponding to a value near to the serviceability limit state pressure and to half and twice that value.

Figure 7(b), in which a logarithmic scale is used for the strain, shows clearly that, for pressures of 100 and 200 kPa, strain values exceed 10^{-3} only in a restricted zone adjacent to the footing.

These results corroborate evidence from other authors (Jardine, Potts, Fourie & Burland, 1986; Burland, 1989; Tatsuoka & Shibuya, 1992; Jardine, 1995; Tatsuoka & Kohata, 1995) of the rather low strain levels involved in a number of soil–structure interaction problems, including shallow foundations, under common working conditions.

A considerable influence of the cemented structure of the soil on the behaviour of the footing is therefore to be expected.

LABORATORY TESTING FOR MECHANICAL CHARACTERIZATION

Note on sampling

Laboratory tests for mechanical characterization were carried out on undisturbed samples taken from large blocks trimmed in open ditches at a depth of 0·5–1 m below the level of the footing base. All the sampling and handling procedures were undertaken with the utmost care in order to preserve the natural structure of the saprolitic soil.

Taking into account, on the one hand, the homogeneity of the ground underneath the footing to depths of about 6 m (see Fig. 2 and Table 1) and, on the other hand, the conclusions on induced strains that have just been presented, the above-mentioned blocks can be taken as representative of the saprolitic soil most intensely affected by the loading of the footing.

Oedometer and triaxial specimens were obtained by driving the rings or split tube samplers into the blocks with the help of a static trust. The requirements of GCO (1990) about the cutting shoe, the wall friction (smooth inox steel faces) and the use of a non-return valve, have been adopted.

It is recognized that tube sampling may significantly reduce stiffness and strength of soils (Hight, Böese, Butcher, Clayton & Smith, 1992). Nevertheless, the very fragile structure of this granitic residual soil did not allow the use of trimming methods to obtain specimens directly from the blocks. The use of these methods seems to be more practicable in residual soils with a clayey matrix, such as gneissic soils from Brazil (Costa-Filho, Döbereiner, De Campos & Vargas, 1989).

Oedometer and isotropic consolidation tests

Figure 8 includes the results of oedometer tests revealing three distinct zones typical of cemented soils (Vaughan, 1988) and corresponding to different states and ranges of the compressibility index: (i) 'stable', $C_r = 0{\cdot}007–0{\cdot}018$; (ii) 'metastable', $C_{cs} = 0{\cdot}129–0{\cdot}289$; and (iii) 'granular or de-structured', $C_c = 0{\cdot}101–0{\cdot}172$. In the first state the soil generally preserves its natural cemented structure, while in the third state this structure is completely destroyed; the intermediate state is associated with progressive de-structuring, resulting from the gradual breakage of interparticle bridges by compression. While the first zone can be likened to an overconsolidated state in transported soils, the second, although apparently similar to a normally-consolidated state, is really a transition state characterized by higher values of compressibility than the same material would have under remoulded

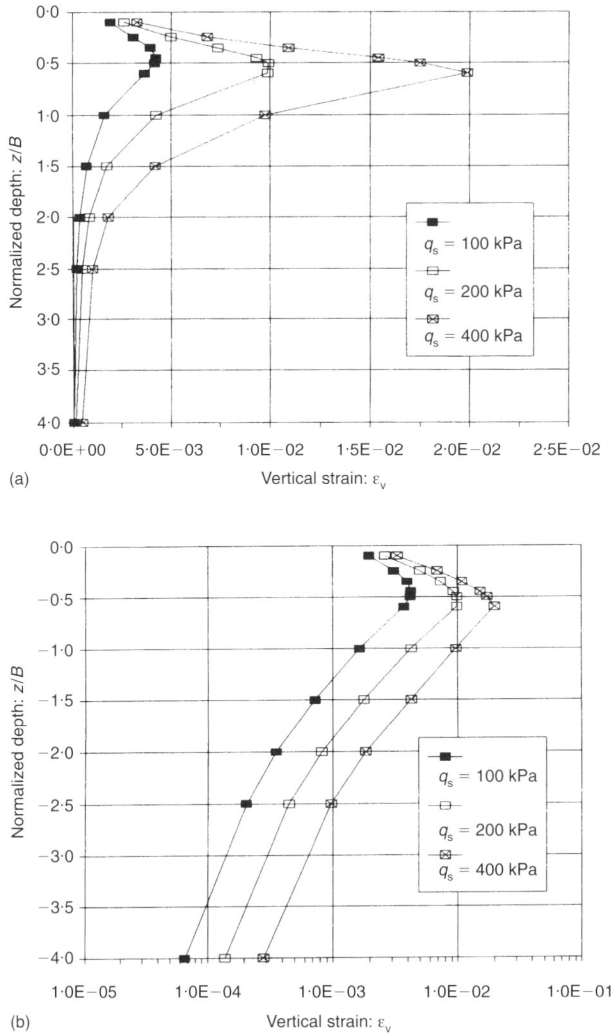

Fig. 7. Maximum principal strain under the centre of the footing from a non-linear elastic simplified analysis: (a) arithmetic scale; (b) logarithmic scale (z = depth; B = footing width)

conditions (granular state). Results for the remoulded soil are also included in the figure for reference.

Therefore, a virtual preconsolidation stress, σ'_{vp}, can be defined as the value which separates the first two zones, where most of the stress states generated by current loading conditions are situated.

In order to evaluate the virtual preconsolidation stress the method proposed by Pacheco e Silva (1970) was used, since it has demonstrated a good reproducibility in residual soils. The values obtained for virtual preconsolidation stress were $\sigma'_{vp} = 85$–140 kPa (see Fig. 8), more than 4 times the natural at-rest vertical effective stress.

It is interesting to note that the pressure of 125 kPa, which corresponds to a transition of behaviour in the footing load test, falls reasonably within the range found for the virtual preconsolidation stress.

Values of C_r and C_c from the tests are reasonably consistent with those commonly reported in the literature for similar soils (Lacerda, Sandroni, Collins, Dias & Prusza, 1985). Nevertheless, the values of the tangent-drained constrained modulus calculated for the same consolidation effective

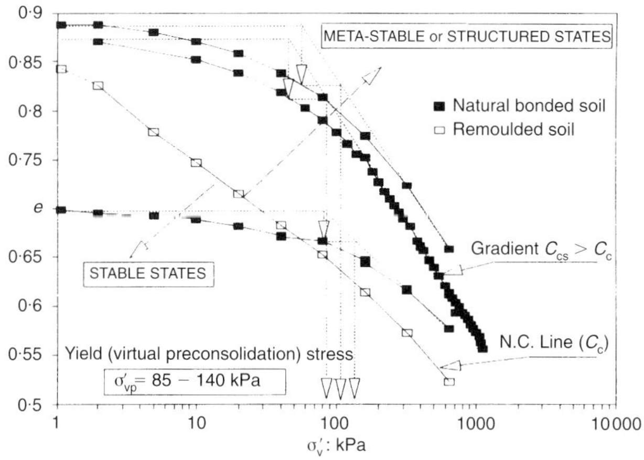

Fig. 8. Oedometer curves – state definition

stresses are substantially lower than the corresponding tangent Young's moduli obtained from triaxial tests (described in the following section) after taking into account the effect of Poisson's ratio. This may be due to the height of these specimens being significantly smaller than that of the triaxial specimens, leading to greater damage during preparation and handling. At small strains, bedding and contact problems may also have contributed to that trend.

Isotropic consolidation tests with local measurement of axial and radial strain provided values of the virtual isotropic preconsolidation stress of $\sigma'_{mp} = 40–60$ kPa. This range is slightly lower than the one deduced from the oedometer tests: considering $K_0 = 0{\cdot}38$ and taking the values of σ'_{vp} indicated above, $\sigma'_{mp} = 50–80$ kPa. This conclusion corroborates the hypothesis of Vaughan (1988) that yield surfaces for volumetric compression are anisotropic and may be centred on the K_0 stress axis (see Fig. 9).

Triaxial compression tests

A large number of triaxial tests were performed with different specimen sizes, consolidation stress conditions and stress-paths.

In this paper are presented just the results from compression tests on specimens of 100 mm diameter with local strain instrumentation. The local strain was measured with a pair of axial strain inclinometers (Ackerley, Hellings & Jardine, 1987).

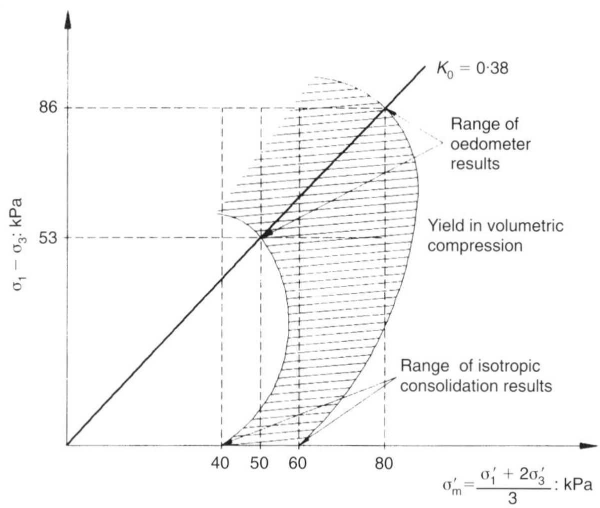

Fig. 9. Yield surfaces for volumetric compression (σ_1 = vertical stress, σ_3 = horizontal stress)

The specimens were previously saturated by applying 10 kPa increments of the back and confining pressures – maintaining an effective isotropic stress of 10 kPa, up to values of the order of 300 kPa to achieve full saturation.

The adopted rate of deformation was 0·002 mm/min, chosen to fulfil the requirements of drained shear conditions. Intermediate unload–reload cycles were performed starting at low to moderate stress levels.

Peak strength parameters obtained from these tests showed a good consistency – ranges of 37–38° for the angle of shearing resistance and 9–12 kPa for the effective cohesion.

In Fig. 10 stress–strain curves from one of the tests obtained by classical (external LVDT) and local strain measurement systems are compared. For these soils, as for many others, the former technique leads to rather unrealistic values of stiffness. The curve for the local instrumentation results is also included in Fig. 11 together with the curves from other tests with similar measurements but different consolidation stresses, σ'_c.

Figure 11 shows that the soil exhibits substantial brittleness for the lowest effective consolidation stress, while for the others the brittle behaviour tends to vanish, probably due to the development of volumetric (collapsible) plastic strains prior to shearing. Thus, for consolidation stresses much higher than the at-rest stress state, the stress–strain response tends to be typical of de-structured materials. For the lower consolidation stresses (10 and 20 kPa) peak values of deviatoric stress are mobilized before the highest value of the dilatancy ratio is reached. This seems to indicate that the mechanical behaviour of the saprolitic soil is controlled by cementation between particles rather than by dilatancy phenomena related to particle interlocking. This latter type of behaviour, typical of dense granular transported soils, is not compatible with the fabric of this soil which exhibits a low to medium density ($e = 0{\cdot}60–0{\cdot}85$) for an essentially sandy soil.

In order to study the dependency of stiffness on the effective consolidation stress, two types of Young's modulus were considered: (i) 'pseudo-elastic', E_{el}, deduced from the initially-linear reload branch of an intermediate unload–reload cycle; (ii) initial tangent, E_{ti}.

Assuming elastic isotropy, the Poisson's ratio ν was determined from the same reload branch as E_{el} (Lade, 1988). The average value obtained was $\nu_{el} = 0{\cdot}26$. It is interesting that this leads to an elastic coefficient of earth pressure at rest of:

$$K_0 = \nu/(1-\nu) = 0{\cdot}35 \quad (1)$$

which is reasonably consistent with the in situ SBPT results.

The analysis of the results, expressed on normalized logarithmic scales in Fig. 12, shows that the dependency of these moduli on the consolidation stress is bilinear, with a noticeable change of slope at stresses of the order of 50–60 kPa. This shows the favourable contribution to the soil stiffness of the natural cemented structure for the lower consolidation stresses.

It is systematically observed that E_{el} values are 2 to 3 times greater than the values of E_{ti}. This discrepancy is believed to be due to slight movements of the inclinometer footings anchored on the wall of the specimens which significantly affect the initial sections of the stress–strain curves and,

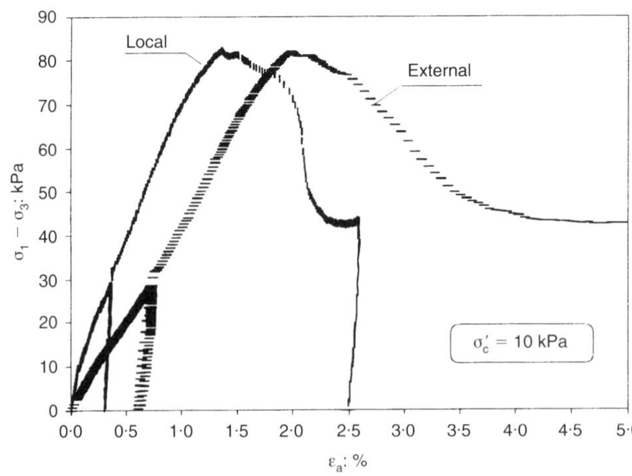

Fig. 10. CID triaxial test ($\sigma'_c = 10$ kPa): comparison of stress–strain relations with local and external measurement of axial strain

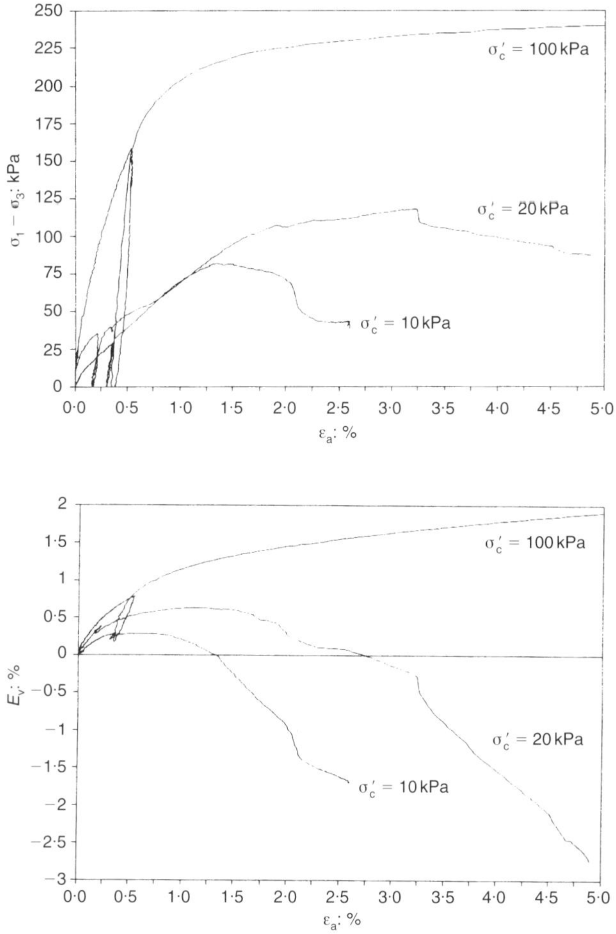

Fig. 11. Results from CID triaxial tests under three distinct consolidation stresses

consequently, the values of E_{ti}.

Several authors (Burland, 1989; Tatsuoka & Shibuya, 1992; Tatsuoka & Kohata, 1995) suggest that the small strain shear moduli evaluated from triaxial tests with local strain instrumentation are similar to those obtained by in situ seismic tests. In the present case, for $\sigma_c' = 10$ kPa, the shear modulus evaluated from E_{el} (assuming $\nu_{el} = 0{\cdot}26$), is $G_{el} = 32{\cdot}8$ MPa. On the other hand, the maximum shear modulus obtained from cross-hole tests for the same consolidation stress is $G_0 = 102$ MPa (see Fig. 2(c)). Therefore, the relation $G_0/G_{el} \cong 3{\cdot}1$ may represent a largely inevitable effect of sample disturbance.

Tatsuoka & Shibuya (1992) suggest a correction method for the laboratory stress–strain curves based on the assumption that the dependency of the deformation moduli on the strain level is similar in laboratory and field tests, whilst the absolute values of laboratory moduli may often be lower than those from in situ testing due to sample disturbance, ground non-uniformity and other effects. The method consists of correcting the modulus from laboratory tests by the above-mentioned factor G_0/G_{el}.

This procedure has been mostly applied in seismic and other dynamic studies but its use for static problems in which strain levels in the ground are small to moderate does not seem to be unreasonable.

The question of the adequacy of CID triaxial tests for the characterization of the stress–strain behaviour of the saprolitic soil, which exhibits K_0 much smaller than 1·0, has been analysed by comparing the tangent moduli evaluated for the same deviatoric stress in CID and CAD triaxial tests with the same effective consolidation radial stress σ'_{3c} (see Table 4).

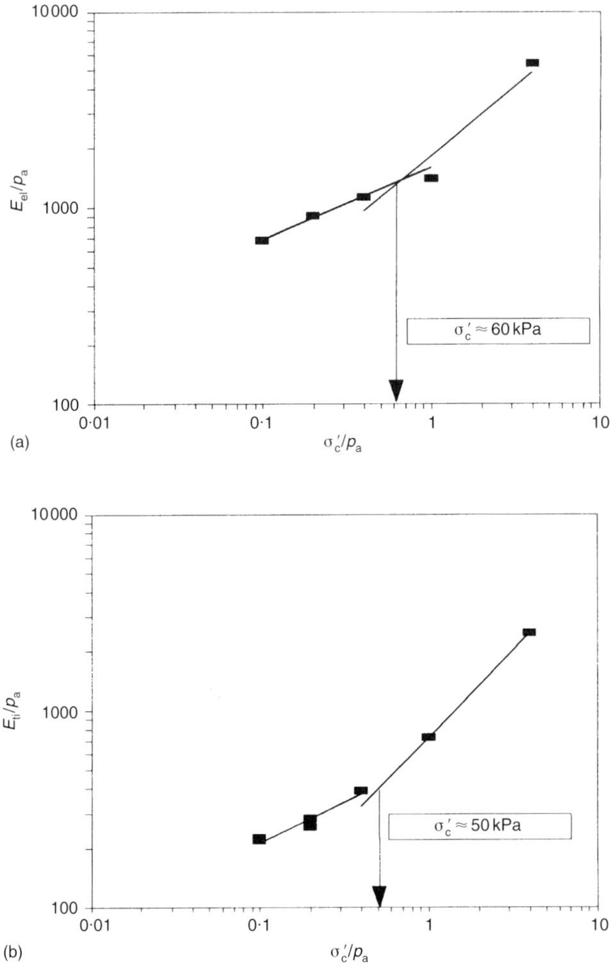

Fig. 12. Young's modulus versus effective consolidation stress, σ'_c, from CID triaxial tests: (a) E_{el}; (b) E_{ti} (p_a = atmospheric pressure)

Table 4. Tangent Young's modulus from CID and CAD ($K_0 = 0{\cdot}5$) triaxial tests, with $\sigma'_{3c} = 10$ kPa

$\sigma_1 - \sigma_3$: kPa	E_t: MPa	
	CID	CAD
10	9·60	10·42
16·3	8·38	7·96
30	7·93	7·57
50	6·68	6·59

The results of these tests, which have all been undertaken with a very low rate of deformation (0·002 mm/min), seem to indicate a fair reproducibility in CID tests of stiffness under initial conditions defined by $K_0 < 1$. Thus, if triaxial specimens of this residual soil are isotropically consolidated inside the volumetric yield surface and if local strain measurement is used, CID tests can serve for stiffness evaluation, thereby avoiding the frequent difficulty of fixing an appropriate value of K_0 to be adopted in a triaxial testing programme.

FOOTING SETTLEMENT EVALUATION BY TRIAXIAL TESTS RESULTS

Linear analysis

Table 5 includes the Young's modulus of the soil for different load stages obtained by back-analysis of the footing load test. The back-analysis assumed a linear elastic layer with constant modulus underlain by a rigid base at 6·0 m depth loaded at the surface by a rigid footing. It should be remembered that the intermediate pressure shown in the table approximately corresponds to the allowable pres-

Table 5. Secant Young's modulus, E_s, evaluated from the footing load test

q_s: kPa	95	190(*)	380
E_s: MPa	23·1	19·4	15·3

(*) Approximately corresponding to the allowable pressure for serviceability limit state design.

sure according to the criterion of Décourt (1992) for residual soils.

Assuming elastic conditions, it is possible to evaluate the position of the so-called settlement centre, z_l (Carrier & Christian, 1973). For circular foundations, its position depends on the variation of the Young's modulus with depth and on the diameter of the footing, B. For the present case it is expected that $z_l \cong B$.

Considering the results of the CID triaxial tests under a consolidation effective stress corresponding to a depth similar to the footing diameter, the tangent Young's modulus for the K_0 shear stress level was found to be $E_{tK0} = 7{\cdot}8$ MPa.

The use of such a modulus in an elastic analysis of the footing load test would lead to a crude overestimation of the observed settlement.

However, if E_{tK0} is multiplied by the above-mentioned factor G_0/G_{el} one obtains 24·3 MPa. This value would provide a good prediction of the settlement for typical working conditions (applied pressure up to about 200 kPa, see Table 5).

This observation suggests that the application of a design methodology which corrects the values of the deformation modulus from triaxial tests by factors referenced to field tests (cross-hole tests or similar) may give good results. This is further explored in the following section.

Simplified non-linear elastic analysis

Taking into account the methodology proposed by Wahls & Gupta (1994), the dependency of the Young's modulus on depth and on the shear stress level in the soil can be evaluated through the results of triaxial tests using the classical equation:

$$E_{tl} = p_a K_l (\sigma'_{ml}/p_a)^{nl} \tag{2}$$

where p_a is the atmospheric pressure and K_l and n_l are dimensionless parameters.

Using the results of several triaxial tests with different consolidation stresses and evaluating, for each test, the tangent modulus, E_{tl}, for distinct shear stress levels, it is possible to compute, for each of these levels, the corresponding values of K_l and n_l. This is reported in Table 6 where q and q_f are the current deviatoric stress and the deviatoric stress at failure, respectively.

A law for the dependency of K_l and n_l on the shear stress level has then to be defined. Since n_l exhibits a modest variation, a constant average value is adopted, $n_l = 0{\cdot}375$. For expressing the dependency of K_l on the stress level, q/q_f (computed from the results of corresponding triaxial tests), the following function is assumed:

$$K_l = C_K \log(q/q_f)_l \tag{3}$$

the evaluation of C_K being outlined on Fig. 13.

Table 6. Values of K_l and n_l as function of the stress level

q/q_f: %	0	10	25	50	75
K_l	520	256	142	88	74
n_l	0·383	0·450	0·488	0·234	0·328

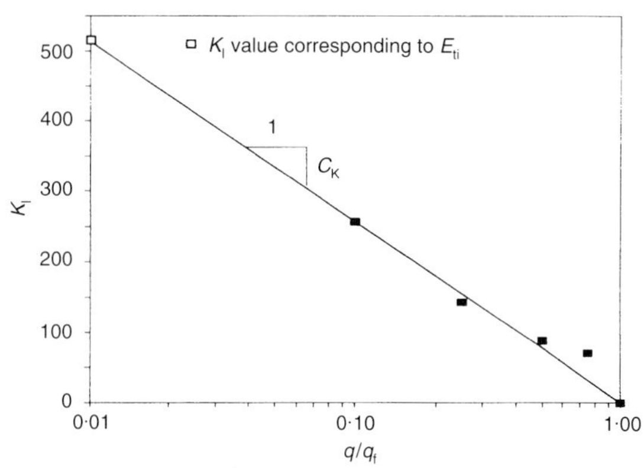

Fig. 13. Variation of K_l with the stress level

These parameters are subsequently used in a simplified non-linear elastic analysis of the footing load test.

The analysis requires the discretization of the ground in several layers ($j = 1, m$) with a thickness of Δh_j and the application of a stepped ($i = 1, n$) loading process. For the applied loading step, Δq_{si}, the vertical and mean effective stress increments in each layer j, $\Delta\sigma_{vij}$ and $\Delta\sigma_{mij}$, are calculated by using, for instance, the I_{vj} and I_{mj} factors proposed by Giroud (1972) for rigid circular foundations:

$$\Delta\sigma_{vij} = I_{vj}\Delta q_{si} \tag{4}$$

$$\Delta\sigma_{mij} = (1+\nu)I_{mj}\Delta q_{si} \tag{5}$$

Stresses at each stage are defined by:

$$\sigma'_{vij} = \sigma'_{v(i-1)j} + \Delta\sigma_{vij} = \sigma'_{v0j} + \Sigma_i\Delta\sigma_{vij} \tag{6}$$

$$\sigma'_{mij} = \sigma'_{m(i-1)j} + \Delta\sigma_{mij} = \sigma'_{m0j} + \Sigma_i\Delta\sigma_{mij} \tag{7}$$

where

$$\sigma'_{m0j} = \sigma'_{v0j}(1+2K_0)/3 \tag{8}$$

The corresponding vertical strain increments are calculated by:

$$\Delta\varepsilon_{vij} = I_{ej}\Delta q_{si}/E_{tij} \tag{9}$$

in which

$$I_{ej} = (1+\nu)[I_{vj} - 3\nu I_{mj}] \tag{10}$$

and E_{tij} represents the tangent Young's modulus. The settlement increment, Δs_i, for each load step can be defined by the equation:

$$\Delta s_i = \Delta q_{si}\Sigma_j[I_{ej}\Delta h_j/E_{tij}] \tag{11}$$

and the total settlement is given by:

$$s_i = \Sigma_i\Delta s_i \tag{12}$$

By means of this incremental process it is possible to select, for each layer, the appropriate deformation modulus. A value of E_{tij} is adopted, for each increment, by considering the average of the values of σ'_m and q/q_f corresponding to the current step (i) and to the previous one ($i-1$). The value of q_f is calculated assuming a Mohr-Coulomb law with the ϕ' and c' parameters deduced from triaxial tests and the value of q given by:

$$q_{ij} = \sigma'_{vij} - \sigma'_{hij} = 1{\cdot}5\ (\sigma'_{vij} - \sigma'_{mij}) \tag{13}$$

The curve obtained by the application of this methodology to the footing load test is compared in Fig. 14(a) with the experimental one. The analytical result is rather poor, as could be easily predicted taking into account the linear back-analysis presented above.

Although the linear elastic assumption for the calculation of the incremental stresses may affect the analytical results (Jardine *et al.*, 1986), it is believed that the main reason for discrepancy is the under-estimation of stiffness by triaxial testing.

Figure 14(b) illustrates a comparison between the observed settlements and the results of the application of the non-linear simplified procedure in which the deformation modulus is now corrected by the factor G_0/G_{el}. The linear diagram obtained by the application of the linear elastic analysis with a constant modulus equal to E_{tK0}. G_0/G_{el} is also shown.

The agreement between the observed results and the analytical curve obtained on the basis of the corrected modulus is very good up to a value twice the allowable pressure. Even the linear elastic analysis gives a good agreement up to loads close to the allowable pressure.

FOOTING SETTLEMENT EVALUATION USING SPT AND CPT TESTING RESULTS

SPT tests

Some empirical methods for the evaluation of settlements of shallow foundations are based on sensitivity analyses of the influencing factors: the applied pressure, q_s, the footing width or diameter, B, and the results from in situ tests like N_{SPT} or $q_{c\text{-}CPT}$.

The expression of Burland & Burbidge (1985), based on over 200 records of settlement of foundations on sands and gravels, is well known:

$$s = \alpha_{BB} q_s B^{0{\cdot}7}/N_{SPT}{}^{1{\cdot}4} \tag{14}$$

with α_{BB} varying between 0·93 and 3·09, the average being 1·71.

The application of this equation to the footing load test provides a settlement two to three times the observed one for working conditions. Rocha Filho (1986) reported ratios of 1·5 to 2·5 after the application of the same equation to Brazilian saprolitic soils.

CPT tests

The semi-empirical method of Schmertmann (1970) assumes a simplified distribution of the influence factor for the vertical strains under the footing, whose values are computed on the basis of a deformation modulus variable with depth in proportion to the CPT cone resistance:

$$E_{si} = \alpha q_{ci} \tag{15}$$

Schmertmann, Hartman & Brown (1978) found $\alpha = 2{\cdot}5$ for axisymetric conditions in granular soils.

It should be remarked that, in spite of using a constant modulus for a given depth, Schmertmann's

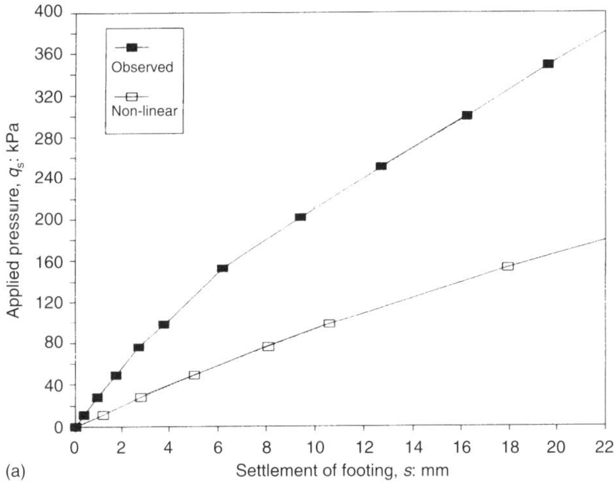

(a)

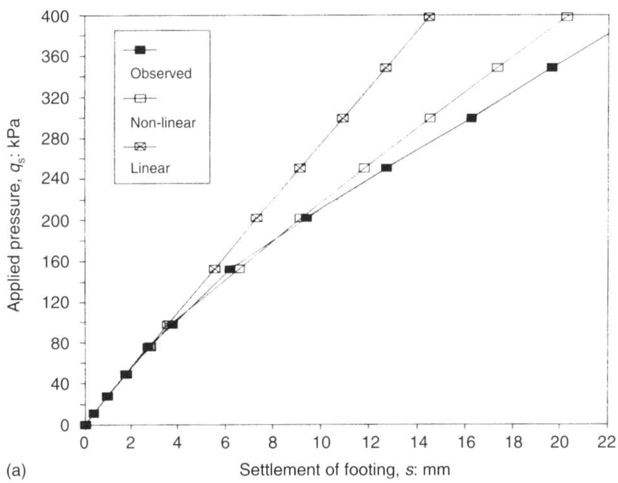

(a)

Fig. 14. Footing load test: comparison of the experimental curve with the one calculated by simplified non-linear analysis: (a) deformation modulus from triaxial tests; (b) deformation modulus corrected by the factor G_0/G_{el}

method provides a non-linear pressure–settlement curve, since the strains depend on the ratio between the incremental pressure and the initial effective vertical stress at foundation level.

This method was applied to the footing load test by adjusting the α parameter in order to fit the observed curve. As shown in Fig. 15, the best agreement was achieved for α values of 4·0 to 4·5. These relatively high values should probably be attributed to the influence of the cemented structure of the saprolitic soil, being situated in the range referred to by Robertson & Campanella (1988) as applicable for aged sands.

CONCLUSIONS

The paper has summarized the work conducted at an experimental site to investigate the behaviour of a saprolitic soil from granite.

The general characterization of the residual soil has been presented in terms of the results of SPT, CPT and CH in situ tests, as well as the grain size distribution, Atterberg limits, physical properties and weathering indices. A carefully-controlled loading test of a circular reinforced concrete footing has been described and the results have been analysed with an emphasis on the pressure–settlement curve for pressures up to the serviceability limit.

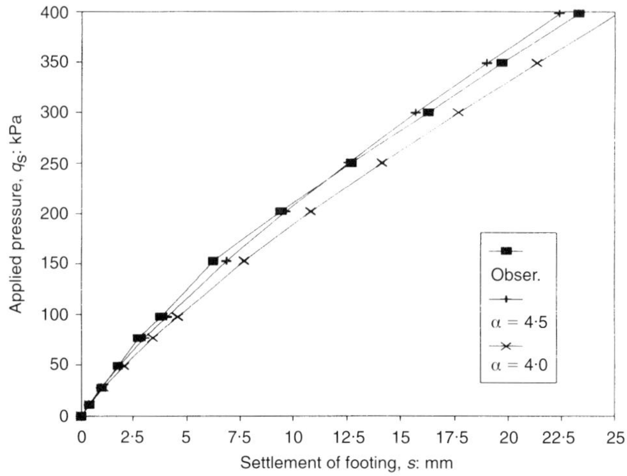

Fig. 15. Back-analysis of the footing load test using CPT results and the semi-empirical method of Schmertmann (1970)

The results presented of oedometer, isotropic compression and triaxial compression tests on high quality block samples using local strain instrumentation can be considered as typical of soils with a cemented structure. In spite of the evidence of such a structure in the laboratory test results and of the careful choice of the parameters expressing the stiffness of the soil (depending on the mean effective stress and on the shear stress level), a simplified non-linear elastic analysis based on these parameters provides a considerable overestimation of the observed settlements.

This discrepancy suggests that, notwithstanding the care taken in the sampling and testing of the soil, the structure of the specimens may have been substantially affected. A substantial part of this damage may have been induced by tube sampling from the blocks.

The evidence of that structural damage is the high value (about 3) reached by the ratio G_0/G_{el} (maximum shear modulus from cross-hole tests and from the reload linear branches of triaxial tests, respectively).

The correction of the stiffness parameters by means of the factor G_0/G_{el} resulted in a very good agreement between calculated and observed results. This methodology, combining results from laboratory and in situ tests in a simple and apparently favourable way, is an approach which has been already explored by other authors.

The interpretation of the footing load test by means of the Burland & Burbidge (1985) equation based on SPT results leads to an overestimation of the observed settlement by a factor of 2–3. If the semi-empirical method of Schmertmann (1970) is applied together with the CPT results the footing settlement is accurately reproduced provided that the value of $\alpha = E/q_c$ is taken in the range of 4·0 to 4·5. Both methods identify this saprolitic soil in the global typology of cemented or granular aged soils.

ACKNOWLEDGEMENTS

The authors are indebted to Dr José Marques and to Dr John Greenfield for their kind suggestions concerning the presentation of the paper.

REFERENCES

Ackerley, S. K., Hellings, J. E. & Jardine, R. J. (1987). Discussion on 'A new device for measuring local axial strains on triaxial specimens' by C. R. Clayton and S. A. Khatrush, 1987, *Géotechnique* **37**, No. 3, 413–417.

ASTM D1194 (1989). Bearing Capacity of Soil for Static Load and Spread Footings. *Annual Book of ASTM Standards* **04.08**, 192–194.

ASTM D2487 (1989). Classification of Soils for Engineering Purposes. *Annual Book of ASTM Standards* **04.08**, 288–297.

Burland, J. B. (1989). Ninth Laurits Bjerrum Memorial Lecture: 'Small is beautiful'. The stiffness of soils at small strains. *Canadian Geotechnical Journal* **26**, 499–516.

Burland, J. B. & Burbidge, M. C. (1985). Settlement of foundations on sand and gravel. *Proc. Institution of Civil Engineers* **78**, 1325–1381.

Carrier III, W. D. & Christian, J. T. (1973). Rigid circular plate resting on a non-homogeneous elastic half-space. *Géotechnique* **23**, No. 1, 67–84.

Collins, R. (1985). Towards characterisation of tropical soil microstructure. *Proc. 1st Int. Conf. on Geome-*

chanics in Tropical Lateritic and Saprolitic Soils, Brasilia **1**, 85–96, and Discussion, **3**, 207–209.

Costa-Filho, L. M., Döbereiner, L., De Campos, T. M. P. & Vargas Jr, E. (1989). Fabric and engineering properties of saprolites and laterites. *General report/Discussion Session 6 – Invited lecture. Proc. 12th ICSMFE, Rio de Janeiro* **4**, 2463–2476.

da Fonseca, A. V. (1996). Geomechanics in Residual Soils from Porto Granite. Criteria for the Design of Shallow Foundations (in Portuguese). *PhD Thesis, University of Porto.*

da Fonseca, A. V., Matos Fernandes, M., Cardoso, A. S. & Martins, J. B. (1994). Portuguese experience on geotechnical characterisation of residual soils from granite. *Proc. 13th Int. Conf. Soil Mec., New Delhi* **1**, 377–380.

Décourt, L. (1992). SPT in non classical material. *Proc. US/Brazil Geotechnical Workshop on Applicability of Classical Soil Mechanics Principles in Structured Soil Belo Horizonte*, 67–100.

GCO (1990). *Guide to Site Investigation*. Reprint. Geotechnical Engineering Office, Hong Kong.

Giroud, J. P. (1972). *Tables pour le calcul des fondations*. Dunod.

GOST 20276 (1985). Soils. Field Methods for Determining Deformation Characteristics. *National Soviet standard, translated from Russian into English in LNEC Technical Information Report ICT – ITG19.*

Hight, D. W., Böese, R., Butcher, A. P., Clayton, C. R. I. & Smith, P. R. (1992). Disturbance of the Bothkennar clay prior to testing. *Géotechnique* **42**, No. 2, 199–217.

Jardine, R. J. (1995). One perspective of the pre-failure deformation characteristics of some geomaterials. *Proc. 1st Int. Conf. on Pre-failure Deformation Characteristics of Geomaterials, Sapporo* **2**, 855–885.

Jardine, R. J., Potts, D. M., Fourie, A. B. & Burland, J. B. (1986). Studies of the influence of non-linear stress-strain characteristics in soil-structure interaction. *Géotechnique* **36**, No. 3, 377–396.

Lacerda, W. A., Sandroni, S. S., Collins, K., Dias, R. D. & Prusza, V. Z. (1985). Compressibility properties of lateritic and saprolitic soils. *Progress Report (1982–1985)*, Com. on Tropical Soils of the ISSMFE. ABMS, Brasília, 85–113.

Lade, P. V. (1988). Model and parameters for the elastic behaviour of soils. *Proc. 6th Int. Conf. Numerical Methods in Geomechanics, Innsbruck* **1**, 359–364.

Lumb, P. (1962). The properties of decomposed granite. *Géotechnique* **12**, No. 3, 226–243.

Pacheco e Silva, F. (1970). A new construction for graphic evaluation of preconsolidation stress of a soil sample (in Portuguese). *Proc. IV COBRA MSFE, Rio de Janeiro* **1**, 225–232.

Robertson, P. K. (1990). Soil classification using the cone penetration test. *Canadian Geotechnical Journal* **27**, 151–158.

Robertson, P. K. & Campanella, R. G. (1988). Guidelines for geotechnical design using CPT and CPTU data. *Report of Federal Highway Administration, Washington, FAWA-PA-87-014-84-24*, **II**.

Rocha Filho, P. (1986). Discussion on 'Settlement of foundations on sand and gravel' by J. B. Burland & M. C. Burbidge, 1985. *Proc. of Institution of Civil Engineers* **79**, 1633–1635.

Rocha Filho, P., Antunes, F. S. & Falcão, M. F. G. (1985). Qualitative influence of the weathering degree upon the mechanical properties of a young gneissic residual soil. *Proc. 1st Int. Conf. on Geomechanics in Tropical Lateritic and Saprolitic Soils, Brasília* **1**, 281–294.

Schmertmann, J. H. (1970). Static cone to compute static settlement over sand. *J. Soil Mech. Fdn Div. Am. Soc. Civ. Engrs* **96**, No. 3, 1011–1043.

Schmertmann, J. H., Hartman, J. P. & Brown, P. R. (1978). Improved strain influence factor diagram. *J. Geot. Eng. Div. Am. Soc. Civ. Engrs* **104**, No. 8, 1131–1135.

Tatsuoka, F. & Shibuya, S. (1992). Deformation characteristics of soils and rocks from field and laboratory tests. *Proc. 9th Asian Regional Conf. Soil Mec., Bangkok* **2**, 101–170.

Tatsuoka, F. & Kohata, Y. (1995). Stiffness of hard soils and soft rocks in engineering applications. *Proc. 1st Int. Conf. on Pre-failure Deformation Characteristics of Geomaterials, Sapporo* **2**, 947–1063.

Vaughan, P. R. (1988). Characterising the mechanical properties of *in situ* residual soils. *Proc. 2nd Int. Conf. on Geomechanics in Tropical Soil, Singapore* **2**, 469–486.

Vaughan, P. R. & Kwan, C. W. (1984). Weathering, structure and *in situ* stress in residual soils. *Géotechnique* **34**, No. 1, 43–59.

Wahls, H. E. & Gupta, M. (1994). Settlement of shallow foundations on sand. *Vertical and Horizontal Deformations of Foundations and Embankments, ASCE Geot. Spec. Pub.* **40**, Vol. 1, 190–260.

Springman, S. M., Balachandran, S. & Jommi, C. (1997). *Géotechnique* **47**, No. 3, 653–663

Modelling pre-failure deformation behaviour of reinforced soil walls

S. M. SPRINGMAN,* S. BALACHANDRAN† and C. JOMMI‡

The dominant form of failure in a wall reinforced by extensible geotextiles is the deformation mode, since it is unlikely that ultimate failure will be critical. A reliable method for displacement prediction is thus the key factor in the analysis and design of reinforced soil structures. Centrifuge modelling was conducted on a series of small-scale, multi-layered reinforced walls, of prototype height 6–15 m, constructed in dry sand and subjected to a vertical surcharge load. Vertical and horizontal displacements of the wall were measured by transducers. Photographic records of markers placed within the fill allowed determination of deformation mechanisms. Two different model reinforcements were used in each of two walls, and all of these were instrumented to measure tension. Numerical back-analyses of these walls provided insight into reinforced wall behaviour, and validated the centrifuge model test data. Significant physical and numerical data are compared and discussed.

KEYWORDS: centrifuge modelling; deformation; geotextiles; numerical modelling and analysis; reinforced soils; retaining walls.

La forme prédominante de rupture dans un mur renforcé de géomatériaux extensibles est le mode de déformation, vu qu'il est peu probable que la ruine soit critique. Il est donc capital d'avoir une méthode fiable de prévision des déplacements pour l'analyse et le dimensionnement des structures en sol renforcé. La modélisation en centrifugeuse a porté sur une série de murs à échelle réduite (correspondant à une hauteur de prototype de 6 à 15 m), renforcés en plusieurs couches, construits sur du sable sec et soumis à une surcharge verticale. Le déplacement vertical et horizontal des murs a été mesuré par des capteurs. Des marqueurs ont été placés dans le remblai pour permettre de photographier une section du modèle afin d'observer les mécanismes de déformation. Deux différents types de renforcements ont été utilisés dans chacun des deux murs. Tous les géomatériaux ont été instrumentés pour la mesure des tensions. Une analyse numérique des murs a permis de mieux comprendre le comportement des murs en sol renforcé et de valider les résultats des essais en centrifuge. Les auteurs comparent et discutent les résultats significatifs des mesures et les données numériques.

INTRODUCTION

Prediction and control of deformation become the dominant factors when considering serviceability limit state design for soil structures with extensible reinforcement. At present, no reliable methods for displacement prediction are available owing to the complicated interaction which controls the reinforced soil behaviour. Pre-failure deformation mechanisms have been investigated using both centrifuge modelling and numerical analysis. While centrifuge models were conceived in order to obtain real information on the behaviour of reinforced walls in the expected working stress range, numerical analyses were performed to validate the model tests.

Centrifuge modelling aims to achieve, at lower cost and under well-controlled experimental conditions, global behaviour which would be consistent with prototype performance. However, prototypes and models will experience different stress paths. In the centrifuge, the model is constructed at 1 *g* and then the gravity level is raised, so that the size of the completed model simply grows; this differs from layer-by-layer construction.

Model walls (Fig. 1), between 6 and 15 m high at prototype scale, reinforced by woven geotextiles with a wrap-round facing, were constructed using dry granular soil. Surcharge loading was applied to the wall and both external and cross-sectional deformations were measured.

Numerical back-analyses were performed to compare the data obtained in the centrifuge with

Manuscript received 16 October 1996; revised manuscript accepted 14 February 1997.
Discussion on this paper closes 1 November 1997.

* Eidgenössiche Technische Hochschule, Zürich; formerly University of Cambridge.
† University of Cambridge.
‡ Politecnico di Milano.

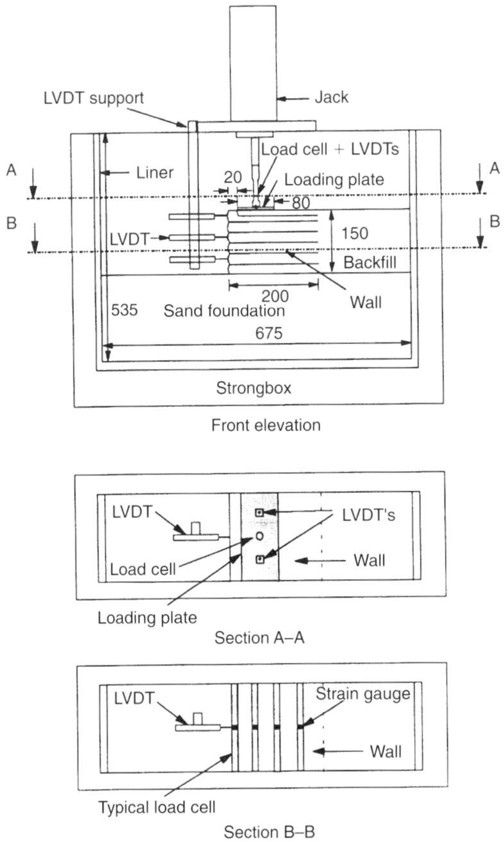

Fig. 1. Centrifuge model test set-up (dimensions in millimetres)

those obtained by numerical representation of the construction procedures and surcharge loading. The analyses were carried out both by modelling the soil and reinforcement separately and, also, by adopting a homogeneous constitutive model for the reinforced soil.

Physical and numerical deformation data of a typical section through walls reinforced with the different geotextiles are compared and discussed.

CENTRIFUGE MODELLING

Small-scale $1/n$ centrifuge models were investigated at enhanced gravity (ng) to provide equivalent full-scale stress conditions, with the multiple advantages of parametric variation, time and cost.

A wrap-round reinforced soil wall (Fig. 1) was constructed in a strongbox at 1g on a founding stratum of fine sand poured to 80% relative density (Balachandran, 1997). Six layers of a woven geotextile especially designed at a reduced scale stiffness, of total length 275 mm and instrumented with full-width strain-gauged tension load cells (Sharma, 1994; Balachandran, 1997), were laid horizontally at 25 mm vertical intervals to a wall height of 150 mm. The 'in-air' load–extension response is shown in Fig. 2 for these 'microweave' (test SB2) and 'bandage' (test SB10) reinforcements (Table 1). Modelling the soil–reinforcement interaction in the centrifuge has been discussed previously by Springman *et al.* (1992).

A greased semi-transparent latex rubber layer was positioned adjacent to the front and rear strongbox boundaries to minimize the effects of side friction. Direct shear tests showed that this reduced the soil–wall adhesion to ~ 5° at the stress levels concerned.

Template blocks were placed to the left of the layer under construction so that the geotextile could be laid on top of the lower sand foundation or layer and supported on the curved boundary. Sand fill, with a median particle size of 0·13 mm, was then poured to 80% relative density, including coloured particles for marking a horizontal layer at each reinforcement level and to give spot markings mid-layer. Cross-sectional displacements of these markers, seen through the latex, were recorded on photographic film during the tests, with subsequent analysis to reveal deformations and strains.

The geotextile–soil interface friction angle was investigated by covering a rigid block with the geotextile, and shearing along the interface with sand in a direct shear box. Peak values ranged from 30 to 35° for 'microweave' (strands 0·45 mm in diameter at 2·38 mm spacing). For the 'bandage' (0·06 mm diameter strands at 1·52 mm spacing), the peak values were 30–32°, at vertical stress levels of 50–300 kPa for both geomaterials. For consistency of method, the sand strength was also tested in this way. The ultimate values were marginally lower (32–34°) than triaxial test data (~ 35°) on the same sand at comparable confining stresses, with dilation angles ranging from 7 to 14°.

A rigid footing was placed on top of the wall with vertical displacement transducers attached, and vertical load was supplied through a load cell by way of a hydraulic jack. Lateral displacements

Table 1. Tensile properties of geomaterials

Test code	Geomaterial	'In-air' ultimate tensile strain: %	'In-air' ultimate tensile strength: kN/m
SB2	Microweave (woven)	16–19	17·5
SB10	Bandage (woven)	3–4	0·8

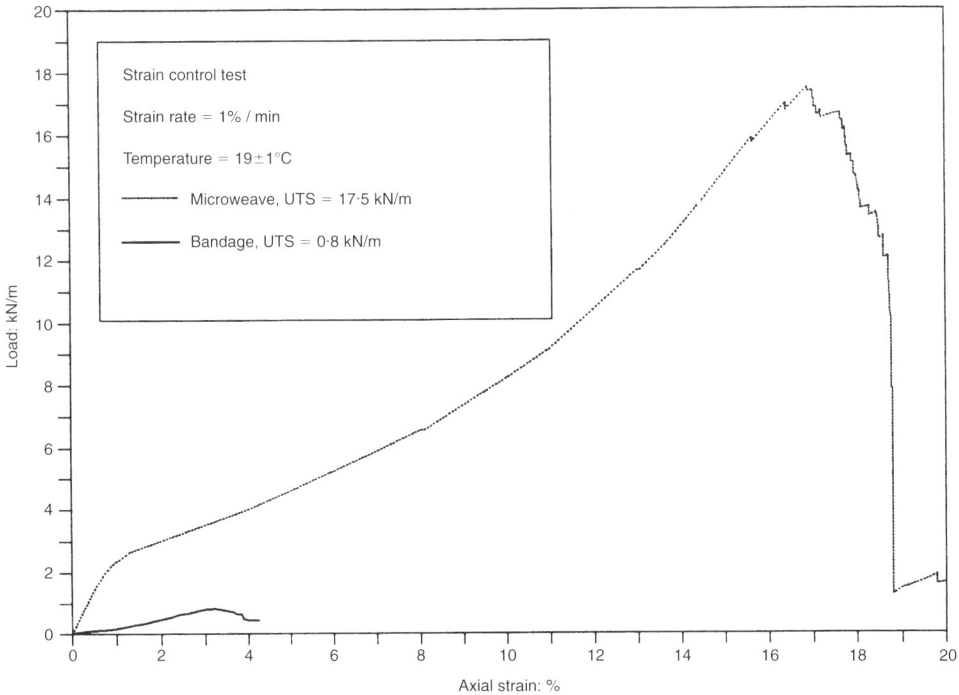

Fig. 2. Tensile test: comparison of microweave and bandage (UTS = ultimate tensile strength)

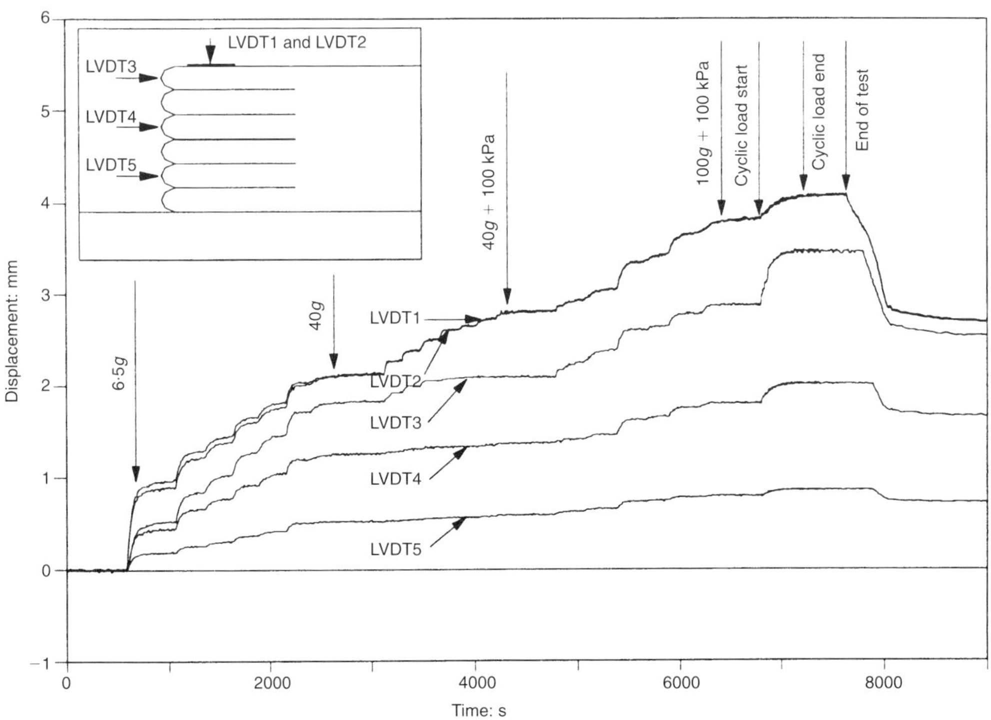

Fig. 3. Horizontal and vertical LVDT measurements in test SB10 (bandage) at model scale

were measured at the wall face (Fig. 3) using linear variable differential transformers (LVDTs).

The gravity level was increased in $5g$ increments up to $40g$, when the wall modelled a 6 m high structure ($40 \times 0{\cdot}15$ m), with 8 m long reinforcement. The self-weight of the smooth footing applied an average 10·5 kPa and this loading was then raised to 100 kPa. Thereafter, the vertical load was kept constant at 100 kPa as the gravity level was increased in $10g$ increments up to $100g$, where the equivalent prototype wall height was 15 m.

Typical externally measured displacement data (at model scale) reflect the stepwise increase in gravity levels (Fig. 3). At $40g$, the scaling factor to the prototype value is 40, and at $100g$ it is 100, so that peak lateral and vertical displacements at the top of the wall, due to this idealized construction method and loading, are 282 and 379 mm respectively for test SB10.

Cross-sectional deformation data is given in Fig. 4 at $40g$ and $100g$, both with 100 kPa surcharge, for the weakest 'bandage' geomaterial (SB10), and in Fig. 5 for 'microweave' (SB2) at $100g$ with 100 kPa. In these figures, the far-field vertical displacement at an equivalent depth, due to sand self-weight increasing with gravity level, has been subtracted from the total displacements to give a clearer picture of the deformation mechanism. Despite the non-linear soil response, the deformation mechanisms are similar at both stress levels and for reinforcements which have different stiffnesses. Deformations are mostly concentrated in a triangular area bounded by a line inclined at $(45 + \psi/2)°$ to the horizontal, in which the net displacement vectors increase in magnitude towards the top corner of the wall. Shear strain ranging from 1·75 to 3%, combined with dilation $5° < \psi < 9°$, can be observed, and these values are lower for the stiffer wall (SB2).

If a triangular simple shear strain γ mechanism (Bransby & Milligan, 1975), with allowance for dilation ψ, is superimposed on the deformations (Figs 4 and 5), then the reinforced soil structure has performed like soil retained behind a rigid wall with wall rotation θ, where $\gamma = 2\theta \sec \psi$. Table 2 summarizes this data. The inclination of displacement vectors with the theoretical zero extension

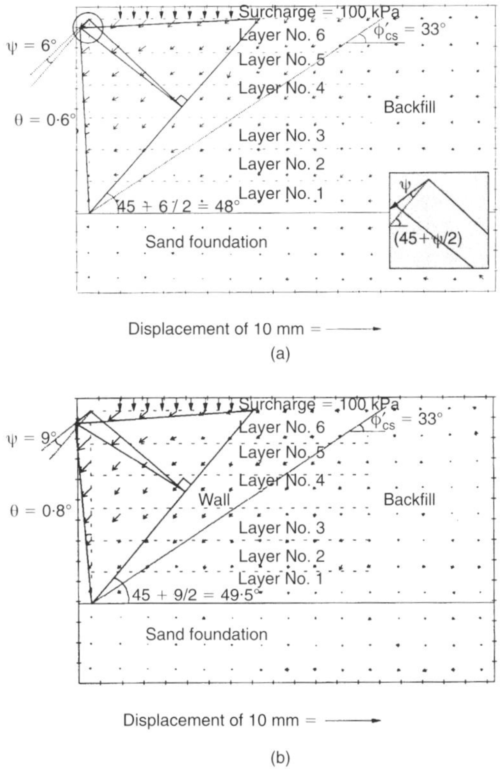

Fig. 4. Displacement mechanism at (a) $40g$ + 100 kPa; (*b*) $100g$ + 100 kPa (test SB10, bandage)

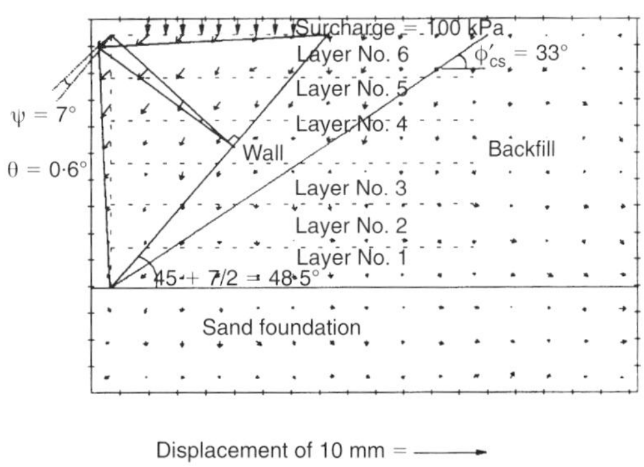

Fig. 5. Displacement mechanism at $40g$ + 100 kPa (test SB2, microweave)

Table 2. Details of simple displacement mechanism

Test code	g level and surcharge load	θ: degrees	ψ: degrees	γ: %
SB2	$40g$ + 100 kPa	0·5	5	1·75
	$100g$ + 100 kPa	0·6	7	2·11
SB10	$40g$ + 100 kPa	0·6	6	2·00
	$100g$ + 100 kPa	0·8	9	2·83

lines coincides almost everywhere with ψ (see insert in Fig. 4(a)).

NUMERICAL MODELLING

The global behaviour of reinforced soil may be analysed by modelling sand and reinforcement separately, but homogeneous constitutive models for the whole reinforced mass have also been proposed (Schweiger & Pande, 1986; Canetta & Nova, 1989; Cividini *et al.* 1994).

Numerical back-analyses of the centrifuge model tests were performed using both homogeneous and non-homogeneous methods with the FLAC finite difference code, to provide further insight into the deformation behaviour of the reinforced soil walls. In the non-homogeneous analysis, two-dimensional sheet elements represented the geomaterials, with a soil–reinforcement interface friction angle of 30°. A homogenization procedure for geomaterial reinforced soil was also used (Jommi *et al.*, 1995). This is based on energy equivalence, and relies on equilibrium and compatibility conditions which must be satisfied at the soil–reinforcement interface.

Sand has been modelled as isotropic linear elastic–perfectly plastic material, with a Mohr–Coulomb shear strength criterion. A non-associated flow rule, defined through the dilatancy angle, has been adopted. The friction angle and dilation angle were derived from the experimental laboratory data presented previously. Poisson's ratio was taken as 0·3. A mean value of the Young's modulus for the wall was calculated using the Ortiz *et al.* (1986) relationship

$$E = E_0(\sigma/\sigma_a)^{\alpha} \quad (1)$$

where σ and σ_a are the vertical stress component and the atmospheric pressure respectively. E_0 and α were taken as 15 MPa and 0·75 respectively, as suggested for a uniform sand; σ was based on the mean stress expected inside the test step (e.g. from 1 g to 40g). This mean value of elastic modulus was increased before applying the surcharge, and then kept constant until the end of the test (Table 3). The stress variation was significant during these test steps, so a stiffness which varied with stress level, stress path and stress history would have been preferable.

As the analyses have been focused on the working, pre-failure range for the reinforced soil, only the nominally perfectly elastic stiffness parameters of the reinforcement, based on the experimental tests (Fig. 2 and Table 3), are relevant.

The geometric characterization of the numerical model reproduced the centrifuge geometry almost exactly. Plane strain analyses were performed on a mesh ~ 1500 elements, assuming smooth lateral mesh boundaries and a rigid, perfectly rough base. After model construction at centrifuge scale, gravity was increased in steps to 40g, followed by

Table 3. Parameters adopted in the numerical analyses

Properties of sand					
	Unit weight: kN/m^3	Bulk modulus: MPa	Shear modulus: MPa	Friction angle: degrees	Dilation angle: degrees
Founding stratum	14·9‡	12·6†–41·2*	5·8†–19·0*	35	5
Reinforced soil and backfill	14·9‡	6·5†–21·0*	3·0†–9·7*	35	10

Properties of geomaterial			
Geomaterial	Thickness: mm	Young's modulus: MPa	Poisson's ratio
Microweave	0·043	7200	0·35
Bandage	0·060	1600	0·35

* Parameter used for analysis after 40g
† Parameter used for analysis up to 40g
‡ Value corrected to allow for vertical sidewall friction

Table 4. Vertical displacement of surcharge load plate: mm (model scale)

Test code	Method used	40g + 10·5 kPa	40g + 100 kPa	100g + 100 kPa
SB2	Centrifuge	2·30	2·82	3·73
	Homogeneous	2·62	3·95	6·30
	Non-homogeneous	2·30	3·16	4·31
SB10	Centrifuge	2·11	2·82	3·79
	Homogeneous	2·68	4·22	6·68
	Non-homogeneous	2·39	3·40	4·80

surcharge loading, and then a second stepped increase in gravity up to 100g to reproduce the centrifuge operational steps.

The experimental vertical displacements of the footing (Table 4) and the horizontal displacements of the facing for cases SB2 and SB10 are compared with the numerical results obtained for both models (Figs 6 and 7). The vertical displacements agree moderately well, as do the mean values of the horizontal displacement of the facing, for the centrifuge and homogeneous analysis. The lateral displacement at the top of the wall is captured better by the non-homogeneous analysis, despite higher net magnitudes in the lower half due to larger lateral displacements at the base. The discrepancies between calculated and experimentally observed total displacements may be largely attributed to the mean stiffness assumed for the wall. To model increasing stress levels in the centrifuge, appropriate parameter variation in the pre-failure range is necessary to reproduce global displacements of the reinforced soil mass.

To validate the numerical analyses effectively, the global experimental deformation pattern should be obtained. Both the experimental data and the non-homogeneous results show almost linear variation of horizontal displacement with height, with the maximum value at the top of the wall. In the homogeneous analysis, the wall bulges at mid-

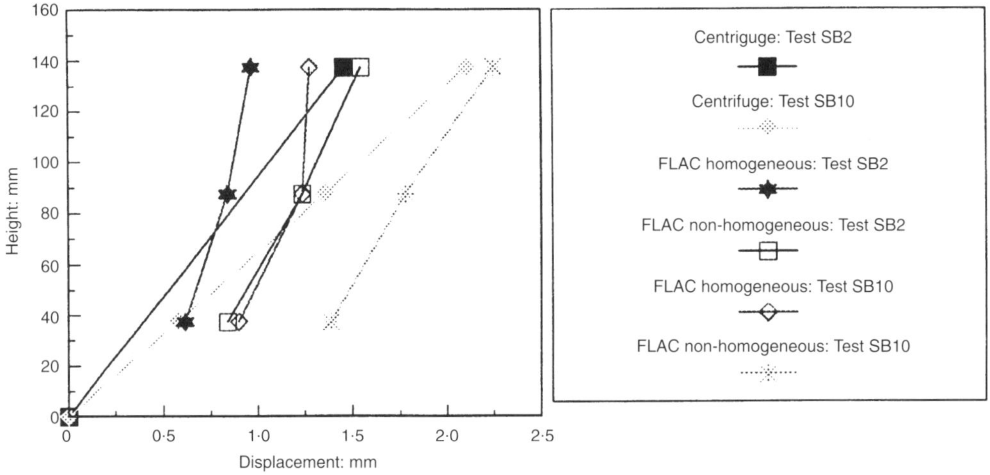

Fig. 6. Comparison of numerical and centrifuge data for horizontal wall displacement (tests SB2, microweave, and SB10, bandage) at 40g + 100 kPa

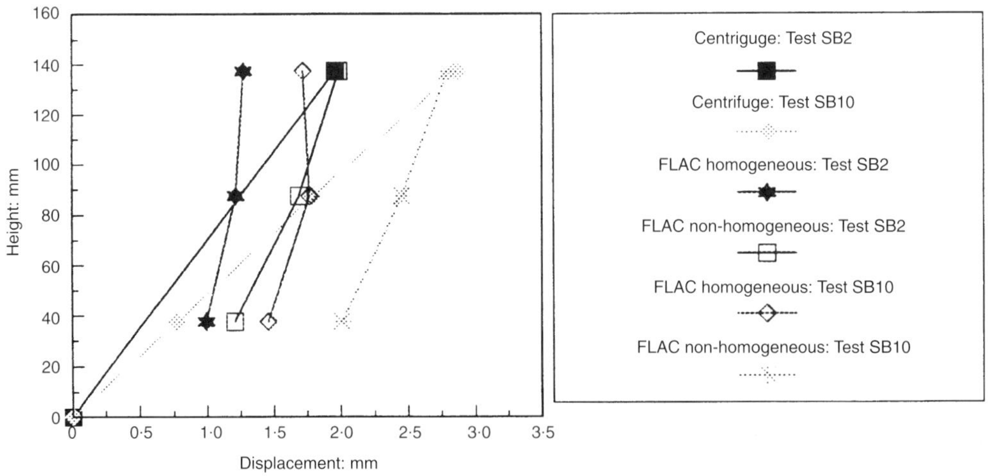

Fig. 7. Comparison of numerical and centrifuge data for horizontal wall displacement (tests SB2, microweave, and SB10, bandage) at 100g + 100 kPa

height, thus suggesting a different deformation mechanism. Vector plots of displacements at $100g + 100$ kPa for both analyses are shown in Figs 8–11, with the vertical far-field displacements subtracted as for the centrifuge data (Figs 4–5). The net displacement distributions obtained from the non-homogeneous analysis for both SB2 and SB10 (Figs 9 and 11) reproduce the deformation mechanism observed experimentally (Figs 4(b) and 5): a clear demarcation is noted between a dilating wedge defined by a line at $(45 + \psi/2)°$ to the horizontal with net displacement inclined at ψ to the theoretical zero extension lines, and the rest of the fill which experiences little, if any, shear

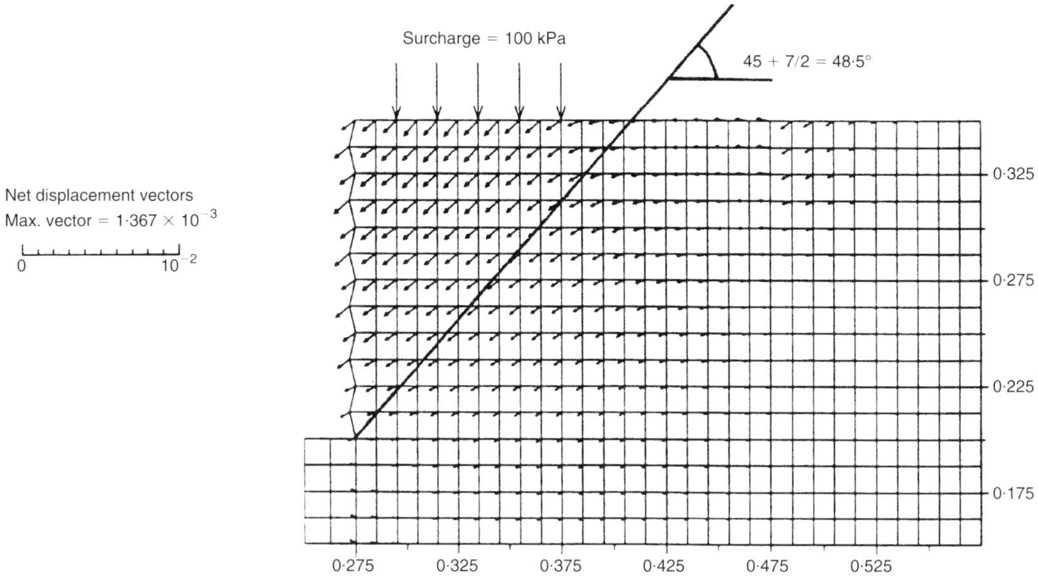

Fig. 8. Displacement pattern of the model wall predicted by homogeneous analysis (test SB2, microweave) at $100g + 100$ kPa

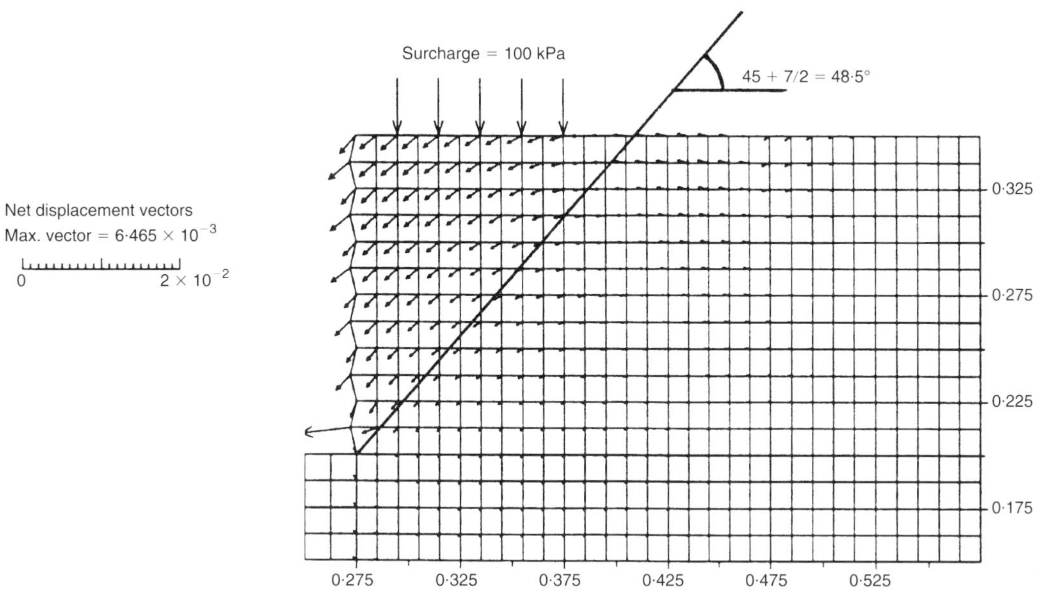

Fig. 9. Displacement pattern of the model wall predicted by non-homogeneous analysis (test SB2, microweave) at $100g + 100$ kPa

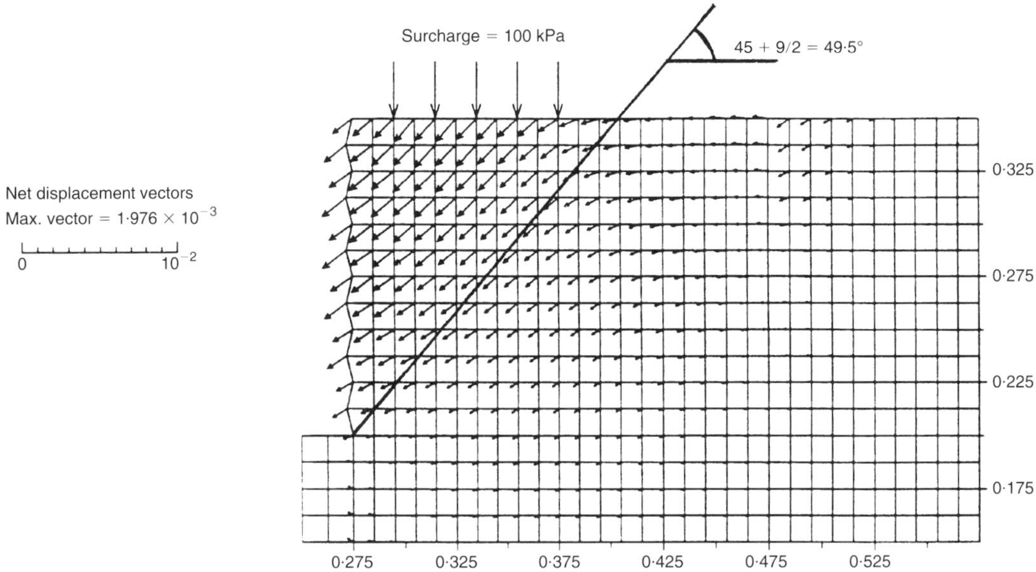

Fig. 10. Displacement pattern of the model wall predicted by homogeneous analysis (test SB10, bandage) at 100*g* + 100 kPa

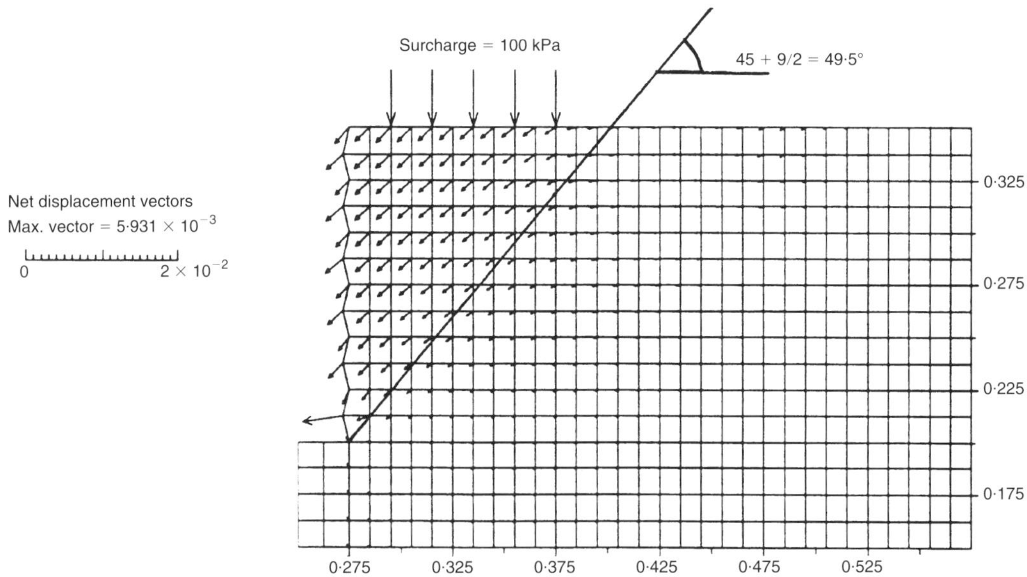

Fig. 11. Displacement pattern of the model wall predicted by non-homogeneous analysis (test SB10, bandage) at 100*g* + 100 kPa

strain. The shear strains are almost constant and approximately 3%; this compares quite well with experimental values (Table 2). The homogeneous displacements (Figs 8 and 10) are distributed more extensively, with ψ varying from 3° to 10° on top of the wall to the front, confirming that a different deformation mechanism is produced. Since the reinforcement is 'smeared' into the entire volume, it offers lateral confinement everywhere, even under zero external horizontal stress. This is unrealistic for widely spaced reinforced soil walls.

The *bulging* deformation mechanism is often observed following step-by-step construction of a prototype wall with a temporary propping of the single layer under compaction. The experimental wall deformation profile reported herein, referred to schematically by Karpurapu & Bathurst (1995), was also reported in the centrifuge at $12g$ by Ho & Rowe (1994). It is likely to be representative of walls constructed by propping over the full height.

The vertical stresses predicted at $100g + 100$ kPa by the two analyses are compared for SB2 (Figs 12 and 13). These are almost exactly replicated for SB10. The vertical stresses due to self-weight and surcharge load spread to the wall facing in the homogeneous model, causing significant bulging at the wall face. For the non-homogeneous case, stresses are mainly transferred behind the soil retained within the wrap-round construction, where most of the tension develops. This is validated by experimental data. Tension distributions from both numerical and experimental results for test SB2 at $100g + 100$ kPa (Fig. 14) are compared. Despite slightly higher values at the face of the wall, tensions obtained in the non-homogeneous analysis reproduce the experimental distribution, while tensions deduced from the homo- geneous analysis are much greater at the wall face (excepting the top layer) and are more widely distributed in the lower four layers.

CONCLUSIONS

The centrifuge data showed a continuous form of deformation mechanism throughout the loading pattern. Constant shear strain and dilation were deduced, similar to the case of externally supported retaining walls.

Numerical analyses, performed with simple constitutive models with mean values for material parameters, lead to better understanding of the deformation mechanism of reinforced soil walls, and highlight the role played by stress history in soil–reinforcement interaction. With equivalent model geometry and construction procedures, and separate discretization of soil and reinforcement, the experimental deformation mechanism was well reproduced. A homogeneous model which smeared the reinforcement into the whole soil mass showed noticeable differences in global deformation behaviour (in spite of comparable values for total displacements) and did not appear to suit such widely spaced reinforced soil walls.

When numerical models validate the physical data effectively, they become valuable for preliminary design. Nevertheless, if an improved prediction of real prototype displacement is needed, advanced constitutive models which represent variation in soil stiffness and strength with stress path and stress history are desirable.

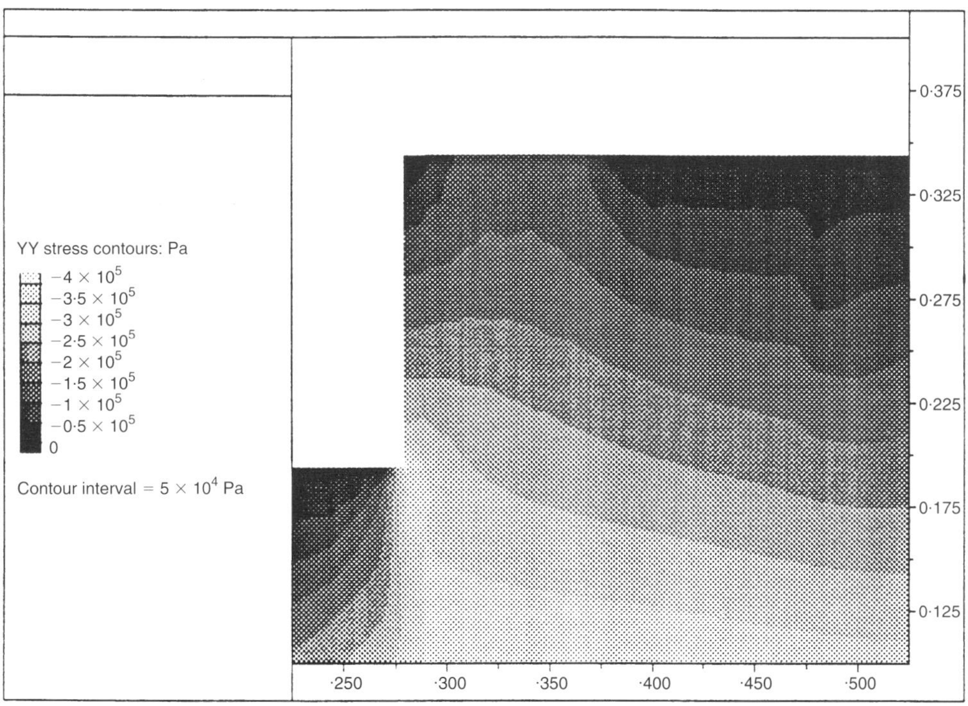

Fig. 12. Vertical stress predicted by homogeneous analysis (test SB2, microweave) at $100g + 100$ kPa

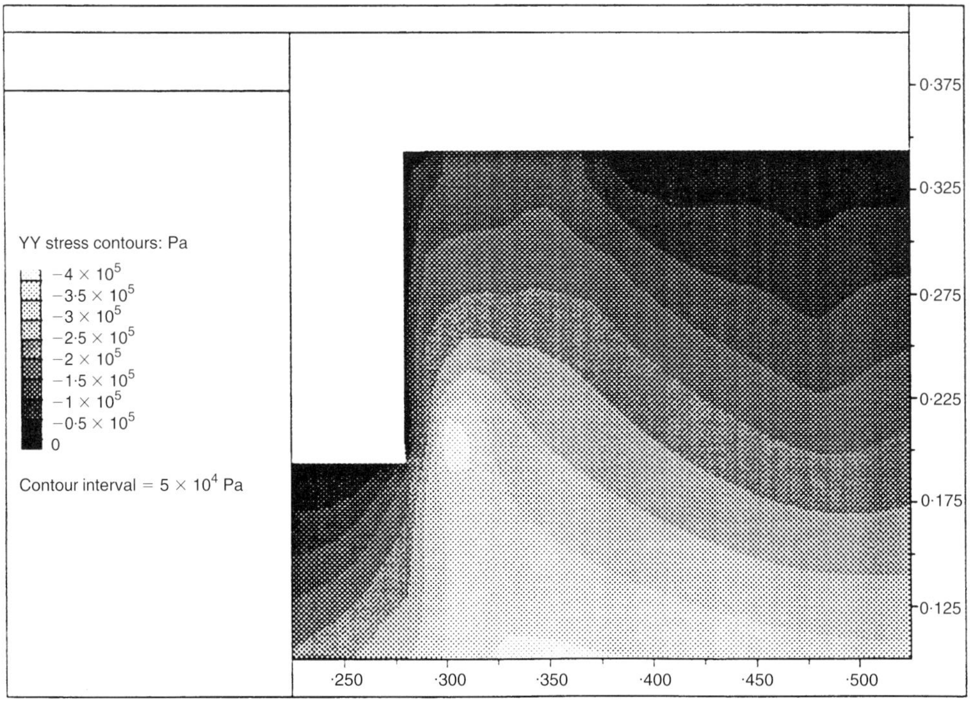

Fig. 13. Vertical stress predicted by non-homogeneous analysis (test SB2, microweave) at 100*g* + 100 kPa

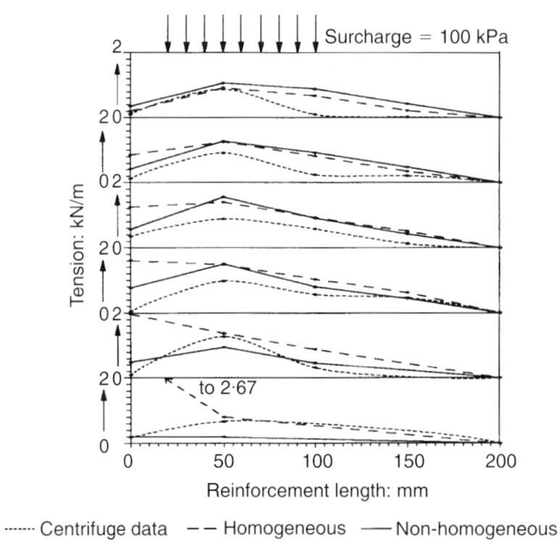

Fig. 14. Tension distribution at 100*g* + 100 kPa (test SB2, microweave)

ACKNOWLEDGEMENTS

The authors wish to thank the technicians and academic staff of the Cambridge Geotechnical Centrifuge Centre for their invaluable help. They also wish to acknowledge financial support from King's College, Cambridge, and EU ALERT Geomaterials.

REFERENCES

Balachandran, S. (1997). *Modelling of geosynthetic reinforced soil walls*. PhD thesis, University of Cambridge.

Bransby, P. L. & Milligan, G. W. E. (1975). Soil deformation near cantilever sheet pile walls. *Géotechnique* **25**, No. 2, 175–195.

Canetta, G. & Nova, R. (1989). A numerical method for the analysis of ground improved by columnar inclusion. *Computers and Geotechnics* **7**, 99–114.

Cividini, A., Donelli, M. & Sterpi, D. (1994). On the mechanical behaviour of sand reinforced with geotextiles. In *Numerical methods in geotechnical engineering* (ed. I. M. Smith), pp. 299–304. Rotterdam: Balkema.

Ho, S. K. & Rowe, K. R. (1994). Predicted behavior of two centrifugal model soil walls. *Proc. ASCE* **120**, No. 10, 1845–1873.

Jommi, C., Nova, R. & Gomis, F. (1995). Numerical analysis of reinforced earth walls via a homogenization method. In *Numerical methods in geomechanics—NUMOG V* (eds Pande and Pietruszczak) pp. 231–236. Rotterdam: Balkema.

Karpurapu, R. & Bathurst, R. J. (1995). Behaviour of

geosynthetic reinforced soil retaining walls using the finite element method. *Computers & Geotechnics* **17**, 279–299.

Ortiz, J. M. R., Serra, J. & Oteo, C. (1986). *Curso aplicado de cimentaciones*, 3rd edn. Madrid: Colegio Oficial de Arquitectos de Madrid.

Schweiger, H. F. & Pande, G. N. (1986). Numerical analysis of stone column supported foundations. *Computers and Geotechnics* **2**, 347–372.

Sharma, J. (1994). *Behaviour of reinforced embankment on soft clay.* PhD thesis, University of Cambridge.

Springman, S. M., Bolton, M. D., Sharma, J. & Balachandran, S. (1992). Modelling and instrumentation of a geotextile in the geotechnical centrifuge. *Proc. Int. Symp. on Earth Reinforcement Practice, Kyushu* 1, pp. 167–172.

Hird, C. C. & Pierpoint, N. D. (1997). *Géotechnique* **47**, No. 3, 665–691

Stiffness determination and deformation analysis for a trial excavation in Oxford Clay

C. C. HIRD* and N. D. PIERPOINT†

Block samples of the heavily overconsolidated Oxford Clay were retrieved from a 10 m deep, instrumented, trial excavation at Elstow, Hertfordshire, and from a nearby brick pit. Triaxial stress path tests were conducted on 100 mm diameter specimens to determine appropriate stiffness parameters for numerical modelling of the excavation. Each test followed a succession of constant p' (mean effective normal stress) and constant q (deviator stress) paths in the extension region. Small axial and radial strains were measured locally using a proximity transducer system, permitting stiffness parameters for the coupled deviatoric and volumetric behaviour to be derived and plotted against changes of strain energy for modelling purposes. The maximum shear stiffness, G_0, was also determined using bender elements. During the multi-stage tests, stiffness behaviour at small strains appeared to be governed by the recent strain history and recent stress history effects were not as prominent as those reported in the literature for other clay soils. Finite element modelling of the excavation was carried out, employing a non-linear cross-anisotropic elastic model in which stress path dependent stiffnesses were specified. Predicted lateral displacements were modestly overestimated by comparison with field measurements, but vertical displacements were generally underpredicted and to a greater degree.

KEYWORDS: clays; excavation; ground movements; laboratory tests; numerical modelling and analysis; stiffness.

Des blocs d'échantillonnage d'argile d'Oxford fortement préconsolidée ont été prélevés dans une fouille de recherche instrumentée, profonde de 10 m, à Elstow dans le Hertfordshire, ainsi que dans une argilière voisine. On a effectué des essais de parcours de contraintes triaxiales sur des échantillons de 100 mm de diamètre pour déterminer les paramètres de rigidité convenant à une modélisation numérique de la fouille. Chaque essai a suivi une succession de parcours de contrainte constante p' (contrainte intergranulaire normale moyenne) et de contrainte constante q (contrainte déviatrice) dans la zone d'allongement. On a mesuré localement les petites contraintes axiales et radiales à l'aide d'un réseau de capteurs de proximité en vue de calculer les paramètres de rigidité pour le comportement dévuatorique et volumétrique jumelé et de les mettre en corrélation avec les variations d'énergie de contrainte à des fins de modélisation. In a également déterminé la rigidité maximale au cisaillement G_0 à l'aide d'éléments de flexion. Dans les essais à étages multiples, la rigidité aux petites contraintes a semblé être dictée par les contraintes subies récemment, et les effets des tensions récentes ne semblent pas aussi importants que ceux dont font état d'autres travaux portant sur d'autres sols argileux. On a effectué une modélisation de la fouille par éléments finis en utilisant un modèle élastique trans-anisotrope non linéaire où l'on a précisé des rigidités tributaires des parcours de contrainte. Les déplacements latéraux prédits ont été légèrement supérieurs aux déplacements mesurés sur le terrain, tandis que les déplacements verticaux ont été généralement sous-estimés et dans de plus grandes proportions.

Manuscript received 23 October 1996; revised manuscript accepted 11 March 1997.
Discussion on this paper closes 1 November 1997.
* University of Sheffield
† Mott MacDonald, Croydon

INTRODUCTION

In the mid-1980s ground investigations were conducted at a number of British sites, with a view to establishing their suitability for the construction of underground repositories for low level nuclear waste. In assessing suitability, an important requirement was the prediction of ground movements due to excavation, which were to be kept within tolerable limits.

One of the sites was located on the Oxford Clay at Elstow, about 5 km south of Bedford in Hertfordshire. At this site the ground investigation included the construction and monitoring of a full scale trial excavation, with the dimensions indicated in Fig. 1. The scale of the excavation was dictated by the need to achieve stress changes in the ground comparable with those produced by the construction of a prototype nuclear waste repository. An extensive array of instrumentation, shown in plan in Fig. 1(a), was installed to monitor horizontal movement, vertical movement and pore water pressure in the surrounding ground. Construction took place in the spring of 1987, as indicated in Fig. 1(b). However, as the process of excavation was nearing completion, the investigation was curtailed and, although monitoring continued for a limited period, most of the field displacement data obtained from the excavation were archived without being analysed or interpreted.

Subsequently, but several years later, a laboratory research programme was undertaken with the aim of investigating the stiffness of the Oxford

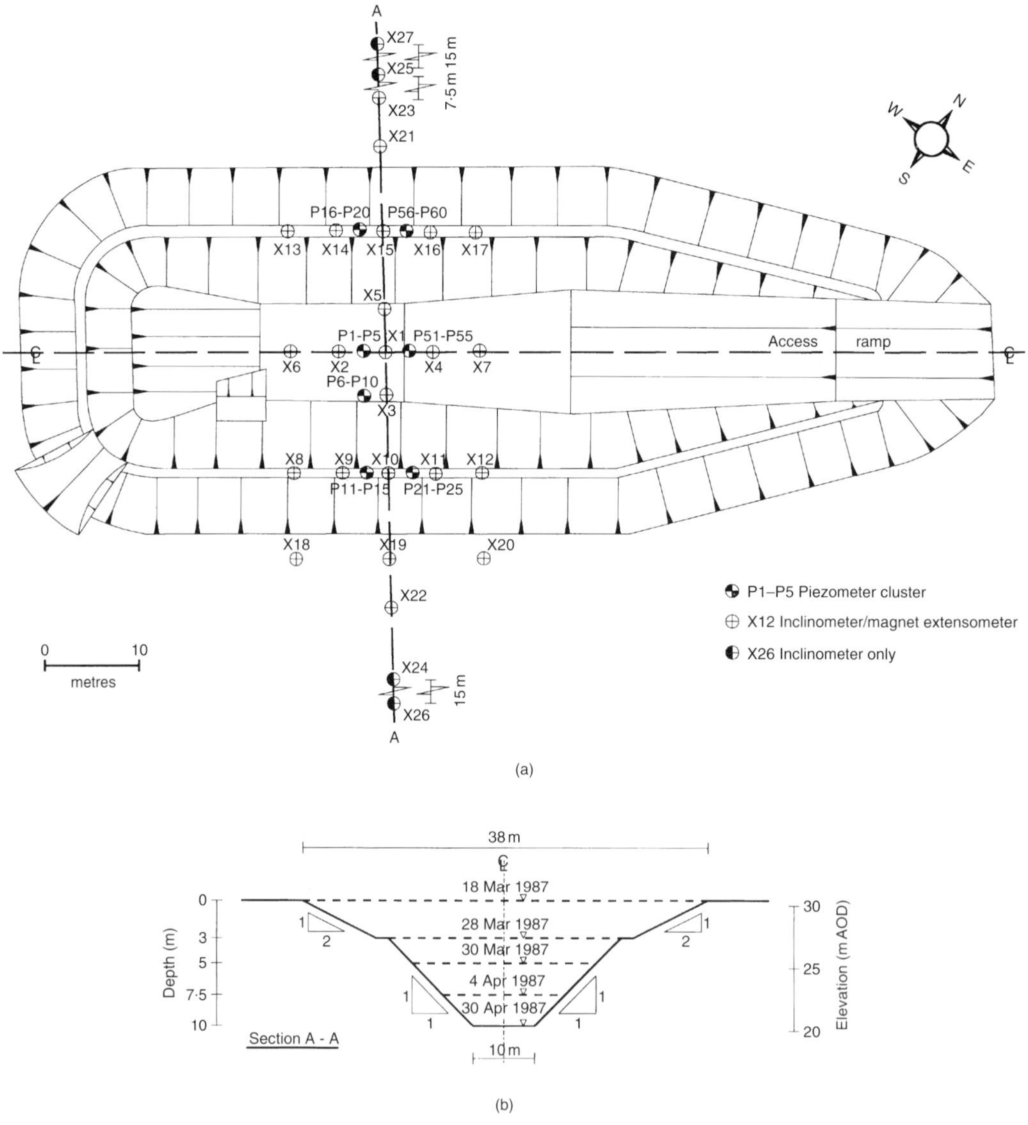

Fig. 1. The Elstow trial excavation: (a) plan; (b) transverse section

Clay and providing parameters for use in numerical models of the excavation. A finite element prediction was then made of the excavation behaviour and compared with the field observations. This study forms the subject of the present paper.

The investigation of Oxford Clay stiffness began in 1992 and was influenced by developments in the understanding of soil behaviour gained over the preceding decade. It was therefore designed to provide information on the non-linearity of stiffness at small strains ($<0{\cdot}1\%$), taking account of the in situ stress state and recent stress history of the soil.

Multi-stage stress path tests were carried out in the triaxial cell on 100 mm diameter specimens trimmed from larger block samples. These tests were intended to permit the maximum amount of information to be gained from a limited number of available samples, and also to provide complete and consistent sets of parameters without the complication of inter-specimen variation. Because the local instrumentation used to measure the strains in the tests could not accurately record very small strains ($<0{\cdot}001\%$), in some tests simultaneous dynamic stiffness measurements at such strain levels were made using bender elements.

Since it was desired to make a Class C prediction (Lambe, 1973) of excavation behaviour, the numerical modelling had to be completed before the field data were taken out of the archive and, in the time available, only a relatively simple model could be employed. Nevertheless, this incorporated both the non-linearity and anisotropy observed in the laboratory tests. Only the single prediction, and one later variation of this analysis, will presently be discussed, although during back-analysis several parametric studies were actually conducted (Pierpoint, 1996).

GROUND CONDITIONS

Soil profile

The stratigraphy at the site of the excavation is summarized in Fig. 2. Beneath an essentially level ground surface, and about 1 m of superficial material (Head), the heavily overconsolidated Oxford Clay extends to a depth of about 14 m. Its upper surface is weathered to a depth of about 3 m. Below the Oxford Clay, the Kellaways beds form a gradual and rather variable transition to generally harder strata.

The geological characteristics of the Oxford Clay are well documented (Callomon, 1968) and investigations of geotechnical properties have been reported by Parry (1972), Jackson & Fookes (1974), Burland *et al.* (1977) and others. In Table 1 the index property information available from the literature is compared with data obtained in a routine manner from the present site. None of the quantities listed in Table 1 showed a significant variation with depth at Elstow, although the plastic and liquid limits tended to reduce slightly in the weathered zone. The average liquidity index decreased with depth from about 0·3 at the top of the weathered zone to about −0·1 at the base of the Oxford Clay.

In situ stresses and stress history

According to Jackson & Fookes (1974), the overconsolidation of the Oxford Clay has been brought about by the removal of approximately 500 m of overburden, although a series of high pressure oedometer tests by Pierpoint (1996) suggested that the overburden thickness at Elstow might have been even greater or that ageing processes might have had an equivalent effect. The resulting profile of overconsolidation ratio (OCR) is shown alongside the profiles of current total vertical stress and pore water pressure, based on field measurements at the site, in Fig. 3(a). A hydrostatic groundwater regime has been assumed, despite evidence of slight underdrainage in the Kellaways beds.

Knowledge of the current lateral stresses is sparse but, given the high degree of overconsolidation, it is likely that much of the Oxford Clay at Elstow is close to a state of passive failure. Limits to the lateral earth pressure coefficient, K_0, based on a range of frictional strengths are shown in Fig. 3(b). Peak angles of shearing resistance ϕ', measured in triaxial compression, varied between 21° and 32° with an average of 26°; average values reported by Parry (1972), Jackson & Fookes (1974) and Burland *et al.* (1977) are 2–4° higher. Also shown in Fig. 3(b) are the estimated reductions of K_0 from the passive limits due to a modest reloading of the Oxford Clay, arising from the deposition of the thin layer of Head (see Fig. 2). These were estimated following the method of Burland *et al.* (1979) and are significant. However, the reloading could have been negated by a 2 m rise in the groundwater level and, in the absence of any information regarding water table fluctuations, it is unknown to what extent reloading has occurred.

EXPERIMENTAL PROCEDURES

Block sampling

During the construction of the excavation, blocks of the Oxford Clay had been retrieved in order to permit laboratory tests to be carried out on relatively undisturbed soil. Cubic blocks, with a side length of 300 mm, were cut using hand tools or, in the stiffer material, a chain saw. The samples were very well sealed in aluminium foil and waxed cheesecloth, and were packed inside a wooden box

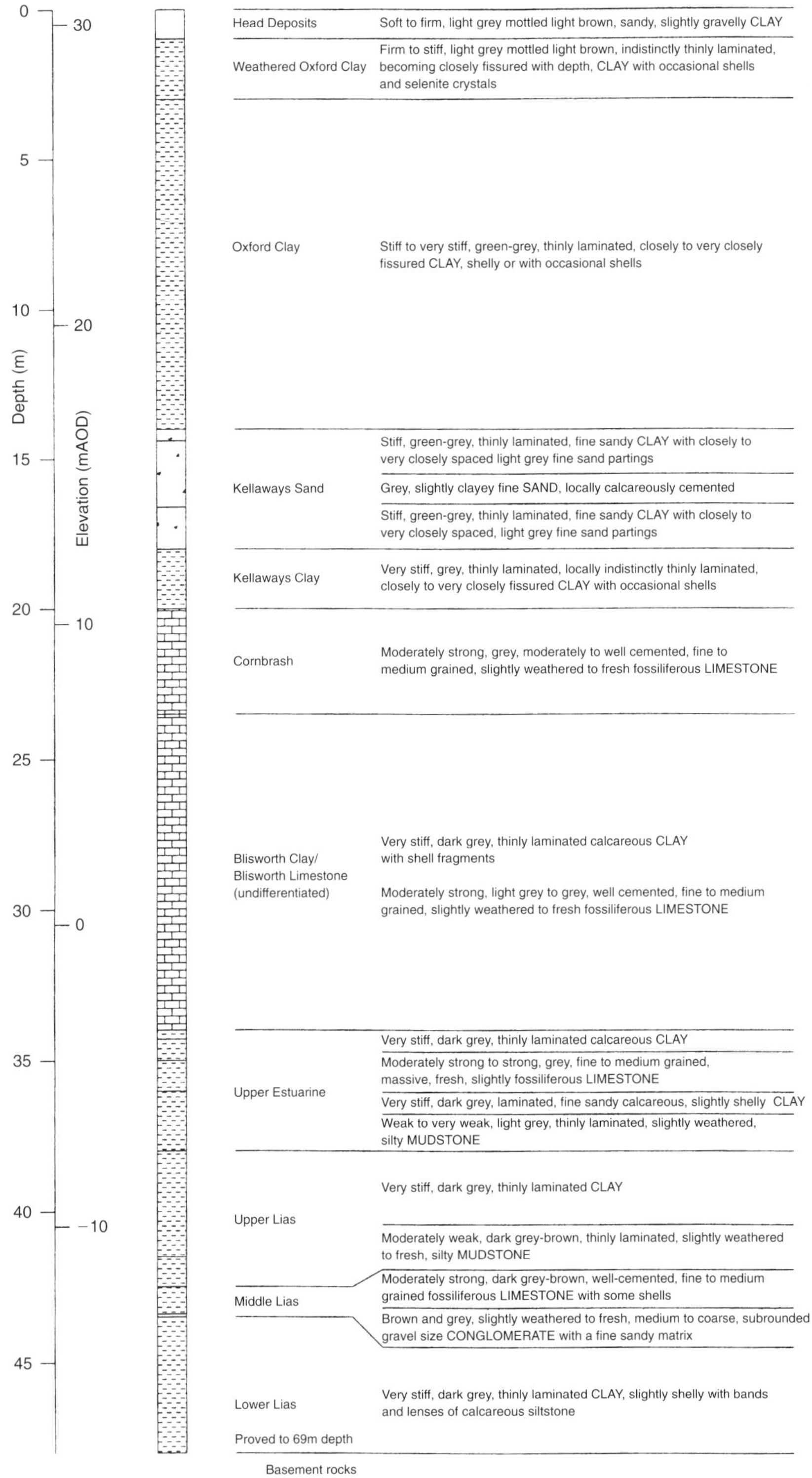

Fig. 2. Stratigraphic profile at Elstow

Table 1. Index properties of Oxford Clay

Source	Liquid limit (%)	Plastic limit (%)	Clay fraction (%)	Activity
Elstow site investigation	52–70	23–33	30–35	0·69–0·91
Parry (1972)	65–75	20–30		
Jackson (1973)	52–69		47–66	0·40–0·50
Jackson & Fookes (1974)	62–70	25 35	50–70	0·50–0·75
Burland *et al.* (1977)			> 55	
Cripps & Taylor (1981)	45–75		30–70	
Crabb & Atkinson (1991)	60	32		

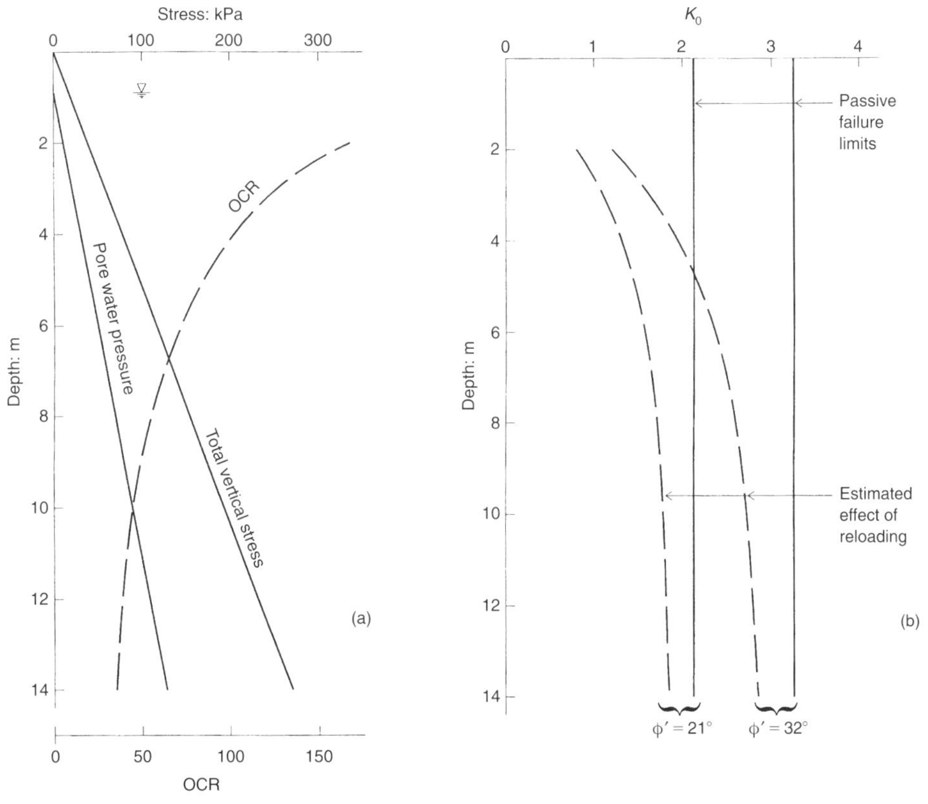

Fig. 3. Profiles of (a) vertical stress, pore pressure and OCR, and (b) K_0 at Elstow

using polyurethane foam. Ten of these samples had remained in storage until the start of the present investigation, although it was never intended that such a long period should elapse before laboratory testing. When the samples were opened for inspection, most showed obvious signs of deterioration. However, three samples appeared to be in good condition and, despite some anxiety over the possible effects of storage, these were used subsequently to prepare triaxial specimens.

To provide further material for testing, in June 1993 an additional two block samples were cut from the side of Kempston brick pit, about 500 m away from the site of the Elstow excavation. These measured about 400 mm by 200 mm by 400 mm deep and were obtained using hand tools only. It had to be accepted that the effects of stress relief for these samples might be different from those for the Elstow samples and also that, at any given elevation, a difference of stratigraphic level of about 7 m was likely between the two sampling locations due to the regional dip (about 1° on average). The Elstow and Kempston materials used in the investigation happened to come from about the same elevation, but no significant lithological differences were observed. Also, the plastic and liquid limits for the Kempston material (32 and 67% respectively) were almost identical to those of the Elstow material (33 and 69%). These observations suggest that differences of material type were minimal.

The 100 mm diameter by 200 mm long triaxial specimens were trimmed either by hand on a soil lathe (early tests) or, much more rapidly and accurately, by using a vertically-mounted band saw (later tests). With the band saw, no additional mechanical disturbance of the specimen was observed and the problems of drying and opening of fissures, which were evident during hand trimming, were minimized.

Triaxial testing

The specimens were tested in a hydraulic stress path apparatus equipped with proximity transducers, to measure both axial and radial strains in the central part of the specimen, and with submersible LVDTs, to measure axial strain between the specimen end caps. The strain instrumentation has been described by Hird & Hajj (1995).

The sequence of stress paths applied to the specimen is shown schematically in Fig. 4(a) in terms of the deviator stress, q, and the mean effective normal stress, p'. Commencing from an isotropic stress state (I), established by applying an initial cell pressure under undrained conditions and then saturating the specimen with the aid of back pressure, the specimen was first brought by way of an approach route IKCO, or IJOCO, to an assumed in situ stress state (O) lying just above the passive failure limit. The final section of the approach path (CO), which had a length of 80–100 kPa, was intended to simulate, approximately, the recent in situ stress history of the soil, i.e. during swelling. In view of the uncertainty regarding reloading of the soil in situ, it was decided not to simulate any such process. Although reloading would generally be potentially significant for the determination of stiffness parameters, in the present case the reloading stress path would have been relatively short (< 20 kPa) so that the influence of the unloading would probably have remained dominant. At the assumed in situ stress state $q/p' = -0{\cdot}75$, corresponding to $K_0 = 2{\cdot}5$; this was as close to the passive limit as it was practicable to go without running the risk of immediately failing the specimen. A range of such states was employed, corresponding to depths of 5–10 m, in order to gain an insight into the variation of stiffness with depth. However, the 'depth' employed in the test did not necessarily equal the depth of the sample from which the specimen was drawn.

From the point O established above, a series of stress probes was carried out along constant p' and constant q directions. Used in combination (Yudhbir & Khan, 1980; Atkinson & Richardson, 1985) these permitted the determination of all four stiffnesses in the constitutive equation

$$\begin{bmatrix} \delta\varepsilon_{\text{v}} \\ \delta\varepsilon_{\text{s}} \end{bmatrix} = \begin{bmatrix} \dfrac{1}{K'} & \dfrac{1}{J'_{\text{qv}}} \\ \dfrac{1}{J'_{\text{ps}}} & \dfrac{1}{3G'} \end{bmatrix} \begin{bmatrix} \delta p' \\ \delta q \end{bmatrix} \tag{1}$$

where G' is the deviatoric stiffness, K' is the bulk stiffness, J'_{ps} and J'_{qv} are the cross-coupling stiffnesses, and ε_{v} and ε_{s} are the volumetric and shear strain parameters (defined so as to correspond to p' and q). As indicated in Fig. 4(a), four probes (OA, OB, OC and OD) would be needed to fully

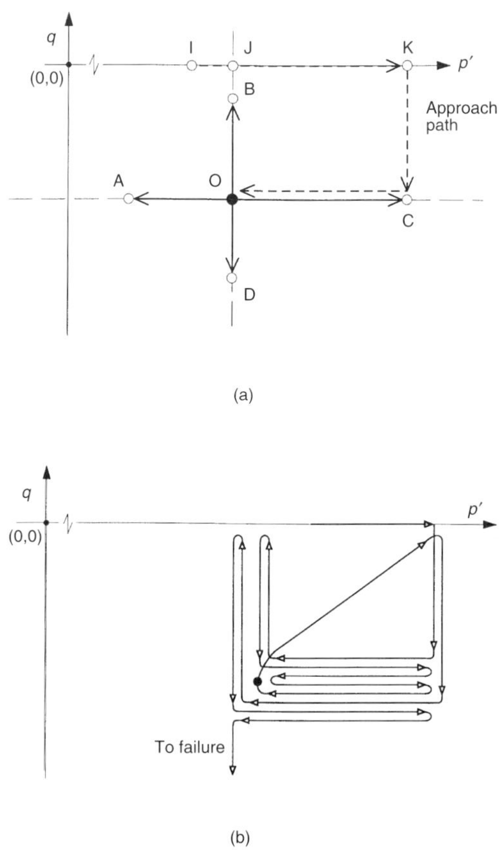

Fig. 4. Schematic triaxial stress paths: (a) probing from estimated in situ stress state; (b) multi-stage procedure

define the stiffness characteristics in triaxial stress space, assuming that the behaviour in each quadrant of the diagram could be defined from the adjacent pair of stress probes. In the present case, probing in directions OA and OD was likely to result in failure of the specimen, and thus effort was concentrated in directions OB and OC. In two tests a diagonal probe between OB and OC was incorporated to assist in the validation of numerical models. The final probe in each test took the specimen to failure along OD.

For reasons stated in the introduction, a multi-stage testing procedure was devised, so that probes OB, OC, OD and, where applicable, the diagonal probe could all be carried out on the same specimen. In order to eliminate a differential recent stress history effect, after each probe had ended point O was regained by returning the stress state to point C and retracing the recent stress history along CO before the next probe was applied. This led to the sequence of stress paths illustrated schematically in Fig. 4(b). In order to check on whether permanent changes were being induced by such a procedure, a repeat of the first probe (OB) was undertaken towards the end of the sequence.

In practice, for various operational reasons (e.g. equipment failures and time constraints), the application of the full sequence of paths shown in Fig. 4(b) was rarely achieved. As explained later, one specimen could only be tested in compression and a different sequence of paths was followed.

The loading rate adopted throughout the test programme was an axial stress change of 1 kPa/hour (implying that q and p' changed at rates of 1·5 and 1 kPa/hour along constant p' and constant q paths, respectively). This was the minimum rate at which the tests could be conducted in the time available – a full multi-stage test took about 4 months – but meant that, even though filter paper side drains were installed, some excess pore water pressures had to be tolerated in the central region of the specimen. An indication of the degree of non-equalization of pore pressure was obtained by mounting a miniature probe (Hight, 1982) between the filter paper strips on the specimen boundary. In most cases the discrepancy between the pore pressures measured at the mid-height and at the base of the specimen was less than 10 kPa; where it was greater than 10 kPa, the test data were rejected.

In investigations of this kind, it is necessary to introduce holding periods between test stages in order to reduce the effects of residual creep strains, which can otherwise be confusing. Ideally, the residual creep rate would be reduced to a negligible level by comparison with the strain rate along the subsequent stress path (Jardine, 1994; Hight & Higgins, 1994). In the present case, it was not possible to eliminate the effect of creep without introducing excessive delays, as the strain rates during the test stages were already small, but holding periods lasting at least 2–3 days were generally employed to reduce the creep to a manageable level.

DYNAMIC STIFFNESS MEASUREMENTS

Measurements of shear wave velocity were made in a few specimens using bender elements installed in the end caps, as described by Bates (1989). Both low frequency (10 Hz) square wave and variable frequency (0·5–50 kHz) sine wave excitations were employed, with an emphasis on the former. The received signals were stacked and an average of 256 consecutive traces was calculated in order to overcome electrical noise. For reasons discussed by Jovičić *et al.* (1996), the interpretation of the square wave results proved problematic and thus only the limited sine wave data, expressed in terms of elastic stiffness, are considered worthy of reporting below.

EXPERIMENTAL RESULTS

Table 2 summarizes the triaxial test programme, after excluding two tests in which equipment failures were experienced, one test in which the soil failed as the assumed in situ stresses were applied and two tests in which the equalization of pore water pressure was insufficient. The stress paths achieved in the four remaining tests are illustrated in Fig. 5. It can be seen that only tests 3 and 6 ran to completion in the intended manner, and therefore the most useful results came from these two

Table 2. Summary of stress path tests

Test no.	Start date	End date	No. of stress path stages	Origin and depth of specimen	p' (kPa) at assumed in situ stress state	Comment
1	12 June 1992	27 July 1992	6	Elstow 4·9 m	100	Trial test
3	1 Nov. 1992	26 Feb. 1993	18	Elstow 5·6 m	150	Full test
6	9 July 1993	10 Nov. 1993	16	Kempston 5·1 m	200	Full test
9	11 Nov. 1993	5 April 1994	7	Kempston 5·1 m	200	Non-typical test

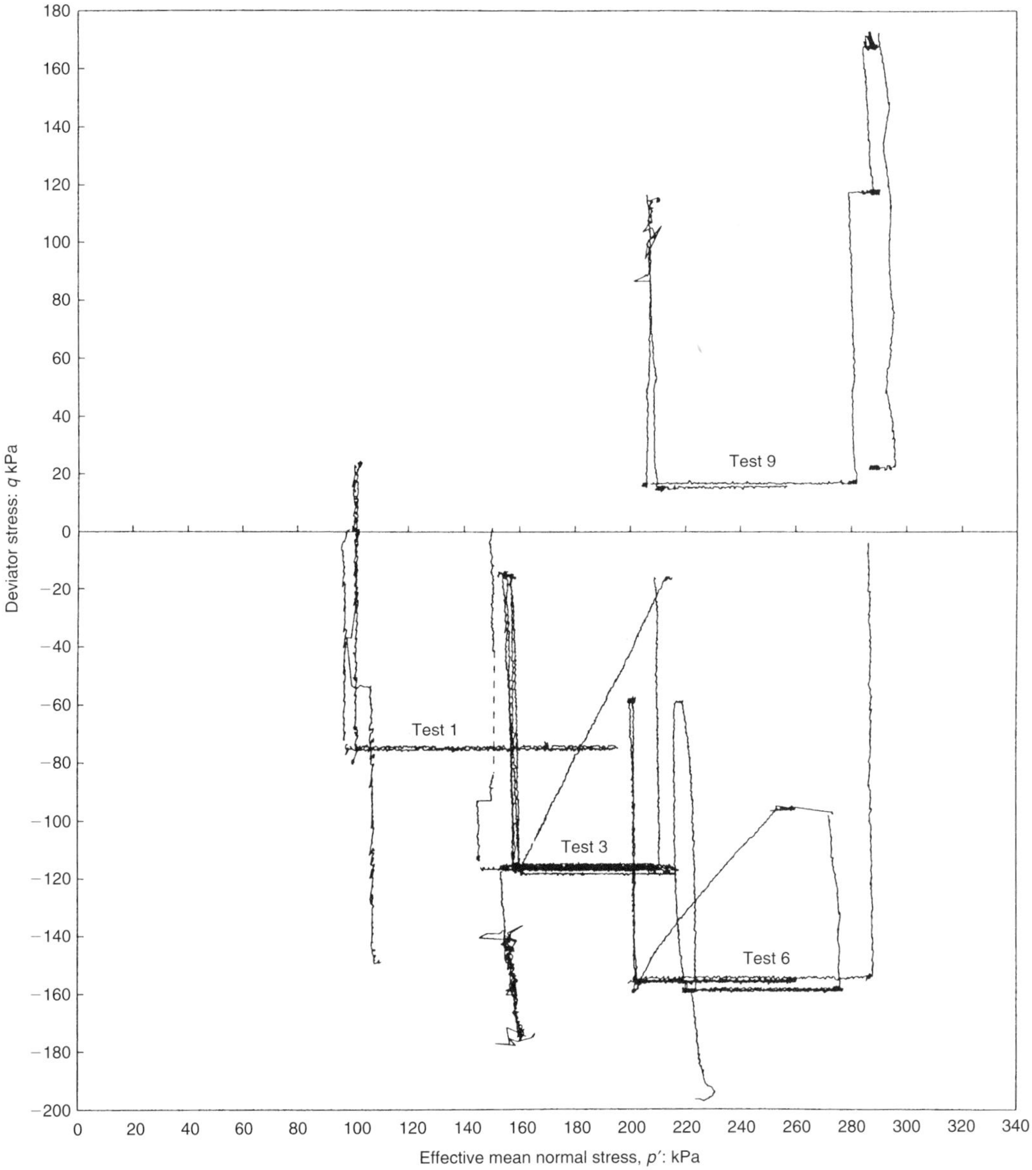

(a)

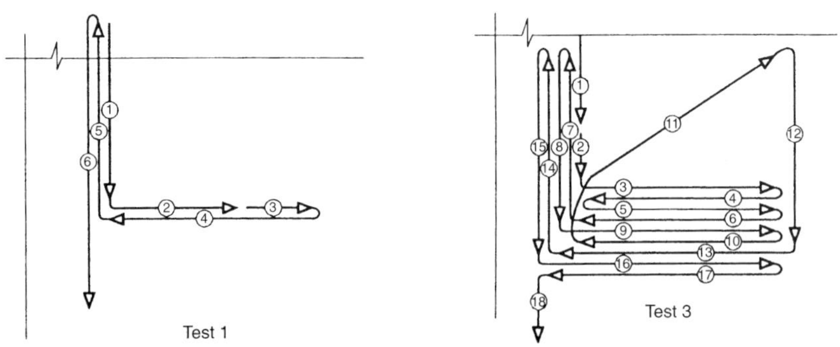

(b) Contd. overleaf

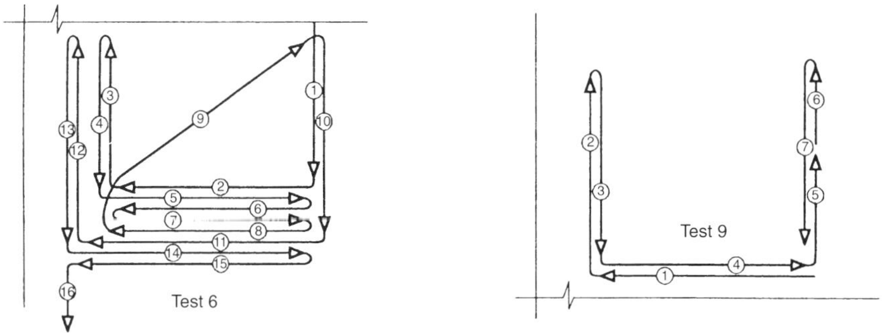

(b)

Fig. 5. (contd.) Stress paths in multi-stage triaxial tests: (a) experimental (b) schematic (showing stage numbers)

tests. In test 6 a control error led to a small translation of some of the constant p' stress paths from their intended position. Also, in this test one of the strain targets for the local measurement of axial strain became detached at a relatively early stage. However, until then there had been excellent agreement, perhaps fortuitous, between the average local and end cap axial strain measurements. It was therefore decided to continue the test and base the interpretation upon the latter. In test 9 the suction connection between the specimen's top cap and the loading ram, employed to permit negative deviator stresses to be applied, could not be made to operate because of slight tilting of the top cap. Consequently, the specimen had to be tested in compression, following a sequence of stress paths which only loosely resembled the usual sequence.

Cumulative strain paths followed during the tests are depicted in Fig. 6, while the changes of strain occurring during the individual stress path stages are shown in Fig. 7. In both these figures the test stage numbers correspond to those identified in Fig. 5. The volumetric strains were calculated from the average axial and radial strains.

Stiffness parameters were extracted from the stress–strain data using a linear spline interpolation technique. This involved performing linear regressions on overlapping sets of 10–20 data points, covering a stress change of less than 5 kPa, to obtain the tangent stiffnesses at the midpoints of the data sets.

Effects of stress history and creep

In both tests 3 and 6 very similar constant q and constant p' paths were followed several times, either forwards or backwards and after either a 90° or a 180° turn in the stress path direction. Changes of stress for these test stage families are plotted against the corresponding changes of strain in Fig. 8 for test 3 and in Fig. 9 for test 6. In Table 3 the residual creep rates at the start of each stage are compared with the subsequent strain rates after 0·01% shear or volume strain, as appropriate, had been recorded. In some cases the residual volumetric creep rate could not be determined reliably, as fluctuations occurred which were of the same order as the mean value; these were attributed to the effect of temperature variation on the instrumentation system.

At the larger strain levels, the data in Figs 8 and 9 are reasonably well grouped and no particular influence of recent stress history differences, including consequential creep effects, or direction of travel along the stress path is evident. The noticeably lower stiffnesses of stage 13 in Fig. 8(b) (left hand diagram) and stage 2 in Fig. 9(b) are an exception and will be discussed below. Also, stage 17 in Fig. 8(b) was severely affected by creep, as the preceding holding period was exceptionally short (only 1·5 h). The spread of results relative to their average tends to increase as the strain level reduces and generally becomes substantial at the 0·01% strain level. Inaccuracies in stress path control and in the measurements of stress or strain must have contributed to the variation, together with the effects of creep. The data in Table 3 suggest that creep may have had an effect on nearly all of the stiffness measurements. However, because of the other factors involved and the uncertainties surrounding some of the volumetric creep rates, the pattern of results in Figs 8 and 9 cannot be correlated with the tabulated creep rates.

The relative absence of systematic recent stress history effects in the above data is in contrast to other research (e.g. Atkinson *et al.*, 1990; Jardine, 1994). However, a tentative explanation can be advanced on the assumption that the recent strain history is of greater importance than the stress history. In Fig. 7 it is seen that the angle between the individual strain paths corresponding to constant q and constant p' stress paths is very acute

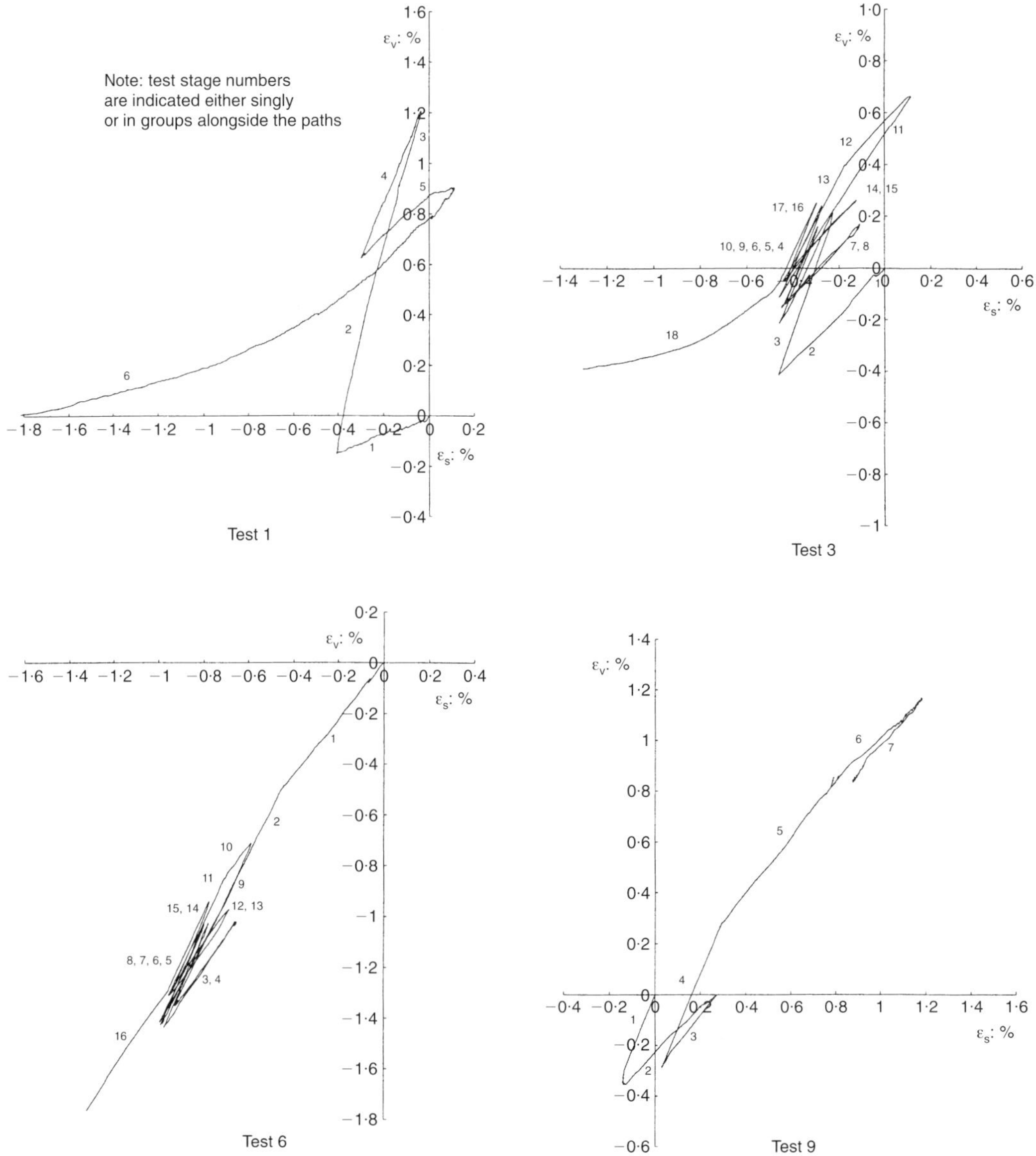

Fig. 6. Cumulative strain paths in multi-stage triaxial tests

(about 20° on average) when both volume strain and shear strain have the same sign. With the sequence of stress paths applied, in most cases a reversal of strain took place when a new test stage commenced. Thus, in terms of recent strain history, there was usually little difference between a stress path turn of 90° and one of 180°, which corresponded respectively to turns of about 160° and 180° in the strain path. In just three cases (stage 13 of test 3 and stages 2 and 11 of test 6), no reversal of strain took place, so that a 90° turn in the stress path corresponded to a turn of only about 20° in the strain path, and in two of these cases a significantly less stiff response was recorded. In the third case (stage 11 of test 6) the reduction of stiffness was less marked, but the length of the stress path before the turn was relatively short.

Further evidence of the predominance of the influence of strain history was obtained in test 9.

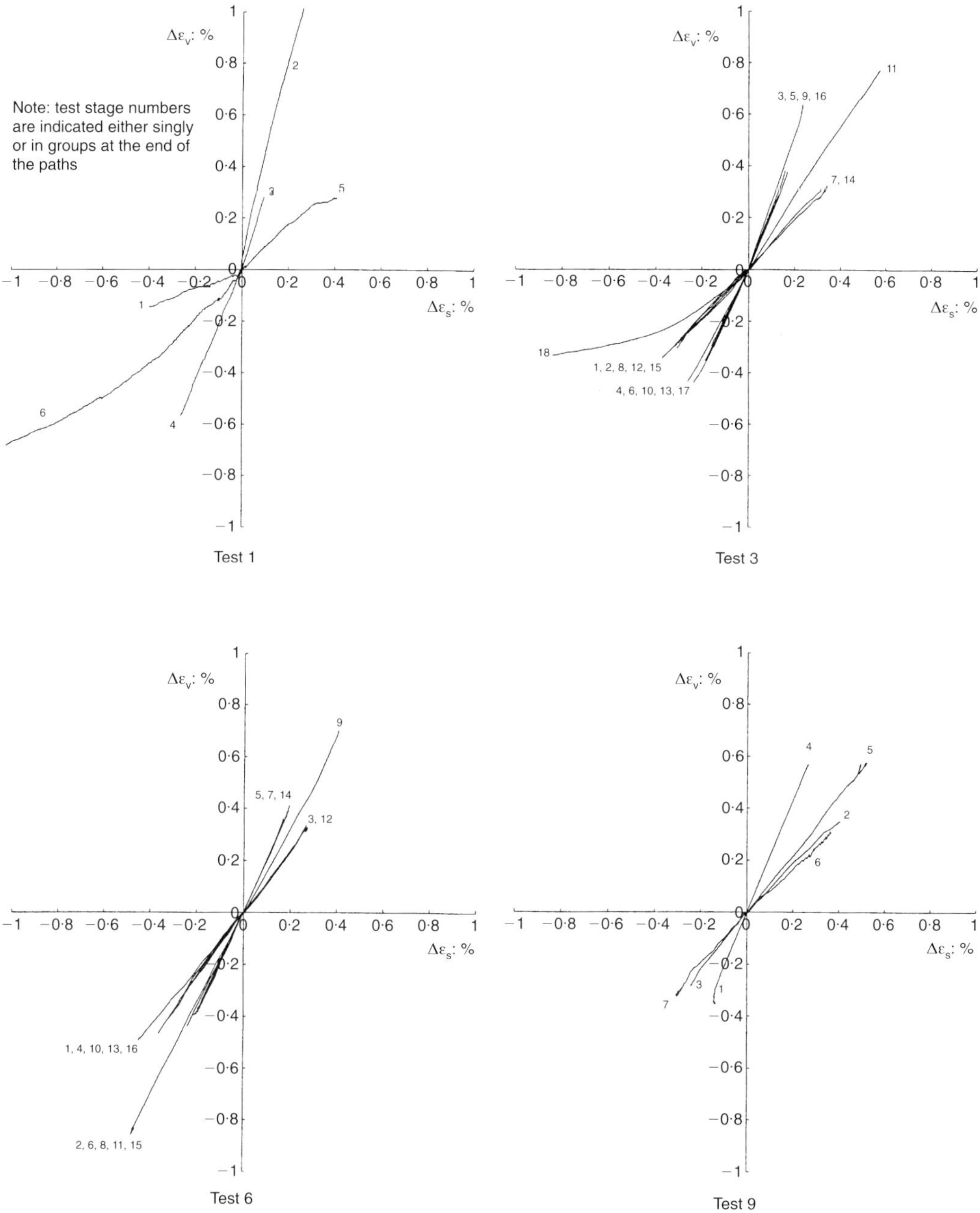

Fig. 7. Strain paths in individual stages of triaxial tests

The relevant results are shown in Fig. 10. Although (unusually) there was no rotation of the stress path from stage 5 to stage 6, the response in stage 6 was as stiff as in stage 5, after a 90° rotation. This result is explained by the similarity of the recent strain history of stages 5 and 6 (see results for stages 4 and 5 in Fig. 7). It also demonstrates the recovery of stiffness achieved during the holding period. As would be expected, the response in stage 7 was stiffer after a 180° rotation in both the

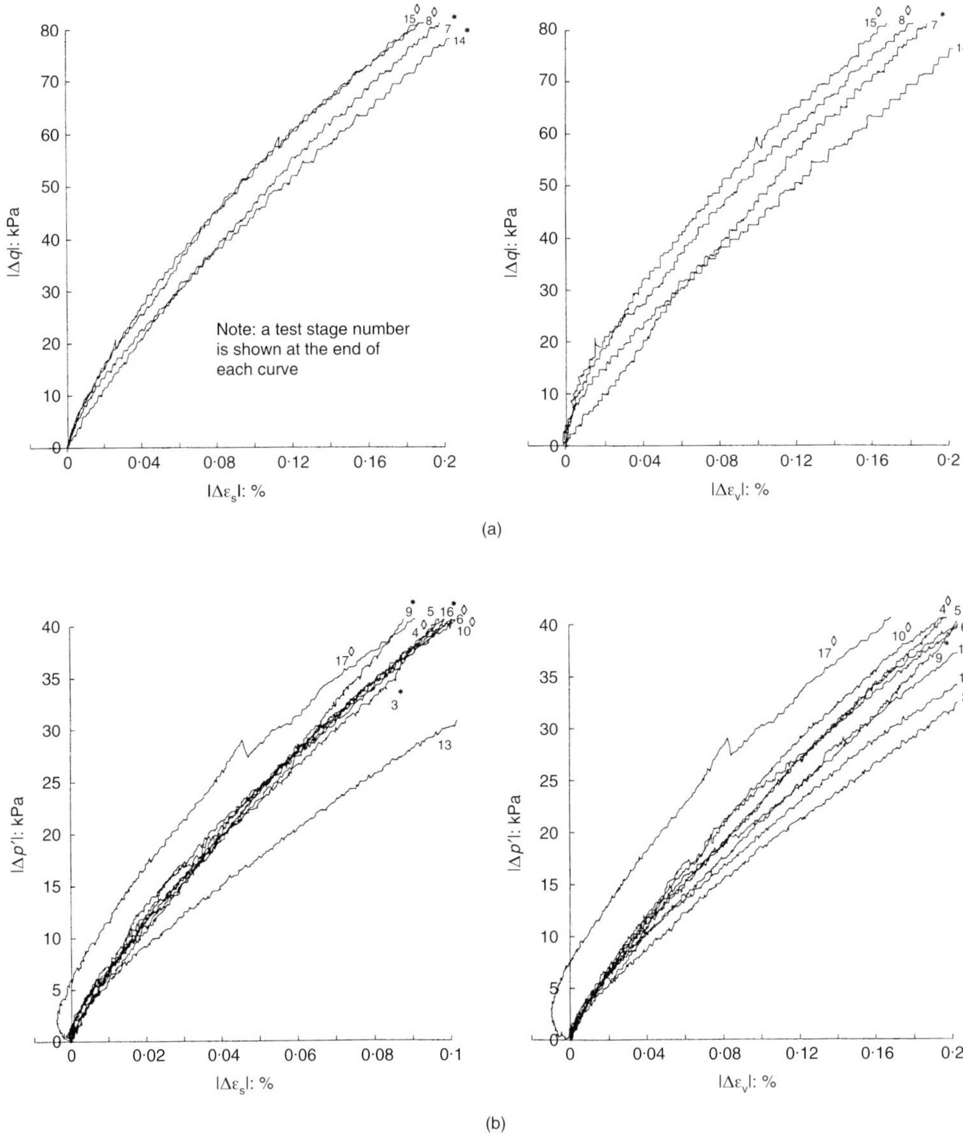

Fig. 8. Stress–strain curves from test 3: (a) constant p' paths, (b) constant q paths

stress and strain paths, the irregularity in the response between 0·02 and 0·06% strain being due to a deterioration in the control of p'.

Validity of multi-stage testing

Contained in Figs 8 and 9 are subsets of results for test stages which had an identical recent stress history but which were conducted at various intervals during the test as a whole. These sets are indicated by asterisks or diamonds placed against the relevant stage numbers on the diagrams. Note that the pair of stage 2 and stage 11 in test 6 is not included because the preceding paths differed greatly in length. Taking account of the comments made in the previous section, the consistency of each set of results indicates that the multi-stage technique was valid.

Effects of confining stress and material variation

For each test the intended stress level at the assumed in situ stress state and the origin of the

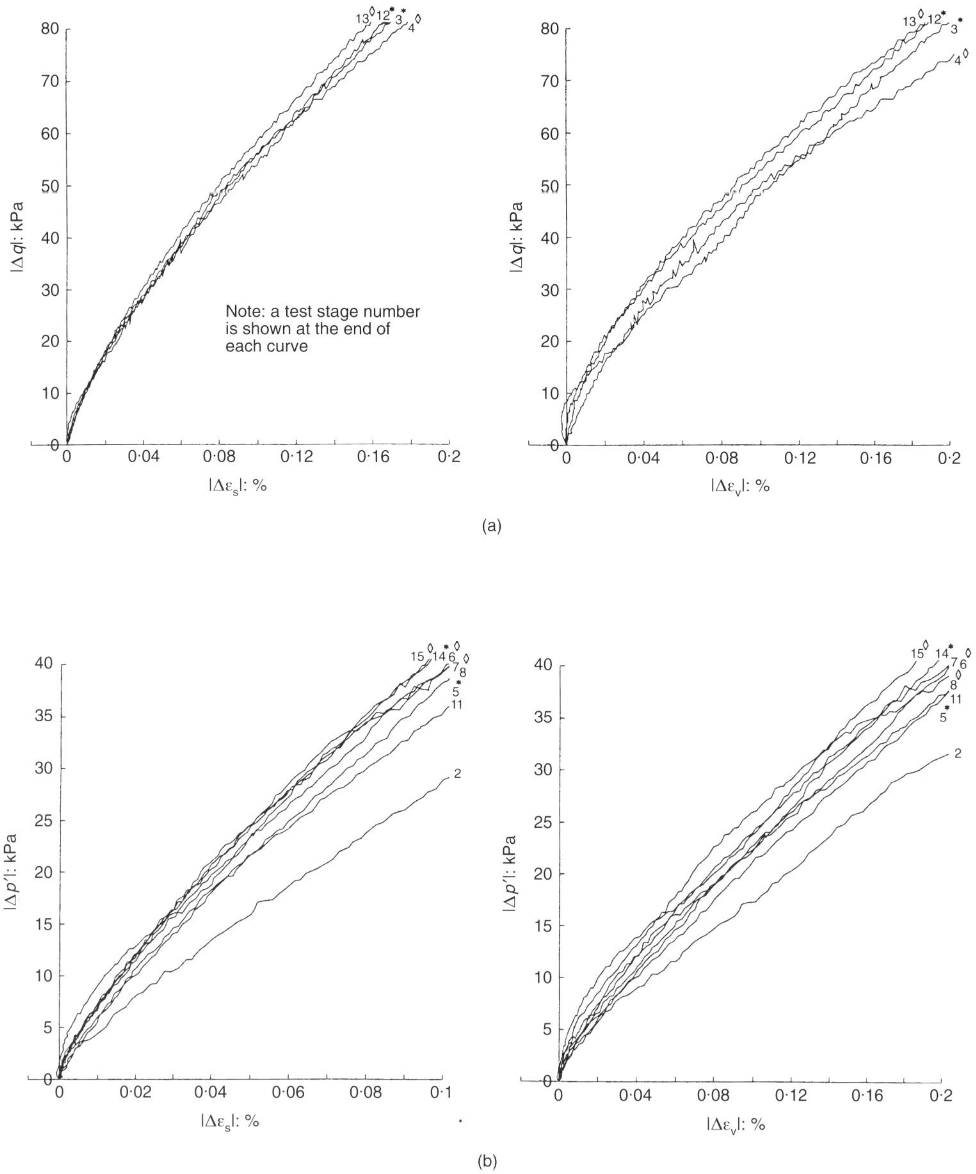

Fig. 9. Stress–strain curves from test 6: (a) constant p' paths, (b) constant q paths

specimen are given in Table 2. In order to permit an assessment to be made of these factors, stress–strain curves from closely corresponding constant p' stages of all four tests are compared in Fig. 11; the average value of p' actually achieved in each case is indicated. The stiffness exhibited towards the end of stage 2 of test 9 reduces as the stress path heads towards the compression failure envelope.

Because the two specimens from Kempston pit were both tested at higher stress levels than the two from Elstow, it is difficult to separate the influence of material source and confining stress. However, on the hypothesis that the effect of material variation was negligible, the results for shear stiffness display the expected hierarchy and, when plotted in the appropriate form (Fig. 12), suggest that G' increases as $(p')^N$ where $N = 0{\cdot}67$

Table 3. Comparison of residual creep rates and strain rates in subsequent test stages

Test number	Stage number	Residual shear strain creep rate, R1 (%/hour × 10^{-3})	Residual volume strain creep rate, R2 (%/hour × 10^{-3})	Shear strain rate after 0·01% shear strain, R3 (%/hour × 10^{-3})	Volume strain rate after 0·01% volume strain, R4 (%/hour × 10^{-3})	Ratio of creep to test stage shear strain rates, R1/R3	Ratio of creep to test stage volume strain rates, R2/R4
3	1			−1·59	−2·24		
	2	−3·00	−2·90	−9·50	−10·80	0·32	0·27
	3	−0·82	−0·69	2·18	3·89	−0·38	−0·18
	4	0·09	0·27	−1·70	−2·84	−0·05	−0·10
	5	−0·25	−0·33	2·17	3·33	−0·12	−0·10
	6	0·03	0·13	−1·78	−3·03	−0·01	−0·04
	7	−0·26	−0·45	2·67	3·20	−0·10	−0·14
	8	0·07	(0·10)	−2·22	−2·12	−0·03	−0·05
	9	−0·24	−(0·10)	1·61	3·65	−0·15	−0·03
	10	0·09	0·37	1·54	−2·37	0·06	−0·16
	11	−0·26	−0·46	2·44	2·28	−0·11	−0·20
	12	0·67	0·74	−2·04	−2·40	−0·33	−0·31
	13	−0·40	−(0·35)	−2·16	−2·92	0·19	0·12
	14	−0·47	−0·69	2·63	3·03	−0·18	−0·23
	15	0·34	0·43	−2·15	−2·14	−0·16	−0·20
	16	−0·44	−0·48	2·07	3·58	−0·21	−0·13
	17	2·64	5·88	−1·62	−3·23	−1·63	−1·82
	18	−0·53	−0·80	−3·90	−4·57	0·14	0·18
6	1			−2·16	−2·17		
	2	−0·23	−(0·48)	−2·17	−2·84	0·11	0·17
	3	−0·49	−(0·57)	2·59	3·71	−0·19	−0·15
	4	0·01	0·01	−1·82	−2·10	−0·01	0·00
	5	−0·77	−0·89	1·84	3·10	−0·42	−0·29
	6	0·17	0·40	−1·76	−2·61	−0·10	−0·15
	7	−0·22	−(0·15)	1·70	2·76	−0·13	−0·05
	8	0·07	(0·33)	−1·62	−2·30	−0·04	−0·14
	9	−0·17	−(0·05)	2·72	3·63	−0·06	−0·01
	10	0·41	1·54	−1·55	−1·76	−0·26	−0·88
	11	−0·12	−(0·16)	−1·87	3·19	0·06	−0·05
	12	−0·28	−0·26	1·79	2·00	−0·16	−0·13
	13	0·24	0·28	−1·86	−1·62	−0·13	−0·17
	14	−0·01	−0·07	1·50	2·46	−0·01	−0·03
	15	0·13	0·34	−1·34	−2·06	−0·10	−0·17
	16	−0·60	−1·03	−2·32	−2·88	0·26	0·36

Note: bracketed values estimated allowing for fluctuations (see text)

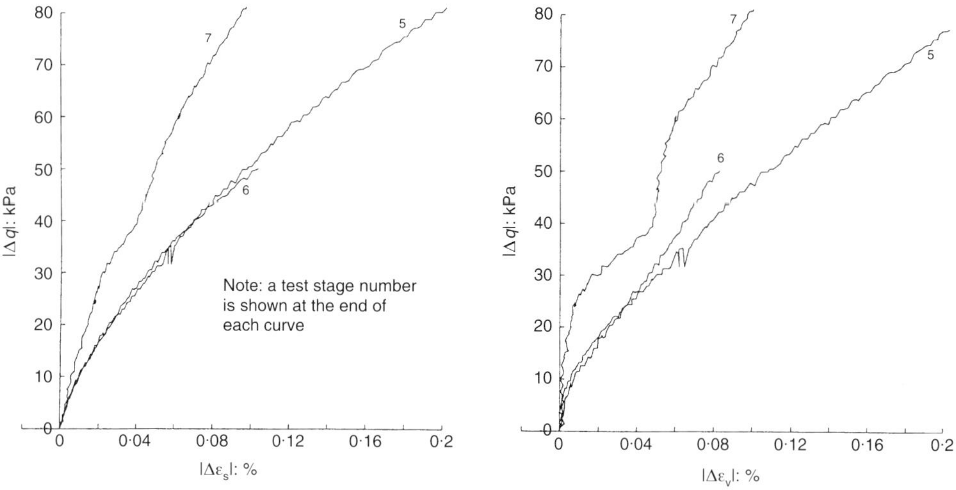

Fig. 10. Stress–strain curves from constant p' paths in test 9

on average. This interpretation, indicated by the trend lines in Fig. 12, is plausible but the expected increase of N with increasing strain level (Viggiani & Atkinson, 1995; Jardine, 1994) is not apparent. This may be because of the limited number of data and their scattered nature.

Interpretation of overall behaviour pattern

It may be argued (Burland & Georgiannou, 1991) that changes in strain energy provide a logical and generalized basis for expressing the non-linearity of soil stiffness. In calculating these changes, only the work associated with the stress changes giving rise to the strains is considered; this quantity, termed 'incremental strain energy', is a much more convenient indicator of stiffness than the total change of strain energy. The incremental strain energy per unit volume during a given triaxial test stage may be calculated as

$$U = \sum_{0}^{\varepsilon_v} \Delta p' \delta\varepsilon_v + \sum_{0}^{\varepsilon_s} \Delta q \delta\varepsilon_s \tag{2}$$

where $\Delta p'$ and Δq are the stress changes from the start of the test stage. The derived contours of incremental strain energy, U, surrounding the assumed in situ stress point are plotted in the plane of the principal stress changes in Figs 13 and 14 for tests 3 and 6 respectively. Axes of stress change are chosen to remove the effect of small differences in the starting points of the various stress paths; the inset diagrams show the stress paths with axes of stress rather than stress change. In each figure, the relative spacing of the contours in the directions of the axes reflects a significant degree of anisotropy, with a stiffer response (larger spacing) evident in the horizontal direction. It can also be seen that the contours close up in the lower part of the diagram as failure in extension is approached.

The variation of the four stiffnesses of equation (1) with incremental strain energy in the same two tests is shown in Figs 15 and 16. For the test stages with either q decreasing or p' increasing (OD and OC in Fig. 4(a)), the initial values of the cross-coupling stiffnesses, J'_{qv} and J'_{ps}, are in reasonable agreement. However, at low incremental strain energies in the test stages with either q or p' increasing (OB and OC in Fig. 4(a)), they are unequal, implying that the initial behaviour is not elastic (Atkinson & Richardson, 1985). Given the unexpected nature of this result, the difficulty of measuring high coupling stiffnesses accurately and the possible effects of residual creep, further experimental work is needed to confirm its validity. In Fig. 15 the erratic increase in J'_{qv} in stage 18 of test 3, when U is about 0·01 kJ/m^3, was caused by an irregularity in the stress path which is visible in Fig. 13.

Comparison of static and dynamic stiffness measurements

Dynamic measurements of the maximum shear stiffness, G_0, were made using sinusoidal pulses in test 6 and in one of the abortive tests not listed in Table 2. In each case, either 11 or 12 individual measurements were made and the confining stress,

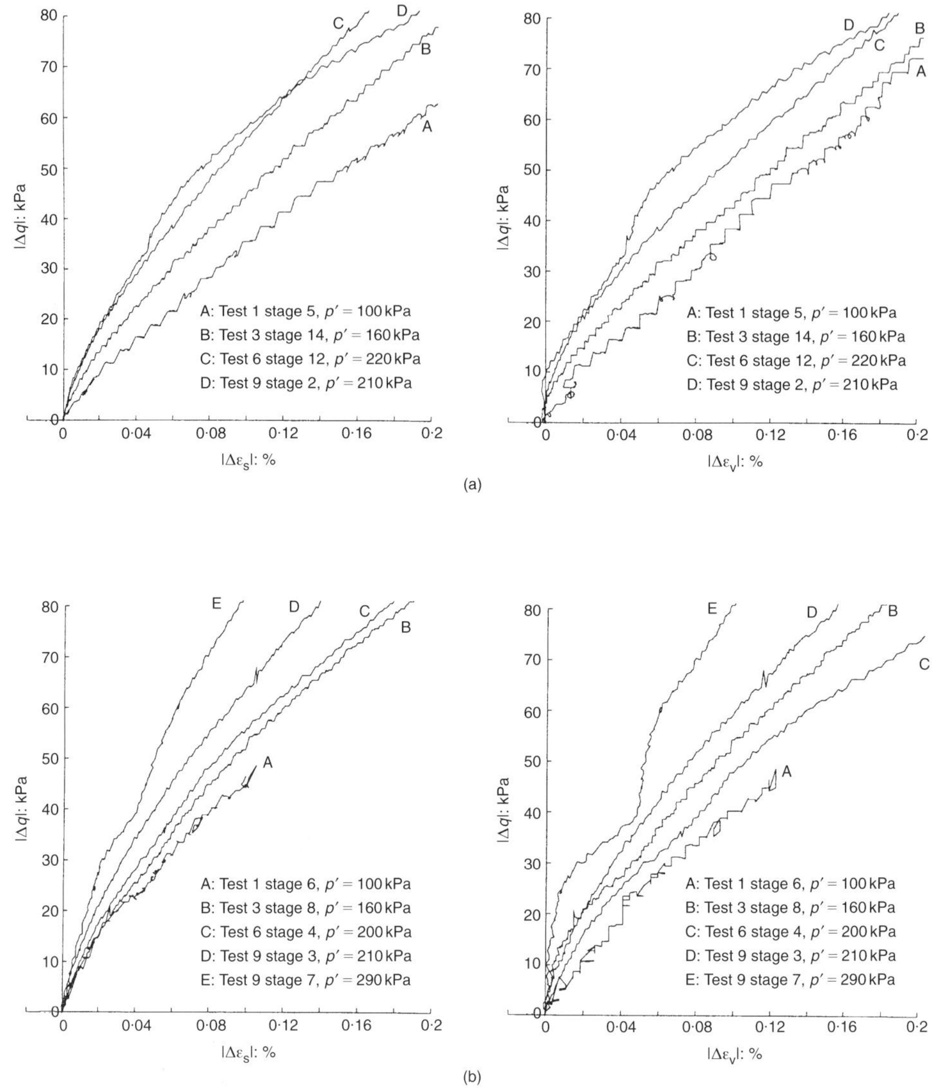

Fig. 11. Comparison of stress–strain curves from constant p' paths of all tests: (a) after a clockwise 90° stress path turn (b) after a 180° stress path turn

p', at the time of the measurements was about 278 kPa. In test 6, on average $G_0 = 75$ MPa and, in the other test, $G_0 = 76$ MPa. While it is evident that these values of G_0 appear relatively high when compared with the data for G' (the deviatoric stiffness) in Figs 15 and 16, a proper comparison of G_0 and G' requires knowledge of the degree of elastic anisotropy and is not attempted here. However, by allowing for the effects of stress level and anisotropy, it is not difficult to reconcile the static and dynamic measurements of soil stiffness (Pierpoint, 1996).

NUMERICAL MODELLING

Constitutive model

The behaviour of the Oxford Clay was modelled using a cross-anisotropic elastic model (Graham & Houlsby, 1983) implemented in terms of effective stress in the finite element program CRISP (Britto & Gunn, 1987). This model requires the parameters G', K' and J' where $J' = J'_{qv} = J'_{ps}$. Non-linearity was incorporated by varying the stiffness parameters with the incremental strain energy as the analysis progressed. It was assumed that the dependence of the stiffnesses on incremental strain energy would be the same under plane strain and axisymmetric conditions. There was also an optional facility in the model to vary the stiffnesses automatically with p'.

The stiffnesses were dependent on the current stress path direction. Because of difficulties that arise in identifying stress path reversals correctly if the deviator stress is defined in terms of principal stresses, the vertical and horizontal stresses were used to calculate a deviator stress defined in the same way as in the triaxial tests and hence to determine whether the soil was deforming in compression or extension. In doing this, the effects of principal stress rotation were ignored. With reference to Fig. 4(a), for compression paths (directed upwards) stiffnesses derived from the triaxial test stages along OB and OC were used, while for extension paths (directed downwards) stiffnesses were based on the test stages along OD and OC. This approach was developed in the absence of any test stages along OA. The single coupling stiffness, J', required by the model was dependent on stress path direction, again assessed crudely using the vertical and horizontal stresses and was scaled linearly with the inclination of the path between $J' = J'_{ps}$ when p' = constant and $J' = J'_{qv}$ when q = constant, with parameters selected for compression or extension as appropriate. When a reversal in stress path direction was detected, stiffnesses were reassigned as if the soil was following a path from its original starting point and the strain energy was reset to zero. If a reversal had occurred after significant compression, the new stiffnesses assigned would have been too low; in the triaxial tests stiffnesses along path BO after a reversal were similar to those along OB and not to the lower ones along OD. However, all the reversals encountered in modelling the excavation were from extension to compression, for which the approach adopted was considered a reasonable approximation. Almost all the

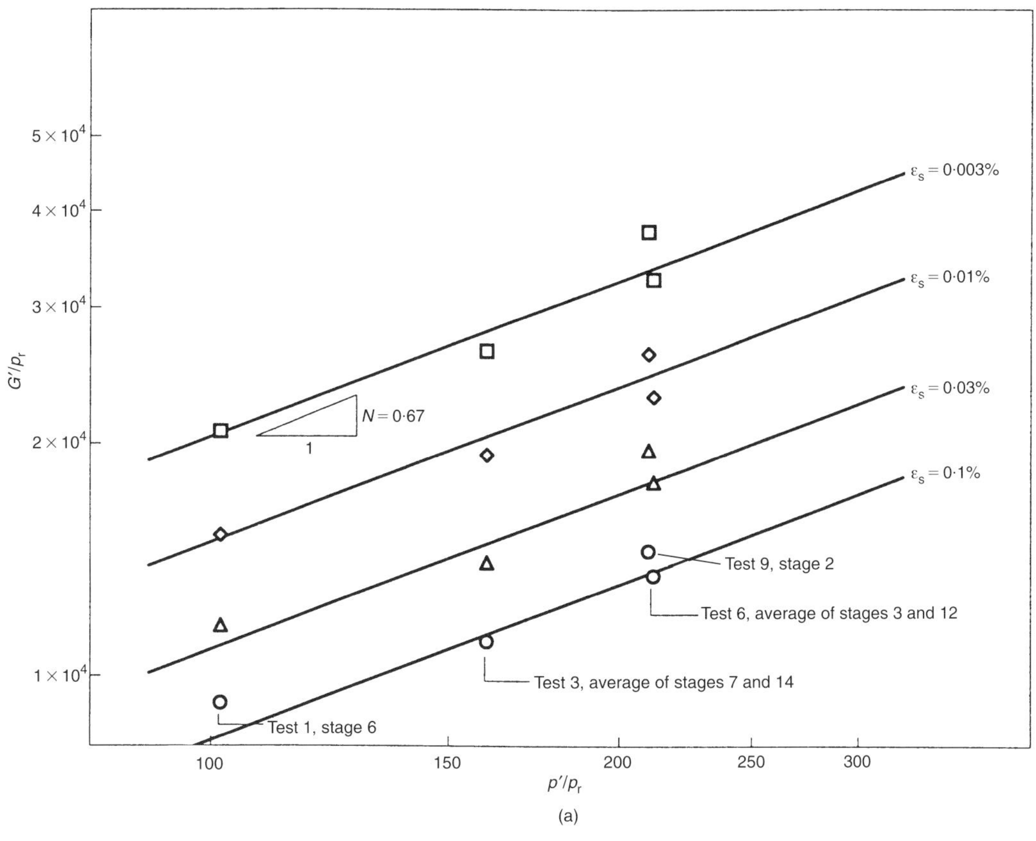

Fig. 12a. caption overleaf

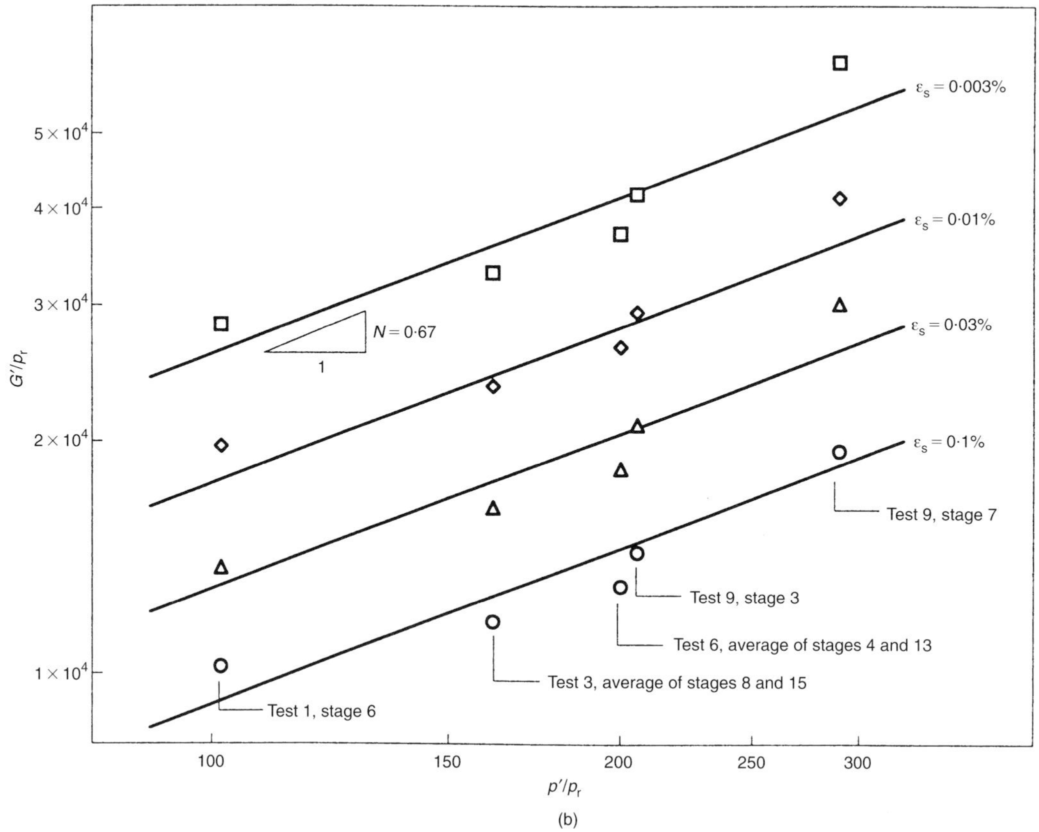

Fig. 12. G' plotted against p' at various strain levels on dual logarithmic scales and normalized using a reference pressure $p_r = 1$ kPa: (a) after a clockwise 90° stress path turn (b) after a 180° stress path turn

soil lying vertically below the excavation sides suffered a stress path reversal.

As shown by Hird & Pierpoint (1994), the model correctly predicted the behaviour observed along a diagonal stress path in the triaxial tests. The quality of the prediction was relatively insensitive to the choice of J' between the limits $J' = J_{ps}$ and $J' = J_{qv}$. However, predictions made without incorporating anisotropy ($J' = \infty$) were greatly inferior.

Parameter selection

Input stiffness parameters for the Oxford Clay are shown varying with incremental strain energy in Fig. 17. These relationships were polynomial fits to the results of stages 7, 12 and 16 of test 6, as presented in Fig. 16, with suitable truncations at high and low strain energies. In the first analysis (the Class C prediction) the initial stiffnesses were assumed to be constant with depth and did not change with stress level during the analysis; in the second analysis (the later variation) all the stiffness parameters were assumed to be proportional to $(p')^{0{\cdot}67}$, as shown for G' in Fig. 12. When $p' = 221$ kPa, corresponding initially to a depth of about 12 m, the parameters were the same as those shown in Fig. 17 and the stiffnesses were therefore lower above this level than in the first analysis. There were no other differences between the two analyses.

The underlying strata (see Fig. 2) were modelled as linear cross-anisotropic materials with the stiffness parameters given in Table 4, which were crudely estimated by making use of geophysical test data.

For simplicity, the initial in situ stresses in all strata were based on a K_0 value of 2·25. This was within the range shown for the Oxford Clay in Fig. 3, although slightly different to the value adopted

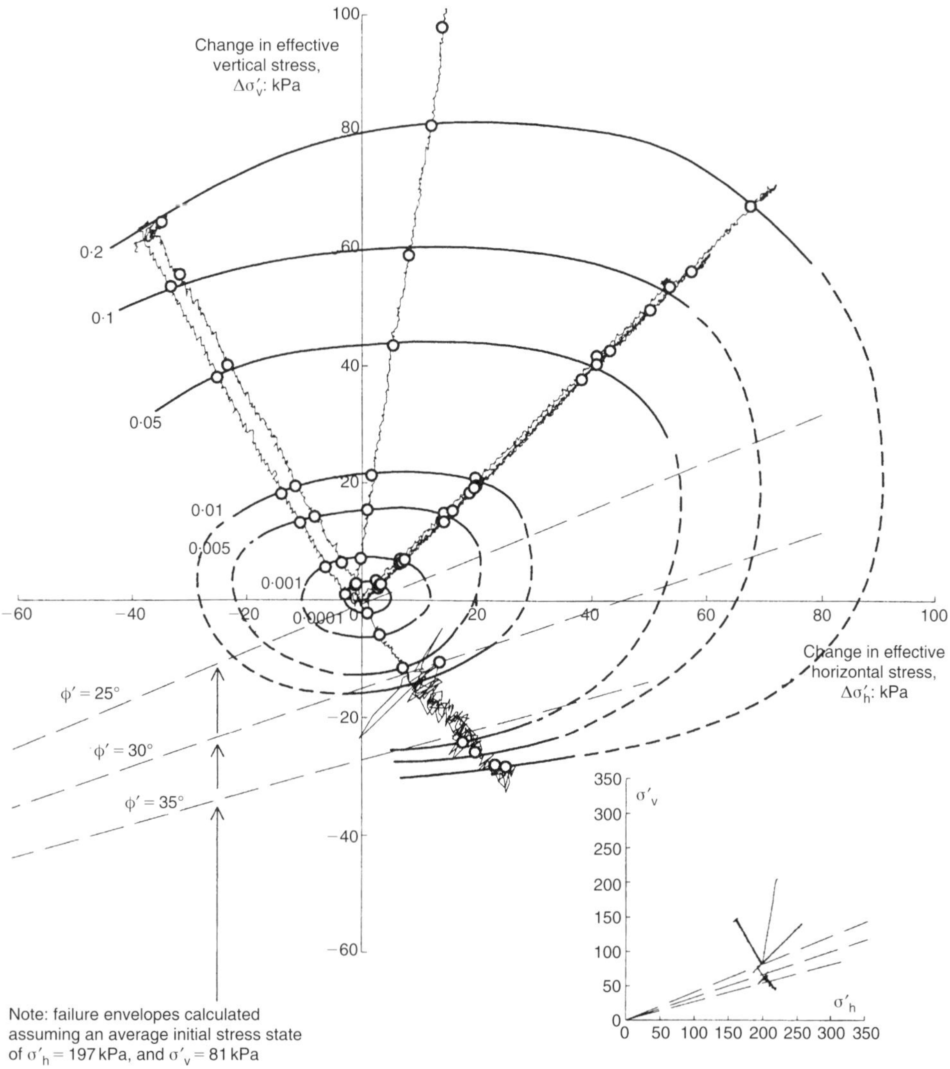

Fig. 13. Contours of incremental strain energy (kJ/m³) from test 3 plotted in principal stress space

in the triaxial tests. The Oxford Clay, Kellaways Clay and Blisworth strata were modelled as undrained, by specifying a relatively high value of bulk stiffness for the pore water (2000 MPa), and the Kellaways Sand and Cornbrash strata as fully drained.

Comparison of observed and predicted displacements

Predicted displacements at the end of construction are compared with field measurements in Figs 18, 19 and 20. The field data were not always consistent and some were considered unreliable. Therefore, where possible, averages of readings from similarly-positioned instruments have been shown and the range of observations at different times during the post-construction monitoring period has been indicated. In Fig. 18 the field data at −13, 0, +13 and +21 m from the centreline (averages from either 3 or 5 instruments) are more dependable than the remainder (values from single instruments). The predictions come from plane strain analyses of one half of the transverse section of the excavation shown in Fig. 1(b). For these analyses the finite element mesh, comprising some

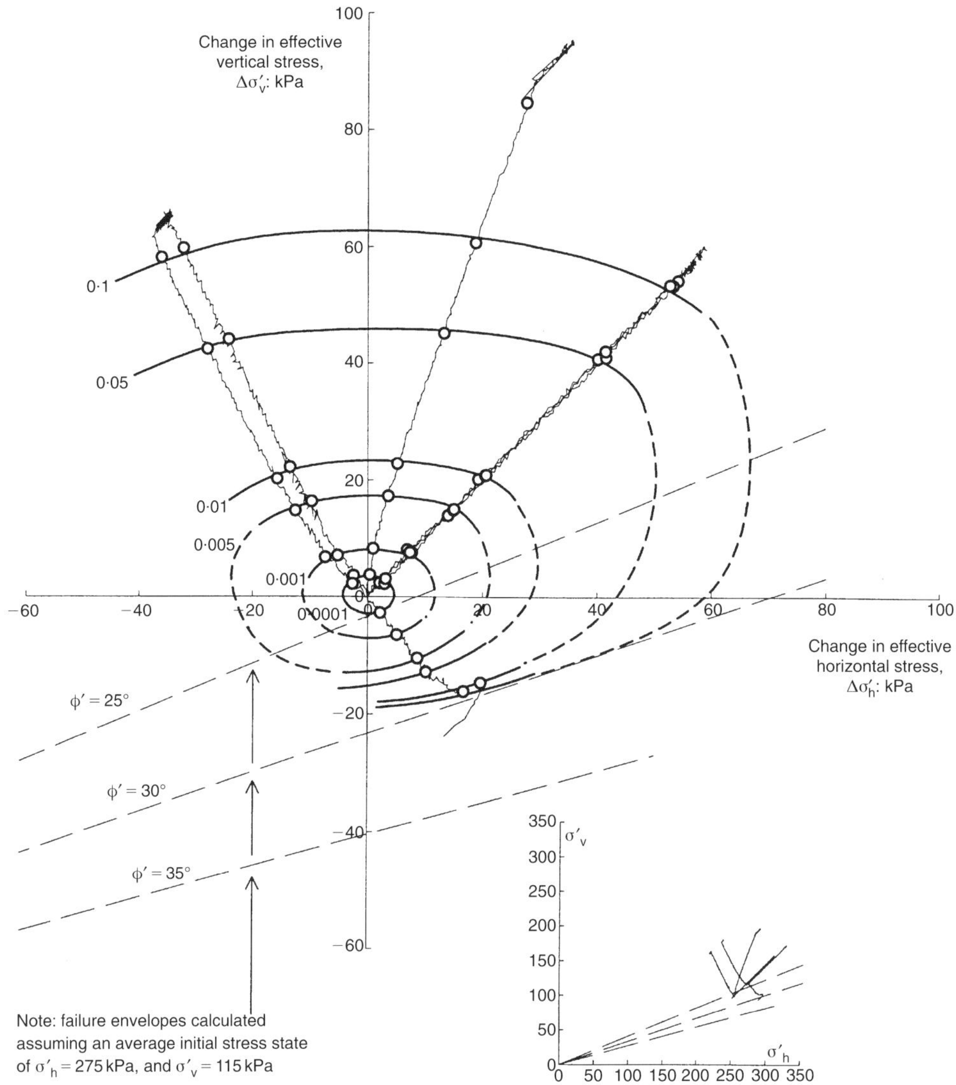

Fig. 14. Contours of incremental strain energy (kJ/m³) from test 6 plotted in principal stress space

1700 quadrilateral and triangular linear strain elements, extended to a distance of 50 m from the centreline (assumed to be a line of symmetry) and a depth of 25 m. The outer vertical boundary was fixed horizontally but not vertically; the base was fixed vertically but not horizontally, although horizontal fixity might have been preferable. Further details of the analysis are given by Pierpoint (1996).

Although the symmetry of the excavation was carefully preserved at all stages, the measured lateral displacements were noticeably larger on one side than the other. In general, these lateral displacements were overpredicted. Fig. 19 shows that the analyses significantly overpredicted the movement below the base of the Oxford Clay and, given the simple way in which the stiffnesses of the lower strata were estimated, this is not surprising. The lack of fixity of the bottom boundary also contributes slightly to the result. If the displacement solely within the Oxford Clay is considered, the overpredictions are relatively modest in relation to the average field response. Although Fig. 18 appears to show that some lateral movements in-

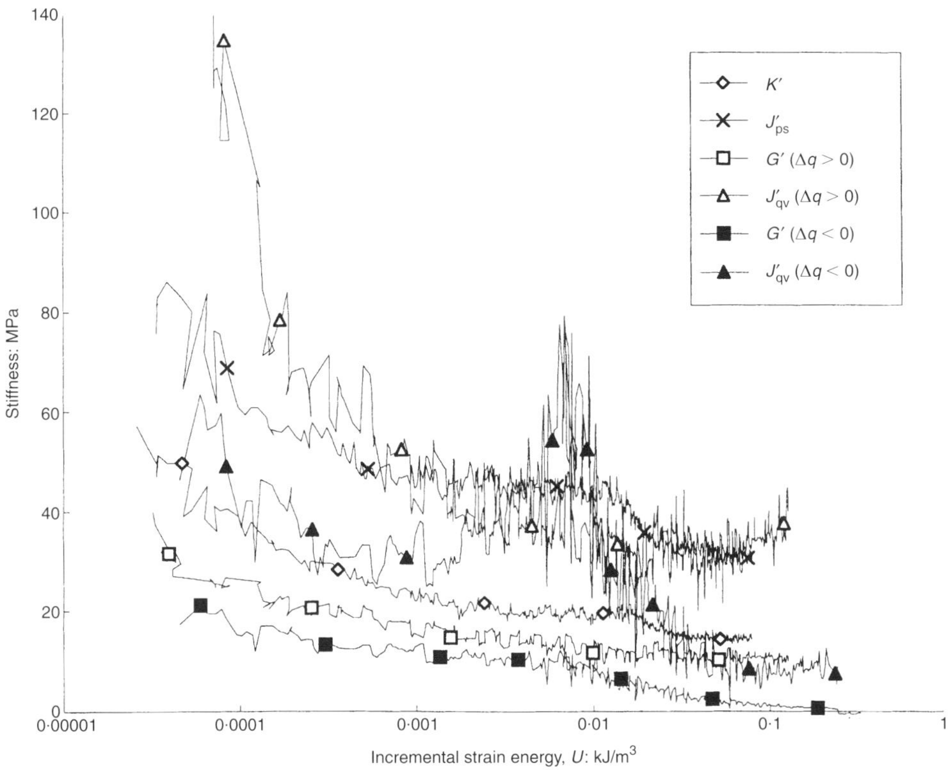

Fig. 15. Variation of stiffness parameters with incremental strain energy in test 3 (stages 5, 14, and 18)

creased during the monitoring period, when all the available data were scrutinized a systematic change of lateral displacement with time after construction was not readily apparent.

In contrast, vertical displacements, both the settlement outside the excavation and the heave within its boundaries, were underpredicted. It is likely that the analyses failed to capture the correct distribution of settlement with distance from the excavation centreline; relatively uniform surface settlement was predicted between the edge of the excavation and the mesh boundary, whereas a more localized settlement trough probably developed. Further analysis showed that moving the boundary outwards improved the form of the settlement results slightly but not sufficiently to give good agreement with the field data. As seen in Fig. 20, the heave of the excavation base increased significantly with time, so that the reasonably good prediction at the end of construction then appears to deteriorate. An analysis of the combination of measured lateral and vertical movements by Pierpoint (1996) points to an expansion of the Oxford Clay in this region during construction, attributable to the opening of fissures, and it is presumed that cavitation took place in the pore water to allow the fissures to open. During the post-construction period, there was no evidence of drainage towards the excavation in the measured pore pressure responses, which remained almost constant (Rudrum, 1990).* Nevertheless, further expansion occurred and it is likely that some swelling took place near the clay surface at the base of the excavation as rainwater infiltrated.

Distribution of shear strain

The distribution of shear strain in the Oxford Clay at the end of the first analysis is shown in Fig. 21; a very similar distribution, with slightly higher magnitudes at the upper levels, was obtained in the second analysis. The predicted strains are generally small, but there is a localized concentration of strain below the junction of the base and the side of the excavation.

* None of the field data reported by Rudrum were seen by the authors before the numerical prediction was made.

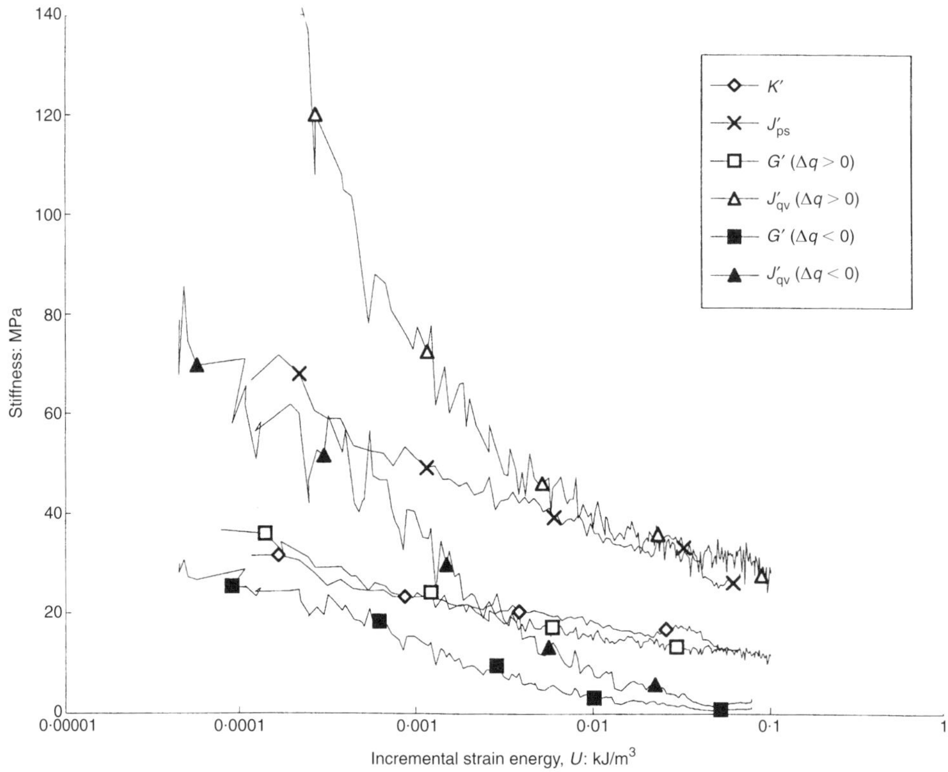

Fig. 16. Variation of stiffness parameters with incremental strain energy in test 6 (stages 7, 12, and 16)

A potentially serious limitation of the constitutive model was that it did not incorporate a failure criterion. Therefore, the computed stresses at the end of the analysis were checked to find out whether an appropriate frictional failure criterion would have been violated. The mobilized angle of shearing resistance was highest in the localized area of strain referred to above, but nowhere did it exceed 31° in either analysis. This was considered satisfactory. However, the results of the numerical analysis, in terms of both stresses and displacements, could have been strongly dependent on the proximity of the assumed in situ stress state to passive failure in the triaxial test used to obtain the stiffness parameters (Test 6). As previously indicated, the actual in situ stress conditions are not accurately known and thus it is also not known how well the assumed condition matched reality.

CONCLUSIONS

The viability of a multi-stage triaxial stress path test for investigating the stiffness of a heavily overconsolidated clay at small strains has been demonstrated. The main advantage of the multi-stage test is that it can extract a large amount of consistent information from a high quality sample; an obvious disadvantage is that the test has a long duration.

In conducting such a test, residual creep at the start of a test stage is, potentially, a significant complicating factor, which may require lengthy holding periods if it is to be reduced to an acceptable level. This problem increases in severity as the loading rates are reduced to permit the intended effective stress paths to be followed more closely.

For the Oxford Clay, recent strain history appeared to have a greater influence than the recent stress history on the subsequent stress–strain behaviour at small strains. Because of the extreme non-orthogonality of the strain paths followed during orthogonal stress paths, and the particular sequence of stress paths applied, relatively few effects of recent stress history were apparent.

The Oxford Clay exhibited a predictably non-linear and a significantly anisotropic behaviour. Although a relatively simple, non-linear, cross-anisotropic, elastic model was fitted to the test data,

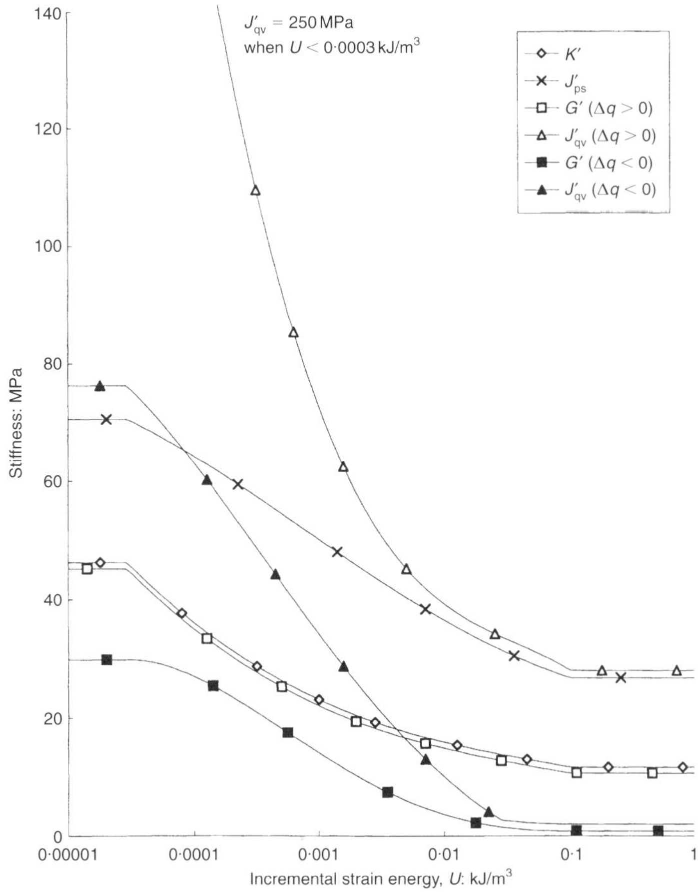

Fig. 17. Stiffnesses employed in the numerical model

Table 4. Stiffness parameters for strata below the Oxford Clay

Stratum	G' (MPa)	K' (MPa)	J' (MPa)
Kellaways Sand	110	220	510
Kellaways Clay	110	220	510
Cornbrash	1370	2810	6420
Blisworth Clay/Limestone	270	540	1240

the calibration of more sophisticated models, such as those proposed by Stallebrass (1990) and Simpson (1992), would be possible.

Despite the fact the Elstow excavation was a well controlled field experiment, supported by an extensive ground investigation, uncertainties still surrounded some of the important input information required for a numerical model. The computed results may have been sensitive to the assumed in situ stress state in the triaxial test from which the stiffness parameters were drawn. In view of this, and the simplifications made in modelling the soil behaviour, it may be fortuitous that the discrepancies between predicted and observed ground deformations were not larger. The predictions of horizontal displacement were more satisfactory than those of vertical displacement.

ACKNOWLEDGEMENTS

The work was supported by EPSRC CASE studentship in collaboration with Mott MacDonald. The technical assistance of Mr P. Osborne, Mr M. Foster and Dr J. Bennell is gratefully acknowledged. The assistance of the London Brick Com-

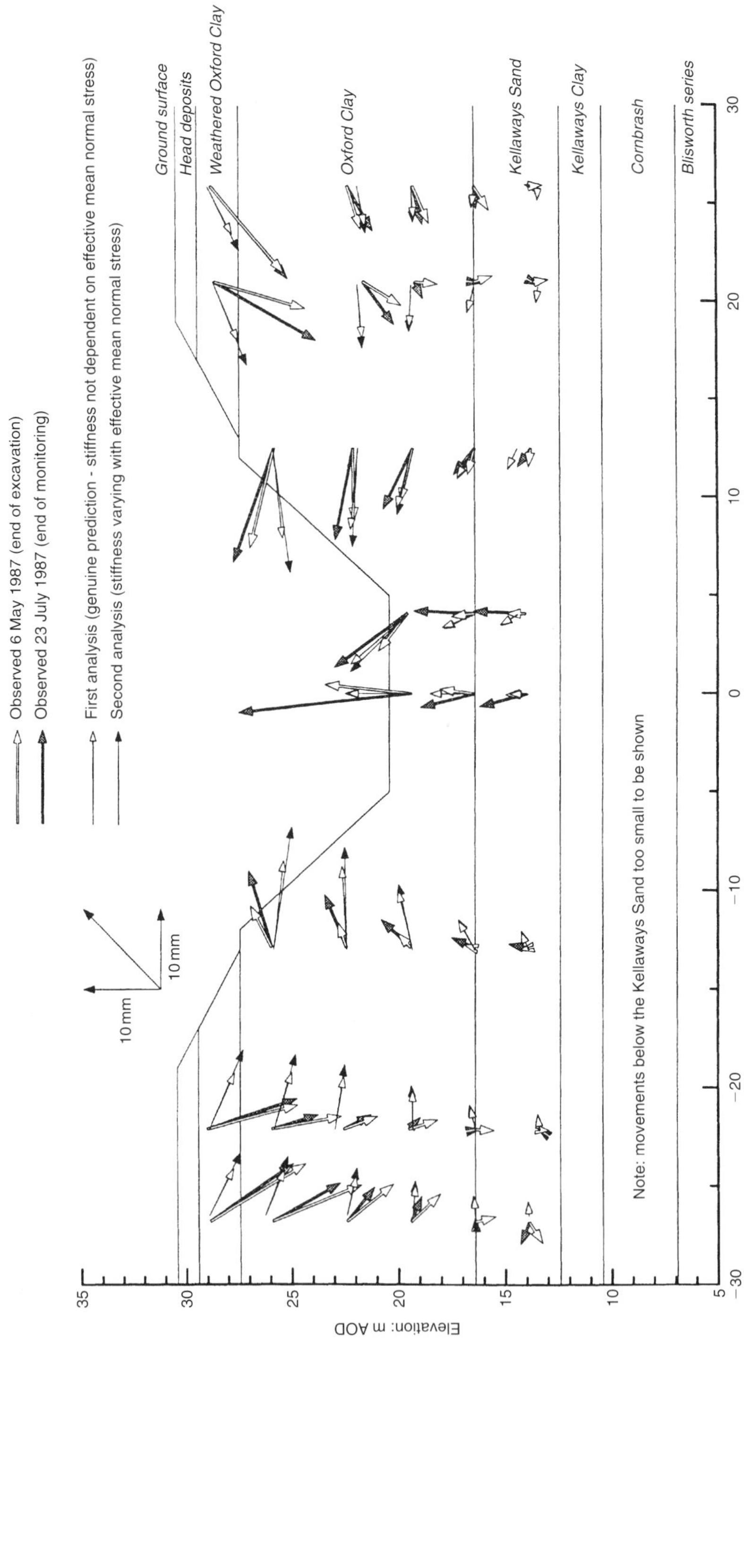

Fig. 18. Comparison of predicted and observed ground movements

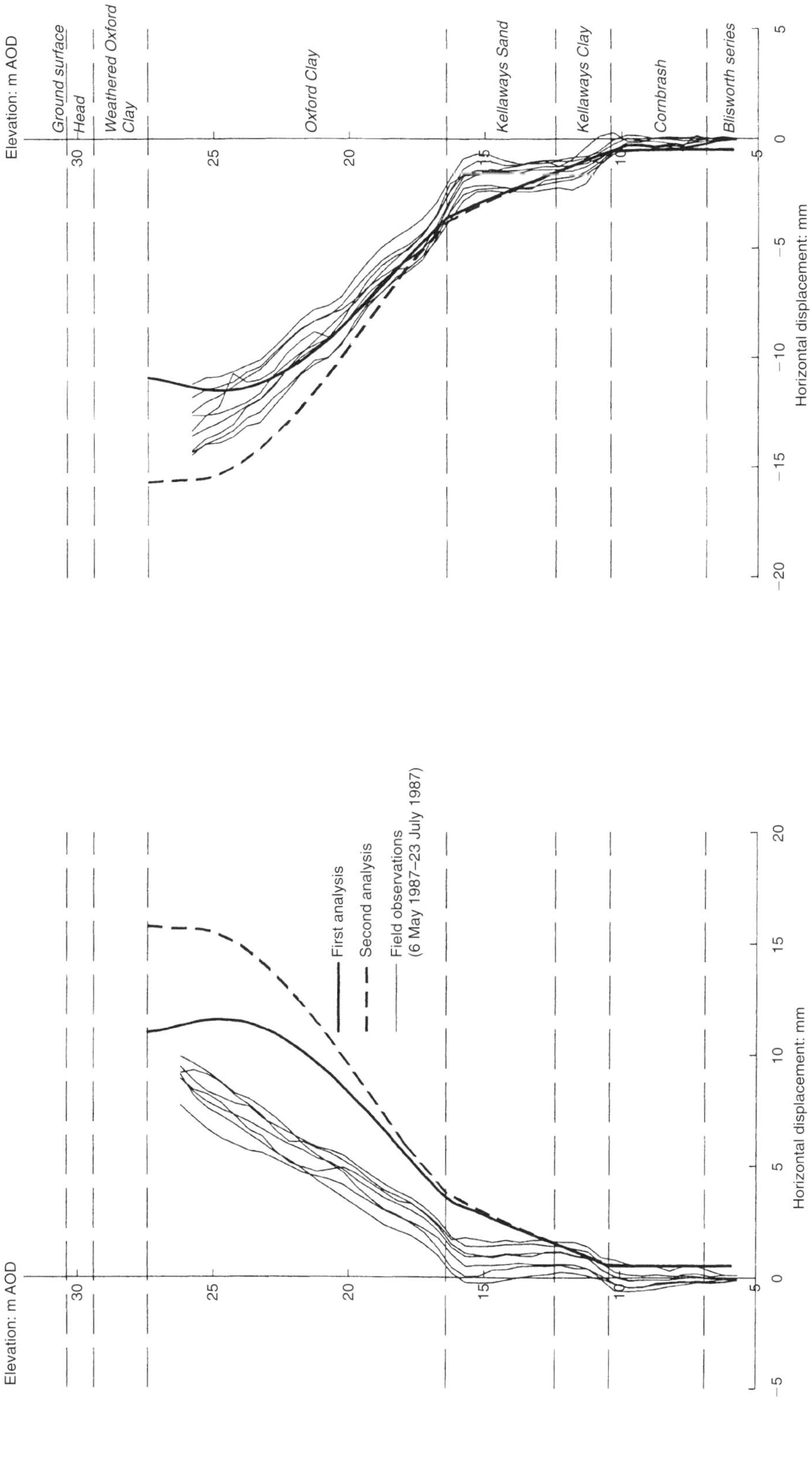

Fig. 19. Comparison of predicted and observed lateral displacements (13 m from excavation centreline)

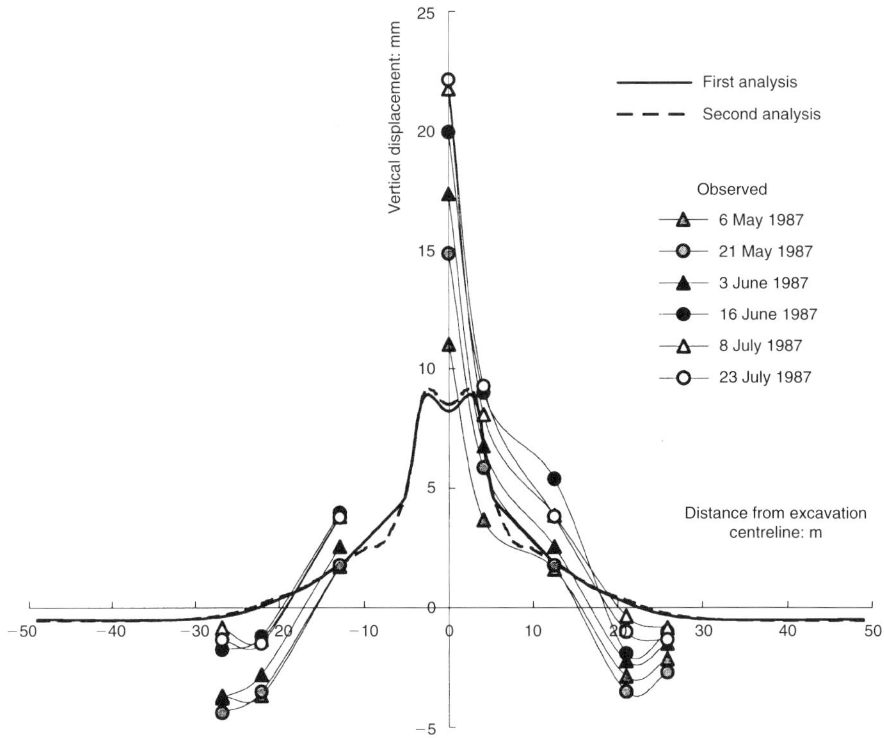

Fig. 20. Comparison of predicted and observed vertical displacements (11 m below ground surface)

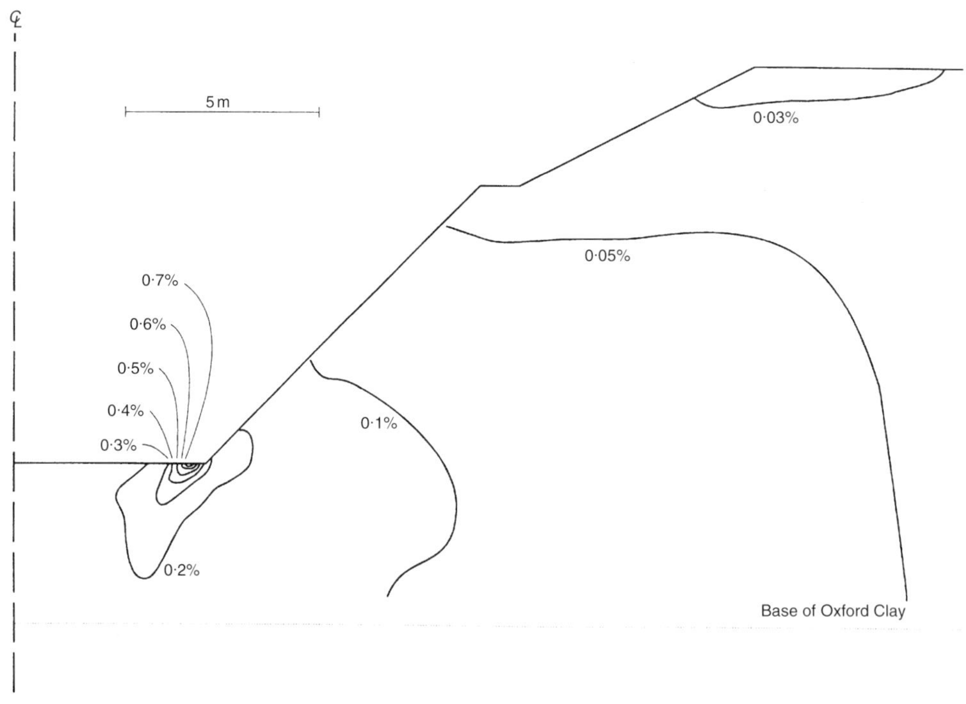

Fig. 21. Predicted shear strains (first analysis)

pany in providing access to Kempston pit is also appreciated.

REFERENCES

Atkinson, J. H. and Richardson, D. (1985). Elasticity and normality in soil – experimental examinations. *Géotechnique* **35**, No. 4, 443–449.

Atkinson, J. H., Richardson, D. & Stallebrass, S. E. (1990). Effect of recent stress history on the stiffness of overconsolidated soil. *Géotechnique* **40**, No. 4, 531–540.

Bates, C. R. (1989). Dynamic soil property measurements during triaxial testing. *Géotechnique* **39**, No. 4, 721–726.

Britto, A. M. & Gunn, M. J. (1987). *Critical State Soil Mechanics via Finite Elements*. Chichester: Ellis Horwood.

Burland, J. B. & Georgiannou, V. N. (1991). Small strain stiffness under generalised stress changes. *Proc. 10th Eur. Conf. Soil Mech., Florence* **1**, 41–44.

Burland, J. B., Longworth, T. I. & Moore, J. F. A. (1977). A study of ground movement and progressive failure caused by a deep excavation in Oxford Clay. *Géotechnique* **27**, No. 4, 557–591.

Burland, J. B., Simpson, B. & St. John, H. D. (1979). Movements around excavation in London Clay. *Proc. 7th Eur. Conf. Soil Mech., Brighton* **1**, 13–30.

Callomon, J. H. (1968). The Kellaways Beds and the Oxford Clay. In: *The Geology of the East Midlands* (eds P. C. Sylvester-Bradley & T. D. Ford), Leicester University Press, 264–290.

Crabb, G. I. & Atkinson, J. H. (1991). Determination of soil strength parameters for the analysis of highway slope failures. In: *Slope Stability Engineering: Developments and Applications* (ed. R. Chandler), London: Thomas Telford, 13–18.

Cripps, J. C. & Taylor, R. K. (1981). The engineering properties of mudrocks. *Quart. J. Eng. Geol.* **14**, 325–346.

Graham, G. & Houlsby, G. T. (1983). Anisotropic elasticity of a natural clay. *Géotechnique* **33**, No. 2, 165–180.

Hight, D. W. (1982). A simple piezometer probe for the routine measurement of pore pressure in triaxial tests on saturated specimens. *Géotechnique* **32**, No. 4, 396–401.

Hight, D. W. & Higgins, K. G. (1994). An approach to the prediction of ground movements in engineering practice: background and application. *Proc. 1st Int. Conf. Pre-failure Deformation Characteristics of Geomaterials, Sapporo (Japan)* **2**, 909–945.

Hird, C. C. & Hajj, A. R. (1995). Simulation of tube sampling effects on the stiffness of clays. *Geotechnical Testing Journal* **18**, No. 1, 3–14.

Hird, C. C. & Pierpoint, N. D. (1994). A non-linear anisotropic elastic model for overconsolidated clay based on strain energy. *Proc. 3rd Eur. Conf. on Numerical Methods in Geotechnical Engineering, Manchester*, 67–74.

Jackson, J. O. (1973). The alteration characteristics of the Lower Oxford Clay. *Clay Minerals* **10**, 113–126.

Jackson, J. O. and Fookes, P. G. (1974). Relationship of the estimated former burial depth of Lower Oxford Clay to some soil properties. *Quart. J. Eng. Geol.* **7**, 137–179.

Jardine, R. J. (1994). One perspective of the pre-failure deformation characteristics of some geomaterials. *Proc. 1st Int. Conf. on Pre-failure Deformation Characteristics of Geomaterials, Sapporo (Japan)* **2**, 855–885.

Jovičić, V., Coop, M. R. & Simic, M. (1996). Objective criteria for determining G_{max} from bender element tests. *Geotechnique* **46**, No. 2, 357–362.

Lambe, T. W. (1973). 13th Rankine Lecture: Predictions in soil engineering. *Geotechnique* **23**, No. 2, 149–202.

Parry, R. H. G. (1972). Some properties of heavily overconsolidated Oxford Clay at a site near Bedford. *Géotechnique* **22**, No. 3, 485–507.

Pierpoint, N. D. (1996). The prediction and back analysis of excavation behaviour in Oxford Clay. University of Sheffield, PhD thesis.

Rudrum, D. M. (1990). The short-term behaviour of a trial excavation in Oxford Clay. Imperial College, London, MSc dissertation.

Simpson, B. (1992). Thirty-second Rankine Lecture: Retaining structures: displacement and design. *Géotechnique* **42**, No. 4, 541–576.

Stallebrass, S. E. (1990). Modelling the effect of recent stress history on the deformation of overconsolidated soils. City University, London, PhD thesis.

Viggiani, G. & Atkinson, J. H. (1995). Stiffness of fine-grained soil at very small strains, *Geotechnique* **45**, No. 2, 249–265.

Yudhbir & Khan, K. K. (1980). Testing for evaluation of stress-strain behaviour of clays. *Geotechnical Testing Journal* **3**, No. 1, 18–29.

Addenbrooke, T. I., Potts, D. M. & Puzrin, A. M. (1997). *Géotechnique* **47**, No. 3, 693–712

The influence of pre-failure soil stiffness on the numerical analysis of tunnel construction

T. I. ADDENBROOKE,* D. M. POTTS* and A. M. PUZRIN†

The use of the finite-element method to analyse tunnels is becoming more widespread, but any prediction is dependent (among other things) on the model adopted for the pre-failure soil behaviour. This paper compares and contrasts plane strain predictions of ground movement for both single- and twin-tunnel excavations in stiff clay modelled as

(*a*) isotropic linear elastic perfectly plastic
(*b*) anisotropic linear elastic perfectly plastic
(*c*) isotropic non-linear elastic perfectly plastic with shear stiffness dependent on deviatoric strain and mean effective stress, and bulk modulus dependent on volumetric strain and mean effective stress
(*d*) anisotropic non-linear elastic perfectly plastic employing the model in (*c*) above
(*e*) isotropic non-linear elastic perfectly plastic with shear and bulk stiffness dependent on deviatoric strain level, mean effective stress, and loading reversals.

The analyses model the geometry of the twin Jubilee Line Extension Project tunnels beneath St James's Park (London, UK), and field data are presented for comparison (Standing *et al.*, 1996). By considering the predicted surface settlement, the study shows the importance of modelling non-linear elasticity, and the effect of introducing a soft independent shear modulus. The differences in subsurface displacements for isotropic and anisotropic models are highlighted. The subsequent modelling of an adjacent tunnel excavation exposes more detailed features of all the models. It is concluded (*a*) that anisotropic parameters appropriate to London Clay do not enhance the plane strain predictions of ground movement as long as non-linear pre-failure deformation behaviour is being modelled; (*b*) that a soft anisotropic shear modulus significantly improves greenfield predictions but not twin-tunnel predictions; (*c*) and that accounting for load reversal effects does influence an analysis of this problem (St James's Park twin tunnels).

L'analyse des tunnels a de plus en plus souvent recours à la méthode des éléments finis, mais toute prévision dépend (entre autres) de la modélisation du comportement des sols avant la rupture. L'article compare les mouvements de sol de tunnels simples et doubles creusés dans de l'argile rigide, sous l'effet de déformations planes, prédits dans les modèles suivants

(*a*) isotrope, linéaire, élastique, parfaitement plastique
(*b*) anisotrope, linéaire, élastique, parfaitement plastique
(*c*) isotrope, non linéaire, élastique, parfaitement plastique, la rigidité au cisaillement variant en fonction des contraintes déviatrices et de la contrainte intergranulaire moyenne, et le module de compression variant en fonction de la contrainte hydrostatique et de la contrainte intergranulaire moyenne
(*d*) anisotrope, non linéaire, élastique, parfaitement plastique, en utilisant le modèle ci-dessus (*c*)
(*e*) isotrope, non linéaire, élastique, parfaitement plastique, la rigidité au cisaillement et la rigidité à la compression variant en fonction des contraintes déviatrices, de la contrainte intergranulaire moyenne et des contraintes subies récemment.

Les analyses modélisent la géométrie des tunnels doubles du prolongement de la ligne de métro Jubilee Line sous St James's Park, et présentent des données relevées sur le terrain à titre de comparaison (Standing *et al.*, 1996). En examinant le tassement de surface prédit, l'étude montre l'importance de la modélisation de l'élasticité non linéaire, et l'effet d'un module de cisaillement tendre indépendant. L'article fait ressortir les différences de déplacements souterrains des modèles isotropes et anisotropes. La modélisation subséquente de l'excavation d'un tunnel adjacent met en évidence des aspects plus détaillés de tous les modèles. L'article conclut (*a*) que les paramètres anisotropes propres à l'argile londonnienne n'améliorent pas les prévisions des mouvements de sol sous l'effet de déformations planes dans la modélisation des déformations non linéaires avant la rupture;

(*b*) qu'un module de cisaillement anisotrope tendre mène à de meilleures prévisions dans le cas de nouveaux chantiers, mais pas celui de tunnels doubles; et (*c*) que l'historique des contraintes récentes compte peu dans l'analyse de ce problème (tunnels doubles de St James's Park).

KEYWORDS: constitutive relations; ground movements; numerical modelling and analysis; settlement; tunnels.

INTRODUCTION

It became evident during the 1970s that the stiffness of London Clay determined in the laboratory was considerably lower than that determined from back analysis of field measurements (Burland, 1989). Fig. 1 demonstrates this by plotting the variation of undrained Young's modulus with mean effective stress derived from a number of sites around London. Also plotted are the ranges of results from the careful testing of specimens from block samples of London Clay retrieved at Ashford Common (Bishop *et al.*, 1965). The values of back-analysed stiffness are significantly greater than the laboratory stiffness. The stiffness determined from deep excavations (5, 6 and 7 in Fig. 1), with dominant horizontal stress changes, is higher that that from vertical loading situations (1, 2, 3 and 4 in Fig. 1), possibly indicating anisotropy of stiffness. At the end of the 1970s, local measurement of strains on samples during triaxial tests revealed much larger stiffness at small strains than previously determined with strain measurement across the end platens. This explains the underprediction of laboratory-determined stiffness seen in Fig. 1. Seismic and ultrasonic tests have recently shown the anisotropy of high stiffness at small strains (e.g. see Porovic, 1995). In 1993, Gunn reported the finite-element developments associated with the tunnelling research programme at Cambridge University in the late 1970s and early 1980s. The finite-element calculations predicted much shallower and wider distributions of surface settlement than was seen in model tests or real tunnelling. The reason was quickly identified as being the elastic part of any constitutive model used to represent the stress–strain behaviour of soil. Gunn (1993) showed the improvement in predicted surface settlement gained with non-linear elastic, rather than linear elastic, behaviour pre-failure.

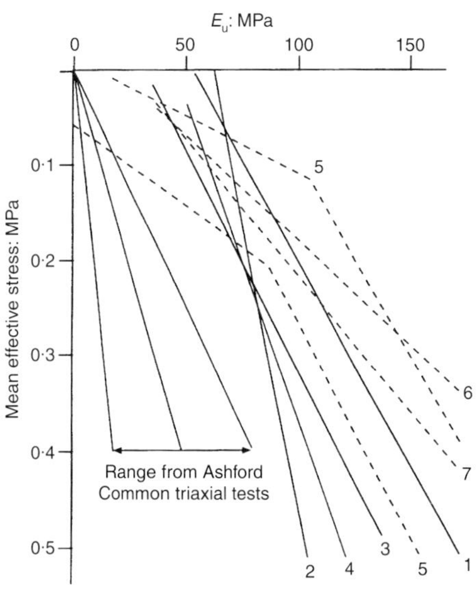

Fig. 1. Variation of undrained Young's modulus with mean normal effective stress derived from various sites on London Clay: 1, Chelsea; 2, Hendon; 3, Hyde Park; 4, Commercial Union; 5, Brittanic House (two lines); 6, New Palace Yard; 7, YMCA. (After Burland, 1989)

Lee & Rowe (1989) carried out linear elastic perfectly plastic finite element analyses of tunnelling specifically to show the influence of cross-anisotropic parameters pre-failure. They concluded that particular attention should be given to the ratio of independent shear modulus to vertical modulus. Their study modelled a soil with an initial horizontal to vertical effective stress ratio (K_0) < 1. Gunn (1993) in referring to Lee & Rowe (1989) stated that, in his experience, the predictions of ground deformations above tunnels in heavily overconsolidated London Clay were not improved by the use of appropriate anisotropic elastic moduli. Simpson *et al.* (1996) concluded from analyses of excavation in London Clay that the predicted surface settlement trough is substantially influenced by the anisotropic shear modulus, but little influenced by non-linearity. These findings appear contradictory.

As numerical analysis of tunnelling has become more widely used, attempts have been made to model the characteristics of the pre-failure defor-

Manuscript received 31 October 1996; revised manuscript accepted 21 March 1997.
Discussion on this paper closes 1 Septemebr 1997.
* Imperial College of Science, Technology and Medicine, London.
† Technion – Israel Institute of Technology, Haifa.

mation behaviour of clays interpreted through back analyses of field data, and observed in the laboratory. The analyses in this paper consider aspects of all the models discussed above, by considering isotropic and anisotropic, linear and non-linear pre-failure deformation behaviour, and also introduces the use of a new non-linear elastic model which accounts for loading reversals (Puzrin & Burland, 1997). The Imperial College Finite Element Programme (ICFEP) has been used to carry out all the analyses presented. The tunnel analyses model the twin tunnels of the Jubilee Line Extension at St James's Park (London, UK), and field data are presented for comparison (Standing *et al.*, 1996).

CONSTITUTIVE MODELS USED TO DEFINE PRE-FAILURE DEFORMATION CHARACTERISTICS

A linear elastic model, if isotropic, requires the specification of two elastic constants. If the soil is cross-anisotropic, then five parameters must be defined: E_v', the vertical Young's modulus; E_h', the horizontal Young's modulus; μ'_{vh}, the Poisson's ratio for the influence of increments of vertical effective stress on horizontal strain; μ'_{hh}, the Poisson's ratio for the influence of increments of horizontal effective stress on the horizontal strain in the orthogonal horizontal direction; and G_{vh}, the shear modulus in the vertical planes (the independent shear modulus). The ratios $n' = E_v'/E_h'$ and $m' = G_{vh}/E_v'$ are often quoted.

One of the models coded in ICFEP which reproduces non-linear behaviour at small strains is outlined by Jardine *et al.* (1986). This model is called model J4 for consistency with Puzrin & Burland (1996). The expressions that describe the variation of shear and bulk moduli in the non-linear region are given in Appendix 1, as are the parameters used in model J4 throughout this paper.

Model J4 has been adapted to model non-linear anisotropy. Model J4 defines the variation of shear and bulk moduli G and K, which give a modulus E' and a Poisson's ratio μ'. The anisotropic elastic stiffness matrix is obtained from constant ratios of μ_{vh}'/μ', μ_{hh}'/μ' E_h'/E', n' and m'. Thus the anisotropic moduli reduce by means of the trigonometric functions defined in Appendix 1. This is referred to as model AJ4.

Puzrin & Burland (1997) have proposed an alternative small strain model which has been coded in ICFEP. The stress space surrounding a stress point, or local origin, is divided into three regions (Fig. 2). The first region is a linear elastic region (LER); in the second region, the small strain region (SSR), the deviatoric and volumetric stress–strain behaviour is defined by a logarithmic curve (Puzrin & Burland, 1996). Each of these regions is bounded by an elliptical boundary surface. In the final region, between the SSR boundary and the plastic yield surface (which is defined independent of the pre-failure model), the soil behaviour is again linear elastic, G and K being associated with the SSR boundary. The model, called model L4 by Puzrin & Burland (1996), is defined in Appendix 2.

Figure 3 shows the variation of secant Young's modulus ($E_u = 3G$) normalized by mean effective stress p', when plotted against percentage axial strain from triaxial test simulations. Both models J4 and L4 compare extremely well with laboratory data for London Clay undergoing undrained triaxial extension (CAU, 'consolidated anisotropically un-

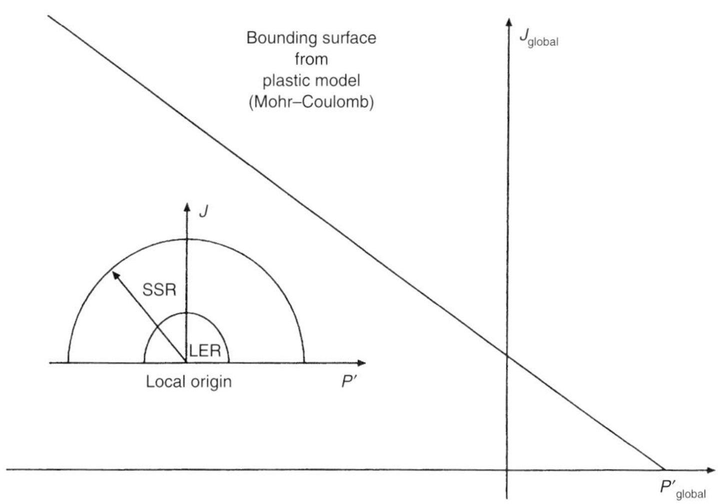

Fig. 2. Three stress regions surrounding a local origin in J–p' space for model L4 (tension positive)

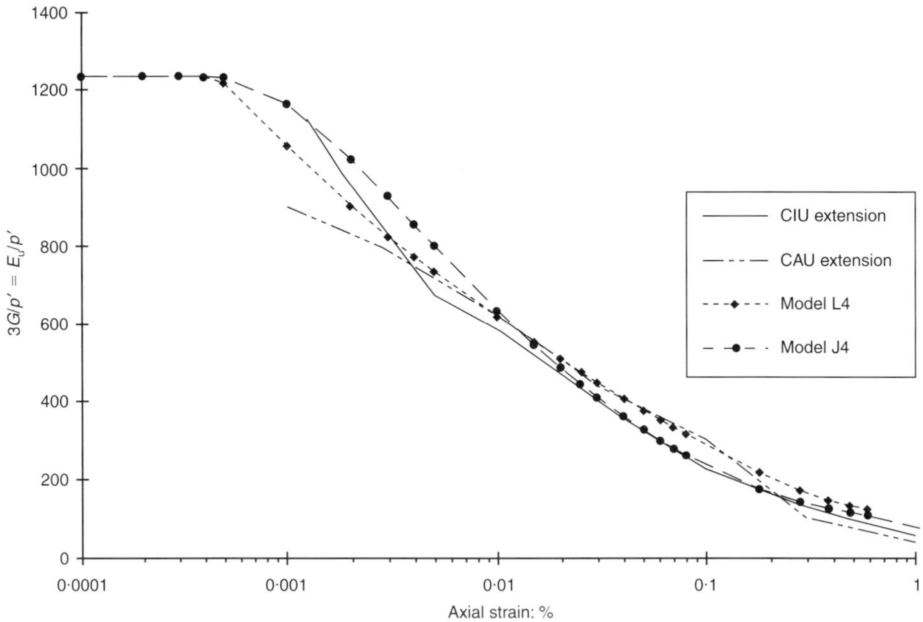

Fig. 3. Secant stiffness–strain curves: comparison between ICFEP models and test data

drained test', and CIU, 'consolidated isotropically undrained test'). Model L4 models a more rapid initial reduction in stiffness in the small strain region, and is softer than model J4 up to 0·01% axial strain; above this strain level, model J4 is softer than model L4.

There are a number of key differences between the three non-linear pre-failure models, J4, L4 and AJ4.

- Models J4 and L4 are isotropic, while model AJ4 is anisotropic.
- Models J4 and AJ4 use a trigonometric function to define the stress–strain behaviour in the non-linear region, while model L4 uses a logarithmic function.
- Models J4 and AJ4 allow only shear stiffness reduction on shear straining with no volumetric strain, and bulk stiffness reduction on volumetric straining with no shear strain, while model L4 forces shear and bulk moduli reduction together if there is only shear or only volumetric strain.
- Model L4 redefines the local origin of stress at a load reversal point, reinvoking the initial high elastic stiffness behaviour. Puzrin & Burland (1997) put forward a criterion for checking whether a change in direction of stress path is associated with a load reversal in a soil element (see Appendix 2). Although acknowledged as a simplified approach to modelling recent stress history effects, no such comparable feature exists within model J4 or AJ4.

ANALYSIS DETAILS

Each of the analyses presented in this paper modelled excavation of the two diagonally oriented running tunnels of the Jubilee Line Extension beneath St James's Park. The geometry is presented in Fig. 4, showing the soil profile, and the finite-element mesh in Fig. 5. The westbound tunnel was excavated first: the ground response is not therefore influenced by any existing surface or subsurface

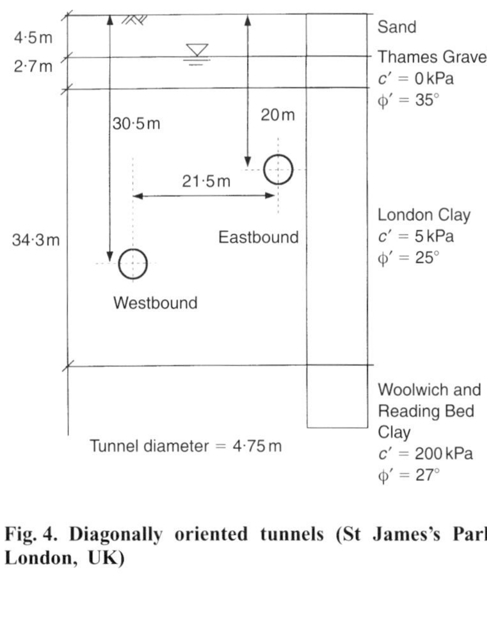

Fig. 4. Diagonally oriented tunnels (St James's Park, London, UK)

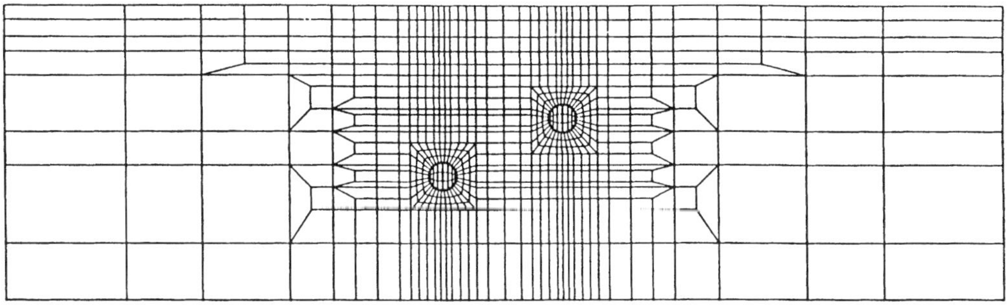

Fig. 5. Finite-element mesh

structures, and is thus termed 'greenfield'. The settlement above such an excavation is the 'greenfield settlement'. The eastbound tunnel is not a greenfield excavation because of the existing westbound tunnel.

A coupled consolidation formulation was employed. Excavation of the tunnels over a simulated time period of 8 h was separated by a rest period of 8·5 months. The sand and gravel were modelled as drained non-consolidating materials. The London Clay was attributed with an anisotropic permeability decreasing with depth. The reduction in horizontal permeability was based on falling head tests carried out near this site at the Palace of Westminster, reported by Burland & Hancock (1977) and shown in Fig. 6(a). The Palace of Westminster site is 300 m from the St James's Park site. The ratio of vertical to horizontal permeability was taken as 1/10 at the top of the London Clay, increasing to 1/2 at the bottom. This profile was adopted as it gave an excellent prediction of the measured pore water pressure profile, as demon-

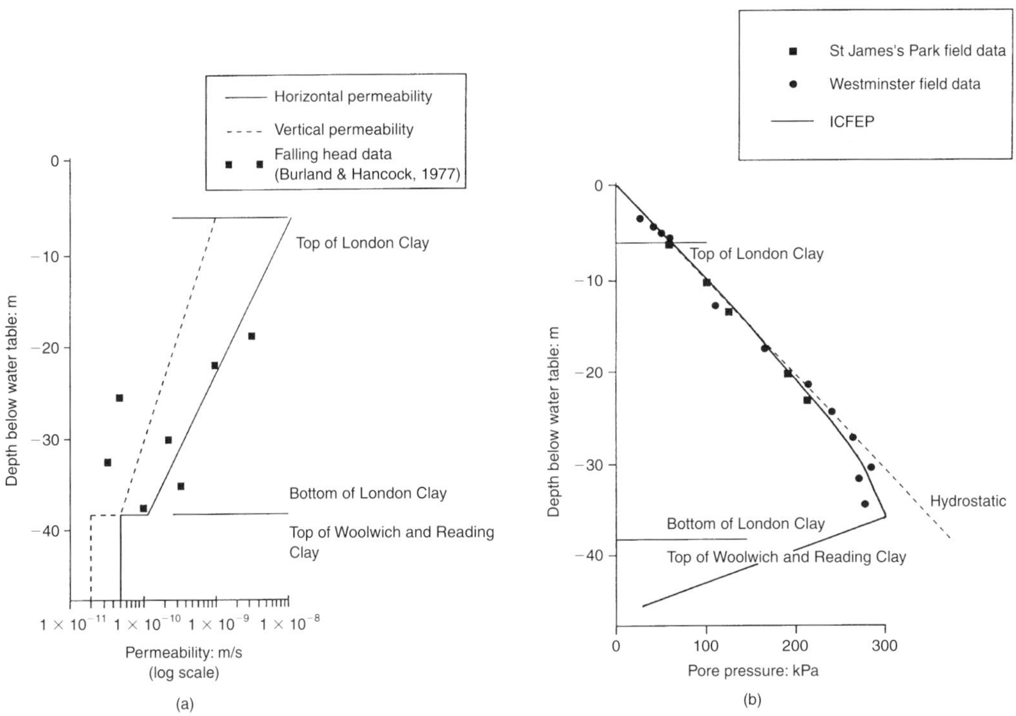

Fig. 6. (a) Permeability profile: (b) underdrained seepage pressure profile, Westminster, London, UK. (ICFEP and field data, Westminster, Higgins *et al.* (1996); St James's Park, Nyren (1995))

strated below. The Woolwich and Reading Clay was attributed with a constant anisotropic permeability.

The initial stresses prescribed an underdrained pore water pressure profile from a water table located 4·5 m below the ground surface (hydrostatic in the Thames Gravel). Fig. 6(b) shows the agreement between the prescribed underdrained pore water pressure profile and field data from St James's Park and Westminster. This prescribed profile is consistent with the anisotropic permeability defined above. An initial effective stress ratio K_0 of 0·5 was prescribed in the sand and gravel, and 1·5 in the London Clay and Woolwich and Reading Clay. In the absence of site measurements of K_0 the value adopted in the London Clay is from the 'upper-bound' profile given by Hight & Higgins (1995), for London Clay between 10 and 30 m below the ground surface. Their profile is based on suction measurements in samples of London Clay using the filter paper technique.

The displacement boundary conditions permitted no horizontal displacement along the two vertical mesh boundaries, and no displacement along the bottom horizontal boundary. Hydraulic boundary conditions dictated that the pore water pressures in the Thames Gravel remained hydrostatic. The initial pore water pressures along the remote boundaries to the clay strata were maintained throughout the analysis. The tunnels were treated as drains using a special boundary condition which prevents water being drawn across the boundary at nodes where suction exists in the adjacent soil, but permits free flow at nodes where the soil water pressures are compressive. The excavation boundary condition simulates the removal of material from the finite-element mesh; the stresses acting at the excavation boundary as a result of the soil elements removed are divided by the number of increments of excavation. These linearly divided stresses are applied in the reverse direction, at that boundary, on each of the increments of excavation. Incremental excavation permits the monitoring of the volume of the developing surface settlement profile, this being a value frequently measured on site. Volume loss is defined as the volume of soil excavated in excess of the theoretical volume of excavation (per unit advance of tunnel), and is quoted as a percentage of the theoretical volume. Rapid excavation in a low-permeability clay results in no volume change, so the volume of the surface settlement profile is equal to the volume loss. The tunnel lining (modelled as a continuous concrete ring: parameters listed in Appendix 5) is constructed once the desired volume loss has been obtained. A volume loss of 3·3% was prescribed for the first tunnel, and 2·9% for the second (calculated from the settlement data at St James's Park; Standing *et al.*, 1996).

SOIL PARAMETERS FOR THE ANALYSES

Made ground was isotropic linear elastic, $E' = 3000$ kPa and $\mu' = 0{\cdot}2$. The Thames Gravel, London Clay and Woolwich and Reading Clay were elastic perfectly plastic with Mohr–Coulomb yield surfaces and plastic potentials (definition and parameters in Appendix 3). The Thames Gravel and the Woolwich and Reading Clay were isotropic linear elastic when the London Clay was linear (parameters in Appendix 4), and isotropic non-linear elastic using model J4 when the London Clay was non-linear (parameters in Appendix 1).

Pre-failure, the London Clay was linear isotropic or anisotropic, or alternatively non-linear isotropic (model J4 or L4), or anisotropic (model AJ4). The linear elastic parameters modelled stiffness increasing with depth, using parameters based on the back-analysed data of Fig. 1 (see Appendix 4). The isotropic non-linear elastic parameters were based on curve fitting to stiffness strain data from locally instrumented triaxial samples (see Appendix 1 for model J4 and Appendix 2 for model L4). Two non-linear anisotropic analyses are presented in this paper, one defines $n' = 0{\cdot}625$ and $m' = 0{\cdot}444$ (model AJ4i), the second reduces m' to 0·2, making the clay very soft in shear (model AJ4ii). Published laboratory data do not give a relationship for the decay of G_{vh} with strain. Based on laboratory and field data for London Clay the initial G_{vh} has been determined by Simpson *et al.* (1996) to be 0·65 G_{hh} (G_{hh}, the shear modulus in a horizontal plane, is dependent on E_h'). With $\mu' = 0{\cdot}2$ and $n' = 0{\cdot}625$, this gives $m' = 0{\cdot}433$. This is very similar to the value adopted in this paper, $m' = 0{\cdot}444$ (assumed by Burland & Kalra, 1986). Lee & Rowe (1989) found an m' value of between 0·2 and 0·25 gave a good surface settlement match with centrifuge and field data (data not for London Clay). The value of m' for model AJ4i is therefore supported by laboratory and field measurements reported by others. The value for model AJ4ii, however, is adopted from the successful work of Lee & Rowe (1989), which is not justified for London Clay by test data.

Model J4 does not automatically reinvoke high stiffness behaviour on loading reversal. For two analyses (isotropic and soft anisotropic), high stiffness behaviour was reinvoked across the entire mesh by zeroing the strains before starting the second excavation. The isotropic analysis is model J40, and the soft anisotropic analysis is model AJ4ii0. This is an approach suggested by Jardine *et al.* (1991) for simulating the kinematic nature of soil stiffness in response to load reversals. Unlike model L4 which detects loading reversals and acts accordingly, model J40 reinvokes high stiffness behaviour at the beginning of a user-defined increment in all the elements in the analysis regardless of the actual location of any reversals.

TRIAXIAL TESTS TO COMPARE THE DIFFERENT MODELS

A single axisymmetric element was used to model ideal undrained extension triaxial tests on London Clay. The initial stresses represented the end of isotropic consolidation to 750 kPa mean stress. The tests were strain controlled by the incremental application of a prescribed displacement at the top. Each non-linear model was used to obtain a comparison of the predicted pre-failure behaviour. The soil was loaded to failure (Mohr–Coulomb yield surface). Fig. 7 shows the stress paths followed, plotted as q against p', where q is the deviator stress and p' the mean stress. Within the yield surface both the isotropic models follow a vertical stress path, and the anisotropic model a negative gradient stress path with increasing p' accompanying the reduction in q.

The stress–strain behaviour and the development of excess pore water pressure with axial strain are compared in Fig. 8. Shear stress is plotted as the change in q. Fig. 8(a) shows the region up to 0·01% axial strain, and Fig. 8(b) up to 0·4% (0·4% axial strain is the limit of non-linear behaviour). The laboratory data for the equivalent test is also plotted. In the very small strain region, the three non-linear models show very similar stress–strain behaviour, and agreement with the laboratory data is excellent (Fig. 8(a)). At larger strains the stress–strain curve for model L4 lies above the others (higher stiffness over the larger strain range shown in Fig. 4), and all the models plot slightly above the laboratory data (Fig. 8(b)). The anisotropic model develops higher excess pore water pressures than the isotropic models.

These analyses show the similarity in the stress–strain behaviour for monotonic loading, and the expected difference in pore water pressure development between isotropic and anisotropic models.

RESULTS FROM TUNNEL ANALYSES

The surface settlement profiles are discussed first, drawing comparisons between the different models and the field data. Following this, the numerical predictions of subsurface ground movements from the different models are compared and contrasted.

Surface settlements

The first series of results presented consider the greenfield surface settlement profiles above the westbound tunnel.

Table 1 shows the consistency of the controlled volume loss with the non-linear soil models, ranging from 3·2 to 3·3%. It is evident, however, that significantly different degrees of unloading were required to achieve this volume loss for the different models. Percentage unloading is obtained from the ratio of the number of increments of excavation to achieve the desired volume loss to the total number of increments of excavation. After 100% unloading the two linear pre-failure models failed to achieve the volume loss measured on site. The surface

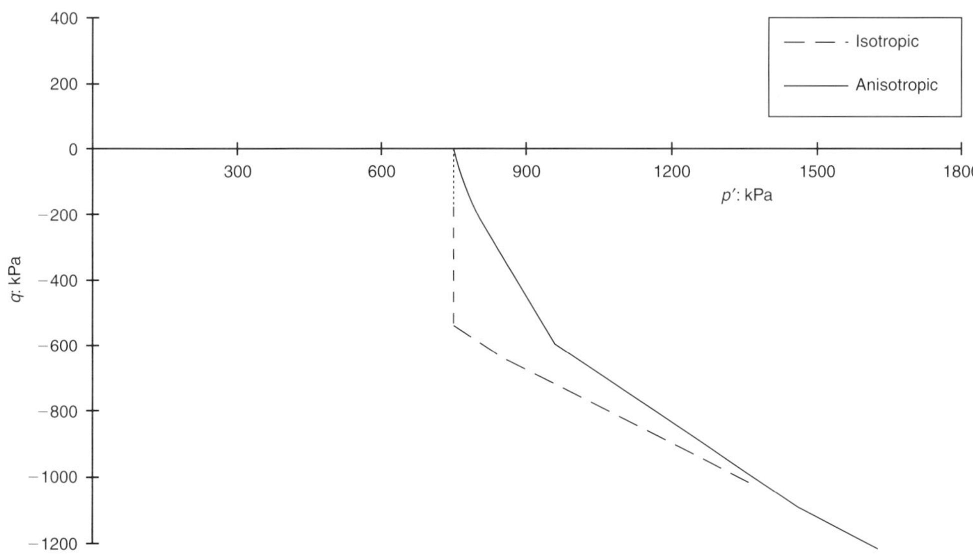

Fig. 7. Stress paths. Isotropically consolidated ($p_0{}' = 750$ kPa) extension test, comparison between isotropic and anisotropic models

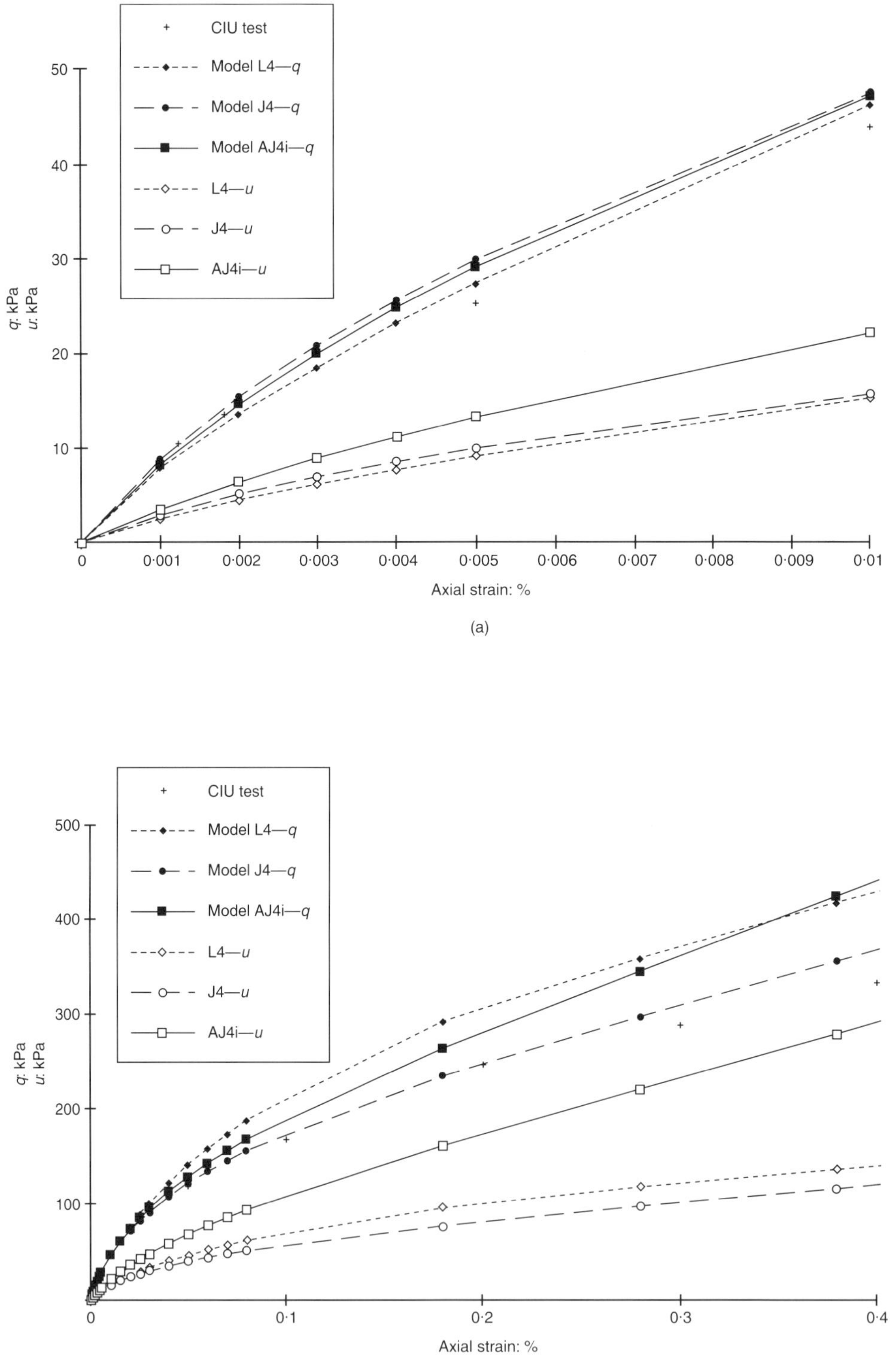

Fig. 8. Stress–strain and excess water pore pressure–strain curves: comparison between ICFEP models and laboratory test data points. Isotropically consolidated ($p_0{}' = 750$ kPa) extension test

Table 1. Percentage unloading of excavation before lining was constructed to achieve the measured volume loss for the westbound excavation of 3·3%

Analysis (London Clay model)	Percentage unloading: %	Volume loss: %
Isotropic linear elastic	100	1·89
Anisotropic linear elastic	100	2·37
Non-linear model J4	75	3·2
Non-linear model L4	80	3·3
Non-linear model AJ4i ($n' = 0{\cdot}625$, $m' = 0{\cdot}444$)	76	3·2
Non-linear model AJ4ii ($n' = 0{\cdot}625$, $m' = 0{\cdot}2$)	57·5	3·2

settlement profiles are presented in Fig. 9(a) for the isotropic soil models. The linear pre-failure model is shown for 100% unloading, and is clearly too wide and shallow, and of the wrong shape. The non-linear analyses predict deeper, narrower profiles. There is little distinction between models J4 and L4. Fig. 9(b) presents the same comparison for the anisotropic soil models. The non-linear pre-failure model predicts deeper, narrower profiles than the linear model. The softer the independent shear modulus (model AJ4ii), the deeper and narrower still the profile. It is clear that non-linear behaviour pre-failure is required to achieve deeper surface settlement profiles closer to the field measurements,

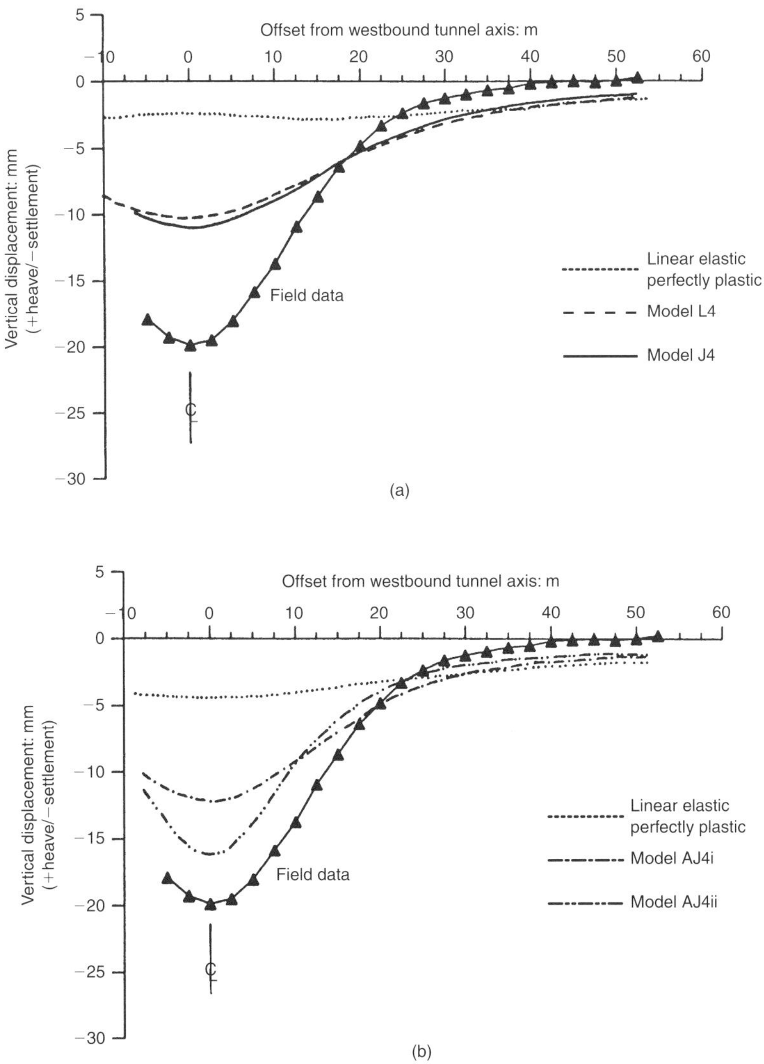

Fig. 9. Surface settlement profiles—first tunnel: (a) isotropic models; (b) anisotropic models. (Models, ICFEP data; field data, Standing *et al.* (1996))

and that the predicted shape and magnitude is improved with a soft anisotropic shear modulus (the maximum predicted settlement is compared with the field data in Table 2).

Having stiffness varying with strain level concentrates the strains at the unloading boundary, so increasing the movements close to the tunnel and reducing them further away to the sides. Adopting a stiffness-strain relationship in model L4 with double the value of shear modulus in the LER (the initial shear modulus, G_0) may be thought to improve this still further. Fig. 10 shows that the stiffness–strain curve adopted still lies within the bounds indicated by the laboratory data for axial strains greater than 0·001% (triaxial test). Carrying out the greenfield tunnel excavation with this stiffer model, however, made no change to the distribution of strain level across the mesh, or to the surface settlement profile at the desired volume loss.

The linear pre-failure models are not considered further as they do not accurately predict the strains at the tunnel boundary in response to excavation.

Figure 11 plots the settlement as a result solely of excavation of the second tunnel (eastbound tunnel), and Table 2 gives the magnitude of maximum predicted settlement. At this stage two new analyses are introduced, models J40 and AJ4ii0, for which high stiffness behaviour was reinvoked across the entire mesh before the second excavation (zeroed strain analyses). Table 3 shows the percentage unloading required to achieve the de-

Table 2. Maximum surface settlement as a result of excavation of each tunnel

Pre-failure model	S_{max} westbound: mm	S_{max} eastbound: mm
J4	11·1	10·8
J40	–	14·3
L4	10·5	12·8
AJ4i	12·5	11·5
AJ4ii	16·6	14·6
AJ4ii0	–	19·9
Field data (Standing *et al.*, 1996)	19·9	23·5

Table 3. Percentage unloading of excavation before lining was constructed to achieve the measured volume loss for the eastbound excavation of 2·9%

Analysis (London Clay model)	Unloading: %	Volume loss: %
Non-linear model J4	65	2·99
Non-linear model J40	75	2·83
Non-linear model L4	75	2·86
Non-linear model AJ4i ($n' = 0{\cdot}625$, $m' = 0{\cdot}444$)	60·5	3·02
Non-linear model AJ4ii ($n' = 0{\cdot}625$, $m' = 0{\cdot}2$)	45	3·08
Non-linear model AJ4ii0 ($n' = 0{\cdot}625$, $m' = 0{\cdot}2$)	57·5	2·94

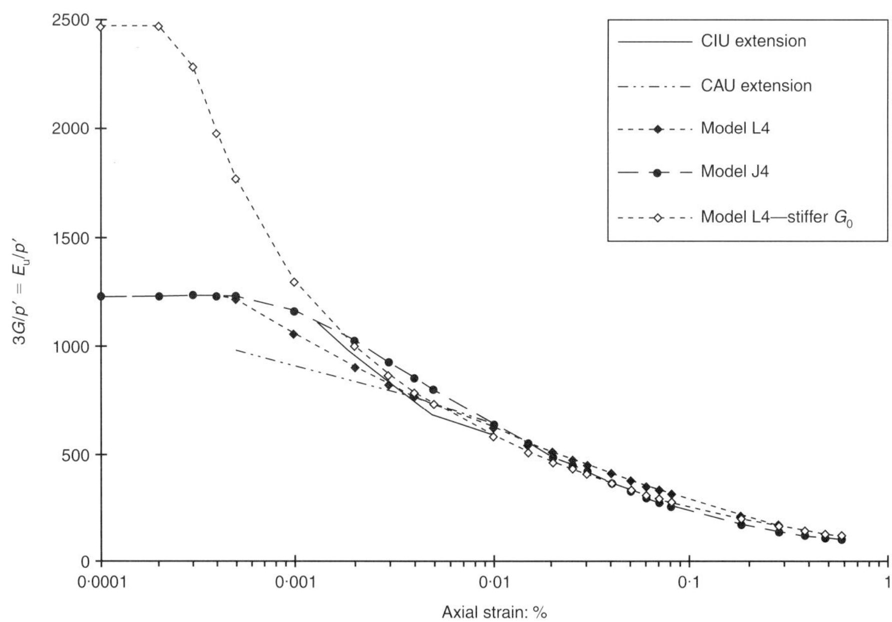

Fig. 10. Secant stiffness–strain curves: comparison between ICFEP models and test data

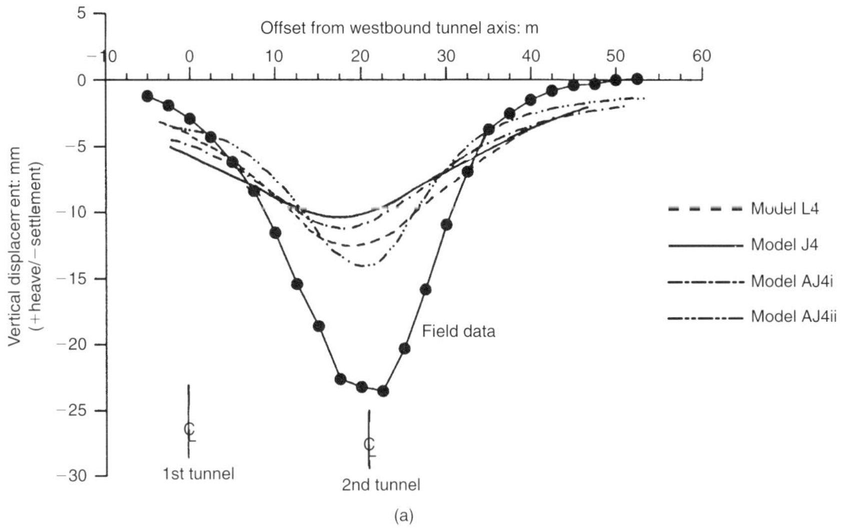

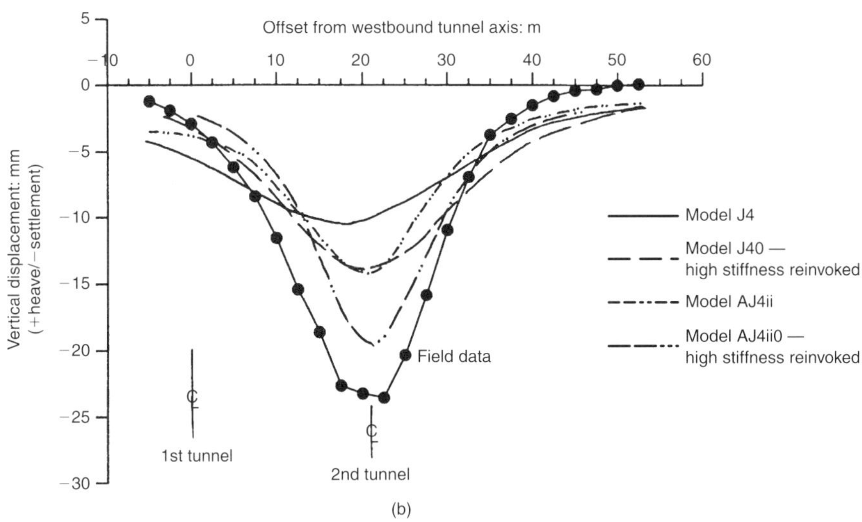

Fig. 11. Surface settlement profiles—second tunnel: (a) non-linear models; (b) non-linear models with and without zeroed strains. (Models, ICFEP data; field data, Standing *et al.* (1996))

sired volume loss, which is held in the range 2·8–3·1%. Fig. 11(a) shows the profiles from the four non-zeroed strain non-linear analyses. It is clear that in each case the predicted surface settlement profile is far wider and shallower than the field data. The similarity of prediction by models J4 and AJ4i implies that employing anisotropic parameters appropriate to London Clay is not a significant factor. The position of maximum surface settlement is pulled towards the first tunnel in the cases of models J4, L4 and AJ4i, but model AJ4ii predicts the centre line settlement as the maximum. The field data exhibit a slight asymmetry, the profile being wider on the side of the existing tunnel than on the opposite side. This feature is displayed by the numerical analyses. Model L4 predicts a deeper, narrower settlement profile than model J4, the two models having predicted very similar profiles for the greenfield tunnel (Fig. 9(a)). The difference here must be attributed to the detection of load reversals in some elements by model L4 and the consequent reinvoking of high stiffness behaviour. Model AJ4ii still predicts the deepest, narrowest profile, but the difference between this model and the others is less than in the greenfield instance (see Table 2).

Figure 11(b) shows the settlement predicted by the two zeroed strain analyses compared with their respective non-zeroed equivalents. In both cases (comparison of model J4 with J40, and of model AJ4ii with AJ4ii0) the magnitude of the predicted settlement is greater, and the trough width narrower. The zeroed strain analyses do, however, lose the asymmetry of the predicted surface settlement profile.

Subsurface ground movements (field data unpublished)

The surface settlement profiles show different results for different pre-failure models for the behaviour of the stiff clay being tunnelled through.

The plots of ground displacement towards the greenfield tunnel excavation, vertical down the axis line, and horizontal along the axis level (Figs 12(a) and 12(b), respectively) show that at the tunnel

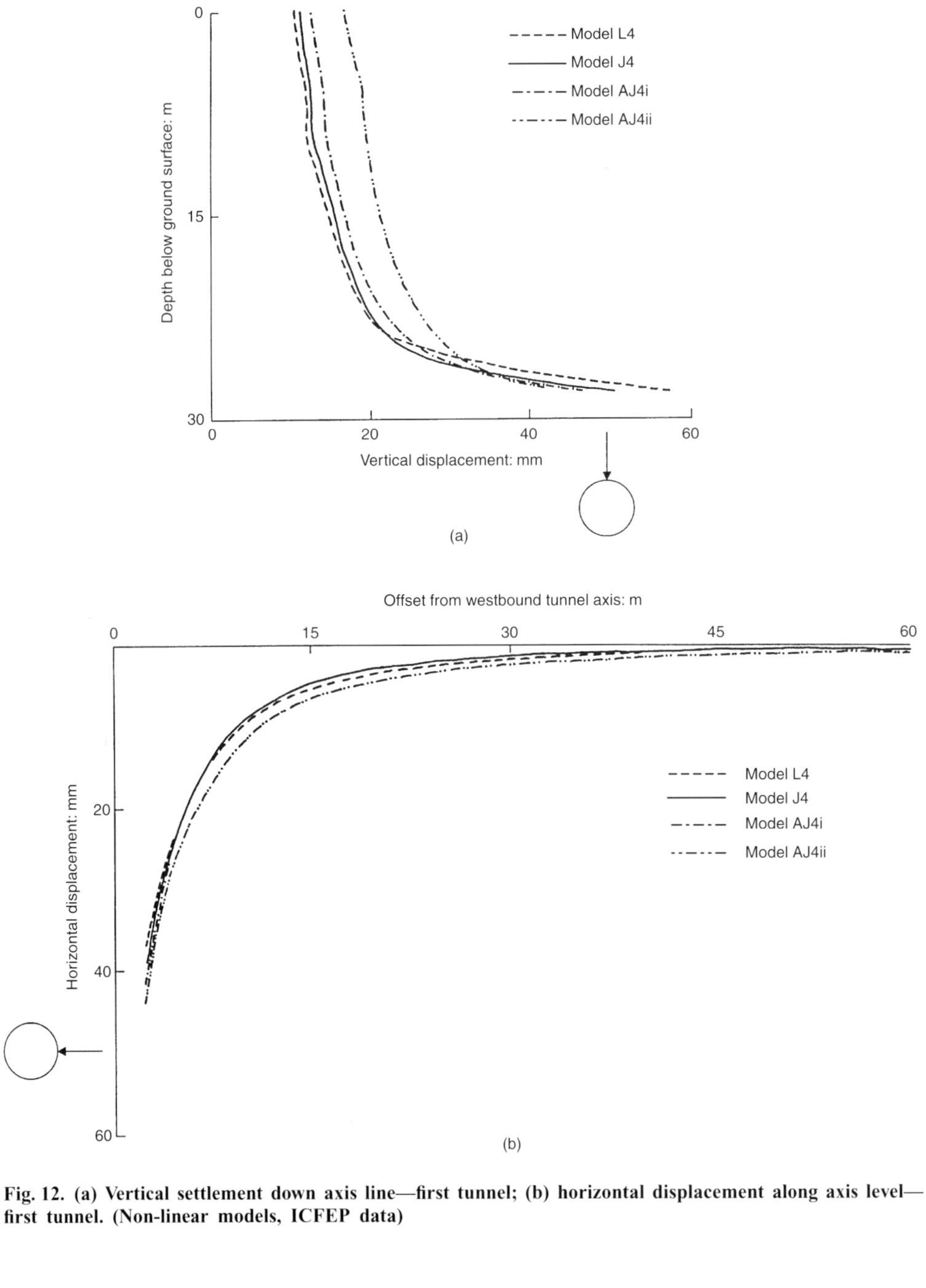

Fig. 12. (a) Vertical settlement down axis line—first tunnel; (b) horizontal displacement along axis level—first tunnel. (Non-linear models, ICFEP data)

boundary the vertical displacement at the crown is greatest for model L4, reducing for model J4, then again for model AJ4i, and the least crown settlement is for model AJ4ii. The relative magnitudes of the horizontal displacement at the springline are in the opposite order (maintaining constant volume loss). From the surface to 1 diameter above the crown the vertical displacement down the axis line is greatest with model AJ4ii, less for models AJ4i and J4 and least for model L4. There is less concentration of displacement within 1 diameter of the crown with model AJ4ii than the other models. The horizontal displacements for models J4, L4 and AJ4i are almost indistinguishable from each other. With model AJ4ii the horizontal displacement is increased at all points away from the tunnel. The deeper, narrower greenfield surface settlement profile is therefore achieved with a soft independent shear modulus by altering the movement at the tunnel boundary itself to give a more uniform pattern of displacement (approximately 45 mm inward displacement at the crown and the springline, compared with nearly 60 mm inward displacement at the crown and 35 mm at the springline with model L4).

Figure 13 shows the development of volume loss with unloading into the second tunnel for models J4 and L4, and plots the equivalent percentage unloading for a greenfield tunnel excavation at the same depth. With model J4 the equivalent volume loss is achieved at a lesser percentage unloading when the tunnel is the second of a pair. This shows the ground in the vicinity of the second tunnel to be softened by excavation of the first. In contrast, model L4 develops volume loss similarly in both greenfield and twin-tunnel instances. The soil around the second tunnel, although softened by excavation of the first tunnel, is clearly experiencing load reversals on commencement of the second excavation. Interrogation of the analysis revealed that soil above and below the second tunnel, and soil directly between the first and second tunnel, experienced load reversals (as defined in model L4 using normalized stresses, see Appendix 2). The evidence from the analyses is that with the same tunnelling technique and standard of workmanship, a greater or equal volume loss is expected in the second tunnel. The field data, with less volume loss into the second tunnel than the first, implies some difference other than soil stiffness for west- and eastbound (for example, the standard of workmanship, or construction sequence at the tunnel face).

The ground movements into the second tunnel are shown in Fig. 14. Both vertical and horizontal displacement patterns are very similar for models J4 and AJ4i, supporting the similarity in their surface settlement predictions. Model L4 (similar to model J4 in the greenfield situation) predicts greater vertical and reduced horizontal displacements. Model AJ4ii, which still gives the least crown settlement, gives more settlement nearer the ground surface. The asymmetry of the ground behaviour, reflected in the surface settlement profiles

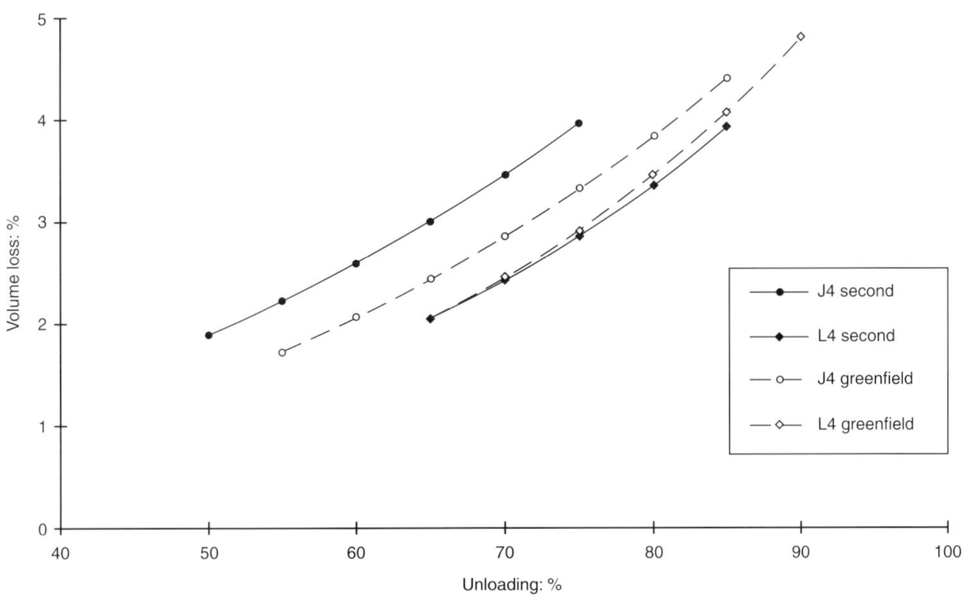

Fig. 13. Development of volume loss with percentage unloading for the non-linear models J4 and L4, comparing the greenfield result with the second tunnel result

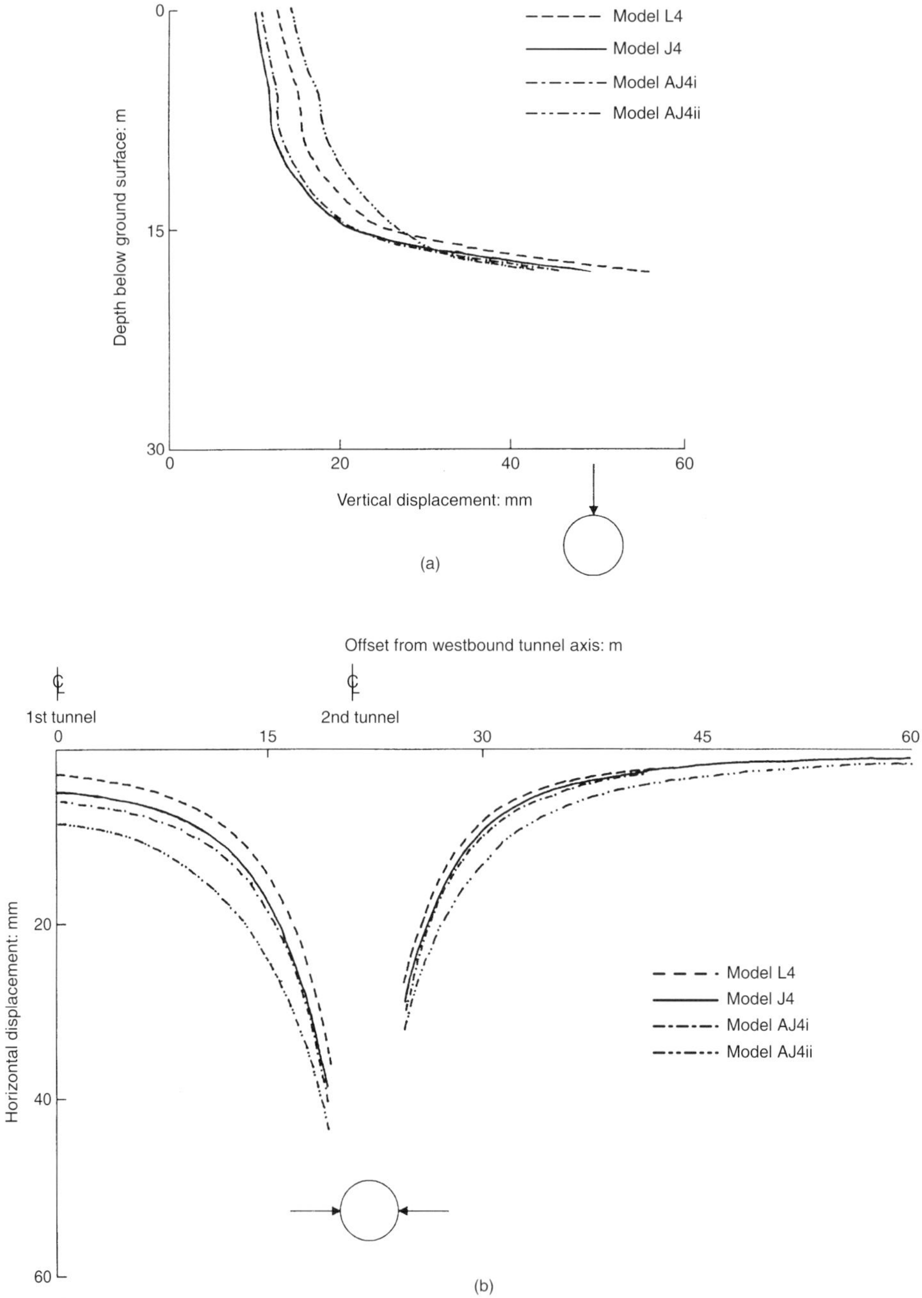

Fig. 14. (a) Vertical settlement down axis line—second tunnel; (b) horizontal displacement along axis level—second tunnel. (Non-linear models, ICFEP data)

of Fig. 11(a), is clear in Fig. 14(b), which shows greater horizontal displacement into the springline on the side of the existing tunnel than on the opposite side for all the models. The difference between horizontal displacement predicted by model AJ4ii and the other models is greater into the second tunnel than into the first (cf. Fig. 11(b)), while the difference in predicted vertical displacement above the crown with the different models is less into the second tunnel (cf. Fig. 11(a)), as reflected at the surface.

The zeroed strain analyses, which improved the

magnitude of the surface settlement prediction, show increased vertical displacements down the axis line when compared with their non-zeroed counterparts (Fig. 15(a)). The crown settlement is also increased. The increased crown settlement is countered by a reduction in horizontal movement at the springline on the side of the existing tunnel (Fig. 15(b)). The horizontal displacements on the opposite side are comparatively similar with and without zeroed strains.

There are two contrasting patterns of subsurface ground movement displayed here, both of which

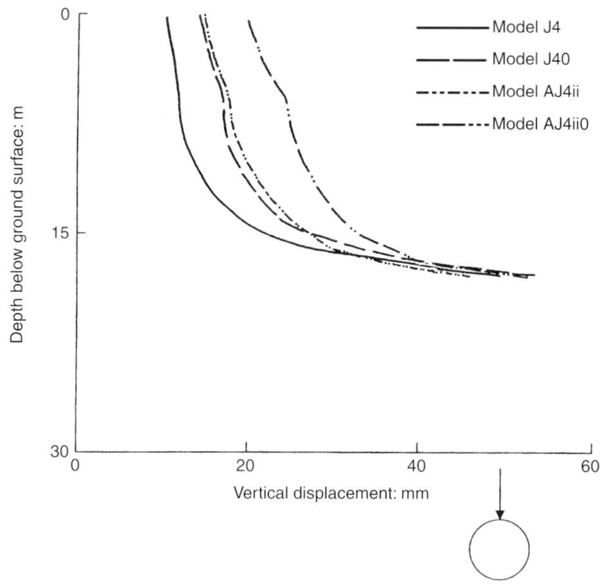

(a)

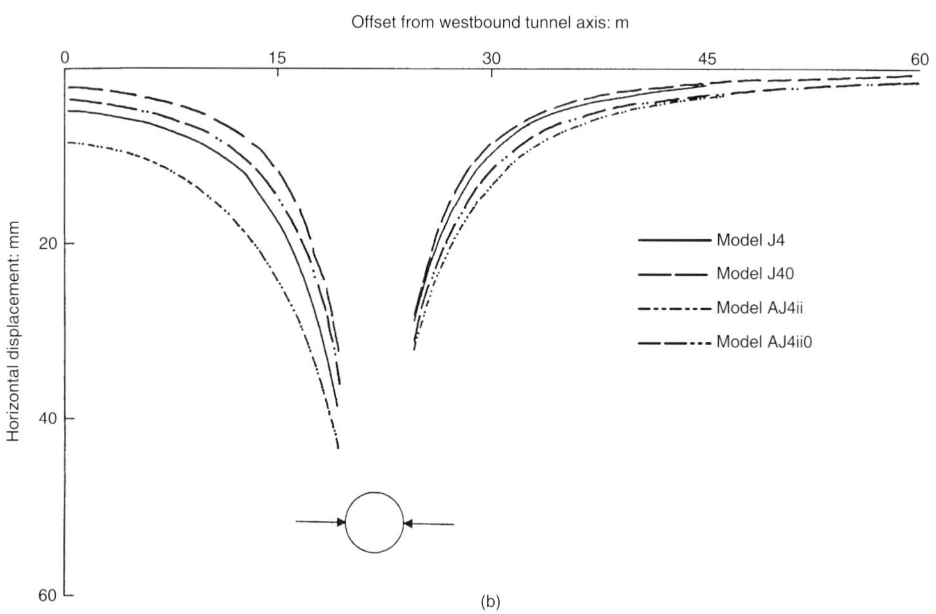

(b)

Fig. 15. (a) Vertical settlement down axis line—second tunnel; (b) horizontal displacement along axis level—second tunnel. (Non-linear models with and without zeroed strains, ICFEP data)

produce deeper, narrower surface settlement profiles, but which are achieved through different variations in the pre-failure deformation model. The greenfield surface settlement profile is improved by the use of non-linear stress–strain behaviour at small strains. The settlement profile above the second adjacent excavation is similarly improved by the reintroduction of non-linear stress–strain behaviour at small strains through the automatic detection of load reversals or through the manual zeroing of the strains across the entire mesh. The influence of such behaviour on the ground movements adjacent to the tunnel is to increase the crown settlement and reduce the springline displacement. The settlement profiles can be further deepened and narrowed, in accordance with the field data, by the use of anisotropic parameters within the non-linear model of stress–strain behaviour at small strains. This has been demonstrated most clearly by the use of a very soft independent shear modulus (model AJ4ii) in this paper. The influence of softening the independent shear modulus within a non-linear anisotropic model is, in contrast to the influence of non-linearity itself, to reduce the crown settlement and increase the springline displacement, producing a more uniform pattern of deformation at the tunnel.

CONCLUSIONS

This paper has looked at the influence of the model adopted for pre-failure soil stiffness on the numerical analysis of tunnel construction. The specific geometry adopted was that of the Jubilee Line Extension tunnels beneath St James's Park in London, UK. The analyses were all plane strain, as the amount of computer resource required for sophisticated three-dimensional analysis is prohibitively large. The influence of three-dimensional effects on numerical predictions of settlement is beyond the scope of this paper, but may contribute to the discrepancy between measured field settlements and the predictions presented.

Greenfield tunnel excavation

Non-linear pre-failure deformation models are necessary to achieve volume loss of the magnitude measured at St James's Park. Isotropic and anisotropic linear pre-failure deformation models fail to predict sufficient straining at the tunnel boundary. The non-linear models achieved the measured volume loss at different increments of unloading. Models J4, L4 and AJ4i required the lining to be constructed after 75–80% unloading. Making the soil softer in shear (model AJ4ii) accelerated the development of volume loss with percentage unloading, and the lining was constructed after 57·5% unloading. The effect of this is to reduce the extent of any zones of yielding soil on completion of excavation, so reducing the magnitude of the drop in pore pressure and the increase in mean effective stress during yielding in a dilatant soil. This influences the stiffness of a stress-level-dependent soil model, and the degree of swelling or consolidation expected in the intermediate term. Constructing the lining earlier in the analysis increases the predicted stresses carried in the lining in the short term.

Non-linear pre-failure models produce considerably deeper, narrower predictions of surface settlement than linear models. Isotropic non-linear models, however, still predict shallower and wider profiles than that measured at St James's Park. Increasing the initial small strain stiffness in the linear elastic region does not improve the settlement prediction at the desired volume loss.

Introducing anisotropic parameters appropriate to London Clay into a non-linear model gives little improvement on the isotropic results. A more reasonable prediction for settlement above a greenfield tunnel excavation can be achieved by employing a very soft independent shear modulus in the anisotropic stiffness matrix. The soft shear modulus causes increased inward displacement at the springline and reduced settlement at the crown, leading to a more uniform distribution round the tunnel than the isotropic models, and increased horizontal subsurface displacement over the full mesh width. Such a soft independent shear modulus cannot, however, be justified from laboratory or field test data for London Clay.

It is apparent that unrealistic soil stiffness is required to achieve an improved prediction of surface settlement above a greenfield tunnel excavation when modelled in plane strain with $K_0 > 1$. If only surface settlement is of interest, then this is a satisfactory device for obtaining good plane strain profiles. The usefulness of such an approach for obtaining information with respect to ground and lining stresses, and pore water pressure distributions, and subsurface ground movements must, however, be questioned.

Adjacent tunnel excavation (non-linear pre-failure deformation models)

If no account is taken of loading reversal effects, the development of volume loss with percentage unloading shows the soil to be softer around the second tunnel than initially around the first. This is not the case when loading reversal effects are accounted for in the fashion of model L4 (a change in direction of the increment of normalized stress reinvoking high stiffness behaviour), in which case the development of volume loss is similar in both greenfield and twin tunnels.

All models predict shallower and wider surface settlement profiles than the measured data. The asymmetry of the measured profile is reproduced, but is overexaggerated above the first tunnel. An isotropic model which captures stress path reversal effects by reinvoking small-strain high-stiffness behaviour (model L4) does improve the settlement prediction above the second excavation. Load reversals were identified above and below the second tunnel, and in the region between the two tunnels. Manually introducing high-stiffness behaviour across the entire mesh prior to excavation of the second tunnel improves the magnitude of the predicted surface settlement by reducing the exaggerated settlement above the existing tunnel. The subsurface displacements show this to be achieved by a reduction in the horizontal axis level movements.

The failure of a soft independent shear modulus to predict the settlement profile above the second tunnel, and its lack of justification from test data, leads further to the conclusion that this approach is a successful device for obtaining the short-term surface settlement above a greenfield excavation alone (if restricted to plane strain analysis). Again, it cannot be justified in general to model soil behaviour accurately around tunnel excavations in high-K_0 stiff clay. Modelling load reversal effects by redefining the local stress origin at a point of normalized deviatoric or mean stress reversal has been shown to be an influencing factor in these analyses (model L4), and can improve predictions. The reinvoking of small-strain high-stiffness behaviour across the entire mesh prior to excavation of the second tunnel proves to be a successful device for achieving a more accurate prediction of the settlement above the second excavation. This approach cannot, however, be justified on the grounds of stress path reversal alone, as the entire soil mass does not experience such reversals when modelled in plane strain.

ACKNOWLEDGEMENTS

The provision of triaxial test data by the Geotechnical Consulting Group is duly acknowledged. The data were obtained from Jubilee Line Extension Project site investigations which were used to establish parameters appropriate to analyses of Westminster Station.

APPENDIX 1. EQUATIONS DEFINING NON-LINEAR ELASTIC MODEL J4

$$\frac{3G}{p'} = C_1 + C_2 \cos\left\{\alpha\left[\log_{10}\left(\frac{E_d}{\sqrt{3}C_3}\right)\right]^{\gamma}\right\}$$

$$\frac{K}{p'} = C_4 + C_5 \cos\left[\delta\left(\log_{10}\frac{(\varepsilon_v)}{C_6}\right)^{\lambda}\right] \quad (1)$$

where G is the secant shear modulus, K is the secant bulk modulus, p' is the mean effective stress, E_d is the deviatoric strain invariant used in ICFEP, ε_v is the volumetric strain, and C_1, C_2, C_3, C_4, C_5, C_6, α, γ, δ and λ are all constants. E_d is related to ε_a (the axial strain observed in undrained triaxial tests) by the expression

$$E_d = \sqrt{3}\varepsilon_a \quad (2)$$

where

$$E_d = 2\sqrt{\tfrac{1}{6}[(\varepsilon_1 - \varepsilon_2)^2 + (\varepsilon_1 - \varepsilon_3)^2 + (\varepsilon_2 - \varepsilon_3)^2]} \quad (3)$$

with ε_1, ε_2 and ε_3 being principal strains.

The constants are obtained from a fit to laboratory data from stress path tests (Jardine *et al.*, 1986). Throughout an analysis the stiffness at a particular point is continually changing. It depends on both the current strain and the current mean effective stress at that point. Until a specified minimum strain $E_{d\,min}$ or $\varepsilon_{v\,min}$ is exceeded, the stiffness varies only with the mean effective stress. This condition also applies once a specified upper strain limit is exceeded, $E_{d\,max}$ or $\varepsilon_{v\,max}$. $E_{d\,min}$, $\varepsilon_{v\,min}$ and $E_{d\,max}$, $\varepsilon_{v\,max}$ are required 'cut-offs' because of the trigonometric nature of the equations (see Fig. 3). The magnitude of the stiffness is prevented from falling below specified minimum values (G_{min} or K_{min}).

Table 4. Small-strain parameters for model J4

Strata	C_1	C_2	C_3: %	α	γ	$E_{d\,min}$: %	$E_{d\,max}$: %	G_{min}: kPa
Thames Gravel	1380·0	1248·0	5·0 × 10^{-4}	0·974	0·940	8·833 46 × 10^{-4}	0·346 41	2000·0
London Clay	1120·0	1016·0	1·0 × 10^{-4}	1·335	0·617	8·660 25 × 10^{-4}	0·692 820	2333·3
Woolwich and Reading Clay	1000·0	1045·0	5·0 × 10^{-4}	1·334	0·591	13·856 40 × 10^{-4}	0·381 050	2666·7
	C_4	C_5	C_6: %	δ	λ	$\varepsilon_{v\,min}$: %	$\varepsilon_{v\,max}$: %	K_{min}: kPa
Thames Gravel	275·0	225·0	2·0 × 10^{-3}	0·998	1·044	2·1 × 10^{-3}	0·20	5000·0
London Clay	549·0	506·0	1·0 × 10^{-3}	2·069	0·420	5·0 × 10^{-3}	0·15	3000·0
Woolwich and Reading Clay	530·0	460·0	5·0 × 10^{-4}	1·492	0·678	1·5 × 10^{-3}	0·16	5000·0

Anisotropic parameters for model AJ4 (London Clay only).
The parameters in Table A1 define the variation of G and K, which define the isotropic Young's modulus and Poisson's ratio, and the following relations form the cross-anisotropic elastic stiffness, matrix, and these relations remain constant throughout the analyses:

$$\mu_{vh}' = \mu_{hh}' = \mu_{isotropic}' \quad E_h' = 1{\cdot}28 E_{isotropic}'$$

For model AJ4i, $n' = 0{\cdot}625$ and $m' = 0{\cdot}444$; for model AJ4ii, $n' = 0{\cdot}625$ and $m' = 0{\cdot}2$. Note: with $n' = 0{\cdot}625$, the average of E_h' and E_v' is very nearly equal to $E_{isotropic}'$ $(0{\cdot}5\ (E_h' + E_v') = 1{\cdot}04 E_{isotropic}')$.

APPENDIX 2. EQUATIONS DEFINING NON-LINEAR ELASTIC MODEL L4

Note that p', J (the mean and deviatoric stress invariants), ε_v and E_d (the volumetric and shear strain invariants) are calculated from the local origin (see Fig. 2)

Elliptical boundary to the LER

$$F_{LER}(p', J) = 1 + n^2\left(\frac{J}{p'}\right)^2 - \left(\frac{a_{LER}}{p'}\right)^2 = 0 \tag{4}$$

where a_{LER} is a parameter defining the size of the LER. It can be determined from an undrained triaxial test as $a_{LER} = nJ_{LER}{}^u$, where $J_{LER}{}^u$ is the value of the deviatoric stress at the LER boundary and $n = \sqrt{(K_{LER}/G_{LER})}$. Within this boundary, the moduli are defined by

$$K_{LER} = \frac{K_{LER}^{ref}}{(p'^{ref})^{\beta}}(p')^{\beta} \quad G_{LER} = \frac{G_{LER}^{ref}}{(p'^{ref})^{\gamma}}(p')^{\gamma} \tag{5}$$

where $K_{LER}{}^{ref}$ and $G_{LER}{}^{ref}$ are values of the bulk and shear elastic moduli, respectively, in the LER at the mean effective stress p'^{ref}.

Elliptical boundary to the SSR (same shape as LER boundary)

$$F_{SSR}(p', J) = 1 + n^2\left(\frac{J}{p'}\right)^2 - \left(\frac{a_{SSR}}{p'}\right)^2 = 0 \tag{6}$$

where a_{SSR} is a parameter defining the size of the SSR. It can be determined from an undrained triaxial test as $a_{SSR} = nJ_{SSR}{}^u$, where $J_{SSR}{}^u$ is the value of the deviatoric stress at the SSR boundary, and n is defined above. Between the LER and the SSR the elastic moduli obey the following logarithmic reductions

$$K = K_{LER}\left[1 - \alpha\left(\frac{(x_v - x_e)R[\ln(1 + x_v - x_e)]^{(R-1)}}{(1 + x_v - x_e)} + [\ln(1 + x_v - x_e)]^R\right)\right] \tag{7}$$

$$G = G_{LER}\left[1 - \alpha\left(\frac{(x_D - x_e)R[\ln(1 + x_D - x_e)]^{(R-1)}}{(1 + x_D - x_e)} + [\ln(1 + x_D - x_e)]^R\right)\right] \tag{8}$$

where

$$\alpha = \frac{x_u - 1}{(x_u - x_e)[\ln(1 + x_u - x_e)]^R} \tag{9}$$

$$R = \left[\left(\frac{(1 - x_e)}{x_u - x_e)} - b\right)\left(\frac{(1 + x_u - x_e)[\ln(1 + x_u - x_e)]}{(x_u - 1)}\right)\right] \tag{10}$$

$$x_u = \frac{2K_{LER}u}{a_{SSR}a_{SSR}} \quad x_v = \frac{|\varepsilon_v|K_{LER}}{p'_{SSR}}$$

$$x_D = \frac{E_d G_{LER}}{J_{SSR}} \quad x_e = \frac{a_{LER}}{a_{SSR}} \tag{11}$$

in which x_u is the normalized strain at the SSR boundary, and x_e the normalized strain at the LER boundary, and b is the ratio of tangent modulus on the SSR boundary to the value within the LER

$$b = \frac{G_{SSR}}{G_{LER}} = \frac{K_{SSR}}{K_{LER}} \tag{12}$$

and u is the incremental strain energy, which can be determined from an undrained triaxial test, being equal to $0{\cdot}5E_{SSR}{}^u J_{SSR}{}^u$, where $E_{SSR}{}^u$ and $J_{SSR}{}^u$ are the deviatoric strain and deviatoric stress on the SSR boundary, respectively.

Outside the SSR the moduli are bulk modulus, $K = bK_{LER}$, and shear modulus, $G = bG_{LER}$.

With this model, a stress path reversal results in the relocation of the local stress origin at the reversal point, so reinvoking the small-strain high-stiffness behaviour. In this respect, a stress path reversal is defined if the increment of normalized shear or mean stress is less than zero; the current stress being normalized by its linear projection on the SSR.

Table 5. Small-strain parameters for model L4

u/p'	$5{\cdot}851\,064 \times 10^{-4}$
b	0·0643
a_{LER}/p'	$2{\cdot}524\,468 \times 10^{-3}$
a_{SSR}/p'	0·191 3149
K_{LER}/p'	214.18
G_{LER}/p'	414.33
β	1.0
γ	1.0
K_{min}	3000.0
G_{min}	2333.3

APPENDIX 3. MOHR–COULOMB YIELD SURFACE AND PLASTIC POTENTIAL AND STRESS INVARIANTS

All the analyses presented in this paper employed the three-dimensional Mohr–Coulomb failure criterion to define the yield of the stiff clay and the gravel. The criterion is specified in terms of the effective stress invariants p', J and θ

$$p' = \frac{(\sigma_1' + \sigma_2' + \sigma_3')}{3} \tag{13}$$

$$J = \sqrt{\tfrac{1}{6}[(\sigma'_1 - \sigma'_2)^2 + (\sigma'_2 - \sigma'_3)^2 + (\sigma'_3 - \sigma'_1)^2]} \quad (14)$$

$$\theta = -\tan^{-1}\left(\frac{(2b-1)}{\sqrt{3}}\right) \quad (15)$$

where $b = (\sigma'_2 - \sigma'_3)/(\sigma'_1 - \sigma'_3)$, such that in triaxial compression ($b = 0$), $\theta = 30°$, and in triaxial extension ($b = 1$), $\theta = -30°$. The yield surface, with a tension positive sign convention, is defined thus

$$F(\sigma) = \frac{J}{(-p' + a)g(\theta)} - 1 = 0 \quad (16)$$

where a is the tensile limit of p' (related to the cohesion c' through $g(\theta)$), and $g(\theta)$ is a function defining the shape of the yield surface in the deviatoric plane and the gradient of the yield surface in the J, p' plane

$$g(\theta) = \frac{\sin\theta}{\cos\theta + (1/\sqrt{3})(\sin\theta\sin\phi')} \quad (17)$$

The plastic potential, $G(\sigma)$, defining plastic straining, has the same equation as the yield function, but with the angle of dilation, ν', defining the gradient of the function in the above equation, rather than ϕ'.

Table 6. Mohr–Coulomb yield surface parameters, plastic potential parameters, and unit weight

	Sand	Gravel	London Clay	Woolwich and Reading Clay
Strength parameters	Linear elastic	$c' = 0$ kPa $\phi' = 35{\cdot}0°$	$c' = 5{\cdot}0$ kPa $\phi' = 25{\cdot}0°$	$c' = 200$ kPa $\phi' = 27{\cdot}0°$
Angle of dilation	Linear elastic	$\nu' = 17{\cdot}5°$	$\nu' = 12{\cdot}5°$	$\nu' = 13{\cdot}5°$
Bulk unit weight: kN/m^3	$\gamma_{dry} = 18$ $\gamma_{sat} = 20$	$\gamma_{sat} = 20$	$\gamma_{sat} = 20$	$\gamma_{sat} = 20$

APPENDIX 4. LINEAR ELASTIC SOIL PARAMETERS

Table 7. Isotropic linear elastic parameters, varying with depth below ground surface, z (in metres)

	Sand	Gravel	London Clay	Woolwich and Reading Clay
Young's modulus, E': kPa	5000	$6000z$	$6000z$	$6000z$
Poisson's ratio, μ'	0·3	0·2	0·2	0·2

Table 8. Anisotropic linear elastic parameters, varying with depth below ground surface, z (in metres)

	E_v'	E_h'	μ_{vh}'	μ_{hh}'	G_{vh}	n'	m'
London Clay	$7500 + 3900z$ kPa	$1{\cdot}6E_v'$	0·125	0·0	$0{\cdot}444E_v'$	0·625	0·444

Burland & Kalra (1986).

APPENDIX 5. TUNNEL LINING PARAMETERS

Table 9. Tunnel lining parameters

Young's modulus	Poisson's ratio	Cross-sectional area	Second moment of area
28×10^6 kN/m^2	0·15	0·168 m^2/m	$3{\cdot}951\,36 \times 10^{-4}$ m^4/m

Unit weight = 24 kN/m^3.

NOTATION

Stiffness parameters

E_u undrained Young's modulus
E' drained Young's modulus
E'_h drained Young's modulus in the horizontal direction
E'_v drained Young's modulus in the vertical direction
G shear modulus
G_0 initial value of G
G_{vh} value of G in the vertical plane
G_{hh} value of G in the horizontal plane
K bulk modulus
m' ratio of G_{vh} to E'_v
n' ratio of E_v' to E_h'
μ' drained Poisson's ratio
μ'_{hh} drained Poisson's ratio in the horizontal plane
μ'_{vh} drained Poisson's ratio in the vertical plane

Stress parameters

J deviatoric stress invariant
K_0 Ratio of initial vertical to horizontal effective stress
p' mean normal effective stress
q deviatoric stress in triaxial space
u pore water pressure
$\sigma'_1, \sigma'_2, \sigma'_3$ principal effective stresses

Strain parameters

E_d	deviatoric strain invariant
ε_v	volumetric strain
ε_a	axial strain
ε_1, ε_2, ε_3	principal strains

Other parameters

c'	cohesive strength
z	depth below ground surface
γ	unit weight
ν'	angle of dilation
ϕ'	angle of shearing resistance

REFERENCES

Bishop, A. W., Webb, D. L. & Lewin, P. I. (1965). Undisturbed samples of London Clay from the Ashford Common shaft: strength–effective stress relationships. *Géotechnique* **15**, No. 1, 1–31.

Burland, J. B. (1989). Small is beautiful – the stiffness of soils at small strains. Ninth Laurits Bjerrum Memorial Lecture. *Can. Geotech. J.* **26**, 499–516.

Burland, J. B. & Hancock, R. J. R. (1977). Underground carpark at the House of Commons, London: geotechnical aspects. *Struct. Eng.* **55**, No. 2, 87–100.

Burland, J. B. & Kalra (1986). Queen Elizabeth II Conference Centre: geotechnical aspects. *Proc. Inst. Civ. Engrs, Part I* **80**, 1479–1503.

Gunn, M. J. (1993). The prediction of surface settlement profiles due to tunnelling. *Predictive soil mechanics, proceedings of the Wroth memorial symposium*, pp. 304–316.

Higgins, K. G., Potts, D. M. & Mair, R. J. (1996). Numerical modelling of the influence of the Westminster Station excavation and tunnelling on Big Ben Clock Tower. *Proceedings of the international symposium on geotechnical aspects of underground construction in soft ground*, London, preprint vol., pp. 447–452.

Hight, D. W. & Higgins, K. G. (1995). An approach to the prediction of ground movements in engineering practice: background and application. In *Pre-failure deformation of geomaterials* (eds S. Shibuya, T. Mitachi & S. Miura), pp. 909–945. Rotterdam: Balkema.

Jardine, R. J., Potts, D. M., Fourie, A. B. & Burland, J. B. (1986). Studies of the influence of non-linear stress–strain characteristics in soil-structure interaction. *Géotechnique* **36**, No. 3, 377–396.

Jardine, R. J., Potts, D. M., StJohn, H. D. & Hight, D. W. (1991). Some practical applications of a non-linear ground model. *Xth Eur. Conf. Soil Mech. Fdn Engng, Florence* **1**, 223–228.

Lee, K. M. & Rowe, R. K. (1989). Deformations caused by surface loading and tunnelling: the role of elastic anisotropy. *Géotechnique* **39**, No. 1, 125–140.

Nyren, R. (1995). Internal communication. Imperial College of Science, Technology and Medicine, London.

Porovic, E. (1995). *Investigations of soil behaviour using a resonant column torsional shear hollow cylinder apparatus*. Imperial College of Science, Technology and Medicine, London, PhD thesis.

Puzrin, A. S. & Burland, J. B. (1996). A logarithmic stress–strain function for rocks and soils. *Géotechnique* **46**, No. 1, 157–164.

Puzrin, A. S & Burland, J. B. (1997). Non-linear model of small strain behaviour of soils. *Géotechnique*, (in press for publication in 1998).

Simpson, B., Atkinson, J. H. & Jovicic, V. (1996). The influence of anisotropy on calculations of ground settlements above tunnels. *Proceedings of the international symposium on geotechnical aspects of underground construction in soft ground*, London, preprint vol., pp. 511–514.

Standing, J. R., Nyren, R. J., Longworth, T. I. & Burland, J. B. (1996). The measurement of ground movements due to tunnelling at two control sites along the Jubilee Line Extension. *Proceedings of the international symposium on geotechnical aspects of underground construction in soft ground*, London, preprint vol., 659–664.

Part two

Additional papers

Pradhan, T. B. S. & Ueno, Y. (1998). Pre-failure deformation behaviour of geomaterials, 329–335

Cyclic deformation characteristics of clay under different consolidation histories

T. B. S. PRADHAN* and Y. UENO†

The main objectives of the present paper are to investigate (*a*) the effect of consolidation stress history on the small to medium strain level stiffness of clay subjected to either monotonic or cyclic loadings, and (*b*) the correlation between the stiffness from anisotropically and isotropically consolidated specimens and from cyclic and monotonic tests. A series of undrained cyclic and monotonic triaxial tests using a reconstituted clay have been carried out on specimens consolidated under different anisotropic stress conditions. The anisotropic stress conditions were specified in such a way that the consolidated vertical stress is changed arbitrarily while keeping the mean effective stress constant and vice versa. Equivalent shear modulus at different strain levels was calculated and the effect of anisotropic consolidated stress ratio is discussed. It was found that small strain stiffness depends largely on stress ratio even though the effective vertical consolidation stress was kept constant. The stiffness of an anisotropically consolidated specimen can be estimated from the conventional isotropically consolidated test when the isotropic stress is made equal to the mean of major and minor principal stresses in anisotropic stress conditions.

KEYWORDS: anisotropic stress; clay; cyclic loading; isotropic stress; monotonic loading; small strain stiffness.

Le principal objectif de cet exposé est d'étudier (a) l'effet de l'histoire de la contrainte de consolidation sur une faible à moyenne rigidité de déformation dans l'argile soumise à des chargements monotoniques ou cycliques et (b) la corrélation entre la rigidité de spécimens consolidés de manière anisotrope et isotrope et celle d'essais cycliques et monotoniques. Nous avons fait une série d'essais triaxiaux cycliques et monotoniques non drainés utilisant une argile reconstituée sur des spécimens consolidés dans différentes conditions de contrainte anisotropes. Les conditions de contrainte anisotropes ont été spécifiées de telle manière que la contrainte consolidée verticale a été changée arbitrairement alors que la contrainte moyenne effective restait constante et vice versa. Nous avons calculé le module de cisaillement équivalent, pour différents niveaux de contrainte, et nous examinons l'effet du taux de contrainte consolidé anisotrope. Nous avons observé qu'une faible rigidité de déformation dépend en grande partie du taux de contrainte même si la contrainte de consolidation verticale effective reste constante. La rigidité du spécimen consolidé anisotrope peut être évaluée grâce à l'essai consolidé isotrope conventionnel quand la contrainte isotrope est rendue égale à la moyenne des contraintes principales majeures et mineures en conditions de contrainte anisotrope.

INTRODUCTION

In order to perform a precise seismic response analysis of clay deposit it is necessary to evaluate accurately the shear modulus and damping ratio of clay at strain level less than about 10^{-5} under an anisotropic stress condition very close to that in the field. However, quite often the conventional method of measuring the shear modulus, damping ratio and its strain level dependency by using a soil specimen which is consolidated under isotropic stress condition in triaxial apparatus is used. Much work has been carried out on sands to investigate the effect of the anisotropic stress conditions on the small strain stiffness and its stress level dependency. However, studies on clay are few and limited.

Lawrence (1965) measured the ultrasonic shear wave velocity in triaxial soil specimens subjected to anisotropic consolidation stresses, and concluded that the shear wave velocity is affected only by the mean effective principal stress and is independent of the shear stress level. Hardin (1966), from resonant column tests on anisotropically consolidated sand specimens, has pointed out that small strain stiffness depends only on the square root of the mean principal stress and is independent of the

Manuscript received 7 February 1997.
* Yokohama National University, Yokohama.
† Taisei Construction Co., Tokyo.

initial static shear stress. Hardin & Black (1968) showed that this conclusion is also applicable to clays. Porovic & Jardine (1994) showed that the small strain stiffness for both K_0 and isotropically consolidated sand specimens depends on the mean effective consolidation stress. On the other hand, some researches show that the small strain stiffness depends on the stress ratio and the stress-paths (e.g. Kuribayashi *et al.*, 1976; Tatsuoka *et al.*, 1979; Yu *et al.*, 1984; Roesler, 1979; Lo Presti *et al.*, 1995). Lo Presti *et al.* (1995), from test results on Pisa clay, showed that the Young's modulus E_0 values at small strain level ($<10^{-5}$) obtained from isotropically and K_0 consolidated specimens are virtually the same when the vertical effective stresses are kept the same. They concluded that E_0 depends solely on the axial consolidation stress, while at larger strain level a greater stiffness is observed for isotropically consolidated specimens. Yu *et al.* (1984), from resonant column tests on dry sands, showed that small strain stiffness depends equally on the static principal stresses in the direction of wave propagation and in the direction of particle motion, while the effect of the third principal stress is unimportant.

In the present study a series of comprehensive tests on clay using triaxial apparatus was carried out on specimens consolidated to different stress ratios in the compression side only. The effect of stress ratio on the shear modulus of clay at small strain to medium strain level is discussed. The stiffness values from the monotonic and cyclic loading tests are compared and discussed from the point of view of the generated excess pore water pressure and loading strain rate.

TESTING MATERIAL

A soil sample from Tokyo Bay was passed through a 75 μm sieve and used as the testing material. The physical properties of the sample thus prepared were: particle density 2·69 Mg/m^3, liquid limit 80%, plasticity index I_p 45. The grain size distribution curve of the soil is shown in Fig. 1. The sample was mixed thoroughly at a water content of twice the liquid limit and then preconsolidated one-dimensionally at a vertical pressure of 50 kPa in a mould of inner diameter 30 cm.

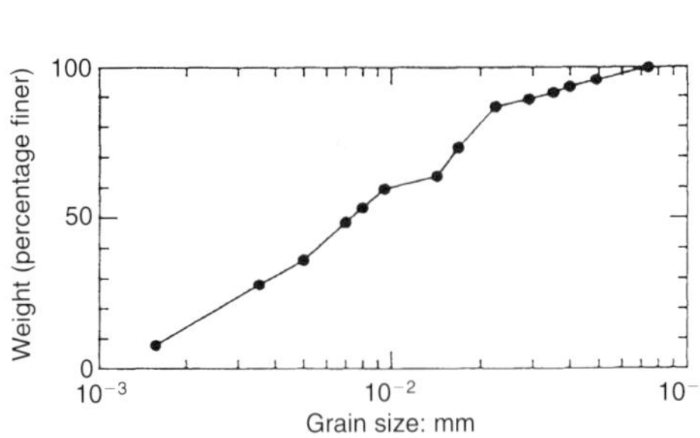

Fig. 1. Grading curve of the clay

TESTING METHODS AND TESTING CONDITIONS

Triaxial specimens of height 10 cm and diameter 5 cm were used. Porous stones together with a non-compressible chemical filter were used at both ends in order to eliminate the bedding error. Axial displacement was measured by non-contact proximeter just above the cap inside the triaxial cell. Axial load was measured inside the triaxial cell by a high-rigidity and high-sensitivity load cell. A computerized pneumatic system to control the axial stress and radial stress was used to apply a prescribed stress-path to the specimen. Pore water pressures at the top and bottom ends and at the midheight periphery of the specimen were measured. Details on the measurement of the pore pressure at the midheight of the specimen are given in Pradhan *et al.* (1995). A back pressure of 200 kPa was applied to ensure high saturation of the specimen. The calibration of the system and the repeatability of the tests were checked in the preliminary test series.

A soil specimen is first consolidated to a prescribed stress-path of constant stress ratio and then subjected to either a cyclic loading test (CLT) or a monotonic loading test (MLT) under the undrained condition. Fig. 2 shows the stress-paths employed in this study. The solid circles show the stress points where shearing tests were carried out. During the consolidation the rate of increase in control pressure was kept constant at 0·5 kPa/min. The consolidation was terminated when the excess pore water pressure at the midheight of the specimen became almost zero. The consolidation period is strictly controlled so as not to measure the stiffness increase due to secondary

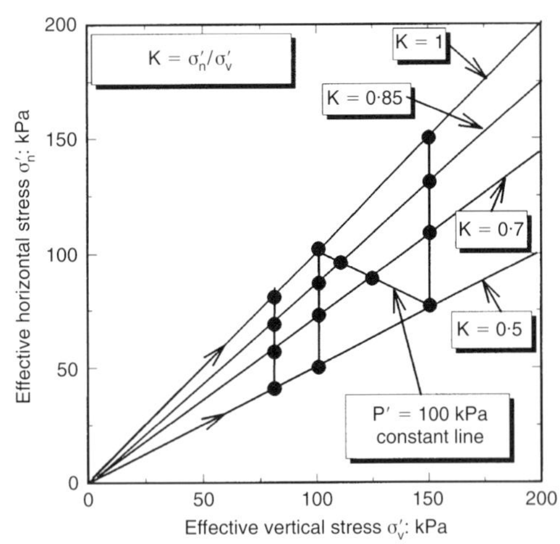

Fig. 2. Effective stress paths used

consolidation (Afifi & Richart, 1973). In case of CLTs, eleven constant shear stress cycles were applied at different stages of strain levels. The equivalent shear modulus and damping ratio were calculated by the ten cycle loop. The loading frequency was 0·1 Hz in all cases. In case of MLTs, axial strain rate was kept constant during shearing and the strain rates were varied to 0·01, 0·5, 0·1, 0·2%/min.

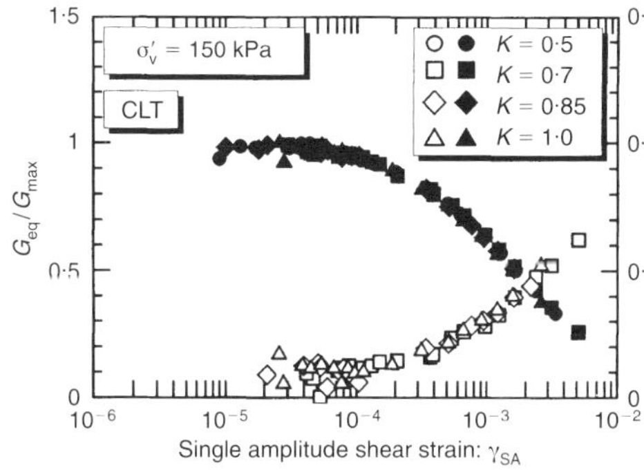

Fig. 4. Relationship between G_{eq}/G_{max}, h and shear strain

TEST RESULTS AND DISCUSSION

In order to take account of the effect of the void ratio after consolidation, a void ratio function $F(e) = e^{-13}$ has been used as proposed by Jamiolkowski *et al.* (1991), where e is the void ratio of the soil specimen after consolidation. A Poisson's ratio of 0·5 is assumed due to the undrained condition. All stiffnesses are expressed in the normalized form using the void ratio function.

Effect of stress ratio during consolidation

Fig. 3 shows a typical relationship between the normalized equivalent shear modulus G_{eq}, damping ratio h and the single amplitude shear strain γ_{SA} for different K $(= \sigma'_h/\sigma'_v$; where σ'_h, σ'_v are respectively the effective horizontal (radial) and vertical consolidation stresses), while σ'_v was kept constant at 150 kPa. It is clear from the figure that the stiffness at small strain of 10^{-5} is larger for the isotropic stress condition ($K = 1$) and lower for smaller values of K. Similar behaviour was observed for other σ'_v values. This tendency contradicts the findings of Lo Presti *et al.* (1995). This result implies that the stiffness is also affected by the lateral stress. However, the damping ratio h seems not to be affected significantly by the lateral stress. G_{eq} is normalized by G_{max}, and plotted against γ_{SA} in Fig. 4. G_{max} is the elastic shear modulus defined at a strain level of 10^{-5}. It can be observed from Fig. 4 that a unique relationship exists between G_{eq}/G_{max} and γ_{SA}, which can be expressed by a hyperbolic function as proposed by Hardin & Drnevich (1972). This implies that the degree of the strain level dependency of the shear modulus is the same for different K values when effective vertical stress is kept constant. The normalized stiffness is plotted against γ_{SA} in Fig. 5 for different K values when the mean effective principal stress p' $(= (\sigma'_v + 2\sigma'_h)/3)$ is kept constant at 100 kPa. It is clear from this figure that the normalized stiffness values at medium to large strain level are almost the same, but G_{max} is slightly different for different K values. In this case, G_{max} seems larger for lower values of K and this tendency is opposite to the one observed in Fig. 3. In order to see more clearly the effect of vertical consolidation stress on the small strain stiffness, normalized G_{max} is plotted against σ'_v in Fig. 6. Solid lines are the fitted lines for constant K values. From this figure, (*a*) normalized G_{max} increases as K increases if σ'_v is kept constant; (*b*) normalized G_{max} increases with σ'_v for constant K (this stress level dependency is almost the same irrespective of the value of K); (*c*) the broken line shows the data points connecting constant

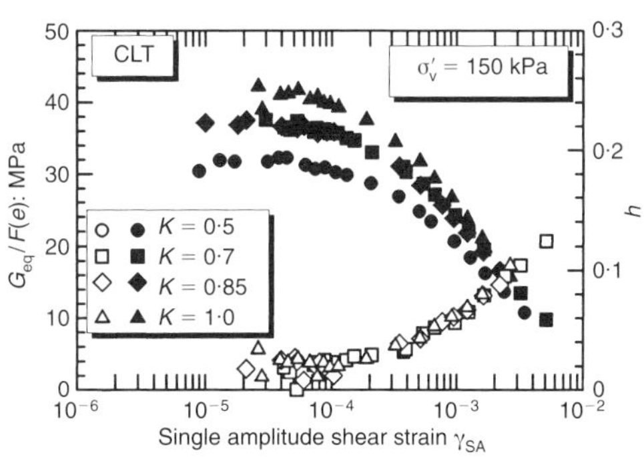

Fig. 3. Relationship between $G_{eq}/F(e)$, h and shear strain

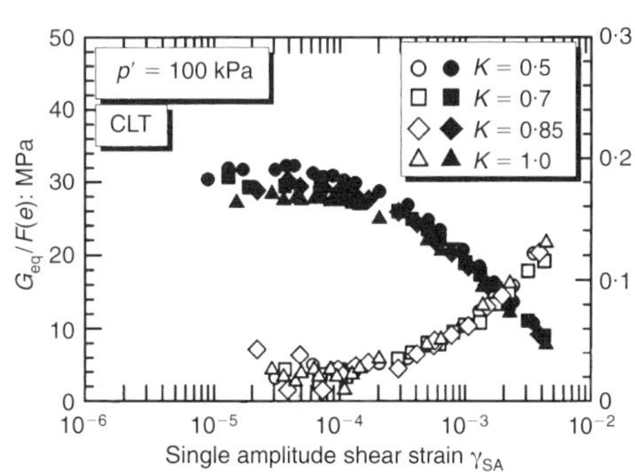

Fig. 5. Relationship between $G_{eq}/F(e)$, h and shear strain

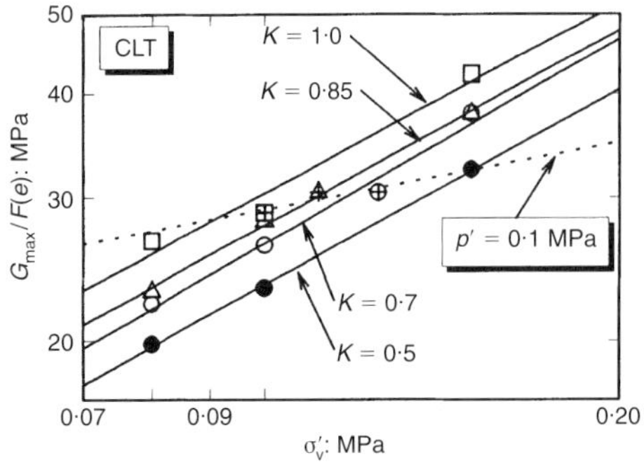

Fig. 6. Relationship between $G_{max}/F(e)$ and σ'_v

$p' = 100$ kPa for different K values. Since the slope of the broken line is smaller than that of the solid lines, it can be concluded that the effect of the K value is smaller when p' is kept constant. If the slope of the broken line is zero, it means that the normalized small strain stiffness is independent of K and σ'_v if p' is kept constant.

Fig. 7 shows the normalized G_{max} plotted against p' for different values of K and σ'_v. Although there is some scatter in data, it can be surmised that $G_{max}/F(e)$ depends uniquely on p' rather than on σ'_v for different values of stress ratio K.

For the $K = 1$ condition, the fitted solid line in Fig. 6 can be expressed by

$$G_{max} = 176\,F(e)\sigma'^{0.76}_v \quad \text{(in MPa)} \tag{1}$$

The modulus exponent in equation (1) is 0·76, which is larger than 0·5 as proposed by Hardin & Black (1968). Recently, Cascante & Santamarina (1996) studied the modulus exponent analytically and experimentally. According to Hertzian contact theory which is based on elastic contact, the modulus exponent is theoretically 1/3, but Cascante & Santamarina showed that for brittle contact, which might occur in clay, the exponent is expected to be higher. The authors believe that this modulus exponent depends largely on the void ratio function and on the compressibility of the soil. Further experimental study in this aspect is needed.

The relationship of $G_{max}/F(e)$ for anisotropically consolidated specimens ($K \neq 1$) normalized by $G_{max}/F(e)$ for isotropically consolidated specimens ($K = 1$) and the stress ratio K is plotted in Fig. 8. $G_{max,v}$ is the stiffness when σ'_v is kept constant; $G_{max,p}$ is that when p' is kept constant. $G_{max,m}$ is the stiffness when the mean of two principal stresses $\sigma'_m(= (\sigma'_v + \sigma'_h)/2)$ is kept constant. $G_{max,m}$ for $K = 1$ is estimated by using equation (1). From this figure, the following points are noted.

(*a*) The normalized stiffness from anisotropically consolidated specimens is smaller than that from isotropically consolidated ($K = 1$) specimens when σ'_v is kept constant in both cases. The difference is larger for lower values of K. It can be inferred from this that the normalized G_{max} is overestimated if stiffness is measured from conventional triaxial tests when consolidated isotropically under the same σ'_v as in the field where K is less than 1.

(*b*) The normalized stiffness from anisotropically consolidated specimens is larger than that from isotropically consolidated ($K = 1$) specimens when p' is kept constant in both cases. This means that the normalized G_{max} is underestimated if stiffness is measured from a triaxial test when consolidated isotropically under the same p' as in the field where K is less than 1.

(*c*) The normalized stiffness from anisotropically consolidated specimens is almost the same as that from isotropically consolidated ($K = 1$) specimens when σ'_m is kept constant in both cases.

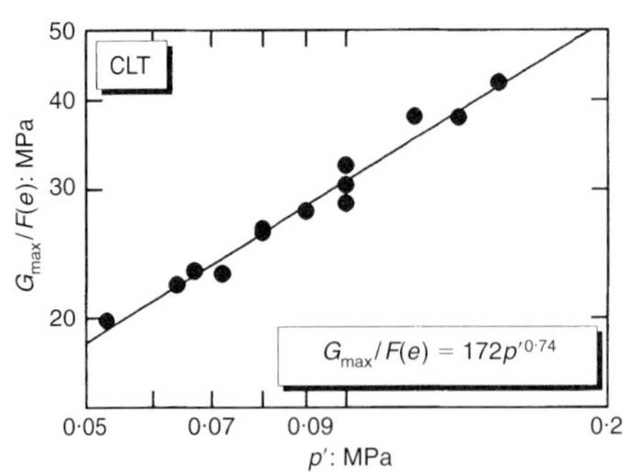

Fig. 7. Relationship between $G_{max}/F(e)$ and p'

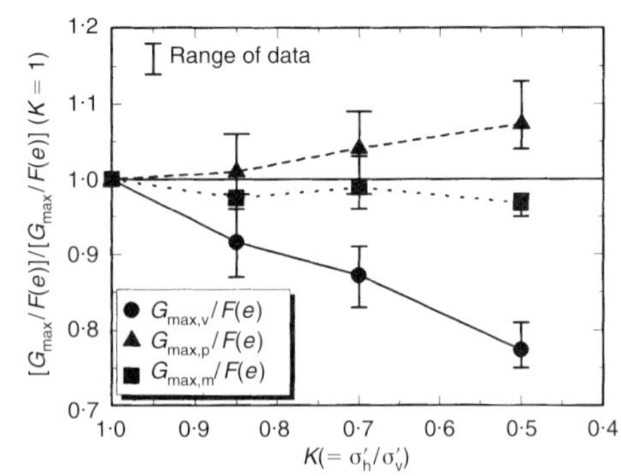

Fig. 8. Relationship between $\{G_{max}/F(e)\}/\{G_{max}/F(e)\}_{K=1}$ and K

Hence it can be concluded that a similar value of the stiffness at a small strain level in the anisotropic stress condition can be obtained from a conventional triaxial test irrespective of the stress ratio K if the specimen is isotropically consolidated under a stress equal to σ'_m.

Correlation between monotonic and cyclic loadings

For undrained MLTs under constant axial strain rate of 0·01%/min, normalized secant Young's modulus E_{sec} is plotted against axial strain in Fig. 9 for constant σ'_v and different K values. It can be observed that E_{sec} at a small strain level of 10^{-5} is different for different K values, even though σ'_v is kept constant. E_{sec} is higher for higher K values. However, the strain level dependency is almost the same. This behaviour is similar to that observed in CLTs.

Normalized E_{sec} values for MLTs under constant mean stress p' but different K values are shown in Fig. 10 by solid and broken lines. G_{max} seems to be independent of the K values if p' is kept constant. In the same figure, data from CLTs are

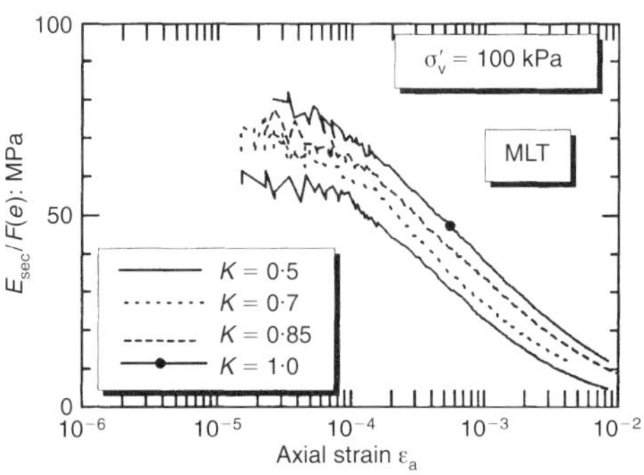

Fig. 9. Relationship between $E_{sec}/F(e)$ and axial strain

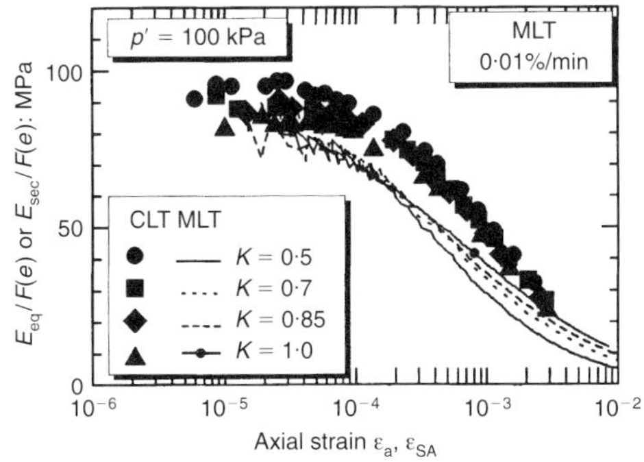

Fig. 10. Relationship between E_{eq} or $E_{sec}/F(e)$ and axial strain

shown by large solid symbols. The following points can be noted from the figure.

(*a*) Small strain stiffness (strain level less than 10^{-5}) is almost the same in CLTs and MLTs for identical consolidated stress ratio. At this strain level, the material behaves as elastic and hence should be independent of loading patterns such as loading frequency and strain rate. Tatsuoka & Shibuya (1991) have shown similar data for sand and soft rock.

(*b*) The degradation of stiffness is higher in MLTs than in CLTs. To clarify this point, the effects of generated excess pore pressure and the strain rate on the stiffness have been investigated. The loading frequency in CLTs is fixed to 0·1 Hz, and the strain rate in MLTs has been varied. Excess pore water pressures Δu generated during CLTs and MLTs are plotted against axial strain in Fig. 11 by large symbols and by lines respectively. Δu is the mean of top, bottom and midheight of the specimen, but the three values are almost the same. It can be observed that Δu is virtually zero up to a strain level of about 5×10^{-5}, and after that it gradually increases. The increase is predominant from a strain level of about 3×10^{-4}. The rate of increase is faster in case of CLTs. From this point of view it can be surmised that the stiffness degradation should be larger in case of CLTs. However, the tendency observed in Fig. 10 is different, and hence it can be concluded that the effect of excess pore water pressure generated during loading on the stiffness degradation is very small.

Calculated values of axial strain rate are plotted against either axial strain ε_a in MLTs or single

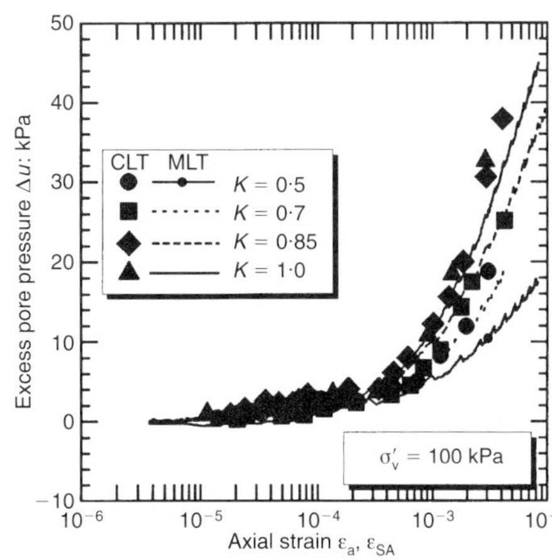

Fig. 11. Relationship between excess pore water pressure and axial strain

amplitude axial strain ε_{SA} in CLTs in Fig. 12. The strain rate in CLTs increases markedly from a strain level of about 5×10^{-5}. In CLTs, due to the higher strain rate induced, the degradation of stiffness from this strain level is smaller than that in MLTs, which might be triggered by some kind of hardening phenomenon due to higher strain rate. Further research is needed to clarify this. MLTs with different axial strain rates with CLT data for constant σ'_v and $K = 1$ are plotted in Fig. 13. It is seen that the stiffness curve is slightly shifted up in the medium strain level as the strain rate increases. However, the stiffness at a strain level of 10^{-5} is not much affected by the strain rate. It should also be stated that almost no significant difference was observed in excess pore water pressure due to the difference in strain rate. The degradation curve in MLTs with a loading strain rate of 0·2%/min is quite similar to that obtained from CLTs at a loading frequency of 0·1 Hz.

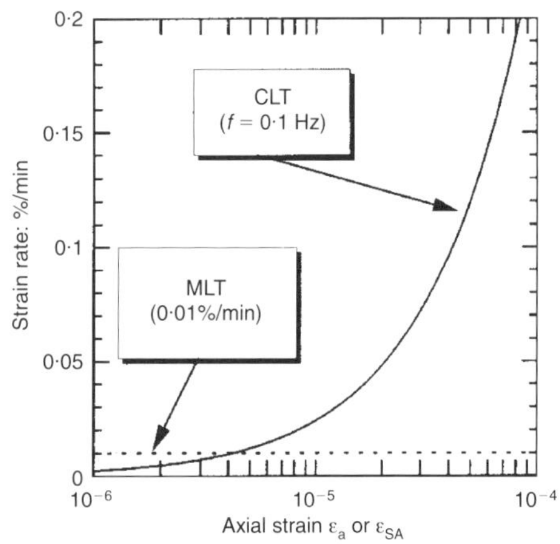

Fig. 12. Relationship between axial strain rate and axial strain

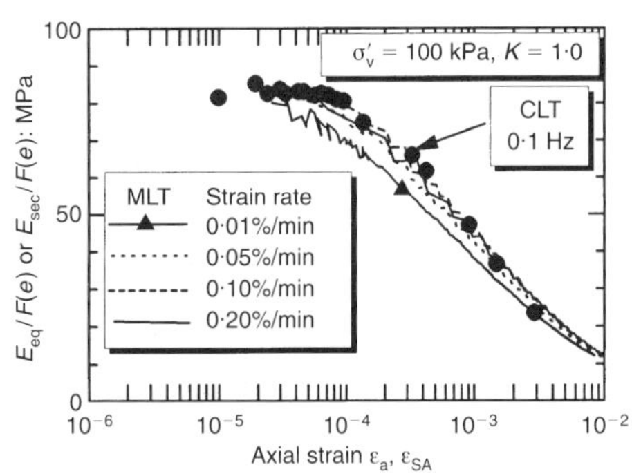

Fig. 13. Relationship between E_{eq} or $E_{sec}/F(e)$ and axial strain

Hence, very similar values of maximum shear modulus G_{max} (at strain level less than 10^{-5}) and similar stiffness degradation curves can be obtained from MLTs and CLTs if an adequate strain rate is taken into consideration based on Fig. 12.

CONCLUSIONS

The following conclusions can be drawn from the limited number of tests performed in the present study.

(*a*) Normalized stiffness G_{max} at small strain increases as the stress ratio K in anisotropically consolidated clay increases if effective vertical consolidation stress σ'_v is kept constant. However, the stress level dependency is almost the same irrespective of the value of K.

(*b*) A similar value of the stiffness at small strain in the anisotropic stress condition can be obtained from conventional triaxial testing irrespective of the stress ratio K if the specimen is consolidated isotropically under a stress equal to the mean values σ'_m of major and minor principal stresses.

(*c*) Similar values of stiffness G_{max} at small strain and similar stiffness degradation curves can be obtained from MLTs and CLTs if an adequate strain rate is considered.

REFERENCES

Afifi, S. S. & Richart, F. E. Jr (1973). Stress-history effects on shear modulus of soils. *Soil Fdns* **13**, No. 1, 77–95.

Cascante, G. & Santamarina, J. C. (1996). Interparticle contact behavior and wave propagation. *J. Geotech. Engng* **122**, No. 10, 831–839.

Hardin, B. O. (1966). Sand stiffness under various triaxial stresses. *J. Soil Mech. Fdn Div. Am. Soc. Civ. Engrs* **92**, SM2, 27–42.

Hardin, B. O. & Black, W. L. (1968). Vibration modulus of normally consolidated clay. *J. Soil Mech. Fdn Div. Am. Soc. Civ. Engrs* **94**, SM2, 353–369.

Hardin, B. O. & Drnevich, V. P. (1972). Shear modulus and damping in soils: design equations and curves. *J. Soil Mech. Fdn Div. Am. Soc. Civ. Engrs* **98**, SM7, 667–692.

Jamiolkowski, M., Leroueil, S. & Lo Presti, D. C. F. (1991). Design parameters from theory to practice. *Proc. Geo-Coast-91*, **2**, 877–917.

Kuribayashi, E., Iwasaki, T. & Tatsuoka, F. (1976). Effects of stress–strain condition on dynamic properties of sands. *Proc. Jpn Soc. Civ. Engrs*, No. 242, 105–114.

Lawrence, F. V. Jr (1965). Ultrasonic shear wave velocities in sand and clay. Report R65-05, Civ. Engng Dept, MIT, Cambridge, MA.

Lo Presti, D. C. F., Jamiolkowski, M. & Pallara, O. (1995). *Proc. 1st Int. Conf. on Earthquake Geotech. Engng*, **1**, 113–118.

Porovic, E. & Jardine, R. J. (1994). Some observations on

the static and dynamic shear stiffness of Ham River sand. *Proc. Int. Conf. on Prefailure Deformation of Geomaterials*, **1**, 25–30.

Pradhan, T. B. S., Imai, G., Ueno, Y. & Dam, L. (1995). Subsequent consolidation of clay subjected to undrained cyclic loading. *Proc. 3rd Int. Conf. on Recent Advances in Geotech. Earthquake Engng and Soil Dynamics*, **1**, St Louis, 61–64.

Roesler, S. K. (1979). Anisotropic shear modulus due to stress anisotropy. *J. Geotech. Engng Div. Am. Soc. Civ. Engrs* **105**, GT7, 871–880.

Tatsuoka, F., Iwasaki, T., Fukushima, S. & Sudo, H. (1979). Stress conditions and stress histories affecting shear modulus and damping of sand under cyclic loading. *Soils Fdns* **19**, No. 2, 29–43.

Tatsuoka, F. & Shibuya, S. (1991). Deformation characteristics of soils and rocks from field and laboratory tests. *Proc. 9th Asian Regional Conf. on SMFE*, **2**, 101–170.

Yu, P. & Richart, F. E. (1984). Stress ratio effects on shear modulus of dry sands. *J. Geotech. Engng, Am. Soc. Civ. Engrs* **110**, No. 3, 331–345.

Chu, J. (1998). Pre-failure deformation behaviour of geomaterials, 337–344

Pre-failure strain softening and pre-failure instability of a granular soil

J. CHU*

Strain softening in the form of decreasing shear resistance and instability, the inability to sustain given loads, can occur before the failure state is attained. This so-called pre-failure strain softening and pre-failure instability has been observed to occur only for very loose sand under an undrained condition. Experimental data are presented in this paper to demonstrate that pre-failure strain softening and pre-failure instability can also occur for dense sand under a strain-path controlled test condition. The factors that affect the prefailure strain softening and prefailure instability of sand are identified. Conditions for the occurrence of pre-failure strain softening and pre-failure instability are established. The relationships between pre-failure strain softening and pre-failure instability are discussed.

KEYWORDS: constitutive relations; failure; laboratory tests; plasticity; sands; shear strength.

L'amollissement de déformation sous forme d'une diminution de la résistance au cisaillement et d'une instabilité dans l'incapacité de soutenir des charges données peut se produire avant que l'état de défaillance ne soit atteint. Les observations ont montré que cet l'amollissement de déformation avant défaillance' et cette 'instabilité avant défaillance' se produisaient uniquement en cas de sable très boulant et en condition de non drainage. Nous présentons dans cet exposé des données expérimentales pour démontrer que l'amollissement de déformation avant défaillance et l'instabilité avant défaillance peuvent aussi se produire dans le cas de sable compact en condition d'essai contrôlé sur la trajectoire de déformation. Nous identifions les facteurs qui affectent l'amollissement de déformation avant défaillance et l'instabilité avant défaillance du sable. Nous établissons les conditions provoquant l'amollissement de déformation avant défaillance et l'instabilité avant défaillance. Nous discutons des relations entre l'amollissement de déformation avant défaillance et l'instabilité avant défaillance.

INTRODUCTION

It is well known that the effective stress-path resulting from an undrained triaxial test on a very loose sand specimen is typically as shown in Fig. 1 (DeGregorio, 1986; Sladen & Oswell, 1988; Chu, 1995). When the specimen is sheared in a deformation controlled mode, the deviator stress q decreases after a peak value has been attained. The behaviour associated with $dq < 0$, or more precisely $d\sigma_{ij}\, d\varepsilon_{ij} < 0$, as defined by Valanis (1985), has been referred to as strain softening (Provest & Hoeg, 1975; Read & Hegemier, 1984; Chu *et al.*, 1992). This type of strain softening occurs before the effective stress-path reaches the failure line, and therefore has been called pre-failure strain softening (Chu *et al.*, 1992). The word 'pre-failure' is used to distinguish this from the type of strain

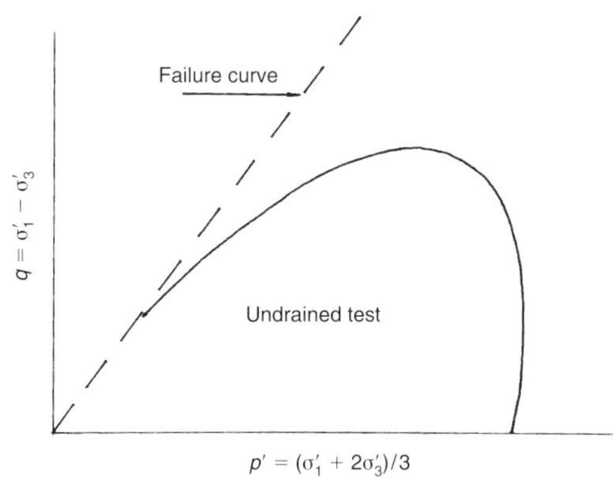

Fig. 1. Typical effective stress-path of undrained test on very loose sand

softening which occurs for dense sand in a drained test when the peak deviator stress is attained. Although it is debatable whether the strain softening that occurs during a drained test is a material

Manuscript received 19 February 1997.

* School of Civil and Structural Engineering, Nanyang Technological University.

behaviour or merely a result of the non-homogeneous deformation (Read & Hegemier, 1984; Chu *et al.*, 1992), there is little doubt that the above-mentioned pre-failure strain softening is a real material behaviour and therefore deserves further investigation.

On the other hand, when a very loose sand specimen is sheared in a stress controlled mode under an undrained condition, a run-away instability in the form of a sudden increase in pore water pressure and axial strain may occur (Lade & Pradel, 1990; Chu, 1991b). A typical type of instability is static liquefaction (Castro, 1969). Static liquefaction normally occurs after the deviator stress exceeds the peak and the effective stress-path followed is similar to that shown in Fig. 1 (Castro, 1969; DeGregorio, 1986). This type of instability has also been called pre-failure instability (Chu, 1991a, 1991b).

Under an undrained condition, pre-failure strain softening and pre-failure instability can occur only for very loose sand, as has been established by experimental data (Castro, 1969; Sladen & Oswell, 1988). The undrained condition inherently implies that void ratio of a sand deposit before and after liquefaction is the same. The adequacy of this assumption has been questioned. Based on the observation of shaking table tests, Seed (1987) has indicated that 'it is possible that there is a redistribution of water content in sand samples in the laboratory and in sand layers in the field'. Therefore, the assumption of an undrained condition may not always be realistic. The pre-failure behaviour of sand should also be studied under other than undrained conditions.

The objectives of this paper are to examine whether the pre-failure strain softening or pre-failure instability behaviour may occur for dense sand and, if so, the factors which control those behaviours. As the pre-failure behaviour of sand is significantly affected by the stress- and strain-paths, the pre-failure strain softening and instability should be studied over a wide range of stress- and strain paths. Chu *et al.* (1992) have indicated that sand may be more susceptible to strain softening under a strain-path controlled condition. In a stress-path test, the variation of stress is controlled, and the specimen may be forced to an inadmissible state upon the onset of strain softening, whereas in a strain-path test, the strain increment ratio is controlled and the decrease in stresses associated with strain softening can therefore develop freely. Experimental results are presented below to illustrate how pre-failure strain softening and pre-failure instability can occur for dense sand under a strain-path controlled condition. The conditions for the occurrence of pre-failure strain softening and pre-failure instability of sand are established. The relationship between pre-failure strain softening and pre-failure instability is discussed.

TEST MATERIAL, ARRANGEMENT AND METHODOLOGY

The material tested was a uniformly graded quartz sand with index properties given in Table 1. The specimens were prepared by following a standardized procedure of depositing dry sand into water. The specimens so prepared had dimensions of 100 mm diameter by 100 mm height and void ratios ranging from 0·599 to 0·621.

The test arrangement for instability tests is shown in Fig. 2. A digital pressure volume controller (DPVC) controlled Bellofram actuator was used to control the axial load. The test arrangement for strain softening tests is similar to that shown in Fig. 2, except that the Bellofram actuator was not used. Free-ends (Rowe & Barden, 1964) with enlarged platens were used to reduce the platen restraint. The bedding and membrane-penetration errors were effectively reduced by a liquid rubber technique (Lo *et al.*, 1989).

A strain-path test was to control the strain increment ratio $d\varepsilon_v/d\varepsilon_1$. This was realized by controlling the volume change of the specimen by way of a DPVC in accordance with the measured axial deformation (Chu, 1991a; Chu *et al.*, 1992). When $d\varepsilon_v/d\varepsilon_1 < 0$ water flows into and when $d\varepsilon_v/d\varepsilon_1 > 0$ water flows out of the specimen. When $d\varepsilon_v/d\varepsilon_1 = 0$, it becomes an undrained test.

PRE-FAILURE STRAIN SOFTENING

Tests under an undrained condition (i.e. $d\varepsilon_v/d\varepsilon_1 = 0$) had been conducted and pre-failure strain softening was observed to occur only for very loose sand (Chu, 1995). To investigate whether pre-failure strain softening will occur under other than undrained conditions, strain-path tests with non-zero $d\varepsilon_v/d\varepsilon_1$ ratios were conducted. The stress–strain curves and the effective stress-paths

Table 1. Index properties of the tested sand

Mean size: mm	Uniformity coefficient	Specific gravity	Void ratio	
			Minimum	Maximum
0·3	1·5	2·65	0·565	0·855

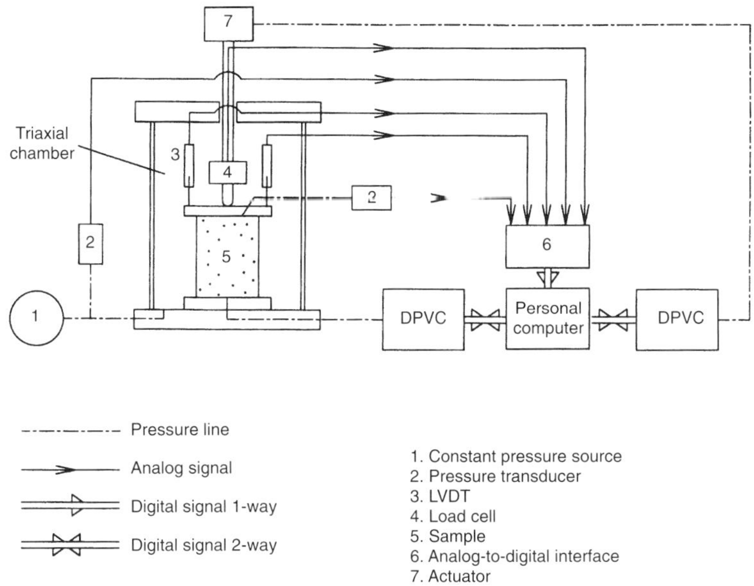

Fig. 2. Test arrangement

of two strain-path tests with $d\varepsilon_v/d\varepsilon_1 = -0{\cdot}67$, $\zeta 1$ and $\zeta 2$, are presented in Figs 3(a) and 3(b). The two tests were conducted on dense sand under effective confining stresses of 200 kPa and 600 kPa respectively. The failure curve as determined by drained conventional triaxial compression (CTC) tests is also plotted in Fig. 3(b). It can be seen that strain softening occurred before the stresses reached failure. Therefore, pre-failure strain softening has occurred for dense sand under a strain-path controlled condition. It should be pointed out that in both tests strain softening occurred at a shear strain as small as 0·3% (Fig. 3(a)). Therefore, the observed pre-failure strain softening could not

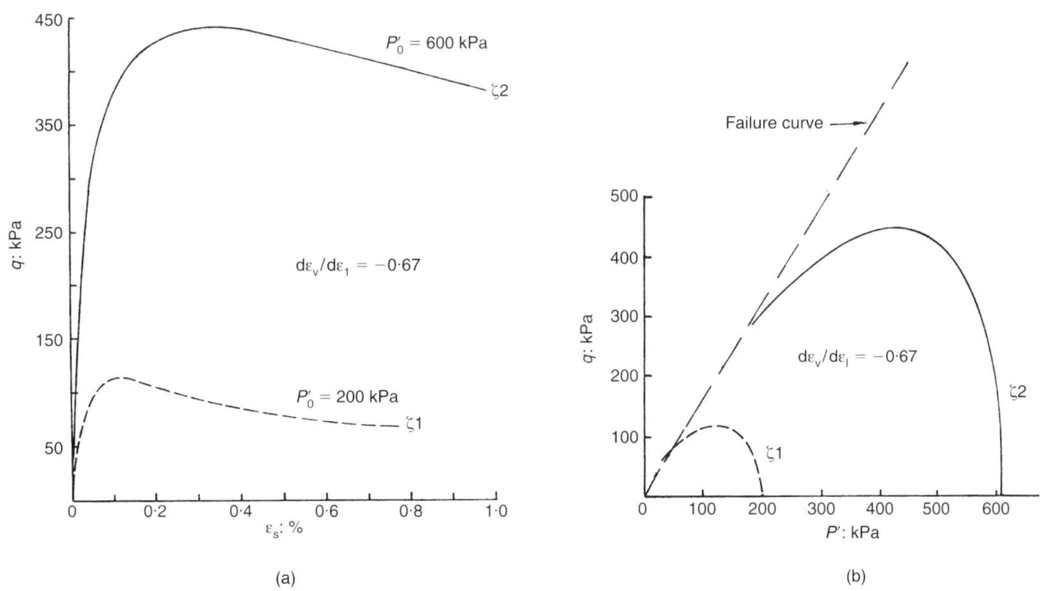

Fig. 3. Pre-failure strain softening behaviour along $d\varepsilon_v/d\varepsilon_1$ path: (a) stress–strain curves; (b) effective stress paths

possibly be due to strain localization, and was a material property.

When duplicated dense specimens were tested under a strain-path of $d\varepsilon_v/d\varepsilon_1 = -0{\cdot}11$, pre-failure strain softening did not occur, as can be seen from Fig. 4, where the effective stress-paths of tests ζ3, ζ4 and ζ5 are presented. It can be concluded from the test results that the occurrence of pre-failure strain softening is affected by the strain-increment ratio. The smaller (in algebra) the $d\varepsilon_v/d\varepsilon_1$, the higher will be the tendency for pre-failure instability to occur.

The pre-failure strain softening behaviour is also affected by the effective confining stress. The effective stress-paths of two strain-path tests with $d\varepsilon_v/d\varepsilon_1 = -0{\cdot}54$, ζ7 and ζ8, are presented in Fig. 5. It can be seen that pre-failure strain softening occurred in the test (ζ8) with a higher effective confining stress, but not in the test (ζ7) with a lower one.

Based on the above testing results, the factors that affect the occurrence of pre-failure strain softening may be identified as

(*a*) *the density of the soil*—the looser the soil, the higher is the tendency for strain softening to occur given that the other conditions are the same
(*b*) *the strain increment ratio*—the smaller the $d\varepsilon_v/d\varepsilon_1$ ratio, the higher is the tendency for strain softening to occur
(*c*) *the effective confining stress*—the higher the confining stress, the lower is the tendency to have strain softening.

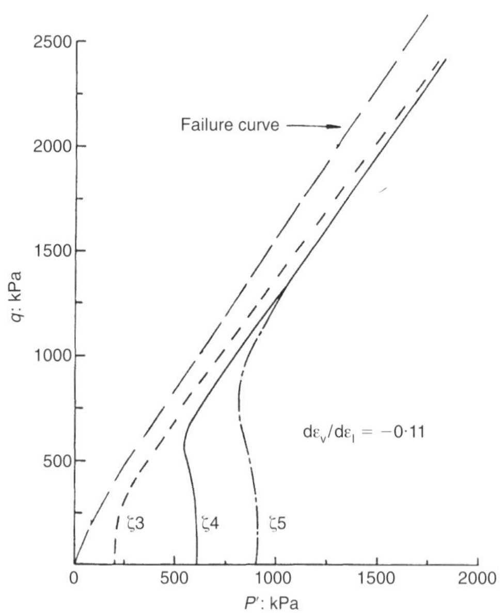

Fig. 4. Influence of $d\varepsilon_v/d\varepsilon_1$ on pre-failure strain softening

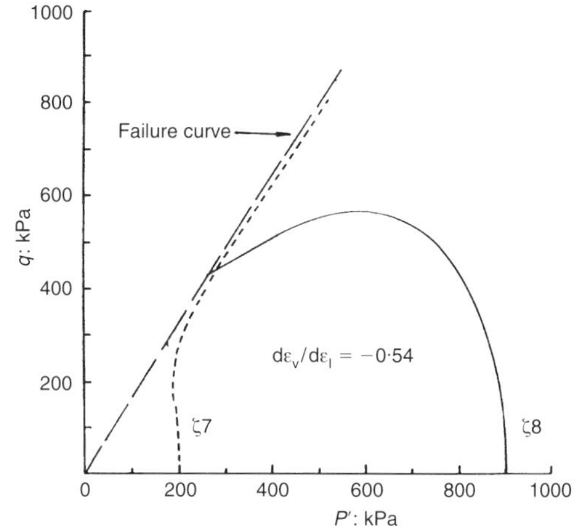

Fig. 5. Influence of σ'_{30} on pre-failure strain softening

An examination of the effective stress-paths of pre-failure strain softening tests has suggested that $dp' < 0$ when strain softening occurs, i.e. the pore water pressure is increasing. The volume change behaviour of a drained test defines the strain increment ratio at which pore water (or back) pressure will not change. The $d\varepsilon_v/d\varepsilon_1$ ratios manifested in some typical drained tests for dense and medium loose specimens are shown in Fig. 6. It can be seen from Fig. 6 that the $d\varepsilon_v/d\varepsilon_1$ ratio decreases with shearing, but approaches a constant when the axial strain exceeds 3% to 4%. For convenience, the plateau value of $d\varepsilon_v/d\varepsilon_1$ will be denoted as $(d\varepsilon_v/d\varepsilon_1)_d$. The $(d\varepsilon_v/d\varepsilon_1)_d$ ratio signifies the strain increment ratio the soil will manifest under a zero pore water pressure generation state. If a $d\varepsilon_v/d\varepsilon_1$ ratio smaller than $(d\varepsilon_v/d\varepsilon_1)_d$ is imposed, the specimen will dilate more (or contract less) and positive pore water pressure will be generated. On the other hand, if a $d\varepsilon_v/d\varepsilon_1$ ratio higher than $(d\varepsilon_v/d\varepsilon_1)_d$ is

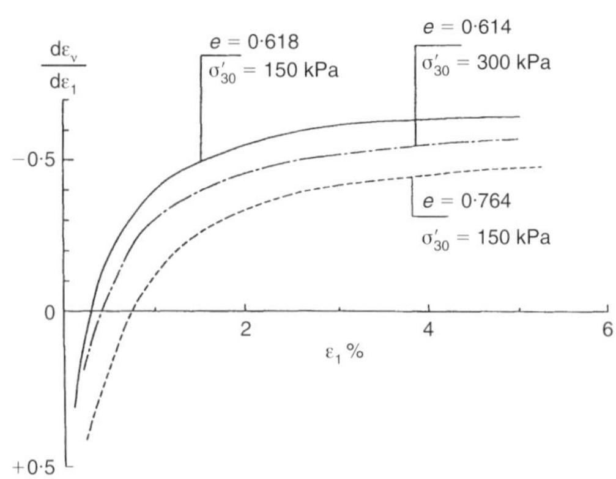

Fig. 6. Strain increment ratio in drained CTC tests

imposed, the specimen will dilate less (or contract more) and negative pore water pressure will be generated. The $(d\varepsilon_v/d\varepsilon_1)_d$ ratio for dense sand is in the range of $-0{\cdot}5$ to $-0{\cdot}6$ (Fig. 6). This explains why the strain softening occurred in tests with $d\varepsilon_v/d\varepsilon_1 = -0{\cdot}67$, but not in tests with $d\varepsilon_v/d\varepsilon_1 = -0{\cdot}11$. Therefore, the conditions for the occurrence of pre-failure instability can be written as

$$\left(\frac{d\varepsilon_v}{d\varepsilon_1}\right)_i - \left(\frac{d\varepsilon_v}{d\varepsilon_1}\right)_d \leqslant 0 \tag{1}$$

$(d\varepsilon_v/d\varepsilon_1)_i$ represents the strain increment ratio imposed on the soil specimen. It can be seen from Fig. 6 that $(d\varepsilon_v/d\varepsilon_1)_d$ ratios are affected by the density of sand and the effective confining stress. The denser the soil or the smaller the σ'_{30}, the smaller the $(d\varepsilon_v/d\varepsilon_1)_d$ ratio will be. This explains how the density and σ'_{30} affect the pre-failure strain softening. For undrained tests, $(d\varepsilon_v/d\varepsilon_1)_i = 0$. Inequality (1) requires $(d\varepsilon_v/d\varepsilon_1)_d \geqslant 0$, i.e. the volumetric change of the specimen has to be contraction. This is possible only for very loose sand. Therefore, only very loose sand can have strain softening under the undrained condition.

The $(d\varepsilon_v/d\varepsilon_1)_d$ ratio can be predicted by a well-established constitutive model. For illustration, Rowe's stress–dilatancy equation (Rowe, 1962, 1971) is used to calculate the $(d\varepsilon_v/d\varepsilon_1)_d$ manifested in a drained CTC test

$$\left(\frac{d\varepsilon_v}{d\varepsilon_1}\right)_d = 1 - \frac{1}{K}\left(\frac{\sigma'_1}{\sigma'_3}\right)_f \tag{2}$$

where K is a soil parameter and $(\sigma'_1/\sigma'_3)_f$ is the failure stress ratio achieved in a drained CTC test. Combining equations (1) and (2) leads to

$$\left(\frac{d\varepsilon_v}{d\varepsilon_1}\right)_i \leqslant 1 - \frac{1}{K}\left(\frac{\sigma'_1}{\sigma'_3}\right)_f \tag{3}$$

It should be pointed out that Rowe's stress–dilatancy equation is established on a 'rigid granular' particle assumption, i.e. the stress–dilatancy behaviour is independent of the effective confining stress. Consequently the influence of the effective confining stress cannot be reflected in equation (2).

PRE-FAILURE INSTABILITY

The term 'instability' is used to describe a runaway deformation or the collapse of a soil specimen which occurs due to its inability to sustain a given load or stress. To study the instability behaviour of soil, tests must be conducted under a load (or stress) controlled mode.

Pre-failure instability can occur for very loose sand under an undrained condition (Lade & Pradel, 1990; Chu, 1991b). To study whether pre-failure instability can occur for dense sand, the stability of dense specimens under constant external load conditions was examined under various strain-path controlled conditions.

Figure 7(a) presents the effective stress-paths followed in three stability tests, Z01, Z02, and Z10, on dense sand. In conducting the tests, the specimens were first sheared along a drained CTC path with $\sigma'_{30} = 150$ kPa to a stress ratio of 3·5, 3·2, and 2·0 for tests Z01, Z02, and Z10 respectively. The external loads, i.e. the axial load and the cell pressure, were then maintained constant to conduct instability checking along a strain-path of $d\varepsilon_v/d\varepsilon_1 = -0{\cdot}67$. Under these conditions, instability occurred for tests Z01 and Z02, but not for test Z10. When instability occurred, the axial strain and the pore water pressure increased suddenly, as shown in Fig. 7(b) for test Z01. Physically it was observed that the specimens collapsed in a split second. The stress states at which instability occurred were below the failure line. Therefore, pre-failure instability had occurred.

Two other tests, Z03 and Z04, were conducted by drained shearing to $\sigma'_1/\sigma'_3 = 3{\cdot}0$ under $\sigma'_{30} = 150$ kPa. Instability checking was then carried out by imposing a constant load condition and a strain path of $d\varepsilon_v/d\varepsilon_1 = -0{\cdot}54$. The results of tests Z03 and Z04 are shown in Figs 8 and 9 respectively. The difference between the two tests was the void ratios. Test Z03 used dense sand with $e = 0{\cdot}60$, and test Z04 medium loose sand with $e = 0{\cdot}764$. Pre-failure instability occurred for test Z04 (Fig. 9), but not for test Z03 (Fig. 8).

It should be pointed out that the instability behaviour presented in Figs 7 and 9 is not caused by controlling a negative strain increment ratio. This is shown by tests Z10 and Z03. Instability occurs only when appropriate conditions are met. The observed instability was not due to strain localization or rate/time effect either, as discussed by Chu (1991a).

The test results presented above have demonstrated that the occurrence of pre-failure instability is affected by

(*a*) the strain increment ratio—for a given condition, instability occurs only if an adequately small strain increment ratio is imposed on the specimen
(*b*) the initial density of the soil—given that the other conditions are the same, the looser the sample, the higher is the tendency for instability to occur.
(*c*) the initial effective stress ratio—instability occurs only when the initial stress state is adequately high.

The effective confining stress σ'_{30} may also affect

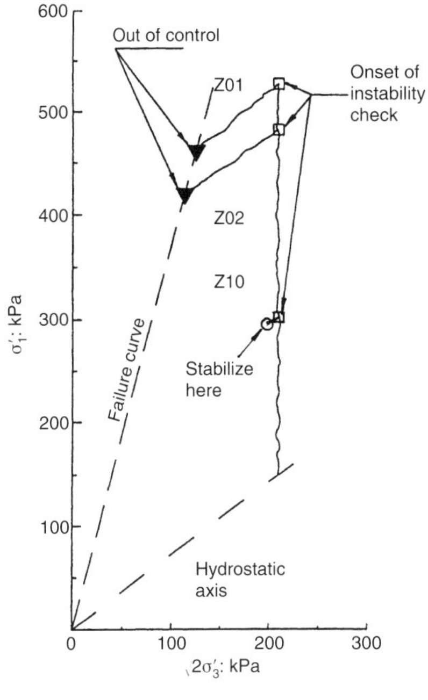

(a)

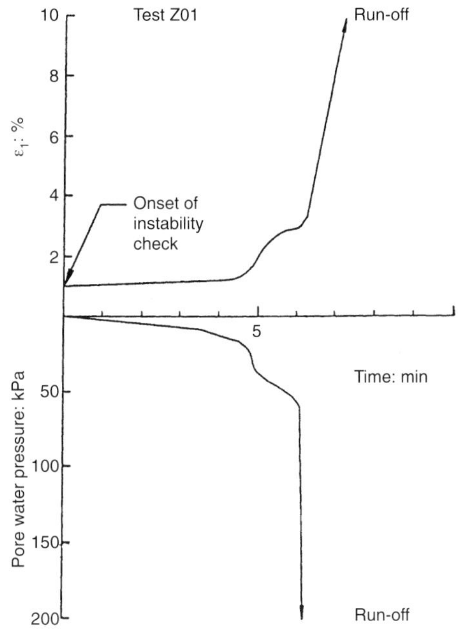

(b)

Fig. 7. Pre-failure instability behaviour along $d\varepsilon_v/d\varepsilon_1$ path: (a) effective stress paths; (b) axial strain and pore water pressure development

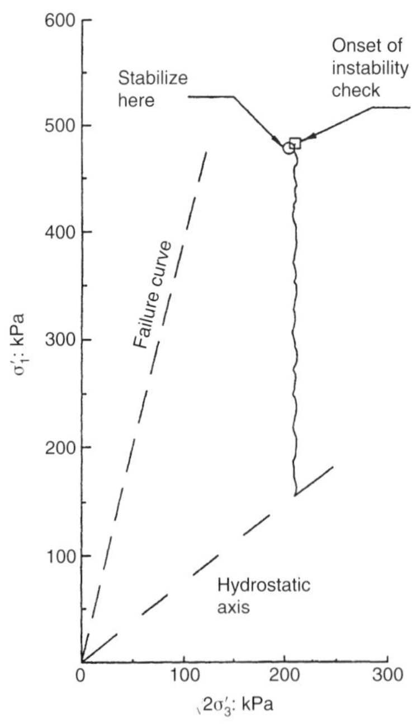

(a)

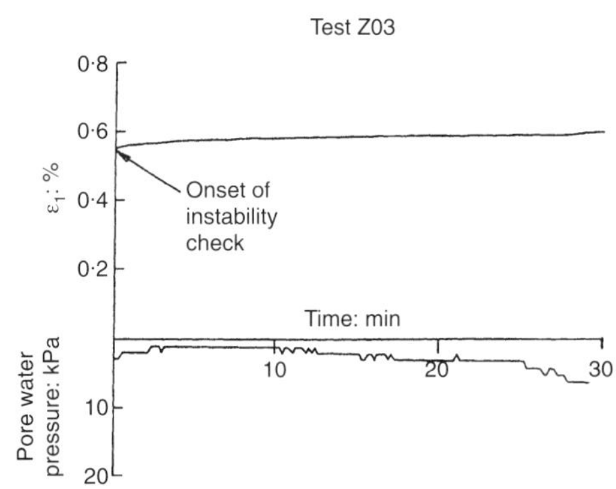

(b)

Fig. 8. Response of test Z03: (a) effective stress path; (b) axial strain and pore water pressure development

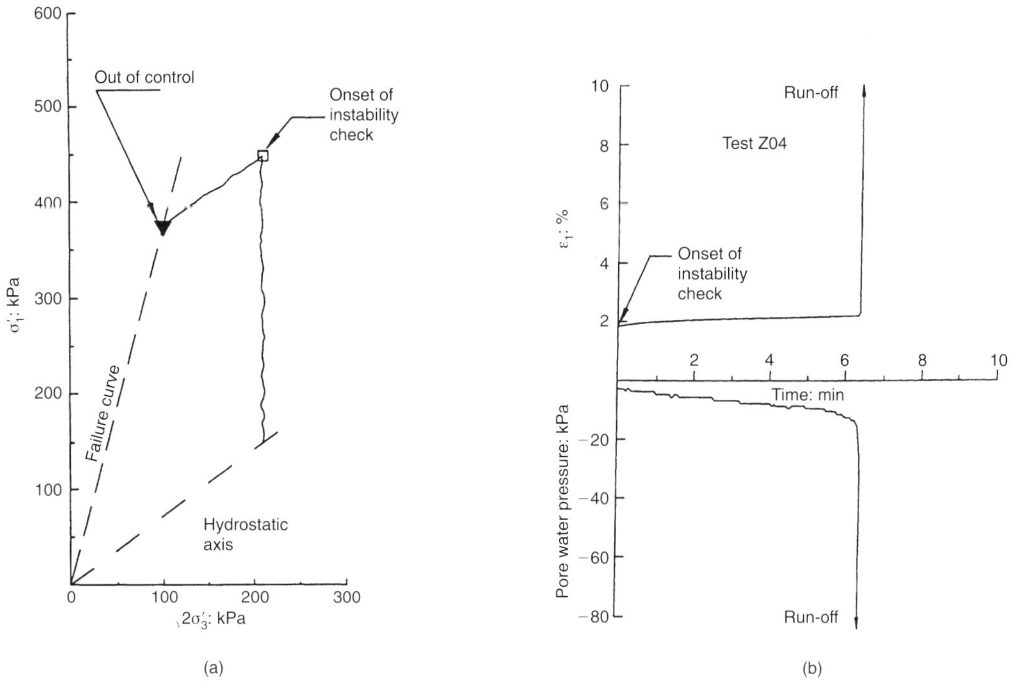

Fig. 9. Response of test Z04: (a) effective stress path; (b) axial strain and pore water pressure development

the occurrence of pre-failure instability, although this effect of σ'_{30} could not be shown as the tests conducted were within a narrow range of σ'_{30}.

Drained tests with deviator stresses maintained at different pre-failure stress levels were also conducted, and instability was not observed in any of the drained tests (Lee, 1997). In fact, all the test results show that the occurrence of pre-failure instability is associated with an increase in pore water pressure. A condition to induce continuous pore water pressure generation is necessary for pre-failure instability to occur. The strain increment ratio and density influence the pore water pressure change in the same way as discussed in the preceding section for pre-failure strain softening. As such, the conditions stipulated in Equation (3) are also applicable to pre-failure instability. These conditions are sufficient for the occurrence of strain softening, but not for the occurrence of pre-failure instability, as the condition on stress level has also to be satisfied.

How high should the initial stress level be in order to induce pre-failure instability? As pre-failure instability is accompanied by a sudden increase in axial strain, a condition to allow continuous plastic flow to develop must be provided. To incur plastic flow, the stress increment direction has to be outwards from the yield surface for the yield surface to expand. A typical yield surface for sand is shown in Fig. 10. During the course of instability, the deviator stress q is either maintained constant or decreasing, i.e. $dq \leqslant 0$ and $dp' < 0$ as pore water pressure is increasing. Consequently, the stress increment direction can only be pointing to a direction shown in Fig. 10. It can be seen from Fig. 10 that only within the hatched zone would it be possible for the stress increment direction to point outwards from the yield surface. This hatched area is characterized by a negative normal direction. Mathematically, this can be expressed as

$$\frac{\partial f}{\partial p'} < 0 \tag{4}$$

where f is the yield surface.

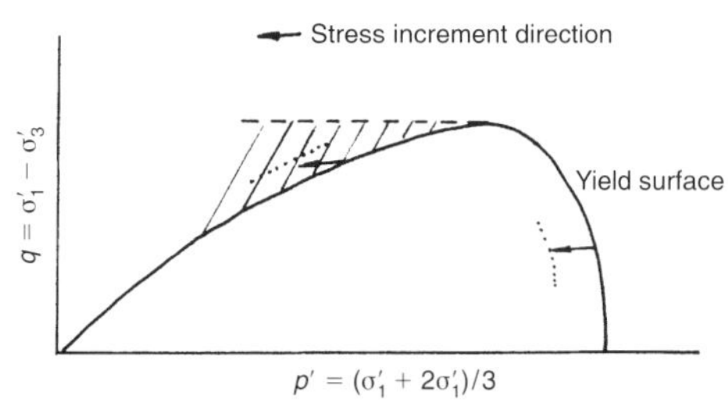

Fig. 10. Admissible zone for instability

In the hatched zone, $dq < 0$, i.e. this is also the zone where pre-failure strain softening is occurring. Therefore, the occurrence of pre-failure strain softening is a necessary condition for the occurrence of pre-failure instability. It should be noted that strain softening occurs under a deformation controlled loading condition, whereas instability occurs under a load controlled loading condition. These two types of behaviour could be different manifestations of the same mechanism under the different loading conditions (Chu, 1991a).

CONCLUSIONS

As the pre-failure behaviour of sand is strongly path-dependent, the pre-failure strain softening and instability behaviour should be studied along different stress- or strain-paths. Under an undrained path ($d\varepsilon_v/d\varepsilon_1 = 0$), pre-failure strain softening and pre-failure instability can occur only for very loose sand. Under a $d\varepsilon_v/d\varepsilon_1$ controlled path, pre-failure strain softening and pre-failure instability can occur for dense sand. The factors that affect the occurrence of pre-failure strain softening are (*a*) the density of soil, (*b*) the effective confining stress and (*c*) the strain increment ratio $d\varepsilon_v/d\varepsilon_1$. These factors also affect the occurrence of pre-failure instability. The necessary and sufficient condition for pre-failure instability to occur is $(d\varepsilon_v/d\varepsilon_1)_i < (d\varepsilon_v/d\varepsilon_1)_d$, i.e. the strain increment ratio imposed should be smaller than that manifested in a drained test. This condition is necessary but not sufficient for the occurrence of pre-failure instability. The initial stress level must also be sufficiently high to have pre-failure instability.

REFERENCES

Castro, G. (1969). *Liquefaction of sands*. Harvard Soil Mechanics Series no. 81, Harvard University, Cambridge, MA.

Chu, J. (1991a). *Strain softening behaviour of granular soils under strain path testing*. PhD thesis, University of New South Wales.

Chu, J. (1991b). Liquefaction of sands under undrained and non-undrained conditions. *Soil dynamics and earthquake engineering V*, pp. 277–291. Karlsruhe, Germany, Ed. ZBF, Computational Mechanics Publication.

Chu, J. (1995). An experimental examination of the critical state and other similar concepts for granular soils. *Can. Geotech. J.* **31**, No. 6, 1065–1075.

Chu, J., Lo, S.-C. R. & Lee, I. K. (1992). Strain softening behaviour of a granular soil in strain path testing. *J. Geotech. Engng Am. Soc. Civ. Engrs* **118**, No. 2, 191–208.

DeGregorio, V. B. (1986). Loading systems, sample preparation, and liquefaction. *J. Geotech. Engng Am. Soc. Civ. Engrs* **116**, No. 5, 805–821.

Lade, P. V. & Pradel, D. (1990). Instability and plastic flow of soils. I: Experimental observations. *J. Engng Mech. Am. Soc. Civ. Engrs* **116**, No. 11, 2532–2550.

Lee, C. W. (1997). *Instability of loose sand. Final year project report*, School of Civil and Structural Engineering, Nanyang Technological University.

Lo, S.-C. R., Chu, J. & Lee, I. K. (1989). A technique for reducing membrane penetration and bedding errors. *Geotech. Test. J. Am. Soc. Test. Mater.* **12**, No. 4, 311–316.

Provest, J.-H. & Hoeg, K. (1975). Soil mechanics and plasticity analysis of strain softening. *Géotéchnique* **25**, No. 2, 279–297.

Read, H. E. & Hegemier, G. A. (1984). Strain softening of rock, soil and concrete—a Review article. *Mech. Mater.* **3**, 271–294.

Rowe, P. W. (1962). The stress dilatancy relationship for static equilibrium of an assembly of particules in contact. *Proc. R. Soc.* **A267**, 500–529.

Rowe, P. W. (1971). Theoretical meaning and observed values of deformation parameters for soil. *Proc. Roscoe Memorial Symp. on Stress–Strain Behaviour of Soils*, Cambridge, 143–194.

Rowe, P. W. & Barden, L. (1964). Importance of free ends in triaxial testing. *J. Soil Mech. Fdn Engng Div. Am. Soc. Civ. Engrs* **90**, No. 1, 1–15.

Seed, H. B. (1987). Design problems in soil liquefaction. *J. Geotech. Engng Am. Soc. Civ. Engrs* **113**, No. 8, 827–845.

Sladen, J. A. & Oswell, J. M. (1988). The behaviour of very loose sand in triaxial compression test. *Can Geotech. J.* **26**, No. 2, 103–113.

Valanis, K. C. (1985). On the uniqueness of solution of the initial value problem in strain softening materials. *J. Appl. Mech. Am. Soc. Mech. Engrs* **52**, 649–653.

Dasari, G. R. & Bolton, M. D. (1998). Pre-failure deformation behaviour of geomaterials, 345–352

Comparison of field and laboratory stiffness of Gault Clay

G. R. DASARI* and M. D. BOLTON*

The stress–strain characteristics of undisturbed Gault Clay soil samples from the Madingley test site at Cambridge were investigated in a triaxial cell. A modified version of the local deformation transducer was developed to measure axial strains on triaxial samples. Bulk and shear stiffnesses were calculated during constant deviatoric and constant mean effective stress-paths respectively. The laboratory measured stiffness was compared with existing field test data.

KEYWORDS: Gault Clay; local strains; pressure meter; Rayleigh wave; stiffness; triaxial tests.

Nous avons étudié les caractéristiques contrainte-déformation de l'argile de Gault non perturbée sur des échantillons provenant du site d'essai de Madingley à Cambridge, dans une cellule triaxiale. Nous avons développé une version modifiée du transducteur de déformation locale pour mesurer les déformations axiales sur des échantillons triaxiaux. Nous avons calculé les rigidités globales et les rigidités de cisaillement pendant, respectivement, une contrainte déviatrice constante et des trajectoires de contrainte effective moyenne constante. Nous avons comparé la rigidité mesurée en laboratoire avec les résultats d'essais déjà obtenus sur le terrain.

INTRODUCTION

The importance of small strain measurement to evaluate the pre-failure stress–strain behaviour of soils and its influence on analysis of boundary value problems has been investigated in the past few years. For example, accurate prediction of ground movements due to a tunnel excavation demands both correct magnitude and variation of stiffness with strain from very small strains. However, 'back-calculated' soil stiffnes was reported to be higher than the values measured in laboratory tests (St John, 1975). Recent research has demonstrated that discrepancies between the back-calculated soil stiffness and the stiffness measured in laboratory tests could mainly be attributed to the magnitude of strain. The objective of this research paper is to investigate the stress–strain characteristics of Gault Clay in the laboratory using internal gauges and to compare these with corresponding field data to ascertain what parameters besides strain have an influence.

Data of the stress–strain behaviour of Gault Clay in Cambridge, UK have been limited to a number of field tests carried out by the Building Research Establishment (Abbiss, 1981; Powell & Butcher, 1991). The field tests included geophysical methods and a self-boring pressuremeter. However, the small strain field stiffnesses measured by various methods differ by a factor of five. In contrast, few laboratory studies of Gault Clay have been reported in the literature. As far as the authors are aware, no laboratory data have been published on the small strain stiffness of undisturbed Gault Clay.

LOCAL STRAIN MEASUREMENT

The principal sources of error in the measurement of axial strain using gauges outside the cell have been identified by, among others, Baldi *et al.* (1988). A large number of methods have been developed to measure strains locally on a triaxial sample. Here, a modified version of the local deformation transducer (LDT) similar to the one used by Goto *et al.* (1991) was developed to measure axial strains on triaxial samples. As shown in Fig. 1, LDTs were connected to soil sample with hinges which were attached to membrane using super glue. The working principle of the LDT is very simple. As the soil sample deforms, the strain gauge bridge fixed at the centre of the LDT measures the bending strain in terms of voltage. Initial calibration of bending strain (voltage) against axial strain is used to evaluate the axial strain.

Modified LDT

Initially LDTs based on the original design by Goto *et al.* (1991) were adopted. However, two major problems were soon encountered. Firstly, there was a lack of fit at the reception corner of the hinge attachment when the strip was attached

Manuscript received 27 May 1997.
* Engineering Department, Cambridge University.

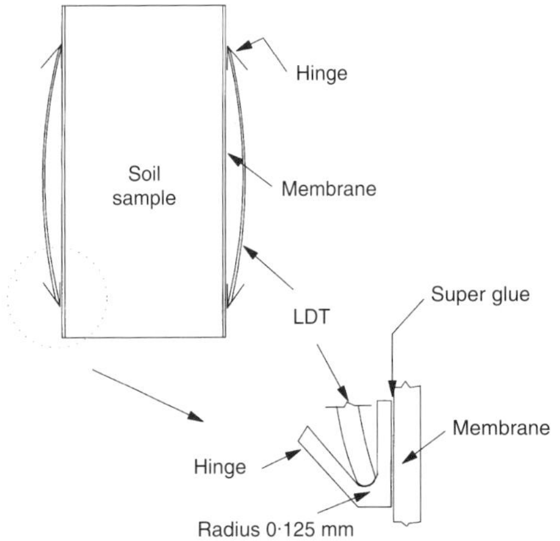

Fig. 1. LDTs connected to soil sample

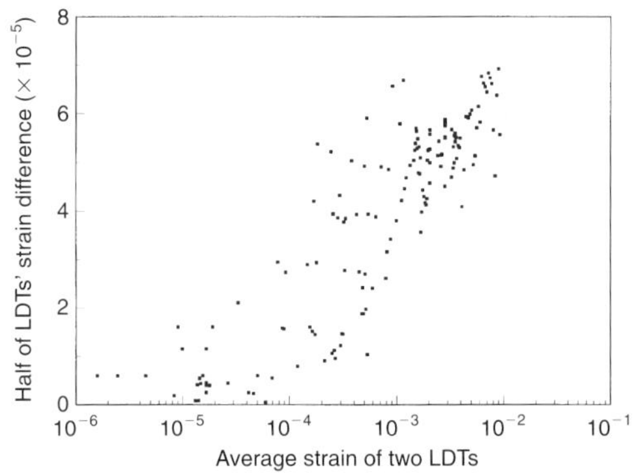

Fig. 2. Average axial strain plotted against the discrepancy between two LDTs

to it. Secondly, the coating applied on the strain gauges was over-heated, causing a permanent shift in the output. These factors were found to be affecting the repeatability of the calibration curves. To solve the first problem, the ends of each strip and the reception corner of each hinge were machined to a radius of 0·125 mm (see Fig. 1). This allows the LDT strip to fit precisely into the supporting hinge attachment. The second problem was solved by increasing the gauge area. These modifications helped to reduce the discrepancy between various calibration curves.

Resolution and accuracy

The resolution of the LDT was as small as 0·01 μm (from a calibration curve). However, the accuracy based on repeatability of several calibration curves of a single LDT was about 2 μm. The considerable difference between resolution and accuracy is due to (*a*) mechanical imperfections at the hinges, (*b*) hysteresis and (*c*) electrical signal noise. The overall accuracy during a triaxial test was exhibited by the discrepancy between two LDT readings. Axial strain is plotted against the discrepancy between two LDTs for a typical test in Fig. 2. As seen, LDTs are very accurate at small strains. However, the difference between LDTs increases with strain. This is due to the reduction of resolution with increasing strain and also possibly to the local deformation of the membrane by the LDTs. But at 0·01 (i.e. 1%) axial strain, a discrepancy in strain of 7×10^{-5} is not critical.

EXPERIMENTAL PROGRAMME

The main objective of the experimental programme was to measure the small strain stiffness of Gault Clay.

Apparatus

The hydraulic triaxial stress path apparatus consists of a Bishop-Wesley cell and three digital pressure controllers (one each for axial stress, cell pressure and pore pressure/back pressure). The stress path cell was fitted with an internal 5 kN load cell to measure deviatoric stress independently. The resolution of the load cell was about 0·05 N, corresponding to 0·05 kPa on a 38 mm dia. sample.

Soil tested

The Gault Clay samples were obtained from the Madingley test site in Cambridge. The soil samples were taken from 2–3 m depth from two bore holes made by hand auger. At the required depth, 38 mm soil samplers were pushed in by the drop hammer technique to obtain 'undisturbed' soil samples. The ends of the samples were sealed with wax immediately and stored at constant temperature.

The basic properties of Gault Clay in the Cambridge area have been reported by Samuels (1975), Worssam & Taylor (1975) and Ng (1992). Gault Clay was laid down in south-east England as a result of widespread marine incursion that spanned the Middle and Upper Albian stages. Following the deposition of the Gault, the Chalk was laid down as the sea water cleared. During the Tertiary and Quaternary epochs, uplift and extensive erosion took place and eventually produced the present landscape. In the Cambridge area, an estimated 200–400 m of overlying Chalk layer had been eroded (Lings *et al.*, 1991). Gault Clay consists of stiff to hard silty grey clay with high plasticity and contains closely spaced fissures and joints. The plastic limit is about 46%, generally above the water content, which indicates that the clay is heavily overconsolidated. Gault Clay is calcareous, containing up to 30% by weight of calcium carbonate (Samuels, 1975). Gault Clay at Cambridge appears to have a clay fraction (<2 μm) of about 68% (Thompson, 1962).

Sample preparation

Soil samples of 38 mm diameter × 90 mm length were carefully extracted. The samples were trimmed with parallel and flat edges and placed on the pedestal of the triaxial cell. Hinges were attached to the sample membrane using super glue. As shown in Fig. 1, two centrally strain gauged LDTs were inserted into the hinges. The distance between the two hinges was slightly shorter than the length of the strip so that, when the strip was mounted, it balanced itself against the fixed attachments by its own elastic force.

Stress paths

Before testing the soil samples were saturated so as to give a B value of 0·75–0·8. For a fully saturated soft clay B is close to 1·0, but for a fully saturated stiff overconsolidated clay B is less than 1·0 (Black & Lee, 1973).

Five tests were conducted on undisturbed samples of Gault Clay (see Table 1). The stress states at the end of saturation, isotropic swelling and shearing are given in Table 1 in terms of mean effective stress p' and deviatoric stress q. The soil samples were saturated at a cell pressure 100 kPa more than the value of p' prior to shearing. The soil samples were then allowed to swell back isotropically. Soil samples were kept 6 to 8 h before the shearing. The soil samples were then sheared in compression at constant p'. Top and bottom drainages were allowed and a very slow rate of 0·5 kPa/h was used.

RESULTS

In each test, bulk stiffness and shear stiffness were derived during isotropic swelling and during constant p' stress paths respectively. The tangent stiffness was obtained as slope of two adjacent stress–strain readings. However, due to the scatter in the data at small strains ($<3 \times 10^{-5}$), it was not possible to obtain tangent stiffness directly from two adjacent measurements. A piece-wise polynomial fitted through the points was used to evaluate the small strain stiffness.

The variation of stiffness ratio, stiffness normalized with p' (K'/p' or G/p'), is computed for each stress path. The bulk stiffness ratio (K'/p') against absolute volumetric strain ($|\varepsilon_v|$) and shear stiffness ratio against deviatoric strain (ε_q) are presented in Figs 3 and 4 respectively. The data suggest that the tests conducted at lower p' values give higher initial stiffness ratios than tests conducted at higher values of p'. The ratio decreased with increasing strain and reached a common value irrespective of the magnitude of p'.

The experimental results on overconsolidated clay indicated that G_{max}/p' was dependent on p' (Houlsby & Wroth, 1991). Viggiani & Atkinson (1995) reported that laboratory tests on reconstituted Speswhite kaolin samples also indicated the dependency of G_{max}/p' on p'. The data of Gault Clay clearly follow a similar behaviour. Note that variation in K'/p' is smaller than that of G/p'.

MODELLING OF STIFFNESS

At very small strains, the tangent shear modulus of soil is constant at G_{max} for a given p' and overconsolidation ratio (OCR) and is assumed to be given by (Houlsby & Wroth, 1991)

$$G_{max} = A(p'/p_r)^{n1}\text{OCR}^{m1} \quad (1)$$

where reference pressure p_r (equal to 1 kPa) is included so that the parameters A and $n1$ are dimensionless but their numerical values depend on the choice of reference pressure. The gradual deterioration of shear stiffness with ε_q can be approximated by

$$G = B(p'/p_r)^{n2}\text{OCR}^{m2}\varepsilon_q^{b2} \quad (2)$$

where A, $n1$, $m1$, B, $n2$, $m2$ and $b2$ are soil para-

Table 1. Details of the tests conducted on Gault Clay

Test no.	Depth of sampling: m	Stresses at end of saturation: kPa		Stresses at end of swelling: kPa		Stresses at end of shearing: kPa		Water content at end of shearing: %
		p'_s	q_s	p'_i	q_i	p'_f	q_f	
1	2·00–2·25	150	0	50	0	50	40	31·7
2	2·50–2·75	200	0	100	0	Slow cyclic test		30·4
3	2·00–2·25	300	0	200	0	200	100	29·5
4	2·50–2·75	500	0	400	0	400	200	28·5
5	3·00–3·25	200	0	100	0	Slow cyclic test		30·6

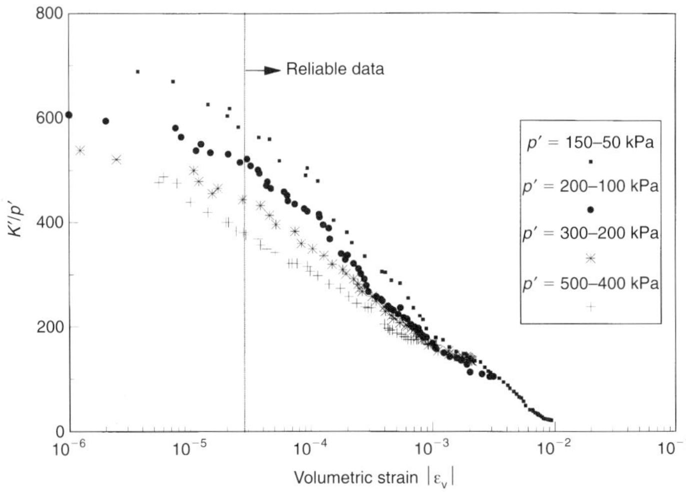

Fig. 3. Bulk stiffness ratio against volumetric strain during swelling (tests 1–4)

meters. Similarly, bulk stiffness is assumed to be in the form of

$$K'_{\max} = C(p'/p_r)^{n3}\,\mathrm{OCR}^{m3} \tag{3}$$

$$K' = D(p'/p_r)^{n4}\,\mathrm{OCR}^{m4}\varepsilon_v^{b4} \tag{4}$$

where C, $n3$, $m3$, D, $n4$, $m4$ and $b4$ are parameters defining bulk modulus.

Analysis of shear stiffness

The $G_{\max}$ values measured in various tests are presented in Fig. 5. The preconsolidation pressure of Gault Clay in Cambridge was taken to be 3500 kPa (Lings *et al.*, 1991). The parameter $(n1 - m1)$ was obtained using equation (1) by plotting $G_{\max}$ against p' on a log–log graph (see Fig. 5). The pressure exponent $n1$ can be obtained if the parameter $m1$ is known. Here, the parameter $m1$ is assumed to be equal to 0·2. This is based on values for other clays, e.g. Speswhite kaolin ($m1 = 0{\cdot}196$) and London Clay ($m1 = 0{\cdot}25$), as quoted by Viggiani & Atkinson (1995). The $G_{\max}$ values for a normally consolidated soil are calculated using equation (1) and are presented in Fig. 5.

The normally consolidated values of shear modulus against p' at various deviatoric strains are

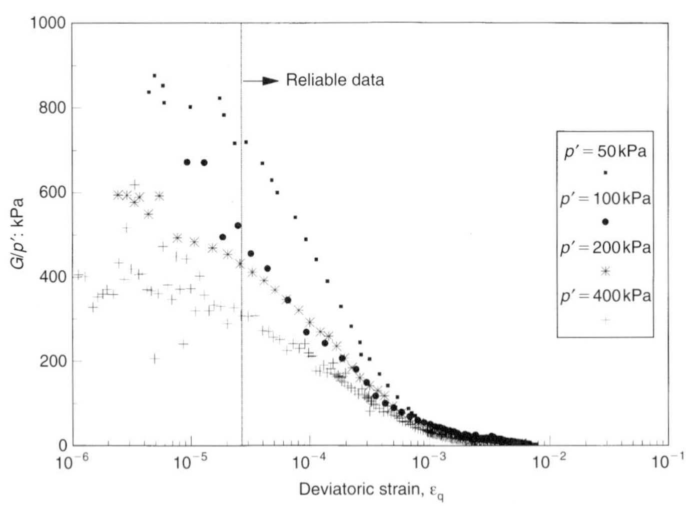

Fig. 4. Shear stiffness ratio against deviatoric strain during constant p' tests (tests 1–4)

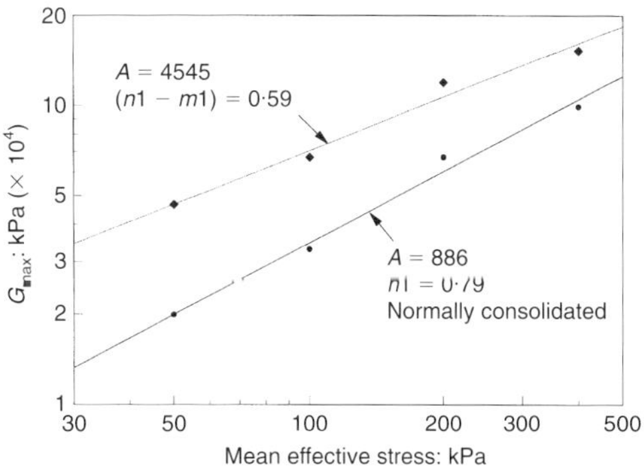

Fig. 5. Variation of G_{max} with p' for Gault Clay

presented in Fig. 6. At a given deviatoric strain, the shear modulus increases with p'. The slope of the line in Fig. 6 gives a value of the pressure exponent $n2$. The pressure exponent $n2$ increased with increasing deviatoric strain. The increase of pressure exponent $n2$ with strain was also observed by Iwasaki *et al.* (1978). The gradual increase of the pressure exponent value towards unity was attributed to development of plastic response of soil by Viggiani & Atkinson (1995). The pressure exponents presented here are higher than the London Clay values measured by Viggiani & Atkinson.

In Fig. 7, the same data are re-plotted on logarithmic axes of deviatoric strain and shear

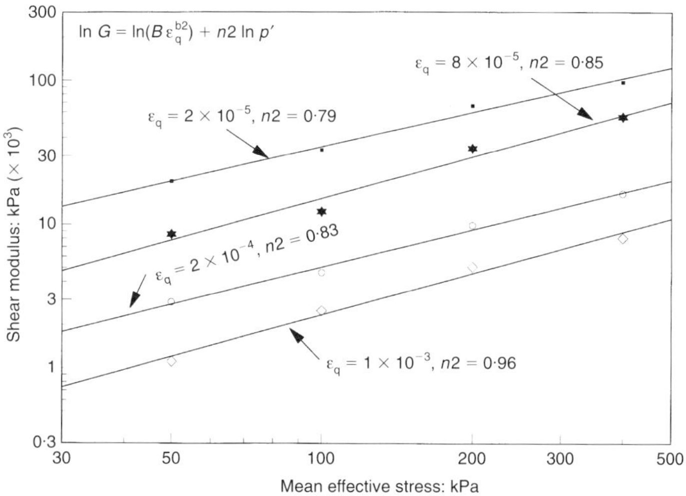

Fig. 6. Variation of G with p' for Gault Clay

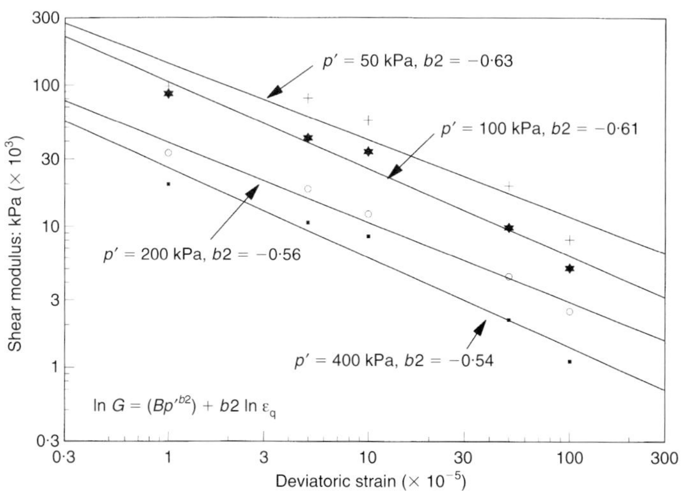

Fig. 7. Variation of G with deviatoric strain ε_q for Gault Clay

modulus. The strain exponent $b2$ increases from $-0{\cdot}63$ at mean effective stress of 50 kPa to $-0{\cdot}54$ at 400 kPa. It appears that stiffness degradation is rapid at lower values of p' (small value of $b2$). A similar analysis for bulk stiffness can be found in Dasari (1996).

COMPARISON OF FIELD AND LABORATORY STIFFNESS

Recent advances in local strain measurement in laboratory tests allowed comparison of field and laboratory stiffness at similar strains. Hird *et al.* (1991) showed good agreement between moduli from small strain triaxial and field plate load tests on stiff clay at the Cowden site in Holderness. They attributed this agreement to the same mode of shear in both tests. Viggiani & Atkinson (1995) reported that G_{max} measured in situ using Rayleigh waves compared well with the corresponding values obtained in the laboratory tests. However, Clayton *et al.* (1994) found that laboratory measured stiffness was much higher than G_{max} obtained using the Rayleigh wave method for highly fractured weak rocks. The above results indicate that the agreement between laboratory and field tests depends on many factors. This issue is investigated for Gault Clay in the following paragraphs.

Abbiss (1981) and Powell & Butcher (1991) measured the secant shear stiffness of Gault Clay using the Rayleigh wave method. These geophysical measurements were estimated to be at shear strains of 10^{-6} to 10^{-5} and are shown in Fig. 8. The secant shear stiffness of Gault Clay obtained from the laboratory tests is also shown in Fig. 8 for comparison. The stiffness from the Rayleigh wave is much smaller than the triaxial data. The Rayleigh wave measures vertical stiffness at shallower depths (lower p'), but this cannot explain why much higher stiffness was found in triaxial data.

Ng *et al.* (1995) reported that the stiffness curve measured in the triaxial test corresponding to $p' = 100$ kPa is close to the stiffness curve derived from the back-analysis of a diaphragm wall constructed in Gault Clay. This tends to support the validity of values measured in triaxial tests. Therefore, the reasons for the much lower stiffness values obtained by the Rayleigh wave method are at present unknown and need further investigation.

Powell & Butcher (1991) presented stiffness–strain relation of Gault Clay at the Madingley site. The stiffness–strain relation is obtained using a self-boring pressuremeter. The stiffness data from the pressuremeter tests and the triaxial tests are presented in Fig. 9. The pressuremeter data are lower and upper bounds of one conventional pressuremeter modulus G_p and other secant moduli G_s obtained from a stress–strain curve following the analysis of Palmer (1972) as suggested by Muir Wood (1990). The stiffness derived from pressuremeter tests lies above the triaxial data. This is expected, as in a pressuremeter the response is governed by horizontal stiffness whereas a triaxial test measures vertical stiffness. For an overconsolidated clay such as Gault Clay, horizontal stiffness is higher than vertical stiffness.

Laboratory measurements on soil samples give spot values under controlled boundary conditions, but sample preparation may have altered their in situ soil structure. The field measurements take place on in situ soils, but the boundary conditions are not clearly known and measurements are over a large volume of soil. Also, the differences between

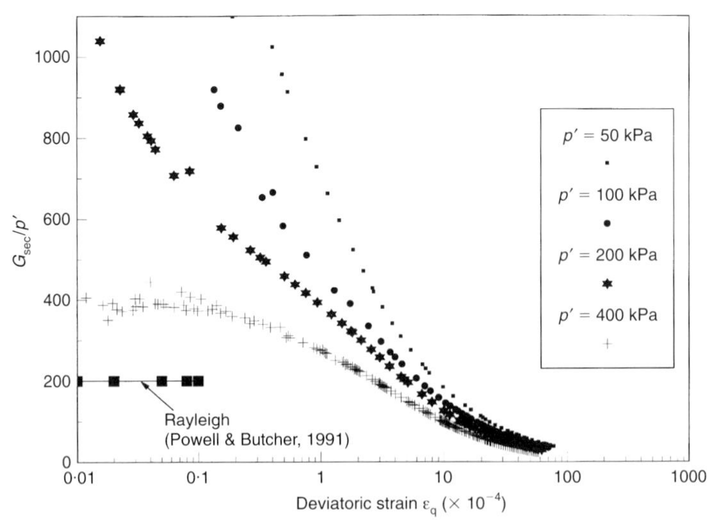

Fig. 8. Triaxial Gault Clay stiffness compared with Rayleigh wave method

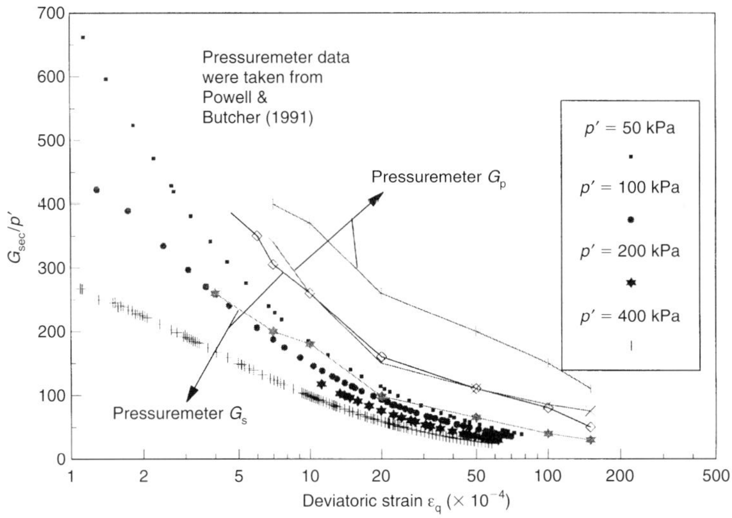

Fig. 9. Triaxial stiffness of Gault Clay compared with pressuremeter data

field and laboratory could be due to other reasons. As shown in Fig. 6, the comparison of stiffness ratio from various test results also depends on the magnitude of p'. Therefore, it is important to know the value of p' in order to compare stiffness ratios from various tests. Bellotti *et al.* (1996) reported that the small strain stiffness is also dependent on individual stress components. The stress component in the laboratory tests is not identical to field stress state, which might have also contributed to the difference.

CONCLUSIONS

Modified LDTs have been used successfully to investigate the stiffness of undisturbed Gault Clay, from small (3×10^{-5}) to large (0·01) strains. The stiffness–strain relations (S-curves) of Gault Clay are presented. The ratios of maximum stiffness normalized with p' (G_{max}/p' and K'_{max}/p') were found to be dependent on the magnitude of p'. The smaller the p' value, the higher is the ratio. Therefore, it is important to know the value of p' to compare stiffness ratios from various tests. Also, in the analysis of problems, soil models should consider the change in stiffness ratios with respect to p'. There is a gradual increase of pressure exponent $n1$ from about 0·6 at 3×10^{-5} towards 1·0 at 0·01 strain.

The comparison of triaxial stiffness to the Rayleigh wave measurements showed that the Rayleigh wave method underestimated Gault Clay stiffness. The reasons for the discrepancy need further investigation. The pressuremeter stiffness was found to be higher than the triaxial data. This could be due to anisotropy of Gault Clay.

REFERENCES

Abbiss, C. P. (1981). Shear wave measurements of the elasticity of the ground. *Géotechnique* **31**, No. 1, 91–104.

Bellotti, R., Jamiolkowski, M., Lo Presti, D. C. F. & O'Neill, D. A. (1996). Anisotropy of small strain stiffness in Ticino sand. *Géotechnique* **46**, No. 1, 115–131.

Black, D. K. & Lee, K. L. (1973). Saturating laboratory samples by back pressure. *J. Soil Mech. Fdn Engng Am. Soc. Civ. Engrs* **99**, SM1, 75–93.

Clayton, C. R. I., Gordon, M. A. & Matthews, M. C. (1994). Measurement of stiffness of soils and weak rocks using small strain laboratory testing and field geophysics. *Pre-failure deformation of geomaterials*, Shibuya *et al.* (eds), Sapporo, pp. 229–234.

Baldi, G., Hight, D. W. & Thomas, G. E. (1988). A re-evaluation of conventional triaxial test methods. *Advanced triaxial testing of soil and rock,* ASTM STP 977, 219–263.

Dasari, G. R. (1996). *Modelling the variation of soil stiffness during sequential construction.* PhD thesis, Cambridge University.

Goto, S., Tatsuoka, F., Shibuya, S., Kim, Y. S. & Sato, T. (1991). A simple gauge for local small strain measurements in the laboratory. *Soils Fdns* **31**, No. 1, 169–180.

Hird, C. C., Powell, J. J. M. & Yung, P. C. Y. (1991). Investigations of the stiffness of a glacial till. *Proc. 10th Eur. Conf. Soil Mech., Florence*, **1**, 107–110.

Houlsby, G. T. & Wroth, C. P. (1991). The variation of shear modulus of a clay with pressure and overconsolidation ratio. *Soils Fdns* **31**, 138–143.

Iwasaki, T., Tatsuoka, F. & Takagi, Y. (1978). Shear moduli of sands under cyclic torsional shear loading. *Soils Fdns* **18**, No. 1, 39–56.

Lings, M. L., Nash, D. F. T., Ng, C. W. W. & Boyce, M. D. (1991). Observed behaviour of a deep excavation in Gault Clay: a preliminary appraisal. *Proc. 10th Eur. Conf. Soil Mech., Florence*, **2**, 467–470.

Muir Wood, D. (1990). Strain dependent moduli and pressuremeter tests. *Géotechnique* **40**, 509–512.

Ng, C. W. W. (1992). *An evaluation of soil–structure interaction associated with a multi-propped excavation*. PhD thesis, University of Bristol.

Ng, C. W. W., Bolton, M. D. & Dasari, G. R. (1995). The small strain stiffness of a carbonate stiff clay. *Soils Fdns* **35**, No. 4, 109–114.

Palmer, A. C. (1972). Undrained plane strain expansion of a cylindrical cavity in clay: a simple interpretation for the pressuremeter test. *Géotechnique* **22**, No. 4, 451–457.

Powell, J. J. M. & Butcher, A. P. (1991). Assessment of ground stiffness from field and laboratory tests. *Proc. 10th Eur. Conf. Soil Mech., Florence*, **1**, 153–156.

Samuels, S. G. (1975). Some properties of the Gault Clay from the Ely–Ouse Essex water tunnel. *Géotechnique* **25**, No. 2, 239–264.

St John, H. D. (1975). *Field and theoretical studies of the behaviour of ground deep excavations in London Clay.* PhD thesis, Cambridge University.

Thompson, W. J. (1962). *Some deformation characteristics of Cambridge Gault Clay.* PhD thesis, University of Cambridge.

Viggiani, G. & Atkinson, J. H. (1995). Stiffness of fine grained soils at very small strains. *Géotechnique* **45**, No. 2, 249–265.

Worssam, B. C. & Taylor, J. H. (1975). *Geology of the country around Cambridge*, 2nd edn. London: HMSO.

Nataatmadja, A. (1998). Pre-failure deformation behaviour of geomaterials, 353–361

Yielding and locking under repeated loading

A. NATAATMADJA*

This paper describes the response of granular materials under repeated load triaxial loading. It is demonstrated that the resilient behaviour of the materials is more complex than can be predicted by the Hertzian contact theory. Repeated load triaxial tests with constant confining pressure will always cause yielding with very high resilient modulus values at initial loading stages. On the other hand, tests with cyclic confining pressure do not always start with yielding. The use of a simple two-parameter model to characterize the resilient modulus variation is demonstrated and the results indicate its accuracy in predicting the dynamic stress–strain relations.

KEYWORDS: aggregate; modelling; repeated load; resilient modulus; triaxial testing.

Cet exposé décrit la réaction des matières granulaires soumises à des charges répétitives sur trois axes. Nous démontrons que le comportement résilient de ces matières est plus complexe que celui pronostiqué par la théorie de contact hertzien. Les essais de charges triaxiales répétitives avec une pression de confinement constante vont toujours causer une élasticité avec de très hautes valeurs de module de résilience lors des premiers stades de chargement. D'autre part, les essais avec pression de confinement cyclique ne commencent pas toujours par causer une élasticité. Nous démontrons comment l'on peut utiliser un modèle simple à deux paramètres pour caractériser la variation de module de résistance; les résultats indiquent qu'on peut prévoir avec exactitude les relations dynamiques contrainte-déformation.

INTRODUCTION

Base course materials have special characteristics. Being coarse (sometimes up to 40 mm diameter) and cohesionless, they function more as discrete particles than as a continuum. Boyce (1980) pointed out that three processes may contribute to strain in a granular material: (*a*) particle deformation around points of contact, (*b*) slip between particles at points of contact, and (*c*) crushing of particles at points of contact. It was suggested that elastic strain is mainly related to the interparticle deformation and crushing may be eliminated as an irreversible process. During cyclic or repeated loading slip between particles may also occur, but some of this slip may be reversed during the return part of each cycle. Hence, it was suggested that the resilient strain observed during repeated loading may be greater than the strain calculated from a purely elastic model. On the other hand, Kalteziotis & Sarsby (1985) indicated that during repeated loading of granular materials ranging from medium silt to coarse sand, the recoverable stress–strain relation could be accurately described by using an empirical modification of the classical elastic theory of Hertz (Timoshenko & Goodier, 1970).

Manuscript received 10 December 1996.
* School of Engineering, Griffith University.

This paper presents the results of an investigation of the stress–strain characteristics of base course materials under repeated loading. The analytical and experimental results demonstrate that under repeated loading coarse-grained materials undergo yielding and locking, producing elastic strains which can be smaller than those calculated from a purely elastic theory.

THE CONTACT THEORY

If elastic is assumed to be solely due to particle deformation, this strain can be calculated in the case of a regular array of spheres in contact. According to the theory of Hertz, the relative displacement d of two spherical elastic particles in contact under a normal force P is given by

$$d = cP^m \tag{1}$$

where c is a function of the radius and elastic properties of the spheres. The value of exponent m depends mainly on the rate of increase of contact area. For the case in which this rate is negligible, i.e. contact between two flat, smooth surfaces, m is equal to unity. If the rate of increase of contact area is similar to that for the point of contact between two spheres, m is equal to two-thirds. Soils, however, have irregular particle curvature and size and indeterminate distribution of normal and shear forces. Hence, m values other than $\frac{2}{3}$ are possible (Ko & Scott, 1967).

Rowe (1971) proposed that the elastic deformation of a particulate mass would be an equation which directly related resultant principal strain of the mass to the corresponding applied principal stress

$$\varepsilon_{1,2,3} = S_{1,2,3}(\sigma_{1,2,3}/E)^m \tag{2}$$

where S and m are elastic coefficients which are functions of the structure of all the particle contacts, and E is the Young's modulus of the parent material. This equation suggests that a single m value may be used to characterize the stress–strain variation of the material for a broad spectrum of loading. There is evidence, however, that for a higher vertical stress the stress–strain curve becomes steeper as the material becomes denser and interparticle sliding becomes increasingly more difficult, and hence the coefficient m may vary with the applied stress.

Ko & Scott (1967), for example, found that the deformation of sand in hydrostatic compression is non-linear but nearly elastic. However, it was indicated that the behaviour of the material was not as predicted by the Hertzian contact theory. The soil becomes stiffer more rapidly than predicted by the theory. It was postulated that the deviation was caused by the increased number of grain contacts continuously being made as the stress is increased. Experimental work with plexiglass cylinders by Travers *et al.* (1987) supports this observation.

Furthermore, Lambe & Whitman (1969) suggested that as load is applied, strain results primarily from deformation of the individual grains. With further loading, grains start to slide relative to one another and the resistance to straining decreases. In a confined compression, the grains subsequently come into a more compact arrangement and interparticle sliding becomes increasingly more difficult and causes grain locking. During isotropic compression, however, the change in the lateral stress is greater than that of confined compression for a given change in axial stress, which results in a smaller vertical strain. On the other hand, static triaxial loading results in yielding, in which the resistance to straining decreases with increasing load. Fig. 1 shows the comparison of stress–strain curves for the three types of compression.

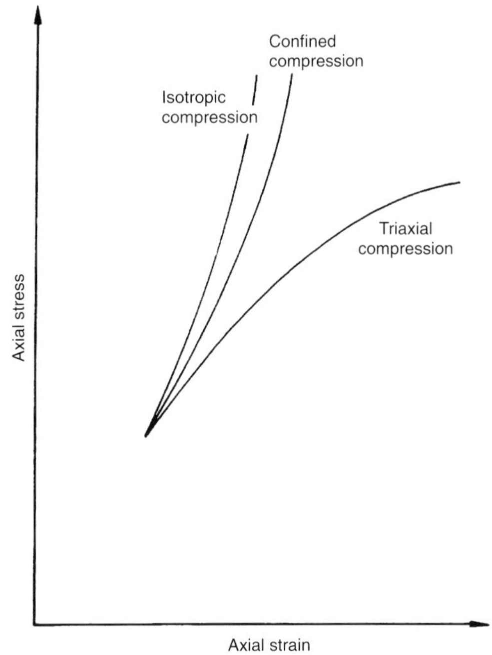

Fig. 1. Comparison of stress–strain curves, after Lambe & Whitman (1969)

THE S-SHAPED STRESS–STRAIN CURVE

Whitman *et al.* (1964) described the results of confined compression tests on a sand, in which the load was applied repeatedly. They demonstrated that the stress–strain curves under the loading had a change of curvature. For a small load the curve is convex upwards, while for a larger load it becomes concave upwards, i.e. having an S-shape (Fig. 2). With increasing axial stress change $\Delta\sigma_A$, the secant modulus first decreases (yielding) and then increases (locking). The range of $\Delta\sigma_A$ for which yielding occurs is related to the initial axial load σ_{AO}. Numerically, this range is something like $\Delta\sigma_A < \frac{1}{2}\sigma_{AO}$. For very small $\Delta\sigma_A$ (say $< \frac{1}{10}\sigma_{AO}$), very large values of secant modulus can be measured (Fig. 3). The behaviour of the material is thus more complicated than has been suggested by the classical Hertzian contact theory.

Although the experimental nature of the repeated load triaxial test is similar to isotropic compression and confined compression, the evidence of an S-shaped curve for base course materials under repeated triaxial loading has rarely been discussed in the literature. However, results of Hicks (1970) and Uzan (1985)—shown in Figs 4 and 5 respectively—indicate that the resilient modulus for a particular confining pressure firstly decreases with increasing deviator stress, but then increases with a further increase in deviator stress (Resilient modulus M_r is defined as the ratio of the repeated deviator stress q_r to the recoverable or resilient strain ε_r). This suggests that S-shaped stress–strain curves do exist.

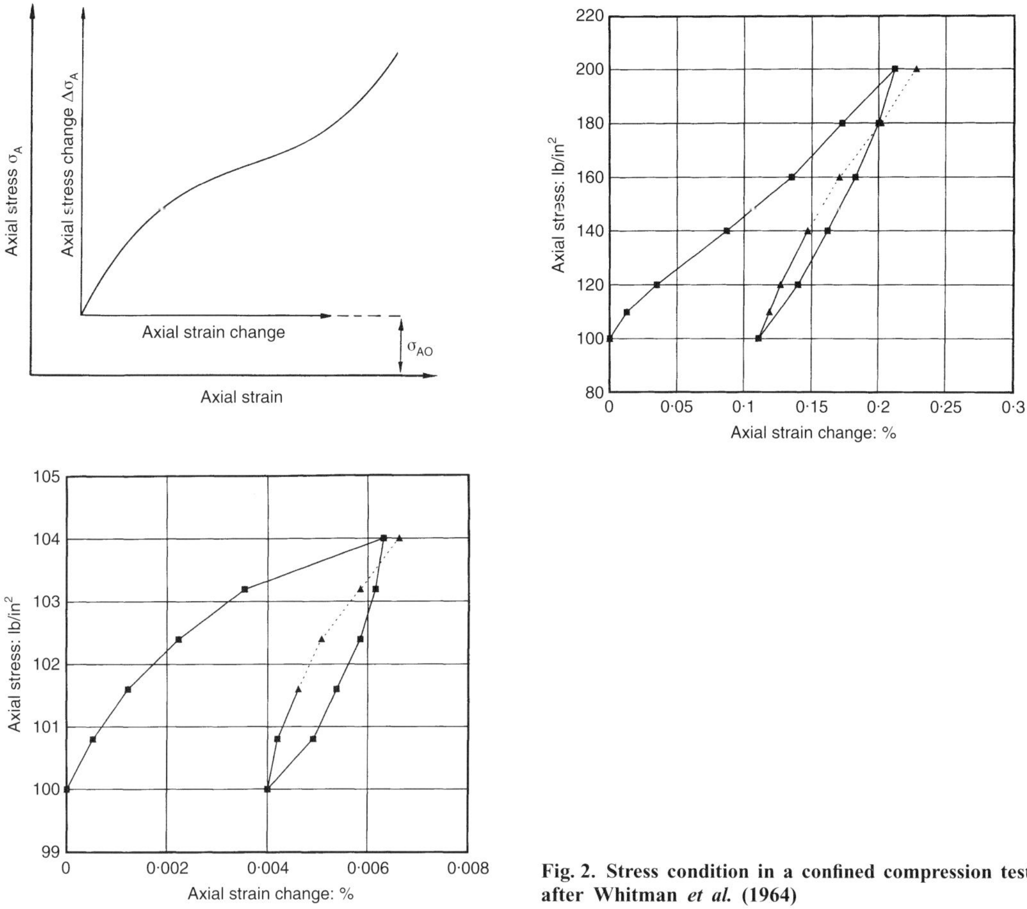

Fig. 2. Stress condition in a confined compression test, after Whitman *et al.* (1964)

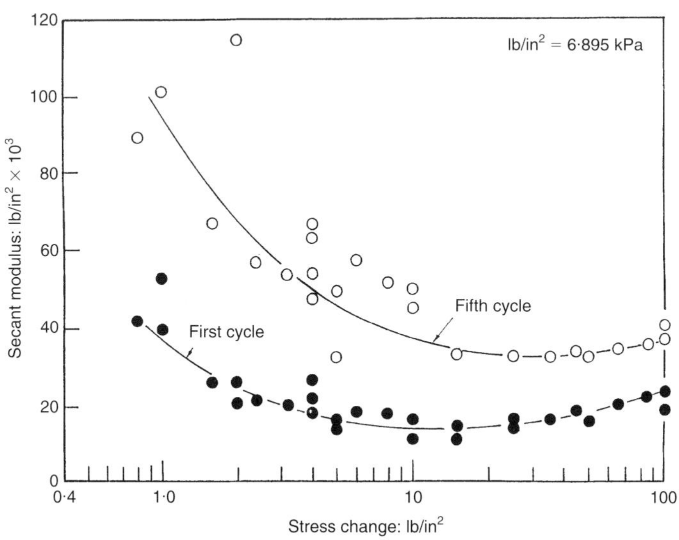

Fig. 3. Relationship between secant modulus and stress change of a sand in a confined compression test, after Whitman *et al.* (1964)

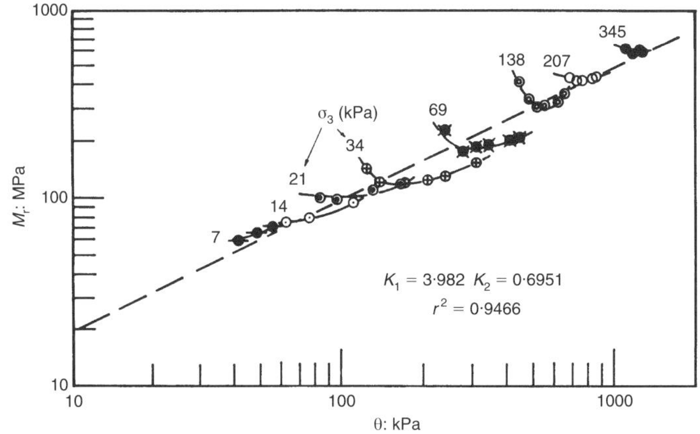

Fig. 4. Variation of resilient modulus with bulk stress for a dry base course material, after Hicks (1970)

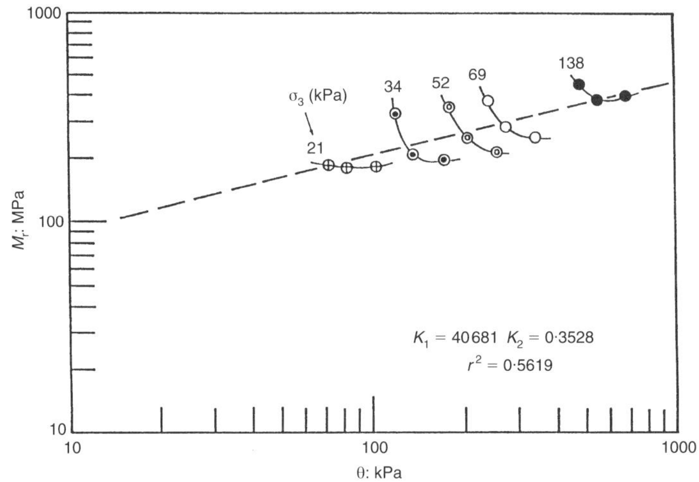

Fig. 5. Variation of resilient modulus with bulk stress for a dense graded aggregate, after Uzan (1985)

REPEATED LOAD TRIAXIAL TEST

To investigate the stress–strain relation of base course materials under repeated loading, an experimental programme has been conducted, including testing under static confining pressure as well as testing with confining pressure cycled in phase with the deviator stress. The triaxial machine, accommodating a specimen size of 400 mm high and 200 mm diameter, is currently the largest of its kind in Australia (Fig. 6). Details of the equipment and experimental procedures can be found in Nataatmadja & Parkin (1988, 1990).

Fig. 6. Repeated load triaxial test equipment at Griffith University

Test with static confining pressure

Figure 7 shows the typical stress–strain curves of the present investigation which are obtained

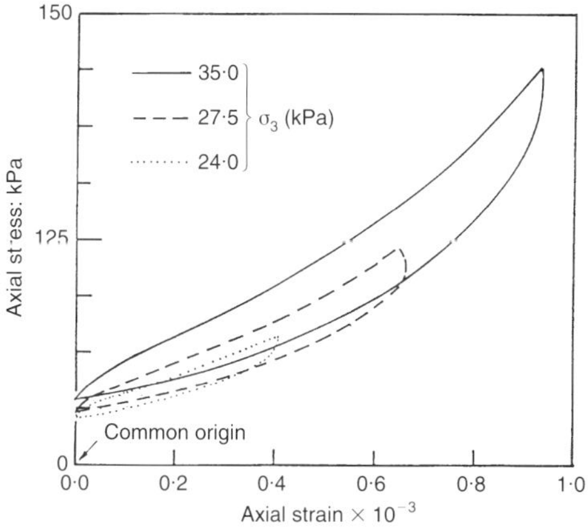

Fig. 7. Typical stress–strain curve of a crushed rhyolite—test with constant confining pressure

from repeated load triaxial testing of a well-graded dry crushed Rhyolite with a maximum size of 19 mm compacted to 95% modified density. It is seen that the curves have an S-shape similar to that described in Whitman *et al.* (1964).

The nature of the stress–strain variation will undoubtedly affect the calculated resilient modulus. In Fig. 8 the resilient modulus is plotted against the sum of principal stresses (first invariant of stress or bulk stress θ) for each confining stress in a double log plot. The linear fitting line on a double log plot is known as the K-theta model (Hicks & Monismith, 1971), which can be expressed as

$$M_r = K_1 \theta^{K_2} \tag{6}$$

where M_r is the resilient modulus, and K_1 and K_2 are material parameters. Although the model has been used in many pavement analyses, it certainly fails to model the variation of the resilient modulus for a broad range of stresses. The coefficient of determination is very low ($r^2 = 0{\cdot}4658$), indicating a poor correlation.

When the resilient modulus is normalized by means of the ratio of the deviator stress to the sum of principal stresses, and plotted against the deviator stress, a significant relationship is obtained (Fig. 9). The equation of the two-parameter model (Nataatmadja & Parkin, 1989) for the fitting line is

$$(M_r q_r)/\theta = A + Bq_r \tag{7}$$

where M_r is the resilient modulus (kPa), q_r is the repeated deviator stress (kPa), θ is the sum of principal stresses or the first stress invariant (kPa), and A (kPa) and B (dimensionless) are material constants.

Note that in this method, the modulus is expressed as a function of θ and q_r, in accordance with developments in the resilient characterization of granular materials (Tam & Brown, 1988; Uzan, 1992; Karasahin *et al.*, 1993). From equation (7), the resilient strain can be calculated by

$$\varepsilon_r = q_r^{\,2}/[(3\sigma_3 + q_r)(A + Bq_r)] \tag{8}$$

Fig. 10 presents the comparison between the calculated and the actual stress–strain curves of Hicks (1970) and Uzan (1985). Fig. 11 shows similar curves obtained from the experimental work of the Author. In general, the agreement is very good and the predicted stress–strain curves are also S-shaped.

Furthermore, from Figs 10 and 11 it can also be deduced that when the applied deviator stress is lower than a certain critical deviator stress q_{crit}, the modulus decreases with increasing deviator stress. This is related to an initially convex upwards curve (yielding). On the other hand, when the applied deviator stress is greater than q_{crit}, a concave

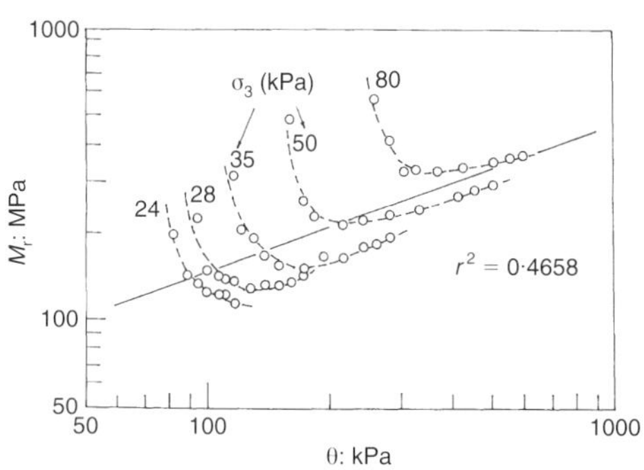

Fig. 8. Variation of resilient modulus with bulk stress and confining pressure for a crushed rhyolite—test with constant confining pressure

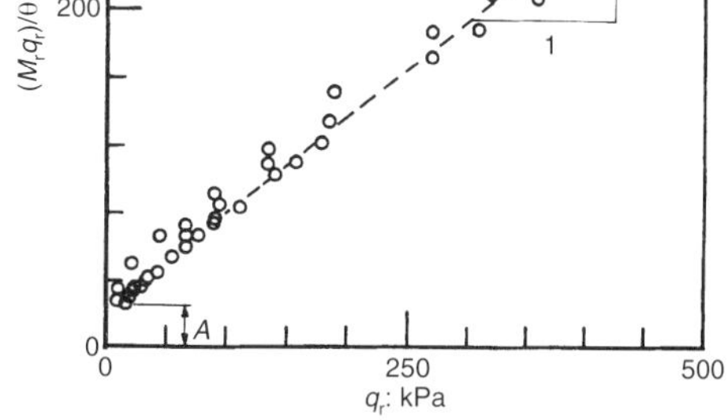

Fig. 9. Plot of two-parameter model for a crushed rhyolite—test with constant confining pressure

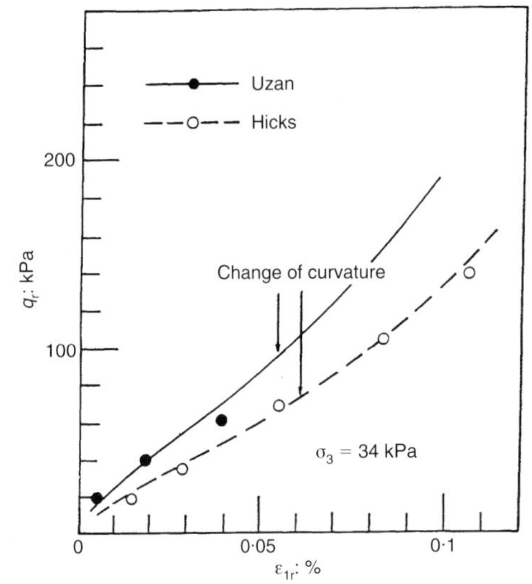

Fig. 10. Comparison between predicted stress–strain curves and experimental results of Hicks (1970) and Uzan (1985)

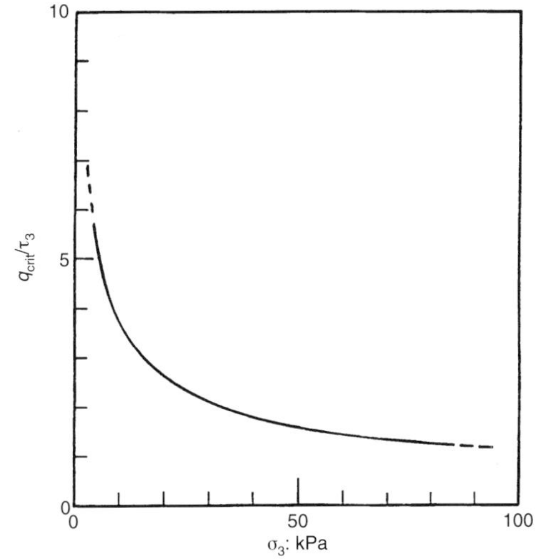

Fig. 12. Calculated variation of critical deviator stress of a crushed rhyolite—test with constant confining pressure

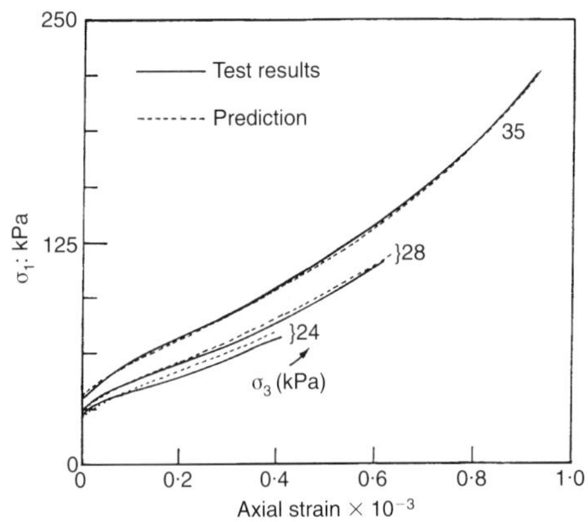

Fig. 11. Comparison between predicted stress–strain curves and experimental results for a crushed rhyolite—test with constant confining pressure

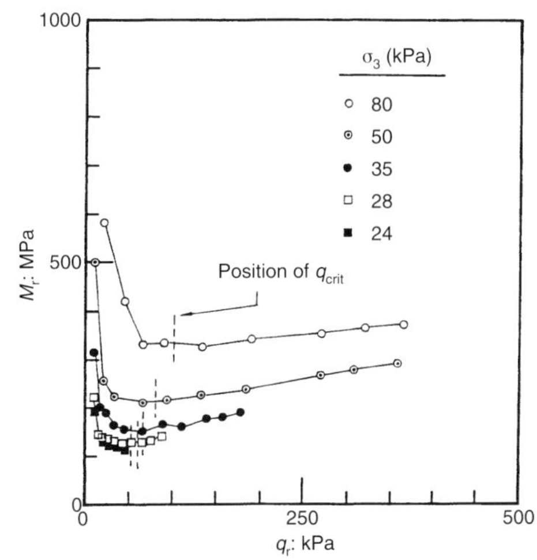

Fig. 13. Plot of resilient modulus against repeated deviator stress for a crushed rhyolite—test with constant confining pressure

upwards curve (locking) occurs and this corresponds to an increase in modulus value with an increase in the applied deviator stress. This critical deviator stress can be obtained from $\mathrm{d}M_r/\mathrm{d}q_r = 0$, which gives

$$q_{crit} = \sqrt{(3A\sigma_3/B)} \tag{9}$$

The result of the calculation of q_{crit} for the present investigation is shown in Fig. 12. It is seen that the ratio q_{crit}/σ_3 is not constant as suggested by Whitman *et al.* (1964), but decreases with increasing σ_3. Fig. 13 shows a plot of the resilient modulus against the repeated deviator stress and the predicted critical deviator stress values are indicated for each confining pressure. Fig. 14 presents the variation of resilient modulus with resilient strain from the experiment and the predicted relations.

Lambe & Whitman (1969) stated that for most engineering problems, the critical deviator stress is probably less than 10 kPa and hence is of no practical significance. However, the present study indicates that for a pavement base course, where the residual confining stress can be as high as 40 kPa (Hicks, 1970), this critical stress may sig-

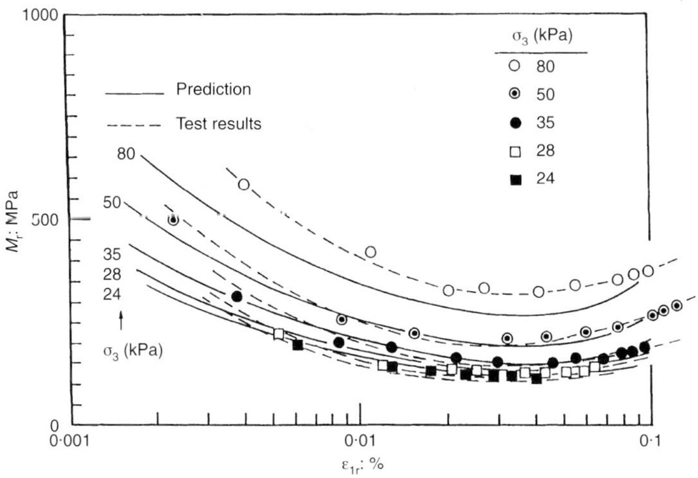

Fig. 14. Variation of resilient modulus with resilient axial strain for a crushed rhyolite—test with constant confining pressure

nificantly affect the resilient modulus, especially in the lower part of a thick base course.

Test with cyclic confining pressure

When cyclic confining pressure is applied in phase with a cyclic deviator stress, the vertical strain will be affected by both the cyclic confining pressure and the initial (static) confining pressure on which the cyclic confining pressure is superimposed. Typical resilient modulus values from the present study are shown in Fig. 15, in which the resilient modulus is plotted against the sum of principal stress in double log plot. Comparing this plot with Fig. 8 (test with static confining pressure), it is seen that the results show a different trend, especially at low deviator stress values. Again it is not possible to fit a meaningful common line for the results.

Modifying equation (7) to include the effects of cyclic confining pressure and initial confining pressure results in

$$(M_r\sigma_1)/\theta = A' + B'q_r \tag{10}$$

where M_r is the resilient modulus (kPa), q_r is the cyclic deviator stress (kPa), θ is the sum of principal stresses (kPa), σ_1 is the total vertical stress (kPa), and A' (kPa) and B' (dimensionless) are

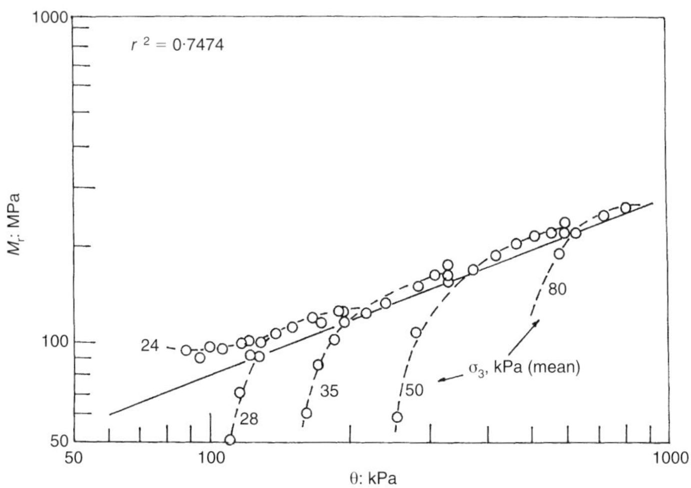

Fig. 15. Variation of resilient modulus with bulk stress and confining pressure for a crushed rhyolite—test with cyclic confining pressure

material constants. Fig. 16 shows the application of the model, which gives a very good correlation ($r^2 = 0{\cdot}9333$).

By analogy to a test with static confining pressure, a critical repeated deviator stress (q_{crit}), defined as the cyclic deviator stress which produces the minimum modulus value for a particular initial and cyclic confining pressure, can be found from $dM_r/dq_r = 0$, which gives

$$B'q_r^2 + 2B'(\sigma_{3i} + \sigma_{3r})q_r + 3B'\sigma_{3r}^2 + 6B'\sigma_{3i}\sigma_{3r} - 2A'\sigma_{3r} + 3B'\sigma_{3i}^2 - 2A'\sigma_{3i} = 0 \quad (11)$$

where σ_{3r} is cyclic confining pressure and σ_{3i} is initial confining pressure.

Figure 17 shows the variation of q_{crit} with total confining pressure σ_3. It is seen that with an increase of cyclic confining pressure, the critical deviator stress decreases until it vanishes for a test with σ_{3r} greater than a critical cyclic confining pressure (σ_{3crit}), found to be 40 kPa for this material. So, when the applied cyclic confining pressure is higher than σ_{3crit}, the modulus will always increase with an increase of deviator stress due to locking.

It is worth noting that Brown & Hyde (1975) suggested that a test with constant confining pressure can produce resilient modulus variation comparable to that of a test with cyclic confining pressure if the constant confining pressure is equal to the mean value of the cyclic confining pressure. However, this cannot be justified in the light of the present results.

With an initial confining pressure of 20 kPa and cyclic confining pressure of 40 kPa, for example, q_{crit} is already zero, while in the test with static confining pressure of 40 kPa (mean value of 40 kPa plus 20 kPa), the corresponding q_{crit} is about 75 kPa. Clearly, under cyclic confining pressure application the stress–strain curve has more tendency to have a concave upwards (locking) shape. Hence, depending on the applied stresses, the stress–strain characteristics in tests with cyclic confining pressure can be significantly different from those obtained in tests with constant confining pressure. This finding underlines the fundamental difference between the two test methods, which may result in significantly different resilient properties. Therefore, test standardization is deemed necessary.

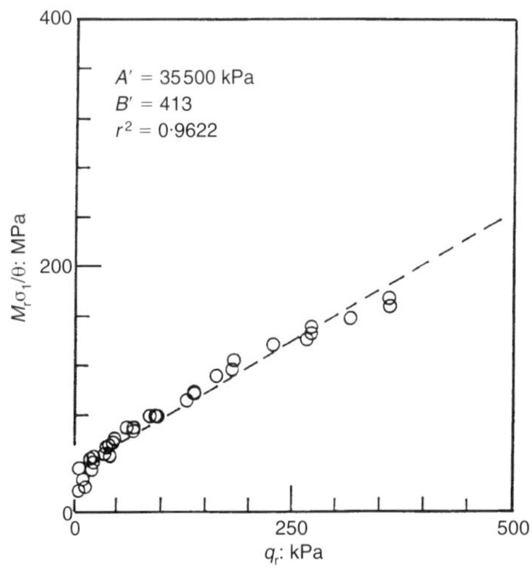

Fig. 16. Plot of two-parameter model for a crushed rhyolite—test with cyclic confining pressure

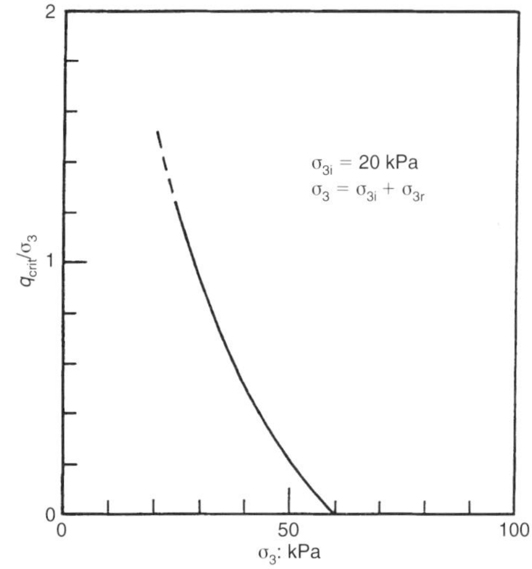

Fig. 17. Calculated variation of critical deviator stress of a crushed rhyolite—test with cyclic confining pressure

CONCLUSION

The behaviour of granular materials under repeated loading is more complex than can be predicted by the Hertzian contact theory, but can be explained by use of the concepts of yielding and locking. Repeated load triaxial tests with constant confining pressure will always cause yielding with very high resilient modulus values at initial loading stages. When the applied deviator stress is higher than a certain critical value, the stress–strain curve will change in curvature and will have a concave upwards shape.

On the other hand, tests with cyclic confining pressure do not always start with yielding. Yielding occurs only when the cyclic confining pressure is less than a certain critical confining pressure and the repeated deviator stress is lower than a certain critical deviator stress.

The use of a simple two-parameter model has been demonstrated and the results indicate its accuracy in predicting the stress–strain relations of base course materials under repeated loading. The model can also explain the behaviour of the material in terms of yielding and locking.

REFERENCES

Boyce, H. R. (1980). A non-linear model for the elastic behaviour of granular materials under repeated loading. *Proc. Int. Symp. on Soils under Cyclic and Transient Loading, Swansea*, 285–294.

Brown, S. F. & Hyde, A. F. L. (1975). Significance of cyclic confining stress in repeated load triaxial testing of granular material. *Transp. Res. Rec.*, No. 537, 49–58.

Hicks, R. G. (1970). *Factors influencing the resilient properties of granular materials*. PhD thesis, University of California, Berkeley.

Hicks, R. G. & Monismith, C. L. (1971). Factors influencing the resilient response of granular materials. *Highway Res. Rec.*, No. 345, 15–31.

Kalteziotis, N. A. & Sarsby, R. W. (1985). Recoverable deformation of particulate soils. *Proc. 11th Int. Conf. on SMFE, San Francisco*, 531–534.

Karasahin, M., Dawson, A. R. & Holden, J. T. (1993). Applicability of resilient constitutive models of granular materials for unbound base layers. *Transp. Res. Rec.*, No. 1406, 98–107.

Ko, H. Y. & Scott, R. F. (1967). Deformation of sand in hydrostatic compression. *J. Soil Mech. Fdn Div. Am. Soc. Civ. Engrs* **93**, SM3, 137–156.

Lambe, T. W. & Whitman, R. V. (1969). *Soil mechanics*, SI edn, pp. 122–136. New York: Wiley.

Nataatmadja, A. & Parkin, A. K. (1988). A large cell for repeated load triaxial testing of base course material. *Proc. 14th ARRB Conf. Canberra*, **14** (7), 85–93.

Nataatmadja, A. & Parkin, A. K. (1989). Characterization of granular materials for pavements. *Can. Geotech. J.* **26**, 725–730.

Nataatmadja, A. & Parkin, A. K. (1990). Axial deformation measurement in repeated load triaxial testing. *Geotech. Test J.* **13**, No. 1, 45–48.

Rowe, P. W. (1971). Theoretical meaning and observed values of deformation parameters for soils. *Roscoe Memorial Symp.*, Cambridge University.

Tam, W. A. & Brown, S. F. (1988). Use of the falling weight deflectometer for in-situ evaluation of granular materials in pavements. *Proc. ARRB Conf., Canberra*, **14** (5), 155–163.

Timoshenko, S. P. & Goodier, J. N. (1970). *Theory of elasticity*, 3rd edn., pp. 409–414. New York: McGraw-Hill.

Travers, T., Ammi, M., Bideau, D., Gervois, A., Messager, J. C. & Troadec, J. P. (1987). Uniaxial compression of 2D packings of cylinder—effect of weak disorder. *Euro-phys. Lett.* **4**, No. 3, 329–332.

Uzan, J. (1985). Characterisation of granular material. *Transp. Res. Rec.*, No. 1022, 52–59.

Uzan, J. (1992). Resilient characterization of pavement materials. *Int. J. Anal. Methods Geomech.* **16**, 453–459.

Whitman, R. V., Miller, E. T. & Moore, P. J. (1964). Yielding and locking of confined sand. *J. Soil Mech Fdn Div. Am. Soc. Civ. Engrs* **90**, SM4, 57–84.

Lo Presti *et al.* (1998). Pre-failure deformation behaviour of geomaterials, 363–372

Assessment of quasi-linear models for sands

D.C.F. LO PRESTI,* O. PALLARA,* V. FIORAVANTE† and M. JAMIOLKOWSKI*

Appropriate modelling of the strain level dependency of soil stiffness is of great importance in order to estimate correctly the displacements of foundations under working load conditions. Soil deformations, which mainly occur in the case of well-designed foundations, are within the range of 0·001% to 1%, therefore the assumed model should be able to reproduce stress–strain curves, especially in this interval. In the light of this observation, quasi-linear stress–strain relationships have been reviewed. This review includes recently proposed models that overcome some intrinsic limitations of the conventional hyperbolic model. Model parameters were determined for a selected database of drained compression loading triaxial tests that refer to four sands with different mineralogical and physical characteristics. The sands were tested in a triaxial apparatus equipped with gauges for the local axial and radial strain measurements, so that the stress–strain curve is accurately defined, especially in the strain interval of interest. A parameter was introduced to evaluate the capacity of each model to fit the experimental results, and especially those concerning the variation of Young's modulus with axial strain. An attempt was also made to find a correlation between the model parameters and test conditions (void ratio, consolidation stresses, consolidation stress ratio, overconsolidation ratio). It was seen that the last two of these factors significantly affect the model parameters. It was also attempted to reproduce stress–strain curves obtained for different stress-paths using these correlations.

KEYWORDS: constitutive relations; laboratory tests; sands; stiffness.

Une bonne maquette du niveau de dépendance vis-à-vis de la déformation de la rigidité du sol est très importante pour évaluer correctement les déplacements de fondations en condition de surcharge. Comme les déformations du sol, qui se produisent principalement dans le cas de fondations bien conçues, sont de l'ordre de 0,001% à 1%, le modèle adopté devrait pouvoir reproduire les courbes contrainte-déformation, surtout dans cet intervalle. À la lumière de cette observation, nous avons révisé les relations quasi linéaires contrainte-déformation. Cette révision tient compte des maquettes récemment proposées qui viennent à bout de certaines des limitations intrinsèques de la maquette hyperbolique conventionnelle. Nous avons déterminé les paramètres de la maquette pour former une base de données choisies provenant d'essais triaxiaux de charge de compression drainée qui se rapportent à quatre sables ayant différentes caractéristiques minéralogiques et physiques. Ces sables ont été testés dans un appareil triaxial équipé de jauges donnant des mesures de déformation axiale et radiale locale afin que la courbe contrainte-déformation soit définie avec exactitude, surtout dans l'intervalle de déformation qui nous intéresse. Un paramètre a été introduit pour évaluer en quoi chaque maquette correspond aux résultats expérimentaux, et surtout à ceux concernant l'écart entre le module de Young et la déformation axiale. Nous avons essayé également de trouver une corrélation entre les paramètres de la maquette et les conditions d'essai (taux de porosité, contraintes de consolidation, rapport contrainte consolidation, taux de surconsolidation). Nous avons découvert que ces deux derniers facteurs affectent de manière significative les paramètres de la maquette. Nous avons essayé également de reproduire les courbes contrainte-déformation obtenues pour différentes trajectoires de contrainte utilisant ces corrélations.

INTRODUCTION

The simplest way of accounting for material non-linearity by means of analytical expressions is to use quasi-linear stress–strain relationships. In principle, the use of such relationships should be limited to materials which exhibit elastic non-linear

Manuscript received 23 December 1996.

* Department of Structural Engineering, Politecnico di Torino.

† Geotechnical Engineer, Ismes Spa of Bergamo.

behaviour; thus, according to Jardine (1992), Tatsuoka & Shibuya (1992) and Jamiolkowski *et al.* (1991, 1995), these relationships should be used, in the case of soils, only for problems which involve small strain levels of less than 0·01%. Indeed, according to those authors, the above limit corresponds, as a first approximation, to the so-called volumetric threshold strain, beyond which, in the case of soils, the observed permanent plastic deformations are no longer negligible. However, in some cases the use of quasi-linear stress–strain relationships is justified even though the expected soil deformations exceed the abovementioned limit. For example, resorting to the quasi-linear approach is reasonable when computing the settlement of foundations under working load conditions. In this case, which does not involve unloading for any soil element, the most relevant problem is that of representing the stiffness decay with the increase of the strain level, which can be very easily accomplished by using quasi-linear relationships.

In the case of granular soils, the stiffness non-linearity can be inferred from the observed settlements of structures and large-scale prototypes (Berardi & Lancellotta, 1991) or from the results of laboratory tests on reconstituted specimens which, in turn, can be correlated to the in situ test results (N_{spt}, q_c). The latter approach can be used confidently in the case of man-made granular fills, well-compacted structural fills and very recent normally-consolidated (NC) cohesionless deposits, but not in the case of aged granular soils. Duncan & Chang (1970) and Duncan *et al.* (1984) have proposed using the hyperbolic model (based on laboratory test results) to take soil non-linearity into account, and they have implemented such a model in a finite-element code for the analysis of earth dams.

The hyperbolic model (Cox, 1850; Kondner, 1963; Kondner & Zelasko, 1963) is quite popular and simple to use. Only two parameters (a and b) are required to define the stress–strain relationship

$$q - q_o = \frac{\varepsilon_a}{a + b \cdot \varepsilon_a} \qquad (1)$$

where q is the deviatoric stress, q_o is the deviatoric stress at the end of consolidation, ε_a is the axial strain, a is the inverse of the initial tangent modulus $(1/E_o)$ and b is the inverse of the limit deviatoric stress $1/(q - q_o)_{lim}$.

However, despite its simplicity, the hyperbolic model has some intrinsic limitations that have already been pointed out by many researchers (Tatsuoka & Shibuya, 1992). The inverse of the secant Young's modulus $1/E_s = \varepsilon_a/(q - q_o)$ is plotted against the axial deformation in Fig. 1, for a reconstituted Toyoura sand specimen subjected to a compression loading triaxial test. As is known, a linear relationship should exist between $(1/E_s)$ and ε_a, according to the hyperbolic model. It can be seen that such a linear relationship does not exist and that the a parameter obtained by regression analysis is much greater than $1/E_o$. Therefore, the stiffness at small and intermediate strains is largely

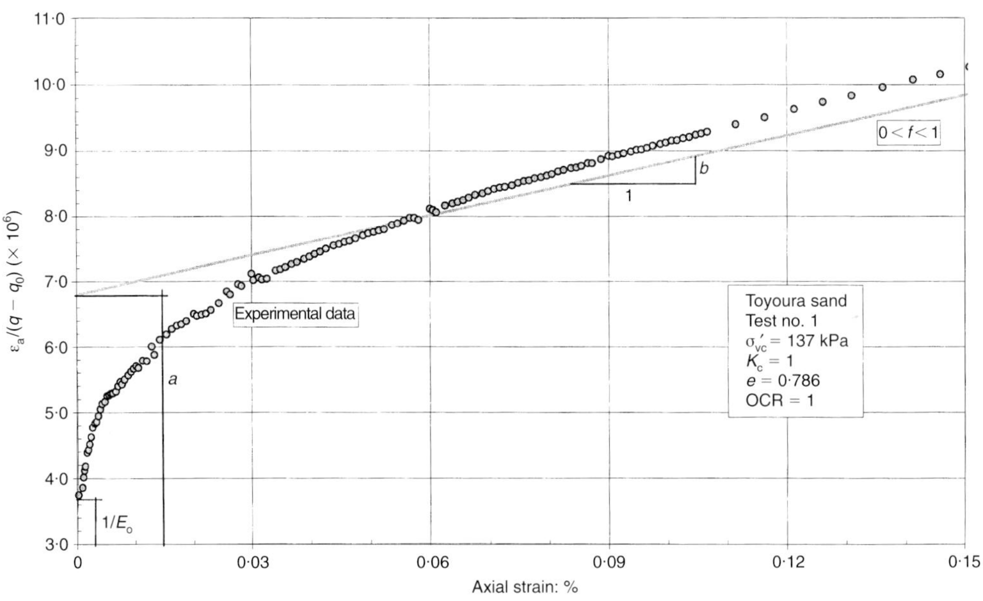

Fig. 1. Determination of hyperbolic model parameters

underestimated by the hyperbolic model when using the parameters obtained by means of linear regression (see Fig. 2).

In principle, it should be possible to assume appropriate values for a and b, but in this case the hyperbolic model generally overestimates the stiffness, as shown in Fig. 3, where the normalized secant Young's modulus (E_s/E_0) is plotted against the mobilization factor

$$f = \frac{q - q_o}{(q - q_o)_{max}}$$

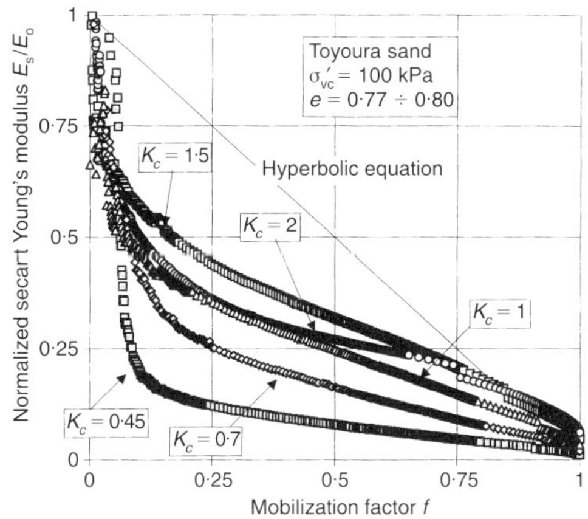

Fig. 3. Normalized secant Young's modulus of Toyoura sand plotted against mobilization factor (Lo Presti *et al.*, 1995)

CONSIDERED EQUATIONS

Tatsuoka & Shibuya (1992) proposed modifying the hyperbolic model in order to overcome the abovementioned limitations. The modification mainly consists of assuming model parameters which depend on the strain level. The equation proposed by Tatsuoka & Shibuya (1992) can be written as

$$q - q_o = \frac{\varepsilon_a}{a(\varepsilon_a) + b(\varepsilon_a).\varepsilon_a} \tag{2}$$

or in terms of normalized stress and strain:

$$\frac{y}{x} = \frac{1}{1/C_1(x) + x/C_2(x)} \tag{3}$$

where $y = (q - q_o)/(q - q_o)_{max}$ is the normalized deviatoric stress, $x = \varepsilon_a/\varepsilon_R$ is the normalized axial strain, $C_1(x)$ and $C_2(x)$ are the model parameters depending on x and $\varepsilon_R = (q - q_o)_{max}/E_o$ is a reference strain which represents the elastic axial strain at failure. In particular, $C_1(x)$ and $C_2(x)$ are periodic functions of x and require the determination of five parameters.

Fahey & Carter (1993) also proposed modifying the hyperbolic model in order to obtain a better fit of the laboratory experimental data. Their equation can be written as

$$\frac{y}{x} = 1 - F.(y)^g \tag{4}$$

where F and g are the model parameters.

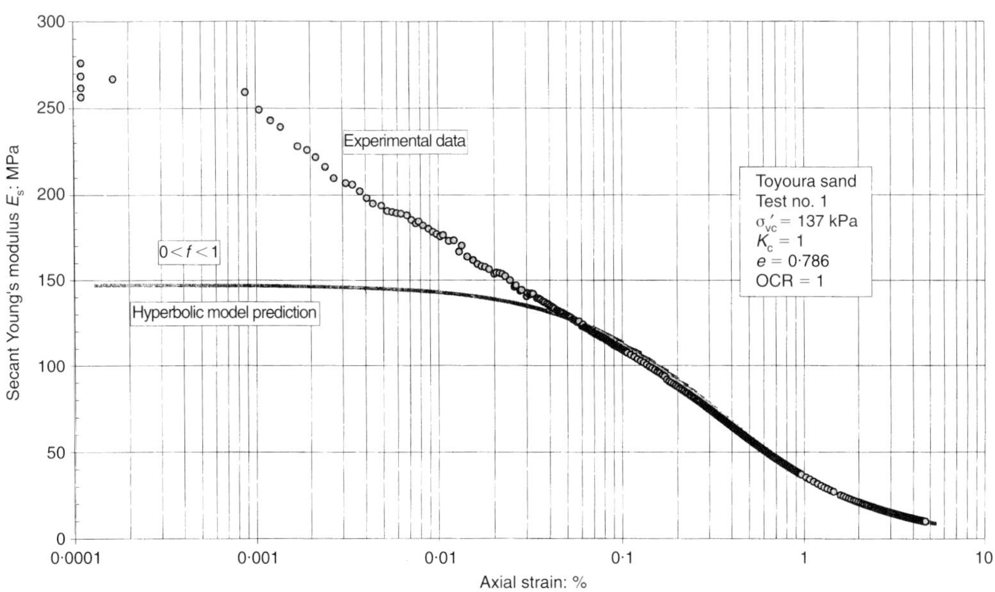

Fig. 2. Comparison of experimental data with hyperbolic model prediction

Even though only two parameters are required to define this model, the proposed equation is more flexible than the original hyperbola and capable of fitting the experimental stress–stain results. However, it should be stressed that this model was developed to simulate the unload–reload loops during a pressuremeter test in cohesionless soils.

Puzrin & Burland (1996) proposed a logarithmic function to fit the experimental stress–strain curves. This function has two analytical expressions

$$y = x - \alpha_s(x - x_e)\cdot[\ln(1 + x + x_e)]^{R_s} \quad (5)$$

for strain level between the elastic limit and the strain which usually results close to the volumetric threshold strain

$$y = \left\{\frac{x - x_l}{1 - y_l} g - \alpha_L \frac{x - x_l}{1 - y_l} g\left[\ln\left(1 + \frac{x - x_l}{1 - y_l} g\right)\right]^{R_L}\right\} \cdot (1 - y_l) + y_l \quad (6)$$

for strain levels greater than the volumetric threshold strain, where x_l and y_l are, as a first approximation, the normalized volumetric threshold strain and the related normalized deviatoric stress, respectively; α_s, α_L, R_s, R_L and g are the other model parameters.

CAPABILITY OF THE MODELS

The parameters of the previously considered functions were determined for a selected database of drained compression loading triaxial tests that refer to four sands whose main characteristics are listed in Table 1 (Fioravante *et al.*, 1994, 1998). The tests were performed in triaxial apparatus equipped with gauges for the local axial and radial strain measurements (Fioravante *et al.*, 1994; Pallara, 1995).

The capability of each model to fit the experimental results was evaluated by means of the following parameter

$$E_R = \frac{E_{mod} - E_{exp}}{E_{exp}} 100 \quad (7)$$

where E_{mod} is Young's modulus inferred from the model, E_{exp} is experimentally determined Young's modulus and E_R is relative error as a percentage.

Different values of E_R can be obtained at different strain levels. The plot of E_R against axial deformation shows the model accuracy over the entire strain interval considered. A synthetic evaluation of model capability for each test is obtained through the parameter

$$M_Q = \sqrt{\left(\frac{\sum E_R^2}{n - 1}\right)} \quad (8)$$

i.e. the standard error where n is the number of data for a given experimental stress–strain curve.

An example of computed E_R and M_Q is shown in Fig. 4. Fig. 5 shows the M_Q parameter determined for each model in the case of Kenya sand. It is possible to see that M_Q in the case of the Tatsuoka & Shibuya (TS) and Puzrin & Burland (PB) equations is very low, generally less than 10%. This also holds for the other sands. The M_Q parameter for tests on Kenya sand, in the case of the Fahey & Carter (FC) equation, is quite large (up to 80%). In particular, the FC equation is not capable of fitting the experimental results of K_o consolidated and overconsolidated specimens very well: moreover, in the case of very crushable sand such as the Quiou sand, the FC equation does not work very well. For this sand, the M_Q parameter, in the case of the FC equation, is extremely large, up to 140%, while for the Toyoura and Ticino sands it is generally less than 30%.

The model parameters are influenced, to a certain extent, by the test conditions. As far as the FC equation is considered, the F parameter is always very close to 1, in the case of carbonate sands. In the case of Toyoura and Ticino sands, the F parameter decreases from 1 to about 0·84 as K_c increases from K_o(NC) to 2. The exponent g of the FC equation is influenced by the consolidation stress ratio K_c and by the mechanical overconsolidation or prestressing. In particular for Toyoura and Ticino sands, g increases from 0·25 to 0·32 as K_c increases from K_o(NC) to 2. On the other hand, in the case of overconsolidated or

Table 1. Main characteristics of the tested sands

Sand	FC (%)	G_s	D_{50} (mm)	U_c	e_{max}	e_{min}	Mineralogy
Toyoura	—	2·650	0·22	1·35	0·985	0·611	quarz
Quiou	2	2·716	0·75	4·40	1·281	0·831	carbonate
Ticino	—	2·685	0·55	1·50	0·930	0·579	silica
Kenya	—	2·785	0·13	1·85	1·777	1·283	carbonate

FC = fine content, G_s = specific gravity, D_{50} = mean grain size, U_c = coefficient of uniformity, e_{max}, e_{min} = maximum and minimum void ratio.

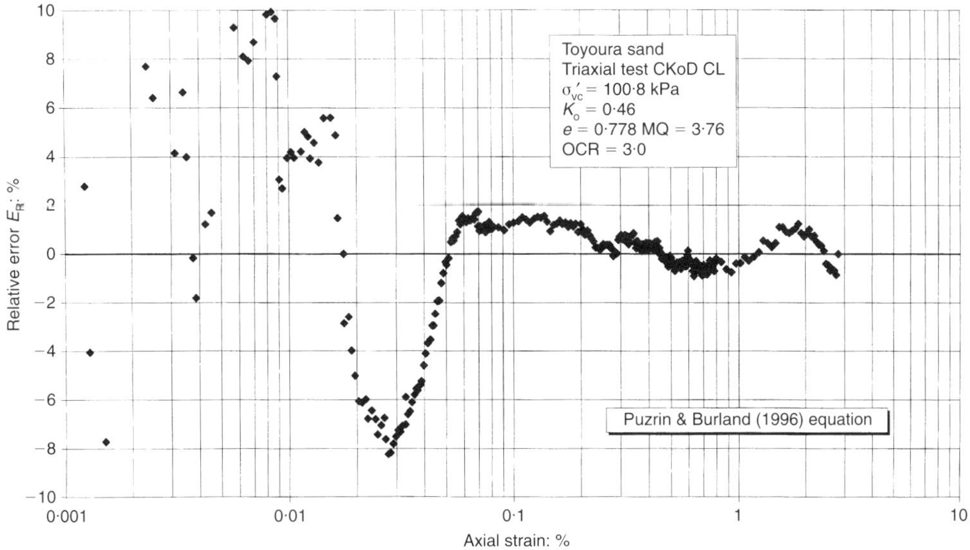

Fig. 4. Example of E_R and MQ computation

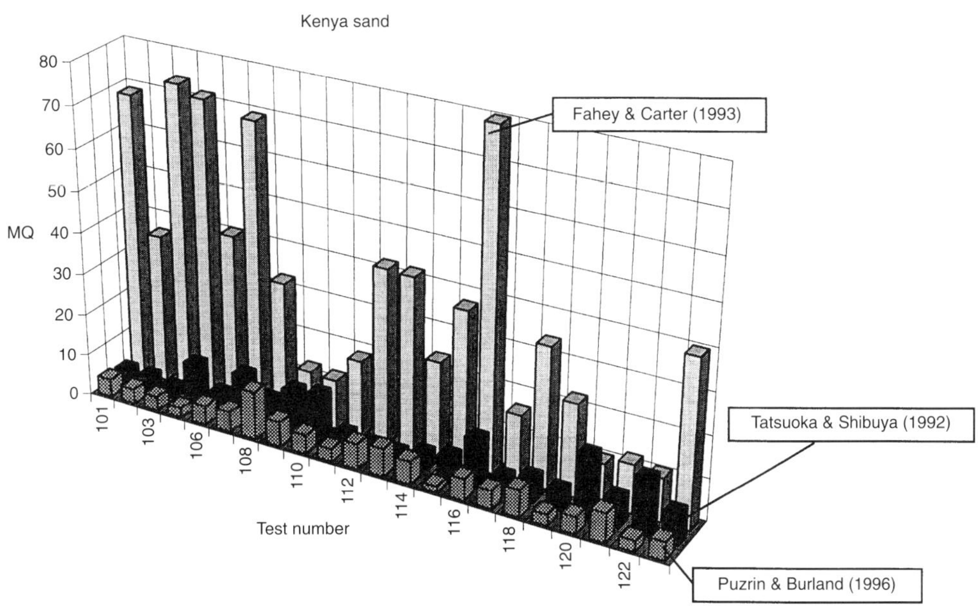

Fig. 5. Capability of quasi-linear relationships

prestressed specimens with overconsolidation ratio (OCR) ≤ 3, g is approximately 0·6 to 0·8. The dependence of g on test conditions for the four considered sands is summarized in Table 2. In the case of the very crushable Quiou sand g increases with an increase of the consolidation pressure.

The parameters of other considered equations are also influenced by K_c and overconsolidation. Detailed information about this influence is given by Pallara (1996).

The dependence of the model parameters on K_c and various overconsolidation mechanisms reflects the experimentally observed influence of these

Table 2. Dependence of the FC equation parameters on test conditions

Sand, test conditions	Value of g	
	NC	OC
Toyoura and Ticino, $K_c = K_o$ to 2, OCR $\leqslant 3$	0·25–0·32	0·6–0·8
Kenya, $K_c = K_o$ to 1, OCR $\leqslant 6$	0·50–0·60	0·7–0·8
Quiou, $K_c = K_o$ to 2, OCR $\leqslant 3$ $\sigma'_{vc} \leqslant 100$ kPa	0·10–0·20	0·5–0·7
Quiou, $K_c = 1$ 250 kPa $\leqslant \sigma'_{vc} \leqslant 450$ kPa	0·45	—

F $\approx$ 1 for Quiou and Kenya sands. 0·84 $\leqslant F \leqslant$ 1·0 for Ticino and Toyoura sands (F increases as K_c decreases from 2 to K_o).

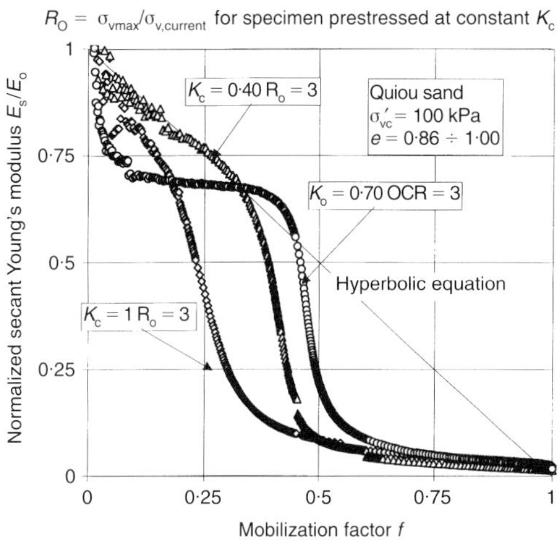

Fig. 6. Normalized secant Young's modulus of Quiou sand plotted against mobilization factor (Lo Presti *et al.*, 1995)

factors on the decay of the sand stiffness with the increase of the stress level (see Figs 3 and 6). The influence of K_c could be qualitatively explained by

(*a*) The relative position of the stress point at the end of consolidation with respect to the failure envelope (compression loading side)

(*b*) the effect of the so-called recent stress history (Jardine, 1992). In particular specimens subjected to consolidation with $K_o(\mathrm{NC}) \leqslant K_c \leqslant 2$ and then sheared under compression loading conditions, a rotation of the current effective stress path occurs with respect to that followed during consolidation.

It was also attempted to simulate the experimental stress–strain curves of specimens subjected to different stress paths. In particular, the following stress paths were considered

(*a*) compression loading triaxial tests at constant horizontal stress (σ_h = constant) performed on Toyoura, Ticino, Quiou and Kenya sand specimens

(*b*) compression loading triaxial tests at s' = constant and p' = constant performed on Ticino sand specimens

(*c*) extension loading triaxial tests at constant vertical stress (σ_v = constant) performed on Ticino sand specimens.

The following procedure was adopted to reproduce the stress–strain curves only from the knowledge of the soil state at the end of consolidation (void ratio and consolidation stresses).

(*a*) The peak angle of shear resistance φ'_p was evaluated for each test. For the Ticino, Kenya and Quiou sands, φ'_p was evaluated by means of equations similar to the one proposed by Bolton (1986) as follows

$$\text{(Ticino sand)} \quad \varphi'_p = 34{\cdot}6 + 3{\cdot}5.I_d.[10 - \ln(p')] - 3{\cdot} \tag{9}$$

$$\text{(Quiou sand)} \quad \varphi'_p = 39{\cdot}8 + 2{\cdot}2.I_d.[7{\cdot}9 - \ln(p')] - 2{\cdot} \tag{10}$$

$$\text{(Kenya sand)} \quad \varphi'_p = 40{\cdot}3 + 4{\cdot}2.I_d.[9{\cdot}5 - \ln(p')] - 4{\cdot} \tag{11}$$

where I_d is the relative density as a fraction of 1 and p' is the mean effective stress at failure, expressed in kPa.

For Toyoura sand, φ'_p was evaluated from the data provided by Fukushima & Tatsuoka (1984) as a function of the void ratio and the confining stress at the end of consolidation. The shear strength $(q - q_o)_{max}$ was evaluated for each test from the φ'_p value.

(*b*) The initial tangent Young's modulus was evaluated by means of the following equations (Hoque *et al.*, 1995, Pallara 1996)

$$\text{(Ticino sand)} \quad E_o = 1510\frac{(2{\cdot}17 - e)^2}{(1 + e)}\sigma'^{\,0{\cdot}53}_{vc}.p_a^{0{\cdot}47} \tag{12}$$

(Toyoura sand) $E_o = 2200\frac{(2{\cdot}17 - e)^2}{(1+e)}\sigma'^{0{\cdot}49}_{vc}.p_a^{0{\cdot}51}$ (13)

(Quiou sand) $E_o = \frac{1610}{e^{1{\cdot}3}}\sigma'^{0{\cdot}62}_{vc}.p_a^{0.38}.OCR^{0{\cdot}2}$ (14)

(Kenya sand) $E_o = \frac{3045}{e^{1{\cdot}55}}\sigma'^{0{\cdot}52}_{vc}.p_a^{0{\cdot}48}.OCR^{0{\cdot}21}$ (15)

(*c*) The E_s/E_o ratio was evaluated by means of the quasi-linear equations as a function of the mobilization factor. The equation parameters were selected on the base of K_c and overconsolidation. It is worth while to remember that only the FC equations explicitly provide E_s/E_o as a function of f. Therefore, it was possible to evaluate the secant Young's modulus for a given stress level.

(*d*) The secant Poisson's ratio of Ticino and Toyoura sand was evaluated as a function of the mobilization factor by means of (Pallara, 1996)

$$\nu = 0{\cdot}15 + 0{\cdot}45f \quad (16)$$

valid for $0{\cdot}15 \leqslant \nu \leqslant 0{\cdot}5$

(*e*) Axial and horizontal deformations were computed by means of elastic relations.

Compression loading triaxial tests at constant horizontal stress (σ_h = constant)

TS and PB equations, when used as a predictive model, are less accurate. The MQ parameter for the TS equation, in the case of Toyoura sand, is shown in Fig. 7. It is seen that MQ increases from less than 10% to about 40% when the TS equation is used as a model. These indications also hold in the case of the PB equation and for the four tested sands.

When the FC equation is used as a predictive model, the MQ values remain more or less unchanged.

Other stress-paths

The considered equations were not able to predict horizontal strains correctly, especially in the case of extension tests, where ε_h was greatly underestimated. The prediction of ε_a was not so bad in the case of compression tests, while, for the extension stress path, the axial deformation was underestimated. Typical examples of experimental against predicted stress–strain curves are shown in Figs 8 and 9 for the extension stress-path and Figs 10 and 11 for the compression stress paths. It is important to point out that axial strain of about 0·5% corresponds to a mobilization factor of approximately 0·7.

Better predictions of strain were obtained using the TS and PB equations only in the case of K_o consolidated or overconsolidated specimens.

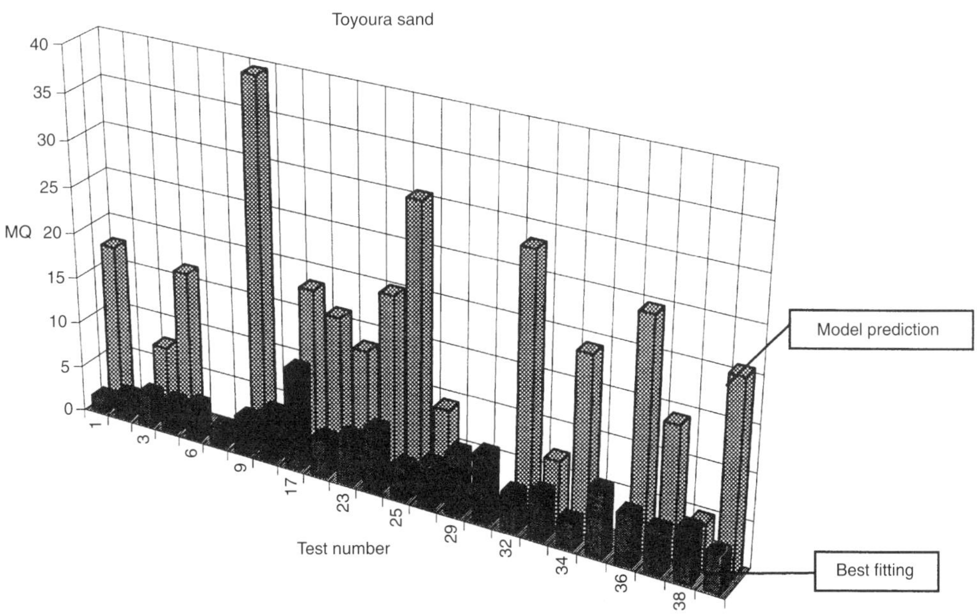

Fig. 7. Model accuracy for compression loading TX test

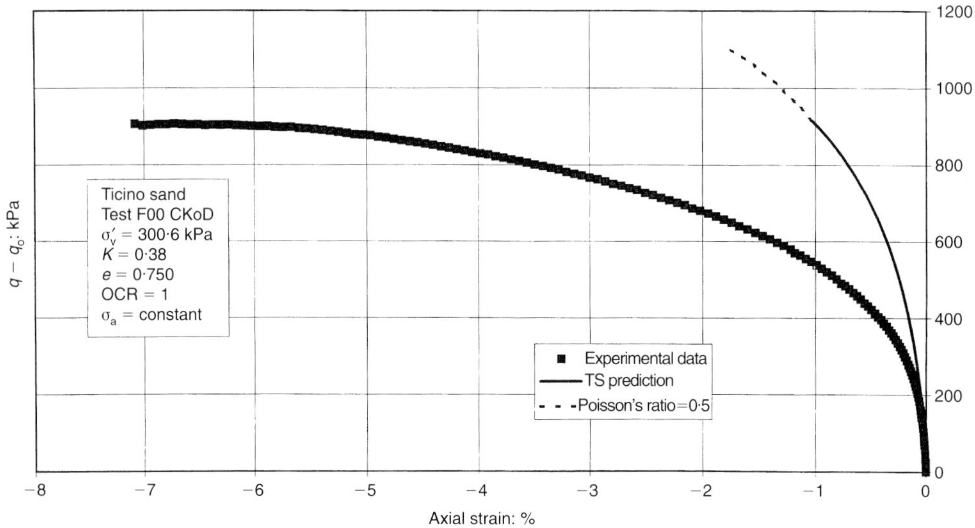

Fig. 8. Model prediction for extension stress path

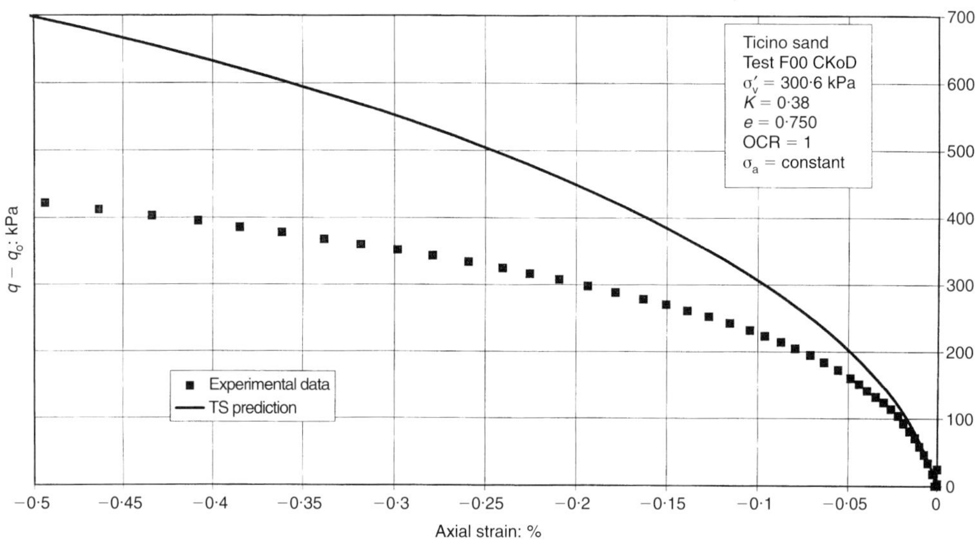

Fig. 9. Model prediction for extension stress path

CONCLUSIONS

The following conclusions can be drawn.

(*a*) The Tatsuoka & Shibuya (1992) and Puzrin & Burland (1996) equations provide an excellent fitting of experimental stress–strain data, but require the determination of at least five parameters. The Fahey & Carter (1993) equation is much simpler to use and requires only two parameters, but the experimental data are fitted less accurately.

(*b*) The use of quasi-linear equations to predict vertical settlements of foundations appears reasonable and justified in the light of the good prediction of axial deformation obtained in the case of compression stress-paths.

(*c*) Quasi-linear equations do not seem capable of predicting horizontal deformations or of giving reasonable prediction of both vertical and horizontal deformations, especially in the case of problems that involve extension stress-paths.

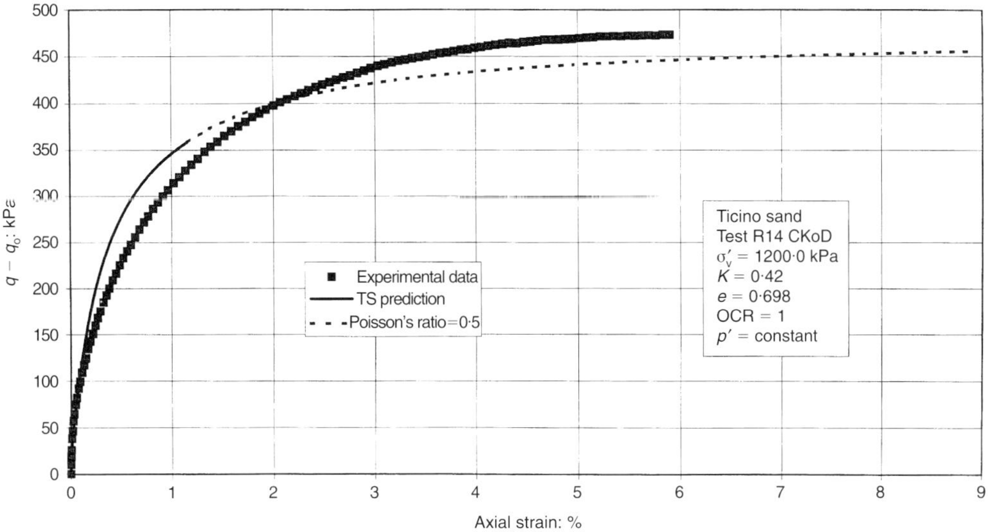

Fig. 10 Model prediction for compression stress path

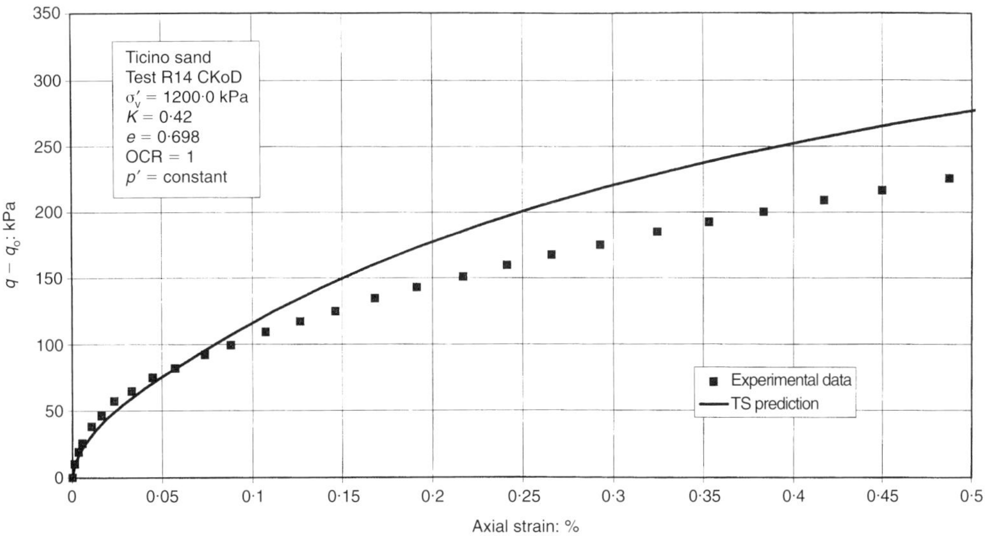

Fig. 11. Model prediction for compression stress path

REFERENCES

Berardi, R. & Lancellotta, R. (1991). Stiffness of granular soils from field performance. *Géotechnique* **41**, No. 1, 149–157.

Bolton, M. D. (1986). The strength and dilatancy of sands. *Géotechnique* **36**, No. 1, 65–78.

Cox, H. (1850). The deflection of imperfectly elastic beams and the hyperbolic law of elasticity. *Trans. Cambridge Philos. Soc.* Part 2, No. 9, 177–190.

Duncan, J. M. & Chang, C.-Y. (1970). Non linear analysis of stress and strain in soils. *J. Soil. Mech. Fdn Div. Am. Soc. Civ. Engrs* **96**, SM5, 1629–1653.

Duncan, J. M., Seed, R. B., Wong, K. S. & Ozawa, Y. (1984). *FEADAM84: a computer program for finite element analysis of dams*. Geotechnical Engineering Research Report SU/GT/84-03, Department of Civil Engineering, Stanford University.

Fahey, M. & Carter, J. P. (1993). A finite element study of the pressuremeter test in sand using a non-linear elastic plastic model. *Can. Geotech. J.* **30**, 348–362.

Fioravante, V., Jamiolkowski, M. & Lo Presti, D. C. F. (1994). Stiffness of carbonatic Quiou sand. *Proc. 13th ICSMFE, New Dehli*, **1**, 163–167.

Fioravante, V., Jamiolkowski, M., Lo Presti, D. C. F., Manfredini, G. & Pedroni, S. (1998). Coefficient of the earth pressure at rest on cohesionless soils from in situ test. *Géotechnique* **48**, No. 3.

Fukushima, S. & Tatsuoka, F. (1984) Strength and deformation characteristics of saturated sand at extremely low pressures. *Soils Fdns* **24**, No. 4, 30–48.

Hoque, E., Tatsuoka, F., Sato, T. & Kohata, Y. (1995) Inherent and stress induced anisotropy in small strain stiffness of granular materials. *Proc. 1st Int. Conf. on Earthquake Geotechnical Engineering, Tokyo*, **1**, 277–282.

Jamiolkowski, M., Leroueil, S. & Lo Presti, D. C. F. (1991). Design parameters from theory to practice. *Proc. Int. Conf. on Geotechnical Engineering for Coastal Development, Geo-Coast 91, Yokohama*, **2**, 877–917.

Jamiolkowski, M., Lo Presti, D. C. F. & Pallara, O. (1995). Role of in-situ testing in geotechnical earthquake engineering. *Proc. 3rd Int. Conf. on Recent Advances in Geotechnical Earthquake Engineering and Soil Dynamics, State of the Art 7*, **3**, 1523–1546.

Jardine, R. J. (1992). Some observations on the kinematic nature of soil stiffness. *Soils Fdns* **32**, 111–124.

Kondner, R. L. (1963). Hyperbolic stress–strain response: cohesive soils. *J. Soil. Mech. Fdn Div. Am. Soc. Civ. Engrs* **89**, SM1, 115–143.

Kondner, R. L. & Zelasko, J. S. (1963). A hyperbolic stress–strain formulation for sands. *Proc. 2nd Pan-American Conf. on SMFE, Rio de Janeiro, Brazil*, **1**, 289–324.

Lo Presti, D. C. F., Jamiolkowski, M., Pallara, O., Pisciotta, V. & Ture, S. (1995). Stress dependence of sand stiffness. *Proc. 3rd Int. Conf. on Recent Advances in Geotechnical Earthquake Engineering and Soil Dynamics*, **1**, 71–76.

Pallara, O. (1995). *Comportamento sforzi–deformazioni di due sabbie soggette a sollecitazioni monotone e cicliche*. PhD thesis, Department of Structural Engineering, Politecnico di Torino (in Italian).

Pallara, O. (1996). *Verifica delle leggi elastiche quasilineari nelle sabbie mediante prove di laboratorio*. Rapporto di ricerca, Department of Structural Engineering, Politecnico di Torino.

Puzrin, A. M. & Burland, J. B. (1996). A logarithmic stress–strain function for rocks and soils. *Géotechnique* **46**, No. 1, 157–164.

Tatsuoka, F. & Shibuya, S. (1992). Deformation characteristics of soil and rocks from field and laboratory tests. Keynote lecture, *Proc. 9th Asian Conf. on SMFE, Bangkok*, **2**, 101–190.

Yasuhara *et al.* (1998). Pre-failure deformation behaviour of geomaterials, 373–382

Cyclic stiffness of plastic silt with an initial drained shear stress

K. YASUHARA,* A. F. L. HYDE,† N. TOYOTA‡ and S. MURAKAMI*

The behaviour of soils beneath buildings and earth structures during earthquakes and beneath offshore structures during storms is influenced by the initial sustained shear stress produced by the self-weight of these structures. Silty materials are often used for reclaimed fill areas in seismic zones, yet little is known of their behaviour under cyclic loading. The cyclic behaviour of cohesive soils can be divided into that occurring during cyclic loading and that occurring after cyclic loading. This paper describes the results from cyclic and post-cyclic triaxial undrained compression tests on a plastic silt and, in particular, the effect of an initial static drained shear stress on those characteristics. The results from a series of undrained cyclic triaxial tests show that the monotonic undrained behaviour of a plastic silt both before and after undrained cyclic loading exhibits greater strain softening as the initial static shear stress is increased. However, it is also concluded that the post-cyclic degradation of both strength and Young's modulus for a plastic silt can be correlated to the excess pore pressure normalized by the initial mean effective confining stress, and that these relationships are not influenced by the magnitude of initial shear stress. The degradation in strength and stiffness of a plastic silt with an initial shear stress can be predicted using previously proposed equations for use with cohesive soils consolidated with no initial shear stress.

KEYWORDS: cyclic stiffness; degradation; initial shear stress; plastic silt.

Le comportement des sols sous les immeubles et les structures terrestres pendant les tremblements de terre et sous les structures en mer pendant les tempêtes est influencé par la tension de cisaillement initiale produite par le poids même de ces structures. On utilise souvent des matières limoneuses pour remplir les zones réhabilitées dans les régions sismiques, bien qu'on ne connaisse pas grand-chose de leur comportement sous charges cycliques. Le comportement cyclique des sols cohésifs se divise en comportement pendant les charges cycliques et en comportement après les charges cycliques; cet exposé décrit les résultats des essais de compression non drainée sur trois axes cycliques et post-cycliques sur un limon plastique et en particulier l'effet qu'une tension de cisaillement drainé statique initiale a sur ces caractéristiques. Les résultats d'une série d'essais triaxiaux cycliques non drainés montrent que le comportement monotonique non drainé d'un limon plastique avant et après le chargement cyclique non drainé présente un amollissement de déformation plus grand quand la tension de cisaillement statique initiale augmente. Pourtant nous concluons aussi que la dégradation post cyclique du module de Young et du module de résistance pour un limon plastique peut être mise en corrélation avec la pression interstitielle excessive normalisée par la contrainte de confinement effective moyenne initiale et que ces relations ne sont pas influencées par la magnitude de la tension de cisaillement initiale. La dégradation de la résistance et de la rigidité d'un limon plastique avec une tension de cisaillement initiale peut être calculée en utilisant les équations proposées précédemment avec les sols cohésifs consolidés sans tension de cisaillement initiale.

INTRODUCTION

It is known that the behaviour of soils beneath buildings and earth structures during earthquakes and that beneath offshore structures during storms is influenced by the initial sustained shear stress produced by the self-weight of those structures. Previous experimental research indicates that the cyclic behaviour of soils depends on the magnitude of this initial shear stress (Ishihara & Takatsu, 1977; Andersen *et al.*, 1980; Vaid & Chern, 1986). According to research by Hyodo *et al.* (1994) in particular, this initial static shear stress influences the cyclic strength of clay in a manner

Manuscript received 1 April 1997.
* Department of Urban and Civil Engineering, Ibaraki University, Japan.
† Department of Civil and Structural Engineering, University of Sheffield, UK.
‡ Tokyo Electrical Power Co. Ltd., Tokyo, Japan.

different from that for sand. Cyclic strength here is defined as the stress amplitude to cause a prescribed strain amplitude after a defined number of cycles, for example 10 or 20 cycles for earthquake loading. It was found that the cyclic strength of clay decreases with an increasing initial drained shear stress, as opposed to that of sand, which increases. This difference is a very important finding not only from a fundamental but also from a practical point of view.

However, since few studies have been carried out to investigate the effect of an initial drained shear stress on the post-cyclic behaviour of silty soils, there is no accepted view as to whether this could be an important parameter in the estimation of earthquake damage to structures which are founded on such soils. This paper, therefore, aims to supplement the data on this point from cyclic triaxial tests on a reconstituted plastic silt. The reason for using a plastic silt was that although Hyde & Brown (1976) and Hyde & Ward (1983, 1985, 1986) have presented work on the cyclic behaviour of a silty clay, no systematic research has been reported on the cyclic and post-cyclic behaviour of silts. This is despite the fact that silty materials are often used for reclaimed fill areas in seismic zones and that these have suffered damage after earthquakes such as at Chibaken-tohooki in 1987 and Hyogoken-nanbu in 1995.

OUTLINE OF CYCLIC TRIAXIAL TESTS

Although post-cyclic behaviour can be investigated using direct-simple shear and torsional shear tests, the cyclic triaxial test was adopted for its simplicity and ease of use. The triaxial equipment used is illustrated in Fig. 1. One of the characteristic features of this equipment was the capability of running stress-controlled and strain-controlled cyclic triaxial tests in sequence, thus enabling the carrying out of stress-controlled cyclic and strain-controlled post-cyclic monotonic tests on a given soil specimen.

The soil used was the silt fraction abstracted from a silty clay (Keuper marl) using a rotary classifier. The powder was mixed with water to a slurry of approximately 80% water content. The soil slurry obtained in this way was poured into a large consolidation vessel, 20 cm in diameter and 40 cm high. A consolidation pressure of 50 kPa was then applied to the slurry to obtain a homogeneous cylinder of soil whose moisture content was approximately 28% to 30%, and which was suitable for trimming to a specimen (50 mm diameter and 100 mm height) for triaxial tests. The index properties of the plastic silt were: $\rho_s = 2{\cdot}69\ \mathrm{Mg/m^3}$, $w_L = 40{\cdot}2\%$ and plasticity index $I_p = 23$. Representative mechanical indices obtained from oedometer and triaxial tests were $C_c = 0{\cdot}23$, $C_s = 0{\cdot}042$, coefficient of earth pres-

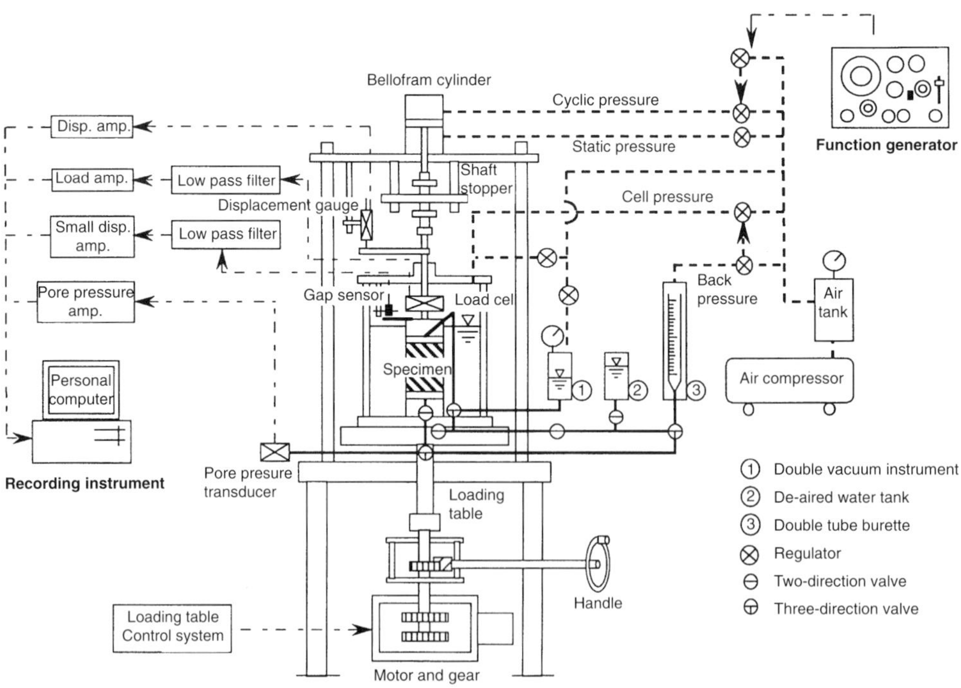

Fig. 1. Triaxial apparatus

sure at rest $K_0 = 0{\cdot}58$ and effective internal friction angle $\phi' = 29{\cdot}4°$.

Each specimen was isotropically preconsolidated in the triaxial cell under a given confining pressure for 24 h and an initial shear stress τ_s was then applied to the specimen in steps under drained conditions until the mean effective confining pressure reached 196 kPa, as shown in Fig. 2. Following this, a sinusoidal 0·1 Hz stress-controlled cyclic stress was applied under undrained conditions. Four normalized levels of initial shear stress given by τ_s/p'_c (where τ_s is the initial sustained shear stress and p'_c is the mean normal effective consolidation stress) were combined with different cyclic shear stress ratios τ_d/p'_c, where τ_d is the shear stress amplitude, as shown in Table 1. The number of cyclic shear stress applications was selected arbitrarily in order to obtain results for a range of cyclic-induced excess pore pressure ratios u/p'_c from 0 to 1·0. This was in order to investigate the use of previous relations proposed by Yasuhara *et al.* (1983, 1992), Yasuhara (1985, 1995), Matsui *et al.* (1992), Yasuhara & Hyde (1997a) and Yasuhara *et al.* (1997c), used for predicting degradation in strength and stiffness in cohesive soils, which were given as a function of the excess pore pressure ratio u/p'_c.

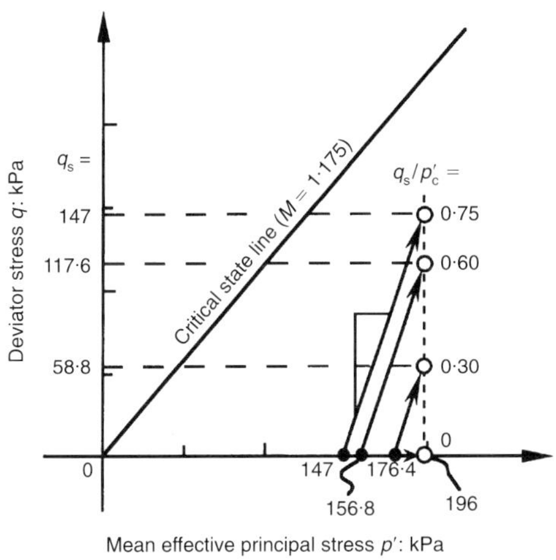

Fig. 2. Consolidation and initial shear stress-paths

After cyclic loading was terminated, each specimen was left for a few hours under undrained conditions with the initial sustained shear stress to attain a homogeneous distribution of cyclic-induced excess pore pressure. Then a monotonic load with a strain rate of 0·1%/min was applied to the specimens, also under undrained conditions. The testing procedure for this series of tests is illustrated in Fig. 3.

Monotonic undrained compression triaxial tests were conducted on similar specimens with and without the initial shear stress, without cyclic

Table 1. Testing parameters

Mean normal effective consolidation stress p'_c: kPa	Consolidation cell pressure σ'_3: kPa	Initial static deviator stress q_s: kPa	Cyclic pore pressure ratio $\Delta u/p'_c$	Cyclic shear stress ratio τ_d/p'_c
196	196	0	0	0
196	196	0	0·829	0·250
196	196	0	0·852	0·170
196	196	0	0·858	0·168
196	196	0	0·533	0·135
196	196	0	0·284	0·115
196	176·4	58·8	0	0
196	176·4	58·8	0·741	0·165
196	176·4	58·8	0·448	0·144
196	176·4	58·8	0·334	0·137
196	175·4	58·8	0·318	0·135
196	176·4	58·8	0·163	0·107
196	176·4	58·8	0·129	0·103
196	156·8	117·6	0	0
196	156·8	117·6	0·467	0·145
196	156·8	117·6	0·354	0·126
196	156·8	117·6	0·0138	0·099
196	147	147	0	0
196	147	147	0·394	0·129
196	147	147	0·311	0·120
196	147	147	0·101	0·087
196	147	147	0·045	0·073

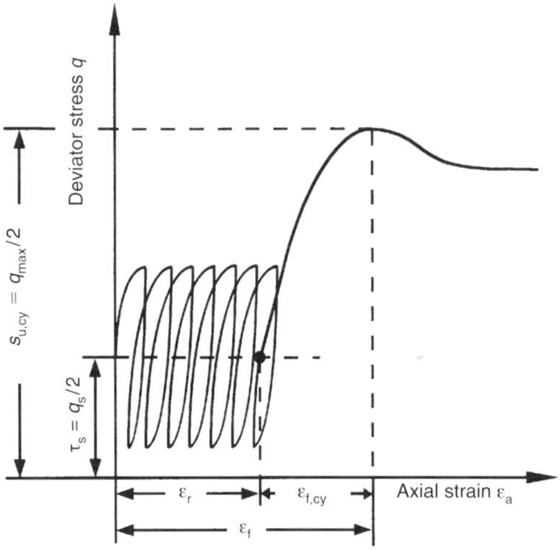

Fig. 3. Schematic testing procedure

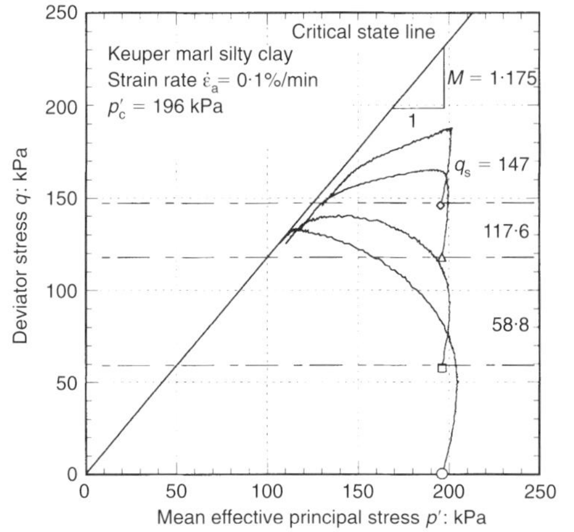

Fig. 4. Monotonic stress paths for non-cycled samples

loading. These were for comparison with the results from post-cyclic monotonic undrained shear tests.

EFFECT OF INITIAL STATIC SHEAR STRESS ON MONOTONIC SHEAR BEHAVIOUR PRIOR TO CYCLIC LOADING

Stress–strain characteristics

Stress–strain relations, strength and deformation modulus were obtained from monotonic undrained triaxial compression tests carried out without any prior undrained cyclic loading. Samples were consolidated to the same combinations of shear and mean normal effective stress as shown in Fig. 2 and then subjected to standard undrained triaxial testing. The strain rate was 0·1%/min and the water content for each specimen w_i was between 28% and 30%.

Figure 4 shows the undrained triaxial effective stress paths in $p'-q$ space for four kinds of specimen with initial shear stress ratio τ_s/p'_c ranging from 0 to 0·375. It can be seen from Fig. 4 that strain softening took place after a local peak deviator stress was reached. The tendency for strain softening became more marked as the initial shear stress increased. However, every path finally fell onto the critical state line which had a gradient of $M = 1{\cdot}175$. By considering the effective stress paths together with the stress–strain curves presented in Fig. 5, it can also be seen that the softening behaviour occurred after a peak strength was reached at fairly small strain levels. At the same time, the stress decrement where softening is exhibited is more rapid as the peak strength increases along with increasing initial static shear stress. These tendencies are in very good agreement with the results obtained by Hyodo *et al.* (1994) from undrained triaxial tests on a highly plastic clay with a plasticity index $I_p = 70$.

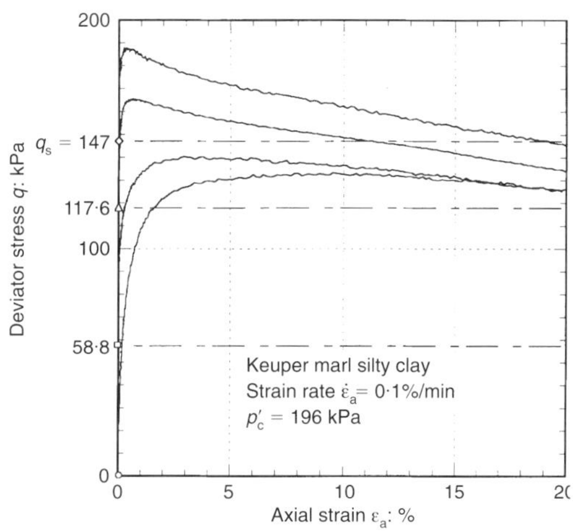

Fig. 5. Stress–strain curves for non-cycled samples

Since shear behaviour is generally governed by the effective stress ratio $\eta = q/p'$ rather than by the deviator stress q, Fig. 6 shows the axial strain ε_a plotted against effective stress ratio η. It can be seen in Fig. 6 that the stress ratio–strain curves all merge at about 6% strain and reach the critical state at between 9% and 10% strain.

Initial secant modulus

The deformation modulus of a plastic silt is defined by the gradient of the straight line which connects the origin and the point corresponding to 50% of the maximum deviator stress difference,

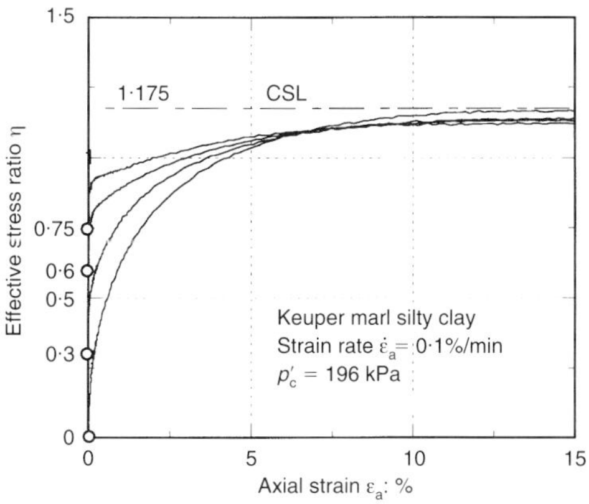

Fig. 6. Effective stress ratio plotted against strain for non-cycled samples

$(q_{max} - q_s)$, in $q{-}\varepsilon_a$ curves, where q_{max} is the peak deviator stress and q_s is the initial drained deviator stress. This is the so-called initial secant modulus E_{50}. Figs 7 and 8 show the variation of initial secant modulus E_{50} and rigidity index E_{50/S_u} with the magnitude of initial static shear stress normalized by the effective confining pressure, τ_s/p'_c. The results from tests on Itsukaichi clay ($\rho_s = 2{\cdot}53$, $w_L = 124\%$, $I_p = 73$) obtained by Sugiyama (1995) are also plotted in these figures for comparison. It is clear from these results that both the secant modulus and the stiffness index for Keuper marl silt with a low plasticity index increase more with increasing initial static shear stress than those of Itsukaichi clay with a higher plasticity index.

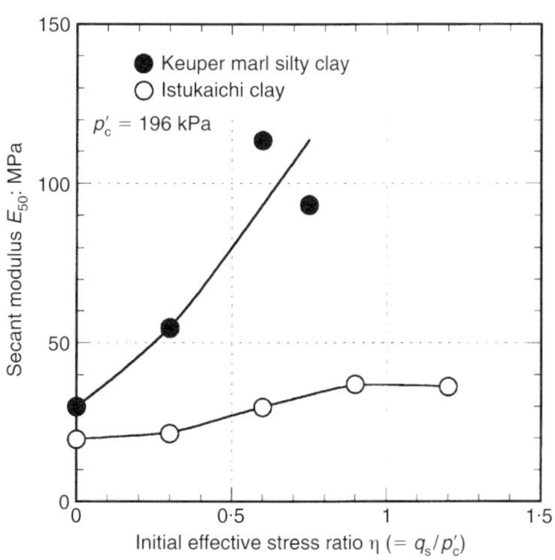

Fig. 7. Secant modulus plotted against initial drained shear stress ratio

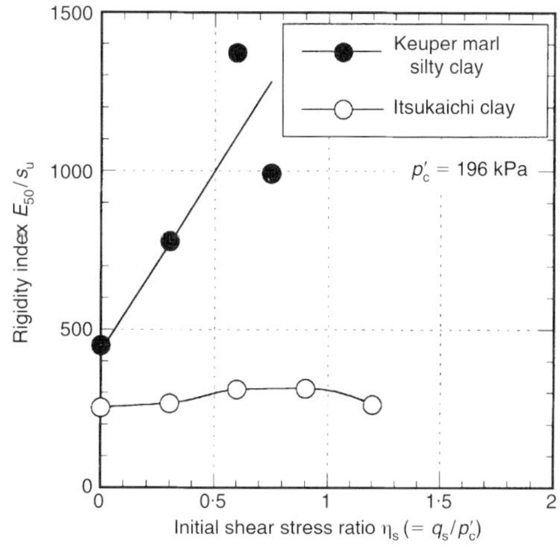

Fig. 8. Rigidity index plotted against drained shear stress ratio

EFFECT OF INITIAL SUSTAINED SHEAR STRESS ON CYCLIC STIFFNESS DEGRADATION

Stiffness degradation with number of load cycles

The effect of cyclic degradation for cohesive soils under strain-controlled conditions can be expressed quantitatively in terms of the degradation index δ_D proposed by Idriss *et al.* (1978) for the triaxial mode of deformation. The degradation parameter can be defined alternatively as

$$\delta_D = \frac{E_{SN}}{E_{S1}} = N^{-d} \tag{1}$$

where E_{S1} and E_{SN} are secant Young's moduli in the first and Nth cycles, N is number of load cycles, and d is the slope of the straight line plotted on a log–log scale for E_{SN}/E_{S1}, against N.

Following the same procedure as proposed by Idriss *et al.* (1978), it was found, however, that the linearity holds better between $G_{eq}/(G_{eq})_{N=1}$ and N if plotted on a semi-log scale, as shown in Fig. 9, rather than a log–log scale. This can then be written as

$$\frac{G_{eq}}{(G_{eq})_{N=1}} = 1 - A \log_{10} N \tag{2}$$

where A is the gradient of the lines in Fig. 9 and is called the stiffness degradation index. This is different to the definition used by Idriss *et al.* (1978) and further used by Vucetic & Dobry (1988). When the values of A obtained from Fig. 9 are plotted against the cyclic shear stress ratio τ_d/p'_c, the relation shown in Fig. 10 is obtained, which can be formulated as

$$A = \alpha\left(\frac{\tau_d}{p'_c}\right)^2 \tag{3}$$

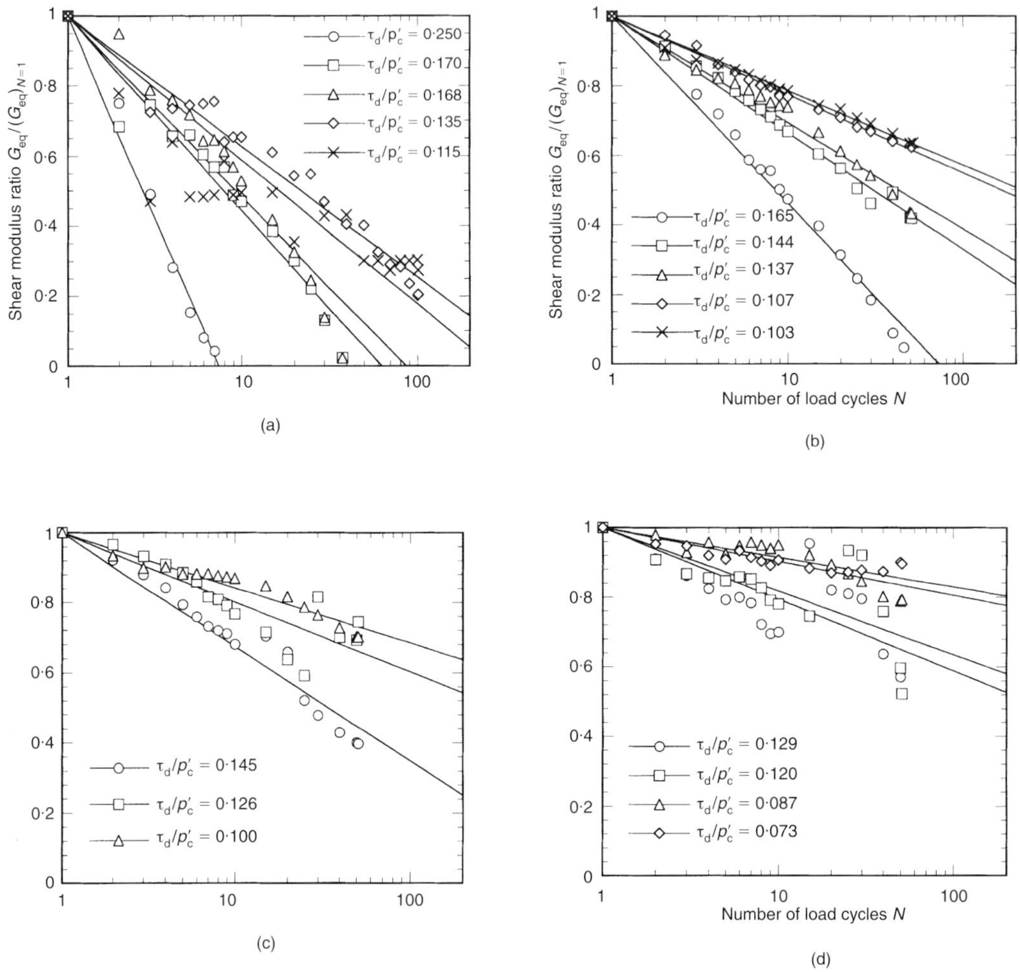

Fig. 9. Degradation of shear modulus ratio: (a) $q_s/p'_c = 0$; (b) $q_s/p'_c = 0{\cdot}3$; (c) $q_s/p'_c = 0$; (d) $q_s/p'_c = 0{\cdot}75$

where α is an experimental constant equal to 17·7 for the Keuper marl silt. The data in Fig. 10 are independent of the initial sustained shear stress τ_s/p'_c, and hence equation (3) is valid for specimens with and without an initial sustained shear stress. Inserting of equation (3) into equation (2) yields

$$\frac{G_{eq}}{(G_{eq})_{N=1}} = 1 - \alpha\left(\frac{\tau_d}{p'_c}\right)^2 \log_{10} N \tag{4}$$

To determine the maximum equivalent shear modulus G_{max}, the equivalent shear modulus in the first cycle is plotted against the cyclic shear stress τ_d/p'_c for sustained initial shear stress ratios $\tau_s/p'_c = 0$, 0·15, 0·3, 0·375. By extrapolating the relation between $(G_{eq})_{N=1}$ and τ_d/p'_c in Fig. 11, one obtains the intercept for $(G_{eq})_{N=1}$ corresponding to $\tau_d/p'_c = 0$, which is a theoretical maximum value for G_{max}. The equation for the lines in Fig. 9 is then

$$(G_{eq})_{N=1} = G_{max} - \beta\frac{\tau_d}{p'_c} \tag{5}$$

where $\beta = 217{\cdot}1$ for Keuper marl silt. Substitution of equation (5) into equation (4) gives

$$G_{eq} = \left[G_{max} - \beta\left(\frac{\tau_d}{p'_c}\right)\right]\left[1 - \alpha\left(\frac{\tau_d}{p'_c}\right)^2 \log_{10} N\right] \tag{6}$$

thus allowing the computation of the equivalent shear modulus G_{eq} at any number of load cycles N.

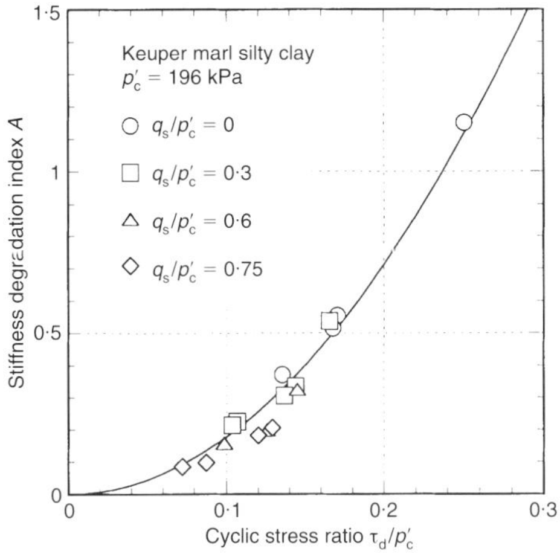

Fig. 10. Stiffness degradation index

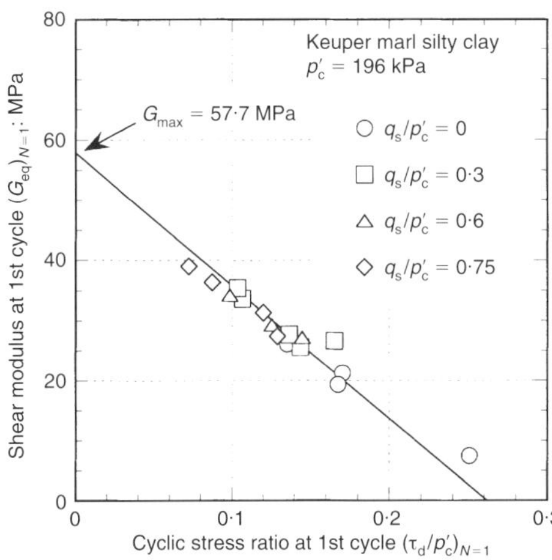

Fig. 11. Determination of maximum shear modulus

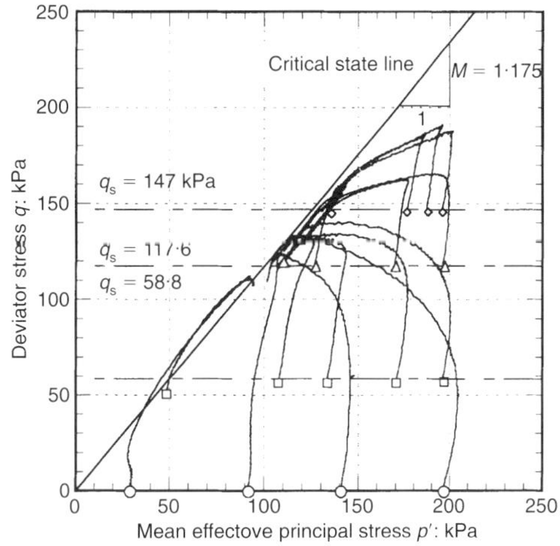

Fig. 12. Effective stress paths for cycled and non-cycled monotonic tests

EFFECT OF INITIAL STATIC SHEAR STRESS ON POST-CYCLIC MONOTONIC UNDRAINED SHEAR DEFORMATION

Effective stress-paths

Figure 12 shows a family of stress-paths for post-cyclic undrained shearing of specimens, each with an initial drained static shear stress. Included with these stress paths are those for samples tested without cyclic loading, i.e. $\Delta u/p'_c = 0$. In each case: (*a*) strain softening occurs after a local peak deviator stress is reached; (*b*) the effective stress decrement due to softening is more rapid with increasing initial static shear stress; (*c*) every path moves onto the projection of the critical state line (CSL), and approaches the critical point.

Stress–strain performance

Figure 13 shows the post-cyclic deviator stress against axial strain curves for specimens with and without an initial static shear stress. It can be seen from Fig. 13 that every curve in q–ε_a space tends to approach the original q–ε_a curve for no initial drained shear stress and ε_a then goes on to values in excess of 20% to 30%. The gradient and hence the stiffness at the beginning of each curve decreases along with the peak deviator stress, with increasing cyclic-induced excess pore pressures. Softening behaviour occurs after the peak strength is reached at relatively small strains. This is also true for the samples shown in Fig. 5, which have not been cyclically loaded. It should be noted that in Fig. 13, ε_a for post-cyclic tests is equal to the

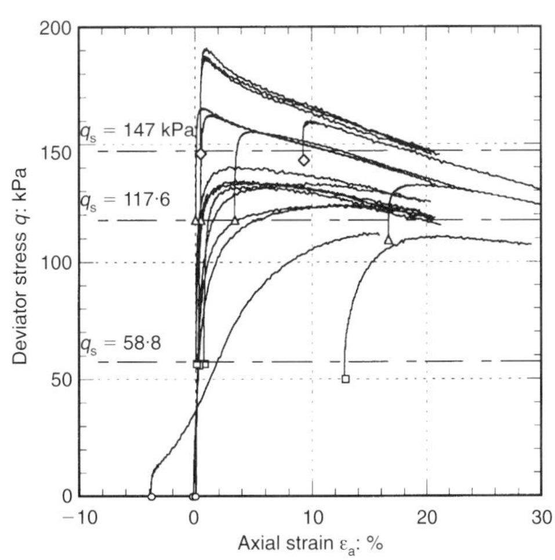

Fig. 13. Non-cyclic and post-cyclic stress–strain curves

summation of the cyclic-induced residual axial strains and the post-cyclic axial strains.

As in Fig. 6 for no cyclic loading, post-cyclic axial strain against effective stress ratio relations are plotted in Fig. 14. In this case the degree of convergence is not as marked as in Fig. 6, but the lines for all samples including those that were heavily overconsolidated seem to be converging on a point just below the critical effective stress ratio.

EFFECT OF INITIAL SHEAR STRESS ON DEGRADATION OF POST-CYCLIC STIFFNESS

Degradation in deformation modulus

Yasuhara *et al.* (1992, 1994, 1997a) and Yasuhara (1994) have shown that the stiffness and strength of plastic soils degrade due to undrained cyclic loading. To illustrate this behaviour quantitatively the post-cyclic secant modulus $E_{50,cy}$ (determined at 50% of the sample peak strength) was determined from Fig. 13. $E_{50,cy}$ was then normalized by $E_{50,NC}$, obtained for given values of q_s/p'_c from samples not subjected to cyclic loading whose stress–strain curves are shown in Fig. 5. The values of stiffness without cyclic loading were determined from Fig. 5. Normalized post-cyclic stiffness for four levels of τ_s/p'_c are plotted in Fig. 15 against the cyclic-induced excess pore pressures during undrained cyclic loading but before monotonic undrained shearing. The line drawn in Fig. 15 was plotted using

$$\frac{E_{cy}}{E_i} = \frac{E_{50,cy}}{E_{50,NC}} = \frac{1 - C\left(\dfrac{1}{1-\lambda}\right)\ln\left(\dfrac{1}{1-\dfrac{\Delta u}{p'_c}}\right)}{\dfrac{1}{1-\dfrac{\Delta u}{p'_c}}} \tag{7}$$

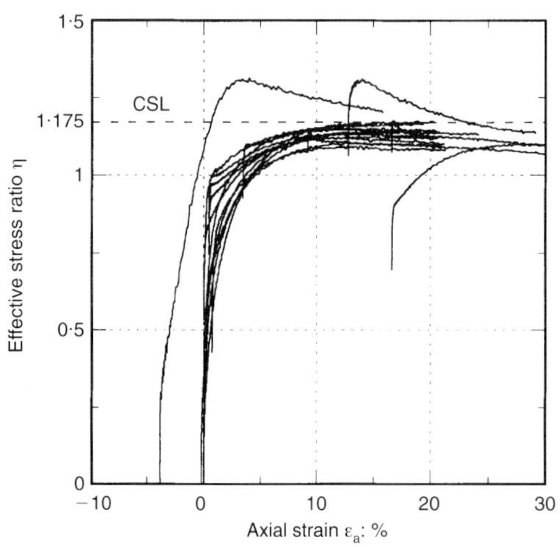

Fig. 14. Post-cyclic effective stress ratio

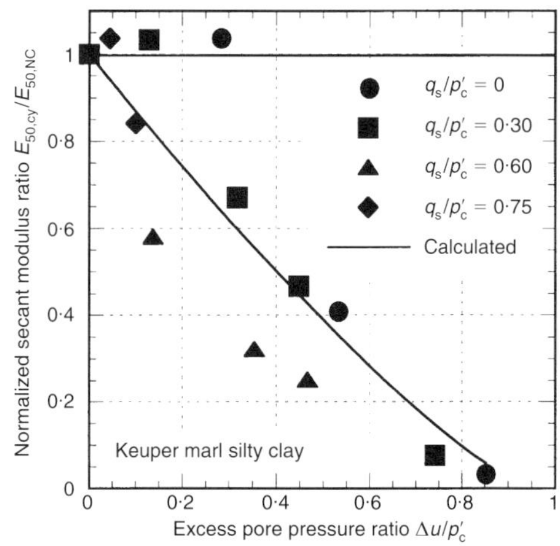

Fig. 15. Degradation of secant modulus

where C_s, C_c are the swelling and compression indices, and $\lambda = C_s/C_c$ and C is given by Wroth and Houlsby (1985) as

$$C = \frac{\left(\dfrac{(E/p')_{OC}}{(E/p')_{NC}}\right) - 1}{\ln \text{OCR}} \tag{8}$$

Although equation (7) was originally proposed for predicting the degradation of stiffness for cohesive soils with no initial shear stress (Yasuhara *et al.*, 1994; Yasuhara & Hyde, 1997a), very good agreement can be seen between the data and the lines drawn using equation (7) in Fig. 15. At the same time, it would appear that the degradation in stiffness with the normalized excess pore pressure is not influenced by the magnitude of the initial drained shear stress. It can be said, therefore, that no matter what magnitude of initial shear stress a soil might be subjected to, the degradation in stiffness is a function of the normalized excess pore pressure u/p'_c.

Degradation in rigidity index

The change in rigidity index $I_r = E/s_u$ is frequently used for estimating the disturbance of cohesive soils due to sampling and pile driving. By combining equation (7) for post-cyclic degradation in stiffness with the following equation for post-cyclic degradation in undrained strength, from Yasuhara (1992) and Yasuhara (1995)

$$\frac{s_{u,cy}}{s_{u,NC}} = \left[\left(\frac{1}{1-\dfrac{\Delta u}{p'_c}}\right)^{1/1-\lambda}\right]^{\Lambda_0 - 1 + \lambda} = \left[\frac{1}{1-\dfrac{\Delta u}{p'_c}}\right]^{\Lambda_0/(1-\lambda)-1} \tag{9}$$

where

$$\Lambda_0 = \frac{\log_{10}\left\{\dfrac{S_u/p'_{OC}}{S_u/p'_{NC}}\right\}}{\log_{10}\mathrm{OCR}} \tag{10}$$

one obtains

$$\frac{I_{r,cy}}{I_{r,NC}} = \frac{1 - \dfrac{C}{\Lambda}\ln[1/(1-\Delta u/p'_c)]}{[1/(1-\Delta u/p'_c)]^{\Lambda_0/\Lambda}} \tag{11}$$

This gives the degradation in rigidity index. Equation (11) is plotted in Fig. 16 and compared with data observed in post-cyclic monotonic undrained compression tests. It is seen from Fig. 16 that equation (11) gives a good prediction of the observed decrease in rigidity index for a plastic silt.

Relationship between degradation in strength and stiffness

When the term for normalized excess pore pressure ratio $\Delta u/p'_c$ is eliminated from equations (7) and (8), one obtains

$$\frac{E_{50,cy}}{E_{50,NC}} = \frac{1 - \dfrac{C}{\Lambda_0 - (1 - C_s/C_c)}\ln\left(\dfrac{s_{u,cy}}{s_{u,NC}}\right)}{\left(\dfrac{s_{u,cy}}{s_{u,NC}}\right)^{(1-C_s/C_c)/\{\Lambda_0-(1-C_s/C_c)\}}} \tag{12}$$

Use of equation (12) enables the degradation in post-cyclic stiffness to be predicted in terms of post-cyclic undrained strength. To confirm this, the results calculated using equation (12) are compared in Fig. 17 with the results observed in cyclic triaxial tests on plastic silts both with and without an initial static shear stress. Good agreement can be seen between the calculated and observed results. It can also be seen in Fig. 17 that the degradation in stiffness is more marked than that for undrained strength, as has already been confirmed for cohesive soils with no initial static shear stress (Yasuhara & Hyde, 1997a).

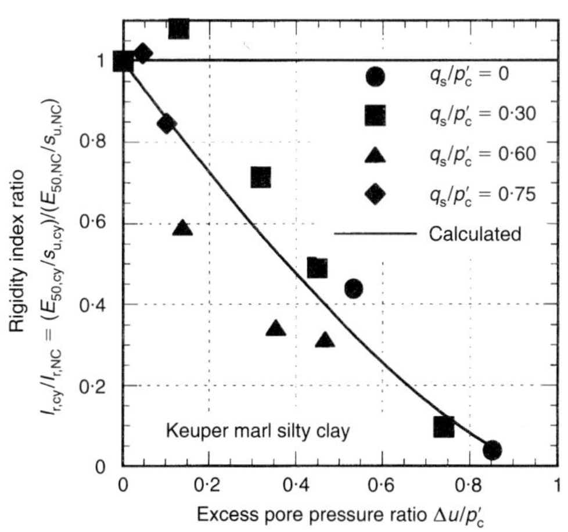

Fig. 16. Degradation of rigidity index

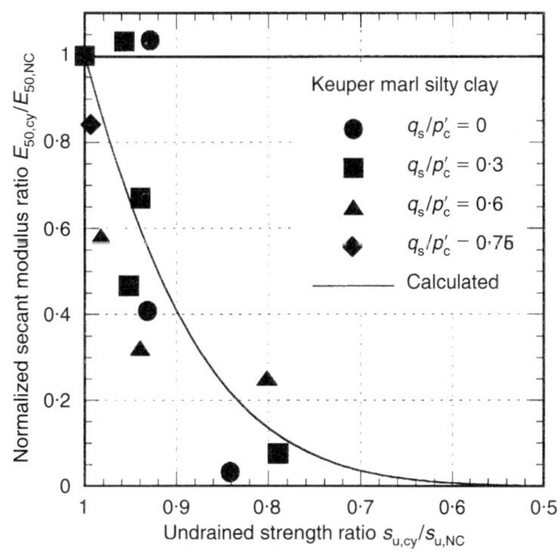

Fig. 17. Relation between secant modulus ratio and undrained strength ratio

CONCLUSIONS

(*a*) The secant modulus of a plastic silt under monotonic loading prior to any cyclic loading increased with increasing initial sustained drained shear stress. However, the variation of the secant modulus with initial shear stress seemed to depend on the plasticity index.

(*b*) The equivalent modulus G_{eq} was dependent on the normalized level of cyclic shear stress and decayed exponentially with the number of cycles of loading. This relation has been empirically formulated, independently of the initial shear stress.

(*c*) The degradation in post-cyclic secant modulus and rigidity index of a plastic silt with initial static drained shear stress was found to be governed by the cyclic-induced excess pore pressure. This tendency applies regardless of the initial shear stress and can be predicted using a method previously proposed for soil with no initial drained shear stress.

REFERENCES

Andersen, K. H., Pool, J. H., Brown, S. F. & Rosenbrand, W. F. (1980). Cyclic and static laboratory tests on Drammen clay. *J. Geotech. Engng Div. Am. Soc. Civ. Engrs* **106**, No. 5, 499–530.

Hyde, A. F. L. & Brown, S. F. (1976). The plastic deformation of a silty clay under creep and repeated loading, *Géotechnique* **26**, No. 1, 173–184.

Hyde, A. F. L. & Ward, S. J. (1983). Yield boundary for cyclic loading of a silty clay. *Conference on Geotechnical Practice in Offshore Engineering*, ASCE, 99–110.

Hyde, A. F. L. & Ward, S. J. (1985). A pore pressure and stability model for a silty clay under repeated loading. *Géotechnique* **35**, No. 2, 113–125.

Hyde, A. F. L. & Ward, S. J. (1986). The effect of cyclic loading on the undrained shear strength of a silty clay. *Marine Geotechnol.* **6**, No. 3, 299–314.

Hyodo, M., Yamamoto, Y. & Sugiyama, M. (1994). Undrained cyclic shear behaviour of normally consolidated clay subjected to initial static shear stress. *Soils Fdns* **34**, No. 54, 1–11.

Idriss, I. M., Dobry, R. & Singh, R. M. (1978). Nonlinear behavior of soft clays during cyclic loading *J. Geotech. Engng Div. Am. Soc. Civ. Engrs* **104**, GT12, 1427–1447.

Ishihara, K. & Takatsu, H. (1977). Pore pressure build-up in initially sheared sand subjected to irregular excitation. *Proc. 6th World Conf. on Earthquake Engng* **3**, 2163–2168.

Matsui, T., Bahr, M. A. & Abe, N. (1992). Estimation of shear characteristics of degradation and stress–strain relation of saturated clays after cyclic loading. *Soils Fdns* **32**, No. 2, 161–172.

Sugiyama, T. (1995). *Undrained cyclic shear characteristics of clays*. PhD thesis, Yamaguchi University (in Japanese).

Vaid, Y. & Chern, J. C. (1983). Effect of static shear on resistance to liquefaction. *Soils Fdns* **23**, No. 1, 47–60.

Vucetic, M. & Dobry, R. (1988). Degradation of marine clays under cyclic loading. *J. Geotech. Engng. Div. Am. Soc. Civ. Engrs* **120**, GT12, 2208–2227.

Wroth, C. P. & Houlsby, G. (1985). Soil mechanics, property characterisation and analysis procedures. *Proc. 11th ICSMFE, San Francisco*, **1**, 1–55.

Yasuhara, K., Hirao, K., Fujiwara, H. & Ue, S. (1983). Undrained shear behaviour of quasi-overconsolidated clay induced by cyclic loading. *Proc. IUTAM Symp. on Seabed Mechanics*, Graham & Trotman, London 17–24.

Yasuhara, K. (1985). Undrained and drained cyclic triaxial tests on a marine clay, *Proc. 11th ICSMFE, San Francisco*, **1**, 1195–1198.

Yasuhara, K. (1992). Effects of cyclic loading on undrained strength and compressibility of clay. *Soils Fdns* **32**, No. 1, 100–116.

Yasuhara, K., Satoh, K. & Hyde, A. F. L. (1994). Postcyclic undrained stiffness for clays. *Proc. Int. Symp. Pre-failure Deformation of Geomaterials* (IS-SAPPORO94), **1**, 483–489.

Yasuhara, K. (1994). Postcyclic undrained strength of cohesive soils, *J. Geotech. Engng Div. Am. Soc. Civ. Engrs* **120**, No. 11, 1961–1979.

Yasuhara, K. (1995). Earthquake-induced settlement in peaty ground. *Proc. Int. Workshop Peat*, **1**, 5.1–5.9.

Yasuhara, K. and Nagano, M. (1995). Post-cyclic behaviour of clay. *Proc. 10th Asian Reg. Conf. SMFE*, **1**, 119–122.

Yasuhara, K. & Hyde, A. F. L. (1997a). Method for estimating postcyclic undrained secant modulus of clays. *J. Geotech. Engng Div. Am. Soc. Civ. Engrs* **123**, No. 3, 204–211.

Yasuhara, K., Murakami, S. & Toyota, N. (1997b). Wave-induced residual settlements in soft seabed soils. *Proc. Int. Conf. BOSS, Delft*, **2**, 211–224.

Yasuhara, K. (1997). Effects of initial static shear stress on post-cyclic degradation of a plastic silt. *Proc. 13th ICSMFE, Hamburg*, **1**, 439–442.

Yasuhara, K., Murakami, S., Yokokawa, S., Toyota, N. & Hyde, A. F. L. (1997c). Cyclic softening and degradation of a plastic silt with initial shear stress. *Proc. Int. Symp. Deformation and Progressive Failure in Geomaterials* (IS-NAGOYA97), Nagoya, **1**, 617–623.

Schanz, T. & Vermeer, P. A. (1998). Pre-failure deformation behaviour of geomaterials, 383–387

On the stiffness of sands

T. SCHANZ* and P. A. VERMEER*

Experimental data are presented on the stiffness of many different sands under initial loading, both for one-dimensional and for triaxial compression. A normalized oedometer modulus and a normalized triaxial modulus are introduced, both of which depend only on density and the grain distribution. On comparing a large number of experimental results it is possible to conclude that the two normalized moduli usually have about the same value. It must be taken into account that the oedometer modulus is normalized by means of the axial stress, whereas the lateral stress is used to normalize the triaxial modulus. In a shearing type of problem, the correlation between the two moduli can be used to convert oedometer moduli, often provided in engineering practice, into appropriate triaxial moduli.

KEYWORDS: compressibility; constitutive relations; elasticity; laboratory tests; sands; stiffness.

Dans cet exposé, nous présentons des données expérimentales sur la rigidité de nombreuses variétés de sable sous charge initiale, à la fois pour une compression unidimensionnelle et pour une compression triaxiale. Nous introduisons un module d'oedomètre normalisé et un module triaxial normalisé, chacun ne dépendant que de la densité et de la distribution des grains. En comparant un grand nombre de résultats expérimentaux, il est possible de conclure que les deux modules normalisés ont en général à peu près la même valeur. Il faut tenir compte du fait que le module d'oedomètre est normalisé au moyen d'une contrainte axiale, alors que la contrainte latérale est utilisée pour normaliser le module triaxial. En ce qui concerne un problème de cisaillement, la corrélation entre les deux modules peut être utilisée pour convertir les modules d'oedomètre, souvent fournis en ingénierie, en modules triaxiaux appropriés.

INTRODUCTION

While stability problems in geotechnical engineering can usually be solved satisfactorily, it is difficult to calculate structure settlements and displacements sufficiently accurately in advance. Despite the considerable growth of knowledge in the past few decades, this is a field in which the displacements measured often turn out to be considerably smaller, and sometimes much larger, than those calculated. The reason for such differences usually lies in the selection of the stiffness parameters used to characterize the deformation behaviour of a soil layer. Most research has probably been completed on determining settlements of shallow foundations and dams or embankment constructions. For such compression-type problems, stiffness moduli are usually obtained from oedometer tests, with application of appropriate pressure ranges.

In the past 20 years, triaxial tests have also come to be widely used to determine soil stiffness. In undrained tests on saturated samples the undrained triaxial (Young's) modulus E_u, which, for example, is used to calculate the immmediate settlement of a foundation is determined. In drained triaxial tests on cohesionless soil the effective modulus E of the soil skeleton is measured, which in turn can be used to estimate pile settlements (Wroth & Houlsby, 1985) or, for example, settlements due to excavations or tunneling. As secant moduli, E_u and E fall rapidly with increasing mobilization of the shear strength, and thus there are several options for determining these moduli. For cyclical loading of structures, the high initial tangent modulus of the test is generally of interest. For routine settlement computations it is more likely that a smaller modulus will be used, although it is still unclear how the modulus is best determined from a triaxial test. This question, at least for sand, is examined in this paper. With reference to test results on several sands it is shown that $E_{oed} \approx 1{\cdot}5E_{50}$, where the subscript 50 signifies that this Young's modulus is measured as the secant modulus corresponding to a stress level of 50% of the peak shear strength. This relation is an empirical correlation which is derived on the basis of test results from several institutes. Finally, advice is given on the realistic choice of stiffness parameters for numerical simulations depending on the type of loading involved. In the future, constitutive models used for calculations might be checked on their capability to simulate this particu-

Manuscript received 30 June 1997.

* Stuttgart University.

lar relation between the stiffness under one-dimensional compression and the shear loading stiffness.

STIFFNESS IN OEDOMETER TESTS

In this section the well-known behaviour of sands in compression tests in which lateral straining is prevented is briefly summarized. Terzaghi (1925) conducted such tests primarily on clay samples from which he identified an almost logarithmic relation; Ohde (1939, 1951) studied silts and sands as well, and derived the pressure-dependent oedometer modulus

$$E_{oed} = E_{oed}^{a}\left(\frac{\sigma_1}{\sigma_a}\right)^m \tag{1}$$

where $\sigma_a = 100$ kPa and σ_1 is the vertical effective stress. In this paper σ_1 always signifies effective stresses ($\sigma_1' = \sigma_1$). It should be noted that Ohde did not introduce a normalized stiffness E_{oed}^{a}. Instead he used the constant v_e such that

$$E_{oed}^{a} = v_e \sigma_a \tag{2}$$

Ohde (1951) reports values between 100 and 750 for the coefficient v_e, giving normalized oedometer moduli of between 10 and 75 MPa. Von Soos (1980) gives very similar values, $v_e = 150$ for a loose fine sand and $v_e = 700$ for a dense coarse sand. This obviously confirms the slightly larger range $10 < E_{oed}^{a} < 75$ MPa.

Figure 1 shows that the oedometer modulus according to Ohde is a tangent modulus, and that for $\sigma_1 = \sigma_a$ it results in the special case $E_{oed} = E_{oed}^{a}$. In practice, however, it can also be measured as a secant modulus, for instance across a stress range of 50 to 150 kPa. Unfortunately Ohde's fundamental research (1939, 1951) remained unknown abroad, where the experimental pressure dependence of the oedometer modulus was first published by Janbu (1963).

Figure 2 shows the values measured by Janbu (1963) and Von Soos (1980) for the exponent m in the range 0·4 to 0·75. The data here are taken from a large number of different sands. Aside from these general data three sands are subject to particular examination in this paper, since for these sands both oedometer results and triaxial results are available. These are Karlsruhe sand (Bauer, 1992), the French Hostun sand (Flavigny *et al.*, 1993) and the Japanese Toyoura sand (Goto & Tatsuoka, 1988; Bauer, 1992). The characteristic basic data of these three sands are listed in Table 1. They are fine sands with mean grain diameters of 0·16, 0·332 and 0·33 mm. All three sands are characterized by a coefficient of uniformity $C_u \approx 1{\cdot}5$. The examined ranges of initial porosity n_0 and relative density I_D are also given in the table.

Figure 3 shows the normalized oedometer moduli as a function of the initial porosity n_0. In this figure the data from Janbu (1963) and Von Soos (1980) lie in the shaded region, and the data for the three special sands are shown separately. Figure 3 clearly gives rise to a much larger modulus for a very dense sand than for a very loose sand, with ratios of 2 to 3 between the two extremes. On closer examination the data are seen to correspond to Von Soos's table in Smoltczyk (1992), in which a coarse sand is given as being up to twice as stiff as a fine sand.

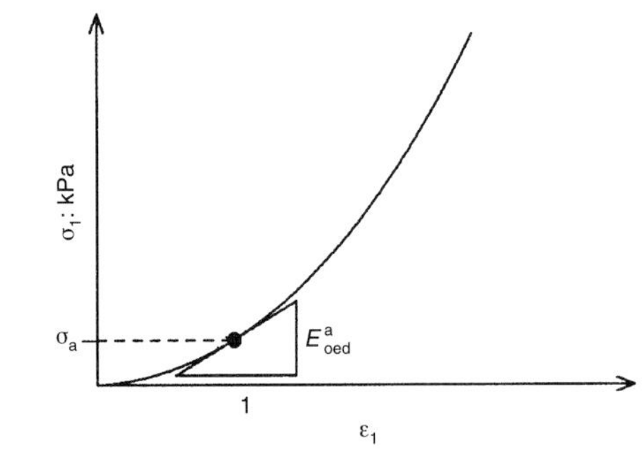

Fig. 1. Behaviour of sand in oedometer tests and definition of the normalized oedometer modulus

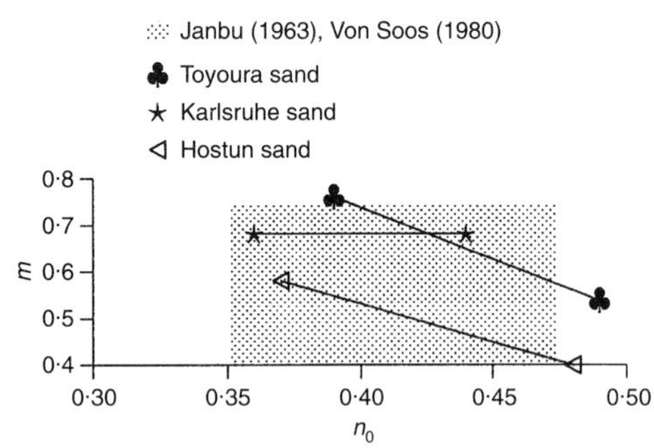

Fig. 2. Values for the exponent *m* determined in oedometer tests

Table 1. Parameters of three sands

	d_{50}: mm	C_u	I_D	n_0
Toyoura sand	0·16	1·46	0·21–0·8	0·40–0·47
Hostun sand	0·32	1·5	0·30–1.15	0·37–0·48
Karlsruhe sand	0·33	1·5	0·16–0·98	0·35–0·43

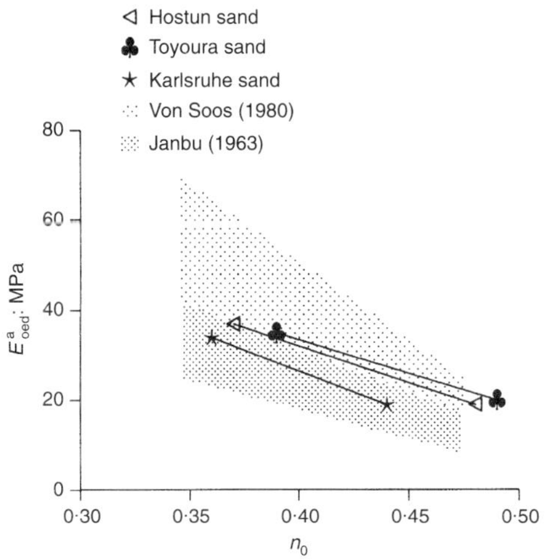

Fig. 3. Normalized stiffness modulus of various sands taken from oedometer tests

STIFFNESS IN TRIAXIAL COMPRESSION TESTS

In this section results of drained triaxial tests on sand samples are examined. Following isotropic consolidation where $\sigma_1 = \sigma_2 = \sigma_3$, shearing is performed under drained conditions by axial compression. In the process, the cell pressure and thus the smallest primary stress $\sigma_2 = \sigma_3$ are kept constant while the vertical stress σ_1 is measured as a function of the axial compression ε_1. Fig. 4 illustrates a characteristic behaviour and also shows the definition of the Young's modulus E_{50} which is a secant modulus corresponding to 50% of the peak shear strength.

While the pressure dependence of the oedometer modulus is formulated with reference to the larger primary stress σ_1, according to Kondner (1963) and Duncan & Chang (1970) in triaxial tests the smaller primary stress σ_3 is generally considered. Apart from this difference, by analogy with the oedometer modulus this results in the empirical relation

$$E_{50} = E^a_{50}\left(\frac{\sigma_3}{\sigma_a}\right)^m \tag{3}$$

The E^a_{50} modulus is again a normalized modulus, but this time not with a largest primary stress σ_1 but with a smallest primary stress $\sigma_3 = 100$ kPa, i.e. the confining pressure. It is additionally influenced by the grain distribution d_{50} and the coefficient of uniformity C_u. Just as in the oedometer test, values for m between 0·4 and 0·7 are found in the triaxial test, as shown in Fig. 5.

In Fig. 6 Young's moduli measured in relation to the porosity are applied. Since they are seldom published directly in the literature, they have been determined from the stress–strain relations of a large number of sands.

Here, only test results with a lateral pressure of $\sigma_3 = 100$ kPa were evaluated. Thus no test results were extrapolated which had cell pressures that differed from the reference pressure σ_a.

Apart from the authors mentioned in Fig. 5, test results from Lee & Seed (1967), Arslan (1980) and Von Wolffersdorff (1992) were also evaluated and are shown in Fig. 6. A large amount of data is available on Hostun sand, partly with the same porosity, because various institutes are conducting tests on it (Schanz & Vermeer, 1996). A conversion was carried out only on the values of Lade & Nelson (1987). In contrast to the data for Hostun sand, these are results from seven different sands and are hence regarded as especially interesting. Extrapolation was necessary here since Lade & Nelson consider only Young's moduli E^a_{ur} for unloading/reloading rather than for initial loading. This modulus practically corresponds to the E-static modulus mentioned by Lorenz and Klein (1980), although those authors do not mention any pressure dependence. The extrapolation consisted in assuming the relation $E^a_{ur} = 5E^a_{50}$. Instead of the factor 5, a factor of 4 might as well have been used.

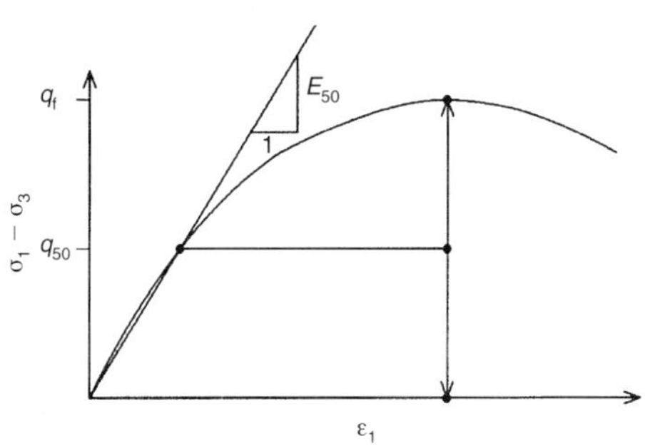

Fig. 4. Behaviour of sand in a triaxial test and definition of E_{50}

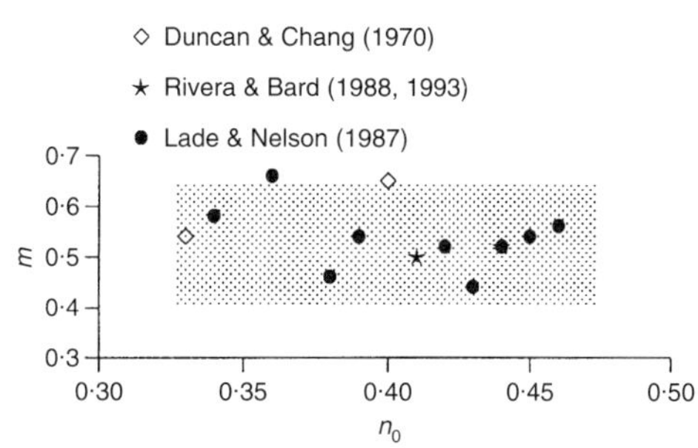

Fig. 5. Values for the exponent m determined in triaxial tests

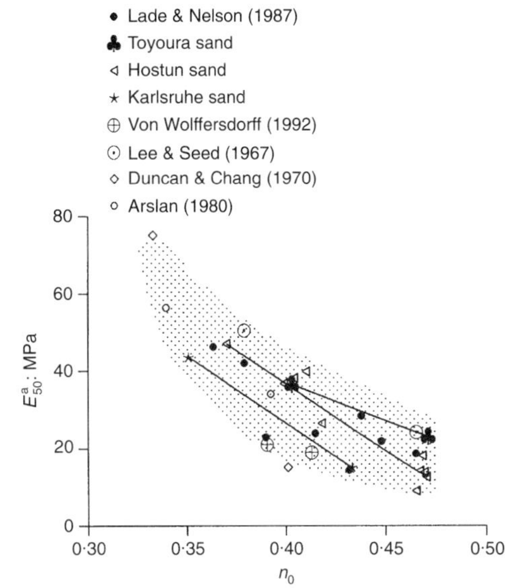

Fig. 6. Normalized Young's modulus for various sands

OEDOMETER MODULUS AGAINST YOUNG'S MODULUS

Now that the results both for the oedometer modulus E_{oed} and for the Young's modulus E_{50} have been put in context, these moduli ought first to be compared in purely empirical terms. At first glance the normalized values appear to be similar already, since both Fig. 3 and Fig. 6 show moduli with values between 10 and 75 MPa, whereby a loose fine sand tends more towards the lower limit and a compact coarse sand tends more towards the upper limit. Both figures display a ratio of about 2 to 7 between the compact coarse sand stiffness and the loose fine sand stiffnesses. Unfortunately, a more precise comparison of the oedometer modulus with the Young's modulus is possible only for the three special sands (Hostun, Karlsruhe and Toyoura sands) because only here are values for the two different moduli available.

The values in Fig. 7 then show that the oedometer moduli E^a_{oed} determined in the oedometer correlate well with the secant moduli E^a_{50} from drained normal triaxial tests, at least after suitable normalization. Here, the different normalization of the two moduli must be taken into consideration. The oedometer modulus was normalized for a vertical stress $\sigma_a = \sigma_1$; the Young's modulus was normalized for a horizontal stress $\sigma_a = \sigma_3$. With anisotropic stresses there is therefore a difference between the current oedometer modulus and the current Young's modulus. According to equations (1) and (3)

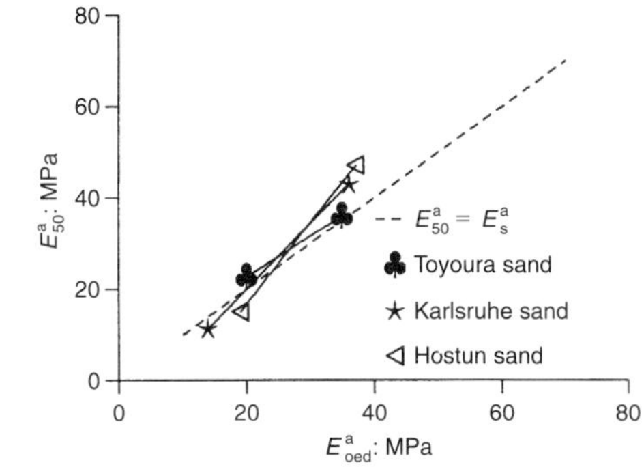

Fig. 7. Comparison of normalized moduli from oedometer and triaxial tests

$$E_{50} = E^a_{50}\left(\frac{\sigma_3}{\sigma_a}\right)^m, \; E_s = E^a_s\left(\frac{\sigma_1}{\sigma_a}\right)^m \tag{4}$$

and from this

$$\frac{E_{oed}}{E_{50}} = \frac{E^a_{oed}}{E^a_{50}}\left(\frac{\sigma_1}{\sigma_3}\right)^m \approx \left(\frac{\sigma_1}{\sigma_3}\right)^m \tag{5}$$

With earth pressure at rest, $\sigma_3 = K_0\sigma_1$ applies, where $K_0 \approx 0{\cdot}4$ for a normally consolidated sand, and it follows that

$$E_{50} = K_0^m E_s \rightarrow E_{oed} \approx 1{\cdot}5 E_{50} \tag{6}$$

It must be stressed that the above relations between the oedometer modulus E_{oed} and the Young modulus's E_{50} are purely empirical correlations which, particularly in non-linear computations, have an important role to play. In a routine settlement computation no stress–strain relation is extracted from triaxial tests, but only the stiffness modulus. Even if closed-form elastic solutions which require a Young's modulus are used, this Young's modulus E is calculated from E_{oed} with the help of the well-known formula from elasticity

$$E = \frac{(1+\nu)(1-2\nu)}{(1-\nu)} E_{oed} \tag{7}$$

For given values of the oedometer modulus E_{oed} and Poisson's ratio ν the Young's modulus is derived from this. Small values for ν are usually assumed in routine settlement computations. If $\nu = 0$ is selected, which is more or less the standard choice, it follows that $E = E_{oed}$ and, in the authors' experience, this means that $E \approx E_{25}$, i.e. a secant modulus of the triaxial test at 25% of the peak shear strength. Although the choice of $\nu = 0$ and $E = E_{oed}$ is customary in normal settlement computation practice, finite element (FE) calculations are difficult to conduct with this choice for Poisson's ratio. If an elastic constitutive equation is used, the resulting earth pressure at rest coefficient is $K_0 = \nu/(1-\nu)$ and if $\nu = 0$ this results in

$K_0 = 0$. This means that the horizontal stresses do not increase and FE calculations would therefore produce very unrealistic stress fields. For normally compacted soils, earth pressure at rest coefficients are $K_0 \approx 0{\cdot}5$ and consistent elasticity calculations therefore require a Poisson's ratio of $\nu \approx \frac{1}{3}$. FE calculations to determine settlements would thus be performed not with $\nu = 0$ and $E = E_{\text{oed}}$ but rather with $\nu = \frac{1}{3}$ and $E = \frac{2}{3}E_{\text{oed}} \approx E_{50}$.

CONCLUSIONS AND OUTLOOK

A normalized oedometer modulus $E^{\text{a}}_{\text{oed}}$ is introduced in this paper which varies with the density and the grain distribution, but not with the stress level. In this context it is necessary to mention that the behaviour of sands under initial loading and therefore a normally consolidated soil are being considered. In addition, a normalized Young's modulus E^{a}_{50} is introduced which can be derived from triaxial tests and, like the normalized oedometer modulus, is dependent only on density and the grain distribution. On comparing a large number of experimental results it was possible to conclude that the two moduli usually have about the same value. Thus there is a correlation between moduli from oedometer and triaxial tests.

The question again arises of which stiffness modulus one should use in numerical calculations: in engineering practice data on E_{oed} are mostly available, and from the test data presented in this paper the empirical relation $E^{\text{a}}_{\text{oed}} \approx E^{\text{a}}_{50}$ is found. It should be realized, however, that the oedometer modulus E_{oed} is normalized by means of the axial stress σ_1, whereas the lateral stress σ_3 is used to normalize the Young modulus from a triaxial test. When this normalization is omitted, one has to account for this difference and obtains $E_{\text{oed}} \approx 1{\cdot}5E_{50}$, at least for a normally consolidated soil element with $\sigma_1/\sigma_3 \approx 0{\cdot}4$.

The above correlation can first of all be used for converting oedometer moduli into triaxial moduli, or vice versa. Such a situation occurs for example when one considers a shearing type of problem, with only oedometer moduli available.

Another application of the relationship is in the field of numerical modelling. Here several models have been developed to simulate sand behaviour, and it would be useful to know whether or not these models match the correlation derived in this paper.

REFERENCES

Arslan, M. U. (1980). Zur Frage des elastoplastischen Verformungsverhaltens von Sand. *Mitteilungen der Versuchsanstalt für Bodenmechanik und Grundbau der TH Darmstadt*, Heft 23.

Bard, E. (1993). *Methode d'analyse des lois de comportement des soils*. Thèse de Doctorat de l'Ecole Centrale de Paris.

Bauer, E. (1992). Zum mechanischen Verhalten granularer Stoffe unter vorwiegend ödometrischer Beanspruchung. *Bericht des Inst. für Bodenmechanik und Felsmechanik*, Heft 130, Univ. Karlsruhe.

Duncan, J. M. & Chang, C. Y. (1970). Nonlinear analysis of stress and strain in soils. *J. Soil Mech. Fdn Div. Am. Soc. Civ. Engrs* **96**, SM5, 1629–1653.

Flavigny, E. *et al.* (1993). *Compilation des essais triaxiaux de révolution sur le sable d'Hostun RF.* Internal report IMG, Grenoble.

Goto, S. & Tatsuoka, F. (1988). Effects of end conditions on triaxial compressive strength for cohesionless soil. *Advanced triaxial testing of soil and rock* (eds Donaghe *et al.*), STP 977, ASTM.

Janbu, N. (1963). Soil compressibility as determined by oedometer and triaxial tests. *Proc. Eur. Conf. Soil Mech. Fdn Engng, Wiesbaden*, 19–25.

Kondner, R. L. & Zelasko, J. S. (1963). A hyperbolic stress strain formulation for sands. *Proc. 2nd Pan-Am. Conf. Soil Mech. Fdn Engng*, Rio de Janeiro, Brazil, **1**, 289–324.

Lade, P. V. & Nelson, R. B. (1987). Modelling the elastic behaviour of granular materials. *Int. J. Numer. Anal. Methods Geomech.* **11**, 521–542.

Lee, K. L. & Seed, H. B. (1967). Drained strength characteristics of sand. *J. Soil Mech. Fdn Engng Div. Am. Soc. Civ. Engrs* SM6, 117–141.

Lorenz, H. & Klein, G. (1980). Bodendynamik und Erdbeben. *Grundbautaschenbuch*, 3 Aufl., Teil 1.

Ohde, J. (1939). Zur Theorie der Druckverteilung im Baugrund. *Bauingenieur* **20**, 451.

Ohde, J. (1951). Grundbaumechanik. *Hütte*, Bd. III, 27 Aufl., 886.

Rivera, R. C. (1988). *Détermination des propriétés mécanique de sables et des argils en régime dynamique et cyclique an faibles déformation.* Thèse de Doctorat de l'Ecole Centrale de Paris.

Schanz, T. & Vermeer, P. A. (1996). Angles of friction and dilatancy of sand. *Géotechnique* **45**, No. 4.

Smoltczyk, U. (1992). *Grundbau-Taschenbuch*, Teil 1, 4 Aufl. Berlin: Ernst & Sohn Verlag.

Terzaghi, K. (1925). *Erdbaumechanik auf bodenphysikalischer Grundlage.* Leipzig, Wien: Franz Deuticke.

Von Soos, P. (1980). Eigenschaften von Boden und Fels, ihre Ermitlung im Labor. *Grundbau-Taschenbuch*, Teil 1, 4 Aufl. Berlin: Ernst & Sohn Verlag.

Von Wolffersdorff, P. A. (1992). Dokumentation zur Probebelastung Hochstetten. *Bericht des Inst. für Bodenmechanik und Felsmechanik*, Univ. Karlsruhe.

Part three

Discussion papers

Kuwano, R. & Jardine, R. J. (1998). Pre-failure deformation behaviour of geomaterials, 391–394

Stiffness measurements in a stress-path cell

R. KUWANO* and R. J. JARDINE*

INTRODUCTION

This short contribution reports the steps that have been taken at Imperial College (*a*) to improve local strain resolution, (*b*) to integrate static stress–strain testing with shear wave velocity measurements and (*c*) to interpret the data obtained at very small strains. Zdravkovic & Jardine (1997) described the strong stiffness anisotropy exhibited by a granular soil over the non-linear strain range ($0{\cdot}003\% < \varepsilon_1 < 5\%$), showing that behaviour was inelastic in almost every respect. In particular, the anisotropy was quite different to the classical cross-anisotropic linear elastic model which is often assumed to apply to geomaterials. However, the transducers employed in Zdravkovic & Jardine's hollow cylinder experiments were not able to resolve behaviour within the linear range where others such as Jiang *et al.* (1997) have observed almost elastic behaviour. In order to address this and other questions, the improved triaxial stress-path system shown in Fig. 1 was developed.

LOCAL STRAIN SENSORS

Electro-level inclinometers have been able to resolve local axial strains to around 0·002% for over 15 years (Jardine *et al.*, 1984). They work by converting changes in distance Δd between two pads attached to the sample into a change of inclination $\Delta\theta$ of the electrolytic capsule, and hence a voltage; their non-linear sensitivity $\Delta\theta/\Delta d$ depends on the geometry adopted. Three changes have been made to improve the resolution of these gauges. Firstly, optimized signal conditioning circuits have been built, comprising the power supply, oscillator unit, demodulator and analogue filters. Secondly, the geometry has been modified to give the best combination of working range and sensitivity for samples of 200 mm height. Thirdly, the mechanical operation has been made as smooth as possible.

To measure radial strains, a submersible linear variable differential transformer (LVDT) with a linear working range of ±5 mm has been adopted. The LVDT is supported by a rectangular frame, and the frame and the rod of the LVDT are glued onto the membrane. In order to minimize the frame weight, metal tube was used and the weight of the frame and the LVDT was counter-balanced by small suspended weights. The radial displacement perpendicular to the surface measured on the membrane might suffer from errors when the confining pressure changes. In order to reduce this problem, target pins were installed inside the membrane (Fig. 1), which are made from flat-head screws with an overall length of 5 mm. In the present tests these ensure that the measurements reflect the movements of more than 200 sand particles which are locked into contact with the thread away from the soil–membrane boundary.

The working range, resolution and scatter of the transducers are shown in Table 1. Further details are given by Kuwano (1997).

SHEAR WAVE VELOCITY MEASUREMENTS

The cell shown in Fig. 1 allows measurements to be made of the velocities of horizontally and vertically propagating shear waves using bender elements. The corresponding shear moduli at very small strains, G_{vh}, G_{hh}, G_{hv}, are obtained by the usual equations. In order to mount bender elements on the side of the specimen, a stainless steel housing was designed, whose small size and weight enable it to be placed directly without additional support (Kuwano *et al.*, 1997).

Table 1. Local strain transducer performance

Strain	Transducer used	Working range	Resolution	Scatter
Axial	Modified electro-level inclinometer	95–105 mm 20%	0·1 μm 10^{-4}%	±0·2 μm $\pm 2 \times 10^{-4}$%
Radial	Submersible LVDT strain belt	±5 mm 10%	0·1 μm 10^{-4}%	±0·1 μm $\pm 1 \times 10^{-4}$%

Manuscript received 20 October 1997.
* Department of Civil Engineering, Imperial College.

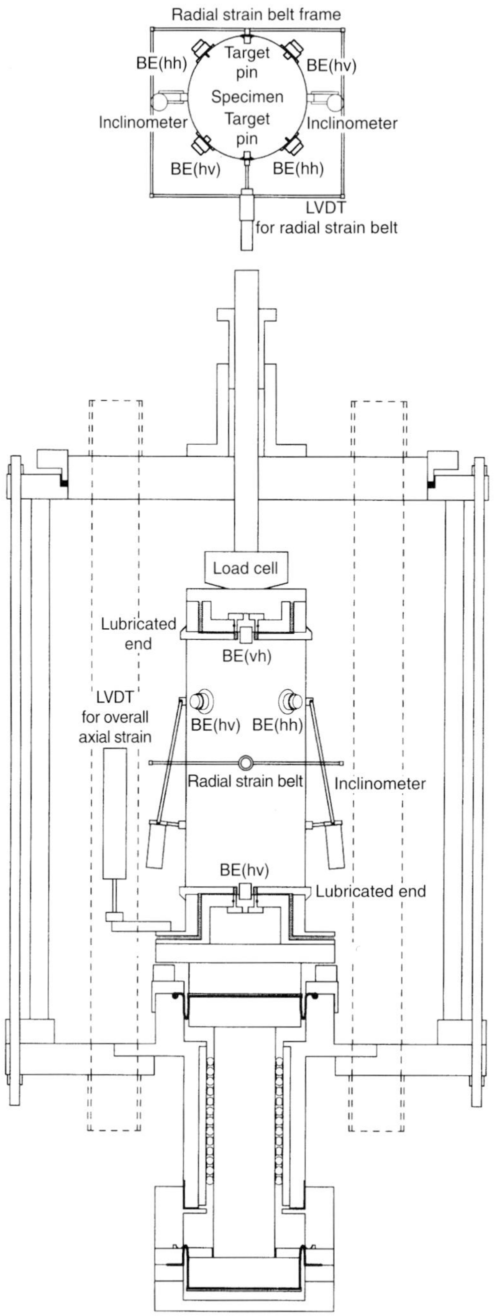

Fig. 1. Triaxial cell and internal transducers (diameter 100 mm, height 200 mm)

PERFORMANCE

The performance of the stress–strain measurement system is illustrated in Fig. 2. A medium-dense Ham River sand sample was consolidated to a vertical (axial) effective stress, $\sigma'_v = 316$ kPa, and a horizontal (radial) effective stress: $\sigma'_h = 142$ kPa. After five hours' ageing, σ'_v was reduced (under drained conditions) by 10 kPa, with σ'_h constant, then increased by 20 kPa before being returned to its original stress state. The resulting strains were small and the equipment was able to resolve smooth stress–strain behaviour at strain levels down to 10^{-4}%. During the initial unload–reload phase, the sample showed a linear elastic response. But on loading past the initial σ'_v level, non-linearity and plastic straining appeared once $\Delta\sigma'_v > 4$ kPa in both axial and radial strains components.

STIFFNESS ANISOTROPY WITHIN THE LINEAR RANGE

If it is assumed that cross-anisotropy applies within the very small linear elastic range, then the triaxial stress–strain relationship is described by

$$\begin{pmatrix} \delta\varepsilon_v \\ \delta\varepsilon_h \end{pmatrix} = \begin{pmatrix} \dfrac{1}{E'_v} & -\dfrac{2\nu'_{hv}}{E'_h} \\ -\dfrac{\nu'_{vh}}{E'_v} & \dfrac{1-\nu'_{hh}}{E'_h} \end{pmatrix} \begin{pmatrix} \delta\sigma'_v \\ \delta\sigma'_h \end{pmatrix} \tag{1}$$

where ε_v is vertical strain, ε_h is horizontal strain, E'_v is Young's modulus in the vertical direction, E'_h is Young's modulus in the horizontal direction, ν'_{vh} is Poisson's ratio for the effect of vertical strain on horizontal strain, ν'_{hv} is Poisson's ratio for the effect of horizontal strain on vertical strain, ν'_{hh} is Poisson's ratio for the effect of horizontal on complementary horizontal strain, σ'_v is effective vertical stress, σ'_h is effective horizontal stress.

If a symmetrical compliance matrix (Love, 1927) is also assumed, then

$$\frac{\nu'_{vh}}{E'_v} = \frac{\nu'_{hv}}{E'_h} \tag{2}$$

From drained axial loading tests ($\Delta\sigma'_h = 0$), E'_v and ν_{vh} can be determined

$$\delta\varepsilon_v = \frac{1}{E'_v}\delta\sigma'_v \tag{3}$$

$$\delta\varepsilon_h = -\frac{\nu'_{vh}}{E'_v}\delta\sigma'_v \tag{4}$$

It is also possible to perform stress-path tests involving purely σ'_h loading ($\Delta\sigma'_v = 0$), leading to axial and radial strains as follows

$$\delta\varepsilon_v = -\frac{2\nu'_{hv}}{E'_h}\delta\sigma'_h \tag{5}$$

$$\delta\varepsilon_h = \frac{1-\nu'_{hh}}{E'_h}\delta\sigma'_h \tag{6}$$

However, without extra data, E'_h, ν'_{hh} cannot be determined. If it is assumed that the shear modulus

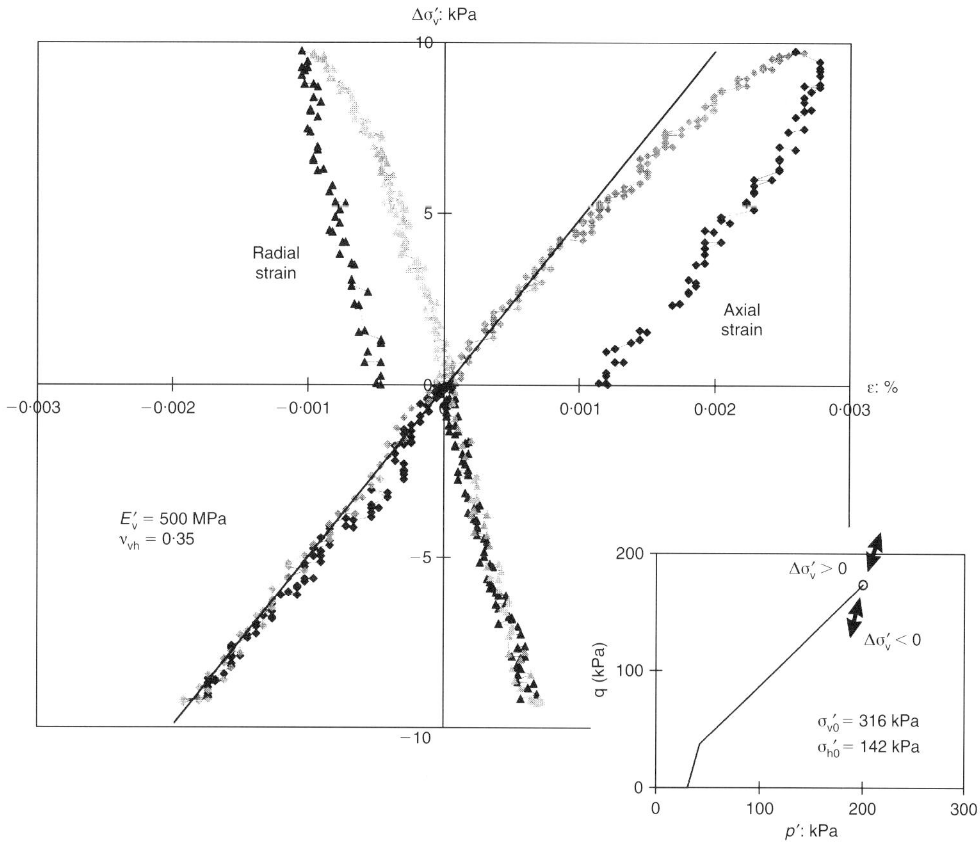

Fig. 2. Performance of static gauges at small strain

in any horizontal plane G_{hh} is completely rate-independent, and can be measured from bender element tests, then

$$G_{hh} = \frac{E'_h}{2(1+\nu'_{hh})} \quad (7)$$

$$E'_h = \frac{4AG_{hh}}{A+2G_{hh}} \quad (8)$$

with

$$\nu_{hh} = \frac{A-2G_{hh}}{A+2G_{hh}} \quad (9)$$

where

$$A = \frac{\delta\sigma'_h}{\delta\varepsilon_h}$$

When the drained axial loading test data shown in Fig. 2 were combined with a lateral loading test and a horizontal bender element measurement obtained at the same stress state, the following cross-anisotropic elastic constants were found, five of which are independent

$$E'_v = 500\text{ MPa} \quad \nu'_{vh} = 0{\cdot}35 \quad G_{vh} = 160\text{ MPa}$$

$$E'_h = 260\text{ MPa} \quad \nu'_{hh} = 0{\cdot}1 \quad G_{hh} = 130\text{ MPa}$$

$$\nu'_{hv} = 0{\cdot}16$$

SUMMARY

A stress-path triaxial cell equipped with sensitive local axial and radial strain transducers and bender elements can allow anisotropic vertical and horizontal stress–strain relationships to be investigated over a wide range of strain. In the limited elastic strain range, the full set of cross-anisotropic constants can be determined by combining static and dynamic stiffness measurements provided that (*a*) the soil's compliance matrix is symmetric and (*b*) the stiffness parameters are rate-independent. Neither assumption holds at larger strains (Zdravkovic & Jardine, 1997), and more complex equipment is required to investigate non-linear anisotropy fully.

ACKNOWLEDGEMENTS

Mr S. Ackerley of Imperial College was responsible for the major improvements described in the signal conditioning system. Mr T. M. Connolly of Imperial College, Mr D. Pennington of Bristol University and Dr V. Jovicic of City University gave valuable help for bender element techniques.

REFERENCES

Jardine, R. J., Symes, M. J. and Burland, J. B. (1984). The measurement of soil stiffness in the triaxial apparatus. *Géotechnique* **34**, No. 3, 323–340.

Jiang, G.-L., Tatsuoka, F., Flora, A. & Koseki, J. (1997). Inherent and stress-state induced anisotropy in very small strain stiffness of a sandy gravel. *Géotechnique* **47**, No. 3, 509–521.

Kuwano, R. (1997). *Anisotropic deformation characteristics of sands.* Internal report, Department of Civil Engineering, Imperial College, London.

Kuwano, R., Connolly, T. M., Jardine, R. J. & Kuwano, J. (1997). Shear wave velocity measurement in the triaxial apparatus using lateral bender elements. *Proc. 32nd Japan National Conf. on Geotechnical Engineering, Kumamoto*, **1**, 809–810.

Love, A. E. H. (1927). *A treatise on mathematical theory of elasticity.* New York: Dover Publications.

Zdravkovic, L. & Jardine, R. J. (1997). Some anisotropic stiffness characteristics of a silt under general stress conditions. *Géotechnique* **47**, No. 3, 407–437.

Jovičić, V. & Coop, M. (1998). Pre-failure deformation behaviour of geomaterials, 395–397

The influence of plastic strains during compression on the inherent anisotropy of G_{max}

V. JOVIČIĆ* and M. COOP†

INTRODUCTION

Much of the discussion during Session 2 addressed the origins of anisotropy of very small strain stiffness in clays, separating out the stress-induced from the inherent component. Pennington (1997) had identified the influence of an anisotropic stress state on G_{max} but, as Professor Wheeler pointed out in his discussion, he seemed not to have addressed the effect of straining on G_{max} anisotropy. Professor Wheeler asked how stable is the inherent anisotropy of G_{max}. He pointed out that when a reconstituted specimen which has been created by one-dimensional compression is transferred to the triaxial apparatus and subjected to an increasing isotropic stress, the plastic strain increments typically become isotropic by the time that the mean effective stress p' has been doubled, and certainly by the time that the isotropic normal compression line is reached. He wondered if this would be also the case for G_{max}. This topic has been addressed in a recently completed research project at City University, and is the subject of this discussion.

In an attempt to reproduce the inherent anisotropy of natural London Clay in the laboratory, a set of five tests was carried out on reconstituted samples. Remoulded natural material was mixed into a slurry and one-dimensionally consolidated in a 100 mm diameter consolidometer. The vertical stress of 400 kPa was chosen to reproduce the in situ stresses on the natural sample which was also tested. Three of the samples, denoted NC, were normally consolidated in the consolidometer to this stress. The other two, denoted OC, were overconsolidated by loading to 1500 kPa and subsequent unloading to a vertical effective stress of 400 kPa so as to better reproduce the geological stress history of the London Clay. After being transferred to the triaxial cell, all the samples were initially consolidated under an isotropic stress of 400 kPa. The OC samples were then compressed isotropically in stages up to 3 MPa as shown in Fig. 1(b). At each stage the samples were allowed to consolidate fully and a bender element test was performed.

The samples tested were cut horizontally or vertically from the cake removed from the consolidometer, and the bender element slots were oriented relative to the principal axis of soil fabric.

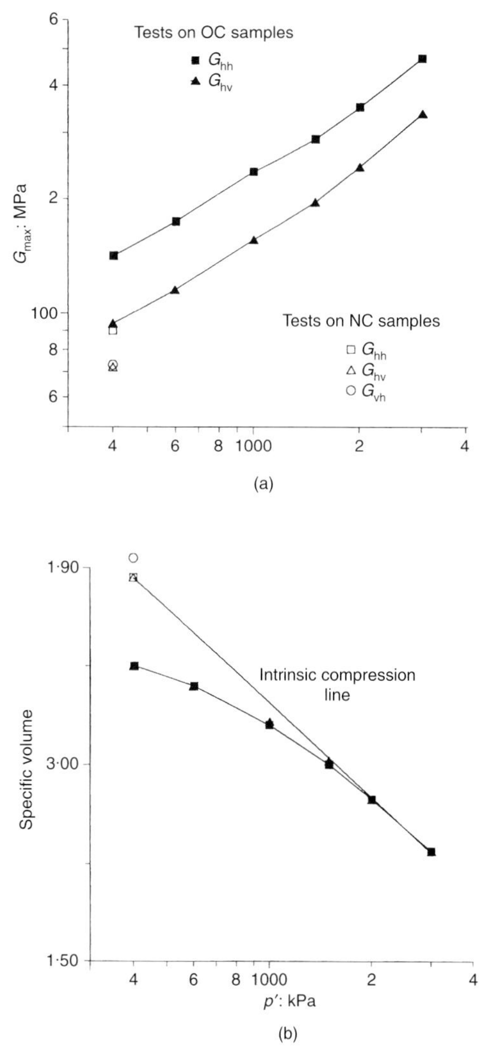

Fig. 1. Bender element tests on oriented samples of reconstituted London Clay (after Jovičić & Coop, 1998)

Manuscript received 16 October 1997
* Arup Geotechnics, formerly City University, London.
† City University, London.

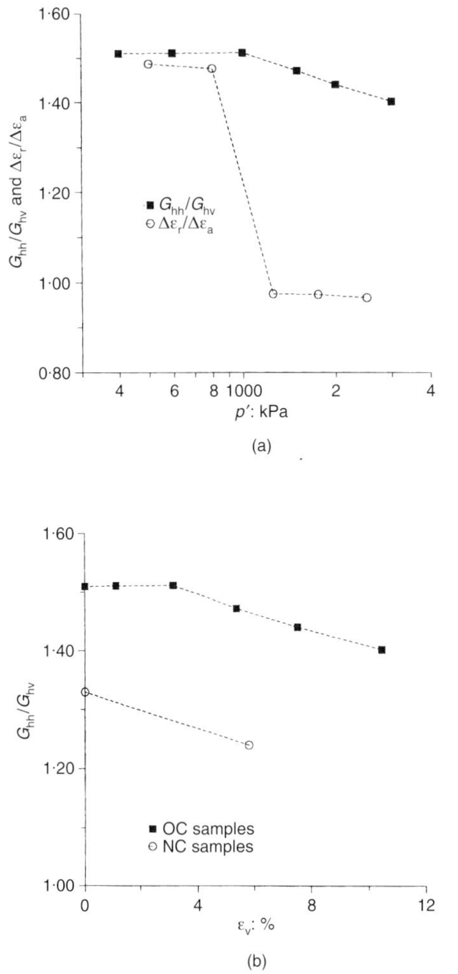

Fig. 2. Variation of the degree of anisotropy and strain increment ratio with (a) mean effective stress and (b) volumetric strain for reconstituted London Clay (after Jovičić & Coop, 1998)

The two OC samples were both cut horizontally and then oriented in the triaxial appraratus so that the bender element would give a shear wave polarized in either the original horizontal or the vertical direction, in order that G_{hh} and G_{hv} could be measured. The use of isotropic stresses throughout triaxial tests ensured that there was no stress-induced effect on the anisotropy measured, so that only the effect of strain-induced anisotropy developed during one-dimensional compression in the consolidometer was seen.

TEST RESULTS

The results of the bender element tests are presented in the form of the variation of G_{max} with p' in Fig. 1(a). At the initial isotropic effective stress of 400 kPa the degree of anisotropy, defined as the ratio G_{hh}/G_{hv}, was found to be 1·5 for the OC samples. This was in good agreement with the degree of anisotropy found in the natural samples of London Clay. The variation of the degree of anisotropy of G_{max} and the strain increment ratio of radial to axial strains are presented in Fig. 2(a). Up to the preconsolidation pressure both ratios remain constant and almost equal at about 1·5. After large strain yield the strain increment ratio immediately becomes isotropic, having a value of about 1·0. In contrast, the degree of anisotropy of G_{max} decreases very gradually so that the strain-induced anisotropy of G_{max} persists long after the strain increments have become isotropic. At the maximum pressure reached the degree of anisotropy was still high at about 1·4. The degree of anisotropy of G_{max} is plotted against volumetric strain in Fig. 2(b). From the data presented it can be concluded that very large strains would be required to obtain an isotropy of G_{max} appropriate to the new stress regime.

Results from the set of tests on the normally consolidated samples are also shown in Fig. 2(b). Before the isotropic consolidation in the triaxial cell, results from bender element tests indicated a degree of anisotropy of 1·33. These measurements were made under a total stress which was isotropic, but with the drainage shut so that no consolidation took place. This degree of anisotropy then reduced to about 1·25 as a result of the isotropic consolidation which followed. The effect observed here is similar to that seen with the overconsolidated samples, again indicating that very large strains are needed to change the degree of inherent anisotropy.

DISCUSSION AND CONCLUSIONS

It was observed from these tests that anisotropy of G_{max} developed within the soil during one-dimensional compression from a slurry. This result demonstrates that anisotropic plastic straining is one means by which an inherent anisotropy of stiffnesses is developed in clays. Anandarajah & Kuganenthira (1995) investigated the strain-induced anisotropy of one-dimensionally compressed kaolin by means of electrical conductivity. They observed that an initially isotropic slurry gradually developed anisotropy during compression. The increasing degree of anisotropy eventually reached a saturation level at about 100 kPa vertical effective stress, after which it did not change. In terms of the anisotropic fabric caused by one-dimensional compression, it therefore appears that there is a limit to the extent to which the development of a fabric influences the material properties of the soil so that continued straining in one particular sense might create a maximum or full degree of anisotropy appropriate to that anisotropic state. This means that where anisotropy has not been implemented by diagenetic process it is possible to reproduce the degree of anisotropy of the natural sample in the laboratory, even if precisely the correct preconsolidation stress is not applied, and this appears to be the case for the London Clay.

It was also shown that plastic straining can reverse the development of an anisotropic fabric so that continued isotropic straining may reduce the degree of anisotropy. The imposition of the strains associated with the new isotropic loading regime tended to erase the existing inherent anisotropy from one-dimensional compression. However, this process was very slow and the inherent anisotropy of G_{max} persisted long after the strain increments reached a ratio appropriate to the new loading condition. Inherent anisotropy in clays can therefore be regarded as a variable factor. In this case it appeared to be related solely to the plastic strain history of the soil up to its current state, and only if there had been very large continuous strains in one sense would the degree of anisotropy have a unique value. It is therefore very difficult to make a complete investigation of the inherent anisotropy of G_{max} in the triaxial apparatus because of the limited strains that may be imposed on the sample.

REFERENCES

Anandarajah, A. & Kuganenthira, N. (1995). Some aspects of fabric anisotropy of soil. *Géotechnique* **45**, No. 1, 69–81.

Jovičić, V. & Coop, M. R. (1998). The measurement of stiffness anisotropy in clays with bender element tests in the triaxial apparatus. *Geotech. Test. J.*, GTJODJ, **21**, No. 1, 3–10.

Pennington, D. (1997). Anisotropy of G_0 shear stiffness in Gault Clay. This volume.

McDermott, R. J. W. (1998). Pre-failure deformation behaviour of geomaterials, 399–401

Variation of elastic moduli with principal effective stress

R. J. W. MCDERMOTT*

This discussion concentrates on the results from three groups of authors—Jiang *et al.*, Kohata *et al.* and Jovičić & Coop—to relate the variation of elastic moduli for granular materials to a power function of the corresponding principal effective stress, and suggests how a generalized form of the Hertz contact equation may lead to a simple model/framework for the collation of such results.

From test results in the very small strain range of less than 0·001%, with due attention to corrections for bedding error and membrane penetration, Jiang *et al.*, on samples of a well graded sandy gravel of a crushed sandstone (called Chiba gravel), and Kohata *et al.*, on samples of eight different sands and gravels (including Chiba gravel), drew the following conclusions:

(*a*) at isotropic stress states the Young's moduli E_v and E_h, in the vertical and horizontal directions respectively, increased in proportion to approximately $(\sigma'_c)^m$ (where σ'_c is the confining pressure). The ratio E^o_v/E^o_h, indicating the inherent anisotropy, was about 2 (Jiang *et al.*), and ranged from 1 to 1·75 for the sands and from 1·75 to 2·3 for the gravels (Kohata *et al.*),

(*b*) for a wide range of anisotropic stress states, E_v was essentially a function of $(\sigma'_v)^m$, while E_h was essentially a function of $(\sigma'_h)^m$, with m equal to about 0·5. Therefore the ratio E_v/E_h increased approximately in proportion to $(\sigma'_v/\sigma'_h)^m$, indicating the stress-state induced anisotropy.

In order to substantiate these two points, I should like to refer to results obtained for pre-peak strains of less than 0·8%, from 100 mm high by 100 mm diameter conventional triaxial samples of a uniform size fraction of fine subangular sand (River Welland (RW) sand), tested between lubricated end platens with careful attention to bedding error and membrane penetration (McDermott, 1982). Comprehensive coverage of this research work has been given by Rowe (1972) and McDermott (1972, 1980). In this discussion the symbols used by the authors have been adopted for ease of comparison. Using the results from cyclic isotropic and anisotropic (CR) tests, analysed by applying a generalized form of the Hertz contact equation, given by equation (1), the writer found that a complete framework could be derived permitting the 'elastic' (recoverable) component of deformation to be assessed in all triaxial compression stress paths

$$\varepsilon^e_{v,h} = S_{v,h}\left(\frac{\sigma'_{v,h}}{E_p}\right)^m \tag{1}$$

where m is the Hertzian exponent, expected to depend on the composition, shape, size and packing of the particles, $S_{v,h}$ are Hertzian elastic coefficients, again dependent on the particle and sample characteristics, and E_p is the Young's modulus of the parent material of the particles.

Equation (1) is a generalized form of the Hertz contact equation for equal spheres in contact, given by the following equation (first used by Wilson & Sutton, 1948).

$$\varepsilon^e = 1{\cdot}55\left(\frac{\sigma}{E_p}\right)^{2/3} \tag{2}$$

By considering the non-linear CR stress–strain curves to be composed of an infinite number of infinitesimal linear increments, the incremental form of equation (1) was applied by the writer to the generalized Hooke's law equations reduced to the axisymmetric case with five compliances A_{ij}, from which the following can be shown to apply

$$A_{11} = 1/E_v = \frac{mS_v}{E^m_p}\sigma'^{\,m-1}_v \tag{3}$$

$$A_{33} = 1/E_h = \frac{mS_h}{E^m_p}\sigma'^{\,m-1}_h \tag{4}$$

It can be seen from equations (3) and (4) that

$$E_v \propto \sigma'^{\,1-m}_v \text{ and } E_h \propto \sigma'^{\,1-m}_h$$

and for the isotropic case with $\sigma'_v = \sigma'_h = \sigma'_c$ for a change in cell pressure $\Delta\sigma'_c$, then

$$\Delta E_v = \Delta E_h \propto \Delta\sigma'^{\,1-m}_c \tag{5}$$

in keeping with the authors' conclusion given as (*a*) above. The ratio E_v/E_h can then be obtained either from equations (3) and (4) or, more simply, from equation (1) using ($E_{v,h} = \sigma'_{v,h}/\varepsilon_{v,h}$), and is given by equation (6) below. In this form E_v/E_h is seen to have greater physical significance because

Manuscript received 17 October 1997.
* Liverpool John Moores University.

it is linked to the Hertzian parameters m and $S_{v,h}$ from equation (1). Equation (6) is in keeping with the authors' conclusion given as (*b*) above, but with their m changed to $1 - m$. The values of E_v and E_h were found by the writer to be of the order of 10^3 less than that for the parent material, E_p.

$$E_v/E_h = S_h/S_v(\sigma'_v/\sigma'_h)^{1-m} \quad (6)$$

The writer's results from the CR stress path data (used in the CR elastic model) for dense fine subangular RW sand showed that the value of m was practically constant and equal to 0·35 for a range of CR values ($1 \leqslant R \leqslant 4{\cdot}7$, $R_{peak} = 5{\cdot}6$) (McDermott, 1981). The value of m was found to increase with increasing particle roundness, as given by comparative tests with dense fine glass ballotini which gave $m \approx 0{\cdot}52$. In addition, m was found to increase with increase in void ratio as given by the results on loose RW sand which gave $m = 0{\cdot}45$. The outline trends from these results are shown in Fig. 1.

From the results of Jiang *et al.* and Kohota *et al.* on crushed Chiba gravel, the corresponding value of m to fit equation (6) would be $1 - m = 0{\cdot}5$ (i.e. $m = 0{\cdot}5$) from the anisotropic tests, and $1 - m = 0{\cdot}52$ (i.e. $m = 0{\cdot}48$) from the isotropic tests. In addition, a measure of the inherent anisotropy is given by $E^o_v/E^o_h = S_h/S_v = 2$ from equation (6), with $\sigma'_v = \sigma'_h$. Similarly, results on the eight different sands and gravels by Kohata *et al.* (their Fig. 8(a)) gave values of $1 - m$ (corresponding to equation (6)) ranging from 0·47 (subrounded SLB sand) to 0·53 (subangular Ticino sand) for the sands, and from 0·51 to 0·57 for the gravels. Using the outline model in Fig. 1, it would appear on first inspection that, since both authors' tests were conducted on dense compacted samples, their m values corresponded more to particles of a rounded nature. The results of the subrounded SLB sand gave the closest fit to the model, with $m = 0{\cdot}53$.

However, there is a possible disparity with the crushed Chiba gravel samples, which one would expect to be more angular. In addition, the well-graded nature of these samples of sandstone would result in a denser packing of the particles, as compared to the more open structure of uniform size particles of fine RW sand, again tending to reduce the value of m. A possible explanation might arise if the intact sandstone were composed of rounded particles which played a more significant role in governing the deformation behaviour. Another possible explanation for the 'high' values of m obtained by both authors is that, although their tests were conducted under isotropic and anisotropic compression, the cyclic parts of the tests, which determined the 'elastic' parameters, were obtained under varying R conditions (vertical cyclic tests followed by lateral cyclic tests). This would have introduced slip (irrecoverable) strains (due to rigid body sliding) into the measurements and affected the resulting value of m. Likewise, it is interesting to note that the exponent n used by Jovičić & Coop in equation (1) of their paper would similarly be given by $n = 1 - m$. From the information given in their Figs 9(a) for Dogs Bay sand (subangular/angular) and 9(c) for Ham River sand (subrounded), they have n values of 0·686 ($m = 0{\cdot}31$) and 0·593 ($m = 0{\cdot}41$) respectively, which give good general agreement with Fig. 1.

It was for the reasons given above that cyclic test data from varying R stress-paths were not used by the writer in formulating an 'elastic' model for sand. Instead, an implicit assumption was made by adopting the CR cyclic test data (to predict the 'elastic' strains in varying R stress-paths) in that the 'elastic' constants are stress-path independent, which is a necessary condition for elastic constants. The resulting 'elastic' strains for $C\sigma'_h$, CP and $C\sigma'_v$ stress-paths, determined from this model, could then be used to correct total strain data to obtain the slip strain data for use in the stress–dilatancy relationship. The model was found to give good consistent results in keeping with the accuracy of the apparatus. However, given the results of Mindlin & Deresiewicz (1953) from ideal assemblies of equal spheres, the CR 'elastic' model could still include local slip over the outer

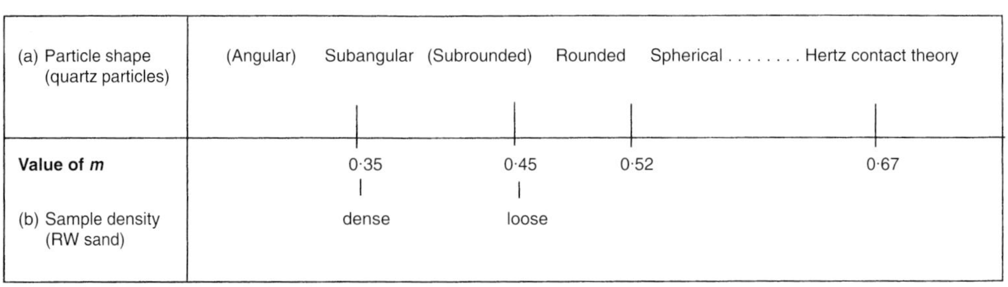

Fig. 1. Variation of the Hertzian parameter *m* with (a) particle shape, (b) sample density

annulus of particle contacts. The good correlation obtained by the authors between the laboratory tests and field seismic data confirms the importance of their work, but in the light of the above, it is suggested that the term 'pseudo-elastic' would be more appropriate.

In summary, this discussion has indicated the importance of relating experimental data to models/frameworks of soils based on known physical constants, for example in this case the Young's modulus of the parent material E_p, and measurable parameters such as the Hertzian-type parameters m, and $S_{v,h}$, which can be linked to the characteristics of the material, e.g. particle composition and shape, sample density. Such models (in keeping with the philosophy of the late Professor Peter Rowe) give physical substantiation to the measured parameters and enable more meaningful qualitative and at best quantitative assessments to be made for the behaviour of the soil under consideration.

REFERENCES

McDermott, R. J. W. (1972). *Deformation characteristics of sand under axisymmetric loading.* PhD thesis, University of Manchester.

McDermott, R. J. W. (1980). The constant stress ratio elastic model for sand. *Soils under cyclic and transient loading* (eds Pande and Zienkiewicz), Vol. 1, pp. 353–376 Rotterdam: Balkema.

McDermott, R. J. W. (1981). Discussion on 'A natural compression law for soils' (Butterfield). *Géotechnique* **31**, No. 4, 564–566.

McDermott, R. J. W. (1982). Discussion on 'Bedding error in triaxial tests on granular media' (Sarsby *et al.*). *Géotechnique* **32**, No. 1, 68–71.

Mindlin, R. D. & Deresiewicz, H. (1953). Elastic spheres in contact under oblique forces. *J. Appl. Mech.* **20**, 327–344.

Rowe, P. W. (1972). Theoretical meaning and observed values of deformation parameters of soil. *Proc. Roscoe Memorial Symp., Cambridge*, 143–194.

Wilson, G. & Sutton, J. L. E. (1948). A contribution to the study of the elastic properties of sand. *Proc. 2nd Int. Conf. Soil Mech.*, **1**, 197–202.

Chan, A. H. C. (1998). Pre-failure deformation behaviour of geomaterials, 403–404

On the use of comprehensive soil models in geotechnical analysis

A. H. C. CHAN*

The discussion leader for the second part of session 2 on Characterization and Modelling raised the following two points for discussion

(*a*) Is it now more important to investigate a wide range of states and classes of soil than to improve measurement techniques further?
(*b*) Is it practical or necessary to include all factors affecting the behaviour of soils in a comprehensive model?

This contribution focuses on the second point, and the following questions are discussed.

(*a*) When should comprehensive models be used?
(*b*) Will the use of simple models with 'conservative' assumptions always lead to safe design?
(*c*) Is a comprehensive model computationally expensive to run?
(*d*) Is parametic identification difficult for comprehensive models?
(*e*) Is a comprehensive model difficult to implement?
(*f*) Is a comprehensive model difficult to understand?
(*g*) Should a comprehensive model be complex?

When should a comprehensive model be used?

A comprehensive model should at least be used in the benchmarking of simpler models to be used in practical applications. The use of comprehensive models will help to identify whether or not the basic assumptions of the simpler models such as elastic unloading would be valid for the application involved. The detail of this benchmarking will no doubt be application-specific, but the use of comprehensive models will help to identify mechanisms that have been neglected in simpler models.

Will the use of simple models with 'conservative' assumptions always lead to safe design?

The answer to this question is that 'conservative' design would result in most static cases involving increase in load, but this may not be the case for unloading and other dynamic-related cyclic events. Under dynamic loading, a 'conservative' design in strength may lead to a reduction of ductility. This reduction could lead to reduction in energy absorption, which may result in brittle failure when a major event occurs.

Soil behaviour near failure is non-linear. Unlike metal, dilatancy during plastic deformation plays an important part in the soil response to loading. This use of a relatively simple associative Mohr–Coulomb model will greatly overestimate the failure load in a confined situation due to its dilative behaviour. A slightly more comprehensive non-associative Mohr–Coulomb model would lead to a more realistic failure load.

In determining whether or not the use of a simple model is adequate, the engineer has to be satisfied that all relevant mechanisms have been included. It is not good enough if separate analysis is performed using different simple models including separate mechanisms, as the combined mechanism may be the one that is critical.

Is a comprehensive model computationally expensive to run?

This is probably still the case for 3-dimensional analysis. However, 2-dimensional static and dynamic analyses can be performed on an Intel-based PC. Developments in computational algorithms, such as the backward Euler method and consistent tangent method, have greatly reduced the computational requirement of non-linear structure, soil and pore-fluid analyses.

Is parametic identification difficult for comprehensive models?

Researchers have put considerable effort into this area. More and more models which can represent complex soil behaviour require a limited number of soil model parameters. Most of these parameters can now be obtained from a relatively limited number of tests. However, there would be a need to develop a third-party-usable procedure so that reasonable parameters can be obtained without substantial experience in mathematical details of the model.

Is a comprehensive model difficult to implement?

In most of the commercially available computer programs such as SAGE-CRISP, LUSAS and ABAQUS, a standard material–model interface is

Manuscript received 21 October 1997.
* School of Civil Engineering, University of Birmingham.

provided. Standard soil model libraries such as SM2D developed by the contributor have greatly simplified the task of material model implementation.

A non-associated Mohr–Coulomb model subroutine has been taken from the library and incorporated into a commercial package. The material model was operational within a few days. The model was then taken through comprehensive validation and checking. Standardization of programming interface and independent Technical Assessment (ITA) should greatly reduce the difficulties and time involved in such implementation.

Is a comprehensive model difficult to understand?

Currently, yes. A 'tower of Babel' is developing in the use of terminology in constitutive modelling. Part of the purpose of developing a soil model library is to collect constitutive models from different authors. The models can then be compared in single element tests and boundary value tests. Through this process, it is hoped that third-party explanations of this vast array of constitutive models can be developed.

Should a comprehensive model be complex?

One possible answer to this question is 'yes, because soil is inherently complex'. But the question has to be asked: are we researching into the right area? As in metal plasticity, ultimate failure is usually associated with strength, i.e. stress criterion. However, for pre-failure deformation and the modelling of anisotropic behaviour, one would question whether a stress-based approach would be the most appropriate. Theories such as the Simpson's brick model and Lensky's strain-based fading memory should be further investigated to establish whether strain-based techniques offer a better and simpler approach to soil behaviour.

In conclusion, a wide range of soil models, from very simple non-linear elasticity to very complex grand unified theory (GUT) models is available. Care must be taken in the choice of models. Comprehensive models should not be dismissed without due consideration and cost–benefit analysis. Lastly, the contributor would like to use a quotation which has been attributed to Einstein to finish this discussion—'As simple as possible, but no simpler'.

Muir Wood, D. (1998). Pre-failure deformation behaviour of geomaterials, 405–407

Constitutive modelling: required complexity?

D. MUIR WOOD*

The scientific method requires that progress in understanding be made by constantly attempting to refute conjectures. So far as the mechanical behaviour of soils is concerned, it is essential that the experimental studies should be performed in association with the development of constitutive models. The hypotheses on which these models are based can then be used to design experimental programmes which will specifically test their validity. The development of appropriate models is then not merely something that may take place after the experiments are finished, but an iterative integrated part of the study.

In the context of applicability of appropriate models, it may well be found that the refutation can be disregarded in certain regions of application where the nature of the problem being studied limits in some way the potential explorations of stress–strain–time space.

It is unfortunate that the language of constitutive modelling and numerical analysis leads to potential confusion over the use of the term '*stiffness*'. Stiffness is associated with the property of an elastic material which, when it is loaded and unloaded, shows entirely recoverable response with no dissipation of energy. Modern experimental techniques indicate that the region of stress space for which the behaviour of real uncemented soils is truly elastic is extremely small.

The term '*stiffness*' is also used to describe the incremental response of an elastic–plastic material, and results of experimental studies are presented through the degradation of instantaneous tangent or secant stiffness with increasing length of strain excursion. This is clearly not an elastic process and, although it may be possible to reproduce certain aspects of the response using a non-linear elastic model, this will almost certainly be unable to describe in a way that is experimentally plausible and thermodynamically acceptable either the response of the soil on non-monotonic loading paths (or paths with corners in stress or strain space) or the volumetric response of the soil. Such a model may be useful for interpolation between experimental data, but provides a theoretically insecure basis for extrapolation.

Manuscript received 10 September 1997.

* University of Bristol.

It is desirable to make use of models which are just sufficiently complex for the intended application in order to minimize the burden of determination of soil parameters. The simplest models with which most engineers are familiar are elastic and elastic–Mohr–Coulomb frictional plastic models. We can develop a framework for introducing a necessary complexity of modelling by considering features of soil response which these models certainly exclude. In so doing we create a series of conjectures which can then be tested with appropriate experiments.

First, it is known that soils show barotropy and pycnotropy—the frictional strength and other mechanical properties of soils are dependent on pressure and density. The framework of critical state soil mechanics and the use of state parameter to describe proximity to critical states are obvious candidates to introduce this type of response.

Second, there is a series of observations—hysteretic stress–strain cycles, asymmetric stiffness or compliance tensors, non-unique stiffness or compliance tensors—which indicate immediately that elastic models stand no chance of success: plasticity is an essential ingredient. Stiffness and compliance tensors for elastic materials (even non-linear elastic materials) are necessarily symmetric.

Third, it is noted that incremental stiffness is influenced by stress history and that soils show significant stress-induced anisotropy. A particular observation that is made regularly is that stiffness increases after every corner in the strain path, with the increase in stiffness being dependent on the sharpness of the corner. Kinematic hardening plasticity provides a ready framework for describing such response.

Fourth, it is observed that incremental stiffness falls continuously with monotonic strain towards failure (and may even become negative if the soil shows strain softening), a character of response that can readily be reproduced using the framework of bounding surface plasticity.

A constitutive model for sand, which is essentially a Mohr–Coulomb elastic–plastic model with these additional features, has been developed by Gajo & Muir Wood (1997a, 1997b). This model, which has been called Severn-Trent sand, is able to reproduce the classical features of volumetric and stress–strain response of sand of various initial densities tested with and without drainage and under different confining pressures (Fig. 1).

In the absence of the intimate links between constitutive modelling and experimental observation that have been advocated, the calibration of a constitutive model is at the mercy of the experimental data available in the published literature. Many of these data come from tests performed in conventional triaxial apparatus, which provide but a limited test of the extrapolatory potential of a model. However, as soon as one extends the set of data to include tests run in different apparatus, with variations of all three principal stresses and rotations of principal axes, then the modeller is confronted with the fallibility of real data. For example, part of the operation of verification of the accuracy of different test devices includes the performance of the same test in different apparatus. 'Triaxial compression' with no rotation of principal axes and with two principal stresses always equal can be performed in conventional triaxial apparatus, true triaxial cubical apparatus, and hollow cylinder apparatus. Typical results for Hostun sand are shown in Fig. 2.

Each test device is a boundary value problem, and the deduction of constitutive parameters from experimental observation becomes a classical deconvolution or inverse problem which may often not be particularly well posed. Individual published data sets frequently come from one device alone, and it may not be simple to deduce the extent to which one can be confident that those data do indeed reflect the ideal response of a single element of homogeneous soil.

A similar incremental approach to the development of constitutive models for natural structured soils works from the widely used modified Cam clay model (Roscoe & Burland, 1968), through the

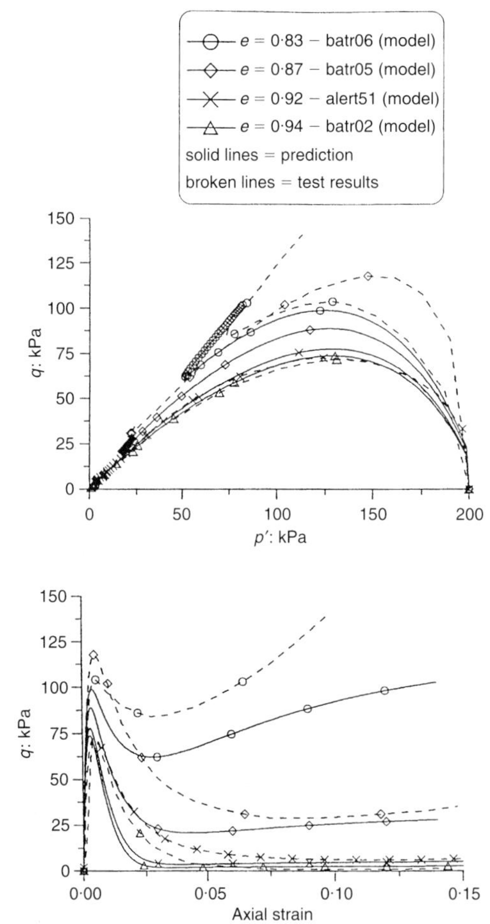

Fig. 1. Undrained triaxial compression tests on Hostun sand with different initial void ratios and predictions using the Severn-Trent sand model (initial mean effective stress $p' = 200$ kPa) (from Gajo & Muir Wood, 1997a)

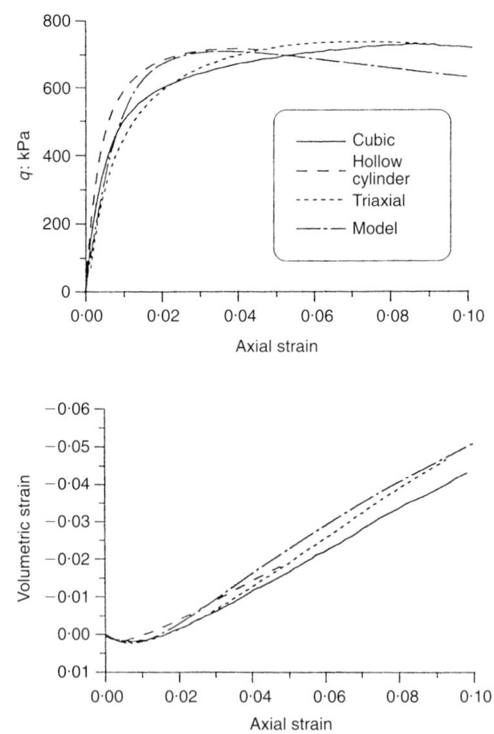

Fig. 2. Drained triaxial compression tests on dense Hostun sand in triaxial apparatus, true triaxial cubical apparatus and hollow cylinder apparatus, and prediction using Severn-Trent sand model (initial void ratio $e = 0{\cdot}61$, initial mean effective stress $p' = 500$ kPa) (from Gajo & Muir Wood, 1997b)

'bubble' kinematic hardening model (Al-Tabbaa & Muir Wood, 1989)—which introduces stress-induced anisotropy and steady stiffness degradation—to a destructuration model (Muir Wood, 1995) which intro-duces a measure of structure which progressively disappears with increasing plastic strain. The models form a hierarchy: without structure the destructuration model degenerates to the 'bubble' model; without kinematic hardening the 'bubble' model degenerates to Cam clay.

REFERENCES

Al-Tabbaa, A & Muir Wood, D. (1989). An experimentally based 'bubble' model for clay. *Numerical models in geomechanics (NUMOG III)* (eds S. Pietruszczak and G. N. Pande), Elsevier Applied Science, Amsterdam, 91–99.

Gajo, A. & Muir Wood, D. (1997a). A kinematic hardening constitutive model for sands: the *q–p* formulation (submitted for publication).

Gajo, A. & Muir Wood, D. (1997b). A kinematic hardening constitutive model for sands: the multiaxial formulation (submitted for publication).

Muir Wood, D. (1995). Kinematic hardening model for structured soil. *Numerical models in geomechanics (NUMOG V)* (eds G. N. Pande and S. Pietruszczak), Balkema, Rotterdam, 83–88.

Roscoe, K. H. & Burland, J. B. (1968). On the generalised stress–strain behaviour of 'wet' clay. *Engineering plasticity* (eds J. Heyman and F. A. Leckie), Cambridge University Press, 535–609.

Higgins, K. G. & Jardine, R. J. (1998). Pre-failure deformation behaviour of geomaterials, 409–411

Experience of the use of non-linear pre-failure constitutive soil models

K. G. HIGGINS* and R. J. JARDINE†

ISSMFE Technical Committee TC-29 is currently compiling a worldwide database of projects where advanced geotechnical testing and non-linear pre-failure constitutive soil models have been combined to make practical predictions of ground movements and soil–structure interaction problems. The committee requested information on the ground conditions, the testing undertaken to derive parameters, and any available field measurements of prototype.

Table 1 provides a minimal summary of the contribution that the authors offered to the TC-29 survey. Over 20 analyses are mentioned, performed over the past 13 years, which employed the non-linear approach set out by Jardine *et al.* (1986) and Jardine & Potts (1988). The list includes three different types of offshore structure and onshore projects involving deep and shallow foundations, embankments, tunnels and excavations. Ground conditions included soft clays, stiff clays, mudstones, inter-bedded sands and clays, sands and cemented sands. The testing undertaken included 'small strain' triaxial stress-path laboratory tests, in situ pressuremeter tests and field geophysics. Most of the analyses were undertaken to aid the design process or assess potential damage. Under these circumstances, engineers have to tend towards conservatism when selecting parameters. Nevertheless, there was generally good agreement between the predictions and field measurements, as is detailed in the cited references. A similar number of unreported analyses have been made for other projects where the clients have made no provision for field measurements. As far as the authors are aware, no practical difficulty arose with any of these schemes as a result of the non-linear numerical analysis.

At the Symposium in Print, David Harris (GCG) presented data from the new Westminster Station excavation in London, which is part of the ongoing Jubilee Line Extension project. He showed very good agreement between the current measured diaphragm wall movements and the most recent predictions, which accounted for revisions in the construction procedure without changing the soil stress–strain parameters. These data were not available when the submission was made to the TC29 committee. Likewise, data from the Al Madinah Mosque excavation have only recently become available. Ground movements appear to have been well predicted while, for reasons not associated with the soil modelling, the measured short-term wall movements ($\sim$ 8 mm) fall below those predicted in 1987 ($\sim$ 14 mm). No 'calibration factor' was applied to the soil parameters, or the numerical results, in any of the cases listed. Where possible, input parameters were interpreted directly from site-specific in situ tests or from laboratory tests on good-quality samples; in other cases, data from nearby sites were used.

In the formulation described by Jardine *et al.* (1986), soil stiffness changes with the mean effective stress and the strain level developed at each finite element Gauss point within the idealized soil mass. Implementation of the model is not straightforward, since the stiffness of the soil needs to be continually updated throughout any analysis. When done correctly, and used in conjunction with an accurate stress point algorithm, this adds considerably to the computer run times. However, the practical benefits of using such procedures have been proved in many applications.

The discussion at the Symposium in Print emphasized that none of the constitutive models currently available is fully representative of soil behaviour and that considerable care is needed when carrying out fully non-linear analysis. In order to make good predictions with the above approach the user needs to be aware of the anisotropic and kinematic nature of soil stiffness. These features are not catered for implicitly and have to be addressed when selecting parameters and deciding whether (and when) to re-invoke the 'small strain' stiffness response during any particular construction sequence. It is also vital (*a*) to make sure that the formation can reproduce the original test data on which it is based before undertaking an analysis, and (*b*) to check the model's behaviour over the effective stress-paths the soil is likely to experience in the analysis. Without these checks the user cannot be confident that the model is behaving reasonably and that inadmissible states are not achieved. This is quite possible if the

Manuscript received 22 October 1997.
* Geotechnical Consulting Group, GCG.
† Imperial College.

Table 1. Projects in which non-linear pre-failure constitutive models have been used for analysis

Project (year of completion)	Location	Type of project	Ground conditions	Tests (lab/field)	Agreement between predictions & measurements	Reference
Magnus (ax. loading) (1982, monitoring 1985)	North Sea	Offshore platform	Stiff to hard clays and dense sands	Small strain triaxial stress path	Good	Jardine & Potts (1992)
Magnus (lat. loading) (1982, monitoring 1985)	North Sea	Offshore platform	Stiff to hard clays and dense sands	Small strain triaxial stress path	Good	Ganendra (1994)
Hutton TLP (1984)	North Sea	Offshore platform	Stiff to hard clays and dense sands	Small strain triaxial stress path	Good	Jardine & Potts (1988)
Brent (ca 1977)	North Sea	Offshore platform	Stiff to hard clays and dense sands	Small strain triaxial stress path	Good (hindcast)	Jardine *et al.* (1991)
Berthierville (ca 1984)	Quebec	Embankment	Soft structured clay	Small strain triaxial stress path	Good (hindcast)	Smith & Jardine (1991)
Hinkley	UK	Nuclear power station foundations	Mudstones	In situ geophysics	See reference	Hight & Higgins (1994)
Sizewell C (not yet built)	UK	Nuclear power station foundations	Cemented sands	In situ geophysics + pressuremeter	Not available	—
NATM Trial Tunnel (1995)	London	NATM tunnel	Stiff clay with inter-bedded stiff clays and sands	—	Good	Kovacevic *et al.* (1996)
Morgan Bank (1990)	London	Excavation	Stiff clay	Small strain triaxial stress path comp. & extn	Good	Jardine *et al.* (1991), St John *et al.* (1992)
Bell Common Tunnel (ca 1983)	London	Cut & cover tunnel	Stiff clay	Small strain triaxial stress path comp. & extn	Good	Fourie *et al.* (1986), Higgins *et al.* (1989)
A12–M11 Link Road (under construction)	London	Cut & cover tunnel	Stiff clay	Pressuremeter	Not yet available	—
London Ring Main (1990)	London	Tunnel	Stiff clay	Small strain triaxial stress path comp. & extn	Good	Skinner (1990)
Westminster Underground Station (under construction)	London	Excavation & tunnels	Stiff clay	Small strain triaxial stress path comp. & extn	Good	Higgins *et al.* (1996)
Canada Water Station (under construction)	London	Excavation	Inter-bedded stiff clays and sands	Small strain triaxial stress path comp. & extn	Not yet available	—
Al Madinah Car Park (1989)	Madinah, Saudi Arabia	Excavation	Inter-bedded soft clays and sands	Small strain triaxial stress path comp. & extn	Fair to good	Puller (1996)
Waterloo Int. Terminal (1991)	London	Excavation & tunnels	Stiff clay	Pressuremeter & small strain triaxial stress path comp. & extn	Good	Hight *et al.* (1992)
Grand Buildings Project (ca 1991)	London	Excavation & tunnels	Stiff clay	—	Not available	—
Angel Underground Station (ca 1993)	London	Piles, excavation & tunnels	Stiff clay with inter-bedded stiff clays and sands	Small strain triaxial stress path comp. & extn	Good	Mair (1993)
Melbury House Development (1996)	London	Piles, excavation & tunnels	Stiff clay	Triaxial tests	Good	—
Finsbury Pavements Development (1996)	London	Excavation & tunnels	Stiff clay	Triaxial tests	Not available	—
Old Street Underground Station (under construction)	London	Tunnels	Inter-bedded stiff clays and sands	—	Fair to good*	Bracegirdle *et al.* (1996)
Poultry development (under construction)	London	Excavation & tunnels	Stiff clay	—	Not available	—

* Measurements at Old Street may have been influenced by other factors such as chemical attack on grout and linings.

model has not been correctly implemented or if it is unstable under certain conditions.

From the papers submitted to this symposium it is apparent that new knowledge gained from laboratory research and developments in constitutive modelling are leading to better models of pre-failure soil behaviour. However, it is important that any such improvements be verified against practical experience, and can be shown to lead to reliable predictions in a suitably wide range of circumstances.

REFERENCES

Bracegirdle, A., Jefferies, S. J., Tedd, P., Crammond, N. J., Chudleigh, I. & Burgess, N. (1996). The investigation of acid generation within the Woolwich and Reading Beds at Old Street and its effect on tunnel linings. *Proc. Int. Symp. on geotechnical aspects of underground construction in soft ground, London*, Balkema.

Fourie, A. B., Potts, D. M. & Jardine, R. J. (1986). The determination of approximate soil stiffness parameters for use in finite element analyses of geotechnical problems. *Proc. 2nd Int. Symp. on Numerical Models in Geomechanics, Ghent.*

Ganendra, D. (1994). *Finite element analysis of laterally loaded piles*. PhD thesis, University of London.

Higgins, K. G., Potts, D. M. & Mair, R. J. (1996). Numerical modelling of the influence of the Westminster Station excavation on the Big Ben clock tower. *Proc. Int. Symp. on geotechnical aspects of underground construction in soft ground*, Balkema.

Higgins, K. G., Potts, D. M. & Symons, I. F. (1989). *Comparison of predicted and measured performance of the retaining walls of the Bell Common Tunnel*. CR 124, Transport & Road Research Laboratory, Crowthorne.

Hight, D. W. & Higgins, K. G. (1994). An approach to the prediction of ground movements in engineering practice: background and application. Keynote lecture, *Proc. IS-Hokkiado*, Balkema, Rotterdam, **2**, pp. 909–946.

Hight, D. W. Higgins, K. G., Jardine, R. J., Potts, D. M., Pickles, A. R. & De Moor, E. K. (1992) Predicted and measured tunnel distortions associated with construction of Waterloo International Terminal. *Proc. Wroth Memorial Symp. on predictive soil mechanics, Oxford*, Thomas Telford, London, pp. 317–338.

Jardine, R. J. & Potts, D. M. (1988). Hutton Tension Leg Platform foundations: an approach to the prediction of driven pile behaviour. *Géotechnique* **38**, No. 2, 231–252.

Jardine, R. J. & Potts, D. M. (1992). Magnus foundations: soil properties and predictions of field behaviour. *Proc. Conf. on large scale pile tests in clay*, Thomas Telford, London, pp. 69–83.

Jardine, R J., Potts, D., M., Fourie, A. B. & Burland, J. B. (1986). Studies of the influence of non-linear stress–strain characteristics in soil–structure interaction. *Géotechnique* **36**, No. 3, 377–396.

Jardine, R. J., St John, H. D., Hight, D. W. & Potts, D. M. (1991). Some practical applications of a non-linear ground model. *Proc. 10th ECSMFE, Florence*, **1**, 223–228.

Kovacevic, N., Edmonds, H. E., Mair, R. J., Higgins, K. G. & Potts, D. M. (1996). Numerical modelling of the NATM and compensation grouting trials at Redcross Way. *Proc. Int. Symp. on geotechnical aspects of underground construction in soft ground, London*, Balkema.

Mair, R. J. (1993). Developments in geotechnical engineering research: application to tunnels and deep excavations. Unwin Memorial Lecture 1992, *Proc. Instn Civ. Engrs, Civ. Engng* **93**, Feb., 27–41.

Puller, M. J. (1996). *Deep excavations: a practical manual.* London: Thomas Telford.

St John, H. D., Potts, D. M., Jardine, R. J. & Higgins, K. G. (1992). Prediction and performance of ground response due to construction of a deep basement at 60 Victoria Embankment. *Wroth Memorial Symp. on predictive soil mechanics, Oxford*, Thomas Telford, London, pp. 581–608.

Skinner, A. E. (1990). *London ring main pipe-jacked water tunnel project*. Personal communication to R. J. Jardine.

Smith, P. R & Jardine, R. J. (1991). The use of triaxial tests for the prediction of settlement. *Geo-coast '91 Int. Conf., Yokohama*, Port and Harbour Research Institute, Yokosuka, pp. 89–92.

Jamiolkowski, M. B. (1998). Pre-failure deformation behaviour of geomaterials, 413–416

Closing remarks

M. B. JAMIOLKOWSKI*

The British Geotechnical Society and the Institution of Civil Engineers are to be praised for the organization of this *Géotechnique* Symposium in Print devoted to the pre-failure behaviour of geomaterials. This event follows the International Symposium organized in 1994 by the Japanese Society of Soil Mechanics and Foundation Engineering devoted to the same topic and held in Sapporo. Both events have gathered researchers working all over the world in the area of a new generation of laboratory tests that, in the past 15 years, have largely contributed to a better understanding of the pre-failure stress–strain–time behaviour of natural soils.

Attempts towards more precise and accurate measurements of mechanical soil properties in the laboratory started with the recognition, in the early 1980s, of the following facts put forward by Professor J. B. Burland in his Laurits Bjerrum Lecture delivered in 1986 (see Burland, 1989).

(*a*) Field evidence: the displacements of *well designed* geotechnical engineered constructions are generally quite small.
(*b*) Predictive capabilities: the conventional computation methods, plus laboratory-determined soil stiffness, overpredict the observed displacements.

These facts have triggered a number of research activities aimed at increasing the reliability of the laboratory-measured soil stiffness, attention being focused on the following crucial issues

(*a*) quality of undisturbed sampling and its engineering quantification
(*b*) importance of the procedures used in the laboratory for the reconsolidation of undisturbed specimens before shearing
(*c*) understanding and overcoming the causes of the lack of precision and accuracy encountered when measuring strains and stresses in conventional triaxial and torsional shear apparatuses (usually yielding unreliable values of stiffness at strains less than 10^{-3})

The most relevant outcomes of those researches were as follows.

(*a*) The recognition of the fundamental differences between the mechanical behaviour of natural soils and soils reconstituted in the laboratory, or substantially disturbed by inappropriate sampling.
(*b*) The implementation of local axial (ε_a) and radial (ε_r) strain measurements (Burland & Symes, 1982; Tatsuoka, 1988; Goto *et al.*, 1991), combined with the insertion of load transducers inside the apparatuses which, together with the careful calibration of the sensors involved now allow the soil stiffness at $\varepsilon_a \approx 5 \times 10^{-6}$ to be measured reliably.

This work leads to a completely new scenario as far as the magnitude of stiffness of $\varepsilon_a < 10^{-3}$ is concerned, generating results similar to those shown in Fig. 1. These results, among the first published in a widely circulated geotechnical journal, showed the great relevance of local strain measurements to the reliable assessment of soil stiffness. Such experimental data have triggered valuable research conducted, world-wide, by many of the individuals who participated in this Symposium, either in person or through written contributions. All of them are deeply involved in advanced laboratory testing, interacting and exchanging their experience, so forming *de facto* a sort of exclusive club whose members have

(*a*) greatly enhanced our potential capability to predict the displacements of structures and the surrounding ground
(*b*) attempted to transfer such a capability to practitioners
(*c*) created a more rational basis for
  (i) programming laboratory and in situ testing
  (ii) evaluating, critically, geotechnical design parameters
  (iii) rationalizing the selection of soil models and computation procedures in relation to the problem in hand.

Within the above framework and based on an examination of the papers presented to this Symposium, the most relevant research results achieved by the members of the 'exclusive club' can be summarized as follows.

(*a*) Relevant to all users:
  (i) acknowledgement of the markedly different behaviour of natural and reconstituted soils

* Department of Structural Engineering, Politecnico di Torino.

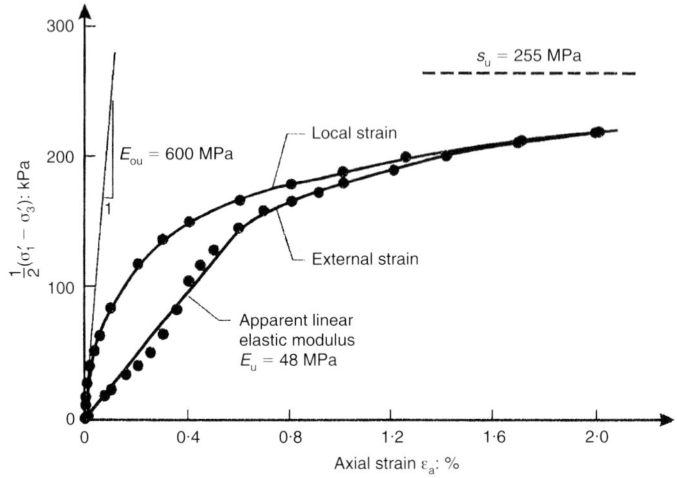

Fig. 1. Triaxial UU compression test on intact lightly OC North Sea clay (Jardine *et al.*, 1984)

- (ii) the real magnitude, hence importance in design, of the stiffness non-linearity in working stress range
- (iii) showing the uniqueness, for a given soil, of the initial stiffness* (E_0, G_0) irrespective of the loading type, i.e. cyclic versus monotonic
- (iv) offering deeper insights into the interpretation of in situ stiffness tests
- (v) rationalizing the role and use of geophysical boreholes tests (DH, CH, seismic CPTU) in geotechnical engineering as means for assessing G_0.

(*b*) Especially relevant for modelling purposes:
- (i) importance of geological and engineering stress histories
- (ii) concept of kinematic yield surfaces located within the state boundary surface
- (iii) role of time and rate effects
- (iv) role of initial and evolving anisotropy.

(*c*) Especially relevant to experimentalists:
- (i) framework for understanding the accuracy and precision necessary in laboratory measurements of strains and stresses
- (ii) criteria for appropriate laboratory reconsolidation procedures
- (iii) recognition of the importance of an adequate data acquisition system and test controlling software.

Despite the important achievements mentioned above, the members of the 'exclusive club' have many important tasks still to fulfil, of which the following appear the most challenging

(*a*) extending the present findings, which are mostly related to triaxial and torsional shear tests, to more general stress conditions (Zdravkovic & Jardine; Hird and Pierpoint)†

(*b*) obtaining a better experimental framing of the description of soil behaviour inside the second kinematic yield surface Y_2 (Jardine, 1985, 1992; Stallebrass, 1990; Stallebrass & Taylor, 1997; Zdravkovic & Jardine)

(*c*) modelling soil stiffness non-linearity far from failure, assessing the role of mathematical functions for fitting stress–strain curves against full constitutive models (Tatsuoka & Shibuya, 1992; Puzrin & Burland, 1996; Stallebrass & Taylor, 1997; Addenbrooke *et al.*; Kohata *et al.*; Hird & Pierpoint)

(*d*) describing stiffness anisotropy beyond the linear threshold strain ε_t^e (Addenbrooke *et al.*; Springman *et al.*; Zdravkovic & Jardine; Tatsuoka *et al.*, 1997)

(*e*) resolving the debate, rehearsed at this Symposium, concerning which principal effective stress components and/or invariants that are best suited to describe the stress dependency of the shear V_s and compression V_p wave velocities as well as of the initial stiffnesses (E_0, G_0) (Roesler, 1979; Yu & Richart, 1984; Lee, 1985; Lo Presti & O'Neill, 1991; Hoque, 1996; Stokoe, 1995; Bellotti *et al.*, 1996; Hight *et al.*; Yiang *et al.*; Kohata *et al.*; Rampello *et al.*; Tatsuoka *et al.*, 1997)

(*f*) developing a clearer picture of the influence of time and rate effects on stiffness and espe-

* Pseudo-elastic Young's and shear modulus respectively, measured at strain ε less than linear threshold strain ε_t^e.

† References without a year are presented in this Symposium.

cially, on material damping (Stokoe *et al.*, 1994; Lo Presti *et al.*, 1996; Hayano *et al.*; Tatsuoka *et al.*, 1997)

(*g*) assessing the influence of soil structure on stiffness and its degradation with strain; investigating the interrelation between soil structure and soil state (Cuccovillo & Coop; Cotecchia & Chandler; Fonseca *et al.*)

(*h*) formulating a simple but rational way to quantify the degree of laboratory sample disturbance (Fonseca *et al.*; Holtz *et al.*, 1986; Lunne *et al.*, 1997; Lo Presti, 1997).

The above issues form the basis for continuing the research described in the present Symposium in the years to come.

However, the most important future goals for the 'exclusive club' are linked to the use of their research for the solution of practical problems, through:

(*a*) the formulation of constitutive models (e.g. Stallebrass, 1990; Simpson, 1993; Stallebrass & Taylor, 1997) incorporating the most important findings from the new generation of laboratory tests

(*b*) the implementation of such models in computer programs and their validation against properly selected benchmark problems

(*c*) with the help of such programs, and back-analyses of well documented case records, validation of the reliability and usefulness of the laboratory-generated data for the solution of everyday design problems.

The above steps are mandatory for a further and definitive consolidation of the remarkable contribution which the members of the 'exclusive club' working in the field of laboratory tests have made in the past 15 years to geotechnical engineering. Without an adequate effort in these directions, a debate will inevitably be raised as to what are the benefits to geotechnical engineers of further increases of precision and accuracy in laboratory stiffness measurements.

REFERENCES

Burland, J. B. (1989). Ninth L. Bjerrum Memorial Lecture: small is beautiful—the stiffness of soils at small strains. *Can. Geotech. J.* **26**, No. 4, 499–516.

Burland, J. B. & Symes, M. J. (1982). A simple axial displacement gauge for use in the triaxial apparatus. *Géotechnique* **32**, No. 1, 62–65.

Goto, S., Tatsuoka, F., Shibuya, S., Kim, Y. S. & Soto, T. (1991). A simple gauge for local small strain measurements in the laboratory. *Soils Fdns* **31**, No. 1, 169–180.

Holtz, R. D., Jamiolkowski, M. & Lancellotta, R. (1986). Lessons from oedometer tests on high quality undisturbed samples. *J. GE* **112**, No. 8, 768–776.

Hoque, E. (1996). *Elastic deformation of sands in triaxial tests.* PhD thesis, University of Tokyo.

Jardine, R. J. (1985). *Investigation of pile–soil behaviour, with special reference to the foundations of offshore structures.* PhD thesis, University of London.

Jardine, R. J. (1992). Some observations on the kinematic nature of soil stiffness. *Soils Fdns* **32**, No. 2, 111–124.

Jardine, B. J., Symes, M. J. & Burland, J. B. (1984). The measurement of stiffness in the triaxial apparatus. *Géotechnique* **34**, No. 3, 323–340.

Lee, S. H. (1985). *Investigation of low-amplitude shear wave velocity in anisotropic material.* PhD thesis, University of Texas at Austin.

Leroueil, S. & Soares, M. E. (1996). SOA Report: *Importance of strain rate and temperature effects in geotechnical engineering.* ASCE Geotechnical Special Publication 61, 1–60.

Lo Presti, D. C. F. (1997). Stress–strain behaviour of undisturbed clays in the laboratory. Panel discussion. Discussion Session 1.1, *Proc. 14th ICSMFE, Hamburg*, in press.

Lo Presti, D. C. F., Jamiolkowski, M., Pallara, O. & Cavallaro, A. (1996). *Rate and creep effect on the stiffness of soils.* ASCE Geotechnical Special Publication 61, 166–180.

Lo Presti, D. C. F. & O'Neill, D. (1991). Laboratory investigation of small strain anisotropy in sand. *Proc. 1st Int. Symp. on Calibration Chamber Testing, Potsdam, NY*, 213–224.

Lunne, T., Berre, T. & Strandvik, S. (1997). Sample disturbance in soft low plasticity Norwegian Clay. *Proc. Symp. on recent developments in soil and pavement mechanics*, Balkema, Rotterdam, 81–102.

Puzrin, A. M. & Burland, J. B. (1996). A logarithmic stress–strain function for rocks and soils. *Géotechnique* **46**, No. 1, 157–164.

Roesler, S. K. (1979). Anisotropic shear modulus due to stress anisotropy. *J. Soil Mech. Fdn Engng Div. Am. Soc. Civ. Engrs* **105**, SM, 871–880.

Simpson, B. (1993). Development and application of a new soil model for prediction of ground movements. *Prediction soil mechanics, Proc. Wroth Memorial Symp., Oxford*, Thomas Telford, London, 628–643.

Stallebrass, S. E. (1990). *The effect of recent stress history on the deformation of overconsolidation soils.* PhD thesis, City University of London.

Stallebrass, S. E. & Taylor, R. N. (1997). The development and evaluation of a constitutive model for the prediction of ground movements in overconsolidated clay. *Géotechnique* **47**, No. 2, 235–254.

Stokoe, K. H., Hwang, S. K., Lee, J. N. K. & Andrus, R. D. (1994). Effects of various parameters on the stiffness and damping of soils at small and medium strains. Keynote lecture, *Proc. IS-Hokkaido*, Balkema, Rotterdam, **2**, 785–816.

Tatsuoka, F. (1988). Some recent developments in triaxial testing systems for cohesionless soils. SOA report, *Advanced Testing of Soils and Rocks*, ASTM STP 977, 7–67.

Tatsuoka, F. & Shibuya, S. (1992). Deformation characteristics of soils and rocks from field and laboratory tests. *Proc. 9th Asian Regional Conf. Soil Mech Fdn Engng, Bangkok*, **2**.

Tatsuoka, F., Jardine, R. J., Lo Presti, D. C. F., Di Benedetto, H. & Kodaka, T. (1997). Characterising the pre-failure deformation properties of geomaterials, theme lecture, Plenary Session I, *Proc. 14th ICSMFE, Hamburg*, in press.

Yu, P. & Richart, F. E. (1984). Stress-ratio effects on shear modulus of dry sands. *J. Geotech. Engng Div. Am. Soc. Civ. Engrs.*

Appendix

Sessions in which papers were presented at the Symposium held on 4 September 1997.

Session 1 Measurements in the laboratory and field

Anisotropy of G_0 shear stiffness in Gault Clay by D. S. Pennington, D. F. T. Nash and M. L. Lings
Laser interferometry to evaluate the performance of local displacement transducers by G. Heymann, C. R. I. Clayton and G. T. Reed
Some anisotropic stiffness characteristics of a silt under general stress conditions by L. Zdravkovic and R. J. Jardine
Deformation characteristics of a sedimentary soft mudstone from triaxial compression tests using rectangular prism specimens by K. Hayano, T. Sato and F. Tatsuoka
Wave velocity and stiffness measurements of the Crag and Lower London Tertiaries at Sizewell by D. W. Hight, J. D. Bennell, B. Chana, P. D. Davis, R. J. Jardine and E. Porovic

Session 2a Characterisation and modelling (A)

Small-strain stiffness of reconstituted clay compressed along constant triaxial effective stress ratio paths by S. Rampello, G. M. B. Viggiani and A. Amorosi
Inherent and stress-state-induced anisotropy in very small strain stiffness of a sandy gravel by G. L. Jiang, F. Tatsuoka, A. Flora and J. Koseki
Modelling of non-linear deformation properties of stiff geomaterials by Y. Kohata, F. Tatsuoka, L. Wang, G. L. Jiang, E. Hoque and T. Kodaka
Shear modulus and damping of soils by D. C. F. Lo Presti, M. Jamiolkowski, O. Pallara, A. Cavallaro and S. Pedroni

Session 2b Characterisation and modelling (B)

Yielding and pre-failure deformation of structured sands by T. Cuccovillo and M. R. Coop
The influence of structure on the pre-failure behaviour of a natural clay by F. Cotecchia and R. J. Chandler
Stiffness of coarse-grained soils at small strains by V. Jovičić and M. R. Coop
Elastic shear modulus of soft clays from shear wave velocity measurement by S Shibuya, S. C. Hwang and T. Mitachi

Session 3 Applications and design/case histories

Displacements of bridge foundations on sedimentary softrock: A case study on small-strain stiffness by K. Izumi, M. Ogihara and H. Kameya
Interpretation of a footing load test on a saprolitic soil from granite by A. J. P. V. da Fonseca, M. A. de Matos Fernandes and A. J. M. S. Cardosa
Modelling pre-failure deformation behaviour of reinforced soil walls by S. M. Springman, S. Balachandran and C. Jommi
Stiffness determination and deformation analysis for a trial excavation in Oxford Clay by C. C. Hird and N. D. Pierpoint
The influence of pre-failure soil stiffness on the numerical analysis of tunnel construction by T. I. Addenbrooke, D. M. Potts and A. M. Puzrin